Selected Papers of
M. Ohya

Selected Papers of

M. Ohya

World Scientific

NEW JERSEY • LONDON • SINGAPORE • BEIJING • SHANGHAI • HONG KONG • TAIPEI • CHENNAI

Published by

World Scientific Publishing Co. Pte. Ltd.

5 Toh Tuck Link, Singapore 596224

USA office: 27 Warren Street, Suite 401-402, Hackensack, NJ 07601

UK office: 57 Shelton Street, Covent Garden, London WC2H 9HE

Library of Congress Cataloging-in-Publication Data

Ohya, Masanori, 1947–
[Papers. Selections]
Selected papers of M. Ohya / edited by N. Watanabe.
p. cm.
Includes bibliographical references.
ISBN-13: 978-981-279-419-2 (hardcover : alk. paper)
ISBN-10: 981-279-419-0 (hardcover : alk. paper)
1. Quantum entropy. 2. Quantum teleportation. 3. Quantum theory. I. Watanabe, N. (Noboru), II. Title.
QC174.85.Q83O3582 2008
530.12--dc22

2007052697

The editor and the publisher would like to thank the following publishers for their assistance and their permission to reproduce the articles found in this volume:

Akadémiai Kiadó
The American Institute of Physics
Elsevier
IEEE Society
Kluwer
Plenum Publishing Co.
Polish Scientific Publishers
The Royal Society of London
Società Italiana di Fisica
Springer
Tokyo Institute of Technology
University of California
Wroclaw University and Wroclaw University of Technology

British Library Cataloguing-in-Publication Data

A catalogue record for this book is available from the British Library.

Printed in Singapore by World Scientific Printers

Masanori Ohya

Preface

1. Introduction

A few years ago 60 years were considered a threshold, just preceding retirement and a gradual disengagements from previous activities.

Contemporary society has changed this situation drastically and the 60 years threshold now marks the entrance of a scientist into the age of synthesis, in which you look back in perspective to the topics that have been at the barycentre of your scientific interest and try to find out a pattern in the development of your interaction with these problems.

In the case of Masanori Ohya, I would describe such a pattern as follows: use of advanced mathematical techniques for the development of quantum information theory and life sciences.

The core inspiration for M. Ohya has undoubtedly been quantum information, of which he can be considered as one of the pioneering figures.

As usual the effort to develop the quantum generalization of a discipline, leads to a new and deeper comprehension of its classical aspects.

This naturally led M. Ohya to consider life science, where the mechanism of elaboration of information plays a crucial role.

Chronologically, the latest development of this intellectual adventure was the merging of these two lines of research into the quantum bio–information program, a bold attempt to investigate the role of quantum mechanics, more specifically of quantum information theory, in the life sciences.

Since quantum mechanics represents the deepest level of development of contemporary physical sciences, it is clear that such role should exist and the role of pioneers is precisely to move the first steps in this direction.

One of the main merits of M. Ohya consists in having understood that such an ambitions scientific program could not be handled by a single individualism: the success of the program required coordination of the energies and enthusiasm of young generations with a solid network of experts in different fields. He succeeded remarkably well in both directions, first of all with an intensive activity as educator which produced hundreds (literally) of students some of whom now have achieved preeminent positions in different branches of science. Second with a careful selection of collaborators from all over the world, which has made Tokyo University of Science one of the main centers of quantum information theory since almost three decades.

One highlight of the national and international network built by M. Ohya is the success of the international journal Open systems and information dynamics, which in a few years has achieved an impact factor higher than many much older scientific journals, and whose foundation was based on an early intuition of M. Ohya anticipating of several years the need for a quantum information journal (in the past few years many such journals have been founded). Another one is the monograph by M. Ohya and D. Petz: Quantum Entropy and his use, now at its second edition at Springer and well established as one of the classics of quantum information theory. In addition the series of annual workshops on Quantum Information Theory and Open Systems, organized by M.Ohya in collaboration with Izumi Ojima since 1992 at the Kyoto Research Institute of Mathematics (RIMS), were the first periodical meeting point for the Japanese researchers interested in the mathematical aspects of quantum information.

These achievements parallel an impressive scientific production, marked by some highly original discoveries as well as the active participation in the editorial board of several international journals, an intense publishing activity including several books in different areas, and in particular the publication of the highly successful Encyclopedia of Information Sciences.

Each of these activities would be more than enough for a normal carrier. Therefore it is truly remarkable how M. Ohya could combine all of them and make them compatible with the additional duties of Dean of the Graduate School of Science and Technology, Director of the Education Research Center for Information Science and Technology, Member of the Committee of the International Institute for Advanced Studies,... which he served during several years, sometimes in coincidence.

Nowadays the pressure of specialization is very high. One of its psychological effects is that many scientists remain captured in a narrow horizon and become existentially unable to appreciate results beyond such an horizon.

There is a sociological counterpart of this effect which makes some people unable to communicate, or to enjoy communication, with people outside a narrowly outlined community. Many scientists, even of very high technical level, do not succeed to avoid these traps.

Therefore those, like M. Ohya, who succeed in maintaining a broad vision of science, while struggling to realize this vision without abdicating the luxury of having a wide human and scientific taste, are rare and precious examples for the future generations.

2. Comments on some of M. Ohya's scientific results

Classical information theory made its transition from a purely engineering and technological level to a well established mathematical theory, based on non trivial theorems, in the 1960's with the work of C. Schannon who, developing earlier intuitions of N. Wiener, introduced the notion of entropy as a measure of information.

The idea of a quantum extension of Shannon's results naturally arose in the context of quantum signal processing and, already in the late 1960's, was discussed by several authors (among whom: Gordon, Lax, Louisell, ...).

Nowadays quantum information, enriched by the engineering appeal of the quantum computer programme, has become a widely practiced scientific discipline and the object of huge investments from all industrialized countries.

But in the early 1980's these investigations were cultivated by a handful of pioneer researchers among whom M. Ohya, who was one following the tracks of his former advisor, Professor Umegaki, the first one to extend the Kulback-Leibler relative information by defining the relative entropy in finite and sigma-finite von Neumann algebras. Later on M. Ohya extended von Neumann's entropy to general C^*–algebras [18].

M. Ohya quantum extension of the notions of mutual entropy and compound channel [Ohya83a], whose classical counterparts play a basic role in Shannon's approach to information theory, is one of the earliest, and still among the most important of his contributions to quantum information theory. A few years later the two notions of compound channel and of transition expectation were combined into the unifying notion of lifting [1] which has now found several applications in different fields.

Inspired by Shannon's work, Kolmogorov introduced entropic type quantities as a measure of complexity and as a tool in approximation theory. These were generalized by M. Ohya to the quantum case [16] and also in the classical domain with the introduction of the notion of chaos degree. This notion found unexpected experimental support in econometrics, (joint papers of M.O. and T. Matsuoka) and in medicine (joint papers of M.O. and K. Sato).

Kolmogorov also introduced a notion of dynamical entropy and proved that it provided the first example of invariants for dynamical systems finer than the spectral invariants introduced by Halmos and von Neumann. The fundamental step in Kolmogorov's construction was to associate a family of finite Markov chains to the given dynamical system. He calculated the entropy of these chains, which can be done explicitly, and taking the supremum over all these chains, he obtained a number depending only on the dynamical system.

Several quantum generalizations of Kolmogorov dynamical entropy have been proposed, but they were lacking this direct and canonical connection with the theory of Markov chains. This motivated the paper [2] where, starting from a quantum dynamical system and extended to this case Kolmogorov's original construction using quantum Markov chains rather than classical. The computation of the entropy of a quantum Markov chain is not so simple as in the classical case, however in several concrete examples it was possible to calculate this entropy or at least to produce a lower bound for it [3]. These results were extended to irreversible dynamical systems by Ohya, Kossakowski and Watanabe [12].

Another anticipating intuition of M. Ohya was the recognition that the theoretical studies on quantum information and communication could find, in quantum computer and quantum cryptography a concrete realization and application. Thus he himself and his group begun to actively work in this field long before it became a fashionable scientific trend.

I consider the channel theoretical formulation of teleportation, that M. Ohya had developed in collaboration with Kei Inoue, Hiroki Suyari and Noboru Watanabe [14],

[17], as the first important contribution of Ohya's group to quantum computer. This has now become the standard mathematical formulation of teleportation

While the original formulation of this notion, due to Bennet and Brassard, looked like a very specific example and did not suggest any insight into its general structure, the channel theoretical formulation of Ohya's group really captured the mathematical essence of teleportation thus opening the way to its realization in arbitrary dimensions.

On this problem several groups were independently working in different parts of the world but, at that moment, it was still open. The first constructive solution of this problem was given in the paper [4] which also included the first proof of the fact that the teleportation scheme must be based on a maximally entangled state in a given basis. Such a result was quite unexpected at that time and was later rediscovered and generalized in various ways by several authors.

Another interesting result, obtained by Ohya in collaboration with Masuda [13], is the possibility to solve the SAT problem in polynomial time by quantum computer. Sharp estimates for the number of steps needed in the Ohya–Masuda algorithm can be found in [5], [6].

However the complete realization of this algorithm still had a problem: one could not exclude the possibility that a positive solution (i.e. one guaranteeing satisfiability) would appear with such a small probability to be indistinguishable from a negative one. The proposal to amplify these probabilities with classical chaos methods, was advanced by Ohya and Volovich. A different, purely quantum method was later proposed in [8] This method is based on the new idea of state–adaptive dynamics applied to the quantum state obtained as output of the Ohya–Masuda algorithm. This is used to construct a physically implementable interaction (hence the name state–adaptive dynamics) capable to drive a system to a stationary state. Thus, by discriminating the limit stationary states, one can discriminate between a positive or negative solution of the SAT problem.

Finally the new notion of degree of entanglement, introduced by M. Ohya and T. Matsuoka [15] and providing a new criterion for entanglement, much easier to verify in concrete cases than previously introduced criteria, was applied in [9] to prove the entangled nature of certain quantum Markov chains.

The above mentioned results represent only a tiny fraction of M. Ohya's scientific production. My selection criterion was very simple: I mentioned only those papers in which either I was directly involved or which inspired some of our joint papers. This kind of joint retrospective is a way to express my pleasure for a tradition of collaboration which is lasting since more that 15 years and which is now evolving in new directions.

Roma, November 2007 **Luigi Accardi**

References

1. Accardi L., Ohya M.: Compound channels, transition expectations and liftings, Applied Mathematics Optimization 39 (1999) 33–59 Volterra preprint N. 75 (1991)
2. Accardi L., Ohya M., Watanabe N.: Note on quantum dynamical entropies, Reports on mathematical physics, 38 N. 3 (1996) 457–469
3. Accardi L., Ohya M., Watanabe N.: Dynamical entropy through quantum Markov chains, Open Systems and Information Dynamics 4 (1997) 71–87
4. Accardi L., Ohya M.: Teleportation of general quantum states, in: Quantum Information, T. Hida, K. Saito (eds.) World Scientific (1999) 59–70 Invited talk to the: International Conference on quantum information and computer, Meijo University 1998 Preprint Volterra N. 354 (1999)
5. Accardi L., Sabbadini R.: On the Ohya–Masuda quantum SAT Algorithm, in: Proceedings International Conference "Unconventional Models of Computations", I. Antoniou, C.S. Calude, M. Dinneen (eds.) Springer (2001) Preprint Volterra, N. 432 (2000)
6. Accardi L., Sabbadini R.: A Generalization of Grover's Algorithm, Proceedings International Conference: Quantum Information III, Meijo University, Nagoya, 27–31 March (2001) World Scientific (2002) qu-phys 0012143 Preprint Volterra, N. 444 (2001)
7. Accardi L., Lu Y.G., Volovich I.: Quantum Theory and its Stochastic Limit, Springer Verlag (2002) Japanese translation Tokyo–Springer, to appear
8. Luigi Accardi, Masanori Ohya: A stochastic limit approach to the SAT problem, Open systems and Information Dynamics, 11 (3) (2004) 219-233
9. L. Accardi, T. Matsuoka, M. Ohya: Entangled Markov chains are indeed entangled, Infinite Dimensional Analysis, Quantum Probability and Related Topics 9 (2006) 379-390
10. Ohya M., Petz D.: Quantum entropy and its use, Springer, Texts and Monographs in Physics (1993)
11. M. Ohya: Mathematical Foundation of Quantum Computer, Maruzen Publ. Company (1998)
12. Ohya M., Kossakowski A., Watanabe N.: Quantum dynamical entropy for completely positive map, Infinite dimensional analysis, quantum probability and related topics, 1 (2) (1999) 267–282 Preprint (1998)
13. Ohya M., Masuda N.: NP problem in quantum algorithm, Open Systems and Information Dynamics, Vol.7, No.1, 33-39, 2000. arXiv:quant–ph/9809074 v2 13 dec (1998)
14. Ohya M., Watanabe N.: On the mathematical treatment of the Fredkin–Toffoli–Milburn gate, Physica D 120 (1998) 206–213
15. M. Ohya, T. Matsuoka: Quantum Entangled State and Its Characterization, Foundation and Probability and Physics-3, AIP Proceedings 750 (2005) 298-306
16. Ohya M.: Complexities and their applications to characterization of chaos, Int. Journ. of Theort. Phy., 37 (1998) 495
17. Ohya M., Inoue K., Suyari H.: Characterization of quantum teleportation processes by nonlinear quantum channel and quantum mutual entropy, SUT Preprint (1997)
18. M.Ohya (1984) Entropy Transmission in C*-dynamical systems, J. Math. Anal.Appl., 100, No.1, 222-235"
19. Ohya M.: On compound state and mutual information in quantum information theory, IEEE Trans. Information Theory, 29 (1983) 770–777

Preface

I am indeed very glad to be able to write a few words as a preface to the collection of selected papers of Professor Ohya. I was in Noda many times on invitation of Prof. Ohya and this gave me many oppotunities of active scientific contacts with him. He is one of the creators and propagators of Information Dynamics as a new branch of modern science. He also came to Torun many times. Open Systems, increases of entropy and entropy as a measure of information, which were the main topics that united our scientific interests and collaboration. Tokyo University of Science was since XIX century, and is at present, an important center of research, especially in mathemetical physics. He is now the editor-in-chief of an international journal "Open Systems and Information Dynamics" which we organized together with Professors Accardi, Kossakowski, Jamiolkowski and others. Professor Ohya was always one of the most active in these fields. On the occasion of his 60th birth anniversary I wish him very cordially the further scientific activity and the best health and greatest satisfaction in science and personal life. His papers give a permanent contribution to Quantum Information, Information Dynamics and Mathematical Physics, but we hope that this will continue for many years in the future. I would like to invite the readers to the interesting papers and reviews of Professor Masanori Ohya as the important steps in modern development of information science.

Torun, June 2007 **Roman. S. Ingarden**

Preface

On the occasion to celebrate the 60th anniversary of Prof. Masanori Ohya's birth, I would like to recollect my personal history of friendship with him. After working on the formulation of quantum fields at finite temperatures (I.O., Ann. Phys. 1981), I became interested in the mutual relations between micro- and macroscopic aspects in quantum theory. So it was natural for me to be led to Prof. Ohya's book, "Quantum Entropy", written (in Japanese) with Prof. Umegaki, immediately after its publication in 1984, which impressed and attracted me very much. This book (which deserves still being translated into international languages) strongly convinced me that certain general mathematical method and framework should be established for treating the problems on the relations between micro- and macroscopic aspects in nature.

Then I decided to meet him to make request for his cooperation in a research project along this line. He kindly accepted my request, and a series of workshops named "Quantum-Field Theoretical Approaches to Evolution Dynamics" started in 1986 at Research Institute for Fundamental Physics (= RIFP and, at present, Yukawa Institute for Theoretical Physics) in Kyoto University. This was really a new interdisciplinary project, covering quantum field theory, measurement theory, quantum theory of information and communications, quantum optics, cosmology and solid state physics, etc., which continued for five years until 1990 at RIFP. After 1991 its host institute was changed to Research Institute for Mathematical Sciences, Kyoto University, and Prof. Ohya served as its chief organizer for thirteen years from 1992 until 2004 when I succeeded the job from him until now.

During this period, of more than twenty years, Prof. Ohya has made essential and important contributions to our science communities, as seen in this volume of his selected papers, including the proposals of such notions as *information dynamics, lifting of channels, chaos degree, adaptive dynamics, degree of entanglement* and so on, which have shed new lights on many fundamental intersecting areas of quantum and of information. If it were not for the fruitful communications and friendship with him, I could never have dared to step into my own new projects based upon"Micro-Macro duality".

Kyoto, November 2007 **Izumi Ojima**

Contents

Introduction

Classical information theory cannot be applied to information communication for micro-scopic objects treated in quantum mechanics. For such study, quantum information theory has been developed. It is based on quantum probability and it is a theory for expressing information by quantum states. Prof. Masanori Ohya studied many topics, for more than thirty years, related to quantum entropy, quantum information, chaos dynamics and life science. His main accomplishments are as follows.

Elucidation of Mathematical Bases of Quantum Channels:

M. Ohya rigorously studied quantum state change including irreversible processes (1970's) in quantum statistical physics and information communication (after 1980) by generalizing the state change to quantum channel. As a by-product of this study, he could succeed to derive the error probability for optical communication processes.

Formulation of Quantum Mutual Information (Entropy):

M. Ohya formulated the quantum mutual entropy (information), which is a natural extension of Shannon's mutual information. Then he generalized it within C*-algebraic framework including several other definitions of both the entropy and the mutual entropy. Thereby, one could define and analyze the channel capacity rigorously. Various studies of quantum entropy including some of the above are discussed in the book "Quantum Entropy and its Use" with D. Petz.

Information Dynamics:

M. Ohya proposed a new concept called Information Dynamics (ID) to integrate various dynamics of the systems with complexities. The mathematical basis and its applications of ID are summarized in the book "Information Dynamics and Open Systems" with R. Ingarden and A. Kossakowski.

Analysis of Quantum Teleportation:

The quantum teleportation model transmitting a quantum state itself was proposed by Bennett et al in early 90's, which is an epoch-making communication method for security. To overcome some difficulties of the previous models, K. Fichtner and M. Ohya made a new model of the quantum teleportation in Bose Fock space, where the incomplete teleportation has been introduced. In 2006, Kossakowski and M. Ohya proposed a new scheme of the teleportation in which the teleportation channel becomes linear and the complete teleportation is possible even for non-maximal entangled states.

Quantum Algorithm:

The quantum computer makes the computing speed extremely high. In 1999-2000, M. Ohya and Masuda wrote the quantum algorithm for the SAT problem, one of the NP-complete problems. Around 2000-2002, M. Ohya and I. Volovich found an algorithm solving the NP-complete problem in polynomial time by combining with the quantum algorithm and the state change in chaotic dynamics. Further, L. Accardi and M. Ohya showed another algorithm for the same problem based on the idea of Adaptive Dynamics in 2005. Recently these algorithms can be written in the language of generalized quantum Turing machine.

Proposal of Adaptive Dynamics:

Existence of substance, more generally of any object, depends on how it is observed. How can one describe this fact mathematically? What is a philosophical basis for the mechanism of the existence? To answer these questions, M. Ohya proposed the Adaptive Dynamics. As the applications of the idea of the Adaptive Dynamics, he succeed to study chaos and quantum algorithms.

Life Science:

M. Ohya started to study Life Science, in particular, Bio-Informatics about 20 years ago. He introduced new measures (entropy evolution rate, a measure of coding structure) to study the evolution (mutation) of species in the level of genome and protein (amino-acid sequence). He and coworkers applied these measures to make phylogenetic trees and to study the change of HIV virus. Moreover, he is now interested in making a model of brain to study its function with K. Fichtner and W. Freudenberg. M. Ohya founded a research center call "Quantum Bio-Informatics (QBIC)", where many researchers from different fields come together and look for new logic (theory) studying quantum information and life sciences.

M. Ohya has been invited from several universities all over the world and he gave many invited talks at international conferences. He has been the editors of several international journals and the representatives of various research projects.

Tokyo, December 2007 **Editors**

Curriculum Vitae

Date/Place of Birth: March 21, 1947, Chiba, Japan
Address/Telephone: Tokyo University of Science, Department of Information Sciences, 2641 Yamazaki, Noda City, Chiba 278-8510, Japan
TEL. +81/471/229357, **FAX.** +81/471/245948
E-Mail. ohya@rs.noda.tus.ac.jp

Education:

1970, University of Tokyo, Dept. of Physics (68-70), Dept. of Mathematics (67-68)
1976, Ph.D., University of Rochester (supervised by Prof. Gerard G. Emch)
D.Sc., Tokyo Institute of Technology (supervised by Prof. Hisaharu Umegaki)

Appointments:

1977-1978 Assistant Professor, Department of Information Sciences, Tokyo University of Science
1978-1982 Junior Associate Professor, Department of Information Sciences, Tokyo University of Science
1982-1987 Associate Professor, Department of Information Sciences, Tokyo University of Science
1987-present Professor, Department of Information Sciences, Tokyo University of Science
2000-2003 Director, Department of Information Sciences, Tokyo University of Science
2001-2005 Director, Frontier Research Center of Computational Science, Tokyo University of Science
2003-2006 Director, Education Reserach Organization for Information Science and Technology, Tokyo University of Science
2004-2006 Dean, Graduate School of Science and Technology, Tokyo University of Science
2002-2006 Committee, International Institute of Advanced Study
2006-present Dean, Science and Technology, Tokyo University of Science
2006-present Director, Quantum Bio-Informatics Center, Tokyo University of Science

Academic Works: (1) He is (or was) the member of the following Societies: Mathematical Society of Japan; The Institute of Electronics, Information and Communication Engeeners (IEICE); The Physical Society of Japan; The Biophysical Society of Japan; American Physical Society (APS); The New York Academy of Sciences. (2) He is (or was); Vice-President of International Society of Information Dynamics; Committee of Information Theory Group of IEICE; Fellow of Mathematical Society of Japan; Chief of Analysis Group in

Mathematical Society of Japan; Representative of Research Projects of RIMS (Research Institute for Mathematical Sciences, Kyoto University); Representative of Research Projects of International Institute of Advanced Study; Special Research Member of International Institute of Advanced Study; Some Committee of the Japanese Ministry of Education, Culture, Sports, Science and Technology. (3) He is (or was) the editors of the following International Journals: Reports on Mathematical Physics; Amino Acid; Infinite Dimensional Analysis, Quantum Probability and Related Topics; Open Systems and Information Dynamics (Chief Editor).

Part Time Professor: Tokyo Institute of Technology, Ochanomizu University, Kyoto University, Hokkaido University, Chiba University, Tsukuba University and others

Visiting Professor: Università di Roma II, Copernicus University, Jena University and others.

Research: Main scientific interests are connected with quantum entropy, quantum information theory, quantum computer, mathematical physics and information genetics.

Selected Papers of M. Ohya

ADAPTIVE DYNAMICS AND ITS APPLICATIONS TO CHAOS AND NPC PROBLEM

MASANORI OHYA

Department of Information Sciences, Tokyo University of Science, 2641, Yamazaki, Noda City, Chiba, Japan

I will discuss the following four (1)-(4) below from both mathematical and philosophical views: (1) What is (or do we mean) the understanding of the existence ? (2) We propose "Adaptive dynamics" to understand the existence. (3) The adaptive dynamics can be used to describe chaos. (4) The adaptive dynamics is applied to the SAT Quantum Algorithm to solve the NP complete problem.

1. Introduction

Natural science is not a copy of nature itself but is a mean to understand several natural phenomena for human beings. Thus it is a sort of a story which we made for recognition of nature, but it is a story beyond each person and personal experience, so that it should have a universality in that sense. Following Wilde, "Nature imitates arts". It is the only way for us to come face to face with nature, which is not our conceit but our limit. After discovery of quantum mechanics, we are forced to face with the facts like the above although there are not many people to feel this difficulty and to look for new description of nature overcoming this difficulty.

In order to understand physical phenomena or other phenomena of human beings, it needs to examine, in various views not only physical but also observational, the ways how object exists and how we can recognize the object. (1) Existence itself, (2) its indicating phenomena and (3) their recognition have been extensively studied by philosophers and some physicists. Explaining (i.e., defining and describing) these three is essentially important for not only philosophy but also physics, information and all other sciences. It is a time for us to explain these three in more rigorous ways beyond usual philosophical and mathematical demonstrations, that is, by finding a method standing on a higher stage made from dialectic mixing of philosophy, mathematics and some others, although its fulfillment is difficult.

First of all, I will mention briefly "phenomenology" of Husserl and "existentialism" of Sartre.

Before Husserl, in the theory of existence by like Kant or Hegel, philosopher could not neglect the existence transcendent (e.g., God), so that they had to distinguish the existence of essence and the existence of phenomenon. An appearance of the essence is a phenomenon and a description of phenomena only is not enough to reach the essence. For instance, Hegel said, "In order to reach the essence it is necessary for mind to develop itself dialectically". In any case, the dualism of the existence of essence and phenomena has been a basis for several philosophies until materialism of Marx and phenomenology of Husserl appeared.

Husserl was against to the idea that the essence of existence is transcendent objects, and he considered that the essence is a chain of phenomena and its integral. The essence is an accurate report of all data (of phenomena) obtained through the stream of consciousness. His consciousness has two characters, "noesis" and "noema". The noesis is the operative part of consciousness to phenomena (objects), in other words, the acting consciousness on objects, and the noema is the object of consciousness experience, i.e., the results obtained by the noesis. His phenomenology is the new dualism of consciousness, but he avoids the existence transcendent, instead, he likes to go to the things themselves.

Under strong influence of Husserl, there appeared several philosophies named "exsitentialism" of Heidegger, of Sartre and of others. Sartre said, "Existence precedes essence". Sartre was affected by "Cogito" of Descartes, and he found two aspects of existence (being) in his famous book "L'Etre et le Neant (Being and Nothingness)", one of which is the "being in itself" and another is the "being for itself". The first one is the being as it is, opaque (nontransparent) being like physical matter itself, being which does not have any connection with another being, being without reason for being, etc. Another one is the being as it is not, transparent being like consciousness, being with cause for being itself, being making any being-in-itself as being, etc. Sartre explained several forms of existence by his new dualism of existence; being-in-itself and being-for-itself. His main concern is being and becoming of human beings, various appearance of emotion, life and ethics, so his expression of philosophy is rather rhetoric and literal. However I will explain that his idea can be applied for the proper interpretation of quantum entropy and dynamics.

1.1. *Entropy, Information in Classical and Quantum World*

Physics is considered as "theory of matter", equivalently, "theory of existence in itself". Information theory (Entropy theory) is considered as "theory of events", so that it will be considered as "theory of changes". Quantum Information can be regarded as a synthesis of these two. The key concept of quantum information bridging between matter and event, so between two modes of being, is "entropy", which was introduced by Shannon in classical systems and by von Neumann in quantum systems. I will discuss how this concept of entropy has a deep connection with the mode of existence considered in the beginning of this section.

According to Shannon, information is related to uncertainty, so it is described by entropy: Information=Uncertainty=Entropy, and dissolution of uncertainty can be regarded as acquisition of information. Historically the concept of entropy was introduced to describe the flow of heat, then it is recognized that the entropy describes chaos or uncertainty of a system. A system is described by a state such as probability measure or density operator, which is a rather abstract concept not belonging to an object (observable) to be measured but a mean to get measured values. Thus the entropy is defined through a state of a system, which implies that the entropy is not an object considered in usual classical physics and it is an existence coming along action of "observation". It is close to (actually more) than a description of chaos which is a mode taken by consciousness to the being-in-itsel. (I will discuss on chaos in Sec.3.) Therefore the entropy can be considered as a representation (formulation) of consciousness involving an observation of a certain object. The concept of entropy is not a direct expression of phenomena associated with a being-in-itself but is a being having an appearance of consciousness to phenomena of a being-in-itself, so that the mode of existence for the entropy is different from two modes of being proposed by Sartre and this third mode is in between being in and for itself. The rigorous (mathematical) study with this third mode of being might be important to solve some problems which we face in several fields.

1.2. *Schematic Expression of Understanding*

Metaphysics, idea, feeling, thought are applied to various existence (series of phenomena), which causes understanding (recognition, theory). To understand a physical system, the usual method, often called "Reductionism", is to divide the system into its elements and to study their relations and combinations, which causes the understanding of the whole system.

Our method is one adding "how to see objects (existence)" to the usual reductionism, so that our method is a mathematical realisation of modern philisophy. The fact "how to see objects" is strongly related to setting the mode for observation, such as selection of phenomena and operation for recognition. Our method is called *"Adaptive dynamics" or "Adaptive scheme"* for understanding the existence.

In this paper, we discuss the conceptual frame of AD and some examples in chaos and quantum algorithm, which are first steps to go to our final aims making complete mahematics for "adaptivity".

2. Adaptive Dynamics-Conceptual Meaning-

The adaptive dynamics has two aspects, one of which is the "observable-adaptivity" and another is the "state-adaptivity".

The idea of observable-adaptivity comes from the papers[38,49,50] studying chaos. We claimed that any observation will be unrelated or even contradicted to mathematical universalities such as taking limits, sup, inf, etc. Observation of chaos is a result due to taking suitable scales of, for example, time, distance or domain, and it will not be possible in the limiting cases.

The idea of the state-adaptivity is implicitly started in constructing a compound state for quantum communication[4,41,42,46]and in Accardi's Chameleon dynamics.[1] This adaptivity can be used to solve a pending problem for more than 30 years whether there exists an algorithm solving a NP complete problem in polynomial time. We found such algorithms first by quantum chaos algorithm[54,55] and secondly by the adaptive dynamics[3] based on quantum algorithm of the SAT[7,52].

We will discuss a bit more on the meaning of the adaptivity for each topics mentioned above.

2.1. *Description of chaos*

There exist several reports saying that one can observe chaos in nature, which are nothing but to report how one could observe the phenomena in specified conditions. It has been difficult to find a satisfactory theory (mathematics) to explain such various chaotic phenomena in a unified way.

An idea describing chaos of a phenomenon is to find some divergence of orbits produced by the dynamics explaining the phenomenon. However to explain such divergence from the differential equation of motion describing the dynamics is often difficult, so that one take (make) a difference equation from that differential equation, for which one has to take a certain time interval τ between two steps of dynamics, that is, one needs a processing discretizing time for observing the chaos. In laboratory, any observation is done in finite size for both time and space, however one believes that natural phenomena do not depend on these sizes how small they are, so that most of mathematics (theory) has been made as free from the sizes taken in laboratory. Therefore mathematical terminologies such as "lim", "sup", "inf" are very often used to define some quantities measuring chaos, and many phenomena showing chaos have been remained unexplained.

In the papers,[38,49,50] we took the opposite position, that is, any observation will be unrelated or even contradicted to such limits. Observation of chaos is a result due to taking suitable scales of, for example, time, distance or domain, and it will not be possible in the limiting cases. In other words, as discussed in Section 1, it is very natural to consider that observation itself plays a similar role of "noesis" of Husserl and the mode of its existence is a "being-for-itself", that is, observation itself can not exist as it is but it exists only through the results (phenomena) of objects obtained by it. Phenomena can not be phenomena without observing them, so to explain the phenomena like chaos it is necessary to find a dynamics with observation.

We claimed that most of chaos are scale-dependent phenomena, so the definition of a degree measuring chaos should depend on certain scales taken and more generally it is important to find mathematics containing the rules (dynamics) of both object and observation, which is "Adaptive dynamics".

Concerning the definition of a criterion measuring chaos, Information Dynamics,[30,47] a scheme to describe many different types of complex systems, can be applied. I introduced a quantity measuring chaos by means of the complexities of ID, and I called it a chaos degree.[30,35,43] Using this degree in adaptive dynamics, we can explain or produce many different types of chaos.

2.2. *Chameleon dynamics*

Accardi considered a problem whether it is possible to explain quantum effects (e.g., EPR(Einstein-Polodoski-Rosen) correlation) by a sort of classical dynamics.[1] He could find a dynamics positively solving the above problem, and he called it "Chameleon dynamics". He considered two systems having their own particles, initially correlated and later separated. After some time, each particle interacts with a measurement apparatus independently.

By the chameleon effect the dynamical evolution of each particle depends on the setting of the nearby apparatus, but not on the setting of the apparatus interacting with the other particle (locality). The he succeeded to reproduce the EPR correlations by this "chameleon dynamics", which is one of surprising results which many physicists were searching for long time. The explicit construction of the dynamics was done in the paper.[2] The interaction between a particle and an apparatus depends on the setting of the apparatus, so that the chameleon dynamics is an adaptive dynamics.

2.3. *Quantum SAT algorithm*

Although the ability of computer is highly progressed, there are several problems which may not be solved effectively, namely, in polynomial time. Among such problems, NP problem and NP complete problem are fundamental. It is known that all NP complete problems are equivalent and an essential question is *"whether there exists an algorithm to solve an NP complete problem in polynomial time"*. They have been studied for decades and for which all known algorithms have an exponential running time in the length of the input so far. The P-problem and NP-problem are those to be considered as follows[10,16,55]

Let us remind what the P-problem and the NP-problem are: Let n be the size of input.

(1)A P-problem is a problem whose time needed for solving the problem is at worst of polynomial time of n. Equivalently, it is a problem which can be recognized in a polynomial time of n by deterministic Turing machine.

(2)An NP-problem is a problem that can be solved in polynomial time by a nondeterministic Turing machine. This can be understood as follows: Let consider a problem to find a solution of $f(x) = 0$. We can check in polynomial time of n whether x_0 is a solution of $f(x) = 0$, but we do not know whether we can find the solution of $f(x) = 0$ in polynomial time of n.

(3) An NP-complete problem is a problem polynomially transformed NP-problem.

There are a lot of NP complete problems, e.g., satisfiable (SAT) problem, travel salesman problem and so on. It is known that all NP complete (NPC for short) problems are equivalent and have been studied for decades, for which all known algorithms have an exponential running time in the length of the input so far. For an essential question to be asked for more than

30 years, that is, *the existence of an algorithm to solve an NP complete problem in polynomial time,* we found two different algorithms.[3,52,54,55]

In[52] we discussed the quantum algorithm of the SAT problem and pointed out that the SAT problem, hence all other NP problems, can be solved in polynomial time by quantum computer if the superposition of two orthogonal vectors $|0\rangle$ and $|1\rangle$ is physically detected. However this detection is considered not to be possible in the present technology. The problem to be overcome is how to distinguish the pure vector $|0\rangle$ from the superposed one $\alpha|0\rangle + \beta|1\rangle$, obtained by our SAT-quantum algorithm, if β is not zero but very small. If such a distinction is possible, then we can solve the NPC problem in the polynomial time.

2.3.1. *Chaos SAT algorithm*

It will be not possible to amplify, by a unitary transformation (usual quantum algorithm), the above small positive $q \equiv |\beta|^2$ into suitable large one to be detected, e.g., $q > 1/2$,with staying $q = 0$ as it is. In the papers[54,55] we proposed to use the output of the quantum computer as an input for another device involving chaotic dynamics, that is, to combine quantum computer with a chaotic dynamics amplifier. We showed that this combination (nonlinear chaos amplifier with the quantum algorithm) provides us with a mathematical algorithm solving NP=P. This algorithm of Ohya and Volovich is going beyond usual (unitary) quantum Turing algorithm, but there exists a generalized quantum Turing machine in which the OV chaos algorithm can be treated.[33,34]

2.3.2. *Adaptive SAT algorithm*

We applied the "adaptive dynamics" to the OM SAT algorithm.[3] That is, the state-adaptive dynamics is applied to the OM SAT algorithm and rescaled the time in the dynamics by the stochastic limit, then we could show that the same amplification (distinction between $q > 0$ and $q = 0$) is possible by unitary adaptive dynamics with the stochastic limit. Its details will be discussed in Section 4. The AO adaptive algorithm can be treated in the frame of generalized quantum Turing machine as a linear TM.

2.4. *Summary*

We summarize our idea on the adaptive dynamics as follows: The mathematical definition of adaptive system proposed was in terms of observables (resp. states).

Two adaptivities are characterized (defined) as follows:

The observable-adaptive dynamics is a dynamics characterized by one of the following two: (1) Measurement depends on how to see an observable to be measured. (2)The interaction between two systems depends on how a fixed observable exists.

The state-adaptive dynamics is a dynamics characterized by one of the following two: (1)Measurement depends on how the state to be used exists. (2)The correlation between two systems interaction depends on the state of at least one of the systems at the instant in which the interaction is switched on.

Examples of the state-adaptivity are seen in compound states[42,48] (or nonlinear liftings[4]) studying quantum communication and in an algorithm solving NP complete problem in polynomial time with stochastic limit.[3]

Examples of the observable-adaptivity are used to understand chaos[38,43] and examine violation of Bell's inequality.[2]

Notice that the definitions of adaptivity make sense both for classical and for quantum systems.

The difference between the property (2) of the state–adaptive system and nonlinear dynamical system should be remarked here:

(i) In nonlinear dynamical systems (such as those whose evolution is described by the Boltzmann equation, or nonlinear Schrödinger equation, etc) the interaction Hamiltonian depends on the state at each time t: $H_I = H_I(\rho_t)$ $(\forall t)$.

(ii) In state–adaptive dynamical systems, the interaction Hamiltonian depends on the state only at each time $t = 0$: $H_I = H_I(\rho_0)$.

The latter class of systems describes the following physical situation: at time $t = -T$ $(T > 0)$ a system S is prepared in a state ψ_{-T} and in the time interval $[-T, 0]$ it evolves according to a fixed (free) dynamics $U_{[-T,0]}$ so that its state at time 0 is $U_{[-T,0]}\psi_{-T} =: \psi_0$ At time $t = 0$ an interaction with another system R is switched on and this interaction depends on the state ψ_0: $H_I = H_I(\psi_0)$. If we interpret the system R as *environment*, we can say that the above interaction describes the response of the environment to the state of the system S. Therefore the adaptive dynamics can be linear and it contains the non-linear dynamics in many occasions.

3. Adaptive Dynamics Describing Chaos

There exist several approaches in the study of chaotic behavior of dynamical systems using the concepts such as (1) entropy and dynamical entropy, (2) Chaitin's complexity, (3) Lyapunov exponent (4) fractal dimension (5) bifurcation (6) ergodicity. However these concepts are rather independently used in each case. In 1991, the present author proposed Information Dynamics[30,45,47] to treat such chaotic behavior of systems from a common standing point, in which a chaos degree to measure the chaos appeared in dynamical systems is defined by means of two complexities in Information Dynamics.[44,45,55] In particular, among several chaos degrees, the entropic chaos degree was introduced in[43] and it has been applied to several dynamical systems.[28,29,43] For instance, semiclassical properties and chaos degree for quantum baker's map have been considered in.[27,28]

3.1. *Information Dynamics*

Information dynamics (ID for short) is a synthesis of the dynamics of state change and the complexity of states. It is a trial to provide a new view for the study of chaotic behavior of systems. We briefly review what ID is.

Let $(\mathcal{A}, \mathfrak{S}, \alpha(G))$ be an input (or initial) system and $(\overline{\mathcal{A}}, \overline{\mathfrak{S}}, \overline{\alpha}(\overline{G}))$ be an output (or final) system. Here $\mathcal{A}$ is a set of some objects to be observed and $\mathfrak{S}$ is a set of some means to get the observed value, $\alpha(G)$ describes a certain evolution of system. Often we have $\mathcal{A} = \overline{\mathcal{A}}$, $\mathfrak{S} = \overline{\mathfrak{S}}$, $\alpha = \overline{\alpha}$. Therefore we claim

[Giving a mathematical structure to input and output triples
≡ Having a theory]

Let $(\mathcal{A}_T, \mathfrak{S}_T, \alpha_T(G_T))$ be the total system of $(\mathcal{A}, \mathfrak{S}, \alpha)$ and $(\overline{\mathcal{A}}, \overline{\mathfrak{S}}, \overline{\alpha})$, and $\mathcal{S}$ be a subset of $\mathfrak{S}$ in which we are measuring observables (e.g., $\mathcal{S}$ is the set of all KMS or stationary states in C*-system).

The dynamics of state change is described by a channel sending a state to another state Λ: $\mathfrak{S} \to \overline{\mathfrak{S}}$ (sometimes $\mathfrak{S} \to \mathfrak{S}$). Moreover ID contains two complexities, which are denoted by C and T. C is the complexity of a state φ measured from a reference system $\mathcal{S}$, in which we actually observe the objects in $\mathcal{A}$ and T is the transmitted complexity associated with a state change $\varphi \to \Lambda\varphi$, both of which should satisfy the following properties:

⟨**Axioms of complexities**⟩

(i) For any $\varphi \in \mathcal{S} \subset \mathfrak{S}$,

$$C^{\mathcal{S}}(\varphi) \geq 0, \ T^{\mathcal{S}}(\varphi; \Lambda) \geq 0$$

(ii) For any disjoint (in a proper sense) bijection $j : ex\mathcal{S} \to ex\mathcal{S}$, the set of all extremal points of $\mathcal{S}$,

$$C^{j(\mathcal{S})}(j(\varphi)) = C^{\mathcal{S}}(\varphi)$$

$$T^{j(S)}(j(\varphi);\Lambda) = T^{S}(\varphi;\Lambda)$$

(iii) For $\Phi \equiv \varphi \otimes \psi \in \mathcal{S}_t \subset \mathfrak{S}_t$, $\psi \in \overline{\mathcal{S}} \subset \overline{\mathfrak{S}}$ (here $\otimes$ is a properly defined product)

$$C^{\mathcal{S}_t}(\Phi) = C^{\mathcal{S}}(\varphi) + C^{\overline{\mathcal{S}}}(\psi)$$

(iv) $0 \le T^{\mathcal{S}}(\varphi;\Lambda) \le C^{\mathcal{S}}(\varphi)$
(v) $T^{\mathcal{S}}(\varphi; id) = C^{\mathcal{S}}(\varphi)$, where "$id$" is an identity map from $\mathfrak{S}$ to $\mathfrak{S}$.

Instead of (iii), when " $\Phi \in \mathcal{S}_T \subset \mathfrak{S}_T$, put $\varphi \equiv \Phi \restriction \mathcal{A}$, $\psi \equiv \Phi \restriction \overline{\mathcal{A}}$ (i.e., the restriction of Φ to $\mathcal{A}$ and $\overline{\mathcal{A}}$, respectively), $C^{\mathcal{S}_t}(\Phi) \le C^{\mathcal{S}}(\varphi) + C^{\overline{\mathcal{S}}}(\psi)$ " is satisfied, C and T is called a pair of strong complexity. Then ID is defined as follows:

Definition 3.1. Information Dynamics is described by

$$\left(\mathcal{A}, \mathfrak{S}, \alpha(G); \overline{\mathcal{A}}, \overline{\mathfrak{S}}, \overline{\alpha}(\overline{G}); \Lambda; C^{\mathcal{S}}(\varphi), T^{\mathcal{S}}(\varphi;\Lambda)\right)$$

and some relations R among them.

Therefore, in the framework of ID, we have to

(i) mathematically determine $\left(\mathcal{A}, \mathfrak{S}, \alpha(G); \overline{\mathcal{A}}, \overline{\mathfrak{S}}, \overline{\alpha}(\overline{G})\right)$
(ii) choose Λ and R, and
(iii) define $C^{\mathcal{S}}(\varphi)$, $T^{\mathcal{S}}(\varphi;\Lambda)$.

3.1.1. *State change and complexities*

ID contains the dynamics of state change as its part. A state change is mathematically described by a unitary evolution, a semi-group dynamics, generally, a channeling transformation (it is simply called "channel"). Let input and output triples $(\mathcal{A}, \mathfrak{S}, \alpha(G))$ and $(\overline{\mathcal{A}}, \overline{\mathfrak{S}}, \overline{\alpha}(\overline{G}))$ be C*-dynamical systems; that is, $\mathcal{A}$ is a C^*-algebra[22] and $\mathfrak{S}$ is its state space and $\alpha(G)$ is an inner evolution of $\mathcal{A}$ with a parameter group G (or semigroup) and so is the output system. Let a channel be a mapping from $\mathfrak{S}(\mathcal{A})$ to $\overline{\mathfrak{S}}(\overline{\mathcal{A}})$.

Although there exist several complexities, one of the most fundamental pairs of C and T in quantum system is the von Neumann entropy and the mutual entropy. Other entropic complexities C and T are ε-entropy, Kolmogorov-Sinai type dynamical entropy, dynamical mutual entropy.[30,40,45]

Here we remind that the quantum entropy and the quantum mutual entropy are examples of our complexities C and T, respectively.

Example 3.1. The entropy S and the mutual entropy I, in both classical and quantum, satisfy the conditions of the complexities C and T of ID. For a density operator ρ in a Hilbert space $\mathcal{H}$ (the case $\mathcal{A} = B(\mathcal{H})$) and a channel Λ, $C(\rho)$ is the entropy $S(\rho)$ and $T(\rho;\Lambda)$ is the mutual entropy $I(\rho;\Lambda)$:

$$C(\rho) = S(\rho) = -tr\rho\log\rho,$$

$$T(\rho;\Lambda) = I(\rho;\Lambda) = \sup\left\{\sum_k \lambda_k S(\Lambda E_k, \Lambda\rho);\ \{E_k\}\right\},$$

where the supremum is taken over all Schatten decompositions $\{E_k\}$ of ρ; $\rho = \sum\limits_k \lambda_k E_k$. In Shannon's communication theory in classical systems, ρ is a probability distribution $p = (p_k) = \sum_k p_k\delta_k$ and Λ is a transition probability matrix $(t_{i,j})$, so that the Schatten decomposition of ρ is unique and the compound state of ρ and its output $\overline{\rho}$ $(\equiv \overline{p} = (\overline{p}_i) = \Lambda p)$ is the joint distribution $r = (r_{i,j})$ with $r_{i,j} \equiv t_{i,j}p_j$. Then the above complexities C and T become the Shannon entropy and mutual entropy, respectively;

$$\begin{aligned} C(p) &= S(p) = -\textstyle\sum_k p_k \log p_k, \\ T(p;\Lambda) &= I(p;\Lambda) = \textstyle\sum_{i,j} r_{i,j}\log \frac{r_{i,j}}{p_j\overline{p}_i}. \end{aligned}$$

We can construct several other types of entropic complexities. For instance, one pair of the complexities is

$$T(\rho;\Lambda) = \sup\left\{\sum_k p_k S(\Lambda\rho_k, \Lambda\rho); \rho = \sum_k p_k\rho_k\right\},\quad C(\rho) = T(\rho; id)$$

where $S(\cdot,\cdot)$ is quantum relative entropy of Umegaki[57] and $\rho = \sum_k p_k\rho_k$ is a finite decomposition of ρ, over all of which the supremum is taken.

Example 3.2. Generalizing the entropy S and the mutual entropy I, we can construct complexities of entropy type: Let $(\mathcal{A}, \mathfrak{S}(\mathcal{A}), \alpha(G))$, $(\overline{\mathcal{A}}, \mathfrak{S}(\overline{\mathcal{A}}), \overline{\alpha}(\overline{G}))$ be C^* systems as before. Let $\mathcal{S}$ be a weak *-compact convex subset of $\mathfrak{S}(\mathcal{A})$ and $M_\varphi(\mathcal{S})$ be the set of all maximal measures μ on $\mathcal{S}$ with the fixed barycenter φ

$$\varphi = \int_{\mathcal{S}} \omega d\mu\,.$$

Moreover let $F_\varphi(\mathcal{S})$ be the set of all measures of finite support with the fixed barycenter φ. The following three pairs C and T satisfy all conditions of the complexities:

$$\begin{aligned}
T^{\mathcal{S}}(\varphi;\Lambda) &\equiv \sup\left\{\int_{\mathcal{S}} S(\Lambda\omega,\Lambda\varphi)d\mu;\mu\in M_{\varphi}(\mathcal{S})\right\}\\
C_T^{\mathcal{S}}(\varphi) &\equiv T^{\mathcal{S}}(\varphi;id)\\
I^{\mathcal{S}}(\varphi;\Lambda) &\equiv \sup\left\{S\left(\int_{\mathcal{S}}\omega\otimes\Lambda\omega d\mu,\varphi\otimes\Lambda\varphi\right);\mu\in M_{\varphi}(\mathcal{S})\right\}\\
C_I^{\mathcal{S}}(\varphi) &\equiv I^{\mathcal{S}}(\varphi;id)\\
J^{\mathcal{S}}(\varphi;\Lambda) &\equiv \sup\left\{\int_{\mathcal{S}} S(\Lambda\omega,\Lambda\varphi)d\mu;\mu\in F_{\varphi}(\mathcal{S})\right\}\\
C_j^{\mathcal{S}}(\varphi) &\equiv J^{\mathcal{S}}(\varphi;id)
\end{aligned}$$

Here, the state $\int_{\mathcal{S}}\omega\otimes\Lambda\omega d\mu$ is the compound state exhibiting the correlation between the initial state and the final state $\Lambda\varphi$, and $S(\cdot,\cdot)$ is quantum relative entropy.[6,53] This compound state was introduced as a quantum generalization of the joint probability measure in CDS (classical dynamical system).[41,42] Note that in the case of $\mathfrak{S}=\mathcal{S}$, $T^{\mathcal{S}}$(resp.$C^{\mathcal{S}}, I^{\mathcal{S}}, J^{\mathcal{S}}$) is denoted by T (resp. C, I, J) for simplicity. These complexities and the mixing $\mathcal{S}$-entropy $S^{\mathcal{S}}(\varphi)$,[40,48] the CNT (Connes-Narnhofer-Thirring) entropy $H_{\varphi}(\mathcal{A})$ satisfy some relations.

We review the definition of the mixing $\mathcal{S}$-entropy here.[44,53] For a state $\varphi\in\mathcal{S}\subset\mathfrak{S}(\mathcal{A})$, put

$$D_{\varphi}(\mathcal{S})\equiv$$

$$\left\{\mu\in M_{\varphi}(\mathcal{S});\exists\{\mu_k\}\subset\mathbf{R}^+ \text{ and } \{\varphi_k\}\subset \mathrm{ex}\mathcal{S} \text{ s.t. } \sum_k\mu_k=1,\mu=\sum_k\mu_k\delta(\varphi_k)\right\},$$

where $\delta(\varphi)$ is the delta measure concentrated on $\{\varphi\}$, and put

$$H(\mu)=-\sum_k\mu_k\log\mu_k$$

for a measure $\mu\in D_{\varphi}(\mathcal{S})$. Then the $\mathcal{S}$-entropy of a state $\varphi\in\mathcal{S}$ is defined as

$$S^{\mathcal{S}}(\varphi)=\begin{cases}\inf\{H(\mu)\ ;\ \mu\in D_{\varphi}(\mathcal{S})\} & \text{when } D_{\varphi}(\mathcal{S})\neq\emptyset\\ +\infty & \text{otherwise}\end{cases},$$

Theorem 3.1. *(1)* $0\le I^{\mathcal{S}}(\varphi;\Lambda)\le T^{\mathcal{S}}(\varphi;\Lambda)\le J^{\mathcal{S}}(\varphi;\Lambda)$. *(2)* $C_I(\varphi)=C_T(\varphi)=C_J(\varphi)=S(\varphi)=H_{\varphi}(\mathcal{A})$. *(3) When* $\mathcal{A}=\overline{\mathcal{A}}=B(\mathcal{H})$, *for any density operator* ρ

$$0\le I^{\mathcal{S}}(\rho;\Lambda)=T^{\mathcal{S}}(\rho;\Lambda)\le J^{\mathcal{S}}(\rho;\Lambda).$$

3.2. *Entropic Chaos Degree*

In quantum systems, if we take $C(\rho) = S(\rho)$ =von Neumann entropy, $T(\rho;\Lambda) = I(\rho;\Lambda)$ =quantum mutual entropy and linear channel Λ, then we have

$$\begin{aligned} D(\rho;\Lambda) &= C(\Lambda\rho) - T(\rho;\Lambda) \\ &= S(\Lambda\rho) - I(\rho;\Lambda) \\ &= S(\Lambda\rho) - \sup\left\{ Tr\left(\sum_n p_n \Lambda E_n (\log \Lambda E_n - \log \Lambda\rho)\right);\ \{E_n\}\right\} \\ &= \inf\left\{\sum_n p_n S(\Lambda E_n);\ \{E_n\}\right\} = \inf\left\{\sum_n p_n C(\Lambda E_n);\ \{E_n\}\right\} \end{aligned}$$

since $S(\Lambda\rho) = -Tr\Lambda\rho\log\Lambda\rho = -Tr\left(\sum_n p_n \Lambda E_n \log \Lambda\rho\right)$ for any Schatten decomposition $\{E_n\}$ of ρ. *Therefore the above quantity $D(\rho;\Lambda)$ can be interpreted as the complexity produced through the channel Λ. We apply this quantity $D(\rho;\Lambda)$ to study chaos even when the channel describing the dynamics is not linear. $D(\rho;\Lambda)$ is called the entropic chaos degree (ECD).*

In order to describe more general dynamics such as in continuous systems, we define the entropic chaos degree in C*-algebraic terminology. This setting will not be used in the sequel application, but for mathematical completeness we will discuss the C*-algebraic setting.

Let $(\mathcal{A}, \mathfrak{S})$ be an input C* system and $(\overline{\mathcal{A}}, \overline{\mathfrak{S}})$ be an output C* system; namely, $\mathcal{A}$ is a C* algebra with unit I and $\mathfrak{S}$ is the set of all states on $\mathcal{A}$. We assume $\overline{\mathcal{A}} = \mathcal{A}$ for simplicity. For a weak* compact convex subset $\mathcal{S}$ (called the reference space) of $\mathfrak{S}$, take a state φ from the set $\mathcal{S}$ and let

$$\varphi = \int_{\mathcal{S}} \omega d\mu_\varphi$$

be an extremal orthogonal decomposition of φ in $\mathcal{S}$, which describes the degree of mixture of φ in the reference space $\mathcal{S}$. In more detail this formula reads

$$\varphi(A) = \int_{\mathcal{S}} \omega(A) d\mu_\varphi(\omega), \quad A \in \mathcal{A}$$

The measure μ_φ is not uniquely determined unless $\mathcal{S}$ is the Schoque simplex, so that the set of all such measures is denoted by $M_\varphi(\mathcal{S})$.

Definition 3.2. The entropic chaos degree with respect to $\varphi \in \mathcal{S}$ and a channel Λ is defined by

$$D^{\mathcal{S}}(\varphi;\Lambda) \equiv \inf\left\{\int_{\mathcal{S}} S^{\mathcal{S}}(\Lambda\omega)\, d\mu; \mu \in M_{\varphi}(\mathcal{S})\right\}$$

where $S^{\mathcal{S}}(\cdot)$ is the mixing $\mathcal{S}$−entropy[42,48] in the reference space $\mathcal{S}$.

When $\mathcal{S} = \mathfrak{S}$, $D^{\mathcal{S}}(\varphi;\Lambda)$ is simply written as $D(\varphi;\Lambda)$. This $D^{\mathcal{S}}(\varphi;\Lambda)$ contains the classical chaos degree and the quantum above. The classical entropic chaos degree is the case that $\mathcal{A}$ is abelian and φ is the probability distribution of a orbit generated by a dynamics (channel) Λ; $\varphi = \sum_k p_k \delta_k$, where δ_k is the delta measure such as $\delta_k(j) \equiv \begin{cases} 1 & (k=j) \\ 0 & (k \neq j) \end{cases}$. Then the classical entropic chaos degree is

$$D_c(\varphi;\Lambda) = \sum_k p_k S(\Lambda\delta_k)$$

with the entropy S.

Summarize that Information Dynamics can be applied to the study of chaos by using more general complexity $C(\varphi)$:

Definition 3.3. (1)ψ is more chaotic than φ if $C(\psi) \geq C(\varphi)$.

(2)When $\varphi \in \mathcal{S}$ changes to $\Lambda\varphi$, the *chaos* degree associated to this state change(dynamics) Λ is given by

$$D^{\mathcal{S}}(\varphi;\Lambda) = \inf\left\{\int_{\mathcal{S}} C^{\mathcal{S}}(\Lambda\varphi)\, d\mu; \mu \in M_{\varphi}(\mathcal{S})\right\}.$$

Definition 3.4. A dynamics Λ produces chaos iff $D^{\mathcal{S}}(\varphi;\Lambda) > 0$.

Remark 3.1. It is important to note here that the dynamics Λ in the definition is not necessarily same as original dynamics (channel) but is one reduced from the original such that it causes an evolution for a certain observed value like orbit. However for simplicity we use the same notation here. In some cases, the above chaos degree $D^{\mathcal{S}}(\varphi;\Lambda)$ can be expressed as

$$D^{\mathcal{S}}(\varphi;\Lambda) = C^{\mathcal{S}}(\Lambda\varphi) - T^{\mathcal{S}}(\varphi;\Lambda).$$

3.3. *Algorithm computing Entropic Chaos Degree*

In order to observe a chaos produced by a dynamics, one often looks at the behavior of orbits made by that dynamics, more generally, looks at the behavior of a certain observed value. Therefore in our scheme we directly compute the chaos degree once a dynamics is explicitly given as a state change of system. However even when the direct calculation does not show a chaos, a chaos will appear if one focuses to some aspect of the state change, e.g., a certain observed value. In the later case, algorithm computing the chaos degree for classical or quantum dynamics consists of the following two cases:

(1) *Dynamics is given by* $\frac{dx}{dt} = F_t(x)$ *with* $x \in I \equiv [a,b]^{\mathbf{N}} \subset \mathbf{R}^{\mathbf{N}}$: First find a difference equation $x_{n+1} = F(x_n)$ with a map F on $I \equiv [a,b]^{\mathbf{N}} \subset \mathbf{R}^{\mathbf{N}}$ into itself, secondly let $A := \{A_i\}$ be a finite partition (i.e., $I \equiv \bigcup_k A_k$, $A_i \cap A_j = \emptyset\ (i \neq j)$). Then the state $\varphi^{(n)}$ at time n of the orbit determined by the difference equation is defined by the probability distribution $\left(p_i^{(n)}\right)$ with a given finite partition $A = \{A_i\}$, that is, $\varphi^{(n)} = \sum_i p_i^{(n)} \delta_i$, where for an initial value $x \in I$ and the characteristic function 1_A

$$p_i^{(n)} \equiv \frac{1}{n+1} \sum_{k=m}^{m+n} 1_{A_i}\left(F^k x\right).$$

Now when the initial value x is distributed due to a measure ν on I after a proper time m, the above $p_i^{(n)}$ is given as

$$p_i^{(n)} \equiv \frac{1}{n+1} \int_I \sum_{k=m}^{m+n} 1_{A_i}\left(F^k x\right) d\nu.$$

The joint distribution $\left(p_{ij}^{(n,n+1)}\right)$ between the time n and $n+1$ is defined by

$$p_{ij}^{(n,n+1)} \equiv \frac{1}{n+1} \sum_{k=m}^{m+n} 1_{A_i}\left(F^k x\right) 1_{A_j}\left(F^{k+1} x\right)$$

or

$$p_{ij}^{(n,n+1)} \equiv \frac{1}{n+1} \int_I \sum_{k=m}^{m+n} 1_{A_i}\left(F^k x\right) 1_{A_j}\left(F^{k+1} x\right) d\nu.$$

Then the channel Λ_n at n is determined by

$$\Lambda_n \equiv \left(\frac{p_{ij}^{(n,n+1)}}{p_i^{(n)}}\right) : \text{transition probability} \Longrightarrow \varphi^{(n+1)} = \Lambda_n \varphi^{(n)},$$

and the entropic chaos degree is given by, for a finite partition $A := \{A_i\}$,

$$D_A(x; F) = D_A\left(p^{(n)}; \Lambda_n\right) = \sum_i p_i^{(n)} S(\Lambda_n \delta_i) = \sum_{i,j} p_{ij}^{(n,n+1)} \log \frac{p_i^{(n)}}{p_{ij}^{(n,n+1)}}. \tag{1}$$

We can judge whether the dynamics causes a chaos or not by the value of D_A for the partition $A = \{A_i\}$ as

$$D_A > 0 \Longleftrightarrow \textit{chaotic}$$
$$D_A = 0 \Longleftrightarrow \textit{stable}.$$

This chaos degree was applied to several dynamical maps such logistic map, Baker's transformation and Tinkerbell map, and it could explain their chaotic characters. This chaos degree has several merits compared with usual measures such as Lyapunov exponent as explained below.

The partition free chaos degree D is defined by the infimum of D_A over all partitions A.Therefore it is said that the dynamics pruduces a chaos in the scale $\{A_k\}$ if D_A is positive.

(2) *Dynamics is given by* $\varphi_t = F_t \varphi_0$ *on a Hilbert space*: Similarly as making a difference equation for (quantum) state, the channel Λ_n at n is first deduced from F_t, which should satisfy $\varphi^{(n+1)} = \Lambda_n \varphi^{(n)}$. By means of this constructed channel, **(i)** we compute the chaos degree D directly according to the definition 3.2 or **(ii)** we take a proper observable X and put $x_n \equiv \varphi^{(n)}(X)$, then go back to the algorithm (1).

The entropic chaos degree for quantum systems has been applied to the analysis of quantum spin system and quantum Baker's type transformation.[27,28,31]

Note that the chaos degree D_A above does depend on a partition A taken, which is somehow different from usual degree of chaos. This is a key point of our understanding of chaos, from which the idea of adaptivity comes, which is discussed in Subsection 3.4.

Example 3.3. Logistic Map

Let us apply the entropy chaos degree (ECD) to logistic map. Chaotic behavior in classical system is often considered as exponential sensitivity to initial condition.

The logistic map is defined by

$$x_{n+1} = a x_n (1 - x_n), x_n \in [0, 1], 0 \leq a \leq 4$$

The solution of this equation bifurcates as Fig. 1.

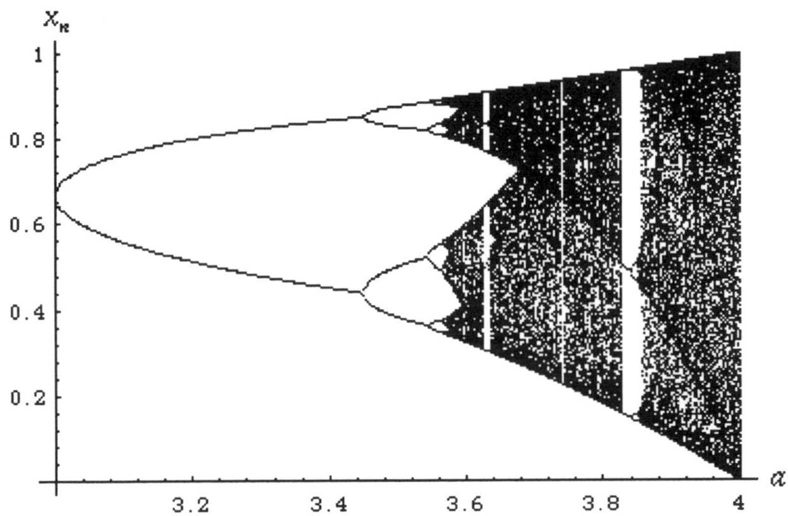

Fig. 1. Logistic map

In order to compare ECD with other measure describing chaos, we take Lyapunov exponent for this comparison: Fig. 2 and Fig. 3.

We computed the entropic chaos degree for various maps in,[29] and it is shown that Lyapunov exponent and chaos degree have clear correspondence. Moreover the ECD resolves some inconvenient points of the Lyapunov exponent as:

(1) Lyapunov exponent takes negative and sometimes $-\infty$, but the ECD is always positive for all $a \geq 0$.
(2) For some map f whose Lyapunov exponent is difficult to compute (e.g., dynamics in $\mathbb{R}^n$ $(n \geq 2)$), the ECD of f is easily computed.
(3) Generally, the algorithm for the ECD is much easier than that for Lyapunov exponent.

3.3.1. *ECD with memory*

Here we generalize the above explained ECD to take the memory effect into account. Although the original ECD is based on the choice of the base space $\Sigma := \{1, 2, \cdots, N\}$, we here take another base; Σ^m, instead of Σ. On

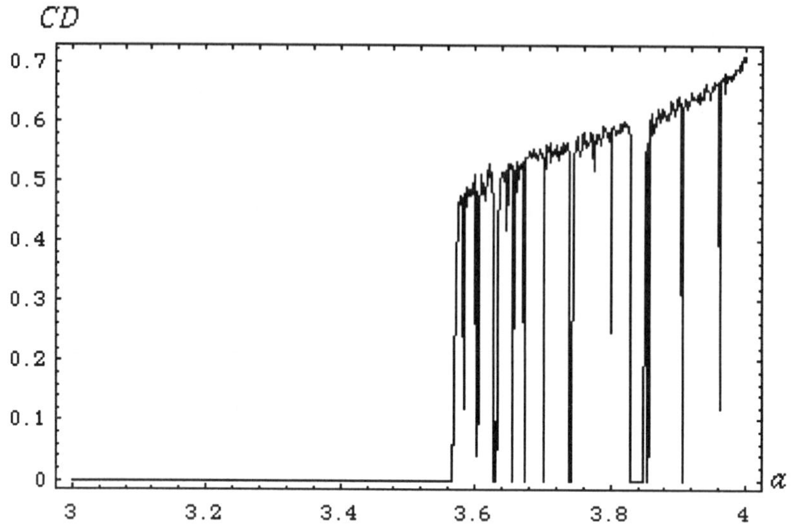

Fig. 2. Entropic Chaos Degree of Logistic Map

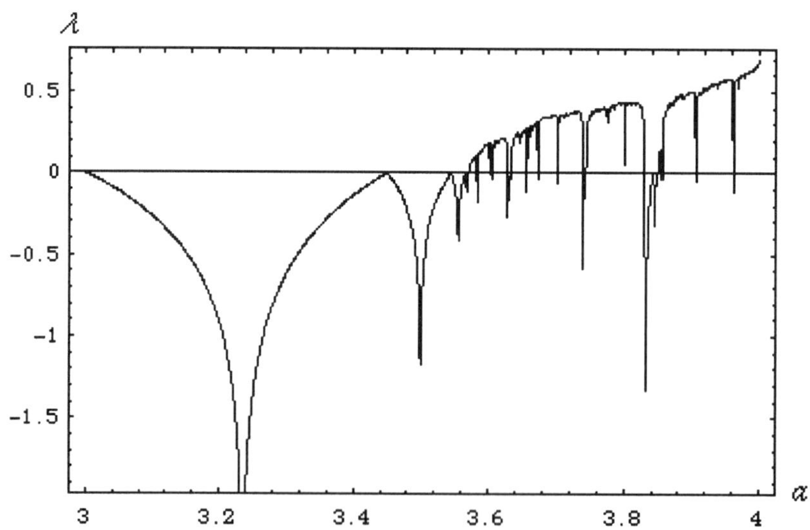

Fig. 3. Lyapunov exponent

this base space, a probability distribution is naturally defined as

$$p_{i_0 i_1 \ldots i_m}^{(n,n+1,\ldots,n+m)} \equiv \frac{1}{n+1} \int_I \sum_{k=m}^{m+n} 1_{A_{i_0}} \left(F^k x\right) 1_{A_{i_1}} \left(F^{k+1} x\right) \cdots 1_{A_{i_m}} \left(F^{k+m} x\right) d\nu.$$

with its mathematical idealization, $p_{i_0i_1\dots i_m} := \lim_{n\to\infty} p^{(n,n+1,\dots,n+m)}_{i_0i_1\dots i_m}$.The channel Λ_m over Σ^m is defined by a transition probability,

$$p_{j_0i_1\dots i_{m+1}}\delta_{i_1j_1}\cdots\delta_{i_mj_m} = p(i_1,i_2,\dots,i_m,i_{m+1}|j_0,j_1,\dots,j_m)p_{j_0,j_1\dots j_m}.$$

Thus it derives the ECD with m-steps memory effect,

$$D^m_A(x;f) = D^m_A(p;\Lambda_m) = \sum_{i_0,i_1,\dots,i_m} p_{i_0i_1\dots i_m}\log\frac{p_{i_0i_1\dots i_{m-1}}}{p_{i_0i_1\dots i_m}}.$$

It notes that this quantity coincides with the original ECD when $m=1$.

This memory effect shows an interesting result, that is, the longer the memory is, the closer the ECD is to the Lyapunov exponent for its positive part.[51]

Theorem 3.2. *For given f, x and A, there exists a probability space (Ω, F, ν) and a random variable g depending on f, x, A such that $\lim_{m\to\infty} D^m_A(x;f) = \int_\Omega g d\nu$ =the positive part of Lyapunov exponent.*

3.4. *Adaptive Chaos Degree*

In adaptive dynamics, it is essential to consider in which states and by which ways we see objects. That is, one has to select phenomena and prepare mode for observation for understanding the whole of a system. Typical adaptive dynamics are the dynamics for state-adaptive and that for observable-adaptive as mentioned in the previous section.

We will discuss how such adaptivities are appeared in dynamics which cause a chaos.

First of all we examine carefully when we say that a certain dynamics produces a chaos. Let us take the logistic map as an example. The original differential equation of the logistic map is

$$\frac{dx}{dt} = ax(1-x), 0 \le a \le 4$$

with initial value x_0 in $[0,1]$. This equation can be easily solved analytically, whose solution (orbit) does not have any chaotic behavior. However once we make the above equation discrete such as

$$x_{n+1} = ax_n(1-x_n), 0 \le a \le 4. \tag{2}$$

This difference equation produces a chaos.

Taking the discrete time is necessary not only to make a chaos but also to observe the orbits drawn by the dynamics. Similarly as quantum mechanics, it is not possible for human being to understand any object without observing it, for which it will not be possible to trace a orbit continuously in time.

Now let us think a finite partition $A = \{A_k; k = 1, \cdots, N\}$ of a proper set $I \equiv [a,b]^{\mathbf{N}} \subset \mathbb{R}^{\mathbf{N}}$ and an equi-partition $B^e = \{B_k^e; k = 1, \cdots, N\}$ of I. Here "equi" means that all elements B_k^e are equivalent. We denote the set of all partitions by $\mathcal{P}$ and the set of all equi-partitions by $\mathcal{P}^e$. Such a partition enables to observe the orbit of a given dynamics, and moreover it provides a criterion for observing chaos. There exist several reports saying that one can observe chaos in nature, which are very much related to how one observes the phenomena, for instance, scale, direction, aspect. It has been difficult to find a satisfactory theory (mathematics) to explain such chaotic phenomena. In the difference equation 2 we take some time interval τ between n and $n+1$, if we take $\tau \to 0$, then we have a complete different dynamics. If we take coarse graining to the orbit of x_t for time during τ; $x_n \equiv \frac{1}{\tau}\int_{(n-1)\tau}^{n\tau} x_t dt$, we again have a very different dynamics. Moreover it is important for mathematical consistency to take the limits $n \to \infty$ or N (the number of equi-partitions)$\to \infty$, i.e., making the partition finer and finer, and consider the limits of some quantities as describing chaos, so that mathematical terminologies such as "lim", "sup", "inf" are very often used to define such quantities. *Let us take the opposite position, that is, any observation will be unrelated or even contradicted to such limits. Observation of chaos is a result due to taking suitable scales of, for example, time, distance or domain, and it will not be possible in the limiting cases.*

It is claimed in[38] that most of chaos are scale-dependent phenomena, so the definition of a degree measuring chaos should depend on certain scales taken. Such a scale dependent dynamics is nothing but adaptive dynamics.

Taking into consideration of this view we modify the definitions of the chaos degree given in the previous section 3.2 as below.

Going back to a triple $(\mathcal{A}, \mathfrak{S}, \alpha(G))$ considered in Section 2 and we use this triple both for an input and an output systems. Let a dynamics be described by a mapping Γ_t with a parameter $t \in G$ from $\mathfrak{S}$ to $\mathfrak{S}$ and let an observation be described by a mapping $\mathcal{O}$ from $(\mathcal{A}, \mathfrak{S}, \alpha(G))$ to a triple $(\mathcal{B}, \mathfrak{T}, \beta(G))$. The triple $(\mathcal{B}, \mathfrak{T}, \beta(G))$ might be same as the original one or its subsystem and the observation map $\mathcal{O}$ may contains several different types of observations, that is, it can be decomposed as $\mathcal{O} = \mathcal{O}_m \cdots \mathcal{O}_1$.Let us list some examples of observations.

For a given dynamics $\frac{d\varphi}{dt} = F(\varphi_t)$, equivalently, $\varphi_t = \Gamma_t \varphi$, one can take several observations.

Example 3.4. Time Scaling (Discretizing): $O_\tau : t \to n$, $\frac{d\varphi}{dt}(t) \to \varphi_{n+1}$, so that $\frac{d\varphi}{dt} = F(\varphi_t) \Rightarrow \varphi_{n+1} = F(\varphi_n)$ and $\varphi_t = \Gamma_t \varphi \Rightarrow \varphi_n = \Gamma_n \varphi$. Here τ is a unit time needed for the observation.

Example 3.5. Size Scaling (Conditional Expectation, Partition): Let $(\mathcal{B}, \mathfrak{T}, \beta(G))$ be a subsystem of $(\mathcal{A}, \mathfrak{S}, \alpha(G))$, both of which have a certain algebraic structure such as C*-algebra or von Neumann algebra. As an example, the subsystem $(\mathcal{B}, \mathfrak{T}, \beta(G))$ has abelian structure describing a macroscopic world which is a subsystem of a non-abelian (non-commutative) system $(\mathcal{A}, \mathfrak{S}, \alpha(G))$ describing a micro-world. A mapping $\mathcal{O}_C$ preserving norm (when it is properly defined) from $\mathcal{A}$ to $\mathcal{B}$ is, in some cases, called a conditional expectation. A typical example of this conditional expectation is according to a projection valued measure

$$\left\{ P_k;\ P_k P_j = P_k \delta_{kj} = P_k^* \delta_{kj} \geqq 0, \sum_k P_k = I \right\}$$

associated with quantum measurement (von Neumann measurement) such that

$$\mathcal{O}_C(\rho) = \sum_k P_k \rho P_k$$

for any quantum state (density operator) ρ. When $\mathcal{B}$ is a von Neumann algebra generated by $\{P_k\}$, it is an abelian algebra isometrically isomorphic to $L^\infty(\Omega)$ with a certain Hausdorff space Ω, so that in this case $\mathcal{O}_C$ sends a general state φ to a probability measure (or distribution) p. Similar example of $\mathcal{O}_C$ is one coming from a certain representation (selection) of a state such as one Schatten decomposition of ρ ;

$$\rho = \mathcal{O}_R \rho = \sum_k \lambda_k E_k$$

by one-dimensional orthogonal projections $\{E_k\}$ associated to the eigenvalues of ρ with $\sum_k E_k = I$. Another important example of the size scaling is due to a finite partition of an underlining space Ω, e.g., space of orbit, defined as

$$\mathcal{O}_P(\Omega) = \left\{ P_k; P_k \cap P_j = P_k \delta_{kj} (k, j = 1, \cdots N),\ \cup_{k=1}^N P_k = \Omega \right\}.$$

3.4.1. *Chaos degree with adaptivity*

We go back to the discussion of the entropic chaos degree. Starting from a given dynamics $\varphi_t = \Gamma_t\varphi$, it becomes $\varphi_n = \Gamma_n\varphi$ after handling the operation $\mathcal{O}_\tau$. Then by taking proper combinations $\mathcal{O}$ of the size scaling operations like $\mathcal{O}_C$, $\mathcal{O}_R$ and $\mathcal{O}_P$, the equation $\varphi_n = \Gamma_n\varphi$ changes to $\mathcal{O}(\varphi_n) = \mathcal{O}(\Gamma_n\varphi)$, which will be written by $\mathcal{O}\varphi_n = \mathcal{O}\Gamma_n\mathcal{O}^{-1}\mathcal{O}\varphi$ or $\varphi_n^{\mathcal{O}} = \Gamma_n^{\mathcal{O}}\varphi^{\mathcal{O}}$. Then our entropic chaos degree is redefined as follows:

Definition 3.5. The entropic chaos degree of Γ with an initial state φ and observation $\mathcal{O}$ is defined by

$$D^{\mathcal{O}}(\varphi;\Gamma) = \int_{\mathcal{O}(\mathfrak{S})} S\left(\Gamma^{\mathcal{O}}\omega^{\mathcal{O}}\right) d\mu^{\mathcal{O}},$$

where $\mu^{\mathcal{O}}$ is the measure operated by $\mathcal{O}$ to a extremal decomposition measure of φ selected by of the observation $\mathcal{O}$ (its part $\mathcal{O}_R$). *The entropic chaos degree of* Γ *with an initial state* φ *is defined by*

$$D(\varphi;\Gamma) = \inf \left\{D^{\mathcal{O}}(\varphi;\Gamma);\mathcal{O}\in\mathcal{SO}\right\},$$

where $\mathcal{SO}$ *is a proper set of observations naturally determined by a given dynamics.*

In this definition , $\mathcal{SO}$ is determined by a given dynamics and some conditions attached to the dynamics, for instance, if we start from a difference equation with a special representation of an initial state, then $\mathcal{SO}$ excludes $\mathcal{O}_\tau$ and $\mathcal{O}_R$.

Then one judges whether a given dynamics causes a chaos or not by the following way.

Definition 3.6. (1) A dynamics Γ is chaotic for an initial state φ in an observation $\mathcal{O}$ iff $D^{\mathcal{O}}(\varphi;\Gamma) > 0$.

(2) A dynamics Γ *is totally chaotic for an initial state* φ *iff* $D(\varphi;\Gamma) > 0$.

The idea introducing in this section to understand chaos can be applied not only to the entropic chaos degree but also to some other degrees such as dynamical entropy, whose applications and the comparison of several degrees will be discussed in.[51]

In the case of logistic map, $x_{n+1} = ax_n(1-x_n) \equiv F(x_n)$, we obtain this difference equation by taking the observation $\mathcal{O}_\tau$ and take an observation $\mathcal{O}_P$ by equi-partition of the orbit space $\Omega = \{x_n\}$ so as to define a state (probability distribution). Thus we can compute the entropic chaos degree in adaptive sense.

As an example, we consider a circle map

$$\theta_{n+1} = f_\nu(\theta_n) = \theta_n + \omega \ (\mathrm{mod}\ 2\pi),$$

where $\omega = 2\pi v$ $(0 < v < 1)$. If v is a rational number N/M, then the orbit $\{\theta_n\}$ is periodic with the period M. If v is irrational, then the orbit $\{\theta_n\}$ densely fills the unit circle for any initial value θ_0; namely, it is a quasiperiodic motion.

Theorem 3.3. *Let* $\mathbf{I} = [0, 2\pi]$ *be partioned into* L *disjoint components with equal length;* $\mathbf{I} = B_1 \cap B_2 \cap \ldots \cap B_L$.

(1) If v *is rational number* N/M *, then the finite equi-partition* $P = \{B_k; k = 1, \cdots, M\}$ *implies* $D^{\mathcal{O}}(\theta_0; f_\nu) = 0$.

(2) If v *is irrational, then* $D^{\mathcal{O}}(\theta_0; f_\nu) > 0$ *for any finite partition* $P = \{B_k\}$.

Note that our entropic chaos degree shows a chaos to quasiperiodic circle dynamics by the observation due to a partition of the orbit, which is different from usual understanding of chaos. However usual belief that quasiperiodic circle dynamics will not cause a chaos is not at all obvious, but is realized in a special limiting case as shown in the following theorem.

Theorem 3.4. *For the above circle map, if* v *is irrational, then* $D(\theta_0; f_\nu) = 0$.

Such a limiting case will not take place in real observation of natural objects, so that we claim that chaos is a phenomenon depending on observations, environment or periphery, which results the adaptive definition of chaos as above. The detailed examination of a map of this type is done in the paper.[13]

Note here that the chaos degree and the adaptivity can be applied to understand quantum dynamics either.[27,28,31]

4. Adaptive Dynamics Solving SAT Problem.

4.1. *SAT problem*

We take the SAT (satisfiable) problem, one of the NP-complete problems, to study whether there exists an algorithm showing NPC=P.

Let $X \equiv \{x_1, \cdots, x_n\}$ be a set. Then x_k and its negation $\bar{x}_k\,(k = 1, 2, \cdots, n)$ are called literals and the set of all such literals is denoted by $X' \equiv \{x_1, \bar{x}_1, \cdots, x_n, \bar{x}_n\}$. The set of all subsets of X' is denoted by $\mathcal{F}(X')$ and an element $C \in \mathcal{F}(X')$ is called a clause. We take a truth assignment to all Boolean variables x_k. If we can assign the truth value to at least one element of C, then C is called satisfiable. When C is satisfiable, the truth value $t(C)$ of C is regarded as true, otherwise, that of C is false. Take the truth values as "true $\leftrightarrow$1, false $\leftrightarrow$0". Then $C$is satisfiable iff $t(C) = 1$.

Let $L = \{0, 1\}$ be a Boolean lattice with usual join $\vee$ and meet $\wedge$, and $t(x)$ be the truth value of a literal x in X. Then the truth value of a clause C is written as $t(C) \equiv \vee_{x \in C} t(x)$.

Moreover the set $\mathcal{C}$ of all clauses $C_j\ (j = 1, 2, \cdots, m)$ is called satisfiable iff the meet of all truth values of C_j is 1; $t(\mathcal{C}) \equiv \wedge_{j=1}^{m} t(C_j) = 1$. Thus the SAT problem is written as follows:

Definition 4.1. SAT Problem: Given a Boolean set $X \equiv \{x_1, \cdots, x_n\}$and a set $\mathcal{C} = \{\mathcal{C}_1, \cdots, \mathcal{C}_m\}$ of clauses, determine whether $\mathcal{C}$ is satisfiable or not.

That is, this problem is to ask whether there exists a truth assignment to make $\mathcal{C}$ satisfiable. It is known in usual algorithm that it is polynomial time to check the satisfiability only when a specific truth assignment is given, but we can not determine the satisfiability in polynomial time when an assignment is not specified.

In[52] we discussed the quantum algorithm of the SAT problem, which was rewritten in[7] with showing that OM SAT-algorithm is combinatric. In[54,55] it is shown that the chaotic quantum algorithm can solve the SAT problem in polynomial time.

Ohya and Masuda pointed out[52] that the SAT problem, hence all other NP problems, can be solved in polynomial time by quantum computer if the superposition of two orthogonal vectors $|0\rangle$ and $|1\rangle$ is physically detected. However this detection is considered not to be possible in the present technology. The problem to be overcome is how to distinguish the pure vector $|0\rangle$ from the superposed one $\alpha\,|0\rangle + \beta\,|1\rangle$, obtained by the OM SAT-quantum algorithm, if β is not zero but very small. If such a distinction is possible, then we can solve the NPC problem in the polynomial time. In[54,55] it is shown that it can be possible by combining nonlinear chaos amplifier with the quantum algorithm, which implies the existence of a mathematical algorithm solving NP=P. The algorithm of Ohya and Volovich is not known to be in the framework of quantum Turing algorithm or not. So the next question is (1) whether there exists a physical realization combining the SAT quantum algorithm with chaos dynamics, or (2) whether there exists another method to achieve the above distinction of two vectors by a suitable unitary evolution so that all process can be discussed by a certain quantum Turing machine (circuits).[32–34] The paper[34] by Iriyama and Ohya of this volume briefly discusses the essence of the quantum SAT algorithm. In this paper, we will discuss the SAT problem with adaptive dynamics based on the work of,[3] which is another method of (2) above.

4.2. *Quantum Algorithm*

The quantum algorithms discussed so far are rather idealized because computation is represented by unitary operations. A unitary operation is rather difficult to realize in physical processes, more realistic operation is one allowing some dissipation like semigroup dynamics. However such dissipative dynamics very much reduces the ability of quantum computation because the ability is based on preserving the entanglement of states and the dissipativity destroys the entanglement. Keeping high ability of quantum computation and good entanglement, it will be necessary to some kinds of amplification in the course of real physical processes in physical devices, which will be similar as amplication processes in quantum communication processes. In this section, to search for more realistic operations in quantum computer, the channel expression will be useful, at least, in the sense of mathematical scheme of quantum computation because the channel is not always unitary and represents many different types of dynamics.

Let $\mathcal{H}$ be a Hilbert space describing input, computation and output (result). As usual, the Hilbert space is $\mathcal{H} = \otimes_1^N \mathbf{C}^2$, and let the basis of $\mathcal{H} = \otimes_1^N \mathbf{C}^2$ be: $e_0\,(= |0\rangle) = |0\rangle \otimes \cdots \otimes |0\rangle \otimes |0\rangle, e_1\,(= |1\rangle) = |0\rangle \otimes \cdots \otimes |0\rangle \otimes |1\rangle, \cdots, e_{2^N-1}\left(= \left|2^N-1\right\rangle\right) = |1\rangle \otimes \cdots \otimes |1\rangle \otimes |1\rangle$.

Any number t $\left(0, \cdots, 2^N-1\right)$ can be expressed by $t = \sum_{k=1}^{N} a_t^{(k)} 2^{k-1}$, $a_t^{(k)} = 0$ *or* 1, so that the associated vector is written by

$$|t\rangle\,(= e_t) = \otimes_{k=1}^N \left|a_t^{(k)}\right\rangle.$$

And applying n-tuples of Hadamrd matrix $A \equiv \frac{1}{\sqrt{2}}\begin{pmatrix}1 & 1\\ 1 & -1\end{pmatrix}$ to the vacuum vector $|0\rangle$, we get $A\,|0\rangle\;(\,= \xi\,(0)) \equiv \otimes_1^N \frac{1}{\sqrt{2}}\,(|0\rangle + |1\rangle)$. Put

$$W(t) \equiv \otimes_{j=1}^{N} \begin{pmatrix} 1 & 0 \\ 0 & \exp(\frac{2\pi it}{2^N} 2^{j-1}) \end{pmatrix}.$$

Then we have

$$\xi(t) \equiv W(t)\xi(0) = \frac{1}{\sqrt{2^N}} \sum_{k=0}^{2^N-1} exp\left(\frac{2\pi itk}{2^N}\right)|k\rangle,$$

which is called Discrete Fourier Transformation. Thus altogether of the above operations, it follows a unitary operator $U_F(t) \equiv W(t) A$ and the vector $\xi(t) = U_F(t)|0\rangle$.

4.2.1. *Channel expression of conventional unitary algorithm*

All conventional unitary algorithms can be written as the following three steps:

(1) Preparation of state: Take a state ρ (e.g., $\rho = |0\rangle\langle 0|$) applying the unitary channel defined by the above $U_F(t) : \Lambda_F = Ad_{U_F(t)}$

$$\Lambda_F = Ad_{U_F} \Longrightarrow \Lambda_F\rho = U_F\rho U_F^*$$

(2) Computation: Let U a unitary operator on $\mathcal{H}$ representing the computation followed by a suitable programming of a certain problem, then the computation is described by a channel $\Lambda_U = Ad_U$ (unitary channel). After the computation, the final state ρ_f will be

$$\rho_f = \Lambda_U\Lambda_F\rho.$$

(3) Register and Measurement: For registeration of the computed result and its measurement we might need an additional system $\mathcal{K}$ (e.g., register), so that the lifting $\mathcal{E}_m$ from $\mathcal{S}(\mathcal{H})$ to $\mathcal{S}(\mathcal{H}\otimes\mathcal{K})$ in the sense of[3] is useful to describe this stage. Thus the whole process is wrtten as

$$\rho_f = \mathcal{E}_m(\Lambda_U\Lambda_F\rho).$$

Finally we measure the state in $\mathcal{K}$: For instance, let $\{P_k; k \in J\}$ be a projection valued measure (PVM) on $\mathcal{K}$

$$\Lambda_m\rho_f = \sum_{k\in J} I\otimes P_k\rho_f I\otimes P_k,$$

after which we can get a desired result by observations in finite times if the size of the set J is small.

4.2.2. *Channel expression of general quantum algorithm*

Since actual physical process is dispative, the above three steps have to be generalized so that the dissipative nature is involved. Such a generalization can be expressed by means of suitable channel, not necessarily unitary, which gives us a basois of the generalized quantum Turing machine.[33,34]

(1) Preparation of state: We may be use the same channel $\Lambda_F = Ad_{U_F}$ in this first step, but if the number of qubits N is large so that it will not be built physically, then Λ_F should be modified, and let denote it by Λ_P.

(2) Computation: This stage is certainly modified to a channel Λ_C reflecting the physical device for computer.

(3) Registering and Measurement: This stage will be remained as aobe. Thus the whole process is written as

$$\rho_f = \mathcal{E}_m \left(\Lambda_C \Lambda_P \rho \right).$$

4.3. *Quantum Algorithm of SAT*

In this subsection, we review fundamentals of quantum computation (see, for instance,[55]) for the SAT problem. Let $\mathbb{C}$ be the set of all complex numbers, and $|0\rangle$ and $|1\rangle$ be the two unit vectors $\binom{1}{0}$ and $\binom{0}{1}$, respectively. Then, for any two complex numbers α and β satisfying $|\alpha|^2 + |\beta|^2 = 1$, $\alpha|0\rangle + \beta|1\rangle$ is called a qubit. For any positive integer N, let $\mathcal{H}$ be the tensor product Hilbert space defined as $\left(\mathbb{C}^2\right)^{\otimes N}$ and let $\left\{|e_i\rangle ; 0 \le i \le 2^{N-1}\right\}$ be the basis as above. For any two qubits $|x\rangle$ and $|y\rangle$, $|x, y\rangle$ and $\left|x^N\right\rangle$ is written as $|x\rangle \otimes |y\rangle$ and $\underbrace{|x\rangle \otimes \cdots \otimes |x\rangle}_{N \text{ times}}$, respectively.

The computation of the truth value can be done by by a combination of the unitary operators on a Hilbert space $\mathcal{H}$, so that the computation is described by the unitary quantum algorithm. The detail of this section is given in the papers,[7,32,52,55] so we will discuss just the essence of the OM algorithm. Throughout this section, let n be the total number of Boolean variables used in the SAT problem. Let $C = \{C_1, \ldots, C_m\}$ be a set of clauses whose cardinality is equal to m. Let $\mathcal{H} = \left(\mathbb{C}^2\right)^{\otimes n+\mu+1}$ be a Hilbert space and $|v_0\rangle$ be the initial state $|v_0\rangle = |0^n, 0^\mu, 0\rangle$, where μ is the number of dust qubits which is determined by the following proposition. Let $U_C^{(n)}$ be a unitary operator for the computation of the SAT:

$$U_C^{(n)}|v_0\rangle = \frac{1}{\sqrt{2^n}}\sum_{i=0}^{2^n-1}|e_i, x^\mu, t_{e_i}(C)\rangle \equiv |v_f\rangle$$

where x^μ denotes the μ strings in the dust bits and $t_{e_i}(C)$ is the truth value of C with e_i. In,[7,34,52] $U_C^{(n)}$ was constructed.

This final state vector $|v_f>$ is also written as

$$|v_f> = \frac{1}{2^{n/2}}\sum_{\varepsilon\in\{0,1\}^n}|\varepsilon, x_{\mu-1}^{\varepsilon}, t_\varepsilon(C)>$$

Theorem 4.1. *C is SAT if and only if,*

$$P_{n+\mu,1}U_C|v\rangle \neq 0$$

where $P_{n+\mu,1}$ denotes the projector

$$P_{n+\mu,1} := 1_{n+\mu-1}\otimes|1><1|$$

onto the subspace of $\mathcal{H}$ spanned by the vectors

$$|\varepsilon_n, \varepsilon_{\mu-1}, 1>,$$

where $\varepsilon_n \in \{0,1\}^n$ and $\varepsilon_{\mu-1}\in\{0,1\}^{\mu-1}$.

According to the standard theory of quantum measurement, after a measurement of the event $P_{n+\mu,1}$, the state $\rho = |v_f><v_f|$ becomes

$$\rho \to \frac{P_{n+\mu,1}\rho P_{n+\mu,1}}{Tr\rho' P_{n+\mu,1}} =: \rho'$$

Thus the solvability of the SAT problem is reduced to check that $\rho' \neq 0$. The difficulty is that the probability of $P_{n+\mu,1}$ is

$$Tr\rho' P_{n+\mu,1} = \|P_{n+\mu,1}\psi\|^2 = \frac{|T(C_0)|}{2^n}$$

where $|TC_0)|$ is the cardinality of the set $T(C_0)$, of all the truth functions t such that $t(C_0) = 1$.

We put $q := \sqrt{\frac{r}{2^n}}$ with $r := |T(C_0)|$ in the sequel. Then if r is suitably large to detect it, then the SAT problem is solved in polynomial time. However, for small r, the probability is very small and this means we in fact don't get an information about the existence of the solution of the equation $t(C_0) = 1$, so that in such a case we need further deliberation.

Let us simplify our notations. After the quantum computation, the quantum computer will be in the state

$$|v_f\rangle = \sqrt{1-q^2}\,|\varphi_0\rangle\otimes|0\rangle + q\,|\varphi_1\rangle\otimes|1\rangle$$

where $|\varphi_1\rangle$ and $|\varphi_0\rangle$ are normalized n qubit states and $q = \sqrt{r/2^n}$. Effectively our problem is reduced to the following 1 qubit problem. We have the state

$$|\psi\rangle = \sqrt{1-q^2}\,|0\rangle + q\,|1\rangle$$

and we want to distinguish between the cases $q = 0$ and $q > 0$(small positive number).

It is argued in[15] that quantum computer can speed up **NP** problems quadratically but not exponentially. The no-go theorem states that if the inner product of two quantum states is close to 1, then the probability that a measurement distinguishes which one of the two it is exponentially small. And one could claim that amplification of this distinguishability is not possible.

At this point we emphasized[55] that we do not propose to make a measurement which will be overwhelmingly likely to fail. What we do it is a proposal to use the output $|\psi\rangle$ of the quantum computer as an input for another device which uses chaotic dynamics.

The amplification would be not possible if we use the standard model of quantum computations with a unitary evolution. However the idea of the papers[54,55] is different. In[54,55] it is proposed to combine quantum computer with a chaotic dynamics amplifier. Such a quantum chaos computer is a new model of computations and we demonstrate that the amplification is possible in the polynomial time.

One could object that we do not suggest a practical realization of the new model of computations. But at the moment nobody knows of how to make a practically useful implementation of the standard model of quantum computing ever. It seems to us that the quantum chaos computer considered in[55] has a potential to be realizable.

4.4. *Quantum chaos algorithm*

Various aspects of classical and quantum chaos have been the subject of numerous studies, see[43] and ref's therein. Here we will argue that chaos can play a constructive role in computations (see[54,55] for the details).

Chaotic behavior in a classical system usually is considered as an exponential sensitivity to initial conditions. It is this sensitivity we would like to use to distinguish between the cases $q = 0$ and $q > 0$ from the previous section.

Using the logistic map

$$x_{n+1} = a x_n (1 - x_n) \equiv f(x), \quad x_n \in [0,1]\,.$$

The properties of the map depend on the parameter a. If we take, for example, $a = 3.71$, the trajectory is very sensitive to the initial value and one has the chaotic behavior. It is important to notice that if the initial value $x_0 = 0$, then $x_n = 0$ for all n.

It is known that any classical algorithm can be implemented on quantum computer. Our quantum chaos computer will be consisting from two blocks. One block is the ordinary quantum computer performing computations with the output $|\psi\rangle = \sqrt{1-q^2}\,|0\rangle + q\,|1\rangle$. The second block is a computer performing computations of the *classical* logistic map. This two blocks should be connected in such a way that the state $|\psi\rangle$ first be transformed into the density matrix of the form

$$\rho = q^2 P_1 + \left(1 - q^2\right) P_0$$

where P_1 and P_0 are projectors to the state vectors $|1\rangle$ and $|0\rangle$. This connection is in fact nontrivial and actually it should be considered as the third block. One has to notice that P_1 and P_0 generate an Abelian algebra which can be considered as a classical system. In the second block the density matrix ρ above is interpreted as the initial data ρ_0, and we apply the logistic map as

$$\rho_m = \frac{(I + f^m(\rho_0)\sigma_3)}{2}$$

where I is the identity matrix and σ_3 is the z-component of Pauli matrix on $\mathbf{C}^2$. To find a proper value m we finally measure the value of σ_3 in the state ρ_m such that

$$M_m \equiv tr\rho_m\sigma_3.$$

We obtain

Theorem 4.2.

$$\rho_m = \frac{(I + f^m(q^2)\sigma_3)}{2}, \quad \text{and } M_m = f^m(q^2).$$

Thus the question is whether we can find such a m in polynomial steps of n satisfying the inequality $M_m \geq \frac{1}{2}$ for very small but non-zero q^2. Here we have to remark that if one has $q = 0$ then $\rho_0 = P_0$ and we obtain $M_m = 0$ for all m. If $q \neq 0$, the stochastic dynamics leads to the amplification of the small magnitude q in such a way that it can be detected as is explained below. The transition from ρ_0 to ρ_m is nonlinear and can be considered as a classical evolution because our algebra generated by P_0 and P_1 is abelian.

The amplification can be done within at most 2n steps due to the following propositions. Since $f^m(q^2)$ is x_m of the logistic map $x_{m+1} = f(x_m)$ with $x_0 = q^2$, we use the notation x_m in the logistic map for simplicity.

Theorem 4.3. *For the logistic map* $x_{n+1} = ax_n(1 - x_n)$ *with* $a \in [0, 4]$ *and* $x_0 \in [0, 1]$, *let* x_0 *be* $\frac{1}{2^n}$ *and a set* J *be* $\{0, 1, 2, \cdots, n, \cdots 2n\}$. *If* a *is* 3.71, *then there exists an integer* m *in* J *satisfying* $x_m > \frac{1}{2}$.

Theorem 4.4. *Let* a *and* n *be the same in the above proposition. If there exists* m_0 *in* J *such that* $x_{m_0} > \frac{1}{2}$, *then* $m_0 > \frac{n-1}{\log_2 3.71 - 1}$.

According to these theorems, it is enough to check the value x_m (M_m) around the above m_0 when q is $\frac{1}{2^n}$ for a large n. More generally, when $q = \frac{k}{2^n}$ with some integer k, it is similarly checked that the value x_m (M_m) becomes over $\frac{1}{2}$ within at most 2n steps.

The complexity of the quantum algorithm for the SAT problem was discussed in Section 3 to be in polynomial time. We have only to consider the number of steps in the classical algorithm for the logistic map performed on quantum computer. It is the probabilistic part of the construction and one has to repeat computations several times to be able to distinguish the cases $q = 0$ and $q > 0$. Thus it seems that the quantum chaos computer can solve the SAT problem in polynomial time.In conclusion of,[54,55] the quantum chaos computer combines the ordinary quantum computer with quantum chaotic dynamics amplifier. It may go beyond the usual quantum Turing algorithm, but such a device can be powerful enough to solve the **NP**-complete problems in the polynomial time.

In the following subsections we will discuss the SAT problem in adaptive dynamics.

Now from the general theory of stochastic limit[9] one knows that, under general ergodicity conditions, an interaction with an environment drives an adaptive dynamical (but not necessarily thermodynamical) equilibrium state which depends on the initial state of the environment and on the interaction Hamiltonian.

If one is able to realize experimentally these state dependent Hamiltonians, one would be able to drive the system S to a pre–assigned dynamical equilibrium state depending on the input state ψ_0.

4.5. *Stochastic limit and adaptive SAT Problem*

We illustrate the general scheme described in the previous section in the simplest case when the state space of the system is $\mathcal{H}_S \equiv \mathbf{C}^2$. We fix an orthonormal basis of $\mathcal{H}_S$ as $\{e_0, e_1\}$.

The unknown state (vector) of the system at time $t = 0$

$$\psi := \sum_{\varepsilon \in \{0,1\}} \alpha_\varepsilon e_\varepsilon = \alpha_0 e_0 + \alpha_1 e_1 \ ; \quad \|\psi\| = 1.$$

In the case of Sec. 3, $\alpha_1$corresponds to q and e_j does to $|j\rangle$ $(j = 0, 1)$. This vector after quantum computation of the SAT problem is taken as input and defines the interaction (*adaptive*) Hamiltonian in an external field

$$\begin{aligned} H_I &= \lambda|\psi\rangle\langle\psi| \otimes (A_g^+ + A_g) \\ &= \sum \lambda \alpha_\varepsilon \overline{\alpha}_{\varepsilon'} |e_\varepsilon\rangle\langle e_{\varepsilon'}| \otimes (A_g^+ + A_g) \end{aligned}$$

where λ is a small coupling constant. Here and in the following summation over repeated indices is understood.

The free system Hamiltonian is taken to be diagonal in the e_ε–basis

$$H_S := \sum_{\varepsilon \in \{0,1\}} E_\varepsilon |e_\varepsilon\rangle\langle e_\varepsilon| = E_0|e_0\rangle\langle e_0| + E_1|e_1\rangle\langle e_1|$$

and the energy levels are ordered so that $E_0 < E_1$.The environment Hamiltonian is

$$H_E := \int \omega\,(k)\, A_k^+ A_k dk,$$

where $\omega(k)$ is a function satisfying the basic analytical assumption of the stochastic limit. Thus the total free Hamiltonian is $H_0 := H_S + H_E$.

The free field evolution is given by

$$e^{itH_0} A^\pm g e^{-itH_0} = A^\pm_{S_t g},$$

where $S_t g(k) = e^{it\omega(k)} g(k)$.

We can distinguish two cases as below, whose cases correspond to two cases of Sec. 3, i.e., $q > 0$ and $q = 0$.

Case (1). If $\alpha_0, \alpha_1 \neq 0$, then, according to the general theory of stochastic limit (i.e., $t \to t/\lambda^2$),[9] the interaction Hamiltonian H_I is in the same universality class as

$$\tilde{H}_I = D \otimes A_g^+ + D^+ \otimes A_g$$

where $D := |e_0\rangle\langle e_1|$. The interaction Hamiltonian at time t is then

$$\tilde{H}_I(t) = e^{-it\omega_0} D \otimes A^+_{S_t g} + \text{ h.c.} = D \otimes A^+(e^{it(\omega(p)-\omega_0)} g) + \text{ h.c.},$$

where $\omega_0 = E_1 - E_0$. The white noise ($\{b_t\}$) Hamiltonian equation associated, via the stochastic golden rule, to this interaction Hamiltonian is

$$\partial_t U_t = i(Db_t^+ + D^+ b_t)U_t$$

Its causally normal ordered form is equivalent to the stochastic differential equation

$$dU_t = (iDdB_t^+ + iD^+ dB_t - \gamma D^+ D dt)U_t,$$

where $dB_t := b_t dt$ and γ is a certain constant. Then we derive the master equation as follows:

$$\frac{d}{dt}\rho_t = L_* \rho_t,$$

where $\rho_t := e^{tL_*}\rho$ and

$$L_* \rho := (Im\gamma)i[\rho, D^+ D] - (\text{Re }\gamma)\{\rho, D^+ D\} + (\text{Re }\gamma)D\rho D^+$$

For $\rho = \rho_0 := |e_0\rangle\langle e_0|$ one has

$$L_* \rho_0 = 0$$

so ρ_0 is an invariant measure. From the Fagnola–Rebolledo criteria,[23] it is the unique invariant measure and the semigroup $\exp(tL_*)$ converges exponentially to it.

Case (2). If $\alpha_1 = 0$, then the interaction Hamiltonian H_I is

$$H_I = \lambda |e_0\rangle\langle e_0| \otimes (A_g^+ + A_g)$$

and, according to the general theory of stochastic limit, the reduced evolution has no damping and corresponds to the pure Hamiltonian

$$H_S + |e_0\rangle\langle e_0| = (E_0 + 1)|e_0\rangle\langle e_0| + E_1 |e_1\rangle\langle e_1|$$

therefore, if we choose the eigenvalues E_1, E_0 to be integers (in appropriate units), then the evolution will be periodic.

Since the eigenvalues E_1, E_0 can be chosen a priori, by fixing the system Hamiltonian H_S, it follows that the period of the evolution can be known a priori. This gives a simple criterium for the solvability of the SAT problem because, by waiting a sufficiently long time one can experimentally detect the difference between a damping and an oscillating behavior.

We used the resulting (flag) state after quantum computation of the truth function of SAT to couple the external field and took the stochastic limit, then our final evolution becomes "linear" for the state ρ described as above. The stochastic limit is historically important to realize macroscopic (time) evolution and it is now rigorously established as explained in,[9] and we gave a general protocol to study the distinction of two cases $\alpha_1 \neq 0$ and $\alpha_1 = 0$ by this rigorous mathematics. The macro-time enables us to measure the behavior of the outcomes practically. Thus we show that it is possible to distinguish two different states, $\sqrt{1-q^2}\,|0\rangle + q\,|1\rangle$ $(q \neq 0)$ and $|0\rangle$ by means of the adaptive dynamics and the stochastic limit. This provides another algorithm solving NPC problem in realistic time.

Acknowledgment
The author thank Monka-Sho for financial support.

References

1. L.Accardi, *Urne e Camaleoni: Dialogo sulla realta, le leggi del caso e la teoria quantistica.* Il Saggiatore, Rome (1997)
2. L.Accardi, K.Imafuku, M.Refoli, On the EPR-Chameleon experiment, Infinite Dimensional Analysis, Quantum Probability and Related Topics Vol. 5, No. 1 (2002) 1-2
3. L.Accardi, M.Ohya, A Stochastic limit approach to the SAT problem", Proceedings of VLSI 2003, and Open systems and Information Dynamics (2004)
4. L.Accardi, M.Ohya, Compound channels, transition expectations, and liftings", Appl. Math. Optim., Vol.39, 33-59 (1999)
5. L.Accardi, M.Ohya, N.Watanabe, Note on quantum dynamical entropies Reports on mathematical physics, vol.38 n.3 457–469 (1996)
6. H.Araki, Relative entropy of states of von Neumann Algebras, Publ.RIMS, Kyoto Univ.Vol.11, 809-833 (1976)
7. L.Accardi, R.Sabbadini, On the Ohya–Masuda quantum SAT Algorithm, in: Proceedings International Conference UMC'01, Springer (2001)
8. K.T.Alligood, T.D.Sauer, J.A.Yorke, Chaos-An Introduction to Dynamical Systems-, Textbooks in Mathematical Sciences, Springer (1996)
9. L.Accardi, Y.G. Lu, I. Volovich: Quantum Theory and its Stochastic Limit. Springer Verlag 2002; Japanese translation, Tokyo–Springer 2003.
10. M. Garey and D. Johnson, *Computers and Intractability - a guide to the theory of NP-completeness*, Freeman, 1979.
11. R.Alicki, Quantum geometry of noncommutative Bernoulli shifts, Banach Center Publications, Mathematics Subject Classification 46L87 (1991)
12. R.Alicki, M.Fannes, Defining quantum dynamical entropy, Lett. Math. Physics, 32, 75-82 (1994)
13. M.Asano, M.Ohya and Y.Togawa, Entropic chaos degree of rotations and log-linear dynamics, QBIC proceedings (this volume), 2007.
14. F. Benatti, Deterministic Chaos in Infinite Quantum Systems, Springer (1993)
15. C. H. Bennett, E. Bernstein, G. Brassard, U. Vazirani, *Strengths and Weaknesses of Quantum Computing*, quant-ph/9701001.
16. R. Cleve, *An Introduction to Quantum Complexity Theory*, quant-ph/9906111.

17. D. Deutsch, *Quantum theory, the Church-Turing principle and the universal quantum computer,* Proc. of Royal Society of London series A, **400**, pp.97-117, 1985.
18. A. Ekert and R. Jozsa, *Quantum computation and Shor's factoring algorithm,* Reviews of Modern Physics, **68** No.3,pp.733-753, 1996.
19. A.Connes, H.Narnhofer, W.Thirring, Dynamical entropy of C*-algebras and von Neumann algebras, Commun.Math.Phys., 112, pp.691-719 (1987)
20. R.L.Devaney, An Introduction to Chaotic dynamical Systems, Benjamin (1986)
21. G.G.Emch, H.Narnhofer, W.Thirring and G.L.Sewell, Anosov actions on non-commutative algebras, J.Math.Phys., **35**, No.11, 5582-5599 (1994)
22. G.G.Emch, Algebraic Methods in Statistical Mechanics and Quantum Field Theory, Wiley (1972)
23. F. Fagnola and R. Rebolledo, On the existence of Stationary States for Quantum Dynamical Semigroups, to appear in J. Math. Phys., 2001.
24. K-H.Fichtner and M.Ohya, Quantum teleportation with entangled states given by beam splittings, *Communications in Mathematical Physics*, **222**, 229 (2001).
25. K-H.Fichtner and M.Ohya, Quantum teleportation and beam splitting, *Communications in Mathematical Physics*, **225**, 67 (2002).
26. K-H.Fichtner, W. Freutenberg and M.Ohya,Teleportation Schemes in Infinite Dimensional Hilbert Spaces, J. Math. Phys. 46, No. 10 (2006).
27. K.Inoue, M.Ohya, I.V.Volovich, Semiclassical properties and chaos degree for the quantum baker's map, J. Math. Phys., 43-2, 734-755 (2002)
28. K.Inoue, M.Ohya, I.V.Volovich, On quantum-classical correspondence for baker's map, quant-ph/0108107 (2001)
29. K.Inoue, M.Ohya and K.Sato, Application of chaos degree to some dynamical systems, Chaos, Soliton & Fractals, 11, 1377-1385 (2000)
30. R.S.Ingarden, A.Kossakowski, M.Ohya, Information Dynamics and Open Systems, Kluwer Publ. Comp. (1997)
31. K.Inoue, M.Ohya, A.Kossakowski, A Description of Quantum Chaos, Tokyo Univ. of Science preprint (2002)
32. S.Iriyama and M.Ohya, Rigorous Estimate for OMV SAT Algorithm, to appear in OSID, 2007.
33. S.Iriyama and M.Ohya, Language Classes Defined by Generalised Quantum Turing Machine, TUS preprint, 2007
34. S.Iriyama and M.Ohya, Review on Quantum Chaos Algorithm and Generalized Quantum Turing Machine, QBIC proceedings (this volume), 2007
35. S.Iriyama, M.Ohya and I.V.Volovich, Generalized Quantum Turing Machine and its Application to the SAT Chaos Algorithm, QP-PQ:Quantum Prob. White Noise Anal., Quantum Information and Computing, 19, World Sci. Publishing, 204-225, 2006.
36. A.Kossakowski, M.Ohya (2006) ; New scheme of quasntum teleportation process, to appear in Infinite Dimensional Analysis and Quantum Probability
37. A.Kossakowski, M.Ohya (2006) Can Non-Maximal Entangled State Achieve a Complete Quantum Teleportation?, Reconsideration of Foundation-3, Amer-

ican Institute of Physics, 810, 211-216.
38. A.Kossakowski, M.Ohya, Y.Togawa, How can we observe and describe chaos? Open System and Information Dynamics, 10(3):221-233 (2003)
39. A.Kossakowski, M.Ohya, N.Watanabe, Quantum dynamical entropy for completely positive maps, Infinite Dimensional Analysis, Quantum Probability and Related Topics, **2**, No.2, 267-282 (1999)
40. N.Muraki, M.Ohya, Entropy functionals of Kolmogorov-Sinai type and their limit theorems, Letter in Mathematical Physics,36, 327-335 (1996)
41. M.Ohya, On compound state and mutual information in quantum information theory, IEEE Trans. Information Theory, 29, No.5, 770-774 (1983)
42. M.Ohya, Some aspects of quantum information theory and their applications to irreversible processes, Rep.Math.Phys., Vol.27, 19-47 (1989)
43. M.Ohya, Complexities and their applications to characterization of chaos, International Journal of Theoretical Physics,Vol.37, No.1, 495-505 (1998)
44. M.Ohya, State change, complexity and fractal in quantum systems, Quantum Communications and Measurement, Plenum Press, New York, 309-320 (1995)
45. M.Ohya, Complexity and fractal dimensions for quantum states, Open Systems and Information Dynamics, 4, 141-157 (1997)
46. M.Ohya, Note on quantum proability, L.Nuovo Cimento, Vol.38, NO1 0, 203-206, (1983)
47. M.Ohya, Information dynamics and its applications to optical communication processes, Lecture Note in Physics, **378**, 81-92 (1991)
48. M.Ohya, Entropy Transmission in C*-dynamical systems, J.Math. Anal.Appl., 100, No.1, 222-235 (1984)
49. M.Ohya (2004) Foundation of Chaos Through Observation, Quantum Information and Complexity edited by T.Hida, K.Saito and Si Si,391-410.
50. M.Ohya (2005): Adaptive Dynamics in Quantum Information and Chaos, in "Stochastic Analysis: Classical and Quantum" ed. by Hida, 127-142.
51. M.Ohya et al, Adaptive dynamics its use in understanding of chaos, TUS preprint
52. M.Ohya, N.Masuda, NP problem in quantum algorithm, Open Systems and Information Dynamics, Vol.7, No.1, 33-39 (2000)
53. M.Ohya, D.Petz, Quantum Entropy and its Use, Springer-Verlag (1993)
54. M.Ohya, I.V.Volovich, New quantum algorithm for studying NP-complete problems, Rep.Math.Phys.,52, No.1,25-33 (2003) and Quantum computing and chaotic amplifier, J.Opt.B (2003)
55. M.Ohya, I.V.Volovich, Mathematical Foundations of Quantum Information and Quantum Computation, to be published in Springer-Verlag
56. P.W. Shor, *Algorithm for quantum computation : Discrete logarithm and factoring algorithm*, Proceedings of the 35th Annual IEEE Symposium on Foundation of Computer Science, pp.124-134, 1994.
57. H.Umegaki, Conditional expectation in operator algebra IV, Kodai Math. Sem. Rep., 14, 59-85 (1962)

JOURNAL OF MATHEMATICAL PHYSICS **46**, 102103 (2005)

Teleportation schemes in infinite dimensional Hilbert spaces

Karl-Heinz Fichtner[a)]
Institut für Angewandte Mathematik, Friedrich-Schiller-Unversität Jena, 07740 Jena, Germany

Wolfgang Freudenberg[b)]
Institut für Mathematik, PF 101344, Brandenburgische Technische Universität Cottbus, 03013 Cottbus, Germany

Masanori Ohya[c)]
Department of Information Science, Tokyo University of Science, Noda City, Chiba 278-8510, Japan

(Received 21 June 2005; accepted 2 August 2005; published online 14 October 2005)

The success of quantum mechanics is due to the discovery that nature is described in infinite dimension Hilbert spaces, so that it is desirable to demonstrate the quantum teleportation process in a certain infinite dimensional Hilbert space. We describe the teleportation process in an infinite dimensional Hilbert space by giving simple examples. © *2005 American Institute of Physics.*
[DOI: 10.1063/1.2044647]

I. INTRODUCTION

In quantum communication theory, we code information by quantum states and send it through a quantum device that is properly designed. If one can send any quantum state from an input system to an output system as it is, that is, if one can find such a method sending an input state without changing it, then it will be an ultimate way for information transmission. It is in quantum teleportation that we can discuss such an ultimate communication system.

The problem of quantum teleportation is whether there exists a physical device and a key (or a set of keys) by which a quantum state attached to a sender (Alice) is completely transmitted and a receiver (Bob) can reconstruct the state sent. Bennett *et al.*[1] showed that such teleportation is possible through a device (channel) made from proper (EPR) entangled states of Bell basis. The basic idea behind their discussion is to divide the information encoded in the state into two parts, classical and quantum, and send them through different channels, a classical channel and an EPR channel. The classical channel is nothing but a simple correspondence between sender and receiver, and the EPR channel is constructed by using a certain entangled state. However the EPR channel is not so stable due to quick decoherence. Fichtner and Ohya[2,3] studied the quantum teleportation by means of general beam splitting processes in Bose Fock space so that it contains the EPR channel as a special case, and they constructed a stable teleportation process with coherent entangled states.

However, all these discussions have been based on finite dimensionality of the Hilbert spaces, attached to Alice and Bob. As is well known, success of quantum mechanics is due to the discovery that nature is described in infinite dimension Hilbert spaces, so that it is desirable to demonstrate the quantum teleportation process in a certain infinite dimensional Hilbert space. This paper is a trial to describe the teleportation process in an infinite dimensional Hilbert space by

[a)]Electronic mail: fichtner@mathematik.uni-jena.de
[b)]Electronic mail: freudenberg@math.tu-cottbus.de
[c)]Electronic mail: ohya@is.noda.tus.ac.jp

giving simple examples. In Sec. II, we fix the notations based on Fock space discussion of the series of papers.[2–4] In Sec. III, the channel expression of the teleportation is reviewed and the entanglement between Alice and Bob is constructed by an isometry operator, on which an operator expression of the teleportation channel is given, and some extreme cases of the teleportation are considered. To be closer to usual teleportation schemes and to get simple and explicit results we consider in Sec. IV the case of product states. In Sec. V, the existence of unitary keys is discussed.

II. BASIC NOTIONS AND NOTATIONS

We consider three complex Hilbert spaces $\mathcal{H}_1$, $\mathcal{H}_2$, and $\mathcal{H}_3$. To Alice there are attached the Hilbert spaces $\mathcal{H}_1$ and $\mathcal{H}_2$. Alice wants to teleportate a state ρ on $\mathcal{H}_1$ to Bob to whom there is attached the Hilbert space $\mathcal{H}_3$. Usually it is assumed that all three Hilbert spaces are finite-dimensional ones. This is also necessary for obtaining perfect teleportation. In the present article we will consider the case of Hilbert spaces being separable but not necessarily finite dimensional. We assume that all three spaces are either infinite-dimensional separable Hilbert spaces or finite-dimensional ones with same finite dimension. The paper continues and generalizes results obtained in Refs. 2–4.

Let us be given orthonormal bases,

$$(\xi_k^1)_{k\in G}, \quad (\xi_k^2)_{k\in G}, \quad (\xi_k^3)_{k\in G}, \tag{1}$$

in $\mathcal{H}_1$, $\mathcal{H}_2$, and $\mathcal{H}_3$ where the at most countable index set G is an abelian (additive) group with operation $\oplus$. An important case is that G is the set of integers $\mathbb{Z}$ where the group operation $\oplus$ will be usual addition. Since we need the structure of a group it is more convenient for our purposes to choose only orthonormal bases consisting of two-sided infinite sequences. To include usual teleportation models (with finite index space G) we consider also the case $G=\{1,\ldots,N\}$ with N belonging to the set $\mathbb{N}$ of natural numbers. In this case the operation $\oplus : G\times G\to G$ is defined by $k\oplus l := (k+l) \bmod N$. The operation inverse to $\oplus$ we denote by $\ominus$. In the latter case $k\ominus l=k-l$ if $k>l$ and $k\ominus l=k-l+N$ if $k\leq l$.

The algebra of all bounded linear operators on a Hilbert space $\mathcal{H}$ we will denote by $\mathcal{B}(\mathcal{H})$. Throughout this article we will assume that all states on a Hilbert space $\mathcal{H}$ are normal states (on $\mathcal{B}(\mathcal{H})$). The set of all normal states we denote by $\mathcal{S}(\mathcal{H})$.

Let $V\in\mathcal{B}(\mathcal{H}_2)$ be an arbitrary unitary operator. Consequently, the sequence $(\eta_n)_{n\in G}$ with $\eta_n := V\xi_n^2, n\in G$ is a second orthonormal basis in $\mathcal{H}_2$. Thus there exists a sequence $(b_{kl})_{k,l\in G}$ such that

$$\eta_k = \sum_{l\in G} b_{kl}\xi_l^2 \quad (k\in G). \tag{2}$$

Obviously, the sequence $(b_{kl})_{k,l\in G}$ has to fulfill

$$\sum_{l\in G} \overline{b_{kl}}b_{ml} = \delta_{k,m} \quad (k,m\in G), \tag{3}$$

where $\delta_{k,m}$ denotes the Kronecker symbol. Observe that for all $m,k\in G$ it holds $\langle \eta_m, \xi_k^2\rangle = \overline{b_{mk}}$. Consequently, we obtain

$$\xi_k^2 = \sum_{m\in G} \overline{b_{mk}}\,\eta_m \quad (k\in G). \tag{4}$$

Since V^* is again unitary and $\xi_n^2 = V^*\eta_n$ we also have

$$\sum_{l\in G} \overline{b_{lk}}b_{lm} = \delta_{k,m} \quad (k,m\in G). \tag{5}$$

Remark: To simplify notations only if there appear ambiguities we will separate multiple indices by "commas," i.e., usually we write b_{kl} instead of $b_{k,l}$.

Further, we consider a sequence $(U_m)_{m\in G}$ of unitary operators on $\mathcal{H}_2$ acting as shift operators on the elements of the (original) basis:

$$U_m\xi_k^2 = \xi_{k\oplus m}^2 \quad (m,k \in G). \tag{6}$$

The Hilbert space $\mathcal{H}_2$ is connected by simple isometries S_1 to $\mathcal{H}_1$ and S_3 to $\mathcal{H}_3$:

$$S_1:\mathcal{H}_2 \to \mathcal{H}_1 \otimes \mathcal{H}_2, \quad S_1(\xi_k^2) = \xi_k^1 \otimes \xi_k^2 \quad (k \in G), \tag{7}$$

$$S_3:\mathcal{H}_2 \to \mathcal{H}_2 \otimes \mathcal{H}_3, \quad S_3(\xi_k^2) = \xi_k^2 \otimes \xi_k^3 \quad (k \in G). \tag{8}$$

Finally, we construct a new basis in $\mathcal{H}_1 \otimes \mathcal{H}_2$ by setting

$$\xi_{km} = (\mathbb{1} \otimes U_m)S_1\eta_k = (\mathbb{1} \otimes U_m)S_1V\xi_k^2 \quad (k,m \in G), \tag{9}$$

where $\mathbb{1}$ denotes the identical operator (from the context it always will be clear on which space $\mathbb{1}$ operates). Observe that for $k,m \in G$

$$\xi_{km} = (\mathbb{1} \otimes U_m)\sum_{l\in G} b_{kl}(\xi_l^1 \otimes \xi_l^2) = \sum_{l\in G} b_{kl}(\xi_l^1 \otimes \xi_{l\oplus m}^2). \tag{10}$$

Proposition 1: The sequence $(\xi_{km})_{k,m\in G}$ is an orthonormal basis in $\mathcal{H}_1 \otimes \mathcal{H}_2$.

Proof: For $n,m,r,s \in G$ we get using (10) and (3)

$$\langle \xi_{nm}, \xi_{rs}\rangle = \sum_{k,l\in G} \overline{b_{nk}}b_{rl}\langle \xi_k^1 \otimes \xi_{k\oplus m}^2, \xi_l^1 \otimes \xi_{l\oplus s}^2\rangle = \sum_{k,l\in G} \overline{b_{nk}}b_{rl}\langle \xi_k^1, \xi_l^1\rangle\langle \xi_{k\oplus m}^2, \xi_{l\oplus s}^2\rangle$$

$$= \sum_{k\in G} \overline{b_{nk}}b_{rk}\langle \xi_{k\oplus m}^2, \xi_{k\oplus s}^2\rangle = \delta_{n,r}\cdot\delta_{m,s}.$$

Thus $(\xi_{km})_{k,m\in G}$ is an orthonormal system in $\mathcal{H}_1 \otimes \mathcal{H}_2$.

To prove completeness we check whether the Parseval equality

$$\|\Phi\|^2 = \sum_{n,m\in G} |\langle \Phi, \xi_{nm}\rangle|^2$$

holds for all $\Phi \in \mathcal{H}_1 \otimes \mathcal{H}_2$. Let

$$\Phi = \sum_{r,s\in G} \alpha_{rs}\xi_r^1 \otimes \xi_s^2. \tag{11}$$

Then

$$\langle \Phi, \xi_{nm}\rangle = \sum_{r,s,k\in G} \overline{\alpha_{rs}}b_{nk}\langle \xi_r^1 \otimes \xi_s^2, \xi_k^1 \otimes \xi_{k\oplus m}^2\rangle = \sum_{r,s\in G} \overline{\alpha_{rs}}b_{nr}\langle \xi_s^2, \xi_{r\oplus m}^2\rangle = \sum_{r\in G} \overline{\alpha_{r,r\oplus m}}b_{nr}.$$

Using (5) this implies

$$\sum_{n,m\in G} |\langle \Phi, \xi_{nm}\rangle|^2 = \sum_{n,m,r,s\in G} \overline{\alpha_{s,s\oplus m}}\alpha_{r,r\oplus m}\overline{b_{nr}}b_{ns} = \sum_{m,r,s\in G} \overline{\alpha_{s,s\oplus m}}\alpha_{r,r\oplus m}\sum_{n\in G} \overline{b_{nr}}b_{ns}$$

$$= \sum_{m,r,s\in G} \overline{\alpha_{s,s\oplus m}}\alpha_{r,r\oplus m}\delta_{r,s} = \sum_{m,r\in G} |\alpha_{r,r\oplus m}|^2 = \sum_{m,r\in G} |\alpha_{r,m}|^2 = \|\Phi\|^2.$$

This ends the proof. □

We denote by $F_{nm} \in \mathcal{B}(\mathcal{H}_1 \otimes \mathcal{H}_2)$ the projection onto ξ_{nm}, i.e.,

$$F_{nm} := |\xi_{nm}\rangle\langle\xi_{nm}| = \langle \xi_{nm}, \cdot\,\rangle\xi_{nm} \quad (n,m \in G). \tag{12}$$

Remark: We will use as well the "scalar product" notation as also the "bra and ket" symbols.

However, using the symbol $|\Phi\rangle$ let us make the convention that the symbol has to denote the normalization of the vector Φ, i.e.,

$$|\Phi\rangle := \frac{\Phi}{\|\Phi\|}. \tag{13}$$

Observe that for $\Phi \in \mathcal{H}_1 \otimes \mathcal{H}_2$ given by (11) and for all $n,m \in G$ one obtains

$$F_{nm}\Phi = \langle \xi_{nm}, \Phi\rangle \xi_{nm} = \sum_{k\in G} \overline{b_{nk}} \langle \xi_k^1 \otimes \xi_{k\oplus m}^2, \Phi\rangle \xi_{nm} = \sum_{k,r,s\in G} \overline{b_{nk}} \alpha_{rs} \langle \xi_k^1 \otimes \xi_{k\oplus m}^2, \xi_r^1 \otimes \xi_s^2\rangle \xi_{nm}$$

$$= \left(\sum_{k\in G} \overline{b_{nk}} \alpha_{k,k\oplus m} \right) \xi_{nm}. \tag{14}$$

Especially,

$$F_{nm}(\xi_r^1 \otimes \xi_s^2) = \delta_{s,r\oplus m} \cdot \overline{b_{nr}} \xi_{nm}. \tag{15}$$

In the subsequent sections we investigate concrete teleportation channels. For this we need an explicit expression for $(F_{nm} \otimes 1)(1 \otimes S_3)$ which maps $\mathcal{H}_1 \otimes \mathcal{H}_2$ into $\mathcal{H}_1 \otimes \mathcal{H}_2 \otimes \mathcal{H}_3$.

Proposition 2: Let $\Phi \in \mathcal{H}_1 \otimes \mathcal{H}_2$ be given by (11). For all $n,m \in G$ it holds

$$(F_{nm} \otimes 1)(1 \otimes S_3)\Phi = \xi_{nm} \otimes \sum_{r\in G} \overline{b_{nr}} \alpha_{r,r\oplus m} \xi_{r\oplus m}^3. \tag{16}$$

Proof: Using definition (8) of the imbedding operator S_3 and (15) we obtain for all $n,m \in G$

$$(F_{nm} \otimes 1)(1 \otimes S_3)\Phi = \sum_{r,s\in G} \alpha_{rs}(F_{nm} \otimes 1)(1 \otimes S_3)\xi_r^1 \otimes \xi_s^2 = \sum_{r,s\in G} \alpha_{rs}(F_{nm} \otimes 1)\xi_r^1 \otimes \xi_s^2 \otimes \xi_s^3$$

$$= \sum_{r,s\in G} \alpha_{rs} \overline{b_{nr}} \xi_{nm} \otimes \xi_s^3 \cdot \delta_{s,r\oplus m} = \sum_{r\in G} \alpha_{r,r\oplus m} \overline{b_{nr}} \xi_{nm} \otimes \xi_{r\oplus m}^3.$$

This proves (16). □

Corollary 3: Let $\Phi \in \mathcal{H}_1 \otimes \mathcal{H}_2$ be given by (11) and assume $\|\Phi\|=1$. For all $n,m \in G$ it holds

$$(F_{nm} \otimes 1)(1 \otimes S_3)|\Phi\rangle\langle\Phi|(1 \otimes S_3^*)(F_{nm} \otimes 1) = F_{nm} \otimes \langle \Psi_{nm}, \cdot\rangle \Psi_{nm} \tag{17}$$

with

$$\Psi_{nm} = \sum_{r\in G} \overline{b_{nr}} \alpha_{r,r\oplus m} \xi_{r\oplus m}^3. \tag{18}$$

Proof: Immediately from Proposition 2 we obtain

$$(F_{nm} \otimes 1)(1 \otimes S_3)|\Phi\rangle\langle\Phi|(1 \otimes S_3^*)(F_{nm} \otimes 1) = \left\langle \sum_{r\in G} \alpha_{r,r\oplus m} \overline{b_{nr}} \xi_{nm} \otimes \xi_{r\oplus m}^3, \cdot \right\rangle \sum_{s\in G} \alpha_{s,s\oplus m} \overline{b_{ns}} \xi_{nm}$$

$$\otimes \xi_{s\oplus m}^3 = |\xi_{nm}\rangle\langle\xi_{nm}|$$

$$\otimes \left\langle \sum_{r\in G} \alpha_{r,r\oplus m} \overline{b_{nr}} \xi_{r\oplus m}^3, \cdot \right\rangle \sum_{s\in G} \alpha_{s,s\oplus m} \overline{b_{ns}} \xi_{s\oplus m}^3.$$

This proves (17). □

Corollary 3 allows us to get explicit formulas for

$$(F_{nm} \otimes 1)(1 \otimes S_3)\varkappa(1 \otimes S_3^*)(F_{nm} \otimes 1),$$

where $\varkappa$ is a normal state on $\mathcal{B}(\mathcal{H}_1 \otimes \mathcal{H}_2)$.

Let $\varkappa$ be given in the form

$$\varkappa = \sum_{u,v \in G} \lambda_{uv} |\Phi_{uv}\rangle\langle\Phi_{uv}| \tag{19}$$

with $(\Phi_{uv})_{u,v \in G}$ being an orthonormal sequence in $\mathcal{H}_1 \otimes \mathcal{H}_2$,

$$\Phi_{uv} = \sum_{r,s \in G} \alpha_{uvrs} \xi_r^1 \otimes \xi_s^2 \tag{20}$$

and $(\lambda_{uv})_{u,v \in G}$ fulfilling

$$\sum_{u,v \in G} \lambda_{uv} = 1, \quad \lambda_{uv} \geqslant 0. \tag{21}$$

Of course, we have for all $u, \tilde{u}, v, \tilde{v} \in G$

$$\sum_{r,s \in G} \overline{\alpha_{uvrs}} \alpha_{\tilde{u}\tilde{v}rs} = \delta_{u,\tilde{u}} \cdot \delta_{v,\tilde{v}}. \tag{22}$$

As an immediate consequence of Corollary 3 we obtain the following result.

Proposition 4: Let $\varkappa$ be given by (19). For all $n, m \in G$ it holds

$$(F_{nm} \otimes 1)(1 \otimes S_3)\varkappa(1 \otimes S_3^*)(F_{nm} \otimes 1) = F_{nm} \otimes \sum_{u,v \in G} \lambda_{uv} \langle \Psi_{uvnm}, \cdot \rangle \Psi_{uvnm}$$

$$= F_{nm} \otimes \sum_{u,v \in G} \lambda_{uv} \|\Psi_{uvnm}\|^2 \cdot |\Psi_{uvnm}\rangle\langle\Psi_{uvnm}| \tag{23}$$

with

$$\Psi_{uvnm} = \sum_{r \in G} \overline{b_{nr}} \alpha_{uvr,r \oplus m} \xi^3_{r \oplus m}. \tag{24}$$

Remark: Observe that the vectors Ψ_{uvnm} usually are not normalized. Further, in general the sequence $(\Psi_{uvnm})_{u,v \in G}$ is not an orthogonal one.

III. THE TELEPORTATION CHANNEL

A. The measurement

Now we apply the model which was used in several papers[2,3,5] for the description of a teleportation scheme. The measurement will be done with the operator

$$F = \sum_{n,m \in G} z_{nm} F_{nm}, \tag{25}$$

where $(z_{nm})_{n,m \in G}$ is a sequence of real numbers and $(F_{nm})_{n,m \in G}$ is the family of orthogonal projections on $\mathcal{H}_1 \otimes \mathcal{H}_2$ introduced in Sec. II (cf. (12) and (10)). In the above-mentioned articles one considers the case of a given state ρ on $\mathcal{H}_1$ (that has to be teleportated to Bob) and an entangled state σ on $\mathcal{H}_2 \otimes \mathcal{H}_3$. Thus the whole system is prepared in the state $\rho \otimes \sigma$. Alice makes a measurement (restricted to $\mathcal{H}_1 \otimes \mathcal{H}_2$) with the operator F given by (25), i.e., the operator $F \otimes 1$ is applied to the system being in the state $\rho \otimes \sigma$. As the result of the measurement Alice obtains a value z_{nm} for some $n, m \in G$. Consequently, after the measurement the whole system will be in the state θ_{nm} on $\mathcal{H}_1 \otimes \mathcal{H}_2 \otimes \mathcal{H}_3$ given by

$$\theta_{nm} := \frac{(F_{nm} \otimes 1)\rho \otimes \sigma(F_{nm} \otimes 1)}{\mathrm{tr}((F_{nm} \otimes 1)\rho \otimes \sigma(F_{nm} \otimes 1))}, \tag{26}$$

where tr denotes the full trace with respect to $\mathcal{H}_1 \otimes \mathcal{H}_2 \otimes \mathcal{H}_3$. Bob who is informed about the result of the measurement controls the state $\tilde{\Lambda}_{nm}(\rho)$ on $\mathcal{H}_3$ being the partial trace of θ_{nm} with respect to $\mathcal{H}_1 \otimes \mathcal{H}_2$:

$$\tilde{\Lambda}_{nm}(\rho) := \mathrm{tr}_{12}(\theta_{nm}) = \frac{\mathrm{tr}_{12}((F_{nm}\otimes 1)\rho\otimes\sigma(F_{nm}\otimes 1))}{\mathrm{tr}((F_{nm}\otimes 1)\rho\otimes\sigma(F_{nm}\otimes 1))}, \tag{27}$$

where tr_{12} denotes the partial trace on $\mathcal{H}_1\otimes\mathcal{H}_2$. Of course we have to assume that the denominator in (26), respectively, in (27) is nonzero. Otherwise, the left-hand side has to be set equal to zero. The mapping $\tilde{\Lambda}_{nm}:\mathcal{S}(\mathcal{H}_1)\to\mathcal{S}(\mathcal{H}_3)$ is called a *teleportation channel.*

The teleportation works perfectly if Bob is able to reconstruct the initial state ρ from $\tilde{\Lambda}_{nm}(\rho)$. We will return to this question in the subsequent sections. Teleportation channels $\tilde{\Lambda}_{nm}$ of the above-mentioned type might be useful also for modeling other transformations of states. For instance in Ref. 4 we proposed an extremely simplified model of certain recognition processes based on a teleportation channel. In this model the spaces $\mathcal{H}_1$, $\mathcal{H}_2$, and $\mathcal{H}_3$ represent the processing part (brain), the memory before and after recognition.

B. The entanglement

Instead of the state $\rho\otimes\sigma$ on $\mathcal{H}_1\otimes\mathcal{H}_2\otimes\mathcal{H}_3$ we will consider now states of the form

$$(1\otimes S_3)\varkappa(1\otimes S_3^*) \tag{28}$$

with $\varkappa$ being a state on $\mathcal{H}_1\otimes\mathcal{H}_2$ and S_3 isometry (8) coupling $\mathcal{H}_1$ to $\mathcal{H}_3$. The simple entanglement between $\mathcal{H}_2$ and $\mathcal{H}_3$ is achieved just by applying S_3. Especially, if $\varkappa$ has the form $\varkappa=\rho^1\otimes\rho^2$ with ρ^j being a state on $\mathcal{H}_j$, $j=1,2$ then using the above cited notations from Refs. 2 and 3 we get $\rho=\rho^1$ and $\sigma=S_3\rho^2S_3^*$. This case of $\varkappa$ being a product state we will discuss in Sec. IV.

We will consider now the channel $\Lambda_{nm}:\mathcal{S}(\mathcal{H}_1\otimes\mathcal{H}_2)\to\mathcal{S}(\mathcal{H}_3)$ given by

$$\Lambda_{nm}(\varkappa) = \frac{\mathrm{tr}_{1,2}((F_{nm}\otimes 1)(1\otimes S_3)\varkappa(1\otimes S_3^*)(F_{nm}\otimes 1))}{\mathrm{tr}((F_{nm}\otimes 1)(1\otimes S_3)\varkappa(1\otimes S_3^*)(F_{nm}\otimes 1))}, \tag{29}$$

where as in (27) tr_{12} denotes the partial trace with respect to $\mathcal{H}_1\otimes\mathcal{H}_2$ and tr the full trace on $\mathcal{H}_1\otimes\mathcal{H}_2\otimes\mathcal{H}_3$. Again we have to assume that the denominator in (29) is greater than zero. Otherwise we set $\Lambda_{nm}(\varkappa)=0$. Observe that for product states $\varkappa=\rho\otimes\sigma$ one has the relation $\tilde{\Lambda}_{nm}(\rho)=\Lambda_{nm}(\rho\otimes\sigma)$.

Remark: Since $\mathrm{tr}_{1,2}((F_{nm}\otimes 1)(1\otimes S_3)\varkappa(1\otimes S_3^*)(F_{nm}\otimes 1))$ is a positive linear functional on $\mathcal{H}_3$ the denominator in (29) can be equal to zero only if this functional is zero. In other words, in this case no information about the measured value z_{nm} can be transmitted to Bob (in a brain model this would mean that no information about the input signal will be stored in the memory). The following result is an immediate consequence of Proposition 4.

Theorem 5: *Let $\varkappa$ be a state on $\mathcal{H}_1\otimes\mathcal{H}_2$ given by (19)–(21). Then for all $n,m\in G$*

$$\Lambda_{nm}(\varkappa) = \frac{\sum_{u,v\in G}\lambda_{uv}\langle\Psi_{uvnm},\cdot\rangle\Psi_{uvnm}}{\mathrm{tr}_3\left(\sum_{u,v\in G}\lambda_{uv}\langle\Psi_{uvnm},\cdot\rangle\Psi_{uvnm}\right)} \tag{30}$$

where $\Psi_{uvnm}\in\mathcal{H}_3$ is given by (24), and $\Lambda_{nm}(\varkappa)$ has to be set equal to zero if the denominator in (30) is equal to zero (tr_3 denotes the trace in $\mathcal{H}_3$).

Sometimes it is more convenient (cf. the second remark in this article) to write (30) in the form

$$\Lambda_{nm}(\varkappa) = \frac{\sum_{u,v\in G}\lambda_{uv}\|\Psi_{uvnm}\|^2|\Psi_{uvnm}\rangle\langle\Psi_{uvnm}|}{\mathrm{tr}\left(\sum_{u,v\in G}\lambda_{uv}\|\Psi_{uvnm}\|^2|\Psi_{uvnm}\rangle\langle\Psi_{uvnm}|\right)}. \tag{31}$$

Of course, mixed states are not necessarily transformed into mixed ones—possibly there is only one pair u,v such that $\|\Psi_{uvnm}\|>0$. However, immediately from (30) we may conclude the following proposition.

Proposition 6: If $\varkappa$ is a pure state on $\mathcal{H}_1 \otimes \mathcal{H}_2$ then $\Lambda_{nm}(\varkappa)$ is a pure state on $\mathcal{H}_3$ or equal to zero.

C. Examples

Example 1: Let $\varkappa = |\Phi\rangle\langle\Phi|$ be a pure state on $\mathcal{H}_1 \otimes \mathcal{H}_2$, $\|\Phi\| = 1$, $\Phi = \Sigma_{r,s\in G}\alpha_{rs}\xi_r^1 \otimes \xi_s^2$. Then we obtain

$$\Lambda_{nm}(\varkappa) = \frac{1}{\|\Psi_{nm}\|^2}\langle\Psi_{nm}, \cdot\,\rangle\Psi_{nm} = |\Psi_{nm}\rangle\langle\Psi_{nm}|$$

with

$$\Psi_{nm} = \sum_{r\in G}\overline{b_{nr}}\alpha_{r,r\oplus m}\xi^3_{r\oplus m},$$

provided $\|\Psi_{nm}\| > 0$. Especially, for $\varkappa = |\Phi\rangle\langle\Phi|$ with $\Phi = \xi_r^1 \otimes \xi_s^2$ we get

$$\Lambda_{nm}(\varkappa) = |\xi^3_{r\oplus m}\rangle\langle\xi^3_{r\oplus m}|$$

provided $b_{n,s\ominus m} \neq 0$.

Let us discuss this result. Measuring the value z_{nm} means that there was made the projection onto $\xi_{nm} = \Sigma_{r\in G} b_{nr}(\xi_r^1 \otimes \xi^2_{r\oplus m})$. Observe that $\Sigma_{r\in G}|b_{nr}|^2|\alpha_{r,r\oplus m}|^2 = \|\Psi_{nm}\|^2 > 0$ if and only if there exists at least one $r \in G$ such that $b_{nr} \neq 0$ and $\alpha_{r,r\oplus m} \neq 0$. Obviously, there exists $r \in G$ such that $b_{nr} \neq 0$. The number $\alpha_{r,r\oplus m}$ is the coefficient of the basis element $\xi_r^1 \otimes \xi^2_{r\oplus m}$ in the expansion of $\varkappa = |\Phi\rangle\langle\Phi|$. So only if $\varkappa$ is receptive for the signal $\xi_r^1 \otimes \xi^2_{r\oplus m}$ for at least one admissible r there will be an output transmitted to Bob.

Example 2: We consider the case $G = \mathbb{Z}$ and fix an $N \in \mathbb{N}$. Let the state $\varkappa$ be a finite mixture of basis elements with equal weights:

$$\varkappa = \sum_{u,v=-N}^{N}\frac{1}{(2N+1)^2}|\xi_u^1 \otimes \xi_v^2\rangle\langle\xi_u^1 \otimes \xi_v^2|. \tag{32}$$

Since in this case $\alpha_{uvrs} = \delta_{u,r}\cdot\delta_{v,s}$ we get from (24)

$$\Psi_{uvnm} = \begin{cases}\delta_{v,u+m}\cdot\overline{b_{nu}}\xi^3_{u+m}, & u,v \in \{-N,\ldots,N\}\\ 0 & \textit{elsewhere}.\end{cases}$$

Especially,

$$\|\Psi_{uvnm}\|^2 = \begin{cases}|b_{nu}|^2, & u,v \in \{-N,\ldots,N\}, v-u=m\\ 0 & \textit{elsewhere}.\end{cases}$$

The conditions $u,v \in \{-N,\ldots,N\}$ and $v-u=m$ imply $m \leq 2N$. So we finally get for $n \in G$ and $m \leq 2N$

$$\Lambda_{nm}(\varkappa) = \frac{1}{C}\sum_{u=-N}^{N-m}|b_{nu}|^2|\xi^3_{u+m}\rangle\langle\xi^3_{u+m}|$$

provided

$$C = \sum_{u=-N}^{N-m}|b_{nu}|^2 > 0.$$

For $m > 2N$ the numerator in (30) is equal to zero. That means that measuring the value z_{nm} with $m > 2N$ it is not possible to send any information to Bob.

In further investigations one should consider more refined measurements than simple projections onto one-dimensional subspaces which are obtained just by a change of the basis and a shifting procedure. The next examples elucidates that the whole procedure becomes trivial if we even do not change the basis.

Example 3: Let us consider the (extreme) case that the second basis in $\mathcal{H}_2$ is the old one, i.e., $\eta_n=\xi_n^2$ for all $n\in G$. This implies $b_{nk}=\delta_{n,k}$, and the projection operator F_{nm} is the projection onto $\xi_{nm}=\xi_n^1\otimes\xi_{n\oplus m}^2$. Let $\varkappa$ be an arbitrary normal state on $\mathcal{H}_1\otimes\mathcal{H}_2$ given in the forms (19)–(21). If $\sum_{u,v}\lambda_{uv}|\alpha_{uvn,n\oplus m}|^2=0$ there will be no output, i.e., $\Lambda_{nm}(\varkappa)=0$. If $\sum_{u,v}\lambda_{uv}|\alpha_{uvn,n\oplus m}|^2>0$ we obtain $\Lambda_{nm}(\varkappa)=|\xi_{n\oplus m}^3\rangle\langle\xi_{n\oplus m}^3|$. Only if the vector $\xi_n^1\otimes\xi_{n\oplus m}^3$ appears in at least one nonzero component Φ_{uv} of the state $\varkappa$ some information will be transmitted and the final state $\Lambda_{nm}(\varkappa)$ will be the pure state $|\xi_{n\oplus m}^3\rangle\langle\xi_{n\oplus m}^3|$.

IV. TRANSFORMATION OF PRODUCT STATES

To be closer to usual teleportation schemes and to get simple and explicit results we want to consider in the sequel the case that the state $\varkappa$ on $\mathcal{H}_1\otimes\mathcal{H}_2$ is a product state:

$$\varkappa=\rho^1\otimes\rho^2 \tag{33}$$

where $\rho^1\in\mathcal{S}(\mathcal{H}_1)$ and $\rho^2\in\mathcal{S}(\mathcal{H}_2)$. Assume $(\Phi_u^1)_{u\in G}$, $(\Phi_u^2)_{u\in G}$ are orthonormal sequences in $\mathcal{H}_1$, respectively, $\mathcal{H}_2$ such that

$$\rho^j=\sum_{u\in G}\lambda_u^j|\Phi_u^j\rangle\langle\Phi_u^j|\quad(j=1,2) \tag{34}$$

with

$$\sum_{u\in G}\lambda_u^j=1,\quad \lambda_v^j\geqslant 0\quad(j=1,2,v\in G) \tag{35}$$

and Φ_u^j having the representation

$$\Phi_u^j=\sum_{r\in G}\alpha_{ur}^j\xi_r^j\quad(j=1,2,u\in G). \tag{36}$$

Using the notations from (19)–(21) we thus get in that case

$$\lambda_{uv}=\lambda_u^1\cdot\lambda_v^2,\qquad \alpha_{uvrs}=\alpha_{ur}^1\cdot\alpha_{vs}^2\quad(u,v,r,s\in G).$$

Obviously, we have

$$\sum_{r\in G}\bar{\alpha}_{ur}^j\alpha_{vr}^j=\delta_{u,v}\quad(j=1,2,u,v\in G).$$

From (24) we conclude

$$\Psi_{uvnm}=\sum_{r\in G}\overline{b_{nr}}\alpha_{ur}^1\cdot\alpha_{v,r\oplus m}^2\xi_{r\oplus m}^3\quad(u,v,n,m\in G) \tag{37}$$

and Theorem 5 can be written in the following form.

Theorem 7: *Let $\varkappa=\rho^1\otimes\rho^2$ be a state on $\mathcal{H}_1\otimes\mathcal{H}_2$ given by* (34)–(36). *Then for all $n,m\in G$*

$$\Lambda_{nm}(\rho^1\otimes\rho^2)=\frac{\sum_{u,v\in G}\lambda_u^1\cdot\lambda_v^2\langle\Psi_{uvnm},\cdot\rangle\Psi_{uvnm}}{\mathrm{tr}_3\left(\sum_{u,v\in G}\lambda_u^1\cdot\lambda_v^2\langle\Psi_{uvnm},\cdot\rangle\Psi_{uvnm}\right)} \tag{38}$$

where $\Psi_{uvnm}\in\mathcal{H}_3$ is given by (37) *and where we have to assume that the denominator in* (30) *is greater than zero* (tr_3 *again denotes the trace in $\mathcal{H}_3$*).

We want to express the channel Λ_{nm} acting from $\mathcal{S}(\mathcal{H}_1\otimes\mathcal{H}_2)$ to $\mathcal{S}(\mathcal{H}_3)$ by a consecutive application of single channels having an intuitive meaning.

First for each $h\in\mathcal{H}_2$ we introduce an operator $A_h:\mathcal{H}_3\to\mathcal{H}_3$ by

$$A_hf=\sum_{r\in G}\langle\xi_r^2,h\rangle\langle\xi_r^3,f\rangle\xi_r^3\quad (f\in\mathcal{H}_3). \tag{39}$$

Especially, for all $r\in G$ we have

$$A_h\xi_r^3=\langle\xi_r^2,h\rangle\xi_r^3 \tag{40}$$

and

$$A_{\xi_r^2}=|\xi_r^3\rangle\langle\xi_r^3|. \tag{41}$$

Obviously,

$$\|A_hf\|^2=\sum_{r\in G}|\langle\xi_r^2,h\rangle\langle\xi_r^3,f\rangle|^2\leq\|h\|^2\|f\|^2,$$

so, for each $h\in\mathcal{H}_2$ it holds $A_h\in\mathcal{B}(\mathcal{H}_3)$.

For each $h\in\mathcal{H}_2$ the adjoint of A_h is given by

$$A_h^*f=\sum_{r\in G}\overline{\langle\xi_r^2,h\rangle}\langle\xi_r^3,f\rangle\xi_r^3\quad (f\in\mathcal{H}_3). \tag{42}$$

On $\mathcal{H}_3$ we define for all $m\in G$ (in the same way as on $\mathcal{H}_2$—cf. (6)) the shift operator $V_m:\mathcal{H}_3\to\mathcal{H}_3$ characterized by

$$V_m\xi_r^3=\xi_{r\oplus m}^3\quad (r\in G) \tag{43}$$

being a unitary operator on $\mathcal{H}_3$.

Finally, let $J_1:\mathcal{H}_1\to\mathcal{H}_3$ and $J_2:\mathcal{H}_2\to\mathcal{H}_3$ be the isomorphisms exchanging the corresponding basis elements, i.e.,

$$J_1\xi_r^1=\xi_r^3,\quad J_2\xi_r^2=\xi_r^3\quad (r\in G). \tag{44}$$

Lemma 8: For all $u,v,n,m\in G$ it holds

$$\Psi_{uvnm}=A_{\Phi_v^2}(V_mA_{\bar{\eta}_n}(J_1\Phi_u^1)), \tag{45}$$

where $\Psi_{uvnm}\in\mathcal{H}_3$ is given by (37), *the vectors Φ_v^j by* (36) *and $\bar{\eta}_n$ by*

$$\overline{\eta_n}=\sum_{l\in G}\overline{b_{nl}}\xi_l^2 \tag{46}$$

(*cf. definition* (2) *of η_n*).

Proof: From

$$J_1\Phi_u^1=\sum_{r\in G}\alpha_{ur}^1\xi_r^3$$

we conclude

$$V_m A_{\overline{\eta}_n}(J_1\Phi_u^1) = V_m\left(\sum_{r\in G} \alpha_{ur}^1 \langle \xi_r^2, \overline{\eta}_n\rangle \xi_r^3\right) = V_m\left(\sum_{r\in G} \alpha_{ur}^1 \overline{b_{nr}} \xi_r^3\right) = \sum_{r\in G} \alpha_{ur}^1 \overline{b_{nr}} \xi_{r\oplus m}^3$$

$$A_{\Phi_v^2} V_m A_{\overline{\eta}_n}(J_1\Phi_u^1) = \sum_{r\in G} \alpha_{ur}^1 \overline{b_{nr}} A_{\Phi_v^2} \xi_{r\oplus m}^3 = \sum_{r\in G} \alpha_{ur}^1 \overline{b_{nr}} \alpha_{v,r\oplus m}^2 \xi_{r\oplus m}^3$$

what proves because of (37) Lemma 8. □

Observe that for all $h \in \mathcal{H}_2$, $f \in \mathcal{H}_3$ and $r \in G$

$$\langle \xi_r^2, h\rangle \langle \xi_r^3, f\rangle = \langle \xi_r^2, J_2^* f\rangle \langle \xi_r^3, J_2 h\rangle .$$

From definition (39) of the operator A_h we thus immediately conclude

Lemma 9: For all $h \in \mathcal{H}_2$, $f \in \mathcal{H}_3$

$$A_h f = A_{J_2^* f}(J_2 h). \tag{47}$$

The following property of the operators A_h will be basic for our description of the channel Λ_{nm}.

Lemma 10: Let τ be a positive trace-class operator on $\mathcal{H}_2$ (possibly also the operator being identically zero) and let us be given two different representations of τ

$$\tau = \sum_{k\in G} \gamma_k^1 |h_k^1\rangle\langle h_k^1| = \sum_{k\in G} \gamma_k^2 |h_k^2\rangle\langle h_k^2| \tag{48}$$

with $\gamma_k^j \geq 0$ and $(h_k^j)_{k\in G}$, $j=1,2$ being orthonormal bases in $\mathcal{H}_2$. Then for each positive trace-class operator α on $\mathcal{H}_3$ we have

$$\sum_{k\in G} \gamma_k^1 A_{h_k^1} \alpha A_{h_k^1}^* = \sum_{k\in G} \gamma_k^2 A_{h_k^2} \alpha A_{h_k^2}^* . \tag{49}$$

Proof: It is sufficient to consider the case

$$\alpha = |f\rangle\langle f| \quad (f \in \mathcal{H}_3, \|f\| = 1).$$

Using Lemma 9 we get for all $g \in \mathcal{H}_3$ and $j=1,2$

$$\sum_{k\in G} \gamma_k^j A_{h_k^j} |f\rangle\langle f| A_{h_k^j}^* g = \sum_{k\in G} \gamma_k^j A_{h_k^j} f \langle A_{h_k^j} f, g\rangle = \sum_{k\in G} \gamma_k^j A_{J_2^* f} J_2 h_k^j \langle A_{J_2^* f} h_k^j, g\rangle = \sum_{k\in G} \gamma_k^j A_{J_2^* f} J_2 |h_k^j\rangle\langle h_k^j| J_2^* A_{J_2^* f}^* g$$

$$= A_{J_2^* f} J_2 \sum_{k\in G} \gamma_k^j |h_k^j\rangle\langle h_k^j| J_2^* A_{J_2^* f}^* g = A_{J_2^* f} J_2 \tau J_2^* A_{J_2^* f}^* g .$$

Consequently, left- and right-hand sides of (49) coincide—the expression in (49) does not depend on the special representation of τ in some basis in $\mathcal{H}_2$.

□

Let us remark that τ may be a finite rank operator since we did not exclude $\gamma_k^j = 0$.

Denote by $\mathcal{T}_j$ the set of all positive trace-class operators on $\mathcal{H}_j$, $j=1,2,3$. For each $\tau \in \mathcal{T}_2$ we define a mapping $K_\tau : \mathcal{T}_3 \to \mathcal{T}_3$ by the ansatz

$$K_\tau(\alpha) = \sum_{k\in G} \gamma_k A_{h_k} \alpha A_{h_k}^* \quad (\alpha \in \mathcal{T}_3), \tag{50}$$

where

$$\tau = \sum_{k\in G} \gamma_k |h_k\rangle\langle h_k| \quad (j=1,2) \tag{51}$$

with $\gamma_k \geq 0$ and $(h_k)_{k\in G}$ being an orthonormal basis in $\mathcal{H}_2$. Lemma 10 guarantees that the mapping K_τ is well defined, i.e., the definition (50) does not depend on the special representation of τ in some basis of $\mathcal{H}_2$.

In the sequel let us make the convention to denote the normalization of a positive trace-class operator $\varkappa$ by $\hat{\varkappa}$, i.e.,

$$\hat{\varkappa} = \frac{\varkappa}{\mathrm{tr}\varkappa} \tag{52}$$

provided $\mathrm{tr}\,\varkappa > 0$.

Now, we are able to express the channel $\Lambda_{nm}(\varkappa)$ as a consecutive application of single channels. This explicit construction will be done only for product states $\varkappa=\rho^1 \otimes \rho^2$. Besides the channel K_τ defined by (50) we still need two other channels—one lifting the state from $\mathcal{H}_1$ to $\mathcal{H}_3$ by the isomorphism J_1 given by (44) and one shifting the states on $\mathcal{H}_3$ with the shift operator V_m defined by (43). We define $K:\mathcal{T}_1 \to \mathcal{T}_3$ by

$$K(\alpha) = J_1 \alpha J_1^* \quad (\alpha \in \mathcal{T}_1) \tag{53}$$

and for $m \in G$ the mapping $K_m:\mathcal{T}_3 \to \mathcal{T}_3$ by

$$K_m(\alpha) = V_m \alpha V_m^* \quad (\alpha \in \mathcal{T}_3). \tag{54}$$

Theorem 11: *For all* $\rho^1 \in \mathcal{S}(\mathcal{H}_1)$, $\rho^2 \in \mathcal{S}(\mathcal{H}_2)$ *and* $n,m \in G$ *it holds*

$$\Lambda_{nm}(\rho^1 \otimes \rho^2) = \hat{K}_{\rho^2} \circ K_m \circ K_{|\overline{\eta_n}\rangle\langle\overline{\eta_n}|} \circ K(\rho^1) \tag{55}$$

provided $\mathrm{tr}\, K_{\rho^2} \circ K_m \circ K_{|\overline{\eta_n}\rangle\langle\overline{\eta_n}|} \circ K(\rho^1) > 0$. *Hereby,* $\overline{\eta_n}$ *is given by* (46).

Proof: Let $\rho^1 \otimes \rho^2$ be given by (34)–(36).

Using (38) in Theorem 7 and the representation (45) of Ψ_{uvnm} we obtain for all $u,v,n,m \in G$

$$\begin{aligned}
\sum_{u,v\in G} \lambda_u^1 \cdot \lambda_v^2 \langle \Psi_{uvnm}, \cdot \rangle \Psi_{uvnm} &= \sum_{u,v\in G} \lambda_u^1 \cdot \lambda_v^2 A_{\Phi_v^2} V_m A_{\overline{\eta_n}} J_1 |\Phi_u^1\rangle\langle \Phi_u^1| J_1^* A_{\overline{\eta_n}}^* V_m^* A_{\Phi_v^2}^* \\
&= \sum_{v\in G} \lambda_v^2 A_{\Phi_v^2} V_m A_{\overline{\eta_n}} J_1(\rho^1) J_1^* A_{\overline{\eta_n}}^* V_m^* A_{\Phi_v^2}^* \\
&= \sum_{v\in G} \lambda_v^2 A_{\Phi_v^2} V_m A_{\overline{\eta_n}} K(\rho^1) A_{\overline{\eta_n}}^* V_m^* A_{\Phi_v^2}^* \\
&= \sum_{v\in G} \lambda_v^2 A_{\Phi_v^2} V_m (K_{|\overline{\eta_n}\rangle\langle\overline{\eta_n}|} \circ K(\rho^1)) V_m^* A_{\Phi_v^2}^* \\
&= \sum_{v\in G} \lambda_v^2 A_{\Phi_v^2} (K_m \circ K_{|\overline{\eta_n}\rangle\langle\overline{\eta_n}|} \circ K(\rho^1)) A_{\Phi_v^2}^* \\
&= K_{\rho^2} \circ K_m \circ K_{|\overline{\eta_n}\rangle\langle\overline{\eta_n}|} \circ K(\rho^1).
\end{aligned}$$

Since $\Lambda_{nm}(\rho^1 \otimes \rho^2)$ is just the normalization of the previous expression using the notation (52) we finally obtain (55). □

The aim of teleportation models is to find methods of transmission of states on $\mathcal{H}_1$ to states on $\mathcal{H}_3$ in such a way that Bob is able to reconstruct perfectly or in a nearly perfect way the emitted state—having only knowledge about the result of a certain measurement. In Sec. III we already gave some examples illustrating the model. For the states considered in these examples it will be impossible to reconstruct all original states. However, one is not interested only in perfect teleportations but also in the deformation of the state, the degree of destruction of the input signal, etc. In Ref. 4 we considered the special case of all spaces being equal and finite dimensional. For examples in this case we refer to this article. Further, one can ask for subsets of $\mathcal{S}(\mathcal{H}_1)$ for which perfect teleportation can be achieved. We continue to illustrate the channels and operations previously introduced by simple examples connected with product states. The aim of the examples below is just to illustrate the mechanism of the procedure.

In what follows, summands for which $\|\Psi_{uvnm}\|=0$ just have to be cancelled. If it is equal to zero no information about the input signal (the measurement of the value z_{nm}) will pass to Bob. Such cases of non-perfect teleportation were considered in Ref. 3.

Example 4: Let us consider the extreme case that $\rho^1=|\xi_k^1\rangle\langle\xi_k^1|$ *for some* $k\in G$. *Then* $K(\rho^1)=|\xi_k^3\rangle\langle\xi_k^3|$, *which implies*

$$K_{|\overline{\eta_n}\rangle\langle\overline{\eta_n}|}\circ K(\rho^1)=|b_{nk}|^2|\xi_k^3\rangle\langle\xi_k^3|.$$

Consequently,

$$K_m\circ K_{|\overline{\eta_n}\rangle\langle\overline{\eta_n}|}\circ K(\rho^1)=|b_{nk}|^2|\xi_{k\oplus m}^3\rangle\langle\xi_{k\oplus m}^3|.$$

From (40) *we know that for each* $h\in\mathcal{H}_2$ *it holds*

$$A_h|\xi_{k\oplus m}^3\rangle\langle\xi_{k\oplus m}^3|A_h^*=|\langle\xi_{k\oplus m}^2,h\rangle|^2|\xi_{k\oplus m}^3\rangle\langle\xi_{k\oplus m}^3|.$$

For arbitrary state ρ^2 *on* $\mathcal{H}_2$ *this leads to*

$$K_{\rho^2}\circ K_m\circ K_{|\overline{\eta_n}\rangle\langle\overline{\eta_n}|}\circ K(\rho^1)=|b_{nk}|^2\mathrm{tr}_2(\rho^2|\xi_{k\oplus m}^2\rangle\langle\xi_{k\oplus m}^2|)\cdot|\xi_{k\oplus m}^3\rangle\langle\xi_{k\oplus m}^3|.$$

Finally, if $|b_{nk}|^2\mathrm{tr}_2(\rho^2|\xi_{k\oplus m}^2\rangle\langle\xi_{k\oplus m}^2|)>0$ *we obtain*

$$\Lambda_{nm}(\rho^1\otimes\rho^2)=|\xi_{k\oplus m}^3\rangle\langle\xi_{k\oplus m}^3|.$$

□

We try to give again a possible interpretation of the result. Measuring z_{nm} we made a projection onto $\xi_{nm}=\Sigma_{r\in G}b_{nr}(\xi_r^1\otimes\xi_{r\oplus m}^2)$. At least one of the coefficients $b_{nr}, r\in G$ has to be different from zero. The state Alice wants to teleportate consists only of the elementary signal ξ_k^1 since $\rho^1=|\xi_k^1\rangle\langle\xi_k^1|$. This implies that only the signal $\xi_k^1\otimes\xi_{k\oplus m}^2$ is able to pass. So it is necessary that $b_{nk}\neq 0$. The information about the index n is not contained in the elementary signal $\xi_k^1\otimes\xi_{k\oplus m}^2$ and will not be contained in the output sent to Bob. However, $b_{nk}\neq 0$ is necessary to get an output at all. Further, for the state ρ^2 given by (34)–(36) one easily checks

$$\mathrm{tr}_2(\rho^2|\xi_{k\oplus m}^2\rangle\langle\xi_{k\oplus m}^2|)=\sum_{u\in G}\lambda_u^2|\alpha_{u,k\oplus m}^2|^2.$$

Consequently, to obtain the output $|\xi_{k\oplus m}^3\rangle\langle\xi_{k\oplus m}^3|$ at least for one Φ_u^2 with $\lambda_u^2>0$ it must hold $\alpha_{u,k\oplus m}^2\neq 0$. If Bob would know that only states of the form $\rho^1=|\xi_k^1\rangle\langle\xi_k^1|$ for some $k\in G$ are sent by Alice he obviously has unitary keys to reconstruct the original state. This of course requires that all $b_{nk}\neq 0$ and that the state ρ^2 is such that for all $k\in G$ there exists at least one $u\in G$ such that $\lambda_u^2>0$ and $\alpha_{uk}^2\neq 0$.

Example 5: Now we consider the other extreme case that ρ^2 *is just an elementary signal, i.e., there exists an* $k\in G$ *such that* $\rho^2=|\xi_k^2\rangle\langle\xi_k^2|$. *Analogous calculation as in Example 4 shows that*

$$\Lambda_{nm}(\rho^1\otimes\rho^2)=|\xi_k^3\rangle\langle\xi_k^3|,$$

where we have to assume that $\mathrm{tr}_1(\rho^1|\xi_{k\ominus m}^1\rangle\langle\xi_{k\ominus m}^1|)>0$. *We already observed in the example above that* ρ^2 *should contain as much information as possible. In the case* $\rho^2=|\xi_k^2\rangle\langle\xi_k^2|$ *Bob obtains no information at all abut the state* ρ^1. *Independently of the result* z_{nm} *of the measurement performed by Alice either* $\Lambda_{nm}(\rho^1\otimes\rho^2)=0$ *or equal to* $|\xi_k^3\rangle\langle\xi_k^3|$. *So we have the most extreme case of loss of information.*

V. EXISTENCE OF UNITARY KEYS

We will consider again the case of product states $\varkappa=\rho^1\otimes\rho^2$ on $\mathcal{H}_1\otimes\mathcal{H}_2$ where $\rho^1\in\mathcal{S}(\mathcal{H}_1)$ and $\rho^2\in\mathcal{S}(\mathcal{H}_2)$. That means that the entangled state σ considered in Refs. 2 and 3 has the form $\sigma=S_3\rho^2S_3^*$. For fixed state ρ^2 on $\mathcal{H}_2$ we consider the channel $\tilde{\Lambda}_{nm}:\mathcal{S}(\mathcal{H}_1)\rightarrow\mathcal{S}(\mathcal{H}_3)$ given by

$$\widetilde{\Lambda}_{nm} := \Lambda_{nm}(\cdot \otimes \rho^2),$$

i.e., (cf. (27) and (29))

$$\widetilde{\Lambda}_{nm}(\rho^1) = \frac{\mathrm{tr}_{1,2}((F_{nm} \otimes 1)(1 \otimes S_3)\rho^1 \otimes \lambda(1 \otimes S_3^*)(F_{nm} \otimes 1))}{\mathrm{tr}((F_{nm} \otimes 1)(1 \otimes S_3)\rho^1 \otimes \lambda(1 \otimes S_3^*)(F_{nm} \otimes 1))}. \tag{56}$$

Let $\mathcal{G} \subseteq \mathcal{S}(\mathcal{H}_1)$ be a fixed set of states on $\mathcal{H}_1$.

Definition 12: We say that the teleportation scheme is perfect if for each $n,m \in G$ there exists a unitary operator $V_{nm}: \mathcal{H}_1 \to \mathcal{H}_3$ such that

$$\widetilde{\Lambda}_{nm}(\rho^1) = V_{nm}\rho^1 V_{nm}^* \quad (\rho^1 \in \mathcal{G}) \tag{57}$$

and

$$\sum_{n,m \in G} \mathrm{tr}((F_{nm} \otimes 1)(1 \otimes S_3)\rho^1 \otimes \rho^2(1 \otimes S_3^*)(F_{nm} \otimes 1)) = 1 \quad (\rho^1 \in \mathcal{G}). \tag{58}$$

The unitary operators V_{nm} are called unitary keys.

Usually, perfect teleportation means that the above-mentioned conditions hold for all $\rho^1 \in \mathcal{S}(\mathcal{H}_1)$ (cf. Refs. 2, 3, and 5). The restriction of the set of states that have to be teleportated enlarges the possibilities for perfect or "nearly" perfect models. Condition (57) means that Bob can reconstruct the original state ρ^1 from the knowledge of the result z_{nm} of the measurement and of the received state $\widetilde{\Lambda}_{nm}(\rho^1)$ where it is assumed that Bob possesses for each $n,m \in G$ the appropriate key V_{nm}. This follows obviously form $\rho^1 = V_{nm}^* \widetilde{\Lambda}_{nm}(\rho^1) V_{nm}$. Finally, (58) means that with probability one Bob will find the proper key, i.e., with certainty there will be a result z_{nm} of measurement for which he has unitary keys.

Let us have a closer look to formula (55). The channel K_m is built with the help of the unitary V_m. The same is true for K since $J_1 J_1^* = J_1^* J_1 = 1$. The channel $K_{|\overline{\eta_n}\rangle\langle\overline{\eta_n}|}$ destroys any hope for unitary keys. An easy calculation shows that for each $f \in \mathcal{H}_3$ it holds

$$A_{\overline{\eta_n}}^* A_{\overline{\eta_n}} f = A_{\overline{\eta_n}} A_{\overline{\eta_n}}^* f = \sum_{r \in G} |b_{nr}|^2 \langle \xi_r^3, f \rangle \xi_r^3. \tag{59}$$

We see that only if G is finite and all $|b_{nk}|$ are equal A will be unitary (up to a constant that vanishes after normalization). This is in accordance with (and another proof of) the fact that perfect teleportation requires finite dimensional spaces and maximal entanglement.

VI. CONCLUDING REMARKS

The aim of this article is to touch the problem of teleportation schemes in infinite-dimensional spaces. The previous results still have to be supplemented by calculations of fidelity and other characteristics. Just to achieve simple explicit expressions we illustrated the model on the most simple sequence of elementary signals (ξ_k^j). To obtain more interesting models one has to refine the above-mentioned models:

(1) consider more complex measurements than simple one-dimensional projections F_{nm},
(2) if the Hilbert space is a symmetric Fock space take *truncated* coherent vectors (exponential vectors with "removed" vacuum part) as basis and beam splittings for the entanglement,
(3) replace the isomorphisms J_1, J_2 and the trivial isometries S_1, S_3 by more complex ones.[6–11]

[1] C. H. Bennett, G. Brassard, C. Crepeau, R. Jozsa, A. Peres, and W. K. Wootters, Phys. Rev. Lett. **70**, 1895 (1993).
[2] K.-H. Fichtner and M. Ohya, Commun. Math. Phys. **222**, 229 (2001).
[3] K.-H. Fichtner and M. Ohya, Commun. Math. Phys. **225**, 67 (2002).
[4] K.-H. Fichtner, W. Freudenberg, and M. Ohya, *Proceedings of the Conference on Quantum Probability and Infinite Dimensional Analysis*, Burg, March 2001, edited by W. Freudenberg (World Scientific, Singapore 2001); *Quantum*

Probability and White Noise Analysis **15**, 85–105 (2003).
[5] L. Accardi and M. Ohya, *Quantum Information* (World Scientific, 1999), pp. 59–70.
[6] L. Accardi and M. Ohya, Appl. Math. Optim. **39**, 33 (1999).
[7] K.-H. Fichtner and W. Freudenberg, Commun. Math. Phys. **137**, 315 (1991).
[8] K.-H. Fichtner, W. Freudenberg, and V. Liebscher, Infinite Dimen. Anal., Quantum Probab., Relat. Top. **1**, 511 (1998).
[9] M. Ohya, *Mathematical Foundation of Quantum Computer* (Maruzen, 1999).
[10] M. Ohya, *Quantum Information II*, edited by T. Hida and K. Saito (World Scientific, Singapore, 2000), pp.149–160.
[11] M. Ohya and I. V. Volovich, *Mathematical Foundation of Quantum Information and Computation* (Springer, New York, in press).

Vol. 55 (2005) *REPORTS ON MATHEMATICAL PHYSICS* No. 1

QUANTUM ALGORITHM FOR SAT PROBLEM AND QUANTUM MUTUAL ENTROPY*

MASANORI OHYA

Department of Information Sciences, Tokyo University of Science,
278 Noda City, Chiba, Japan
(e-mail: Ohya@is.noda.tus.ac.jp)

(*Received November 4, 2004*)

It is von Neumann who opened the window for today's information epoch. He defined quantum entropy including Shannon's information more than 20 years ahead of Shannon, and he explained what computation means mathematically.

In this paper I discuss two problems studied recently by me and my coworkers. One of them concerns a quantum algorithm in a generalized sense solving the SAT problem (one of NP complete problems) and another concerns quantum mutual entropy properly describing quantum communication processes.

Keywords: quantum mutual entropy, quantum algorithm.

1. Introduction

This paper consists of two parts, one (Sections 2 and 3) of them is about quantum algorithm solving the SAT problem, it is based on a series of papers [30, 7, 32, 33, 4]. The other one (Section 4) is about quantum mutual entropy applying quantum communication processes, it is based on [27, 28, 35].

Although the ability of computers is quickly progressing, there are several problems which may not be solved effectively, namely in polynomial time. Among such problems, nonpolynomial (NP) problems and NP complete problems are fundamental. It is known that all NP complete (NPC for short) problems are equivalent and have been studied for decades, for which all known algorithms have an exponential running time in the length of the input so far. An essential question asked for more than 30 years has been *whether there exists an algorithm solving an NP complete problem in polynomial time*. We found two different algorithms solving the NPC problems in polynomial time [30, 32, 33, 3]. In first two sections of the present paper we discuss the essence of these algorithms.

After von Neumann introduced the concept of quantum entropy [23] for density operators, many concepts on various quantum entropies have appeared [23, 31], among which quantum mutual entropy plays an important role. That is, the mutual

*Lecture given at the von Neumann Centennial Conference, Budapest, October 15–20, 2003.

entropy expresses the amount of information sent from input to output, so that it is a basic quantity measuring the ability of a communication channel. I defined the quantum mutual entropy for density operators in 1983 [27] by using Umegaki's relative entropy [40] and I extended it to general C*-dynamical systems by means of Araki's or Uhlmann's relative entropy [6, 39, 31]. Recently several types of quantum mutual entropies have been introduced [38, 8, 10], and they are used to discuss communication processes. In Section 4 of this paper, we compare these mutual entropies from the point of view of information communication based on [35].

2. Quantum chaos algorithm of SAT

Let us remind what the P-problem and the NP-problem are [16, 13]: Let n be the size of input.

(1) A P-problem is a problem in which time needed for solving the problem is at worst of polynomial time of n. Equivalently, it is a problem which can be recognized in polynomial time of n by deterministic Turing machine.

(2) An NP-problem is a problem that can be solved in polynomial time by a nondeterministic Turing machine. This can be understood as follows: Let us consider a problem to find a solution of $f(x) = 0$. We can check in polynomial time of n whether x_0 is a solution of $f(x) = 0$, but we do not know whether we can find the solution of $f(x) = 0$ in polynomial time of n.

(3) An NP-complete problem is a problem of polynomially transformed NP-problem.

We take the SAT (satisfiable) problem, one of the NP-complete problems, to study whether there exists an algorithm showing that NPC becomes P. It is known that the SAT problem is equivalent to any other NPC problem.

Let $X \equiv \{x_1, \ldots, x_n\}$ be a set. Then x_k and its negation $\bar{x}_k$ $(k = 1, 2, \ldots, n)$ are called *literals* and the set of all such literals is denoted by $\tilde{X} \equiv \{x_1, \bar{x}_1, \ldots, x_n, \bar{x}_n\}$. The set of all subsets of X' is denoted by $\mathcal{F}(\tilde{X})$ and an element $C \in \mathcal{F}(\tilde{X})$ is called a *clause*. We take a truth assignment to all Boolean variables x_k. If we can assign the truth value to at least one element of C, then C is called *satisfiable*. When C is satisfiable, the truth value $t(C)$ of C is regarded as true, otherwise, that of C is false. Take the truth values as "true $\leftrightarrow$ 1, false $\leftrightarrow$ 0". Then C is satisfiable iff $t(C) = 1$.

Let $L = \{0, 1\}$ be a Boolean lattice with usual join $\vee$ and meet $\wedge$, and $t(x)$ be the truth value of a literal x in X. Then the truth value of a clause C is written as $t(C) \equiv \vee_{x \in C} t(x)$.

Moreover the set $\mathcal{C}$ of all clauses C_j $(j = 1, 2, \cdots, m)$ is called satisfiable iff the meet of all truth values of C_j is 1; $t(\mathcal{C}) \equiv \wedge_{j=1}^{m} t(C_j) = 1$. Thus the SAT problem is written as follows:

DEFINITION 1. *SAT Problem*: Given a Boolean set $X \equiv \{x_1, \cdots, x_n\}$ and a set $\mathcal{C} = \{C_1, \cdots, C_m\}$ of clauses, determine whether $\mathcal{C}$ is satisfiable or not.

That is, this problem is to ask whether there exists a truth assignment to make $\mathcal{C}$ satisfiable. It is known for usual algorithms that it is polynomial time to check the satisfiability only when a specific truth assignment is given, but we cannot determine the satisfiability in polynomial time when an assignment is not specified.

In [30] we discussed the quantum algorithm of the SAT problem, which was rewritten in [7] with showing that the OM SAT-algorithm is combinatoric. Ohya and Masuda pointed out [30] that the SAT problem, and hence all other NP problems, can be solved in polynomial time by quantum computer if the superposition of two orthogonal vectors $|0\rangle$ and $|1\rangle$ is physically detected. However, this detection is considered not to be possible in the present technology. The problem to be overcome is how to distinguish the pure vector $|0\rangle$ from the superposed one $\alpha|0\rangle + \beta|1\rangle$, obtained by the OM SAT-quantum algorithm, if β is not zero but very small. If such a distinction is possible, then we can solve the NPC problem in polynomial time.

In [32, 33] it is shown that it can be possible by combining nonlinear chaos amplifier with the quantum algorithm, which implies the existence of a mathematical algorithm solving NP = P. The algorithm of Ohya and Volovich is going beyond usual (unitary) quantum Turing algorithm. So the next question is (1) whether there exists more general Turing machine scheme combining the unitary quantum algorithm with chaos dynamics, or (2) whether there exists another method to achieve the above distinction of two vectors by a suitable unitary evolution. In [4], we showed that the stochastic limit, recently extensively studied by Accardi and coworkers [2], can be used to find another method of (2).

In this paper, we review mathematical frame of quantum algorithm in Section 2 and the OV-chaos algorithm. In Section 3, based on the idea of quantum adaptive dynamics [1, 25, 4], we discuss how it can be used to solve the problem NP = P.

2.1. Quantum algorithm

The quantum algorithms discussed so far are rather idealized because computation is represented by unitary operations. A unitary operation is rather difficult to realize in physical processes, more realistic operation is the one allowing some dissipation like semigroup dynamics. For such a realization, we have to generalize the concept of quantum Turing machine so that the generalized one contains nonunitary operations. This work has been done in [4, 21, 5], about which we will not discuss here. In this paper we will explain the algorithms solving the SAT problem in polynomial time.

First we remind the procedure of usual quantum algorithm which is needed to compute the truth value $t(\mathcal{C})$ of the SAT.

Let $\mathcal{H}$ be a Hilbert space describing input, computation and output (result). As usual, the Hilbert space is $\mathcal{H} = \otimes_1^N \mathbb{C}^2$, and let the basis of $\mathcal{H} = \otimes_1^N \mathbb{C}^2$ be: $e_0 (= |0\rangle) = |0\rangle \otimes \cdots \otimes |0\rangle \otimes |0\rangle, e_1 (= |1\rangle) = |0\rangle \otimes \cdots \otimes |0\rangle \otimes |1\rangle, \ldots, e_{2^N-1} (= |2^N - 1\rangle) = |1\rangle \otimes \cdots \otimes |1\rangle \otimes |1\rangle$.

Any number t $\left(0,\ldots,2^N-1\right)$ can be expressed by $t=\sum_{k=1}^{N} a_t^{(k)}2^{k-1},\ a_t^{(k)}=0$ *or* 1, so that the associated vector is written by

$$|t\rangle\,(=e_t)=\otimes_{k=1}^{N}|a_t^{(k)}\rangle.$$

And applying n times the Hadamard matrix $H\equiv\frac{1}{\sqrt{2}}\begin{pmatrix}1&1\\1&-1\end{pmatrix}$ to the vacuum vector $|0\rangle$, we get $H\,|0\rangle\,(\,=\xi\,(0))\equiv\otimes_1^N\frac{1}{\sqrt{2}}\,(|0\rangle+|1\rangle)$. Put

$$W\,(t)\equiv\otimes_{j=1}^{N}\begin{pmatrix}1&0\\0&\exp\left(\frac{2\pi it}{2^N}2^{j-1}\right)\end{pmatrix}.$$

Then we have

$$\xi\,(t)\equiv W\,(t)\,\xi\,(0)=\frac{1}{\sqrt{2^N}}\sum_{k=0}^{2^N-1}\exp\left(\frac{2\pi itk}{2^N}\right)|k\rangle\,,$$

which is called discrete Fourier transformation. Thus from all of the above operations it follows a unitary operator $U_F\,(t)\equiv W(t)H$ and the vector $\xi\,(t)=U_F\,(t)\,|0\rangle$.

All conventional unitary algorithms can be written as the following three steps by means of certain channels on the state space in $\mathcal{H}$ (i.e. a channel is a map sending state to another state):

(1) *Preparation of state*: Take a state ρ (e.g. $\rho=|0\rangle\,\langle 0|$) applying the unitary channel defined by the above $U_F\,(t):\Lambda_F^*=\mathrm{Ad}_{U_F(t)}$,

$$\Lambda_F^*=\mathrm{Ad}_{U_F}\Longrightarrow\Lambda_F^*\rho=U_F\rho U_F^*.$$

(2) *Computation*: Let U be a unitary operator on $\mathcal{H}$ representing the computation followed by a suitable programming of a certain problem, then the computation is described by a channel $\Lambda_U^*=\mathrm{Ad}_U$ (unitary channel). After the computation, the final state ρ_f will be

$$\rho_f=\Lambda_U^*\Lambda_F^*\rho.$$

(3) *Registering and measurement*: For registration of the computed result and its measurement we might need an additional system $\mathcal{K}$ (e.g. register), so that the lifting $\mathcal{E}_m^*$ from $\mathcal{S}\,(\mathcal{H})$ to $\mathcal{S}\,(\mathcal{H}\otimes\mathcal{K})$ in the sense of [3] is useful to describe this stage. Thus the whole process is written as

$$\rho_f=\mathcal{E}_m^*\left(\Lambda_U^*\Lambda_F^*\rho\right).$$

Finally we measure the state in $\mathcal{K}$: For instance, let $\{P_k;\,k\in J\}$ be a projection-valued measure (PVM) on $\mathcal{K}$,

$$\Lambda_m^*\rho_f=\sum_{k\in J}I\otimes P_k\rho_f I\otimes P_k,$$

after which we can get a desired result by observations in finite times whether the size of the set J is small.

REMARK 1. When dissipation is involved, the above three steps have to be generalized so that dissipative nature is involved. Such a generalization can be expressed by means of a suitable channel, not necessarily unitary.

(1) Preparation of state: We may use the same channel $\Lambda_F^* = \mathrm{Ad}_{U_F}$ in this first step, but if the number of qubits N is large so that it will not be built physically, then Λ_F^* should be modified, and let us denote it by Λ_P^*.

(2) Computation: This stage is certainly modified to a channel Λ_C^* reflecting the physical device for computer.

(3) Registering and measurement: This stage will remains as above. Thus the whole process is written as

$$\rho_f = \mathcal{E}_m^* \left(\Lambda_C^* \Lambda_P^* \rho \right).$$

2.2. Quantum algorithm of SAT

We explain the algorithm of the SAT problem which has been introduced by Ohya and Masuda [30] and developed by Accardi and Sabbadini [7]. This quantum algorithm is described by a combination of the unitary operators discussed in the previous section on a Hilbert space $\mathcal{H}$. The detail of this section is given in the papers [30, 7, 33], so we will discuss just the essence of the OM algorithm. Throughout this subsection, let n be the total number of Boolean variables used in the SAT problem.

Let 0 and 1 of the Boolean lattice L be denoted by the vectors $|0\rangle \equiv \begin{pmatrix} 1 \\ 0 \end{pmatrix}$ and $|1\rangle \equiv \begin{pmatrix} 0 \\ 1 \end{pmatrix}$ in the Hilbert space $\mathbb{C}^2$, respectively. That is, the vector $|0\rangle$ corresponds to falseness and $|1\rangle$ to truth.

As we have explained in the previous section, an element $x \in X$ can be denoted by 0 or 1, so by $|0\rangle$ or $|1\rangle$. In order to describe a clause C with at most n length by a quantum state, we need the n-fold tensor product Hilbert space $\mathcal{H} \equiv \otimes_1^n \mathbb{C}^2$. For instance, in the case of $n = 2$, given $C = \{x_1, x_2\}$ with an assignment $x_1 = 0$ and $x_2 = 1$, then the corresponding quantum state vector is $|0\rangle \otimes |1\rangle$, so that the quantum state vector describing C is generally written by $|C\rangle = |x_1\rangle \otimes |x_2\rangle \in \mathcal{H}$ with $x_k = 0$ or 1 $(k = 1, 2)$.

Once $X \equiv \{x_1, \ldots, x_n\}$ and $\mathcal{C} = \{C_1, C_2, \ldots, C_m\}$ are given, the SAT is to find the vector

$$|t(\mathcal{C})\rangle \equiv \wedge_{j=1}^m \vee_{x \in C_j} t(x),$$

where $t(x)$ is $|0\rangle$ or $|1\rangle$ when $x = 0$ or 1, respectively, and $t(x) \wedge t(y) \equiv t(x \wedge y)$, $t(x) \vee t(y) \equiv t(x \vee y)$.

For any two qubits $|x\rangle$ and $|y\rangle$, $|x, y\rangle$ and $|x^N\rangle$ are defined as $|x\rangle \otimes |y\rangle$ and $\underbrace{|x\rangle \otimes \cdots \otimes |x\rangle}_{N \text{ times}}$, respectively. The usual (unitary) quantum computation can be formulated mathematically as the multiplication by unitary operators. Let U_{NOT},U_{CN} and U_{CCN} be the three unitary operators defined as

$$U_{\text{NOT}} \equiv |1\rangle\langle 0| + |0\rangle\langle 1|,$$

$$U_{\text{CN}} \equiv |0\rangle\langle 0| \otimes I + |1\rangle\langle 1| \otimes U_{\text{NOT}},$$

$$U_{\text{CCN}} \equiv |0\rangle\langle 0| \otimes I \otimes I + |1\rangle\langle 1| \otimes |0\rangle\langle 0| \otimes I + |1\rangle\langle 1| \otimes |1\rangle\langle 1| \otimes U_{\text{NOT}}.$$

U_{NOT}, U_{CN} and U_{CCN} are often called NOT-gate, Controlled-NOT gate and Controlled-Controlled-NOT gate, respectively. For any $k \in \mathbb{N}$, $U_H^{(N)}(k)$ denotes the k-fold Hadamard transformation on $(\mathbb{C}^2)^{\otimes N}$ defined as

$$U_H^{(N)}(k)\,|0^N\rangle = \frac{1}{2^{k/2}}(|0\rangle + |1\rangle)^{\otimes k}\,|0^{N-k}\rangle = \frac{1}{2^{k/2}}\sum_{i=0}^{2^k-1}|e_i\rangle \otimes |0^{N-k}\rangle.$$

The above unitary operators can be extended to the unitary operators on $(\mathbb{C}^2)^{\otimes N}$:

$$U_{\text{NOT}}^{(N)}(u) \equiv I^{\otimes u-1} \otimes (|0\rangle\langle 1| + |1\rangle\langle 0|)\, I^{\otimes N-u-1},$$

$$\begin{aligned} U_{\text{CN}}^{(N)}(u, v) \equiv{}& I^{\otimes u-1} \otimes |0\rangle\langle 0| \otimes I^{\otimes N-u-1} + I^{\otimes u-1} \otimes |1\rangle\langle 1| \\ &\otimes I^{\otimes v-u-1} \otimes U_{\text{NOT}} \otimes I^{\otimes N-v-1}, \end{aligned}$$

$$\begin{aligned} U_{\text{CCN}}^{(N)}(u, v, w) ={}& I^{\otimes u-1} \otimes |0\rangle\langle 0| \otimes I^{\otimes N-u-1} + I^{\otimes u-1} \otimes |1\rangle\langle 1| \\ &\otimes I^{\otimes v-u-1} \otimes |0\rangle\langle 0| \otimes I^{\otimes N-v-1} + I^{\otimes u-1} \otimes |1\rangle\langle 1| \\ &\otimes I^{\otimes v-u-1} \otimes |1\rangle\langle 1| \otimes I^{\otimes w-t-1} \otimes U_{\text{NOT}} \otimes I^{\otimes N-w-1}, \end{aligned}$$

where u, v and w are positive integers satisfying $1 \le u < v < w \le N$.

Furthermore we have the following three unitary operators U_{AND}, U_{OR} and U_{COPY}, called the logical gates (see [7]),

$$\begin{aligned} U_{\text{AND}} \equiv{}& \sum_{\varepsilon_1,\varepsilon_2\in\{0,1\}} \{|\varepsilon_1, \varepsilon_2, \varepsilon_1 \wedge \varepsilon_2\rangle\langle \varepsilon_1, \varepsilon_2, 0| + |\varepsilon_1, \varepsilon_2, 1 - \varepsilon_1 \wedge \varepsilon_2\rangle\langle \varepsilon_1, \varepsilon_2, 1|\} \\ ={}& |0,0,0\rangle\langle 0,0,0| + |0,0,1\rangle\langle 0,0,1| + |1,0,0\rangle\langle 1,0,0| + |1,0,1\rangle\langle 1,0,1| \\ &+ |0,1,0\rangle\langle 0,1,0| + |0,1,1\rangle\langle 0,1,1| + |1,1,1\rangle\langle 1,1,0| + |1,1,0\rangle\langle 1,1,1|, \end{aligned}$$

$$\begin{aligned} U_{\text{OR}} \equiv{}& \sum_{\varepsilon_1,\varepsilon_2\in\{0,1\}} \{|\varepsilon_1, \varepsilon_2, \varepsilon_1 \vee \varepsilon_2\rangle\langle \varepsilon_1, \varepsilon_2, 0| + |\varepsilon_1, \varepsilon_2, 1 - \varepsilon_1 \vee \varepsilon_2\rangle\langle \varepsilon_1, \varepsilon_2, 1|\} \\ ={}& |0,0,0\rangle\langle 0,0,0| + |0,0,1\rangle\langle 0,0,1| + |1,0,1\rangle\langle 1,0,0| + |1,0,0\rangle\langle 1,0,1| \\ &+ |0,1,1\rangle\langle 0,1,0| + |0,1,0\rangle\langle 0,1,1| + |1,1,1\rangle\langle 1,1,0| + |1,1,0\rangle\langle 1,1,1|, \end{aligned}$$

$$\begin{aligned}U_{\mathrm{COPY}} &\equiv \sum_{\varepsilon_1\in\{0,1\}} \{|\varepsilon_1,\varepsilon_1\rangle\langle\varepsilon_1,0| + |\varepsilon_1,1-\varepsilon_1\rangle\langle\varepsilon_1,1|\} \\ &= |0,0\rangle\langle 0,0| + |0,1\rangle\langle 0,1| + |1,1\rangle\langle 1,0| + |1,0\rangle\langle 1,1| .\end{aligned}$$

Here ε_1 and ε_2 take the value 0 or 1. We call $U_{\mathrm{AND}}, U_{\mathrm{OR}}$ and U_{COPY}, the AND gate, OR gate and COPY gate, respectively, whose extensions to $(\mathbb{C}^2)^{\otimes N}$ are denoted by $U_{\mathrm{AND}}^{(N)}, U_{\mathrm{OR}}^{(N)}$ and $U_{\mathrm{COPY}}^{(N)}$, which are expressed as

$$\begin{aligned}U_{\mathrm{AND}}^{(N)}(u,v,w) = &\sum_{\varepsilon_1,\varepsilon_2\in\{0,1\}} I^{\otimes u-1}\otimes|\varepsilon_1\rangle\langle\varepsilon_1| I^{\otimes v-u-1}\otimes|\varepsilon_2\rangle\langle\varepsilon_2| \\ &+ I^{\otimes w-v-u-1}\otimes|\varepsilon_1\wedge\varepsilon_2\rangle\langle 0| I^{\otimes N-w-v-u} \\ &+ I^{\otimes u-1}\otimes|\varepsilon_1\rangle\langle\varepsilon_1| I^{\otimes v-u-1} \\ &\otimes|\varepsilon_2\rangle\langle\varepsilon_2| I^{\otimes w-v-u-1}\otimes|1-\varepsilon_1\wedge\varepsilon_2\rangle\langle 1| I^{\otimes N-w-v-u}, \\ U_{\mathrm{OR}}^{(N)}(u,v,w) \equiv &\sum_{\varepsilon_1,\varepsilon_2\in\{0,1\}} I^{\otimes u-1}\otimes|\varepsilon_1\rangle\langle\varepsilon_1| I^{\otimes v-u-1}\otimes|\varepsilon_2\rangle\langle\varepsilon_2| \\ &+ I^{\otimes w-v-u-1}\otimes|\varepsilon_1\vee\varepsilon_2\rangle\langle 0| I^{\otimes N-w-v-u} \\ &+ I^{\otimes u-1}\otimes|\varepsilon_1\rangle\langle\varepsilon_1| I^{\otimes v-u-1}\otimes|\varepsilon_2\rangle\langle\varepsilon_2| \\ &+ I^{\otimes w-v-u-1}\otimes|1-\varepsilon_1\vee\varepsilon_2\rangle\langle 1| I^{\otimes N-w-v-u}, \\ U_{\mathrm{COPY}}^{(N)}(u,v) \equiv &\sum_{\varepsilon_1\in\{0,1\}} I^{\otimes u-1}|\varepsilon_1\rangle\langle\varepsilon_1| I^{\otimes v-u-1}|\varepsilon_1\rangle\langle 0| I^{\otimes N-v-u} \\ &+ I^{\otimes u-1}|\varepsilon_1\rangle\langle\varepsilon_1| I^{\otimes v-u-1}|1-\varepsilon_1\rangle\langle 1| I^{\otimes N-v-u}.\end{aligned}$$

where u, v and w are positive integers satisfying $1 \le u < v < w \le N$. These operators can be written, in terms of elementary gates, as

$$\begin{aligned}U_{\mathrm{OR}}^{(N)}(u,v,w) &= U_{\mathrm{CN}}^{(N)}(u,w)\cdot U_{\mathrm{CN}}^{(N)}(v,w)\cdot U_{\mathrm{CCN}}^{(N)}(u,v,w), \\ U_{\mathrm{AND}}^{(N)}(u,v,w) &= U_{\mathrm{CCN}}^{(N)}(u,v,w), \\ U_{\mathrm{COPY}}^{(N)}(u,v) &= U_{\mathrm{CN}}^{(N)}(u,v).\end{aligned}$$

Let $\mathcal{C}$ be a set of clauses whose cardinality is equal to m. Let $\mathcal{H} = (\mathbb{C}^2)^{\otimes n+\mu+1}$ be a Hilbert space and $|v_0\rangle$ be the initial state $|v_0\rangle = |0^n, 0^\mu, 0\rangle$, where μ is the number of dust qubits (the details can be seen in [19]). Let $U_{\mathcal{C}}^{(n)}$ be a unitary operator for the computation of the SAT,

$$U_{\mathcal{C}}^{(n)}|v_0\rangle = \frac{1}{\sqrt{2^n}}\sum_{i=0}^{2^n-1}\left|e_i, x^\mu, t_{e_i}(\mathcal{C})\right\rangle \equiv \left|v_f\right\rangle,$$

where x^μ denotes a μ strings of binary symbols and $t_{e_i}(\mathcal{C})$ is a truth value of $\mathcal{C}$ with e_i.

Let $\{s_k; k = 1, \ldots, m\}$ be the sequence defined as

$$s_1 = n + 1,$$
$$s_2 = s_1 + \mathrm{card}\,(C_1) + \delta_{1,\mathrm{card}(C_1)} - 1,$$
$$s_i = s_{i-1} + \mathrm{card}\,(C_{i-1}) + \delta_{1,\mathrm{card}(C_{i-1})}, \qquad 3 \le i \le m,$$

where $\mathrm{card}\,(C_i)$ means the cardinality of a clause C_i. Take a value s as

$$s = s_m - 1 + \mathrm{card}\,(C_m) + \delta_{1,\mathrm{card}(C_m)}.$$

Note that the number m of the clause is at most $2n$. Then we have [19]: The total number of dust qubits μ is

$$\begin{aligned}\mu &= s - 1 - n \\ &= \sum_{k=1}^{m} \mathrm{card}\,(C_k) + \delta_{1,\mathrm{card}(C_k)} - 2\end{aligned}$$

for $m \geq 2$. In order to construct $U_C^{(n)}$ concretely, we use the following unitary gates for this concrete expression [30, 7]:

$$U_{\mathrm{AND}}^{(x)}(k) = \begin{cases} U_{\mathrm{AND}}^{(x)}(s_{k+1} - 1, s_{k+2} - 2, s_{k+2} - 1), & 1 \le k \le m-2, \\ U_{\mathrm{AND}}^{(x)}(s_m - 1, s_f - 1, s_f), & k = m-1, \end{cases}$$

$$\begin{aligned} U_{\mathrm{OR}}^{(x)}(k) = \bar{U}_{\mathrm{OR}}^{(x)}&(l_4, s_k - \mathrm{card}\,(C_k) - 1, s_k - \mathrm{card}\,(C_k) - 2) \\ &\cdots\cdot \bar{U}_{\mathrm{OR}}^{(x)}(l_3, s_k, s_k + 1)\, \bar{U}_{\mathrm{OR}}^{(x)}(l_1, l_2, s_k), \end{aligned}$$

$$\bar{U}_{\mathrm{OR}}^{(x)}(u, v, w) = \begin{cases} U_{\mathrm{OR}}^{(x)}(u, v, w), & x_u \in C_k, \\ U_{\mathrm{NOT}}^{(x)}(u) \cdot U_{\mathrm{OR}}^{(x)}(u, v, w) \cdot U_{\mathrm{NOT}}^{(x)}(u), & \bar{x}_u \in C_k, \\ U_{\mathrm{NOT}}^{(x)}(u) \cdot U_{\mathrm{NOT}}^{(x)}(v) \cdot U_{\mathrm{OR}}^{(x)}(u, v, w) \cdot U_{\mathrm{NOT}}^{(x)}(u) \cdot U_{\mathrm{NOT}}^{(x)}(v), & \\ & \bar{x}_u, \bar{x}_v \in C_k \end{cases}$$

where l_1, l_2, l_3, l_4 are positive integers such that $x_z \in C_k$ or $\bar{x}_z \in C_k$ $(z = l_1, \ldots, l_4)$.

THEOREM 1. *The unitary operator* $U_C^{(n)}$ *is represented as*

$$\begin{aligned} U_C^{(n)} = U_{\mathrm{AND}}^{(n+\mu+1)}&(m-1) \cdot U_{\mathrm{AND}}^{(n+\mu+1)}(m-2) \cdots U_{\mathrm{AND}}^{(n+\mu+1)}(1) \\ &\cdot U_{\mathrm{OR}}^{(n+\mu+1)}(m) \cdot U_{\mathrm{OR}}^{(n+\mu+1)}(m-1) \cdots U_{\mathrm{OR}}^{(n+\mu+1)}(1) \cdot U_H^{(n+\mu+1)}(n). \end{aligned}$$

Applying the above unitary operator to the initial state, we obtain the final state ρ. The result of the computation is registered as $|t(C)\rangle$ in the last section of the final vector, which will be taken out by a projection $P_{n+\mu,1} \equiv I^{\otimes n+\mu} \otimes |1\rangle\langle 1|$ onto the subspace of $\mathcal{H}$ spanned by the vectors $|\varepsilon^n, \varepsilon^\mu, 1\rangle$.

The following theorem is easily seen.

THEOREM 2. *$\mathcal{C}$ is SAT if and only if*

$$P_{n+\mu,1}U_C^{(n)}\,|v_0\rangle \neq 0.$$

According to the standard theory of quantum measurement, after a measurement of the event $P_{n+\mu,1}$, the state $\rho = |v_f\rangle\langle v_f|$ becomes

$$\rho \to \frac{P_{n+\mu,1}\rho P_{n+\mu,1}}{\operatorname{Tr}\rho P_{n+\mu,1}} =: \overline{\rho}.$$

Thus the solvability of the SAT problem is reduced to checking that $\rho' \neq 0$. The difficulty is that the probability

$$\operatorname{Tr}\overline{\rho}P_{n+\mu,1} = \|P_{n+\mu,1}\left|v_f\right\rangle\|^2 = \frac{|T(\mathcal{C}_0)|}{2^n}$$

is very small in some cases, where $|T(\mathcal{C}_0)|$ is the cardinality of the set $T(\mathcal{C}_0)$, of all the truth functions t such that $t(\mathcal{C}_0) = 1$.

We put $q \equiv \sqrt{\frac{r}{2^n}}$ with $r \equiv |T(C_0)|$. Then if r is suitably large to detect it, then the SAT problem is solved in polynomial time. However, for small r, the probability is very small so that we in fact do not get any information about the existence of the solution of the equation $t(C_0) = 1$, hence in such a case we need further discussion.

Let go back to the SAT algorithm. After computation, the quantum computer will be in the state

$$\left|v_f\right\rangle = \sqrt{1-q^2}\,|\varphi_0\rangle \otimes |0\rangle + q\,|\varphi_1\rangle \otimes |1\rangle,$$

where $|\varphi_1\rangle$ and $|\varphi_0\rangle$ are normalized n $(= n + \mu)$ qubit states and $q = \sqrt{r/2^n}$. Effectively our problem is reduced to the following 1 qubit problem: The above state $\left|v_f\right\rangle$ is reduced to the state

$$|\psi\rangle = \sqrt{1-q^2}\,|0\rangle + q\,|1\rangle,$$

and we want to distinguish between the cases $q = 0$ and $q > 0$ (small positive number).

It will not be possible to amplify, by a unitary transformation, the above small positive q into suitable large one to be detected, e.g. $q > 1/2$, having $q = 0$ as it is. The amplification would be not possible if we use the standard model of quantum computations with a unitary evolution. What we did in [32, 33] was to propose to use the output $|\psi\rangle$ of the quantum computer as an input for another device involving chaotic dynamics. That is, it was proposed to combine quantum computer with a chaotic dynamics amplifier in [32, 33]. Such a quantum chaos computer is a new model of computations and we could demonstrate that the amplification was possible in the polynomial time.

2.3. Chaos algorithm of SAT

Here we will argue that chaos can play a constructive role in computations (see [32, 33] for the details). Chaotic behaviour in a classical system usually is considered as an exponential sensitivity to initial conditions. We would like to use this sensitivity to distinguish between the cases $q = 0$ and $q > 0$ mentioned in the previous section.

Consider the so-called logistic map

$$x_{n+1} = ax_n(1 - x_n) \equiv g(x), \qquad x_n \in [0, 1].$$

The properties of the map depend on the parameter a. If we take, for example, $a = 3.71$, then the Lyapunov exponent is positive, the trajectory is very sensitive to the initial value and one has the chaotic behaviour [26]. It is important to notice that if the initial value $x_0 = 0$, then $x_n = 0$ for all n.

It is known [14] that any classical algorithm can be implemented on quantum computer. Our quantum chaos computer will consist of two blocks. One block is the ordinary quantum computer performing computations with the output $|\psi\rangle = \sqrt{1-q^2}\,|0\rangle + q\,|1\rangle$. The second block is a computer performing computations of the *classical* logistic map. These two blocks should be connected in such a way that the state $|\psi\rangle$ first be transformed into the density matrix of the form

$$\rho = q^2 P_1 + \left(1 - q^2\right) P_0,$$

where P_1 and P_0 are projectors to the state vectors $|1\rangle$ and $|0\rangle$. This connection is in fact nontrivial and actually it should be considered as the third block. One has to notice that P_1 and P_0 generate an abelian algebra which can be considered as a classical system. In the second block the above density matrix ρ is interpreted as the initial data ρ_0, and we apply the logistic map as

$$\rho_m = \frac{(I + g^m(\rho_0)\sigma_3)}{2},$$

where I is the identity matrix and σ_3 is the z-component of Pauli matrix on $\mathbb{C}^2$. To find a proper value m we finally measure the value of σ_3 in the state ρ_m such that

$$M_m \equiv \operatorname{tr} \rho_m \sigma_3.$$

THEOREM 3.

$$\rho_m = \frac{(I + g^m(q^2)\sigma_3)}{2}, \qquad \textit{and} \qquad M_m = g^m(q^2).$$

Thus the question is whether we can find such an m in polynomial steps of n satisfying the inequality $M_m \geq \frac{1}{2}$ for very small but nonzero q^2. Here we have to remark that if one has $q = 0$ then $\rho_0 = P_0$ and we obtain $M_m = 0$ for all m. If $q \neq 0$, the stochastic dynamics leads to the amplification of the small magnitude q in such a way that it can be detected as is explained below. The transition from ρ_0 to ρ_m is nonlinear and can be considered as a classical evolution because our algebra generated by P_0 and P_1 is abelian. The amplification can be done within at most $2n$ steps due to the following propositions. Since $g^m(q^2)$ is x_m of the logistic map $x_{m+1} = g(x_m)$ with $x_0 = q^2$, we use the notation x_m in the logistic map for simplicity.

THEOREM 4. *For the logistic map* $x_{n+1} = ax_n(1 - x_n)$ *with* $a \in [0,4]$ *and* $x_0 \in [0,1]$, *let* x_0 *be* $\frac{1}{2^n}$ *and a set* J *be* $\{0,1,2,\ldots,n,\ldots,2n\}$. *If* a *is* 3.71, *then there exists an integer* m *in* J *satisfying* $x_m > \frac{1}{2}$.

THEOREM 5. *Let* a *and* n *be the same as in the above theorem. If there exists* m_0 *in* J *such that* $x_{m_0} > \frac{1}{2}$, *then* $m_0 > \frac{n-1}{\log_2 3.71 - 1}$.

According to these theorems, it is enough to check the value x_m (M_m) around the above m_0 when q is $\frac{1}{2^n}$ for a large n. More generally, when $q = \frac{k}{2^n}$ with some integer k, it is similarly checked that the value $x_m(M_m)$ becomes over $\frac{1}{2}$ within at most $2n$ steps.

The complexity of the quantum algorithm for the SAT problem was discussed in Section 3, it was of polynomial time. We have only to consider the number of steps in the classical algorithm for the logistic map performed on quantum computer. It is the probabilistic part of the construction and one has to repeat computations several times to be able to distinguish the cases $q = 0$ and $q > 0$. Thus it seems that the quantum chaos computer can solve the SAT problem in polynomial time.

In conclusion of [33], the quantum chaos computer combines the ordinary quantum computer with quantum chaotic dynamics amplifier. It may go beyond the usual quantum Turing algorithm, but such a device can be powerful enough to solve the NP-complete problems in polynomial time. The detail estimation of the complexity of the SAT algorithm is discussed in [19].

In the next two sections we will discuss the SAT problem in a different view, that is, we will show that the same amplification is possible by unitary dynamics defined in the stochastic limit.

3. Quantum adaptive algorithm of SAT

The idea to develop a mathematical approach to adaptive systems, i.e. to systems whose properties are in part determined as responses to an environment [1, 25], was born in connection with some problems of quantum measurement theory and chaos dynamics.

The mathematical definition of adaptive system is in terms of observables, namely: *an adaptive system is a composite system* whose interaction depends on a fixed observable (typically in a measurement process, this observable is the observable one wants to measure). Such systems may be called *observable-adaptive.*

In [4] we extended this point of view by introducing another natural class of adaptive systems which, in a certain sense, is the dual to the above defined one, namely the class of *state-adaptive* systems. These are defined as follows: a state-adaptive system is a composite system whose interaction depends on the state of at least one of the subsystems at the instant in which the interaction is switched on. We applied the state-adaptivity to quantum computation.

The difference between state-adaptive systems and nonlinear dynamical systems should be emphasized:

(i) In nonlinear dynamical systems (such as those whose evolution is described by the Boltzmann equation, or nonlinear Schrödinger equation, ...,) the interaction Hamiltonian depends on the state at each time t: $H_I = H_I(\rho_t)$; $\forall t$.

(ii) In state-adaptive dynamical systems (such as those considered in the present paper) the interaction Hamiltonian depends on the state only at time $t = 0$: $H_I = H_I(\rho_0)$.

Now, from the general theory of stochastic limit [2] one knows that, under general ergodicity conditions, interaction with an environment drives the system to a dynamical (but not necessarily thermodynamical) equilibrium state which depends on the initial state of the environment and on the interaction Hamiltonian.

Therefore, if one is able to realize experimentally these state dependent Hamiltonians, one would be able to drive the system S to a pre-assigned dynamical equilibrium state depending on the input state ψ_0.

In the following subsection we will substantiate the general scheme described above with an application to the SAT problem described in the previous sections.

3.1. Stochastic limit and SAT problem

We illustrate the general scheme described in the previous section in the simplest case when the state space of the system is $\mathcal{H}_S \equiv \mathbb{C}^2$. We fix an orthonormal basis of $\mathcal{H}_S$ as $\{e_0, e_1\}$.

The unknown state (vector) of the system at time $t = 0$ is

$$\psi := \sum_{\varepsilon \in \{0,1\}} \alpha_\varepsilon e_\varepsilon = \alpha_0 e_0 + \alpha_1 e_1; \qquad \|\psi\| = 1.$$

In Section 3, α_1 corresponds to q and e_j to $|j\rangle$ $(j = 0, 1)$. This vector is taken as input and defines the interaction Hamiltonian in an external field

$$\begin{aligned} H_I &= \lambda |\psi\rangle\langle\psi| \otimes (A_g^+ + A_g) \\ &= \sum \lambda \alpha_\varepsilon \overline{\alpha}_\varepsilon |e_\varepsilon\rangle\langle e_{\varepsilon'}| \otimes (A_g^+ + A_g), \end{aligned}$$

where λ is a small coupling constant. Here and in the following summation over repeated indices is understood.

The free system Hamiltonian is taken to be diagonal in the e_ε-basis,

$$H_S := \sum_{\varepsilon \in \{0,1\}} E_\varepsilon |e_\varepsilon\rangle\langle e_\varepsilon| = E_0 |e_0\rangle\langle e_0| + E_1 |e_1\rangle\langle e_1|$$

and the energy levels are ordered so that $E_0 < E_1$. Thus there is a single Bohr frequency $\omega_0 := E_1 - E_0 > 0$. The 1-particle field Hamiltonian is

$$S_t g(k) = e^{it\omega(k)} g(k),$$

where $\omega(k)$ is a function satisfying the basic analytical assumption of the stochastic limit. Its second quantization is the free field evolution

$$e^{itH_0} A_g e^{-itH_0} = A_{S_t g}.$$

We can distinguish two cases as below, which correspond to two cases of Section 3, i.e. $q > 0$ and $q = 0$.

Case 1. If $\alpha_0, \alpha_1 \neq 0$, then, according to the general theory of stochastic limit (i.e. $t \to t/\lambda^2$) [2], the interaction Hamiltonian H_I is in the same universality class as

$$\tilde{H}_I = D \otimes A_g^+ + D^+ \otimes A_g,$$

where $D := |e_0\rangle\langle e_1|$. The interaction Hamiltonian at time t is then

$$\tilde{H}_I(t) = e^{-it\omega_0} D \otimes A_{S_t g}^+ + \text{ h.c.} = D \otimes A^+(e^{it(\omega(p)-\omega_0)}g) + \text{ h.c.},$$

and the white noise ($\{b_t\}$) Hamiltonian equation associated, via the stochastic golden rule, to this interaction Hamiltonian is

$$\partial_t U_t = i(Db_t^+ + D^+ b_t)U_t.$$

Its causally normal ordered form is equivalent to the stochastic differential equation

$$dU_t = (iDdB_t^+ + iD^+ dB_t - \gamma_- D^+ D dt)U_t,$$

where $dB_t := b_t dt$.

The causally ordered inner Langevin equation is

$$\begin{aligned} dj_t(x) &= dU_t^* x U_t + U_t^* x dU_t + dU_t^* x dU_t \\ &= U_t^*(-iD^+ x dB_t - iDx dB_t^+ - \overline{\gamma}_- D^+ D x dt + ix D dB_t^+ \\ &\quad + ix D^+ dB_t - \gamma_- x D^+ D dt + \gamma_- D^+ x D dt)U_t \\ &= ij_t([x, D^+])dB_t + ij_t([x, D])dB_t^+ \\ &\quad -(\text{Re } \gamma_-) j_t(\{D^+ D, x\})dt + i(Im\gamma_-) j_t([D^+ D, x])dt \\ &\quad + j_t(D^+ x D)(\text{Re } \gamma_-)dt, \end{aligned}$$

where $j_t(x) := U_t^* x U_t$. Therefore the master equation is

$$\begin{aligned} \frac{d}{dt} P^t(x) &= (\text{Im}\,\gamma)i[D^+ D, P^t(x)] - (\text{Re}\gamma_-)\{D^+ D, P^t(x)\} \\ &\quad +(\text{Re } \gamma_-) D^+ P^t(x) D, \end{aligned}$$

where $D^+ D = |e_1\rangle\langle e_1|$ and $D^+ x D = \langle e_0, x e_0\rangle |e_1\rangle\langle e_1|$.

The dual Markovian evolution P_*^t acts on density matrices, and its generator is

$$L_* \rho = (\text{Im}\,\gamma_-)i[\rho, D^+ D] - (\text{Re } \gamma_-)\{\rho, D^+ D\} + (\text{Re } \gamma_-) D\rho D^+.$$

Thus, if $\rho_0 = |e_0\rangle\langle e_0|$ one has

$$L_* \rho_0 = 0,$$

so ρ_0 is an invariant measure. From the Frigerio–Fagnola–Rebolledo criteria, it is the unique invariant measure and the semigroup $\exp(tL_*)$ converges exponentially to it.

Case 2. If $\alpha_1 = 0$, then the interaction Hamiltonian H_I is

$$H_I = \lambda |e_0\rangle\langle e_0| \otimes (A_g^+ + A_g)$$

and, according to the general theory of stochastic limit, the reduced evolution has no damping and corresponds to the pure Hamiltonian

$$H_S + |e_0\rangle\langle e_0| = (E_0 + 1)|e_0\rangle\langle e_0| + E_1|e_1\rangle\langle e_1|.$$

Therefore, if we choose the eigenvalues E_1, E_0 to be integers (in appropriate units), then the evolution will be periodic.

Since the eigenvalues E_1, E_0 can be chosen a priori, by fixing the system Hamiltonian H_S, it follows that the period of the evolution can be known a priori. This gives a simple criterion for the solvability of the SAT problem because, by waiting a sufficiently long time one can experimentally detect the difference between damping and an oscillating behaviour.

A precise estimate of this time can be achieved either by theoretical methods or by computer simulation. Both methods will be analyzed in the complete paper [5].

CONCLUSION 1. We pointed out that it was possible to distinguish two different states, $\sqrt{1-q^2}\,|0\rangle + q\,|1\rangle$ $(q \neq 0)$ and $|0\rangle$ by means of the adaptive dynamics with the stochastic limit.

CONCLUSION 2. Finally we remark that our algorithm can be described by a deterministic generalized quantum Turing machine [21, 5].

4. Comparison of various quantum mutual type entropies

There exist several different types of quantum mutual entropy. The classical mutual entropy was introduced by Shannon to discuss the transmission of information from an input system to an output system [20]. Then Kolmogorov [24], Gelfand and Yaglom [17] gave a measure theoretic expression for the mutual entropy by means of the relative entropy defined by Kullback and Leibler. Shannon's expression for mutual entropy was generalized for the finite-dimensional quantum (matrix) case by Holevo [18, 22]. Ohya took the measure theoretic expression of KGY and defined quantum mutual entropy by means of quantum relative entropy [27, 28]. Recently Shor [38] and Bennett et al. [10] took the coherent information and defined new types of mutual entropy in order to discuss Shannon's coding theorem. In this section, we compare these types of mutual entropies.

The most general form of quantum mutual entropy defined by Ohya, generalizing the KGY measure theoretic mutual entropy, is given as

$$I_1(\varphi; \Lambda) = \sup\left\{ \int_{\mathcal{S}} S^{AU}(\Lambda\omega, \Lambda\varphi)\, d\mu;\ \ \mu \in M_\varphi(\mathcal{S}) \right\}.$$

Here $\mathcal{S}$ is the set of all states in a certain C*-algebra (or von Neumann algebra) describing a quantum system, $S^{AU}(\cdot,\cdot)$ is the relative entropy of Araki [6] or Uhlmann [39] and μ is a measure decomposing the state φ into extremal orthogonal states, i.e. $\varphi = \int_{ex\mathcal{S}} \omega d\mu$, in $\mathcal{S}$, whose set is denoted by $M_\varphi(\mathcal{S})$.

In the case that the C*-algebra is $\mathbf{B}(\mathcal{H})$ and $\mathcal{S}$ is the set of all density operators, the above mutual entropy goes to

$$I_1(\rho;\Lambda) = \sup\left\{\sum_n S^U(\Lambda E_n, \Lambda\rho);\ \rho = \sum_n \lambda_n E_n\right\},$$

where ρ is a density operator (state), $S^U(\cdot,\cdot)$ is Umegaki's relative entropy and $\rho = \sum_n \lambda_n E_n$ is the Schatten–von Neumann (one-dimensional spectral) decomposition. The SN decomposition is not always unique unless $\mathcal{S}$ is Choque simplex, so we take the supremum over all possible decompositions. It is easy to show that we can take orthogonal decomposition instead of the SN decomposition [29].

These quantum mutual entropies are completely quantum, namely, they describe the transmission of information from a quantum input to a quantum output. When the input system is classical, the state ρ is a probability distribution and the Schatten–von Neumann decomposition is unique with delta measures δ_n such that $\rho = \sum_n \lambda_n \delta_n$. In this case we need to code the classical state ρ by a quantum state, whose process is a quantum coding described by a channel Γ such that $\Gamma\delta_n = \sigma_n$ (quantum state) and $\sigma \equiv \Gamma\rho = \sum_n \lambda_n \sigma_n$. Then the quantum mutual entropy $I_1(\rho;\Lambda)$ becomes Holevo's one, that is,

$$I_1(\rho;\Lambda\Gamma) = S(\Lambda\sigma) - \sum_n \lambda_n S(\Lambda\sigma_n)$$

when $\sum_n \lambda_n S(\Lambda\sigma_n)$ is finite.

Let us discuss the entropy exchange [8]. For a state ρ, a channel Λ is defined by an operator-valued measure $\{A_j\}$ such as $\Lambda(\cdot) \equiv \sum_j A_j^* \cdot A_j$. Then define a matrix $W = (W_{ij})$ with $W_{ij} = \dfrac{\operatorname{tr} A_i^* \rho A_j}{\operatorname{tr} \Lambda\rho}$, by which the entropy exchange is defined by

$$S_e(\rho,\Lambda) = -\operatorname{tr} W \log W.$$

Using the above entropy exchange, two types of mutual entropies are defined as below and they are applied to the study of the quantum version of Shannon's coding theorem [8, 38, 10]. The first one is called the coherent information $I_2(\rho;\Lambda)$ and the second one is $I_3(\rho,\Lambda)$, which are defined by

$$I_2(\rho,\Lambda) \equiv S(\Lambda\rho) - S_e(\rho,\Lambda),$$
$$I_3(\rho,\Lambda) \equiv S(\rho) + S(\Lambda\rho) - S_e(\rho,\Lambda).$$

By comparing these mutual entropies for information communication processes, we have the following theorem [35].

THEOREM 6. *When* $\{A_j\}$ *is a projection-valued measure and* $\dim(\operatorname{ran} A_j) = 1$, *for arbitrary state* ρ *we have* (1) $I_1(\rho,\Lambda) \le \min\{S(\rho), S(\Lambda\rho)\}$, (2) $I_2(\rho,\Lambda) = 0$, (3) $I_3(\rho,\Lambda) = S(\rho)$.

From this theorem, the entropy $I_1(\rho,\Lambda)$ only satisfies the inequality held for classical systems, so that only this entropy can be a candidate as the quantum extension of the classical mutual entropy. The other two entropies can describe a sort of entanglement between input and output, such a correlation can be also described by quasi-mutual entropy, some generalization of $I_1(\rho, \Lambda)$, discussed in [29, 9].

Acknowledgements

The author thanks IIAS and SCAT for financial support of this work.

REFERENCES

[1] L. Accardi and K. Imafuku: Control of quantum states by decoherence, to appear in *Open Systems and Information Dynamics*, 2003.

[2] L. Accardi, Y. G. Lu and I. Volovich: *Quantum Theory and its Stochastic Limit*, Springer, Berlin 2002; Japanese translation, Tokyo–Springer 2003.

[3] L. Accardi and M. Ohya: Compound channels, transition expectations, and liftings, *Appl. Math. Optim.* **39** (1999), 33–59.

[4] L. Accardi and M. Ohya: A stochastic limit approach to the SAT problem, to appear.

[5] L. Accardi and M. Ohya: Generalized quantum Turing machine and stochastic limit for the SAT problem, in preparation.

[6] H. Araki: Relative entropy of states of von Neumann algebras, *Publ. RIMS, Kyoto Univ.* **11** (1976), 809–833; Relative entropy for states of von Neumann algebras II, *Publ. RIMS, Kyoto Univ.* **13** (1977), 173–192.

[7] L. Accardi and R. Sabbadini: On the Ohya–Masuda quantum SAT Algorithm, *Proceedings Intern. Conf. "Unconventional Models of Computations"*, I. Antoniou, C.S. Calude, M. Dinneen (eds.) Springer 2001; Preprint Volterra, N. 432, 2000.

[8] H. Barnum, M. A. Nielsen and B. W. Schumacher: Information transmission through a noisy quantum channel, *Phys. Rev. A* **57** (1998), 4153-4175.

[9] V. P. Belavkin and M. Ohya: Quantum entropy and information in discrete entangled states, infinite dimensional analysis, quantum probability and related topics, Vol. 4, No. 2 (2001), 137–160; Quantum entanglements and entangled mutual entropy, *Proc. R. Soc. Lond. A.* **458** (2002), 209–231.

[10] C. H. Bennett, P. W. Shor, J. A. Smolin and A. V. Thapliyalz: Entanglement-assisted capacity of a quantum channel and the reverse Shannon Theorem, quant-ph/0106052.

[11] E. Bernstein and U. Vazirani: Quantum complexity theory, *Proc. of the 25th Annual ACM Symposium on Theory of Computing*, ACM, New York, pp. 11–22 (1993), *SIAM Journal on Computing* **26** (1997), 1411.

[12] C. H. Bennett, E. Bernstein, G. Brassard and U. Vazirani: *Strengths and Weaknesses of Quantum Computing*, quant-ph/9701001.

[13] R. Cleve: *An Introduction to Quantum Complexity Theory*, quant-ph/9906111.

[14] D. Deutsch: Quantum theory, the Church-Turing principle and the universal quantum computer, *Proc. Royal Society of London series A*, **400** (1985), 97–117.

[15] A. Ekert and R. Jozsa: Quantum computation and Shor's factoring algorithm, *Rev. Mod. Phys.* **68** (1996), 733–753.

[16] M. Garey and D. Johnson: *Computers and Intractability—a Guide to the Theory of NP-completeness*, Freeman, 1979.

[17] I. M. Gelfand and A. M. Yaglom: Calculation of the amount of information about a random function contained in another such function, *Amer. Math. Soc. Transl.* **12** (1959), 199–246.

[18] A. S. Holevo, Some estimates for the amount of information transmittable by a quantum communication channel (in Russian), *Problemy Peredachi Informacii* **9** (1973), 3–11.

[19] S. Iriyama and S. Akashi: Complexity of Ohya-Masuda-Volovich algorithm, to appear.

[20] R. S. Ingarden, A. Kossakowski and M. Ohya: *Information Dynamics and Open Systems*, Kluwer 1997.

[21] S. Iriyama, M. Ohya and I. Volovich: Generalized quantum Turing machine and its application to the SAT chaos algorithm, TUS (Tokyo University of Science) preprint, 2003.
[22] R. S. Ingarden: Quantum information theory, *Rep. Math. Phys.* **10** (1976), 43–73.
[23] J. von Neumann: Die mathematischen Grundlagen der Quantenmechanik, Springer, Berlin, 1932.
[24] A. N. Kolmogorov, Theory of transmission of information, *Amer. Math. Soc. Translation*, Ser. 2, **33** (1963), 291–321.
[25] A. Kossakowski, M. Ohya and Y. Togawa: How can we observe and describe chaos?, *Open System and Information Dynamics* **10(3)** (2003), 221–233.
[26] M. Ohya, Complexities and their applications to characterization of chaos, *Int. J. Theor. Phys.* **37** (1998), 495.
[27] M. Ohya: On compound state and mutual information in quantum information theory, *IEEE Trans. Information Theory* **29** (1983), 770–777.
[28] M. Ohya: Some aspects of quantum information theory and their applications to irreversible processes, *Rep. Math. Phys.* **27** (1989), 19–47.
[29] M. Ohya: Fundamentals of quantum mutual entropy and capacity, *Open Systems and Information Dynamics* **6** (1999), 69–78.
[30] M. Ohya and N. Masuda: NP problem in quantum algorithm, *Open Systems and Information Dynamics* **7** (2000), 33–39.
[31] M. Ohya and D. Petz: *Quantum Entropy and its Use*, Springer 1993.
[32] M. Ohya and I. Volovich: Quantum computing, NP-complete problems and chaotic dynamics, Quantum Information II, eds. T. Hida and K. Saito, World Scientific 2000; quant-ph/9912100 and *J. Opt. B*, **5** (2003), 639–642.
[33] M. Ohya and I. Volovich: New quantum algorithm for studying NP-complete problems, *Rep. Math. Phys.* **52** (2003), 25-33.
[34] M. Ohya and I. Volovich: *Quantum Information, Computation, Cryptography and Teleportation*, Springer (to appear).
[35] M. Ohya and N. Watanabe: Remarks on quantum mutual entropy, TUS preprint.
[36] D. Petz and M. Mosonyi: Stationary quantum source coding, *J. Math. Phys.* **42** (2001), 4857-4864.
[37] P. Shor, Algorithm for quantum computation, Discrete logarithm and factoring algorithm, *Proceedings of the 35th Annual IEEE Symposium on Foundation of Computer Science*, pp. 124–134, 1994.
[38] P. Shor: The quantum channel capacity and coherent information, Lecture Notes, *MSRI Workshop on Quantum Computation*, 2002.
[39] A. Uhlmann: Relative entropy and the Wigner–Yanase–Dyson–Lieb concavity in interpolation theory, *Commun. Math. Phys.* **54** (1977), 21–32.
[40] H. Umegaki: Conditional expectations in an operator algebra IV (entropy and information), *Kodai Math. Sem. Rep.* **14** (1962), 59–85.

Open Sys. & Information Dyn. **11**: 219–233, 2004

A Stochastic Limit Approach to the SAT Problem

Luigi Accardi

Centro V. Volterra
Università di Roma Torvergata
Via Orazio Raimondo, 00173 Roma, Italia
e-mail: accardi@volterra.mat.uniroma2.it

Masanori Ohya

Department of Information Sciences
Tokyo University of Science
Noda City, Chiba 278–8510, Japan
e-mail: ohya@is.noda.tus.ac.jp

(Received: January 26, 2004)

Abstract. There exists an important problem whether there exists an algorithm to solve an NP-complete problem in polynomial time. In this paper, a new concept of quantum adaptive stochastic systems is proposed, and it is shown that it can be used to solve the problem above.

1. Introduction

Although the performance of computers is highly progressed, there are several problems which may not be solved effectively, namely, in polynomial time. Among such problems, so-called NP-problems and NP-complete problems are fundamental. It is known that all NP-complete problems are equivalent and an essential question is *whether there exists an algorithm to solve an NP complete problem in polynomial time*. Problems of this kind have been studied for decades and so far all known algorithms have an exponential running time in the length of the input. The standard definition of P- and NP-problems is the following [14, 17, 20]:

DEFINITION 1 Let n be the size of input.

(1) A P-problem is a problem such that the number of elementary steps needed to solve it is polynomial in n. Equivalently, it is a problem which can be recognized in time which is polynomial in n by a deterministic Turing machine.

(2) An NP-problem is a problem which can be solved in polynomial time by a nondeterministic Turing machine.

This can be understood as follows: Let us consider a problem to find a solution of $f(x) = 0$. We can check in time polynomial in n whether x_0 is a solution of $f(x) = 0$, but we do not know whether we can find the solution of $f(x) = 0$ in such time.

DEFINITION 2 An NP-complete problem is a problem to which any other NP-problem can be polynomially transformed.

We take the SAT (satisfiability) problem, one of the NP-complete problems, to study whether there exists an algorithm showing NPC = P. Our aim of this paper and the previous ones [10, 12, 13] is to find a quantum algorithm solving the SAT problem in polynomial time in the size of data.

Let $X \equiv \{x_1, \ldots, x_n\}$ be a set. Then x_k and its negation $\bar{x}_k$ $(k = 1, 2, \ldots, n)$ are called literals and the set of all such literals is denoted by $X' \equiv \{x_1, \bar{x}_1, \ldots, x_n, \bar{x}_n\}$. The set of all subsets of X' is denoted by $\mathcal{F}(X')$ and an element $C \in \mathcal{F}(X')$ is called a clause. We take a truth assignment to all Boolean variables x_k. If we can assign the truth value to at least one element of C, then C is called satisfiable. When C is satisfiable, the truth value $t(C)$ of C is regarded as true, otherwise, it is is false. Taking the truth values as "true $\leftrightarrow$1, false $\leftrightarrow$0". Then C is satisfiable iff $t(C) = 1$.

Let $L = \{0, 1\}$ be a Boolean lattice with usual join $\vee$ and meet $\wedge$ operations, and $t(x)$ be the truth value of a literal x in X. Then the truth value of a clause C is written as $t(C) \equiv \vee_{x \in C} t(x)$.

Moreover the set $\mathcal{C}$ of all clauses C_j $(j = 1, 2, \ldots, m)$ is called satisfiable iff the meet of all truth values of C_j is 1; $t(\mathcal{C}) \equiv \wedge_{j=1}^{m} t(C_j) = 1$. Thus the SAT problem is written as follows:

DEFINITION 3 *SAT Problem: Given a Boolean set $X \equiv \{x_1, \ldots, x_n\}$ and a set $\mathcal{C} = \{C_1, \ldots, C_m\}$ of clauses, determine whether $\mathcal{C}$ is satisfiable or not.*

That is, this problem is to ask whether there exists a truth assignment which makes $\mathcal{C}$ satisfiable. It is known that one needs polynomial time to check the satisfiability when a specific truth assignment is given, but we cannot determine the satisfiability in polynomial time when an assignment is not specified.

In [10] we have discussed the quantum algorithm of the SAT problem, which was rewritten in [18] and we have showed that OM SAT-algorithm is combinatoric. In [12, 13] it is shown that the chaotic quantum algorithm can solve the SAT problem in polynomial time.

Ohya and Masuda pointed out [10] that the SAT problem, and hence all other NP problems, can be solved in polynomial time by a quantum computer if the superposition of two orthogonal vectors $|0\rangle$ and $|1\rangle$ can be physically detected. However this detection is considered impossible with the present day technology. The problem to be overcome is how to distinguish the pure vector $|0\rangle$ from the superposed one $\alpha|0\rangle + \beta|1\rangle$, obtained by the OM SAT-quantum algorithm, if β is not zero but very small. If such a distinction is possible, then we can solve the NPC problem in the polynomial time. In [12, 13] it is shown that it can be possible by combining nonlinear chaos amplifier with the quantum algorithm, which would imply the existence of a mathematical algorithm solving NP = P. It is not known if the algorithm of Ohya and Volovich lies in the framework of quantum Turing algorithms or not. So the next question is (1) whether there exists a physical realization combining the SAT quantum algorithm with chaos dynamics, or (2) whether there exists another method to achieve the above distinction of two vectors by a suitable unitary evolution so that all process can be modeled by a certain quantum Turing machine (circuits).

In this paper, we argue that the stochastic limit, recently extensively studied by Accardi and coworkers [1], can be used to find another method of (2) above. In Sect. 2, we review mathematical frame of quantum algorithm and the OM SAT-algorithm following the representation of Accardi and Sabaddini [18] with a quick review of OV-chaos algorithm in Sect. 3. In Sect. 4, a new concept — quantum adaptive stochastic system — is proposed, and in Sect. 5, we show that it can be used to solve the problem NP = P.

2. Quantum Algorithm

The quantum algorithms discussed so far are rather idealized because computation is represented by unitary operations. A unitary operation is rather difficult to realize in physical processes, a more realistic operation is the one allowing some dissipation like semigroup dynamics. However such dissipative dynamics destroys the entanglement and hence they essentially reduce the ability of quantum computation to preserve the entanglement of states. In order to keep the power of quantum computation and good entanglement, it will be necessary to introduce some kind of amplification in the course of real physical processes in physical devices, which will be similar to the amplication processes in quantum communication. In this section, to look for more realistic operations in a quantum computer, the channel expression will be used, at least, in the sense of mathematical scheme of quantum computation because a channel is not always unitary and represents many different types of dynamics.

Let $\mathcal{H}$ be a Hilbert space describing the input, computation and the output (result). As usual, the Hilbert space is $\mathcal{H} = \otimes_1^N \mathbb{C}^2$, and let the basis of $\mathcal{H} = \otimes_1^N \mathbb{C}^2$ be:

$$\begin{aligned} e_0 &\equiv |0\rangle = |0\rangle \otimes \cdots \otimes |0\rangle \otimes |0\rangle , \\ e_1 &\equiv |1\rangle = |0\rangle \otimes \cdots \otimes |0\rangle \otimes |1\rangle , \\ \cdots & \quad \cdots \\ e_{2^N-1} &\equiv \left|2^N - 1\right\rangle = |1\rangle \otimes \cdots \otimes |1\rangle \otimes |1\rangle . \end{aligned}$$

Any number $t \in \{0, \ldots, 2^N - 1\}$ can be expressed by

$$t = \sum_{k=1}^{N} a_t^{(k)} 2^{k-1},$$

$a_t^{(k)} = 0$ or $a_t^{(k)} = 1$, so that the associated vector is written by

$$|t\rangle \, (= e_t) = \bigotimes_{k=1}^{N} \left| a_t^{(k)} \right\rangle .$$

Applying n-tuples of the Hadamard matrix $A \equiv \frac{1}{\sqrt{2}}\begin{pmatrix} 1 & 1 \\ 1 & -1 \end{pmatrix}$ to the vacuum vector $|0\rangle$, we get

$$A\,|0\rangle\,(\,=\xi\,(0)) \;\equiv\; \bigotimes_{1}^{N} \frac{1}{\sqrt{2}}\,(|0\rangle + |1\rangle)\,.$$

Put

$$W\,(t) \;\equiv\; \bigotimes_{j=1}^{N} \begin{pmatrix} 1 & 0 \\ 0 & \exp(\frac{2\pi i t}{2^N}2^{j-1}) \end{pmatrix}.$$

Then we have

$$\xi\,(t) \;\equiv\; W\,(t)\,\xi\,(0) \;=\; \frac{1}{\sqrt{2^N}}\sum_{k=0}^{2^N-1} \exp\left(\frac{2\pi i t k}{2^N}\right)|k\rangle\,,$$

which is called Discrete Fourier Transformation. The combination of the above operations gives a unitary operator $U_F\,(t) \equiv W\,(t)\,A$ and the vector $\xi\,(t) = U_F\,(t)\,|0\rangle$.

2.1. Channel expression of conventional unitary algorithm

All conventional unitary algorithms can be written as a combination of the following three steps:

(1) Preparation of state: Take a state ρ (e.g., $\rho = |0\rangle\langle 0|$) and apply the unitary channel defined by the above $U_F\,(t) : \Lambda_F^* = \mathrm{Ad}_{U_F(t)}$

$$\Lambda_F^* \;=\; \mathrm{Ad}_{U_F} \quad\Longrightarrow\quad \Lambda_F^*\rho \;=\; U_F\rho U_F^*.$$

(2) Computation: Let U be a unitary operator on $\mathcal{H}$ representing the computation followed by a suitable programming of a certain problem, then the computation is described by a channel $\Lambda_U^* = \mathrm{Ad}_U$ (unitary channel). After the computation, the final state ρ_f will be

$$\rho_f \;=\; \Lambda_U^*\Lambda_F^*\rho\,.$$

(3) Registration and Measurement: For the registration of the computed result and its measurement we may need an additional system $\mathcal{K}$ (e.g., register), so that the lifting $\mathcal{E}_m^*$ from $\mathcal{S}\,(\mathcal{H})$ to $\mathcal{S}\,(\mathcal{H}\otimes\mathcal{K})$ in the sense of [2] is useful to describe this stage. Thus the whole process is wrtten as

$$\rho_f \;=\; \mathcal{E}_m^*\,(\Lambda_U^*\Lambda_F^*\rho)\,.$$

Finally, we measure the state in $\mathcal{K}$: For instance, let $\{P_k; k\in J\}$ be a projection valued measure (PVM) on $\mathcal{K}$

$$\Lambda_m^*\rho_f \;=\; \sum_{k\in J} I\otimes P_k\rho_f I\otimes P_k\,,$$

after which we can get a desired result by observations in finite times if the size of the set J is small.

2.2. Channel expression of the general quantum algorithm

When dissipation is involved the above three steps have to be generalized. Such a generalization can be expressed by means of suitable channel, not necessarily unitary.

(1) Preparation of state: We may use the same channel $\Lambda_F^* = \mathrm{Ad}_{U_F}$ in this first step, but if the number of qubits N is large so that it will not be built physically, then Λ_F^* should be modified; let us denote it by Λ_P^*.
(2) Computation: This stage is certainly modified to a channel Λ_C^* reflecting a physical device realizing it.
(3) Registration and Measurement: This stage is the the same as above. Thus the whole process is written as

$$\rho_f = \mathcal{E}_m^* \left(\Lambda_C^* \Lambda_P^* \rho \right).$$

3. Quantum Algorithm of SAT

Let 0 and 1 of the Boolean lattice L be denoted by the vectors $|0\rangle \equiv \binom{1}{0}$ and $|1\rangle \equiv \binom{0}{1}$ in the Hilbert space $\mathbb{C}^2$, respectively. That is, the vector $|0\rangle$ represents false and $|1\rangle$ truth. This section is based on [10, 18, 3].

As we explained in the previous section, an element $x \in X$ can be denoted by 0 or 1, i.e. by $|0\rangle$ or $|1\rangle$ in the present context. In order to describe a clause C of length at most n by a quantum state, we need the n-tuple tensor product Hilbert space $\mathcal{H} \equiv \otimes_1^n \mathbb{C}^2$. For instance, in the case of $n = 2$, given $C = \{x_1, x_2\}$ with an assignment $x_1 = 0$ and $x_2 = 1$, the corresponding quantum state vector is $|0\rangle \otimes |1\rangle$, so that the quantum state vector describing C is generally written as $|C\rangle = |x_1\rangle \otimes |x_2\rangle \in \mathcal{H}$ with $x_k = 0$ or 1 $(k = 1, 2)$.

The quantum computation is performed by a unitary gate constructed from several fundamental gates such as "Not" gate, "Controlled-Not" gate, "Controlled-Controlled" Not gate [22, 11]. Once $X \equiv \{x_1, \ldots, x_n\}$ and $\mathcal{C} = \{C_1, C_2, \ldots, C_m\}$ are given, the SAT is to find the vector

$$|t(\mathcal{C})\rangle \equiv \bigwedge_{j=1}^{m} \bigvee_{x \in C_j} t(x),$$

where $t(x)$ is $|0\rangle$ or $|1\rangle$ when $x = 0$ or 1, respectively, and $t(x) \wedge t(y) \equiv t(x \wedge y)$, $t(x) \vee t(y) \equiv t(x \vee y)$.

3.1. Logical negation

DEFINITION 4 Let X be a set. A *negation* on X is an involution without fixed points, i.e. a map $X \ni x \mapsto x' \in X$ such that $(x')' = x;\ x \neq x'\ \forall x \in X$. x' is called the *negation* of x.

PROPOSITION 1 *Given a nonempty set X with a negation $(x \mapsto x')$ and denoting*

$$I' := \{x' \in X : x \in I\},$$

for $I \subseteq X$, there exists a set $I \subseteq X$ such that $X = I \cup I'$.

Thus a finite set with a negation must be even. Let X be a finite set with $2n$ elements and with a negation $(x \mapsto x')$. A partition $X = I \cup I'$, $|I| = n$ can be constructed with an n-step algorithm. Not all n-step algorithms are equivalent.

DEFINITION 5 Given a set X with a negation $x \mapsto x'$, a "clause" is a subset of X. A minimal clause is a subset $I \subseteq X$ such that $I \cap I' = \emptyset$ (i.e. if I contains x, it does not contain the negation of x).

In any set X of cardinality $2n$ there are 2^n minimal clauses. Given a set $\widehat{\mathcal{C}}_0$ of clauses, if there are non-minimal clauses in it, then we can eliminate them from $\hat{\mathcal{C}}_0$ because any truth function must be identically zero on a non-minimal clause.

However, to eliminate the non-minimal clauses from $\widehat{\mathcal{C}}_0$, one has to "read" all its elements. Their number can be of order 2^n.

3.2. Truth functions

The set $\{0,1\}$ is a Boolean algebra with the operations

$$\varepsilon \vee \varepsilon' := \max\{\varepsilon, \varepsilon'\}, \qquad \varepsilon \wedge \varepsilon' := \min\{\varepsilon, \varepsilon'\}, \quad \varepsilon, \varepsilon' \in \{0,1\}.$$

A clause truth function on the clauses on the set $X = \{x_1, \ldots, x_n, x_1', \ldots, x_n'\}$ is a boolean algebra homomorphism

$$t : X \to \{0,1\}$$

with the property (*principle of the excluded third*):

$$t(x_j) \vee t(x_j') = 1, \quad \forall j = 1, \ldots, n. \tag{1}$$

Because of (1), such a function is uniquely determined by values $\{t(x_1), \ldots, t(x_n)\}$, hence the number of such functions is 2^n. For this reason, in the following we will simply say *truth function on* $\{x_1, \ldots, x_n\}$ meaning by this a truth function on the clauses of the set $\{x_1, \ldots, x_n, x_1', \ldots, x_n'\}$. Conversely given any n-tuple $\varepsilon = (\varepsilon_1, \ldots, \varepsilon_n) \in \{0,1\}^n$, there exists only one truth function on $\{x_1, \ldots, x_n\}$, with the property that

$$t(x_j) = \varepsilon_j, \quad \forall j = 1, \ldots, n.$$

In what follows, given a truth function t, we denote the string in $\{t(x_1), \ldots, t(x_n)\}$ uniquely associated to that function by ε_t.

Let $\mathcal{T}$ be the set of truth functions on $\{x_1, \ldots, x_n\}$. The function

$$t \in \mathcal{T} \mapsto |t(x_1), \ldots, t(x_n)\rangle \in \otimes^n \mathbb{C}^2$$

defines a one-to-one correspondence between $\mathcal{T}$ and the set $\{0,1\}$, that is, a one-to-one correspondence between truth functions and vectors of the computational basis of $\otimes^n \mathbb{C}^2$

PROPOSITION 2 *Let $C \subseteq X$ be a clause and I, I' the sets associated to it through the procedure explained in Sect. 1. Let t be a truth function on $\{x_1, \ldots, x_n\}$. Then*

$$t(C) = \left[\bigvee_{i\in I} t(x_i)\right] \vee \left[\bigvee_{j\in I'} (1 - t(x_j))\right].$$

Therefore, as stated in Introduction, a set of clauses $\mathcal{C}_0$ is said to be SAT if there exists a truth function t, on $\{x_1, \ldots, x_n\}$ such that

$$t(\mathcal{C}_0) := t\Big(\bigwedge_{C\in\mathcal{C}_0} C\Big) = \prod_{C\in\mathcal{C}_0} t(C) = 1.$$

3.3. Quantum algorithm for the SAT Problem

We review here a technique developed in [10], which shows that the SAT problem can be solved in polynomial time by a quantum computer.

Given a set of clauses $\mathcal{C}_0 = \{C_1, \ldots, C_m\}$ on X, Ohya and Masuda constructed a Hilbert space $\mathcal{H} = \otimes^{n+\mu}\mathbb{C}^2$, where μ is a number that can be chosen linear in mn, and a unitary operator $U_{\mathcal{C}_0} : \mathcal{H} \to \mathcal{H}$ with the property that, for any truth function t,

$$U_{\mathcal{C}_0}|\varepsilon_t, 0_\mu\rangle = |\varepsilon_t, x^{\varepsilon_t}_{\mu-1}, t(\mathcal{C}_0)\rangle,$$

where ε_t is the vector of the computational basis of $\otimes^n\mathbb{C}^2$ corresponding to t, and 0_μ (resp. $x^{\varepsilon}_{\mu-1}$) is a string of μ zeros (resp. a string of $(\mu - 1)$ binary symbols depending on ε).

Furthermore $U_{\mathcal{C}_0}$ is a product of *gates*, namely of unitary operators that act at most on two qubits at a time.

Let $\mathcal{C}_0$ and $U_{\mathcal{C}_0}$ be as above and, for every $\varepsilon \in \{0,1\}^n$, let t_ε be the corresponding truth function. Applying the unitary operator $U_{\mathcal{C}_0}$ to the vector

$$|v\rangle := \frac{1}{2^{n/2}} \sum_{\varepsilon\in\{0,1\}^n} |\varepsilon, 0_\mu\rangle$$

one obtains the final state vector

$$|v_f\rangle := U_{\mathcal{C}_0}|v\rangle = \frac{1}{2^{n/2}} \sum_{\varepsilon\in\{0,1\}^n} |\varepsilon, x^{\varepsilon}_{\mu-1}, t_\varepsilon(\mathcal{C}_0)\rangle.$$

THEOREM 1 *$\mathcal{C}_0$ is satisfiable if and only if*

$$P_{n+\mu,1}U_{\mathcal{C}_0}|v\rangle \neq 0,$$

where $P_{n+\mu,1}$ denotes the projector

$$P_{n+\mu,1} := 1_{n+\mu-1} \otimes |1\rangle\langle 1|$$

on the subspace of $\mathcal{H}$ spanned by the vectors

$$|\varepsilon_n, \varepsilon_{\mu-1}, 1\rangle.$$

According to the standard theory of quantum measurement, after a measurement of the event $P_{n+\mu,1}$, the state $\rho = |v_f\rangle\langle v_f|$ becomes

$$\rho \longrightarrow \frac{P_{n+\mu,1}\rho P_{n+\mu,1}}{\mathrm{Tr}\,\rho' P_{n+\mu,1}} =: \rho' .$$

Thus the solvability of the SAT problem is reduced to check that $\rho' \neq 0$. The difficulty is that the probability of $P_{n+\mu,1}$ is

$$\mathrm{Tr}\,\rho' P_{n+\mu,1} = \|P_{n+\mu,1}\psi\|^2 = \frac{|T(\mathcal{C}_0)|}{2^n},$$

where $|T(\mathcal{C}_0)|$ is the cardinality of the set $T(\mathcal{C}_0)$, of all the truth functions t such that $t(\mathcal{C}_0) = 1$.

We put $q := \sqrt{r/2^n}$ with $r := |T(\mathcal{C}_0)|$ in the sequel. Then if r is suitably large to detect it, then the SAT problem is solved in polynomial time. However, for small r, the probability is very small and this means we in fact don't get an information about the existence of the solution of the equation $t(C_0) = 1$, so that in such a case we need further deliberation.

Let us simplify our notations. After the quantum computation, the quantum computer will be in the state

$$|v_f\rangle = \sqrt{1-q^2}\,|\varphi_0\rangle \otimes |0\rangle + q\,|\varphi_1\rangle \otimes |1\rangle ,$$

where $|\varphi_1\rangle$ and $|\varphi_0\rangle$ are normalized n qubit states and $q = \sqrt{r/2^n}$. Effectively our problem is reduced to the following 1 qubit problem. We have the state

$$|\psi\rangle = \sqrt{1-q^2}\,|0\rangle + q\,|1\rangle$$

and we want to distinguish between the cases $q = 0$ and $q > 0$ (small positive number).

It is argued in [16] that quantum computer can speed up NP problems quadratically but not exponentially. The no-go theorem states that if the inner product of two quantum states is close to 1, then the probability that a measurement distinguishes then is exponentially small. And one could claim that amplification of this distinguishability is not possible.

At this point we emphasized [13] that we do not propose to make a measurement which will be overwhelmingly likely to fail. What we do it is a proposal to use the output $I\,|\psi\rangle$ of the quantum computer as an input for another device which uses chaotic dynamics.

The amplification would be not possible if we use the standard model of quantum computations with a unitary evolution. However the idea of the paper [12, 13] is different. In [12, 13] it is proposed to combine quantum computer with a chaotic dynamics amplifier. Such a quantum chaos computer is a new model of computations and we demonstrate that the amplification is possible in the polynomial time.

One could object that we do not suggest a practical realization of the new model of computations. But at the moment nobody knows of how to make a practically useful implementation of the standard model of quantum computing ever. It seems to us that the quantum chaos computer considered in [13] deserves an investigation and has a potential to be realizable.

Here we mention two works on non-linear quantum evolution to study NP-problems done by Abrams-Lloyd [8] and Czachor [9]. The former was based on the Weinberg model of nonlinear quantum mechanics and the latter was done by means of the Polchinski type description. Czachor's work looks similar to our approach (stochastic limit). Their works are very artificial and conceptually different from ours.

3.4. Chaotic dynamics

Various aspects of classical and quantum chaos have been the subject of numerous studies, see [19] and references therein. Here we will argue that chaos can play a constructive role in computations (see [12, 13] for the details).

Chaotic behaviour in a classical system is usually considered as an exponential sensitivity to initial conditions. It is this sensitivity we would like to use to distinguish between the cases $q = 0$ and $q > 0$ from the previous section.

Consider the so called logistic map which is given by the equation

$$x_{n+1} = ax_n(1 - x_n) \equiv f(x), \qquad x_n \in [0, 1].$$

The properties of the map depend on the parameter a. If we take, for example, $a = 3.71$, then the Lyapunov exponent is positive, the trajectory is very sensitive to the initial value and one has the chaotic behaviour [19]. It is important to notice that if the initial value $x_0 = 0$, then $x_n = 0$ for all n.

It is known [21] that any classical algorithm can be implemented on a quantum computer. Our quantum chaos computer will consist of two blocks. One block is an ordinary quantum computer performing computations with the output $|\psi\rangle = \sqrt{1 - q^2}\,|0\rangle + q\,|1\rangle$. The second block is a computer performing computations of the *classical* logistic map. This two blocks should be connected in such a way that the state $|\psi\rangle$ first be transformed into the density matrix of the form

$$\rho = q^2 P_1 + (1 - q^2) P_0 ,$$

where P_1 and P_0 are projectors to the state vectors $|1\rangle$ and $|0\rangle$. This connection is in fact nontrivial and actually it should be considered as the third block. One has to notice that P_1 and P_0 generate an Abelian algebra which can be considered as a classical system. In the second block the density matrix ρ above is interpreted as the initial data ρ_0, and we apply the logistic map as

$$\rho_m = \frac{(I + f^m(\rho_0)\sigma_3)}{2},$$

where I is the identity matrix and σ_3 is the z-component Pauli matrix on $\mathbb{C}^2$. To find the proper value m we finally measure the value of σ_3 in the state ρ_m such that

$$M_m \equiv \mathrm{Tr}\rho_m\sigma_3 .$$

We obtain

THEOREM 2

$$\rho_m = \frac{(I + f^m(q^2)\sigma_3)}{2} \quad \text{and} \quad M_m = f^m(q^2).$$

Thus the question is whether we can find such an m in polynomial number steps in n satisfying the inequality $M_m \geq \frac{1}{2}$ for very small but non-zero q^2. Here we have to remark that if one has $q = 0$ then $\rho_0 = P_0$ and we obtain $M_m = 0$ for all m. If $q \neq 0$, the stochastic dynamics leads to the amplification of the small magnitude q in such a way that it can be detected as is explained below. The transition from ρ_0 to ρ_m is nonlinear and can be considered as a classical evolution because our algebra generated by P_0 and P_1 is abelian. The amplification can be done within at most 2n steps due to the following propositions. Since $f^m(q^2)$ is x_m of the logistic map $x_{m+1} = f(x_m)$ with $x_0 = q^2$, we use the notation x_m in the logistic map for simplicity.

THEOREM 3 *For the logistic map* $x_{n+1} = ax_n(1 - x_n)$ *with* $a \in [0, 4]$ *and* $x_0 \in [0, 1]$, *let* x_0 *be* $1/2^n$ *and the set* J *be* $\{0, 1, 2, \ldots, n, \ldots 2n\}$. *If* a *is* 3.71, *then there exists an integer* m *in* J *satisfying* $x_m > 1/2$.

THEOREM 4 *Let* a *and* n *be the same as in the above proposition. If there exists* m_0 *in* J *such that* $x_{m_0} > 1/2$, *then* $m_0 > (n - 1)/\log_2 3.71$.

According to these theorems, it is enough to check the value x_m (M_m) around the above m_0 when q is $1/2^n$ for a large n. More generally, when $q = k/2^n$ with some integer k, it is similarly checked that the value x_m (M_m) becomes over $1/2$ within at most $2n$ steps.

The complexity of the quantum algorithm for the SAT problem was discussed in Sect. 3 to be polynomial in time. We have only to consider the number of steps in the classical algorithm for the logistic map performed on a quantum computer. It is the probabilistic part of the construction and one has to repeat computations several times to be able to distinguish the cases $q = 0$ and $q > 0$. Thus it seems that the quantum chaos computer can solve the SAT problem in polynomial time.

In conclusion of [12, 13], the quantum chaos computer combines the ordinary quantum computer with quantum chaotic dynamics amplifier. It may go beyond the usual quantum Turing algorithm, but such a device can be powerful enough to solve the NP-complete problems in the polynomial time. The detailed estimation of the complexity of the SAT algorithm is discussed in [23].

In the next two sections we will discuss the SAT problem from a different point of view, that is, we will show that the same amplification is possible by unitary dynamics defined in the stochastic limit.

4. Quantum Adaptive Systems

The idea to develop a mathematical approach to adaptive systems, i.e. those systems whose properties are in part determined as responses to an environment

[7, 25], was born in connection with some problems of quantum measurement theory and chaos dynamics.

The mathematical definition of an adaptive system is in terms of observables, namely: *an adaptive system is a composite system whose interaction depends on a fixed observable* (typically in a measurement process, this observable is the observable one wants to measure). Such systems may be called *observable-adaptive.*

In the present paper, we want to extend this point of view by introducing another natural class of adaptive systems which, in a certain sense, is the dual to the one defined above, namely the class of *state-adaptive* systems. These are defined as follows: *a state-adaptive system is a composite system whose interaction depends on the state of at least one of the sub-systems at the instant in which the interaction is switched on.*

Notice that both definitions make sense both for classical and for quantum systems. Since in this paper we will be interested in an application of adaptive systems to quantum computation, we will discuss only quantum adaptive systems, but one should keep in mind that all the considerations below apply to classical systems as well.

The difference between state-adaptive systems and nonlinear dynamical systems should be emphasized:

(i) in nonlinear dynamical systems (such as those whose evolution is described by the Boltzmann equation, or nonlinear Schrödinger equation, etc.) the interaction Hamiltonian depends on the state at each time t, i.e. $H_I = H_I(\rho_t)$ $\forall\, t$.

(ii) in state-adaptive dynamical systems (such as those considered in the present paper) the interaction Hamiltonian depends on the state only at each time $t = 0$, i.e. $H_I = H_I(\rho_0)$.

The latter class of systems describes the following physical situation: at time $t = -T$ $(T > 0)$ the system S is prepared in the state ψ_{-T} and in the time interval $[-T, 0]$ it evolves according to a fixed (free) dynamics $U_{[-T,0]}$ so that its state at time 0 is $U_{[-T,0]}\psi_{-T} =: \psi_0$. At time $t = 0$ an interaction with another system R is switched on and this interaction depends on the state ψ_0, i.e. $H_I = H_I(\psi_0)$.

If we interpret the system R as *environment*, we can say that the above interaction describes the response of the environment to the state of the system S.

Now from the general theory of stochastic limit [1] one knows that, under general ergodicity conditions, an interaction with an environment drives the system to a dynamical (but not necessarily thermodynamical) equilibrium state which depends on the initial state of the environment and on the interaction Hamiltonian.

Therefore, if one is able to realize experimentally these state dependent Hamiltonians, one would be able to drive the system S to a pre-assigned dynamical equilibrium state depending on the input state ψ_0.

In the following section we will substantiate the general scheme described above with an application to the quantum computer approach to the SAT problem described in previous sections.

5. Stochastic Limit and SAT Problem

We illustrate the general scheme described in the previous section in the simplest case when the state space of the system is $\mathcal{H}_S \equiv \mathbb{C}^2$. We fix an orthonormal basis of $\mathcal{H}_S$ as $\{e_0, e_1\}$.

The unknown state (vector) of the system at time $t = 0$

$$\psi := \sum_{\varepsilon\in\{0,1\}} \alpha_\varepsilon e_\varepsilon = \alpha_0 e_0 + \alpha_1 e_1 , \quad \|\psi\| = 1 .$$

In the case of Sect. 3, α_1 corresponds to q and e_j does to $|j\rangle$ $(j = 0, 1)$. This vector is taken as input and defines the interaction Hamiltonian in an external field

$$\begin{aligned} H_I &= \lambda|\psi\rangle\langle\psi| \otimes (A_g^+ + A_g) \\ &= \sum \lambda\alpha_\varepsilon\overline{\alpha}_\varepsilon |e_\varepsilon\rangle\langle e_{\varepsilon'}| \otimes (A_g^+ + A_g) , \end{aligned}$$

where λ is a small coupling constant. Here and in the following summation over repeated indices is understood.

The free system Hamiltonian is taken to be diagonal in the e_ε-basis

$$H_S := \sum_{\varepsilon\in\{0,1\}} E_\varepsilon |e_\varepsilon\rangle\langle e_\varepsilon| = E_0|e_0\rangle\langle e_0| + E_1|e_1\rangle\langle e_1|$$

and the energy levels are ordered so that $E_0 < E_1$. Thus there is a single Bohr frequency $\omega_0 := E_1 - E_0 > 0$. The one-particle field Hamiltonian is

$$S_t g(k) = e^{it\omega(k)} g(k) ,$$

where $\omega(k)$ is a function satisfying the basic analytical assumption of the stochastic limit. Its second quantization is the free field evolution

$$e^{itH_0} A^{\pm} g e^{-itH_0} = A^{\pm}_{S_t g}$$

We can distinguish two cases as below, which correspond to two cases of Sect. 3, i.e., $q > 0$ and $q = 0$.

Case 1

If $\alpha_0, \alpha_1 \neq 0$, then, according to the general theory of stochastic limit (i.e., $t \to t/\lambda^2$) [1], the interaction Hamiltonian H_I is in the same universality class as

$$\widetilde{H}_I = D \otimes A_g^+ + D^+ \otimes A_g ,$$

where $D := |e_0\rangle\langle e_1|$ (this means that the two interactions have the same stochastic limit). The interaction Hamiltonian at time t is then

$$\widetilde{H}_I(t) = e^{-it\omega_0} D \otimes A^+_{S_t g} + \text{ h.c.} = D \otimes A^+(e^{it(\omega(p)-\omega_0)} g) + \text{ h.c.}$$

and the white noise ($\{b_t\}$) Hamiltonian equation associated, via the stochastic golden rule, to this interaction Hamiltonian is

$$\partial_t U_t = i(Db_t^+ + D^+ b_t)U_t \,.$$

Its causally normal ordered form is equivalent to the stochastic differential equation

$$dU_t = (iDdB_t^+ + iD^+ dB_t - \gamma_- D^+ D dt)U_t \,,$$

where $dB_t := b_t dt$.

The causally ordered inner Langevin equation is

$$\begin{aligned} dj_t(x) &= dU_t^* x U_t + U_t^* x dU_t + dU_t^* x dU_t \\ &= U_t^*(-iD^+ x dB_t - iDx dB_t^+ - \overline{\gamma}_- D^+ D x dt + ixD dB_t^+ \\ &\quad + ixD^+ dB_t - \gamma_- x D^+ D dt + \gamma_- D^+ x D dt)U_t \\ &= ij_t([x, D^+])dB_t + ij_t([x, D])dB_t^+ \\ &\quad -(\mathrm{Re}\ \gamma_-)j_t(\{D^+ D, x\})dt + i(\mathrm{Im}\,\gamma_-)j_t([D^+ D, x])dt \\ &\quad +j_t(D^+ x D)(\mathrm{Re}\ \gamma_-)dt \,, \end{aligned}$$

where $j_t(x) := U_t^* x U_t$. Therefore the master equation is

$$\begin{aligned} \frac{d}{dt} P^t(x) &= (\mathrm{Im}\,\gamma)i[D^+ D, P^t(x)] - (\mathrm{Re}\,\gamma_-)\{D^+ D, P^t(x)\} \\ &\quad + (\mathrm{Re}\ \gamma_-)D^+ P^t(x) D \,, \end{aligned}$$

where $D^+ D = |e_1\rangle\langle e_1|$ and $D^+ x D = \langle e_0, x e_0\rangle |e_1\rangle\langle e_1|$.

The dual Markovian evolution P_*^t acts on density matrices and its generator is

$$L_* \rho = (\mathrm{Im}\,\gamma_-)i[\rho, D^+ D] - (\mathrm{Re}\ \gamma_-)\{\rho, D^+ D\} + (\mathrm{Re}\ \gamma_-)D\rho D^+ \,.$$

Thus, if $\rho_0 = |e_0\rangle\langle e_0|$ one has

$$L_* \rho_0 = 0$$

so ρ_0 is an invariant measure. From the Fagnola-Rebolledo criteria [26], it is the unique invariant measure and the semigroup $\exp(tL_*)$ converges exponentially to it.

Case 2

If $\alpha_1 = 0$, then the interaction Hamiltonian H_I is

$$H_I = \lambda |e_0\rangle\langle e_0| \otimes (A_g^+ + A_g)$$

and, according to the general theory of stochastic limit, the reduced evolution has no damping and corresponds to the pure Hamiltonian

$$H_S + |e_0\rangle\langle e_0| = (E_0 + 1)|e_0\rangle\langle e_0| + E_1 |e_1\rangle\langle e_1|$$

therefore, if we choose the eigenvalues E_1, E_0 to be integers (in appropriate units), then the evolution will be periodic.

Since the eigenvalues E_1, E_0 can be chosen a priori, by fixing the system Hamiltonian H_S, it follows that the period of the evolution can be known a priori. This gives a simple criterion for the solvability of the SAT problem because, by waiting a sufficiently long time one can experimentally detect the difference between damping and an oscillatory behaviour.

The precise estimate of this time can be achieved either by theoretical methods or by computer simulation. Both methods will be analyzed in the expanded paper [3].

Czachor [9] gave an example of a nonlinear Schrödinger equation to distinguish two cases, similar to $\alpha_1 \neq 0$ and $\alpha_1 = 0$ given above, in a certain oracle computation. We used the resulting (flag) state after quantum computation of the truth function of SAT to couple the external field and took the stochastic limit, then our final evolution becomes "linear" for the state ρ described as above. The stochastic limit is historically important to realize macroscopic (time) evolution and it is now rigorously established as explained in [1], and we gave a general protocol to study the distinction of two cases $\alpha_1 \neq 0$ and $\alpha_1 = 0$ by this rigorous mathematics. The macro-time enables us to measure the behavior of the outcomes practically. Thus our approach is conceptually different from Czachor's. Moreover Czachor discussed that some expectation value is constant for the case $\alpha_1 = 0$ and oscilating for $\alpha_1 \neq 0$, and ours gives the detail behavior of the state w.r.t the macro-time; damping ($\alpha_1 \neq 0$ case) and oscilating ($\alpha_1 = 0$ case)

6. Conclusion

We showed in [10, 12, 13] that we can find an algorithm solving the SAT problems in polynomial number of steps by combining a quantum algorithm with chaotic dynamics. We used the logistic map there, however it is possible to use other chaotic maps if they can amplify one of two coefficients. In this short paper we pointed out that it is possible to distinguish two different states, $\sqrt{1-q^2}\,|0\rangle + q\,|1\rangle$ ($q \neq 0$) and $|0\rangle$ by means of an adaptive dynamics and the stochastic limit. Finally we remark that our algorithms can be described by deterministic general quantum Turing machine [24, 4], whose result is based on the general quantum algorithm mentioned in Sect. 2.

Acknowledgment

The authors thank SCAT for financial support of this joint work. We thank the referee for informing us about the paper of Czachor.

Bibliography

[1] L. Accardi, Y. G. Lu, I. Volovich, *Quantum Theory and its Stochastic Limit*, Springer Verlag 2002, Japanese translation, Tokyo-Springer, 2003.

[2] L. Accardi and M. Ohya, *Compound channels, transition expectations, and liftings*, Appl. Math. Optim. **39**, 33 (1999).

[3] L. Accardi and M. Ohya, *A stochastic limit approach to the SAT problem*, in preparation.

[4] L. Accardi and M. Ohya, *Generalized Quantum Turing machine and stochastic limit for the SAT problem*, TUS preprint.

[5] L. Accardi, R. Sabbadini, *On the Ohya-Masuda quantum SAT Algorithm*, in: Proceedings Intern.Conf. "Unconventional Models of Computations", I. Antoniou, C. S. Calude, M. Dinneen, eds., Springer 2001; Preprint Volterra, N. 432, 2000

[6] L. Accardi, R.Sabbadini, *A Generalization of Grover's Algorithm*, Proceedings Intern. Conf.: Quantum Information III, Meijo University, Nagoya, 27-31 March, 2001; World Scientific 2002; qu-phys 0012143; Preprint Volterra, N. 444, 2001.

[7] L. Accardi and K. Imafuku, *Control of Quantum States by Decoherence*, Volterra Center Preprint No. 542.

[8] D. S. Abrams and S. Lloyd, *Nonlinear quantum mechanics implies polynomial time solution for NP-complete and #P problem*, Phys. Rev. Lett. **81**, 3992 (1998).

[9] M. Czachor, *Notes on nonlinear quantum algorithm*, Acta Phys. Slov. **48**, 157 (1998).

[10] M. Ohya and N. Masuda, *NP problem in Quantum Algorithm*, Open Sys. Information Dyn. **7**, 33 (2000).

[11] M. Ohya, *Mathematical Foundation of Quantum Computer*, Maruzen Publ. Company, 1998.

[12] M. Ohya and I. V. Volovich, *Quantum computing, NP-complete problems and chaotic dynamics*, in: *Quantum Information II*, eds. T.Hida and K.Saito, World Sci. 2000; quant-ph/9912100 and J. Opt. B **5**, 639 (2003).

[13] M. Ohya and I. V. Volovich, *New quantum algorithm for studying NP-complete problems*, Rep. Math. Phys.**52**, 25 (2003).

[14] M. Garey and D. Johnson, *Computers and Intractability — a guide to the theory of NP-completeness*, Freeman, 1979.

[15] P. W. Shor, *Algorithm for quantum computation: Discrete logarithm and factoring algorithm*, Proceedings of the 35th Annual IEEE Symposium on Foundation of Computer Science, pp. 124–134, 1994.

[16] C. H. Bennett, E. Bernstein, G. Brassard, U. Vazirani, *Strengths and Weaknesses of Quantum Computing*, quant-ph/9701001.

[17] R. Cleve, *An Introduction to Quantum Complexity Theory*, quant-ph/9906111.

[18] L. Accardi, R. Sabbadini, *On the Ohya-Masuda quantum SAT Algorithm*, in: Proceedings International Conference "Unconventional Models of Computations", I. Antoniou, C. S. Calude, M. Dinneen, eds., Springer, 2001.

[19] M. Ohya, *Complexities and Their Applications to Characterization of Chaos*, Int. Journ. of Theort. Phys. **37**, 495 (1998).

[20] M. Ohya and I. V. Volovich, *Quantum information, computation, cryptography and teleportation*, Springer, to appear.

[21] D. Deutsch, *Quantum theory, the Church-Turing principle and the universal quantum computer*, Proc. of Royal Society of London series A, **400**, pp. 97–117, 1985.

[22] A. Ekert and R. Jozsa, *Quantum computation and Shor's factoring algorithm*, Reviews of Modern Physics **68**, 733 (1996).

[23] S. Iriyama and S. Akashi, *Complexity of Ohya-Masuda-Volovich algorithm*, to appear.

[24] S. Iriyama and M. Ohya, *On generalized Turing machine*, TUS (Tokyo University of Science) preprint, 2003.

[25] A. Kossakowski, M. Ohya and Y. Togawa, *How can we observe and describe chaos?*, Open Sys. Information Dyn. **10**, 221 (2003).

[26] F. Fagnola and R. Rebolledo, *On the existence of Stationary States for Quantum Dynamical Semigroups*, to appear in J. Math. Phys., 2001.

Vol. 52 (2003) *REPORTS ON MATHEMATICAL PHYSICS* No. 1

NEW QUANTUM ALGORITHM FOR STUDYING NP-COMPLETE PROBLEMS

MASANORI OHYA

Tokyo University of Science, Department of Information Sciences,
Noda City, Chiba 278-8510, Japan
(e-mail: ohya@is.noda.tus.ac.jp)

and

IGOR V. VOLOVICH

Steklov Mathematical Institute, Gubkin St. 8, 117966 Moscow, Russia
(e-mail: volovich@mi.ras.ru)

(*Received December 6, 2002*)

Ordinary approach to quantum algorithm is based on quantum Turing machine or quantum circuits. It is known that this approach is not powerful enough to solve NP-complete problems. In this paper we study a new approach to quantum algorithm which is a combination of the ordinary quantum algorithm with a chaotic dynamical system. We consider the satisfiability problem as an example of NP-complete problems and argue that the problem, in principle, can be solved in polynomial time by using our new quantum algorithm.

Keywords: Quantum algorithm, NP-complete problem, chaotic dynamics.

1. Introduction

Ordinary approach to quantum algorithm is based on quantum Turing machine or quantum circuits [1–3]. It is known that this approach is not powerful enough to solve NP-complete problems [4, 5]. In [6] we have proposed a new approach to quantum algorithm which goes beyond the standard quantum computation paradigm. This new approach is a sort of combination of the ordinary quantum algorithm and a chaotic dynamics. This approach was based on the results obtained in the paper [7].

There are important problems such as the knapsack problem, the travelling salesman problem, the integer programming problem, the subgraph isomorphism problem, the satisfiability problem that have been studied for decades and for which all known algorithms have a running time that is exponential in the length of the input. These five and many other problems belong to the set of NP-complete problems [4].

Many NP-complete problems have been identified, and it seems that such problems are very difficult and probably exponential. If so, solutions are still needed, and in this paper we consider an approach to these problems based on quantum computers and chaotic dynamics as mentioned above.

As in the previous papers [7, 6], we again consider the satisfiability problem as an example of NP-complete problems and argue that the problem, in principle, can be solved in polynomial time by using our new quantum algorithm.

It is widely believed that quantum computers are more efficient than classical computers. In particular, Shor [8, 9] gave a remarkable quantum polynomial-time algorithm for the factoring problem. However, it is known that this problem is not NP-complete but is NP-intermediate.

Since the quantum algorithm of the satisfiability problem (SAT for short) has been considered in [7], Accardi and Sabbadini showed that this algorithm is combinatoric one and they discussed its combinatoric representation [10]. It was shown in [7] that the SAT problem can be solved in polynomial time by using a quantum computer under the assumption that a special superposition of two orthogonal vectors can be physically detected. The problem one has to overcome here is that the output of computations could be a very small number and one needs to amplify it to a reasonable large quantity.

In this paper we construct a new model (representation) of computations which combine ordinary quantum algorithm with a chaotic dynamical system and prove that one can solve the SAT problem in polynomial time.

For a recent discussion of computational complexity in quantum computing see [11–14]. Mathematical features of quantum computing and quantum information theory are summarized in [15].

2. SAT Problem

Let $X \equiv \{x_1, \ldots, x_n\}$ be a set. Then x_k and its negation $\overline{x}_k$ $(k = 1, 2, \ldots, n)$ are called *literals* and the set of all such literals is denoted by $X' = \{x_1, \overline{x}_1, \ldots, x_n, \overline{x}_n\}$. The set of all subsets of X' is denoted by $\mathcal{F}(X')$ and an element $C \in \mathcal{F}(X')$ is called a *clause*. We take a truth assignment to all variables x_k. If we can assign the truth value to at least one element of C, then C is called *satisfiable*. When C is satisfiable, the truth value $t(C)$ of C is regarded as true, otherwise, that of C is false. Take the truth values as true “1”, false “0”. Then

$$C \text{ is satisfiable iff } t(C) = 1.$$

Let $L = \{0, 1\}$ be a Boolean lattice with usual join $\vee$ and meet $\wedge$, and let $t(x)$ be the truth value of a literal x in X. Then the truth value of a clause C is written as

$$t(C) \equiv \vee_{x \in C} t(x).$$

Further, the set $\mathcal{C}$ of all clauses C_j $(j = 1, 2, \ldots m)$ is called satisfiable iff the meet of all truth values of C_j is 1,

$$t(\mathcal{C}) \equiv \wedge_{j=1}^{m} t(C_j) = 1.$$

Thus the SAT problem is defined as follows.

DEFINITION 1. *SAT Problem*: Given a set $X \equiv \{x_1, \dots, x_n\}$ and a set $C = \{C_1, C_2, \dots, C_m\}$ of clauses, determine whether C is satisfiable or not.

That is, this problem is to ask whether there exists a truth assignment to make C satisfiable.

It is known [4] for usual algorithm that the time to check the satisfiability is polynomial only when a specific truth assignment is given, but we cannot determine the satisfiability in polynomial time when an assignment is not specified.

Note that a formula made by the product (AND $\wedge$) of the disjunction (OR $\vee$) of literals is said to be in the *product of sums* (POS) form. For example, the formula

$$(x_1 \vee \overline{x}_2) \wedge (\overline{x}_1) \wedge (x_2 \vee \overline{x}_3)$$

is in the POS form. Thus a formula in the POS form is said to be *satisfiable* if there is an assignment of values to variables so that the formula has value 1. Therefore, the SAT problem can be regarded as determining *whether or not a formula in the POS form is satisfiable.*

The following analytical formulation of the SAT problem is useful. We define a family of Boolean polynomials $f_{\mathcal{A}}$, indexed by the following data. Let $\mathcal{A}$ be a set

$$\mathcal{A} = \{S_1, \dots, S_N, T_1, \dots, T_N\},$$

where $S_i, T_i \subseteq \{1, \dots, n\}$, and $f_{\mathcal{A}}$ be defined as

$$f_{\mathcal{A}}(x_1, \dots, x_n) = \prod_{i=1}^{N} \left(1 + \prod_{a \in S_i}(1 - x_a) \prod_{b \in T_i} x_b\right).$$

We assume here the addition modulo 2. The SAT problem now is to determine whether or not there exists a value of $\mathbf{x} = (x_1, \dots, x_n)$ such that $f_{\mathcal{A}}(\mathbf{x}) = 1$.

3. Quantum algorithm

Although the quantum algorithm of the SAT problem is needed to add the dust bits to the input n bits, the number of dust bits is of the order of n [7, 10]. Therefore for simplicity we will work in this paper in the $(n+1)$-tuple tensor product Hilbert space $\mathcal{H} \equiv \otimes_1^{n+1} \mathbb{C}^2$ with the computational basis

$$|x_1, \dots, x_n, y\rangle = \otimes_{i=1}^{n} |x_i\rangle \otimes |y\rangle,$$

where $x_1, \dots, x_n,\ y = 0$ or 1. We denote $|x_1, \dots, x_n, y\rangle = |\mathbf{x}, y\rangle$. The quantum version of the function $f(\mathbf{x}) := f_{\mathcal{A}}(\mathbf{x})$ is given by the unitary operator $U_f |\mathbf{x}, y\rangle = |\mathbf{x}, y + f(\mathbf{x})\rangle$. We assume that the unitary matrix U_f can be build in the polynomial time, see [7]. Now let us use the usual quantum algorithm:

(i) using the Fourier transform produce from $|\mathbf{0},0\rangle$ the superposition

$$|v\rangle := \frac{1}{\sqrt{2^n}} \sum_{\mathbf{x}} |\mathbf{x}, 0\rangle,$$

(ii) use the unitary matrix U_f to calculate $f(\mathbf{x})$,

$$|v_f\rangle = U_f\,|v\rangle = \frac{1}{\sqrt{2^n}} \sum_{\mathbf{x}} |\mathbf{x}, f(\mathbf{x})\rangle.$$

Now if we measure the last qubit, i.e., apply the projector $P = I \otimes |1\rangle\langle 1|$ to the state $|v_f\rangle$, then we obtain that the probability to find the result $f(\mathbf{x}) = 1$ is $\| P\,|v_f\rangle \|^2 = r/2^n$, where r is the number of roots of the equation $f(\mathbf{x}) = 1$. If r is suitably large to detect it, then the SAT problem is solved in polynomial time. However, for small r, the probability is very small and this means that in fact we do not get any information about the existence of the solution of the equation $f(\mathbf{x}) = 1$, so that in such a case we need further discussion.

Let us simplify our notation. After the step (ii), the quantum computer will be in the state

$$|v_f\rangle = \sqrt{1-q^2}\,|\varphi_0\rangle \otimes |0\rangle + q\,|\varphi_1\rangle \otimes |1\rangle,$$

where $|\varphi_1\rangle$ and $|\varphi_0\rangle$ are normalized n qubit states and $q = \sqrt{r/2^n}$. Effectively our problem is reduced to the following 1-qubit problem. We have the state

$$|\psi\rangle = \sqrt{1-q^2}\,|0\rangle + q\,|1\rangle$$

and we want to distinguish between the cases $q = 0$ and $q > 0$ (a small positive number).

It is argued in [5] that quantum computer can speed up NP problems quadratically but not exponentially. The no-go theorem states that if the inner product of two quantum states is close to 1, then the probability that a measurement distinguishes which one of the two occurs is exponentially small. And one could claim that amplification of this distinguishability is not possible.

At this point we emphasize that we do not propose to make a measurement (not read) which will be overwhelmingly likely to fail. What we do is a proposal to use the output $|\psi\rangle$ of the quantum computer as an input for another device which uses chaotic dynamics in the sequel.

The amplification would not be possible if we used the standard model of quantum computations with a unitary evolution. However, the idea of our paper is different. We propose to combine quantum computer with a chaotic dynamics amplifier. Such a quantum chaos computer is a new model of computations going beyond usual scheme of quantum computation and we demonstrate that the amplification is possible in the polynomial time.

One could object that we do not suggest a practical realization of the new model of computations. But at the moment nobody knows how to make a practically useful implementation of the standard model of quantum computing. Quantum circuit or quantum Turing machine is a mathematical model, though a convincing one. It seems to us that the quantum chaos computer considered in this paper deserves investigation and has a potential to be realizable.

In this paper we propose a mathematical model of computations for solving SAT problem by refining our previous paper [6]. A possible specific physical implementation of quantum chaos computations with some error correction will be discussed in a separate paper [16], which is somehow related to the recently proposed atomic quantum computer [17].

4. Chaotic dynamics

Various aspects of classical and quantum chaos have been the subject of numerous studies, see [18] and references therein. The investigation of quantum chaos by using quantum computers has been proposed in [19–21]. Here we will argue that chaos can play a constructive role in computations.

Chaotic behaviour in a classical system usually is considered as an exponential sensitivity to initial conditions. It is this sensitivity we would like to use to distinguish between the cases $q = 0$ and $q > 0$ from the previous section.

Consider the so-called logistic map which is given by the equation

$$x_{n+1} = ax_n(1 - x_n) \equiv g(x), \qquad x_n \in [0, 1].$$

The properties of the map depend on the parameter a. If we take, for example, $a = 3.71$, then the Lyapunov exponent is positive, the trajectory is very sensitive to the initial value and one has the chaotic behaviour [18]. It is important to notice that if the initial value $x_0 = 0$, then $x_n = 0$ for all n.

It is known [2] that any classical algorithm can be implemented on quantum computer. Our quantum chaos computer will consist of two blocks. One block will be the ordinary quantum computer performing computations with the output $|\psi\rangle = \sqrt{1-q^2}\,|0\rangle + q\,|1\rangle$. The second block will be a computer performing computations of the *classical* logistic map. These two blocks should be connected in such a way that the state $|\psi\rangle$ should first be transformed into the density matrix of the form

$$\rho = q^2 P_1 + \left(1 - q^2\right) P_0,$$

where P_1 and P_0 are projectors to the state vectors $|1\rangle$ and $|0\rangle$. This connection would in fact be nontrivial and actually should be considered as the third block. One has to notice that P_1 and P_0 generate an abelian algebra which can be considered as a classical system. In the second block the density matrix ρ above is interpreted as the initial data ρ_0, and we apply the logistic map as

$$\rho_m = \frac{(I + g^m(\rho_0)\sigma_3)}{2},$$

where I is the identity matrix and σ_3 is the z-component of Pauli matrix on $\mathbb{C}^2$. This expression is different from that of our first paper [6]. To find a proper value m we finally measure the value of σ_3 in the state ρ_m such that

$$M_m \equiv \operatorname{tr} \rho_m \sigma_3.$$

After simple computation we obtain

$$\rho_m = \frac{(I + g^m(q^2)\sigma_3)}{2}, \quad \text{and } M_m = g^m(q^2).$$

Thus the question is whether we can find such m in polynomial steps of n satisfying the inequality $M_m \geqq \frac{1}{2}$ for very small but nonzero q^2. Here we have to remark that if one has $q = 0$ then $\rho_0 = P_0$ and we obtain $M_m = 0$ for all m. If $q \neq 0$, the stochastic dynamics leads to the amplification of the small magnitude q in such a way that it can be detected as is explained below. The transition from ρ_0 to ρ_m is nonlinear and can be considered as a classical evolution because our algebra generated by P_0 and P_1 is abelian. The amplification can be done within at most $2n$ steps due to the following propositions. Since $g^m(q^2)$ is x_m of the logistic map $x_{m+1} = g(x_m)$ with $x_0 = q^2$, we use the notation x_m in the logistic map for simplicity.

PROPOSITION 2. *For the logistic map* $x_{n+1} = ax_n(1 - x_n)$ *with* $a \in [0, 4]$ *and* $x_0 \in [0, 1]$, *let* x_0 *be* $\frac{1}{2^n}$ *and the set* J *be* $\{0, 1, 2, \ldots, n, \ldots, 2n\}$. *If* a *is* 3.71, *then there exists an integer* m *in* J *satisfying* $x_m > \frac{1}{2}$.

Proof: Suppose that there does not exist such m in J. Then $x_m \leq \frac{1}{2}$ for any $m \in J$. The inequality $x_m \leq \frac{1}{2}$ implies

$$x_m = 3.71(1 - x_{m-1})x_{m-1} \geq \frac{3.71}{2}x_{m-1}.$$

Thus we have

$$\frac{1}{2} \geq x_m \geq \frac{3.71}{2}x_{m-1} \geq \cdots \geq \left(\frac{3.71}{2}\right)^m x_0 = \left(\frac{3.71}{2}\right)^m \frac{1}{2^n},$$

from which we get

$$2^{n+m-1} \geq (3.71)^m .$$

According to the above inequality, we obtain

$$m \leq \frac{n-1}{\log_2 3.71 - 1}.$$

Since $\log_2 3.71 \doteqdot 1.8912$, we have

$$m \leq \frac{n-1}{\log_2 3.71 - 1} < \frac{5}{4}(n-1),$$

which is definitely less than $2n-1$ and it is contradictory to the statement "$x_m \leq \frac{1}{2}$ for any $m \in J$". Thus there exists m in J satisfying $x_m > \frac{1}{2}$. □

PROPOSITION 3. *Let a and n be the same as in the above proposition. If there exists m_0 in J such that $x_{m_0} > \frac{1}{2}$, then $m_0 > \frac{n-1}{\log_2 3.71}$.*

Proof: Since $0 \leq x_m \leq 1$, we have

$$x_m = 3.71(1 - x_{m-1})x_{m-1} \leq 3.71x_{m-1},$$

which reduces to

$$x_m \leq (3.71)^m x_0.$$

For m_0 in J satisfying $x_{m_0} > \frac{1}{2}$, it holds

$$x_0 \geq \frac{1}{(3.71)^{m_0}} x_{m_0} > \frac{1}{2 \times (3.71)^{m_0}}.$$

It follows from $x_0 = \frac{1}{2^n}$ that

$$\log_2 2 \times (3.71)^{m_0} > n,$$

which implies

$$m_0 > \frac{n-1}{\log_2 3.71}.$$

□

According to these propositions, it is enough to check the value x_m (M_m) around the above m_0 when q is $\frac{1}{2^n}$ for a large n. More generally, when $q=\frac{k}{2^n}$ with some integer k, it is easily checked that the above two propositions hold and the value x_m (M_m) becomes over $\frac{1}{2}$ around the m_0 above.

One can think about various possible implementations of the idea of using chaotic dynamics for computations, which is an open and very interesting problem. For this problem, realization of nonlinear quantum gates will be essential; it will be discussed in [16].

Finally, we show in Fig. 1 how we can easily amplify the small q in several steps.

5. Conclusion

The complexity of the quantum algorithm for the SAT problem has been considered in [7] where it was shown that one can build the unitary matrix U_f in the polynomial time. We have also to consider the number of steps m in the classical algorithm for the logistic map performed on quantum computer. It is the probabilistic part of the construction and one has to compute several times to be able to distinguish the cases $q = 0$ and $q > 0$. Thus we conclude that the quantum chaos algorithm can solve the SAT problem in polynomial time according to the above propositions.

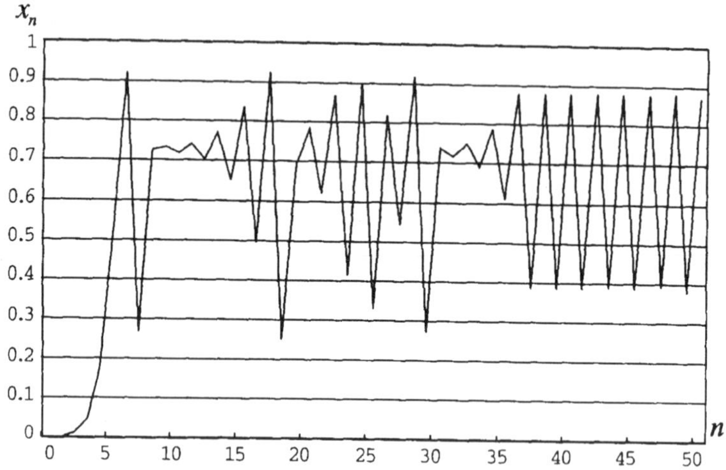

Fig. 1. Change of x_n w.r.t. time n

In conclusion, in this paper the quantum chaos algorithm is proposed. It combines the ordinary quantum algorithm with quantum chaotic dynamics amplifier. We argue that such an algorithm can be powerful enough to solve the NP-complete problems in the polynomial time. Our proposal is to show the existence of algorithm to solve NP-complete problem. The physical implementation of this algorithm is another question and it is strongly desirable to study it further.

REFERENCES

[1] D. Bouwmeester, A. Ekert and A. Zeilinger: *The Physics of Quantum Information*, Springer, Berlin 2001.
[2] D. Deutsch: Quantum theory, the Church–Turing principle and the universal quantum computer, *Proc. Royal Soc. London series A*, **400** (1985), 97–117.
[3] E. Bernstein and U. Vazirani: Quantum Complexity Theory, in *Proc. the 25th Annual ACM Symposium on Theory of Computing*, ACM Press, New York 1993, 11–20.
[4] M. Garey and D. Johnson: *Computers and Intractability—a Guide to the Theory of NP-completeness*, Freeman, 1979.
[5] C. H. Bennett, E. Bernstein, G. Brassard and U. Vazirani: Strengths and Weaknesses of Quantum Computing, quant-ph/9701001
[6] M. Ohya and I. V. Volovich: Quantum computing, NP-complete problems and chaotic dynamics, in T. Hida and K. Saito (eds.), *Quantum Information II*, World Scientific, Singapore 2000; quant-ph/9912100.
[7] M. Ohya and N. Masuda: NP problem in Quantum Algorithm, *Open Systems and Information Dynamics* 7 No.1 (2000), 33–39.
[8] P. W. Shor: Algorithm for quantum computation: Discrete logarithm and factoring algorithm, *Proceedings of the 35th Annual IEEE Symposium on Foundation of Computer Science*, 1994, 124–134.
[9] A. Ekert and R. Jozsa: Quantum computation and Shor's factoring algorithm, *Rev. Mod. Phys.* **68**, No.3 (1996), 733–753.
[10] L. Accardi, R. Sabbadini: On the Ohya–Masuda quantum SAT Algorithm, in *Proceedings International Conference "Unconventional Models of Computations"*, I. Antoniou, C. S. Calude, M. Dinneen (eds.), Springer 2001.

Open Sys. & Information Dyn. **10**: 221–233, 2003

How Can We Observe and Describe Chaos?[†]

Andrzej Kossakowski

Institute of Physics
N. Copernicus University, Grudziądzka 5, 87–100 Toruń, Poland

Masanori Ohya and Yoshio Togawa

Department of Information Sciences
Tokyo University of Science, Noda City, Chiba 278–8510, Japan

(Received: January 31, 2003)

Abstract. We propose a new approach to define chaos in dynamical systems from the point of view of Information Dynamics. Observation of chaos in reality depends upon how to observe it, for instance, how to take the scale in space and time. Therefore it is natural to abandon taking several mathematical limiting procedures. We take account of them, and chaos degree previously introduced is redefined in this paper.

1. Introduction

There exist several attempts to describe chaos appearing in classical or quantum dynamical systems [1 – 16]. One of the present authors introduced Information Dynamics (ID for short) [18] as a frame to discuss complexity and chaos appearing in various fields, in which he tried to find a common basis by synthesizing the state change (dynamics) and the complexity associated with dynamical systems. Since then ID has been applied to several different topics [9, 19], among which chaos degree, a quantity measuring the degree of chaos associated with a dynamics, was introduced by means of the complexities in ID and its entropic version (called Entropic Chaos Degree (ECD for short)) has been computed numerically for rather famous chaotic dynamics such as logistic map, baker's transformation, Tinkerbel map. It is surprising that the result of the ECD exactly matches that of Lyapunov exponent in the case when the later can be computed. Moreover, the algorithm computing the ECD is much easier than that of Lyapunov exponent, so that the ECD is almost always computable even when the Lyapunov exponent is not. However, there are some unclear points, both conceptual and mathematical, why the ECD should be so successful for computational experiments. In this paper we study these points and propose a new description of chaos.

In Sect. 2, we briefly review information dynamics and chaos degree, and in Sect. 3 the entropic chaos degree and its algorithm are recalled with a computational result. In Sect. 4, a new way of detecting chaos from a given dynamics is

[†]This is an invited paper for the anniversary 10th volume of OSID.

discussed based on the ECD, that is, we propose a new approach to define chaos in dynamical systems.

2. Information Dynamics and Chaos Degree

We briefly review what ID is. Let $(\mathcal{A}, \mathfrak{S}, \alpha(G))$ be an input (or initial) system and $(\overline{\mathcal{A}}, \overline{\mathfrak{S}}, \overline{\alpha}(\overline{G}))$ be an output (or final) system. Here $\mathcal{A}$ is the set of objects to be observed and $\mathfrak{S}$ is the set of means to get the observed value, $\alpha(G)$ describes evolution of system with a parameter g in a certain set G. Often we have $\mathcal{A} = \overline{\mathcal{A}}$, $\mathfrak{S} = \overline{\mathfrak{S}}$, $\alpha = \overline{\alpha}$, $G = \overline{G}$. Therefore it can be said that:

giving a mathematical structure to input and output triples $\equiv$ having a theory

The dynamics of state change is described by a channel, that is, a map Λ^*: $\mathfrak{S} \to \overline{\mathfrak{S}}$ (sometimes $\mathfrak{S} \to \mathfrak{S}$). The fundamental point of ID is that ID contains two complexities in itself. Let $(\mathcal{A}_t, \mathfrak{S}_t, \alpha^t(G^t))$ be the total system of $(\mathcal{A}, \mathfrak{S}, \alpha)$ and $(\overline{\mathcal{A}}, \overline{\mathfrak{S}}, \overline{\alpha})$, and $\mathcal{S}$ be a subset of $\mathfrak{S}$ in which we are measuring observables (e.g., $\mathcal{S}$ is the set of all KMS or stationary states in C*-system). The two complexities are denoted by C and T. C is the complexity of a state φ measured from a reference system $\mathcal{S}$, in which we actually observe the objects in $\mathcal{A}$ and T is the transmitted complexity associated with the state change $\varphi \to \Lambda^*\varphi$, both of which should satisfy the following properties:

Axioms of complexities

(i) For any $\varphi \in \mathcal{S} \subset \mathfrak{S}$,

$$C^{\mathcal{S}}(\varphi) \geq 0, \qquad T^{\mathcal{S}}(\varphi; \Lambda^*) \geq 0.$$

(ii) For any orthogonal bijection $j : \mathrm{ex}\,\mathfrak{S} \to \mathrm{ex}\,\mathfrak{S}$, the set of all extremal points of $\mathfrak{S}$,

$$\begin{aligned} C^{j(\mathcal{S})}(j(\varphi)) &= C^{\mathcal{S}}(\varphi) \\ T^{j(\mathcal{S})}(j(\varphi); \Lambda^*) &= T^{\mathcal{S}}(\varphi; \Lambda^*). \end{aligned}$$

(iii) For $\Phi \equiv \varphi \otimes \psi \in \mathcal{S}_t \subset \mathfrak{S}_t$,

$$C^{\mathcal{S}_t}(\Phi) = C^{\mathcal{S}}(\varphi) + C^{\overline{\mathcal{S}}}(\psi).$$

(iv) $0 \leq T^{\mathcal{S}}(\varphi; \Lambda^*) \leq C^{\mathcal{S}}(\varphi)$

(v) $T^{\mathcal{S}}(\varphi; \mathrm{id}) = C^{\mathcal{S}}(\varphi)$, where "id" is the identity map from $\mathfrak{S}$ to $\mathfrak{S}$.

Instead of (iii), when

(iii') $\Phi \in \mathcal{S}_t \subset \mathfrak{S}_t$, put $\varphi \equiv \Phi \upharpoonright \mathcal{A}$ (i.e., the restriction of Φ to $\mathcal{A}$), $\psi \equiv \Phi \upharpoonright \overline{\mathcal{A}}$, $C^{\mathcal{S}_t}(\Phi) \leq C^{\mathcal{S}}(\varphi) + C^{\overline{\mathcal{S}}}(\psi)$

is satisfied, C and T is called a pair of strong complexity. Therefore ID is defined as follows:

DEFINITION 2.1 Information dynamics is described by

$$(\mathcal{A}, \mathfrak{S}, \alpha(G); \overline{\mathcal{A}}, \overline{\mathfrak{S}}, \overline{\alpha}(\overline{G}); \Lambda^*; C^{\mathcal{S}}(\varphi), T^{\mathcal{S}}(\varphi; \Lambda^*))$$

and some relations R among them.

In the framework of ID, we have to

(i) mathematically determine $\mathcal{A}, \mathfrak{S}, \alpha(G); \overline{\mathcal{A}}, \overline{\mathfrak{S}}, \overline{\alpha}(\overline{G})$,
(ii) choose Λ^* and R, and
(iii) define $C^{\mathcal{S}}(\varphi)$, $T^{\mathcal{S}}(\varphi; \Lambda^*)$.

In ID, several different topics can be treated on a common footing so that we can find a new clue bridging several fields.

We assume $\overline{\mathcal{A}} = \mathcal{A}$ for simplicity in the sequel. For a certain subset $\mathcal{S}$ (called the reference space) of $\mathfrak{S}$ and a state $\varphi \in \mathcal{S}$, there exists a decomposition of the state φ into a mixture of extreme (pure) states such that

$$\varphi = \int_{\mathcal{S}} \omega \, d\mu \, .$$

This extremal decomposition of φ describes the degree of mixture of φ in the reference space $\mathcal{S}$. The measure μ is not always unique, so that the set of all such measures is denoted by $M_{\varphi}(\mathcal{S})$.

For instance, when $(\mathcal{A}, \mathfrak{S})$ and is a C*-system containing both classical and quantum systems, that is, $\mathcal{A}$ is a C* algebra and $\mathfrak{S}$ is the set of all states on $\mathcal{A}$, the reference space $\mathcal{S}$ is a weak* compact convex subset of $\mathfrak{S}$ and the measure μ is not uniquely determined unless $\mathcal{S}$ is the Choquet simplex. In this paper we will not go to the details of such general mathematical discussion.

A measure of chaos produced by the dynamics Λ^* is defined in [21, 22]:

DEFINITION 2.2
(1) ψ is more chaotic than φ if $C(\psi) \geq C(\varphi)$.
(2) When $\varphi \in \mathcal{S}$ changes to $\Lambda^*\varphi$, the *chaos* degree associated to this state change (dynamics) Λ^* is given by

$$D^{\mathcal{S}}(\varphi; \Lambda^*) = \inf \left\{ \int_{\mathcal{S}} C^{\mathcal{S}}(\Lambda^* \omega) \, d\mu \, ; \; \mu \in M_{\varphi}(\mathcal{S}) \right\}.$$

DEFINITION 2.3 The dynamics Λ^* produces chaos iff $D^{\mathcal{S}}(\varphi; \Lambda^*) > 0$.

It is important to note here that the dynamics Λ^* in the definition is not necessarily the same as the original dynamics (channel) but is the one reduced from the original one such that it causes an evolution for a certain observed value like an orbit. However for simplicity we often use the same notation in this paper. In some cases, the above chaos degree $D^{\mathcal{S}}(\varphi; \Lambda^*)$ can be expressed as

$$D^{\mathcal{S}}(\varphi; \Lambda^*) = C^{\mathcal{S}}(\Lambda^* \varphi) - T^{\mathcal{S}}(\varphi; \Lambda^*) \, .$$

3. Entropic Chaos Degree and its Algorithm

Although there exist several complexities [20], one of the most useful examples of C and T are Shannon's entropy and mutual entropy in classical systems (von Neumann entropy and quantum mutual entropy in quantum systems [23]), respectively.

The concept of entropy was introduced and developed to study the topics such as irreversible behaviour, symmetry breaking, amount of information transmission, so that it originally describes a certain chaotic property of state.

Let us recall the simplest case of C and T, that is, Shannon's entropy and mutual entropy. In classical communication systems, an input state φ is a probability distribution $p = (p_k) = \sum_k p_k \delta_k$ and a channel Λ^* is a transition probability $(t_{i,j})$, so that the compound state of φ and its output $\overline{\varphi}$ $(\equiv \overline{p} = (\overline{p}_i) = \Lambda^* p)$ is the joint distribution $r = (r_{i,j})$ with $r_{i,j} \equiv t_{i,j} p_j$. Then the complexities C and T are given as

$$
\begin{aligned}
C(p) &= S(p) = -\sum_k p_k \log p_k \,, \\
T(p;\Lambda^*) &= I(p;\Lambda^*) = \sum_{i,j} r_{i,j} \log \frac{r_{i,j}}{p_j \overline{p}_i} \,.
\end{aligned}
$$

Thus the entropic chaos degree of the channel Λ^* becomes

DEFINITION 3.1

$$
D(p;\Lambda^*) = S(\Lambda^* p) - I(p;\Lambda^*) \,.
$$

Quantum version of the above entropic chaos degree was discussed in [10, 22], which we will briefly review here in the case of usual Hilbert space formulation. Let ρ be a quantum state, namely, a density operator on a Hilbert space $\mathcal{H}$, and Λ^* be a channel sending the set $\mathfrak{S}$ of all states on $\mathcal{H}$ into itself. Then the entropic chaos degree is defined by

$$
D(\rho;\Lambda^*) = \inf \left\{ \sum_k \lambda_k S(\Lambda^* E_k) \,;\ \{E_k\} \in \mathcal{E} \right\},
$$

where $\mathcal{E}$ is the set of all Schatten decompositions (i.e., one-dimensional spectral decompositions) of the state $\rho := \sum_k \lambda_k E_k$, and S is the von Neumann entropy.

3.1. An Algorithm Computing Chaos Degree

In order to observe chaos produced by a dynamics, one often looks at the behaviour of orbits produced by that dynamics, more generally, looks at the behaviour of a certain observed value. Therefore, in our scheme we directly compute the chaos degree once the dynamics is explicitly given as a state change of the system. However, even when the direct calculation does not show chaos, it will appear if one focuses at some aspect of the state change, e.g., a certain observed value which may be called an orbit, as usual. The algorithm computing the chaos degree for a dynamics falls into the following two cases [21, 22, 12, 10]:

1. The dynamics is given by $dx/dt = F_t(x)$ with $x \in I \equiv [a,b]^{\mathbb{N}} \subset \mathbb{R}^{\mathbb{N}}$: First find a difference equation $x_{n+1} = F(x_n)$ with the map F on $I \equiv [a,b]^{\mathbb{N}} \subset \mathbb{R}^{\mathbb{N}}$ into itself, secondly let $I \equiv \bigcup_k A_k$ be a finite partition with $A_i \cap A_j = \emptyset$ $(i \neq j)$. Then the state $\varphi^{(n)}$ of the orbit determined by the difference equation is defined by the probability distribution $(p_i^{(n)})$, that is, $\varphi^{(n)} = \sum_i p_i^{(n)} \delta_i$, where for a given initial value $x \in I$ and the characteristic function 1_A

$$p_i^{(n)} \equiv \frac{1}{n+1} \sum_{k=m}^{m+n} 1_{A_i}(F^k x).$$

Now, when the initial value x is distributed according to a measure ν on I, the above $p_i^{(n)}$ is given as

$$p_i^{(n)} \equiv \frac{1}{n+1} \int_I \sum_{k=m}^{m+n} 1_{A_i}(F^k x)\, d\nu.$$

The joint distribution $p_{ij}^{(n,n+1)}$ between the times n and $n+1$ is defined by

$$p_{ij}^{(n,n+1)} \equiv \frac{1}{n+1} \sum_{k=m}^{m+n} 1_{A_i}(F^k x) 1_{A_j}(F^{k+1} x)$$

or

$$p_{ij}^{(n,n+1)} \equiv \frac{1}{n+1} \int_I \sum_{k=m}^{m+n} 1_{A_i}(F^k x) 1_{A_j}(F^{k+1} x)\, d\nu.$$

Then the channel Λ_n^* at n is determined by

$$\Lambda_n^* \equiv \left(\frac{p_{ij}^{(n,n+1)}}{p_i^{(n)}} \right) \implies \varphi^{(n+1)} = \Lambda_n^* \varphi^{(n)},$$

and the entropic chaos degree is given by the Definition 3.1;

$$D_A(x;F) = D_A\left(p^{(n)};\Lambda_n^*\right) = \sum_i p_i^{(n)} S(\Lambda_n^* \delta_i) = \sum_{i,j} p_{ij}^{(n,n+1)} \log \frac{p_i^{(n)}}{p_{ij}^{(n,n+1)}}. \quad (1)$$

We can judge whether the dynamics is chaoting or not by the value of D as in the Definition 2.2:

$$D > 0 \iff \text{chaotic}$$
$$D = 0 \iff \text{stable.}$$

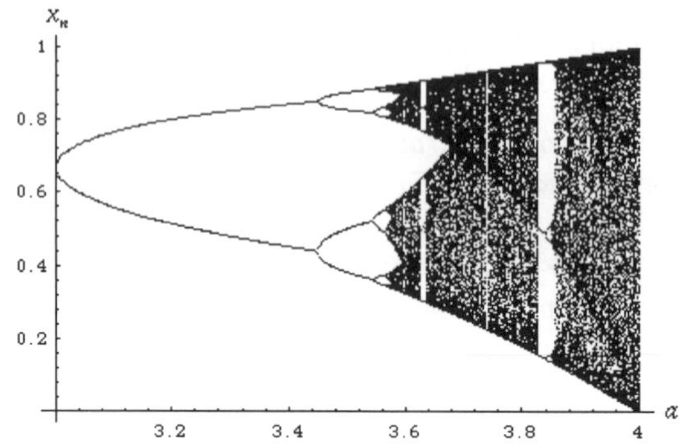

Fig. 1: The bifurcation diagram for logistic map

This chaos degree was applied to several dynamical systems such as logistic map, Baker's transformation and Tinkerbel map, and it could explain their chaotic characters. This chaos degree has several advantages when compared with the usual measures, such as Lyapunov exponents, as explained below.

2. The dynamics is given by $\varphi_t = F_t^* \varphi_0$ on a Hilbert space: Similarly as converting it into the difference equation of state, the channel Λ_n^* at n is first deduced from F_t^*, which should satisfy $\varphi^{(n+1)} = \Lambda_n^* \varphi^{(n)}$. By means of this constructed channel (α) we compute the chaos degree D directly according to the Definition 3.2 or (β) we take a proper observable X and put $x_n \equiv \varphi^{(n)}(X)$, then go back to the algorithm (1).

Note that the chaos degree D does depend on a partition A taken, which is somehow different from usual degree of chaos (cf., dynamical entropy [1, 4, 3, 14]). This is a key point of our understanding of chaos, which will be discussed in the next section.

3.2. Logistic Map

Let us explain how well the entropy chaos degree (ECD) describes the chaotic behaviour of the logistic map.

The logistic map is defined by

$$x_{n+1} = a x_n (1 - x_n) , \qquad x_n \in [0,1] , \qquad 0 \le a \le 4 .$$

The bifurcation diagram for this equation is shown in Fig. 1.

In order to compare ECD with other measures describing chaos, we choose Lyapunov exponents.

Definition of Lyapunov exponent

(1) Let f be a map on $\mathbb{R}$ and let $r_0 \in \mathbb{R}$. Then the Lyapunov exponent $\lambda_{\mathcal{O}}(f)$

for the orbit $\mathcal{O} \equiv \{f^n(x_0)\,;\ n = 0, 1, 2, \ldots\}$ is defined by

$$\lambda_{\mathcal{O}}(f) = \lim_{n\to\infty} \lambda_{\mathcal{O}}^{(n)}(f), \qquad \lambda_{\mathcal{O}}^{(n)}(f) = \frac{1}{n}\log\left|\frac{df^n}{dx}(x_0)\right|.$$

(2) Let $f = (f_1, f_2, \ldots, f_m)$ be a map on $\mathbb{R}^m$ and let $x_0 \in \mathbb{R}^m$. The Jacobi matrix $J_n = Df^n(r_0)$ at r_0 is defined by

$$J_n = Df^n(r_0) = \begin{pmatrix} \frac{\partial f_1^n}{\partial x_1}(r_0) & \cdots & \frac{\partial f_1^n}{\partial x_m}(r_0) \\ \vdots & & \vdots \\ \frac{\partial f_m^n}{\partial x_1}(r_0) & \cdots & \frac{\partial f_m^n}{\partial x_m}(r_0) \end{pmatrix}.$$

Then, the Lyapunov exponent $\lambda_{\mathcal{O}}(f)$ of f for the orbit

$$\mathcal{O} \equiv \{f^n(x_0)\,;\ n = 0, 1, 2, \ldots\}$$

is defined by

$$\lambda_{\mathcal{O}}(f) = \log \tilde{\mu}_1, \qquad \tilde{\mu}_k = \lim_{n\to\infty} (\mu_k^n)^{1/n}, \qquad k = 1, \ldots, m.$$

Here, μ_k^n is the k-th largest square root of the m eigenvalues of the matrix $J_n^T J_n$,

$$\begin{aligned} \lambda_{\mathcal{O}}(f) &> 0 \quad \Rightarrow \quad \text{orbit } \mathcal{O} \text{ is chaotic.} \\ \lambda_{\mathcal{O}}(f) &\le 0 \quad \Rightarrow \quad \text{orbit } \mathcal{O} \text{ is stable.} \end{aligned}$$

The properties of the logistic map depend on the parameter a. If we take a particular constant a, for example, $a = 3.71$, then the Lyapunov exponent and the entropic chaos degree are positive, the trajectory is very sensitive to the initial value and one has the chaotic behaviour.

From the above example and some other maps (see [11]), Lyapunov exponent and the entropic chaos degree have clear correspondence, but the ECD can resolve some inconvenient properties of the Lyapunov exponent as follows:

(1) Lyapunov exponent can take negative values and sometimes even $-\infty$, but the ECD is always positive for any $a \ge 0$.

(2) It is difficult to compute the Lyapunov exponent for some maps like the Tinkerbell map f because it is difficult to compute f^n for large n. On the other hand, the ECD of f is easily computed.

(3) Generally, the algorithm for calculating the ECD is much easier than that for the Lyapunov exponent.

4. New Description of Chaos

First of all, we have to examine carefully when a certain dynamics produces chaos. Let us take the logistic map as an example. The original differential equation of the logistic map is

$$\frac{dx}{dt} = ax(1-x), \qquad 0 \le a \le 4 \tag{2}$$

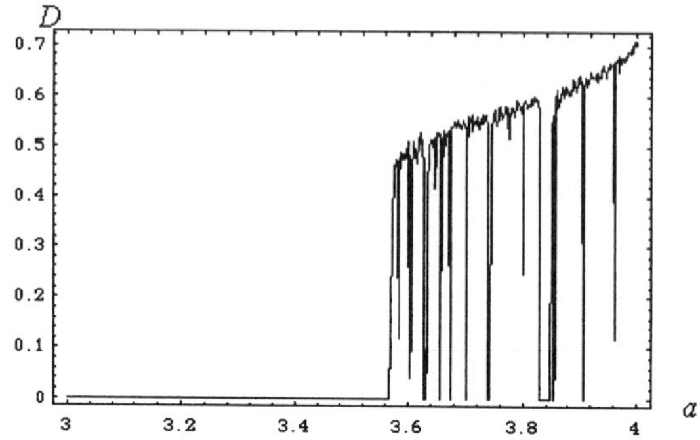

Fig. 2: Chaos degree for logistic map

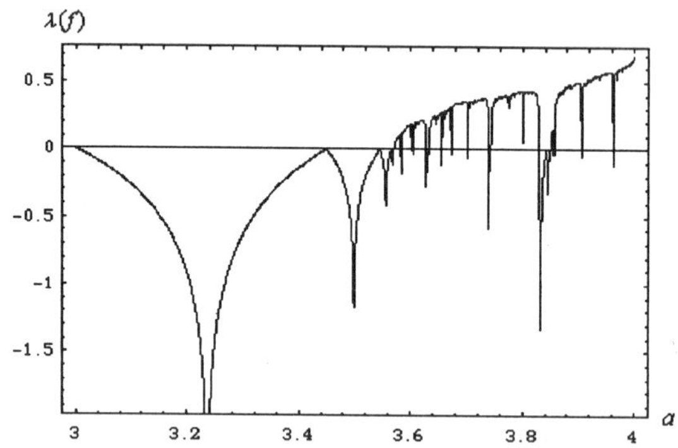

Fig. 3: Lyapunov exponent for logistic map

with initial value x_0 in $[0,1]$. This equation can be easily solved analytically, whose solution (orbit) does not have any chaotic behaviour. However, once we convert it into a discrete equation such as

$$x_{n+1} = ax_n(1-x_n), \qquad 0 \leq a \leq 4, \tag{3}$$

this difference equation produces chaos. Taking the time discrete is necessary not only for observing chaos but also for visualising orbits drawn by the dynamics. Similarly as in quantum mechanics, it is not possible for a human being to understand any object without observing it, for which it will not be possible to trace an orbit continuously in time.

Now let us think about finite partition $A = \{A_k\,;\ k = 1, \ldots, N\}$ of a proper set $I \equiv [a,b]^{\mathrm{N}} \subset \mathbb{R}^{\mathrm{N}}$ and a equi-partition $B^e = \{B_k^e\,;\ k = 1, \ldots, N\}$ of I. Here "equi" means that all elements B_k^e are identical. We denote the set of all partitions by $\mathcal{P}$

and the set of all equi-partitions by $\mathcal{P}^e$. In Sect. 3, we specify a special partition, in particular, an equi-partition for computer experiment calculating the ECD. Such a partition enables us to observe the orbit of a given dynamics, and moreover it provides a criterion for observing chaos. There exist several reports saying that one can observe chaos in nature, which are very much related to how one observes the phenomena, for instance, scale, direction, aspect. It has been difficult to find a satisfactory theory (mathematics) to explain such chaotic phenomena. In the difference equation we can take some time interval τ between n and $n+1$. If we take $\tau \to 0$, then we have a completely different dynamics. If we take coarse graining of the orbit x_t for time interval τ,

$$x_n \equiv \frac{1}{\tau} \int_{(n-1)\tau}^{n\tau} x_t \, dt \,,$$

we again have a very different dynamics. Moreover it is important for mathematical consistency to take the limits $n \to \infty$ or N (the number of equi-partitions)$\to \infty$, i.e., making the partition finer and finer, and to consider the limits of some quantities describing chaos, so that mathematical operations such as "lim", "sup", "inf" are used very often in definitions of such quantities. In this paper we take the opposite position, that is, any observation will be unrelated or even in contradiction with such limits. Detection of chaos is a result of assuming suitable scales of, for example, time, distance or domain, and it will not be possible in the limiting cases.

We claim in this paper that most of chaotic phenomena are scale-dependent, so the definition of a degree measuring chaos should depend on certain scales taken.

Taking this view into consideration we modify the definitions of the chaos degree given in the previous sections as follows.

We go back to the triple $(\mathcal{A}, \mathfrak{S}, \alpha(G))$ considered in Sect. 2 and we use this triple both for an input and for an output systems. Let the dynamics be described by the mapping Γ_t from $\mathfrak{S}$ to $\mathfrak{S}$ with a parameter $t \in G$ and let an observation be described by the mapping $\mathcal{O}$ from $(\mathcal{A}, \mathfrak{S}, \alpha(G))$ to the triple $(\mathcal{B}, \mathfrak{T}, \beta(G))$. The triple $(\mathcal{B}, \mathfrak{T}, \beta(G))$ might be the same as the original one or the one of its subsystem and the observation map $\mathcal{O}$ may contains several different types of observations, that is, it can be decomposed as $\mathcal{O} = \mathcal{O}_m \cdots \mathcal{O}_1$. Let us list some examples of observations.

For a given dynamics $d\varphi/dt = F(\varphi_t)$, equivalently, $\varphi_t = \Gamma_t^* \varphi$, one can take several observations.

EXAMPLE 1 Time Scaling (Discretizing): $\mathcal{O}_\tau : t \to n$, $d\varphi/dt\,(t) \to \varphi_{n+1}$, so that $d\varphi/dt = F(\varphi_t) \Rightarrow \varphi_{n+1} = F(\varphi_t)$ and $\varphi_t = \Gamma_t^* \varphi \Rightarrow \varphi_n = \Gamma_n^* \varphi$. Here τ is the unit of time needed for the observation.

EXAMPLE 2 Size Scaling (Conditional Expectation, Partition): Let $(\mathcal{B}, \mathfrak{T}, \beta(G))$ be a subsystem of $(\mathcal{A}, \mathfrak{S}, \alpha(G))$, both of which have a certain algebraic structure such as C*-algebra or von Neumann algebra. As an example, the subsystem $(\mathcal{B}, \mathfrak{T}, \beta(G))$ has abelian structure describing macroscopic world which is a subsystem of a non-abelian (non-commutative) system $(\mathcal{A}, \mathfrak{S}, \alpha(G))$ describing micro-world. A mapping $\mathcal{O}_C$ preserving norm (when it is properly defined) from $\mathcal{A}$ to $\mathcal{B}$ is, in some cases, called a conditional expectation. A typical example of this conditional expectation is according to a projection valued measure $\{P_k;\ P_kP_j = P_k\delta_{kj} = P_k^*\delta_{kj} \geq 0, \sum_k P_k = I\}$ associated with a quantum measurement (von Neumann measurement) such that $\mathcal{O}_C(\rho) = \sum_k P_k\rho P_k$ for any quantum state (density operator) ρ. When $\mathcal{B}$ is a von Neumann algebra generated by $\{P_k\}$, it is an abelian algebra isometrically isomorphic to $L^\infty(\Omega)$ with a certain Hausdorff space Ω, so that in this case $\mathcal{O}_C$ sends a general state φ to a probability measure (or distribution) p. Similar example of $\mathcal{O}_C$ is one coming from a certain representation (selection) of state such as e.g. Schatten decomposition of $\rho; \rho = \mathcal{O}_R\rho = \sum_k \lambda_k E_k$ by one-dimensional orthgonal projections $\{E_k\}$ associated to the eigenvalues of ρ with $\sum_k E_k = I$. Another important example of size scaling is due to a finite partition of an underlining space Ω, e.g., space of orbit, defined as

$$\mathcal{O}_P(\Omega) = \left\{P_k;\ P_k \cap P_j = P_k\delta_{kj}\,, \quad k, j = 1, \ldots N\,, \quad \bigcup_{k=1}^{N} P_k = \Omega\right\}.$$

We go back to the discussion of the entropic chaos degree. Starting from a given dynamics $\varphi_t = \Gamma_t^*\varphi$, it becomes $\varphi_n = \Gamma_n^*\varphi$ after handling the operation $\mathcal{O}_T$. Then by taking proper combinations $\mathcal{O}$ of the size scaling operations like $\mathcal{O}_C$, $\mathcal{O}_R$ and $\mathcal{O}_P$, the equation $\varphi_n = \Gamma_n^*\varphi$ changes to $\mathcal{O}(\varphi_n) = \mathcal{O}(\Gamma_n^*\varphi)$, which will be written by $\mathcal{O}\varphi_n = \mathcal{O}\Gamma_n^*\mathcal{O}^{-1}\mathcal{O}\varphi$ or $\varphi_n^{\mathcal{O}} = \Gamma_n^{*\mathcal{O}}\varphi^{\mathcal{O}}$. Then our entropic chaos degree is redifined as follows:

DEFINITION 4.1 The entropic chaos degree of Γ^* with an initial state φ and observation $\mathcal{O}$ is defined by

$$D^{\mathcal{O}}(\varphi; \Gamma^*) = \int_{\mathcal{O}(\mathfrak{S})} S\left(\Gamma^{*\mathcal{O}}\omega^{\mathcal{O}}\right) d\mu^{\mathcal{O}},$$

where $\mu^{\mathcal{O}}$ is the measure induced by the operation $\mathcal{O}$ when applied to an extremal decomposition measure of φ selected by the observation $\mathcal{O}$ (its part $\mathcal{O}_R$).

DEFINITION 4.2 The entropic chaos degree of Γ^* with an initial state φ is defined by $D(\varphi; \Gamma^*) = \inf\left\{D^{\mathcal{O}}(\varphi; \Gamma^*); \mathcal{O} \in \mathcal{SO}\right\}$, where $\mathcal{SO}$ is a proper set of observations natually determined by a given dynamics.

Then one judges whether a given dynamics is chaotic or not in the following way.

DEFINITION 4.3

(1) The dynamics Γ^* is chaotic for an initial state φ in an observation $\mathcal{O}$ if and only if $D^{\mathcal{O}}(\varphi; \Gamma^*) > 0$.

(2) Γ^* is totally chaotic for an initial state φ if and only if $D(\varphi; \Gamma^*) > 0$.

In Definition 4.2 $\mathcal{SO}$ is determined by a given dynamics and some conditions attached to the dynamics, for instance, if we start from a difference equation with a special representation of an initial state, then $\mathcal{SO}$ excludes $\mathcal{O}_\tau$ and $\mathcal{O}_R$.

The idea introduced in this paper to understand chaos can be applied not only to the entropic chaos degree but also to some other degrees such as dynamical entropy, whose applications and comparison with several other chaos indicators will be discussed in a forthcoming paper.

In the case of logistic map, $x_{n+1} = ax_n(1 - x_n) \equiv F(x_n)$, we obtain this difference equation by taking the observation $\mathcal{O}_\tau$ and an observation $\mathcal{O}_P$ by equipartition of the orbit space $\Omega = \{x_n\}$ so as to define a state (probability distribution). Thus we can compute the entropic chaos degree as discussed in Sect. 3.

It is important to notice here that the chaos degree does depend on the choice of observations. As an example, we consider the circle map

$$\theta_{n+1} = f_\nu(\theta_n) = \theta_n + \omega \pmod{2\pi}, \tag{4}$$

where $\omega = 2\pi\nu(0 < \nu < 1)$. If ν is a rational number N/M, then the orbit $\{\theta_n\}$ is periodic with the period M. If ν is irrational, then the orbit $\{\theta_n\}$ densely fills the unit circle for any initial value θ_0; namely, it is a quasiperiodic motion.

We proved in [10] the following theorem.

THEOREM 1 *Let $I = [0, 2\pi]$ be partitioned into L disjoint components with equal length; $I = B_1 \cap B_2 \cap \ldots \cap B_L$.*

(1) If ν is a rational number equal to N/M, then the finite equi-partition $P = \{B_k; k = 1, \ldots, M\}$ implies $D^{\mathcal{O}}(\theta_0; f_\nu) = 0$.

(2) If ν is irrational, then $D^{\mathcal{O}}(\theta_0; f_\nu) > 0$ for any finite partition $P=\{B_k\}$.

Note that our entropic chaos degree indicates chaos for the quasiperiodic circle dynamics by the observation according to a partition of the orbit, which is different from the usual understanding of chaos. However, the usual belief that quasiperiodic circle dynamics will not cause chaos is not at all obvious, but is realized in a special limiting case as shown in the following proposition.

PROPOSITION 1 *For the above circle map, if ν is irrational, then $D(\theta_0; f_\nu) = 0$.*

Proof. Let us take an equipartition $P = \{B_k\}$ as

$$B_k \equiv \left\{x;\ 2\pi\frac{k-1}{l} \le x < 2\pi\frac{k}{l}\pi\right\}, \qquad k = 1, 2, \ldots, l,$$

where l is a certain integer and $B_{k+l} = B_k$. When ν is irrational, put $\nu_0 \equiv [l\nu]$ with Gaussian $[\cdot]$. Then $f_\nu(B_k)$ intersects only two intervals $B_{k+\nu_0}$ and $B_{k+\nu_0+1}$, so that we can denote by $(1-s) : s$ the ratio of the Lebesgue measure of $f_\nu(B_k) \cap B_{k+\nu_0}$ and that of $f_\nu(B_k) \cap B_{k+\nu_0+1}$. This s is equal to $l\nu - [l\nu]$ and the entropic chaos degree becomes

$$D^P = -s\log s - (1-s)\log(1-s).$$

Take the continued fraction expansion of ν and denote its j-th approximate by b_j/c_j. Then we have

$$\left|\nu - \frac{b_j}{c_j}\right| \leq \frac{1}{c_j^2}.$$

For the above equi-partition $B = \{B_k\}$ with $l = c_j$, we find

$$|l\nu - b_j| \leq \frac{1}{k}$$

and

$$[l\nu] = \begin{cases} b_j & \text{when} \quad \nu - \frac{b_j}{c_j} > 0 \\ b_{j-1} & \text{when} \quad \nu - \frac{b_j}{c_j} < 0. \end{cases}$$

It implies

$$D^P \fallingdotseq \frac{\log c_j}{c_j},$$

which goes to 0 as $j \to \infty$. Hence $D = \inf\{D^P; P\} = 0$. □

Such a limiting case will not appear in real observation of natural objects, so we claim that chaos is a phenomenon depending on observations, which results in the definition of chaos as above.

In the forthcoming paper [24], we will discuss how obtain chaotic dynamics starting from general differential dynamics in both classical and quantum systems. That is, it is demonstrated how we can get chaotic dynamics by considering observations introduced in this paper, and we calculate the entropic chaos degrees in each case.

Acknowledgment

The authors thank JSPS and SCAT for financial support.

Bibliography

[1] L. Accardi, M. Ohya, and N. Watanabe, *Dynamical entropy through quantum Markov chain*, Open Sys. Information Dyn. **4**, 71 (1997).

[2] K. T. Alligood, T. D. Sauer, and J. A. Yorke, *Chaos-An Introduction to Dynamical Systems*, Textbooks in Mathematical Sciences, Springer, 1996.

[3] R. Alicki and M. Fannes, *Defining quantum dynamical entropy*, Lett. Math. Physics **32**, 75 (1994).

[4] F. Benatti, *Deterministic Chaos in Infinite Quantum Systems*, Springer, 1993.

[5] R. L. Devaney, *An Introduction to Chaotic dynamical Systems*, Benjamin, 1986.

[6] G. Casati and B. Chirikov, *Quantum Chaos: Between Order and Disorder*, Cambridge University Press, 1995.

[7] G. G. Emch, H. Narnhofer, W. Thirring, and G. L. Sewell, *Anosov actions on noncommutative algebras*, J. Math. Phys. **35**, No. 11, 5582 (1994).

[8] H. Hasegawa, *Dynamical formulation of quantum level statistics*, Open Sys. Information Dyn. **4**, 359 (1997).

[9] R. S. Ingarden, A. Kossakowski, and M. Ohya, *Information Dynamics and Open Systems*, Kluwer Academic Publishers, 1997.

[10] K. Inoue, M. Ohya, and A. Kossakowski, *A Description of Quantum Chaos*, preprint, 2002.

[11] K. Inoue, M. Ohya, and K. Sato, *Application of chaos degree to some dynamical systems*, Chaos, Solitons & Fractals **11**, 1377-1385 (2000).

[12] K. Inoue, M. Ohya, and V. Volovich, *Semiclassical properties and chaos degree for the quantum baker's map*, Journal of Mathematical Physics **43**, 734 (2002).

[13] K. Inoue, M. Ohya, and I. V. Volovich, *On quantum-classical correspondence for baker's map*, quant-ph/0108107.

[14] A. Kossakowski, M. Ohya, and N. Watanabe, *Quantum dynamical entropy for completely positive maps*, Infinite Dimensional Analysis, Quantum Probability and Related Topics **2**, 267 (1999).

[15] W. A. Majewski, *Does quantum chaos exitst? A quantum Lyapunov exponents approach*, quant-ph/9805068.

[16] N. Muraki and M. Ohya, *Entropy functionals of Kolmogorov Sinai type and their limit theorems*, Lett. Math. Phys. **36**, 327 (1996).

[17] M. Ohya, *Some aspects of quantum information theory and their applications to irreversible processes*, Rep. Math. Phys. **27**, 19 (1989).

[18] M. Ohya, *Information dynamics and its applications to optical communication processes*, Lecture Note in Physics **378**, 81 (1991).

[19] M. Ohya, *State change, complexity and fractal in quantum systems, Quantum Communications and Measurement*, Plenum Press, New York, 1995, pp. 309–320.

[20] M. Ohya, *Complexity and fractal dimensions for quantum states*, Open Sys. Information Dyn. **4**, 141 (1997).

[21] M. Ohya, *Complexities and their applications to characterization of chaos*, International Journal of Theoretical Physics **37**, 495 (1998).

[22] M. Ohya, *Complexity in Quantum System and its Application to Brain Function* in: *Quantum Information II*, T. Hida and K. Saito, ed., World Scientific, 2000.

[23] M. Ohya and D. Petz, *Quantum Entropy and Its Use*, Springer-Verlag, TMP, (1993).

[24] A. Kossakowski, M. Ohya and Y. Togawa, in preparation.

Infinite Dimensional Analysis, Quantum Probability
and Related Topics
Vol. 6, No. 2 (2003) 301–310

World Scientific
www.worldscientific.com

ON QUANTUM CAPACITY AND ITS BOUND

MASANORI OHYA

*Department of Information Sciences, Science University of Tokyo,
278-8510, Noda City, Chiba, Japan
ohya@is.noda.tus.ac.jp*

IGOR V. VOLOVICH

*Steklov Mathematical Institute, Gubkin St.8,
GSP-1 117966, Moscow, Russia
volovich@mi.ras.ru*

Received 3 December 2002

The quantum capacity of a pure quantum channel and that of classical-quantum-classical channel are discussed in detail based on the fully quantum mechanical mutual entropy. It is proved that the quantum capacity generalizes the so-called Holevo bound.

Keywords: Quantum mutual entropy; quantum capacity; Holevo bound.

AMS Subject Classification: 94A15, 81P99

1. Introduction

Measure theoretic formulation of the mutual entropy (information) in classical systems was done by Kolmogorov[1] and Gelfand, Yaglom,[2] which enabled one to define the capacity of information channel. In quantum systems, there have been several definitions of the mutual entropy for classical input and quantum output.[3–5] In 1983, Ohya defined[6] the fully quantum mechanical mutual entropy, i.e. for quantum input and quantum output, by means of the relative entropy of Umegaki,[7] and he extended it[8] to general quantum systems by using the relative entropy of Araki[9] and Uhlmann.[10] In this short note, we prove that the quantum capacity[11] of a quantum channel derived from the fully quantum mechanical mutual entropy generalizes the so-called Holevo bound.

2. Mutual Entropy

The quantum mutual entropy was introduced in Ref. 6 for a quantum input and quantum output, namely, for a purely quantum channel, and it was generalized for a general quantum system described by C^*-algebraic terminology.[8] We briefly review the mutual entropy in the usual quantum system described by a Hilbert space.

Let $\mathcal{H}$ be a Hilbert space for an input space, $B(\mathcal{H})$ be the set of all bounded linear operators on $\mathcal{H}$ and $\mathcal{S}(\mathcal{H})$ be the set of all density operators on $\mathcal{H}$. An output space is described by another Hilbert space $\tilde{\mathcal{H}}$, but often $\mathcal{H} = \tilde{\mathcal{H}}$. A channel from the input system to the output system is a mapping Λ^* from $\mathcal{S}(\mathcal{H})$ to $\mathcal{S}(\tilde{\mathcal{H}})$.[12] A channel Λ^* is said to be completely positive if the dual map Λ satisfies the following condition: $\sum_{k,j=1}^{n} A_k^* \Lambda(B_k^* B_j) A_j \geq 0$ for any $n \in \mathbb{N}$ and any $A_j \in B(\mathcal{H})$, $B_k \in B(\tilde{\mathcal{H}})$.

An input state $\rho \in \mathcal{S}(\mathcal{H})$ is sent to the output system through a channel Λ^*, so that the output state is written as $\tilde{\rho} \equiv \Lambda^* \rho$. Then it is important to ask how much information of ρ is sent to the output state $\Lambda^* \rho$. This amount of information transmitted from input to output is expressed by the quantum mutual entropy.

The quantum mutual entropy was introduced on the basis of the von Neumann entropy ($S(\rho) \equiv -\mathrm{tr}\, \rho \log \rho$) for purely quantum communication processes. The mutual entropy depends on an input state ρ and a channel Λ^*, so it is denoted by $I(\rho; \Lambda^*)$, which should satisfy the following conditions:

(i) The quantum mutual entropy is well-matched to the von Neumann entropy. Furthermore, if a channel is trivial, i.e. $\Lambda^* =$ identity map, then the mutual entropy equals to the von Neumann entropy: $I(\rho; \mathrm{id}) = S(\rho)$.
(ii) When the system is classical, the quantum mutual entropy reduces to classical one.
(iii) Shannon's fundamental inequality[13] $0 \leq I(\rho; \Lambda^*) \leq S(\rho)$ is held.

To define such a quantum mutual entropy extending Shannon's and Gelfand–Yaglom's classical mutual entropy, we need the quantum relative entropy and the joint state (it is called "compound state" in the sequel) describing the correlation between an input state ρ and the output state $\Lambda^* \rho$ through a channel Λ^*. A finite partition of measurable space in classical case corresponds to an orthogonal decomposition $\{E_k\}$ of the identity operator I of $\mathcal{H}$ in quantum case because the set of all orthogonal projections is considered to make an event system in a quantum system. It is known[14] that the following equality holds:

$$\sup \left\{ -\sum_k \mathrm{tr}\, \rho E_k \log \mathrm{tr}\, \rho E_k ; \{E_k\} \right\} = -\mathrm{tr}\, \rho \log \rho$$

and the supremum is attained when $\{E_k\}$ is a Schatten decomposition[15] of ρ. Therefore the Schatten decomposition is used to define the compound state and the quantum mutual entropy following the formulation of the classical mutual entropy by Kolmogorov, Gelfand and Yaglom.[2]

The compound state σ_E (corresponding to joint state in CS) of ρ and $\Lambda^*\rho$ was introduced in Refs. 6 and 16, which is given by

$$\sigma_E = \sum_k \lambda_k E_k \otimes \Lambda^* E_k \,, \tag{2.1}$$

where E stands for a Schatten decomposition $\{E_k\}$ of ρ, so that the compound state depends on how we decompose the state ρ into basic states (elementary events), in other words, how to observe the input state.

The relative entropy for two states ρ and σ is defined by Umegaki[7] and Lindblad,[17] which is written as

$$S(\rho,\sigma) = \begin{cases} \operatorname{tr} \rho(\log\rho - \log\sigma) & \text{when } \overline{ran\rho} \subset \overline{ran\sigma}\,, \\ \infty & \text{otherwise}\,. \end{cases} \tag{2.2}$$

Then we can define the mutual entropy by means of the compound state and the relative entropy,[6] i.e.

$$I(\rho;\Lambda^*) = \sup\{S(\sigma_E, \rho\otimes\Lambda^*\rho); E = \{E_k\}\}\,, \tag{2.3}$$

where the supremum is taken over all Schatten decompositions because this decomposition is not unique unless every eigenvalue is not degenerated. Some computations reduce it to the following form[6]:

$$I(\rho;\Lambda^*) = \sup\left\{\sum_k \lambda_k S(\Lambda^* E_k, \Lambda^*\rho); E = \{E_k\}\right\}. \tag{2.4}$$

This mutual entropy satisfies all conditions (i)–(iii) mentioned above.

It is important to note here that the Schatten decomposition of ρ is unique when the input system is classical. That is, when an input state ρ is given by a probability distribution or a probability measure. For the case of probability distribution; $\rho = \{\lambda_k\}$, the Schatten decomposition is uniquely given by

$$\rho = \sum_k \lambda_k \delta_k \,, \tag{2.5}$$

where δ_k is the delta measure;

$$\delta_k(j) = \delta_{k,j} = \begin{cases} 0\,, & k \neq j \\ 1 & k = j \end{cases} \quad \forall\, j\,. \tag{2.6}$$

Therefore for any channel Λ^*, the mutual entropy becomes

$$I(\rho;\Lambda^*) = \sum_k \lambda_k S(\Lambda^*\delta_k, \Lambda^*\rho)\,, \tag{2.7}$$

which equals to the following usual expression of Shannon when the minus is well-defined:

$$I(\rho;\Lambda^*) = S(\Lambda^*\rho) - \sum_k \lambda_k S(\Lambda^*\delta_k)\,. \tag{2.8}$$

The above equality has been taken as the definition of the mutual entropy for a classical-quantum channel.[3–5]

Note that the definition (2.3) of the mutual entropy is written as

$$I(\rho;\Lambda^*) = \sup\left\{\sum_k \lambda_k S(\Lambda^*\rho_k,\Lambda^*\rho);\rho=\sum_k \lambda_k\rho_k \in F_o(\rho)\right\},$$

where $F_o(\rho)$ is the set of all orthogonal finite decompositions of ρ. The proof of the above equality is given in Ref. 18 by means of the fundamental properties of the quantum relative entropy.

3. Communication Processes

We discuss communication processes in this section.[4,14] Let $A = \{a_1, a_2, \ldots, a_n\}$ be a set of certain alphabets and Ω be the infinite direct product of $A: \Omega = A^Z \equiv \prod_{-\infty}^{\infty} A$ calling a message space. In order to send an information written by an element of this message space to a receiver, we often need to transfer the message into a proper form for a communication channel. This change of a message is called a coding. In other words, a coding is a measurable one-to-one map ξ from Ω to a proper space X.

Let $(\Omega, \mathcal{F}_\Omega, P(\Omega))$ be an input probability space and X be the coded input space. This space X may be a classical object or a quantum object. For instance, X is a Hilbert space $\mathcal{H}$ of a quantum system, then the coded input system is described by $(B(\mathcal{H}), \mathcal{S}(\mathcal{H}))$.

An output system is similarly described as the input system: The coded output space is denoted by $\tilde{X}$ and the decoded output space is $\tilde{\Omega}$ made by another alphabets. A transmission (map) from X to $\tilde{X}$ is described by a channel reflecting all properties of a physical device, which is denoted by γ here. With a decoding $\tilde{\xi}$, the whole information transmission process is written as

$$\Omega \xrightarrow{\ \xi\ } X \xrightarrow{\ \gamma\ } \tilde{X} \xrightarrow{\ \tilde{\xi}\ } \tilde{\Omega}\,. \tag{3.9}$$

That is, a message $\omega \in \Omega$ is coded to $\xi(\omega)$ and it is sent to the output system through a channel γ, then the output coded message becomes $\gamma \circ \xi(\omega)$ and it is decoded to $\tilde{\xi} \circ \gamma \circ \xi(\omega)$ at a receiver.

This transmission process is mathematically set as follows: M messages are sent to a receiver and the kth message $\omega^{(k)}$ occurs with the probability λ_k. Then the occurrence probability of each message in the sequence $(\omega^{(1)}, \omega^{(2)}, \ldots, \omega^{(M)})$ of M messages is denoted by $\rho = \{\lambda_k\}$, which is a state in a classical system. If ξ is a classical coding, then $\xi(\omega)$ is a classical object such as an electric pulse. If ξ is a quantum coding, then $\xi(\omega)$ is a quantum object (state) such as a coherent state. Here we consider such a quantum coding, so that $\xi(\omega^{(k)})$ is a quantum state, and we denote $\xi(\omega^{(k)})$ by σ_k. Thus the coded state for the sequence $(\omega^{(1)}, \omega^{(2)}, \ldots, \omega^{(M)})$

is written as

$$\sigma = \sum_k \lambda_k \sigma_k \,. \tag{3.10}$$

This state is transmitted through a channel γ. This channel is expressed by a completely positive mapping Γ^*, in the sense of Sec. 1, from the state space of X to that of $\tilde{X}$, hence the output coded quantum state $\tilde{\sigma}$ is $\Gamma^*\sigma$. Since the information transmission process can be understood as a process of state (probability) change, when Ω and $\tilde{\Omega}$ are classical and X and $\tilde{X}$ are quantum, the process (3.9) is written as

$$P(\Omega) \xrightarrow{\Xi^*} \mathcal{S}(\mathcal{H}) \xrightarrow{\Gamma^*} \mathcal{S}(\tilde{\mathcal{H}}) \xrightarrow{\tilde{\Xi}^*} P(\tilde{\Omega}) \,, \tag{3.11}$$

where Ξ^* (resp. $\tilde{\Xi}^*$) is the channel corresponding to the coding ξ (resp. $\tilde{\xi}$) and $\mathcal{S}(\mathcal{H})$ (resp. $\mathcal{S}(\tilde{H})$) is the set of all density operators (states) on $\mathcal{H}$ (resp. $\tilde{\mathcal{H}}$).

We have to be careful in studying the objects in the above transmission process (3.9) or (3.11). Namely, we have to make clear which object is going to be studied. For instance, if we want to know the information capacity of a quantum channel $\gamma(=\Gamma^*)$, then we have to take X so as to describe a quantum system like a Hilbert space and we need to start the study from a quantum state in quantum space X not from a classical state associated to a message. If we like to know the capacity of the whole process including a coding and a decoding, which means the capacity of a channel $\tilde{\xi}\circ\gamma\circ\xi(=\tilde{\Xi}^*\circ\Gamma^*\circ\Xi^*)$, then we have to start from a classical state. In any case, when we are concerned with the capacity of channel, we only have to take the supremum of the mutual entropy $I(\rho;\Lambda^*)$ over a quantum or classical state ρ in a proper set determined by what we like to study with a channel Λ^*. We explain this more precisely in the next section.

4. Channel Capacity

We discuss two types of channel capacity in communication processes, namely, the capacity of a quantum channel Γ^* and that of a classical (classical-quantum-classical) channel $\tilde{\Xi}^*\circ\Gamma^*\circ\Xi^*$.

(1) *Capacity of quantum channel:* The capacity of a quantum channel is the ability of information transmission of a quantum channel itself, so that it does not depend on how to code a message being treated as a classical object and we have to start from an arbitrary quantum state and find the supremum of the mutual entropy. One often makes a mistake at this point. For example, one starts from the coding of a message and compute the supremum of the mutual entropy and he says that the supremum is the capacity of a quantum channel, which is not correct. Even when his coding is a quantum coding and he sends the coded message to a receiver through a quantum channel, if he starts from a classical state, then his capacity is not the capacity of the quantum channel itself. In his case, usual Shannon's theory is applied because he can easily compute the conditional distribution by the usual (classical) way. His supremum is the capacity of a classical-quantum-classical channel, and it is in the second category discussed below.

Let $\mathcal{S}_0$ ($\subset \mathcal{S}(\mathcal{H})$) be the set of all states prepared for expression of information. Then the capacity of the channel Γ^* with respect to $\mathcal{S}_0$ is defined as:

Definition 1. The capacity of a quantum channel Γ^* is

$$C^{\mathcal{S}_0}(\Gamma^*) = \sup\{I(\rho;\Gamma^*);\rho\in\mathcal{S}_0\}\,. \tag{4.12}$$

Here $I(\rho;\Gamma^*)$ is the mutual entropy given in (2.3) or (2.4) with $\Lambda^* = \Gamma^*$.

When $\mathcal{S}_0 = \mathcal{S}(\mathcal{H})$, $C^{\mathcal{S}(\mathcal{H})}(\Gamma^*)$ is denoted by $C(\Gamma^*)$ for simplicity. In Refs. 8, 19 and 18, we also considered the pseudo-quantum capacity $C_p(\Gamma^*)$ defined by (4.12) with the pseudo-mutual entropy $I_p(\rho;\Gamma^*)$ where the supremum is taken over all finite decompositions instead of all orthogonal pure decompositions:

$$I_p(\rho;\Gamma^*) = \sup\left\{\sum_k \lambda_k S(\Gamma^*\rho_k,\Gamma^*\rho);\rho=\sum_k\lambda_k\rho_k\,,\quad \text{finite decomposition}\right\}\,. \tag{4.13}$$

However the pseudo-mutual entropy is not well-matched to the conditions explained in Sec. 2, and it is difficult to compute numerically.[20] From the monotonicity of the mutual entropy,[14] we have

$$0\le C^{\mathcal{S}_0}(\Gamma^*)\le C_p^{\mathcal{S}_0}(\Gamma^*)\le \sup\{S(\rho);\rho\in\mathcal{S}_0\}\,.$$

(2) *Capacity of classical-quantum-classical channel:* The capacity of C-Q-C channel $\tilde{\Xi}^*\circ\Gamma^*\circ\Xi^*$ is the capacity of the information transmission process starting from the coding of messages, therefore it can be considered as the capacity including a coding (and a decoding). As is discussed in Sec. 3, an input state ρ is the probability distribution $\{\lambda_k\}$ of messages, and its Schatten decomposition is unique as (2.5), so the mutual entropy is written by (2.7):

$$I(\rho;\tilde{\Xi}^*\circ\Gamma^*\circ\Xi^*) = \sum_k \lambda_k S(\tilde{\Xi}^*\circ\Gamma^*\circ\Xi^*\delta_k,\tilde{\Xi}^*\circ\Gamma^*\circ\Xi^*\rho)\,. \tag{4.14}$$

If the coding Ξ^* is a quantum coding, then $\Xi^*\delta_k$ is expressed by a quantum state. Let us denote the coded quantum state by σ_k and put $\sigma = \Xi^*\rho = \sum_k\lambda_k\sigma_k$. We denote the set of such quantum codings by $\mathcal{C}$. Then the above mutual entropy becomes

$$I(\rho;\tilde{\Xi}^*\circ\Gamma^*\circ\Xi^*) = \sum_k \lambda_k S(\tilde{\Xi}^*\circ\Gamma^*\sigma_k,\tilde{\Xi}^*\circ\Gamma^*\sigma)\,. \tag{4.15}$$

This is the expression of the mutual entropy of the whole information transmission process starting from a coding of classical messages. Hence the capacity of C-Q-C channel is as follows:

Definition 2. The capacity of C-Q-C channel is

$$C^{P_0}(\tilde{\Xi}^*\circ\Gamma^*\circ\Xi^*) = \sup\{I(\rho;\tilde{\Xi}^*\circ\Gamma^*\circ\Xi^*);\rho\in P_0\}\,, \tag{4.16}$$

where $P_0(\subset P(\Omega))$ is the set of all probability distributions prepared for input (*a priori*) states (distributions or probability measures).

Moreover the capacity for coding free in $\mathcal{C}$ is found by taking the supremum of the mutual entropy (4.15) over all probability distributions in P_0 and all codings in $\mathcal{C}$:

$$C_c^{P_0}(\tilde{\Xi}^* \circ \Gamma^*) = \sup\left\{I(\rho; \tilde{\Xi}^* \circ \Gamma^* \circ \Xi^*); \rho \in P_0, \Xi^* \in \mathcal{C}\right\}. \tag{4.17}$$

There are several ways to decode quantum states such as quantum measurements, so that such decodings and denoted by $\mathcal{D}$. The capacity for decoding free in $\mathcal{D}$ is

$$C_d^{P_0}(\Gamma^* \circ \Xi^*) = \sup\left\{I(\rho; \tilde{\Xi}^* \circ \Gamma^* \circ \Xi^*); \rho \in P_0, \tilde{\Xi}^* \in \mathcal{D}\right\}. \tag{4.18}$$

The last capacity is for both coding and decoding free and it is given by

$$C_{cd}^{P_0}(\Gamma^*) = \sup\left\{I(\rho; \tilde{\Xi}^* \circ \Gamma^* \circ \Xi^*); \rho \in P_0, \Xi^* \in \mathcal{C}, \tilde{\Xi}^* \in \mathcal{D}\right\}. \tag{4.19}$$

These capacities $C_c^{P_0}$, $C_d^{P_0}$, $C_{cd}^{P_0}$ do not measure the ability of the quantum channel Γ^* itself, but measure the ability of Γ^* through the coding and decoding. The above three capacities C^{P_0}, $C_c^{P_0}$, $C_{cd}^{P_0}$ satisfy the following inequalities:

$$0 \le C^{P_0}(\tilde{\Xi}^* \circ \Gamma^* \circ \Xi^*) \le C_c^{P_0}(\tilde{\Xi}^* \circ \Gamma^*),$$
$$C_d^{P_0}(\Gamma^* \circ \Xi^*) \le C_{cd}^{P_0}(\Gamma^*) \le \sup\{S(\rho); \rho \in P_0\},$$

where $S(\rho)$ is not the von Neumann entropy but the Shannon entropy: $\sum \lambda_k \log \lambda_k$.

Remark that if $\sum_k \lambda_k S(\Gamma^* \sigma_k)$ is finite, then (4.15) becomes

$$I(\rho; \tilde{\Xi}^* \circ \Gamma^* \circ \Xi^*) = S(\tilde{\Xi}^* \circ \Gamma^* \sigma) - \sum_k \lambda_k S(\tilde{\Xi}^* \circ \Gamma^* \sigma_k). \tag{4.20}$$

Further, if ρ is a probability measure having a density function $f(\lambda)$ and each λ corresponds to a quantum coded state $\sigma(\lambda)$, then $\sigma = \int f(\lambda)\sigma(\lambda) d\lambda$ and

$$I(\rho; \tilde{\Xi}^* \circ \Gamma^* \circ \Xi^*) = S(\tilde{\Xi}^* \circ \Gamma^* \sigma) - \int f(\lambda) S(\tilde{\Xi}^* \circ \Gamma^* \sigma(\lambda)) d\lambda, \tag{4.21}$$

which is less than

$$S(\Gamma^* \sigma) - \int f(\lambda) S(\Gamma^* \sigma(\lambda)) d\lambda. \tag{4.22}$$

This bound is computed in several cases.[19,21] This bound is a special one of the following inequality

$$I(\rho; \tilde{\Xi}^* \circ \Gamma^* \circ \Xi^*) \le I(\rho; \Gamma^* \circ \Xi^*),$$

which comes from the monotonicity of the relative entropy. When the decoding is not taken into account, we only have to consider the mutual entropy $I(\rho; \Gamma^* \circ \Xi^*)$ above.

Let us define an extension of the functional of the relative entropy. If A and B are two positive Hermitian operators (not necassarily the states, i.e. not necessarily with unit traces) then we set

$$S(A,B) = \mathrm{tr}\, A(\log A - \log B)\,.$$

There is the following *Bogoliubov inequality.*

$$S(A,B) \geq \mathrm{tr}\, A(\log \mathrm{tr}\, A - \log \mathrm{tr}\, B)\,.$$

The following theorem gives us the bound of the mutual entropy $I(\rho;\Gamma^* \circ \Xi^*)$.

Theorem 1. *For a probability distribution* $\rho = \{\lambda_k\}$ *and a quantum coded state* $\sigma = \Xi^*\rho \equiv \sum_k \lambda_k \sigma_k$, $\lambda_k \geq 0$, $\sum_k \lambda_k = 1$, *one has the following inequality for any quantum channel decomposed as* $\Gamma^* = \Gamma_1^* \circ \Gamma_2^*$ *such that* $\Gamma_1^*\sigma \equiv \sum_i E_i \sigma E_i$ *by a projection valued measure* $\{E_i\}$:

$$\begin{aligned}
&\sum_k \lambda_k S(\sigma_k, \sigma) \\
&\quad \geq I(\rho;\Gamma^*\circ\Xi^*) = \sum_k \lambda_k S(\Gamma^*\sigma_k, \Gamma^*\sigma) \\
&\quad \geq \sum_i \left[-\mathrm{tr}(\Gamma_2^*\sigma E_i)\log \mathrm{tr}(\Gamma_2^*\sigma E_i) + \sum_k \lambda_k\, \mathrm{tr}(\Gamma_2^*\sigma_k E_i)\log \mathrm{tr}(\Gamma_2^*\sigma_k E_i)\right].
\end{aligned} \tag{4.23}$$

Proof. The equality $I(\rho;\Gamma^*\circ\Xi^*) = \sum_k \lambda_k S(\Gamma^*\sigma_k, \Gamma^*\sigma)$ is the case of the equality (15), and the first inequality comes from the monotonicity of the relative entropy. Furthermore, by applying again the monotonicity of the relative entropy, we have

$$\begin{aligned}
&\sum_k \lambda_k S(\Gamma^*\sigma_k, \Gamma^*\sigma) \\
&\quad \geq \sum_k \sum_i \lambda_k S(E_i\Gamma_2^*\sigma_k E_i, E_i\Gamma_2^*\sigma_k E_i) \\
&\quad \geq \sum_{k,i} \lambda_k\, \mathrm{tr}(E_i\Gamma_2^*\sigma_k E_i)(\log \mathrm{tr}(E_i\Gamma_2^*\sigma_k E_i) - \log \mathrm{tr}(E_i\Gamma_2^*\sigma E_i)) \\
&\quad = \sum_i \left(-\mathrm{tr}(\Gamma_2^*\sigma E_i)\log \mathrm{tr}(\Gamma_2^*\sigma E_i) + \sum_k \lambda_k\, \mathrm{tr}(\Gamma_2^*\sigma_k E_i)\log \mathrm{tr}(\Gamma_2^*\sigma_k E_i)\right).
\end{aligned}$$

Here the second inequality is due to the Bogoliubov inequality. □

In the case that the channel Γ_2^* is trivial; $\Gamma_2^*\sigma = \sigma$, the above inequality reduces to the bound obtained by Holevo[3]:

$$\begin{aligned}\sum_k \lambda_k S(\sigma_k, \sigma) &= -\mathrm{tr}\,\sigma \log \sigma + \sum_k \lambda_k \,\mathrm{tr}\,\sigma_k \log \sigma_k \\ &\geq I(\rho; \Gamma^* \circ \Xi^*) = I(\rho; \Gamma_1^* \circ \Xi^*) \\ &\geq \sum_i \left[-\mathrm{tr}(\sigma E_i) \log\, \mathrm{tr}(\sigma E_i) + \sum_k \lambda_k \,\mathrm{tr}(\sigma_k E_i) \log\, \mathrm{tr}(\sigma_k E_i) \right]\end{aligned}$$

Remark that the right-hand side in the inequality is sometimes called the *accessible information.*

Using the above upper and lower bounds of the mutual entropy, we can compute these bounds of the capacity in many different cases.

Acknowledgments

The authors thank SCAT for finacial support of this work.

References

1. A. N. Kolmogorov, *Theory of transmission of information*, *Amer. Math. Soc. Trans.* Ser. 2 **33** (1963) 291–321.
2. I. M. Gelfand and A. M. Yaglom, *Calculation of the amount of information about a random function contained in another such function*, *Amer. Math. Soc. Trans.* **12** (1959) 199–246.
3. A. S. Holevo, *Some estimates for the amount of information transmittable by a quantum communication channel* (in Russian), *Prob. Pered. Infor.* **9** (1973) 3–11.
4. R. S. Ingarden, *Quantum information theory*, *Rep. Math. Phys.* **10** (1976) 43–73.
5. L. B. Levitin, *Physical information theory for 30 years: basic concepts and results*, *Springer Lect. Note in Phys.*, *Vol. 378* (Springer, 1991) pp. 101–110.
6. M. Ohya, *On compound state and mutual information in quantum information theory*, *IEEE Trans. Inf. Theory* **29** (1983) pp. 770–777.
7. H. Umegaki, *Conditional expectations in an operator algebra IV (entropy and information)*, *Kodai Math. Sem. Rep.* **14** (1962) 59–85.
8. M. Ohya, *Some aspects of quantum information theory and their applications to irreversible processes*, *Rep. Math. Phys.* **27** (1989) 19–47.
9. H. Araki, *Relative entropy for states of von Neumann algebras*, *Publ. RIMS Kyoto Univ.* **11** (1976) 809–833.
10. A. Uhlmann, *Relative entropy and the Wigner–Yanase–Dyson–Lieb concavity in interpolation theory*, *Commun. Math. Phys.* **54** (1977) 21–32.
11. M. Ohya, *Fundamentals of quantum mutual entropy and capacity*, *Open Systems Infor. Dyn.* **6** (1999) 69-78.
12. M. Ohya, *Quantum ergodic channels in operator algebras*, *J. Math. Anal. Appl.* **84** (1981) 318–327.
13. C. E. Shannon, *Mathematical theory of communication*, *Bell System Tech. J.* **27** (1948) 379–423.
14. M. Ohya and D. Petz, *Quantum Entropy and Its Use* (Springer, 1993).
15. R. Schatten, *Norm Ideals of Completely Continuous Operators* (Springer, 1970).
16. M. Ohya, *Note on quantum probability*, *L. Nuovo Cimento* **38** (1983) 402–406.
17. G. Lindblad, *Entropy, Information and quantum measurements*, *Commun. Math. Phys.* **33** (1973) 111–119.

18. N. Muraki, M. Ohya and D. Petz, *Note on entropy of general quantum systems*, *Open Systems Infor. Dyn.*, **1**, (1992) 43–56.
19. M. Ohya, D. Petz and N. Watanabe, *On capacities of quantum channels*, *Probab. Math. Statist.* **17** (1997) 179–196.
20. M. Ohya, D. Petz and N. Watanabe, *Numerical computation of quantum capacity*, *Int. J. Theor. Phys.* **38** (1998) 507–510.
21. H. P. Yuen and M. Ozawa, *Ultimate information carrying limit of quantum systems*, *Phys. Rev. Lett.* **70** (1993) 363–366.

10.1098/rspa.2001.0867

THE ROYAL SOCIETY

Entanglement, quantum entropy and mutual information

By Viacheslav P. Belavkin[1] and Masanori Ohya[2]

[1]*Department of Mathematics, University of Nottingham, Nottingham NG7 2RD, UK*

[2]*Department of Information Sciences, Science University of Tokyo, 278 Noda City, Chiba, Japan*

Received 18 May 2001; accepted 24 May 2001; published online 29 November 2001

The operational structure of quantum couplings and entanglements is studied and classified for semi-finite von Neumann algebras. We show that the classical–quantum correspondences, such as quantum encodings, can be treated as diagonal semi-classical (d-) couplings, and the entanglements, characterized by truly quantum (q-) couplings, can be regarded as truly quantum encodings. The relative entropy of the d-compound and entangled states leads to two different types of entropy for a given quantum state: the von Neumann entropy, which is achieved as the maximum of mutual information over all d-entanglements, and the dimensional entropy, which is achieved at the standard entanglement (true quantum entanglement) coinciding with a d-entanglement only in the case of pure marginal states. The d- and q-information of a quantum noisy channel are, respectively, defined via the input d- and q-encodings, and the q-capacity of a quantum noiseless channel is found to be the logarithm of the dimensionality of the input algebra. The quantum capacity may double the classical capacity, achieved as the supremum over all d-couplings (or encodings) bounded by the logarithm of the dimensionality of a maximal Abelian subalgebra.

Keywords: entanglements; compound states; quantum entropy and information

1. Introduction

The entanglements, specifically quantum correlations first considered by Schrödinger (1935), are now used to study quantum information processes, in particular, quantum computations, quantum teleportation and quantum cryptography (Bennett *et al.* 1993; Ekert 1993; Jozsa & Schumacher 1994). There have been mathematical studies of the entanglements in Werner (1989, 1998), Bennett *et al.* (1996) and Schumacher (1993*a*,*b*), in which the entangled state of two quantum systems is defined as a compound state which is a convex combination $\sum_n \varrho_n \otimes \varsigma_n p(n)$ with some states ϱ_n and ς_n on the corresponding algebras $\mathcal{A}$ and $\mathcal{B}$. However, it is obvious that there exist several types of correlated states, written as 'separable' forms above. Such correlated, or classically entangled, states have also been discussed in several contexts in quantum probability, such as quantum measurement and filtering (Belavkin 1980, 1994), quantum compound states (Ohya 1983*a*,*b*) and lifting (Accardi & Ohya 2002).

In this paper, we study the mathematical structure of classical–quantum and quantum–quantum couplings to provide a finer classification of quantum separable and entangled states. We also discuss the informational degree of entanglement and entangled quantum mutual entropy and quantum capacity. The latter are treated here solely as quantities arising in certain maximization problems for quantum mutual information which is generalized here for arbitrary semi-finite algebras.

The term entanglement was introduced by Schrödinger in 1935 out of the need to describe correlations of quantum states not captured by mere classical, statistical correlations which are always the convex combinations of non-correlated states. In this spirit, the by now standard definition (Werner 1989) of the entanglement in physics is the state of a compound quantum system 'which cannot be prepared by two separated devices with only correlated classical data as their inputs'. We show that the entangled states can be achieved by quantum (q-) encodings, and that the non-separable couplings of states, in the same way as the separable states, can be achieved by classical (c-) encodings.

The compound states, called o-coupled, are defined by orthogonal decompositions of their marginal states. This is a particular case of a so-called diagonal (d-compound) state of a compound system which is achieved by the classical–quantum correspondences, called encodings. The d-compound states, as convex combinations of the special product states, are most informative among c-compound states, in the sense that the maximum of the mutual entropy over all c-couplings of probe systems $\mathcal{A}$ to the quantum system $\mathcal{B}$, with a given normal state ς, is achieved on the extreme d-coupled (even o-coupled) states. This maximum is the von Neumann entropy, which is bound by the rank capacity $\ln \operatorname{rank} \mathcal{B}$, the supremum of $\mathsf{S}(\varsigma)$ over all ς. The rank $\operatorname{rank} \mathcal{B}$ of the algebra $\mathcal{B}$ is a topological characteristic of $\mathcal{B}$ defined as the dimensionality of the maximal Abelian subalgebra $\mathcal{A} \subseteq \mathcal{B}$ (in the case of the simple $\mathcal{B}$ it coincides with the dimensionality $\dim \mathcal{H}$ of the Hilbert space $\mathcal{H}$ of representation for $\mathcal{B}$). The von Neumann capacity defined as the maximal von Neumann entropy, i.e. as the maximum $\mathsf{C}_{\mathrm{c}} = \ln \operatorname{rank} \mathcal{B}$ of mutual entropy over all c-couplings of the classical probe systems $\mathcal{A}$ to the quantum system $\mathcal{B}$, is finite only if $\operatorname{rank} \mathcal{B} < \infty$. Due to $\dim \mathcal{B} \leqslant (\operatorname{rank} \mathcal{B})^2$ (the equality is only for the simple algebras $\mathcal{B}$), it is achieved on the normal tracial density operator $\sigma = (\operatorname{rank} \mathcal{B})^{-1} I$ only in the case of finite-dimensional $\mathcal{B}$.

We prove that the truly entangled compound states are most informative, in the sense that the maximum of the mutual entropy over all couplings including entanglements of the quantum probe systems $\mathcal{A}$ to the quantum system $\mathcal{B}$ is achieved on a non-separable q-compound state. It is given by the standard entanglement, an extreme entanglement of $\mathcal{A} = \tilde{\mathcal{B}}$ with the marginal state $\varrho = \tilde{\varsigma}$, where $(\tilde{\mathcal{B}}, \tilde{\varsigma})$ is the transposed (time-inversed) system to $(\mathcal{B}, \varsigma)$. The maximal information gained for such extreme q-compound states defines another type of entropy, the q-entropy $\mathsf{H}(\varsigma)$, which is bigger than the von Neumann entropy $\mathsf{S}(\varsigma)$ in the case of mixed ς. The maximum of the q-entropy $\mathsf{H}(\varsigma)$ over all states ς defines the dimensional capacity $\ln \dim \mathcal{B}$. The dimensionality $\dim \mathcal{B}$ of the algebra $\mathcal{B}$ is the major topological characteristic of $\mathcal{B}$, and it gives true quantum capacity of $\mathcal{B}$ achieved at the standard entanglement with the maximal chaotic ς. Thus, the true quantum capacity is the maximum $\mathsf{C}_{\mathrm{q}} = \ln \dim \mathcal{B}$ of the mutual entropy over all, not only classical–quantum, couplings of the probe systems $\mathcal{A}$ to the quantum system $\mathcal{B}$, and it is finite only for the finite-dimensional algebra $\mathcal{B}$. The q-entropy $H(\varsigma)$, also called the dimensional entropy, can be considered as the true quantum entropy, in contrast to the von Neumann entropy $S(\varsigma)$, called also rank-entropy, or c-entropy (semi-classical entropy) as the supremum of mutual entropy over couplings with only classical probe systems $\mathcal{A}$. The capacity C_{q} coincides with C_{c} only in the classical case of the Abelian $\mathcal{B}$, and it is strictly larger than the semi-classical capacity $C_{\mathrm{e}} = \ln \operatorname{rank} \mathcal{B}$ for any noiseless quantum channel. We shall show that the capacity $C_{\mathrm{q}} = \ln \dim \mathcal{B}$ is achieved as the supremum of the quantum Shannon information for the noiseless channel over the entanglements as q-encodings similar to the capacity C_{c}, which is achieved as the supremum over c-encodings described by the classical–quantum correspondences $\mathcal{A} \to \mathcal{B}$.

In this paper we consider the case of semi-finite quantum systems which are described by the von Neumann algebras $\mathcal{A}$ and $\mathcal{B}$ with normal, faithful semi-finite trace. Such quantum systems include all simple quantum systems described by full operator algebras as well as all classical systems as the commutative case. The particular cases of simple and discrete decomposable algebras are considered in Belavkin & Ohya (1998, 2000).

2. Pairings, couplings and entanglements

In this section, we give mathematical characterizations of entanglement in terms of quantum coupling, which is described in terms of transpose-completely positive operations extending individual states to a compound state of a composed quantum system. We show how any normal compound state can be achieved in this way, and introduce the standard entanglement as an operation giving rise to the standard entangled compound state.

Let $\mathcal{H}$ denote the Hilbert space of a quantum system, and $\mathcal{B} = \mathcal{L}(\mathcal{H})$ be the algebra of all linear bounded operators on $\mathcal{H}$. Note that $\mathcal{B}$ consists of all operators $A : \mathcal{H} \to \mathcal{H}$ having the adjoints $A^\dagger$ on $\mathcal{H}$. A linear functional ς on $\mathcal{B}$ with complex-values $\varsigma(B) \in \mathbb{C}$ is called a state on $\mathcal{B}$ if it is positive (i.e. $\varsigma(B) \geqslant 0$ for any positive operator $B = A^\dagger A$ in $\mathcal{B}$) and normalized (i.e. $\varsigma(I) = 1$ for the identity operator I in $\mathcal{A}$). A *normal* state can be expressed as

$$\varsigma(B) = \operatorname{tr} \varkappa^\dagger B \varkappa \equiv \langle B, \sigma \rangle, \qquad B \in \mathcal{B}, \tag{2.1}$$

where $\varkappa$ is a linear Hilbert–Schmidt operator from $\mathcal{H}$ to (another) Hilbert space $\mathcal{G}$, and $\varkappa^\dagger$ is the adjoint operator from $\mathcal{G}$ to $\mathcal{H}$. Here, tr stands for the usual trace in $\mathcal{G}$ (in cases of ambiguity it will also be denoted as $\operatorname{tr}_{\mathcal{G}}$). This $\varkappa$ is called the *amplitude operator*, which can always be considered on $\mathcal{G} = \mathcal{H}$ as the square root of the operator $\varkappa\varkappa^\dagger$ (it is called simply the amplitude if $\mathcal{G}$ is the one-dimensional space $\mathbb{C}$, $\varkappa = \eta \in \mathcal{H}$ with $\varkappa^\dagger \varkappa = \|\eta\|^2 = 1$, in which case $\varkappa^\dagger$ is the functional $\eta^\dagger$ from $\mathcal{H}$ to $\mathbb{C}$).

We can always equip $\mathcal{H}$ (and will equip all auxiliary Hilbert spaces, e.g. $\mathcal{G}$) with an isometric involution $J = J^\dagger$, $J^2 = I$ having the properties of complex conjugation

$$J \sum \lambda_j \eta_j = \sum \bar{\lambda}_j J \eta_j, \quad \forall \lambda_j \in \mathbb{C}, \quad \eta_j \in \mathcal{H},$$

and denote by $\langle B, \sigma \rangle$ the tilde-pairing $\operatorname{tr} B \tilde{\sigma}$ of $\mathcal{B}$ with the trace class operators $\sigma \in \mathcal{T}(\mathcal{H})$ such that $\tilde{\sigma} = J \sigma^\dagger J$. We shall call $\sigma = J \varkappa \varkappa^\dagger J = \tilde{\varkappa}^\dagger \tilde{\varkappa}$ the probability density of the state (2.1) with respect to this pairing and assume that the support E_σ of σ

is the minimal projector $E = E^\dagger \in \mathcal{B}$ for which $\varsigma(E) = 1$, i.e. that $\overline{\widetilde{E_\sigma}} := JE_\sigma J = E_\sigma$. The latter can also be expressed as the symmetricity property $\widetilde{E_\sigma} = E_\sigma$ with respect to the tilde operation (transposition) $\tilde{B} = JB^\dagger J$ on $\mathcal{L}(\mathcal{H})$. One can always assume that J is the standard complex conjugation in an eigenrepresentation of σ such that $\bar{\sigma} = \varkappa\varkappa^\dagger = \tilde{\sigma}$ coincides with σ as the real element of the invariant maximal Abelian subalgebra $\mathcal{A} \subset \mathcal{L}(\mathcal{H})$ of all diagonal (and thus symmetric) operators in this basis.

The auxiliary Hilbert space $\mathcal{G}$ and the amplitude operator in (2.1) are not unique; however, $\varkappa$ is defined uniquely up to a unitary transform $\varkappa^\dagger \mapsto U\varkappa^\dagger$ in $\mathcal{G}$. $\mathcal{G}$ can always be taken to be minimal by identifying it with the support $\mathcal{H}_\sigma = E_\sigma\mathcal{H}$ for σ, defined as the closure of $\sigma\mathcal{H}$ (E_σ is the minimal orthoprojector in $\mathcal{B}$ such that $\sigma E = \sigma$). In general, $\mathcal{G}$ is not one dimensional, the dimensionality $\dim \mathcal{G}$ must not be less than $\operatorname{rank} \varkappa^\dagger = \operatorname{rank} \sigma$, and the dimensionality of the $\operatorname{ran} \rho = \operatorname{ran} \varkappa^\dagger$ of ρ coincides with the support $\mathcal{G}_\rho$ for $\rho = \varkappa^\dagger\varkappa \simeq \tilde{\sigma}$.

Given the amplitude operator $\varkappa : \mathcal{G} \to \mathcal{H}$, one can define not only the state ς but also the normal state

$$\varrho(A) = \operatorname{tr} \tilde{\varkappa}^\dagger A \tilde{\varkappa} \equiv \langle A, \rho \rangle, \quad A \in \mathcal{A}, \tag{2.2}$$

on $\mathcal{A} = \mathcal{L}(\mathcal{G})$, as the marginal of the *pure compound* state,

$$\omega(A \otimes B) = \operatorname{tr} \tilde{A}\varkappa^\dagger B \varkappa = \operatorname{tr} \tilde{\varkappa}^\dagger A \tilde{\varkappa} \tilde{B},$$

where ω is defined on the algebra $\mathcal{A} \otimes \mathcal{B}$ of all bounded operators on the Hilbert tensor product space $\mathcal{G} \otimes \mathcal{H}$.

Indeed, the defined bilinear form, with $\tilde{A} = JA^\dagger J$, is uniquely extended to such a state given on $\mathcal{L}(\mathcal{G} \otimes \mathcal{H})$ by the amplitude $\psi = \varkappa'$, where $\varkappa'$ is uniquely defined by $(\zeta \otimes \eta)^\dagger \varkappa' = \eta^\dagger \varkappa J\zeta$, for all $\zeta \in \mathcal{G}$, $\eta \in \mathcal{H}$.

This pure compound state ω is the so-called *entangled state* (Schrödinger 1935) unless its marginal state ς (and ϱ) is pure corresponding to a rank-one operator $\varkappa^\dagger = \zeta\eta^\dagger$, in which case $\omega = \varrho \otimes \varsigma$ is given by the amplitude $v = \zeta \otimes \eta$. The amplitude operator $\varkappa$ corresponding to mixed states on $\mathcal{A}$ and $\mathcal{B}$ will be called the *entangling operator* of $\rho = \varkappa^\dagger\varkappa$ to $\sigma = \tilde{\varkappa}^\dagger\tilde{\varkappa}$.

As follows from the next theorem, any pure entangled state

$$\omega(A \otimes B) = \psi^\dagger (A \otimes B)\psi, \quad A \otimes B \in \mathcal{L}(\mathcal{G} \otimes \mathcal{H})$$

given by an amplitude $\psi \in \mathcal{G} \otimes \mathcal{H}$, can be described by a unique entanglement $\varkappa$ to the algebra $\mathcal{A} = \mathcal{L}(\mathcal{G})$ of the marginal state ς on $\mathcal{B} = \mathcal{L}(\mathcal{H})$.

Before formulating this theorem in the generality required for further considerations, let us introduce the following notation.

Let $\mathcal{A}$ be a $*$-algebra on $\mathcal{G}$ with a normal, faithful and semi-finite trace μ; let $\mathcal{A}'$ denote the commutant $\{A' \in \mathcal{L}(\mathcal{G}) : [A', A] = 0, \ \forall A \in \mathcal{A}\}$ of $\mathcal{A}$; and let $(\tilde{\mathcal{A}}, \tilde{\mu})$ denote the transposed algebra of the operators $\tilde{A}$ with $\tilde{\mu}(A) = \mu(\tilde{A})$, which may not coincide with $(\mathcal{A}, \mu)$ (nor with $\mathcal{A}'$). We can always assume that $\tilde{A} = JA^\dagger J$, with respect to an involution J on $\mathcal{G}$ representing $\tilde{\mathcal{A}}$ on the same Hilbert space $\mathcal{G}$ and, in most cases, $\tilde{\mathcal{A}} = \mathcal{A}$ and $\tilde{\mu} = \mu$, but not in the *standard representation*, unless $\mathcal{A}$ is Abelian algebra. We denote by $\mathcal{A}_\mu \subseteq \mathcal{A}$ the space of all operators $A \in \mathcal{A}$ in the form $x^\dagger z$, where $x, z \in \mathfrak{a}_\mu$, with $\mathfrak{a}_\mu = \{x \in \mathcal{A} : \mu(x^\dagger x) < \infty\}$. $(\mathcal{G}_\mu, \iota, J_\mu)$ denotes the standard representation $\iota : \mathcal{A} \to \mathcal{L}(\mathcal{G}_\mu)$ given by the left multiplication $\iota(A)x = Ax$ on $\mathfrak{a}_\mu$, with the standard isometric involution $J_\mu : x \mapsto x^\dagger$ defining the representation

$\tilde{\iota}(\tilde{A}) = J_\mu \iota(A^\dagger) J_\mu$ of $\tilde{A}$ on the completion $\mathcal{G}_\mu$ of the module $\mathfrak{a}_\mu$ with respect to the inner product $(x \mid z)_\mu = \mu(x^\dagger z)$. We recall that the von Neumann algebra $\mathcal{A}$ defined by $\mathcal{A}'' = \mathcal{A}$ is anti-isomorphic to $\iota(\mathcal{A})' = J_\mu \iota(\mathcal{A}) J_\mu$ and thus $\tilde{\mathcal{A}} \simeq \iota(\mathcal{A})'$ and that $\tilde{\mathcal{A}} = \mathcal{A}_\mu^*$ as the space of all continuous functionals $\tilde{A} : \phi \mapsto \langle \phi, \tilde{A} \rangle_\mu$ with respect to the $*$-norm $\|\phi\|_* = \sup\{|\mu(A\phi)| : \|A\| \leqslant 1\}$ on $\mathcal{A}_\mu$ and the pairing

$$\langle x^\dagger z, \tilde{A} \rangle_\mu = \mu(zAx^\dagger) = \langle A, \widetilde{x^\dagger z} \rangle_\mu, \quad x^\dagger z \in \mathcal{A}_\mu, \quad \tilde{A} \in \tilde{\mathcal{A}}.$$

The completion of $\mathcal{A}_\mu$ with respect to the norm $\|\cdot\|_*$ is the pre-dual Banach space, denoted as $\mathcal{A}_*$ (if $\mu = \tau \mid \mathcal{A}$ is the usual trace $\tau = \mathrm{tr}_\mathcal{G}$ on $\mathcal{A}$, then $\mathcal{A}_\mu$ coincides with $\mathcal{A}_*$ as the class $\mathcal{A}_\tau = \mathcal{A} \cap \mathcal{T}(\mathcal{G})$ of trace operators $\mathcal{T}(\mathcal{G}) = \{x^\dagger z : x, z \in \mathcal{S}(\mathcal{G})\}$, where $\mathcal{S}(\mathcal{G}) = \{x \in \mathcal{L}(\mathcal{G}) : \mathrm{tr}_\mathcal{G}\, x^\dagger x < \infty\}$).

If $\mathcal{A}$ is not the algebra of all operators $\mathcal{L}(\mathcal{G})$, the density operator ρ for a normal state (2.2) is not unique with respect to $\tau = \mathrm{tr}_\mathcal{G}$. However, it is uniquely defined as the bounded *probability density* $\rho = Jx^\dagger xJ = \bar{x}^\dagger \bar{x}$ with respect to the restriction $\mu = \tau \mid \mathcal{A}$ (i.e. as the density operator with respect to μ), describing this state as $\langle A, \rho \rangle_\mu = \mu(x\widetilde{Ax^\dagger})$ by the additional condition $\varkappa = \bar{x} \in \tilde{\mathcal{A}}_\mu$. Note that each probability density $\rho \in \tilde{\mathcal{A}}_\mu$, describing the normal state $\varrho(A) = \langle A, \rho \rangle_\mu$ on $\mathcal{A} \ni A$, is positive and normalized as $\langle I, \rho \rangle_\mu = 1$. However, the pre-dual space $\tilde{\mathcal{A}}_*$, as the $*$-completion of $\tilde{\mathcal{A}}_\mu$, may consist not only of the bounded densities with respect to μ (although each $\rho \in \tilde{\mathcal{A}}_*$ can always be approximated by the bounded $\rho_n \in \tilde{\mathcal{A}}_\mu$).

In the following formulation, $\mathcal{B}$ can also be the more general von Neumann algebra, rather than $\mathcal{L}(\mathcal{H})$, with a normal, faithful and semi-finite trace $\nu : \mathcal{B}_\nu \mapsto \mathbb{C}$, defining the pairing $\langle B, u^\dagger u \rangle_\nu = \nu(\tilde{u}^\dagger B \tilde{u})$, where $u \in \tilde{\mathfrak{b}}_\nu$ ($\mathcal{B}_\nu = \mathfrak{b}_\nu^\dagger \mathfrak{b}_\nu$ coincides with $\mathcal{B}_*$ in the case of the standard trace $\nu(B\tilde{\sigma}) = \mathrm{tr}\, B\tilde{\sigma} = \langle B, \sigma \rangle_\nu$, when $\mathfrak{b}_\nu$ is the space of Hilbert–Schmidt operators $y \in \mathcal{B}$ and $\tilde{\mathcal{B}} = \mathcal{B}$).

Theorem 2.1. *Let $\omega : \mathcal{A} \otimes \mathcal{B} \to \mathbb{C}$ be a normal compound state*

$$\omega(A \otimes B) = \tau(v(\widetilde{A \otimes B})v^\dagger) := \langle A \otimes B, v^\dagger v \rangle, \tag{2.3}$$

described by an amplitude operator $v : \mathcal{G} \otimes \mathcal{H} \to \mathcal{E} \otimes \mathcal{F}$ on the tensor product of Hilbert spaces $\mathcal{E}$ and $\mathcal{F}$, satisfying the condition

$$v^\dagger v \in \tilde{\mathcal{A}} \otimes \tilde{\mathcal{B}}, \quad \tau(vv^\dagger) = 1,$$

where $\tau \simeq \tilde{\mu} \otimes \tilde{\nu}$ is the trace $\tau(vv^\dagger) = \langle I \otimes I, v^\dagger v \rangle$, defined in (2.3) by the pairing for $\mathcal{A} \otimes \mathcal{B}$ with respect to $\mu \otimes \nu$. Then this state is achieved by an entangling operator $\varkappa : \mathcal{G} \otimes \mathcal{F} \to \mathcal{E} \otimes \mathcal{H}$ as

$$\langle A, \nu(\varkappa^\dagger (I \otimes B) \varkappa) \rangle_\mu = \omega(A \otimes B) = \langle B, \mu(\tilde{\varkappa}^\dagger (A \otimes I) \tilde{\varkappa}) \rangle_\nu, \tag{2.4}$$

for all $A \in \mathcal{A}$ and $B \in \mathcal{B}$, such that

$$\nu(\varkappa^\dagger (I \otimes B) \varkappa) \subseteq \tilde{A}, \quad \mu(\tilde{\varkappa}^\dagger (A \otimes I) \tilde{\varkappa}) \subseteq \tilde{\mathcal{B}}.$$

The operator $\varkappa$ together with $\tilde{\varkappa} = J\varkappa^\dagger J$ is uniquely defined by $v = U\varkappa'$, where

$$(\xi \otimes \eta')^\dagger \varkappa' (\zeta \otimes J\eta) = (\xi \otimes \eta)^\dagger \varkappa (\zeta \otimes J\eta'), \quad \xi \in \mathcal{E}, \quad \eta' \in \mathcal{F}, \quad \zeta \in \mathcal{G}, \quad \eta \in \mathcal{H}, \tag{2.5}$$

up to a unitary transformation U of the minimal subspace space $\operatorname{ran} v \subseteq \mathcal{E} \otimes \mathcal{F}$.

Proof. Without loss of generality, we can assume that $\mathcal{E} = \mathcal{G}_\rho$, $\mathcal{F} = \mathcal{H}_\sigma$ and $v^\dagger = v(E_\rho \otimes E_\sigma)$ as the support $(\mathcal{G} \otimes \mathcal{H})_{v^\dagger v} = \operatorname{ran} v^\dagger$ for $v^\dagger v$ is contained in $\mathcal{G}_\rho \otimes \mathcal{H}_\sigma$. Due to $v^\dagger v \in (\widetilde{\mathcal{A}'} \otimes \widetilde{\mathcal{B}'})'$, the range of v is invariant under the action

$$(A \otimes B)v = v(AE_\rho \otimes BE_\sigma), \quad \forall A \in \widetilde{\mathcal{A}'}, \quad B \in \widetilde{\mathcal{B}'}$$

of the commutant $(\tilde{\mathcal{A}} \otimes \widetilde{\mathcal{B}})' = \widetilde{\mathcal{A}'} \otimes \widetilde{\mathcal{B}'}$. Let us equip $\mathcal{G}$ and $\mathcal{H}$ with the involutions J leaving invariant $\mathcal{G}_\rho = E_\rho \mathcal{G}$ and $\mathcal{H}_\sigma = E_\sigma \mathcal{H}$, denoting $J_\rho = E_\rho J$, $J_\sigma = E_\sigma J$, and $\mathcal{E} \otimes \mathcal{F} = \mathcal{G}_\rho \otimes \mathcal{H}_\sigma$ with the induced involution $J(\zeta \otimes \eta) = J_\rho \zeta \otimes J_\sigma \eta$. It is easy to check for such v and $\varkappa = v'$, defined by $v = \varkappa'$ in (2.5), that for any $A \in \mathcal{A}'$ and $B \in \widetilde{\mathcal{B}'}$

$$\begin{aligned}(\tilde{A}\xi \otimes \eta)^\dagger \varkappa(\zeta \otimes \bar{B} J \eta') &= (\tilde{A}\xi \otimes B\eta')^\dagger v(\xi \otimes J\eta) = (\xi \otimes \eta')^\dagger v(\bar{A}\xi \otimes J\tilde{B}\eta) \\ &= (\xi \otimes \tilde{B}\eta)^\dagger \varkappa(\bar{A}\zeta \otimes J\eta'),\end{aligned}$$

where $\bar{A} = JAJ \in \widetilde{\mathcal{A}'}$, $\bar{B} = JBJ \in \mathcal{B}'$. Hence, for any $B \in \mathcal{B}$,

$$(A \otimes B')\varkappa^\dagger (I \otimes B)\varkappa = \varkappa^\dagger (A \otimes B'B)\varkappa = \varkappa^\dagger (I \otimes B)\varkappa(A \otimes B'),$$

where $A \in \widetilde{\mathcal{A}'_\rho} := \widetilde{\mathcal{A}'} E_\rho$, $B' \in \mathcal{B}'_\sigma := \mathcal{B}' E_\sigma$, and, for any $A \in \mathcal{A}$,

$$(A' \otimes B)\tilde{\varkappa}^\dagger (A \otimes I)\tilde{\varkappa} = \varkappa(A'A \otimes B)\varkappa^\dagger = \tilde{\varkappa}^\dagger (A \otimes I)\tilde{\varkappa}(A' \otimes B),$$

where $A' \in \mathcal{A}'$ and $B \in \widetilde{\mathcal{B}'}$. Thus, for all $A \in \mathcal{A}$ and $B \in \mathcal{B}$,

$$\varkappa^\dagger (I \otimes B)\varkappa \in (\widetilde{\mathcal{A}'_\rho} \otimes \mathcal{B}'_\sigma)', \qquad \tilde{\varkappa}^\dagger (A \otimes I)\tilde{\varkappa} \in (\mathcal{A}'_\rho \otimes \widetilde{\mathcal{B}'_\sigma})'.$$

Moreover, due to $\mathcal{A}''_\rho = E_\rho \mathcal{A} E_\rho \equiv \mathcal{A}_\rho$ and $\mathcal{B}''_\sigma = E_\sigma \mathcal{B} E_\sigma \equiv \mathcal{B}_\sigma$,

$$\begin{aligned}\varkappa^\dagger (I \otimes B)\varkappa &\subseteq J_\rho \mathcal{A}_\mu J_\rho \otimes E_\sigma \mathcal{B}_\nu E_\sigma := (\widetilde{\mathcal{A}_\rho} \otimes \mathcal{B}_\sigma)_{\tilde{\mu} \otimes \nu}, \\ \tilde{\varkappa}^\dagger (A \otimes I)\tilde{\varkappa} &\subseteq E_\rho \mathcal{A}_\mu E_\rho \otimes J_\sigma \mathcal{B}_\nu J_\sigma := (\mathcal{A}_\rho \otimes \widetilde{\mathcal{B}_\sigma})_{\mu \otimes \tilde{\nu}},\end{aligned}$$

as bounded by $\|B\|\varkappa^\dagger \varkappa$ and by $\|A\|\tilde{\varkappa}^\dagger \tilde{\varkappa}$, respectively. The partial traces ν and μ on these reduced algebras are defined as

$$\nu(\varkappa^\dagger (I \otimes B)\varkappa) = \langle B, v^\dagger v\rangle_\nu, \qquad \mu(\tilde{\varkappa}^\dagger (A \otimes I)\tilde{\varkappa}) = \langle A, v^\dagger v\rangle_\mu, \tag{2.6}$$

according to $\langle A, \langle B, v^\dagger v\rangle_\nu\rangle_\mu = \langle A \otimes B, v^\dagger v\rangle = \langle B, \langle A, v^\dagger v\rangle_\mu\rangle_\nu$, where

$$\langle B, v^\dagger v\rangle_\nu = \nu(\widetilde{(I \otimes B)}\widetilde{v^\dagger v}), \qquad \langle A, v^\dagger v\rangle_\mu = \mu(\widetilde{(A \otimes I)}\widetilde{v^\dagger v}).$$

In particular,

$$\nu(\varkappa^\dagger \varkappa) = \tilde{\nu}(v^\dagger v) = \rho, \qquad \mu(\tilde{\varkappa}^\dagger \tilde{\varkappa}) = \tilde{\mu}(v^\dagger v) = \sigma.$$

Any other choice of v with the minimal $\mathcal{E} \otimes \mathcal{F} \simeq \mathcal{G}_\rho \otimes \mathcal{H}_\sigma$ is unitary equivalent to $\varkappa'$. ■

Note that the entangled state (2.3) is written in (2.4) as

$$\langle B, \varpi(A)\rangle_\nu = \omega(A \otimes B) = \langle A, \varpi^\intercal (B)\rangle_\mu,$$

in terms of the mutually adjoint maps $\varpi : \mathcal{A} \to \mathcal{B}_*$ and $\varpi^{\intercal} : \mathcal{B} \to \mathcal{A}_*$. These maps are given in (2.6) as

$$\varpi(A) = \langle A, \upsilon^\dagger \upsilon \rangle_\mu = \widetilde{\pi^*(A)}, \qquad \varpi^{\intercal}(B) = \langle B, \upsilon^\dagger \upsilon \rangle_\nu = \widetilde{\pi(B)}, \tag{2.7}$$

where the linear map $\pi : \mathcal{B} \to \mathcal{A}_\mu$ and the adjoint $\pi^* : \mathcal{A} \to \mathcal{B}_\nu$ are defined as partial traces

$$\pi(B) = \nu((I \otimes B)\widetilde{\upsilon^\dagger \upsilon}), \qquad \pi^*(A) = \mu((A \otimes I)\widetilde{\upsilon^\dagger \upsilon}).$$

The linear normal map ϖ in (2.6) is written in the Kraus–Stinespring form (Stinespring 1955) and thus is completely positive (CP). It is not unital but normalized to the density operators $\sigma = \omega(I)$ with respect to the weight ν.

A linear map $\pi : \mathcal{B} \to \mathcal{A}_*$ is called *tilde-positive* if the map $\pi^{\tilde{}}$ defined as $\pi^{\tilde{}}(B) := J\pi(B)^\dagger J$ is positive for any positive (and thus Hermitian) operator $B \geqslant 0$ in the sense of non-negative definiteness of B. It is called *tilde-completely positive* (TCP) if the operator-matrix $\pi^{\tilde{}}(\boldsymbol{B}) = J\pi(\boldsymbol{B})^\dagger J$ is positive for every positive operator-matrix $\boldsymbol{B} = [B_{ik}] = \boldsymbol{B}^*$, where $\boldsymbol{A}^\dagger = [A^\dagger_{ik}]$, $\boldsymbol{B}^* = [B^\dagger_{ki}]$ (and, thus, $\boldsymbol{A}^\dagger = [A_{ki}]$ for $\boldsymbol{A} = [A_{ik}] \geqslant 0$, and $\boldsymbol{B}^* = \boldsymbol{B}$ for $\boldsymbol{B} \geqslant 0$). Obviously, every tilde-positive and tilde-completely positive π is positive, since positive is $\tilde{A} = JA^\dagger J$ for every positive A, but it is not necessarily completely positive unless $\tilde{A} = A$, for all $A \in \mathcal{A}$, in which case $\mathcal{A}$ is Abelian (or the Abelian is $\mathcal{B}$).

The map π defined in equation (2.8) as a TCP †-map, $\pi(B^\dagger) = \pi(B)^\dagger$, is obviously transpose-CP in the sense of positivity of $\pi(\boldsymbol{B})^\dagger = [\pi(B_{ki})] = \pi(\boldsymbol{B}^\dagger)$ for any $\boldsymbol{B} \geqslant 0$, but it is in general not CP. Because every transpose-CP map can be represented as tilde-CP, there might be a positive-definite matrix $\boldsymbol{B}$ for which $\pi(\boldsymbol{B})$ is not positive. Note that the adjoint map $\pi^* = \bar{\pi}^{\intercal}$ is also TCP, as well as the maps $\tilde{\pi} = \bar{\pi}$ and $\pi^{\intercal} = \bar{\pi}^*$, where $\bar{\pi}(B) = J\pi(\bar{B})J$, obtained from (2.6) as partial tracings,

$$\bar{\pi}(B) = \nu(\varkappa^\dagger (I \otimes \tilde{B})\varkappa), \qquad \pi^{\intercal}(A) = \mu(\tilde{\varkappa}^\dagger (\tilde{A} \otimes I)\tilde{\varkappa}). \tag{2.8}$$

In these terms, the compound state (2.4) is written as

$$\langle A | \pi(B) \rangle_\mu = \omega(A^\dagger \otimes B) = \langle \pi^*(A) | B \rangle_\nu,$$

where $\langle x | y \rangle = \langle y, \bar{x} \rangle$ defines an inner product which coincides in the case of traces with the GNS product $(x \mid y)$.

In the following definition, the pre-dual space $\mathcal{B}_{\intercal} = \tilde{\mathcal{B}}_*$ (as well as $\mathcal{A}_{\intercal} = \tilde{\mathcal{A}}_*$) is identified by the pairing $\langle B, \sigma \rangle_\nu = \varsigma(B)$ with the space of generalized density operators σ, which are thus uniquely defined as selfadjoint (unbounded) operators in $\mathcal{H}$. Note that $\mathcal{B}_{\intercal} = \mathcal{B}_\nu$ if $\mathcal{B} = \tilde{\mathcal{B}}$ and $\nu = \mathrm{tr}_{\mathcal{H}} = \tilde{\nu}$.

Definition 2.2. A TCP map $\pi : \mathcal{B} \to \mathcal{A}_*$ (or $\mathcal{B} \to \mathcal{A}_\mu \subseteq \mathcal{A}_*$) normalized as $\mu(\pi(I)) = 1$ and having an adjoint with $\pi^*(\mathcal{A}) \subseteq \mathcal{B}_*$ ($\pi^*(\mathcal{A}) \subseteq \mathcal{B}_\nu$) is called normal coupling (bounded coupling) of the state $\varsigma = \mu \circ \pi$ on $\mathcal{B}$ to the state $\varrho = \nu \circ \pi^*$ on $\mathcal{A}$. The CP map $\varpi : \mathcal{A} \to \mathcal{B}_{\intercal}$ (or $\mathcal{A} \to \widetilde{\mathcal{B}_\nu} \subseteq \mathcal{B}_{\intercal}$), normalized to the probability density $\sigma = \varpi(I)$ of ς with

$$\varpi^{\intercal}(I) \in \mathcal{B}_* \quad (\varpi^{\intercal}(I) \in \widetilde{\mathcal{A}_\mu}),$$

will be called normal entanglement (bounded entanglement) of the system $(\mathcal{A}, \varrho)$ with the probability density $\rho = \varpi^{\intercal}(I)$ to $(\mathcal{B}, \varsigma)$. The coupling π (entanglement ϖ)

is called truly quantum if it is not CP (not TCP). The self-adjoint entanglement $\varpi_q = \varpi_q^*$ on $(\mathcal{A}, \varrho) = (\tilde{\mathcal{B}}, \tilde{\varsigma})$ (or symmetric coupling $\pi_q = \pi_q^\intercal$ into $\mathcal{A}_* = \mathcal{B}_\intercal$) is called standard for the system $(\mathcal{B}, \varsigma)$ if it is given by

$$\varpi_q(A) = \sigma^{1/2} A \sigma^{1/2}, \qquad \pi_q(B) = \sigma^{1/2} \tilde{B} \sigma^{1/2}. \tag{2.9}$$

Note that the standard entanglement is true as soon as the reduced algebra $\mathcal{B}_\sigma = E_\sigma \mathcal{B} E_\sigma$ on the support $\mathcal{H}_\sigma = E_\sigma \mathcal{H}$ of the state ς is not Abelian, i.e. is not one dimensional in the case $\mathcal{B} = \mathcal{L}(\mathcal{H})$, corresponding to a pure normal ς on $\mathcal{B} = \mathcal{L}(\mathcal{H})$. Indeed, π^q, restricted to $\mathcal{B}_\sigma$, is the composition of the non-degenerated multiplication $\mathcal{B}_\sigma \ni B \mapsto \tilde{\sigma}^{1/2} B\, \tilde{\sigma}^{1/2}$ (which is CP) and the transposition $\tilde{B} = JB^\dagger J$ on $\mathcal{B}_\sigma$ (which is TCP but not CP if $\dim \mathcal{H}_\sigma > 1$).

The standard entanglement in the purely quantum case $\mathcal{B} = \mathcal{B}(\mathcal{H}) = \tilde{\mathcal{B}}$, $\nu = \mathrm{tr} = \tilde{\nu}$ corresponds to the pure *standard compound state*

$$\mathrm{tr}\, A\sigma^{1/2} \tilde{B} \sigma^{1/2} = \omega_q(A \otimes B) = \mathrm{tr}\, B \tilde{\sigma}^{1/2} \tilde{A} \tilde{\sigma}^{1/2} \tag{2.10}$$

on the algebra $\mathcal{B} \otimes \mathcal{B}$. It is given by the amplitude $v' \simeq |\sigma^{1/2}) \equiv \psi$, with $|\sigma^{1/2})^\dagger = \varkappa' \equiv (\sigma^{1/2}|$ defined in (2.5) as $\varkappa'(\zeta \otimes J\eta) = \eta^\dagger \varkappa \zeta$ for $\varkappa = \sigma^{1/2}$.

Any entanglement on $\mathcal{A} = \mathcal{L}(\mathcal{G})$, $\mu = \mathrm{tr}$ corresponding to a pure compound state is true if $\operatorname{rank} \rho = \operatorname{rank} \sigma$ is not one. If the space $\mathcal{G}$ is also minimal, $\mathcal{G} = \mathcal{G}_\rho$, $\pi^\intercal$ is unitary equivalent to the standard one π_q. Indeed, $\varpi(A) = \tilde{\varkappa}^\dagger A \tilde{\varkappa}$ can be decomposed as

$$\varpi(A) = \sigma^{1/2} U^\dagger A U \sigma^{1/2} = \varpi_q(U^\dagger A U),$$

where $U : \sigma^{1/2}\eta \mapsto \tilde{\varkappa}\eta$ is a unitary operator from $\mathcal{H}_\sigma$ onto the support $\mathcal{G}_\rho$ of $\rho = U\sigma U^\dagger$ with non-Abelian $\mathcal{A}_\rho = \mathcal{L}(\mathcal{G}_\rho)$ and $\mathcal{B}_\sigma = U^\dagger \mathcal{A}_\rho U = \mathcal{L}(\mathcal{H}_\sigma)$.

Note that the compound state (2.4) with $\tilde{\varkappa} = \sigma^{1/2}$, corresponding to the standard $\varpi = \varpi_q$, can always be extended to a vector state on $\tilde{\mathcal{B}} \vee \mathcal{B}$ in the standard representation $(\mathcal{H}_\nu, \iota, J_\nu)$ of $\mathcal{B} \equiv \iota(\mathcal{B})$, when $\tilde{\mathcal{B}} = J_\nu \mathcal{B} J_\nu = \mathcal{B}'$. However, it cannot be extended to a normal state on $\tilde{\mathcal{B}} \otimes \mathcal{B}$ in the case of non-atomic $\mathcal{B}$. If $\mathcal{B}$ is a factor, this state is pure, given in the standard representation $\tilde{\mathcal{B}} \vee \mathcal{B} = \mathcal{L}(\mathcal{H}_\nu)$ by the unit vector $y = \tilde{\sigma}^{1/2} \in \mathcal{H}_\nu$. However, it is not normal on $\tilde{\mathcal{B}} \otimes \mathcal{B}$ unless $\mathcal{B}$ is type I: $\mathcal{B} \simeq \mathcal{L}(\mathcal{H})$.

3. c-, d- and o-couplings and encodings

In this section, we discuss the operational meaning of couplings corresponding to different types of encodings, which are treated here solely in terms of coupling maps on input of a quantum physical system. We hope that this mathematical treatment will provide a new physical insight for the corresponding asymptotic problems of quantum information.

The compound states play the role of joint input–output probability measures in classical information channels and can be pure in the quantum case, even if the marginal states are mixed. The pure compound states achieved by an entanglement of mixed input and output states exhibit new, non-classical type correlations, which are responsible for the EPR-type paradoxes in the interpretation of quantum theory (Werner 1989). However, mixed, so-called *separable*, states on $\mathcal{A} \otimes \mathcal{B}$, defined as convex product combinations

$$\omega_c(A \otimes B) = \sum_n \varrho_n(A) \varsigma_n(B) p(n),$$

which we refer as the *c-compound states*, do not exhibit such paradoxical behaviour. Here, $p(n) > 0$, $\sum p_n = 1$ is a probability distribution, and $\varrho_n : \mathcal{A} \to \mathbb{C}$, $\varsigma_n : \mathcal{B} \to \mathbb{C}$ are usually normal states defined by the product densities $\rho_n \otimes \sigma_n \in \mathcal{A}_{\mathrm{T}} \otimes \mathcal{B}_{\mathrm{T}}$ of $\omega_n = \varrho_n \otimes \sigma_n$. Such compound states are achieved by *c-couplings* $\pi_{\mathrm{c}} : \mathcal{B} \to \mathcal{A}_*$, given by $\pi_{\mathrm{c}} = \varpi_{\mathrm{c}}^{\mathrm{T}}$, where

$$\varpi_{\mathrm{c}}(A) = \sum_n \varrho_n(A)\sigma_n p(n), \qquad \varpi_{\mathrm{c}}^{\mathrm{T}}(B) = \sum_n \varsigma_n(B)\rho_n p(n).$$

Here, $\rho_n \in \mathcal{A}_*$ and $\sigma_n \in \mathcal{B}_*$ are the probability densities for ϱ_n and ς_n, with respect to given traces μ and ν on $\mathcal{A}$ and $\mathcal{B}$. Note that the *c-entanglement* ϖ_{c}, being the convex combinations of the primitive CP-TCP maps $\varpi_n(A) = \varrho_n(A)\sigma_n \in \mathcal{B}_{\mathrm{T}}$, is not truly quantum.

The separable states of the particular form

$$\omega_{\mathrm{d}}(A \otimes B) = \sum_n \langle n|A|n\rangle \varsigma(n, B), \tag{3.1}$$

where $\varrho_n(A) = \langle n|A|n\rangle$, are pure states on $\mathcal{A} = \mathcal{L}(\mathcal{G}) = \tilde{\mathcal{A}}$, given by an orthonormal system $\{|n\rangle\} \subset \mathcal{G}$, and $\varsigma(n, B) = \langle B, \sigma(n)\rangle_\nu$ with $\sigma(n) = \sigma_n p(n)$, are usually considered as the proper candidates for the input–output states in the communication channels involving the classical–quantum (c–q) encodings. Such a separable state was introduced by Ohya (1983*a*, 1989) using a Schatten decomposition $\rho = \sum |n\rangle\langle n|p(n)$ of the input density operator $\rho \in \mathcal{T}(\mathcal{G})$ into the orthogonal one-dimensional projectors $\rho_n = |n\rangle\langle n|$. Here, we note that such a state is the mixture of the classical–quantum correspondences $n \mapsto |n\rangle\langle n| \otimes \sigma_n$, which can be described as the composition of quantum channelling $|n\rangle\langle A01\rangle$ and the errorless *encodings* $n \mapsto |n\rangle\langle n|$, in the sense that they can be inverted by the measurements $|n\rangle\langle n| \mapsto n$ as input *decodings*. We shall call such separable states *d-compound* as they are achieved by the diagonal couplings $\pi_{\mathrm{d}} = \varpi_{\mathrm{d}}^{\mathrm{T}}$ (*d-couplings*) to the subalgebra $\mathcal{A}_{\mathrm{d}} \subseteq \mathcal{A}$ of the diagonal operators $A = \sum a(n)|n\rangle\langle n|$, where

$$\varpi_{\mathrm{d}}(A) = \sum_n \langle n|A|n\rangle \sigma(n), \qquad \varpi_{\mathrm{d}}^{\mathrm{T}}(B) = \sum_n \varsigma(n, B)|n\rangle\langle n|, \tag{3.2}$$

with respect to the standard transposition $\langle n|\tilde{A}|m\rangle = \langle m|A|n\rangle$ in the eigenbasis of ρ.

Actually, Ohya obtained the compound states ω_{d} as the result of the composition

$$\omega_{\mathrm{d}}(A \otimes B) = \omega_{\mathrm{o}}(A \otimes \Lambda(B))$$

of quantum channels as normal unital CP maps $\Lambda : \mathcal{B} \to \mathcal{A}$ and the special, *o-compound* states

$$\omega_{\mathrm{o}}(A \otimes B) = \sum_n \langle n|A|n\rangle p(n)\langle n|B|n\rangle, \tag{3.3}$$

corresponding to the orthogonal decompositions

$$\varpi_{\mathrm{o}}(A) = \sum_n \langle n|A|n\rangle p(n)|n\rangle\langle n| = \varpi_0^{\mathrm{T}}(A), \tag{3.4}$$

such that

$$\varsigma_n(B) = \langle n|\Lambda(B)|n\rangle, \qquad \sigma_n = \Lambda^{\mathrm{T}}(|n\rangle\langle n|),$$

where

$$\langle B, \Lambda^{\intercal}(\rho)\rangle_{\nu} = \mathrm{tr}_{\mathcal{G}}\, \Lambda(B)\tilde{\rho}.$$

Assuming that $\langle A, \rho\rangle = \mathrm{tr}_{\mathcal{G}}\, A\tilde{\rho}$, we can extend this construction to any discretely decomposable algebra $\mathcal{A} = \tilde{\mathcal{A}}$ on the Hilbert sum $\mathcal{G} = \oplus\mathcal{G}_i$ with invariant components $\mathcal{G}_i$ under the standard complex conjugation J in the eigenbasis of the density operator $\tilde{\rho} = J\rho J = \rho$. In particular, the von Neumann algebra $\mathcal{A}$ might be Abelian, as it is in the case $\tilde{A} = A$, for all $A \in \mathcal{A}$, e.g. when $\mathcal{A} = \tilde{\mathcal{A}}$ is the diagonal algebra of pointwise multiplications $Ag = ag = \tilde{A}g$ by the bounded functions $n \mapsto a(n) \in \mathbb{C}$ on the functional Hilbert space $\mathcal{G} = \ell^2 \ni g$ with the standard complex conjugation $Jg = \bar{g}$. In this case, the densities $\rho \in \mathcal{A}_*$ are given by the summable functions $p \in \ell^1$ with respect to the standard trace $\mu(\rho) = \sum p(n)$, and any compound state has the separable form with $\varrho_n(A) = a(n)$, corresponding to the Kronecker δ-densities $\rho_n \simeq \delta_n$. The normal states on the $\mathcal{A} \simeq \ell^\infty$ are described by the probability densities $p(n) \geqslant 0$, $\sum p(n) = 1$, with respect to the standard pairing

$$\langle A, \rho\rangle_\mu = \sum a(n)p(n), \quad p \in \ell^1, \quad a \in \ell^\infty$$

of $\mathcal{A}_\mu = \mathcal{A}_*$ with the commutative algebra $\mathcal{A}$. Every normal compound state ω on $\mathcal{A} \otimes \mathcal{B}$ is defined by

$$\omega_{\mathrm{c}}(A \otimes B) = \sum_n a(n)\langle B, \sigma(n)\rangle_\nu,$$

where $\sigma(n) = \sigma_n p(n)$ is the function with positive values $\sigma(n) \in \mathcal{B}_{\intercal}$ normalized to the probability density $p(n) = \langle I, \sigma(n)\rangle_\nu$. Thus, all normal compound states on $\ell^\infty \otimes \mathcal{B}$ are achieved by *c-couplings* $\pi_{\mathrm{c}} = \varpi_{\mathrm{c}}^{\intercal} : \mathcal{B} \to \ell^1$ with $\pi_{\mathrm{c}}^{\intercal} = \varpi_{\mathrm{c}}$ given by convex combinations of the primitive CP (and TCP) maps $\varpi_n(a) = a(n)\sigma_n \in \mathcal{B}_*$,

$$\varpi_{\mathrm{c}}(A) = \sum_n a(n)\sigma(n), \varpi_{\mathrm{c}}^{\intercal}(B) = \sum_n \varsigma(n, B)\delta_n,$$

where $\varsigma(n, B) = \langle B, \sigma(n)\rangle_\nu$.

Note that any d-coupling can be regarded as quantum-classical c-coupling, achieved by the identification $a(n) = \langle n|A|n\rangle$ of $\ell^\infty \ni a$ and the reduced diagonal algebra $\mathcal{A}^0 = \{\sum |n\rangle a(n)\langle n| : A \in \mathcal{A}\}$. This follows simply from the commutativity of the density operators $\rho = \sum |n\rangle\langle n|p(n)$ for the induced states $\varrho(A) = \omega_{\mathrm{d}}(A \otimes I)$ identified with $p \in \ell^1$.

In the case $\mathcal{A} = \mathcal{L}(\mathcal{G})$ and pure elementary states ω_n described by probability amplitudes $v_n = \chi_n \otimes \psi_n$, where $\tilde{\chi}_n \equiv |\chi_n\rangle \in \mathcal{G}$, $\tilde{\psi}_n \equiv |\psi_n\rangle \in \mathcal{H}$, we have density operators $\rho_n = \chi_n^\dagger \chi_n$ and $\sigma_n = \psi_n^\dagger \psi_n$ of rank one. The total compound amplitude is obviously $v = \sum |n\rangle v(n)$, where $v(n) = \chi_n \otimes \psi_n p(n)^{1/2}$ are the amplitude operators $\mathcal{G} \otimes \mathcal{H} \to \ell^2$ satisfying the orthogonality relations

$$v(n)^\dagger v(m) = \rho_n \otimes \sigma_n p(n)\delta_n^m$$

corresponding to the decomposition $v^\dagger v = \sum \rho_n \otimes \sigma_n p(n)$. The 'entangling' operator for the separable state $\varkappa$ can be chosen as either $\varkappa = \sum |n\rangle \varkappa(n)$ or as $\varkappa = \sum \varkappa(n)\langle n|$, or even as $\varkappa = \sum |n\rangle\varkappa(n)\langle n|$ with $\varkappa(n) = \chi_n \otimes \tilde{\psi}(n)$, where $\tilde{\psi}_n(n) = \tilde{\psi}_n p(n)^{1/2}$. In particular, a d-entangling operator $\varkappa$ corresponding to d-encodings (3.2) is diagonal, $\varkappa = \sum |n\rangle\tilde{\psi}(n)\langle n|$ on $\mathcal{G} = \ell^2$, corresponding to the orthogonal $\tilde{\chi}_n = |n\rangle$. Thus, we have proved theorem 3.1, below, in the case of pure states ς_n and ϱ_n. But, before formulating this theorem in a natural generality, let us introduce the following notations.

The general c-compound states on $\mathcal{A} \otimes \mathcal{B}$ are defined as integral convex combinations

$$\omega(A \otimes B) = \int \varrho_x(A)\varsigma_x(B)p(\mathrm{d}x),$$

given by a probability distribution p on the product states $\varrho_x \otimes \varsigma_x$. Such compound states are achieved by convex combinations of the primitive CP (and TCP) maps $\pi_x^- = \varpi_x^\intercal$ with $\varpi_x(A) = \varrho_x(A)\sigma_x$:

$$\varpi_{\mathrm{c}}(A) = \int \varrho_x(A)\sigma_x p(\mathrm{d}x), \varpi_{\mathrm{c}}^\intercal(B) = \int \varsigma_x(B)\rho_x p(\mathrm{d}x). \tag{3.5}$$

This is always the case when the von Neumann algebra $\mathcal{A}$ is Abelian, and, thus, can be identified with the diagonal algebra of multiplications $(Ag)(x) = a(x)g(x)$ by the functions $a \in L_\mu^\infty$ on the functional Hilbert space $\mathcal{G} = L_\mu^2$ with respect to a (not necessarily finite) measure μ on X. It defines trace μ on $\mathcal{A}_\mu \simeq L_\mu^1 \cap L_\mu^\infty$ as the integral $\mu(\rho) = \int p(x)\mu(\mathrm{d}x)$ for the bounded multiplication densities $(\rho g)(x) = p(x)g(x)$. The normal states on $\mathcal{A}$ are given by the probability densities $p \in L_\mu^1$ with respect to the standard pairing

$$\langle A, \rho\rangle_\mu = \int a(x)p(x)\mu(\mathrm{d}x), \quad p \in L_\mu^1, \quad a \in L_\mu^\infty$$

of $\mathcal{A}_* = \mathcal{A}_\intercal \simeq L_\mu^1$ and $\mathcal{A} = \tilde{\mathcal{A}} \simeq L_\mu^\infty$ corresponding to the trivial transposition $\tilde{a} = a$. Any normal compound state ω on $\mathcal{A} \otimes \mathcal{B} \simeq L_\mu^\infty(X \to \mathcal{B})$ is the c-compound state, defined on the diagonal algebra $\mathcal{A}$ by

$$\omega_{\mathrm{d}}(A \otimes B) = \int a(x)\varsigma(x, B)\mu(\mathrm{d}x), \tag{3.6}$$

where $\varsigma(x, B) = \langle B, \sigma(x)\rangle_\nu$ is an absolutely integrable function with density operator values $\sigma(x) = \sigma_x p(x)$ normalized to the probability density $p(x) = \langle I, \sigma(x)\rangle_\nu = \varsigma(x, I)$. It corresponds to d-couplings $\pi_{\mathrm{d}} = \varpi_{\mathrm{d}}^\intercal = \pi_{\mathrm{d}}^-$ with $\pi_{\mathrm{d}}^\intercal = \varpi_{\mathrm{d}}$ decomposing into $\varpi(x, A) = a(x)\sigma(x)$:

$$\varpi_{\mathrm{d}}(A) = \int a(x)\sigma(x)\mu(\mathrm{d}x), \qquad \varpi_{\mathrm{d}}^\intercal(B) = \int \varsigma(x, B)\delta_x\mu(\mathrm{d}x), \tag{3.7}$$

where δ_x is the (generalized) density operator of the Dirac state $\varrho_x(A) = \langle A, \delta_x\rangle_\mu = a(x)$ on the diagonal algebra $\mathcal{A}$.

Theorem 3.1. *Let $\omega_{\mathrm{c}} : \mathcal{A} \otimes \mathcal{B} \to \mathbb{C}$ be a normal c-compound state given as*

$$\omega_{\mathrm{c}}(A \otimes B) = \int \mu_x(\tilde{\chi}_\cdot^\dagger A\tilde{\chi}_\cdot)\nu_x(\tilde{\psi}_\cdot^\dagger B\tilde{\psi}_\cdot)p(\mathrm{d}x), \tag{3.8}$$

where $\chi_x : \mathcal{G} \to \mathcal{E}_x$, $\psi_x : \mathcal{H} \to \mathcal{F}_x$ are linear operators having bounded transpose $\tilde{\chi}_\cdot = J\chi_\cdot^\dagger J_\cdot$, $\tilde{\psi}_\cdot = J\psi_\cdot^\dagger J_\cdot$ on Hilbert spaces

$$\mathcal{E}_\cdot = \int^\oplus \mathcal{E}_x p(\mathrm{d}x), \qquad \mathcal{F}_\cdot = \int^\oplus \mathcal{F}_x p(\mathrm{d}x)$$

with respect to pointwise involution $J_\cdot = J_\cdot^\dagger$. *We also assume that*

$$\chi_x^\dagger \chi_x \in \tilde{\mathcal{A}}, \quad \psi_x^\dagger \psi_x \in \tilde{\mathcal{B}}, \quad \mu_x(\tilde{\chi}_\cdot^\dagger \tilde{\chi}_\cdot) = 1 = \nu_x(\tilde{\psi}_\cdot^\dagger \tilde{\psi}_\cdot)$$

with respect to the traces

$$\mu_x(\tilde{\chi}_\cdot^\dagger \tilde{\chi}_\cdot) = \langle I, \chi_x^\dagger \chi_x \rangle_\mu, \qquad \nu_x(\tilde{\psi}_\cdot^\dagger \tilde{\psi}_\cdot) = \langle I, \psi_x^\dagger \psi_x \rangle_\nu. \tag{3.9}$$

Then this state is achieved by a decomposable entangling operator $\varkappa = \int^\oplus \chi_x \otimes \tilde{\psi}_x p(\mathrm{d}x)$ *defining c-entanglement (3.5) with*

$$\varrho_x(A) = \mu_x(\tilde{\chi}_\cdot^\dagger A \tilde{\chi}_\cdot), \qquad \varsigma_x(B) = \nu_x(\tilde{\psi}_\cdot^\dagger A \tilde{\psi}_\cdot), \tag{3.10}$$

corresponding to the probability densities $\rho_x = \chi_x^\dagger \chi_x$, $\sigma_x = \psi_x^\dagger \psi_x$. *In particular, every d-compound state (3.6) corresponding to* $p(\mathrm{d}x) = p(x)\mu(\mathrm{d}x)$ *with the Abelian algebra* $\mathcal{A}$ *can be achieved by the orthogonal sum of entangling operators* $\varkappa_x = \delta_x \otimes \tilde{\psi}_x$ *defining d-entanglement (3.7) with*

$$\sigma(x) = \psi_x^\dagger \psi_x p(x), \qquad \varsigma(x, B) = \nu_x(\tilde{\psi}_\cdot^\dagger A \tilde{\psi}_\cdot) p(x).$$

Proof. The amplitude operator $v = \int^\otimes v_x p(\mathrm{d}x)$, corresponding to c-compound state (3.8), is defined as the orthogonal sum of $v_x = \chi_x \otimes \psi_x$ on $\mathcal{G} \otimes \mathcal{H}$ into $\int^\oplus \mathcal{E}_x \otimes \mathcal{F}_x p(\mathrm{d}x)$. Without loss of generality, we can assume that $\mathcal{E}_x = \mathcal{G}_\rho$, $\mathcal{F}_x = \mathcal{H}_\sigma$ and $v_x^\dagger = v_x(E_\rho \otimes E_\sigma)$ because the support $(\mathcal{G} \otimes \mathcal{H})_{v_x^\dagger v_x} = \operatorname{ran} v_x^\dagger$ for

$$v_x^\dagger v_x = \chi_x^\dagger \chi_x \otimes \psi_x^\dagger \psi_x = \rho_x \otimes \sigma_x$$

is in $\mathcal{G}_\rho \otimes \mathcal{H}_\sigma$. Due to

$$\chi_x^\dagger \chi_x \in \widetilde{\mathcal{A}'}, \qquad \psi_x^\dagger \psi_x \in \widetilde{\mathcal{B}}'$$

for almost all x, the operators χ_x and ψ_x commute with $A \in \widetilde{\mathcal{A}'}$ and $B \in \widetilde{\mathcal{B}}'$, respectively, and $\tilde{\psi}_x$ commutes with $B \in \mathcal{B}'$ for almost all x. Thus,

$$\tilde{\chi}_x^\dagger A \tilde{\chi}_x \in \mathcal{A}, \qquad \tilde{\psi}_x B \tilde{\psi}_x \in \mathcal{B},$$

which defines the traces (3.9) on $L_p^\infty \otimes \mathcal{A}$ and $L_p^\infty \otimes \mathcal{B}$ for almost all x. The rest of the proof is a repetition of the proof of theorem 2.1 for each x, with the addition that $\varkappa_x$ is the product $v_x' = \chi_x \otimes \tilde{\psi}_x$ for each x. The total entangling operator $\varkappa : \mathcal{G} \otimes \mathcal{F}_\cdot \to \mathcal{E}_\cdot \otimes \mathcal{H}$ acts componentwise as $\varkappa_x(\zeta \otimes \eta_\cdot) = \chi_x \zeta \otimes \tilde{\psi}_x \eta_x$.

In the case of d-compound state (3.6), one should take $\mathcal{G} = L_\mu^2$, $\mathcal{E}_x = \mathbb{C}$, and $\chi_x g = g(x)$. Thus, the entangling operator in this case is given as

$$\varkappa(g \otimes \eta_\cdot) = \int^\otimes g(x) \tilde{\psi}_x \eta_x \mu(\mathrm{d}x), \quad \forall g \in L_\mu^2, \quad \eta_\cdot = \int^\oplus \eta_x \mu(\mathrm{d}x) \in \mathcal{F}_\cdot.$$

∎

Note that c-entanglements ϖ_c in (3.5) are both CP and TCP and are thus not true quantum entanglements. The map $\varpi_c : \mathcal{A} \to \mathcal{B}_\mathrm{T}$ with Abelian algebra $\mathcal{A}$ in (3.7) is described by a $\mathcal{B}_\mathrm{T}$-valued measure $\sigma(\mathrm{d}x) = \sigma(x)\mu(\mathrm{d}x)$ normalized to the input probability measure as $p(\mathrm{d}x) = \langle I, \sigma(\mathrm{d}x) \rangle_\nu$. This gives the concise form for the description of random classical–quantum state correspondences $x \mapsto \sigma_x$ with the given probability measure p, called *encodings* of $\sigma = \int \sigma(\mathrm{d}x)$.

Definition 3.2. Let both algebras $\mathcal{A}$ and $\mathcal{B}$ be non-Abelian. The map $\varpi : \mathcal{A} \to \mathcal{B}_{\intercal}$ is called a *c-encoding* of $(\mathcal{B}, \varsigma)$ if it is a convex combination of the primitive maps $\sigma_n \varrho_n$ given by the probability densities $\sigma_n \in \mathcal{B}_{\intercal}$ and normal states $\varrho_n : \mathcal{A} \to \mathbb{C}$. It is called *d-encoding* if it has the diagonalizing form (3.2) on $\mathcal{A}$, and it is called *o-encoding* if all density operators σ_n are mutually orthogonal: $\sigma_m \sigma_n = 0$, for all $m \neq n$ as in (3.4). The entanglement which is described by non-separable CP map $\varpi : \mathcal{A} \to \mathcal{B}_{\intercal}$ will be called *q-encoding*.

Note that, due to the commutativity of the operators $A \otimes I$ with $I \otimes B$ on $\mathcal{G} \otimes \mathcal{H}$, one can treat the encodings as non-demolition measurements (Belavkin 1994) in $\mathcal{A}$ with respect to $\mathcal{B}$. The corresponding compound state is the state prepared for such measurements on the input $\mathcal{G}$. It coincides with the mixture of the states corresponding to those after the measurement, without reading the message sent. The set of all d-encodings for a Schatten decomposition of the input state ρ on $\mathcal{A}$ is obviously convex with the extreme points given by the pure output states ς_n on $\mathcal{B}$, corresponding to the not necessarily orthogonal (not Schatten) decompositions $\sigma = \sum \sigma(n)$ into the one-dimensional density operators $\sigma(n) = p(n)\sigma_n$.

The Schatten decompositions $\sigma = \sum_n q(n)\sigma_n$ correspond to o-encodings, the extreme d-encodings $\sigma_n = \eta_n \eta_n^{\dagger}$, $p(n) = q(n)$ characterized by the orthogonality $\sigma_m \sigma_n = 0$, $m \neq n$. For each Schatten decomposition of σ they form a convex subset of d-encodings with mixed commuting σ_n.

4. Quantum versus von Neumann entropy.

As we have seen in the previous section, the encodings $\varpi : \mathcal{A} \to \mathcal{B}_{\intercal}$, which are described in (3.7), usually with a discrete Abelian $\mathcal{A}$, correspond to the case (3.2) when the generalized entanglement (2.7) is *d-encoding*, with the diagonal coupling $\pi = \varpi^{\intercal}$ in the eigenrepresentation of a discrete probability density ρ on non-Abelian $\mathcal{A}$. The true quantum entanglements with non-Abelian $\mathcal{A}$ cannot be achieved by d-, or more generally, c-encodings even in the case of discrete $\mathcal{A}$. The non-separable, true entangled states ω called *q-compound* states in Ohya (1989), can be achieved by *q-encodings*, the quantum–quantum non-separable correspondences (2.6) which are not diagonal in the eigenrepresentation of ρ.

As we shall prove in this section, the self-dual standard true entanglement $\varpi_{\mathrm{q}} = \varpi_{\mathrm{q}}^{\intercal}$ to the probe system $(\mathcal{A}^0, \varrho_0) = (\tilde{\mathcal{B}}, \tilde{\varsigma})$, which is defined in (2.9), is the most informative for a quantum system $(\mathcal{B}, \varsigma)$, in the sense that it achieves the maximal mutual information in the coupled system $(\mathcal{A} \otimes \mathcal{B}, \omega)$ when $\omega = \omega_{\mathrm{q}}$, is given in (2.10).

Let us consider entangled mutual information and quantum entropies of states by means of the above three types of compound states. To define the quantum mutual entropy we need to apply a quantum version of the relative entropy to compound states on the algebra $\mathcal{M} = \mathcal{A} \otimes \mathcal{B}$, also called the information divergency of the state ω with respect to a reference state φ on $\mathcal{M}$. The relative entropy was defined in Lindblad (1973), Araki (1976) and Umegaki (1962), even for the most general von Neumann algebra $\mathcal{M}$, but, for our purposes, we need the following explicit formulation.

Let $\mathcal{M}$ be a semi-finite algebra with normal states ω and φ having the density operator $v^{\dagger}v$ and $\phi \in \widetilde{\mathcal{M}}$ with respect to the pairing

$$\langle M, v^{\dagger}v \rangle = \tau(v\tilde{M}v^{\dagger}), \quad M \in \mathcal{M}, \quad vv^{\dagger} \in \widetilde{\mathcal{M}},$$

given by a normal, faithful trace τ on the transposed algebra $\widetilde{\mathcal{M}} = J\mathcal{M}J$ (not necessary decomposable as $\tau = \tilde{\mu} \otimes \tilde{\nu}$ in (2.3) in the case of $\mathcal{M} = \mathcal{A} \otimes \mathcal{B}$). Then the *relative entropy* $R(\omega; \varphi)$ of the state ω, with respect to φ, is given by the formula

$$R(\omega : \varphi) = \tau(v(\ln v^\dagger v - \ln \phi)v^\dagger) = \tau(\omega(\ln \omega - \ln \phi)). \tag{4.1}$$

(For notational simplicity here and below we identify the state ω with its density operator $v^\dagger v$.) It has a positive value $R(\omega : \varphi) \in [0, \infty]$ if the states are equally normalized, say (as usual) $\tau(\omega) = 1 = \tau(\phi)$, and it can be finite only if the state ω is absolutely continuous with respect to the reference state φ, i.e. if $\omega(E) = 0$ for the maximal null-orthoprojector $E \in \mathcal{M}$, $E\phi = 0$. This definition does not depend on the choice of the semi-finite trace τ, and it can also be extended to the arbitrary normal ω and φ with unbounded self-adjoint density operators $v^\dagger v$ and ϕ.

The most important property of the information divergence R is its monotonicity property (Lindblad 1973; Uhlmann 1977), i.e. non-increase in the divergency $R(\omega_0 : \varphi_0)$ after the application of the pre-dual of a normal, completely positive unital map $\mathrm{K} : \mathcal{M} \to \mathcal{M}^0$ to the states ω_0 and φ_0 on a von Neumann algebra $\mathcal{M}^0$:

$$(\omega = \omega_0 \mathrm{K}, \varphi = \varphi_0 \mathrm{K}) \Rightarrow R(\omega : \varphi) \leqslant R(\omega_0 : \varphi_0). \tag{4.2}$$

The *mutual information* $I(\pi) = I(\pi^*)$ in a compound state ω, achieved by a coupling $\pi : \mathcal{B} \to \mathcal{A}_*$, or by $\pi^* : \mathcal{A} \to \mathcal{B}_*$ with the marginals

$$\varrho(A) = \omega(A \otimes I) = \langle A, \rho \rangle_\mu, \varsigma(B) = \omega(I \otimes B) = \langle B, \sigma \rangle_\nu,$$

is defined by the relative entropy

$$I(\pi) = \tau(\omega(\ln \omega - \ln(\rho \otimes I) - \ln(I \otimes \sigma))) = R(\omega : \varrho \otimes \varsigma) \tag{4.3}$$

of the state ω on $\mathcal{M} = \mathcal{A} \otimes \mathcal{B}$ with respect to the product state $\varphi = \varrho \otimes \varsigma$ for $\tau = \tilde{\mu} \otimes \tilde{\nu}$. This quantity, generalizing the classical mutual information corresponding to the case of Abelian $\mathcal{A}$, $\mathcal{B}$, describes an information gain in a quantum system $(\mathcal{A}, \varrho)$ via the entanglement $\varpi^\intercal = \pi^{\tilde{}}$, or in $(\mathcal{B}, \varsigma)$ via an entanglement $\varpi : \mathcal{A} \to \mathcal{B}_\intercal$. It is naturally treated as a measure of the strength of the generalized entanglement having zero value only for completely disentangled states $\omega = \varrho \otimes \varsigma$.

Proposition 4.1. *Let $(\mathcal{A}^0, \mu_0)$ be a quantum system with a normal, faithful semi-finite trace, and $\pi_0 : \mathcal{A}^0 \to \mathcal{B}_*$ be a normal coupling of the state $\varrho_0 = \nu \circ \pi_0$ on $\mathcal{A}^0$ to $\varsigma = \mu \circ \pi$, defining an entanglement $\varpi = \pi^{*\tilde{}}$ of $(\mathcal{A}, \varrho)$ to $(\mathcal{B}, \varsigma)$ by the composition $\pi^* = \pi_0 \mathrm{K}$ with a normal completely positive unital map $\mathrm{K} : \mathcal{A} \to \mathcal{A}^0$. Then $I(\pi) \leqslant I(\pi^0)$, where $\pi^0 = \pi_0^*$. In particular, for each normal c-coupling given by (3.5), such as $\pi^{\tilde{}} = \varpi_{\mathrm{c}}^\intercal$, there exists a not less informative d-coupling $\pi^0 = \varpi_0^\intercal$ with Abelian $\mathcal{A}^0$ corresponding to the encoding $\varpi_0 = \pi_0^{\tilde{}}$ of $(\mathcal{B}, \varsigma)$, and the standard q-coupling $\pi^0 = \pi_{\mathrm{q}}$, $\pi_{\mathrm{q}}(B) = \sigma^{1/2}\tilde{B}\sigma^{1/2}$ to $\varrho_0 = \tilde{\varsigma}$ on $\mathcal{A}^0 = \tilde{\mathcal{B}}$ is the maximal coupling in this sense.*

Proof. The first follows from the monotonicity property (4.2) applied to the extension $\mathrm{K}(A \otimes B) = \mathrm{K}(A) \otimes B$ of the CP map K from $\mathcal{A} \to \mathcal{A}^0$ to $\mathcal{A} \otimes \mathcal{B} \to \mathcal{A}^0 \otimes \mathcal{B}$. The compound state $\omega_0(\mathrm{K} \otimes \mathrm{I})$ (I denotes the identity map $\mathcal{B} \to \mathcal{B}$) is achieved by the entanglement $\varpi = \varpi_0 \mathrm{K}$ and $\varphi_0(\mathrm{K} \otimes \mathrm{I}) = \varrho \otimes \varsigma$, $\varrho = \varrho_0 \mathrm{K}$, corresponding to $\varphi_0 = \varrho_0 \otimes \varsigma$. It corresponds to the coupling $\pi = \mathrm{K}^* \pi_0$, which is defined by $\mathrm{K}^* : \mathcal{A}_*^0 \to \mathcal{A}_*$ as $\mathrm{K}^* \tilde{\rho}_0 = J(\mathrm{K}^\intercal \rho_0)^\dagger J$, where

$$\langle A, \mathrm{K}^\intercal \rho_0 \rangle_\mu = \langle \mathrm{K}A, \rho_0 \rangle_{\mu_0}, \quad \forall A \in \mathcal{A}, \quad \rho_0 \in \mathcal{A}_\intercal^0.$$

This monotonicity property proves, in particular, that, for any separable compound state (3.8) on $\mathcal{A}\otimes\mathcal{B}$, which is prepared by the c-entanglement $\pi_{\mathrm{c}}^{-}=\varpi_{\mathrm{c}}^{\intercal}$, there exists a d-entanglement $\varpi_{0}^{\intercal}=\pi_0$ with $(\mathcal{A}^0,\varrho_0)$ having the same, or even larger information gain (4.3). One can even take a classical system $(\mathcal{A}^0,\varrho_0)$, say the diagonal subalgebra $\mathcal{A}^0\simeq L_p^\infty$ on $\mathcal{G}_0=L_p^2$ with the state ϱ_0, induced by the measure $\mu=p$, and consider the classical–quantum correspondence (encoding)

$$\varpi_0(A^0)=\int a(x)\sigma_x p(\mathrm{d}x),\quad A^0=\int^{\oplus}a(x)p(\mathrm{d}x),\quad a\in L_p^\infty,$$

assigning the states $\varsigma_x(B)=\langle B,\sigma_x\rangle_\nu$ to the letters x with the probabilities $p(\mathrm{d}x)$. In this case, the state ϱ is described by the density $\rho=I$, the multiplication by identity function in L_p^2, ω is multiplication by $\sigma_{\cdot}(x)=6x$ in $L_p^2\otimes\mathcal{H}$ and the mutual information (4.3) is given as

$$\mathsf{I}(\pi^0)=\int\tilde{\nu}_x(\sigma_x(\ln\sigma_x-\ln\sigma))p(\mathrm{d}x)=\mathsf{S}(\sigma)-\int\mathsf{S}(\sigma_x)p(\mathrm{d}x),\tag{4.4}$$

where $\mathsf{S}(\sigma)=-\tilde{\nu}(\sigma\ln\sigma)$. The achieved information gain $\mathsf{I}(\pi^0)$ is larger than $\mathsf{I}(\pi)$, corresponding to $\omega=\int\rho_x\otimes\sigma_x p(\mathrm{d}x)$, because the c-entanglement ϖ_{c} in (3.5) is represented as the composition $\varpi_0\mathrm{K}$ of the encoding $\varpi_0:\mathcal{A}^0\to\mathcal{B}_{\intercal}$ with the CP map

$$\mathrm{K}(A)=\int^{\oplus}\varrho_x(A)p(\mathrm{d}x),\quad A\in\mathcal{A}$$

given by $a(x)=\varrho_x(A)$ for each $A\in\mathcal{A}$. Hence,

$$\pi^*(A)=\widetilde{\varpi(A)}=\widetilde{\varpi_0\mathrm{K}A}=\pi_0(\mathrm{K}A),\quad\forall A\in\mathcal{A},$$

where $\pi_0=\varpi_0^{-}$, and thus, $\mathsf{I}(\pi^0)\geqslant\mathsf{I}(\mathrm{K}^*\pi^0)=\mathsf{I}(\pi)$, where $\pi^0=\pi_0^*=\varpi_0^{\intercal}$.

The inequality (4.2) can also be applied to the standard entanglement corresponding to the compound state (2.10) on $\widetilde{\mathcal{B}}\otimes\mathcal{B}$. Indeed, any normal entanglement $\varpi(A)=\mu(\tilde{\varkappa}^\dagger(A\otimes I)\tilde{\varkappa})$ on $\mathcal{A}$ into $\mathcal{B}_{\intercal}$, described by a CP map $\mathcal{A}\to\widetilde{\mathcal{B}_\nu}$, can be decomposed as

$$\mu(\tilde{\varkappa}^\dagger(A\otimes I)\tilde{\varkappa})=\sigma^{1/2}\mu(X^\dagger(A\otimes I)X)\sigma^{1/2}=\varpi_0(\mathrm{K}A),$$

where $\mathrm{K}A=\mu(X^\dagger(A\otimes I)X)$ is a normal unital CP map $\mathcal{A}\to\tilde{\mathcal{B}}$. It is uniquely given by an operator $X:\mathcal{E}\otimes\mathcal{H}\to\mathcal{G}\otimes\mathcal{F}$ with $\mathcal{E}=\mathcal{G}_\rho$, $\mathcal{H}=\mathcal{F}_\sigma$ satisfying the condition $X(I\otimes\sigma)^{1/2}=\tilde{\varkappa}$, and thus, $X\in\mathcal{A}\otimes\mathcal{B}'$, due to the commutativity of $\tilde{\varkappa}$ with $\mathcal{A}'\otimes\mathcal{B}$ and σ with $\mathcal{B}$. Moreover, the partial trace μ of $X^\dagger X$ is well defined by $\mu(\tilde{\varkappa}^\dagger\tilde{\varkappa})=\sigma$ as $\mu(X^\dagger X)=I$. Thus, $\varpi=\varpi_{\mathrm{q}}\mathrm{K}$ and $\pi=\mathrm{K}^*\pi_{\mathrm{q}}$, where K is a normal unital CP map $\mathcal{A}\to\widetilde{\mathcal{B}}$, and $\mathrm{K}^*:\mathcal{B}_{\intercal}=\widetilde{\mathcal{B}}_*\to\mathcal{A}_*$. Hence, the standard entanglement (coupling) (2.9) corresponds to the maximal mutual information, $\mathsf{I}(\pi_{\mathrm{q}})\geqslant\mathsf{I}(\mathrm{K}^*\pi_{\mathrm{q}})=\mathsf{I}(\pi)$. ■

Note that the mutual information (4.3) is written as

$$\mathsf{I}(\pi)=\mathsf{S}(\rho)+\mathsf{S}(\sigma)-\mathsf{S}(\omega/\varphi),$$

where $\varphi=\mu\otimes\nu$, $\mathsf{S}(\rho)=\mathsf{S}(\varrho/\mu)$, $\mathsf{S}(\sigma)=\mathsf{S}(\varsigma/\nu)$ and

$$\mathsf{S}(\omega/\varphi)=-\tilde{\varphi}(\upsilon(\ln\upsilon^\dagger\upsilon)\upsilon^\dagger)\equiv-\tilde{\varphi}(\upsilon^\dagger\upsilon\ln\upsilon^\dagger\upsilon)\tag{4.5}$$

denotes the *entropy of the density operator* $\upsilon^\dagger\upsilon \in \widetilde{\mathcal{M}}$ of the state ω, with respect to the trace φ on $\mathcal{M}$. Note that the entropy $S(\omega/\varphi)$, coinciding with $-R(\omega : \varphi)$ (see with (4.1) in the case $\tau = \tilde{\varphi}$), is not generally positive, and may not even be bounded from below as a function of ω. However, in the case of irreducible $\mathcal{M}$ it can always be made positive by the choice of the standard trace $\tau = \mathrm{tr}$ on $\mathcal{M}$, in which case it is called the von Neumann entropy of the state ω $(= \upsilon^\dagger\upsilon)$, denoted simply as $S(\omega)$:

$$S(\omega/\tau) = -\operatorname{tr}\omega\ln\omega \equiv S(\omega). \tag{4.6}$$

In the following, we shall assume that $\mathcal{B}$ is a discrete decomposition of the irreducible $\mathcal{B}_i = \mathcal{L}(\mathcal{H}_i) = \widetilde{\mathcal{B}}_i$ with the trace $\nu = \mathrm{tr}_{\mathcal{H}} = \tilde{\nu}$ induced on $\mathcal{B}_* = \mathcal{B}_\tau$. The entropy $S(\sigma) = S(\varsigma/\nu)$ of the density operator σ, for the normal state ς on $\mathcal{B}$, can be found in this case as the maximal information $S(\varsigma) = \sup I(\pi_{\mathrm{c}})$ achieved via all c-encodings $\varpi : \mathcal{A} \mapsto \mathcal{B}_\tau$ of the system $(\mathcal{B}, \varsigma)$, such that $\varpi(I) = \sigma$, $\varpi^\intercal = \pi_{\mathrm{c}}^{\sim}$. Indeed, as follows from the proposition above, it is sufficient to find the maximum of $I(\pi)$ over all d-couplings $\pi = \varpi^\intercal$, mapping $\mathcal{B}$ into Abelian $\mathcal{A}$ with fixed $\varpi(I) = \sigma$, i.e. to find maximum of (4.4) under the condition $\int \sigma_x p(\mathrm{d}x) = \sigma$. Due to positivity of the *d-conditional entropy*

$$S(\pi_{\mathrm{d}}) = -\int \operatorname{tr}(\sigma_x \ln\sigma_x)p(\mathrm{d}x) = \int S(\sigma_x)p(\mathrm{d}x), \tag{4.7}$$

the information $I(\pi^0) = I(\pi_{\mathrm{d}})$ has the maximum $S(\sigma)$, which is achieved on an extreme d-coupling π_{d}^0 when almost all $S(\sigma_x)$ are zero, i.e. when almost all σ_x are one-dimensional projectors $\sigma_x^0 = P_x$ corresponding to pure states ς_x. One can take, for example, the maximal Abelian subalgebra $\mathcal{A}^0 \subseteq \mathcal{B}$ generated by $P_n = |n\rangle\langle n| \in \mathcal{B}$ for a Schatten decomposition $\sigma = \sum_n |n\rangle\langle n|p(n)$ of $\sigma \in \mathcal{B}_\tau$. The maximal value $\ln\operatorname{rank}\mathcal{B}$ of the von Neumann entropy is defined by the dimensionality $\operatorname{rank}\mathcal{B} = \dim\mathcal{A}^0$ of the maximal Abelian subalgebra of the decomposable algebra $\mathcal{B}$, i.e. by $\dim\mathcal{H}$.

However, if π is not c-coupling, the difference $S(\pi) = S(\sigma) - I(\pi)$ can achieve the negative value and may not serve as a measure of conditional entropy in such a case.

Definition 4.2. The supremum of the mutual information

$$H(\varsigma) = \sup\{I(\pi) : \mu \circ \pi = \varsigma\} = I(\pi_{\mathrm{q}}), \tag{4.8}$$

which is achieved on $\mathcal{A} = \tilde{\mathcal{B}}$ for a fixed state $\varsigma(B) = \mathrm{tr}_{\mathcal{H}} B\sigma$ by the standard q-coupling $\pi_{\mathrm{q}}(B) = \sigma^{1/2}\tilde{B}\sigma^{1/2}$, is called *q-entropy* of the state ς. The maximum

$$S(\varsigma) = \sup\{I(\pi_{\mathrm{c}}) : \mu \circ \pi_{\mathrm{c}} = \varsigma\} = I(\pi_{\mathrm{d}}^0)$$

over all c-couplings π_{c} corresponding to c-encodings (3.5), which is achieved on an extreme d-coupling π_{d}^0, is called the *c-entropy* of the state ς. The differences,

$$H(\pi) = H(\varsigma) - I(\pi), \qquad S(\pi) = S(\varsigma) - I(\pi),$$

are called, respectively, the *q-conditional entropy* on $\mathcal{B}$ with respect to $\mathcal{A}$ and the *(degree of) disentanglement* for the coupling $\pi : \mathcal{B} \to \mathcal{A}$. A compound state is said to be *essentially entangled* if $S(\pi) < 0$, and $S(\pi) \geqslant 0$ for a c-coupling $\pi = \pi_{\mathrm{c}}$ (this is called the *c-conditional entropy* on $\mathcal{B}$, with respect to $\mathcal{A}$).

Obviously, $H(\varsigma)$ and $S(\varsigma)$ are both positive and (unlike $S(\sigma) = S(\varsigma/\nu)$) do not depend on the choice of the faithful trace ν on $\mathcal{B}$, and obey the inequality $H(\varsigma) \geqslant S(\varsigma)$. The same is true for the conditional entropies $H(\pi)$ and $S(\pi)$, where $S(\pi)$ always has a positive value

$$S(\pi) \geqslant S(\pi^0) \geqslant 0,$$

in the case of a c-coupling $\pi = \pi_{\mathrm{c}}$, due to $\pi_{\mathrm{c}}^* = \pi_{\mathrm{d}}^* \mathrm{K}$ for a normal unital CP map $\mathrm{K}: \mathcal{A} \to \mathcal{A}^0$, where $\pi^0 = \pi_{\mathrm{d}}$ is a d-coupling with Abelian $\mathcal{A}^0$. But the disentanglement $S(\pi)$ can also achieve the negative value

$$\inf\{S(\pi) : \mu \circ \pi = \varsigma\} = S(\varsigma) - H(\varsigma) = -\sum_i \varkappa(i) S(\sigma_i), \tag{4.9}$$

as the following theorem states in the case of discrete $\mathcal{B}$. Here, the $\sigma_i \in \mathcal{L}(\mathcal{H}_i)$ are the density operators of the normalized factor-states $\varsigma_i = \varkappa(i)^{-1}\varsigma \mid \mathcal{L}(\mathcal{H}_i)$ with $\varkappa(i) = \varsigma(I^i)$, where I^i are the orthoprojectors onto $\mathcal{H}_i$. Note that $H(\varsigma) = S(\varsigma)$ if the algebra $\mathcal{B}$ is completely decomposable, i.e. Abelian. In this case, the maximal value $\ln \operatorname{rank} \mathcal{B}$ of $S(\varsigma)$ can be written as $\ln \dim \mathcal{B}$. The disentanglement $S(\pi)$ is always positive in this case, and $S(\pi) = H(\pi)$ as in the case of Abelian $\mathcal{A}$.

Theorem 4.3. *Let $\mathcal{B}$ be a discrete decomposable algebra on $\mathcal{H} = \oplus_i \mathcal{H}_i$, with a normal state given by the density operator $\sigma = \oplus \sigma(i)$, with respect to the trace $\mu = \mathrm{tr}_{\mathcal{H}}$ on $\mathcal{B}$, and let $\mathcal{C} \subseteq \mathcal{B}$ be its centre with the state $\varkappa = \varsigma \mid \mathcal{C}$ induced by the probability distribution $\varkappa(i) = \mathrm{tr}\,\sigma(i)$. Then the c-entropy $S(\varsigma)$ is given as the von Neumann entropy (4.6) of the density operator σ and the q-entropy (4.8) is given by the formula*

$$H(\varsigma) = \sum_i (\varkappa(i) \ln \varkappa(i) - 2\,\mathrm{tr}_{\mathcal{H}_i} \sigma(i) \ln \sigma(i)). \tag{4.10}$$

This can be written as $H(\varsigma) = H_{\mathcal{B}|\mathcal{C}}(\varsigma) + H_{\mathcal{C}}(\varsigma)$, where $H_{\mathcal{C}}(\varsigma) = -\sum_i \varkappa(i) \ln \varkappa(i)$, and

$$H_{\mathcal{B}|\mathcal{C}}(\varsigma) = -2 \sum_i \varkappa(i)\,\mathrm{tr}_{\mathcal{H}_i} \sigma_i \ln \sigma_i = 2 S_{\mathcal{B}|\mathcal{C}}(\varsigma),$$

with $\sigma_i = \sigma(i)/\varkappa(i)$. $H(\varsigma)$ is finite if $S(\varsigma) < \infty$, and if $\mathcal{B}$ is finite dimensional, it is bounded, with maximal value $H(\varsigma^\circ) = \ln \dim \mathcal{B}$, achieved for $\sigma^\circ = \oplus \sigma_i^\circ \varkappa^\circ(i)$:

$$\sigma_i^\circ = (\dim \mathcal{H}_i)^{-1} I^i, \qquad \varkappa^\circ(i) = \dim \mathcal{B}(i) / \dim \mathcal{B},$$

where $\dim \mathcal{B}(i) = (\dim \mathcal{H}_i)^2$, $\dim \mathcal{B} = \sum_i \dim \mathcal{B}(i)$.

Proof. We have already proved that $S(\varsigma) = S(\sigma)$, where

$$S(\sigma) = -\sum_i \mathrm{tr}_{\mathcal{H}_i} \sigma(i) \ln \sigma(i) = S_{\mathcal{C}}(\varsigma) + S_{\mathcal{B}|\mathcal{C}}(\varsigma),$$

with $S_{\mathcal{C}}(\varsigma) = H_{\mathcal{C}}(\varsigma)$, $S_{\mathcal{B}|\mathcal{C}}(\varsigma) = \sum \varkappa(i) S(\sigma_i) = \frac{1}{2} H_{\mathcal{B}|\mathcal{C}}(\varsigma)$.

The q-entropy $H(\varsigma)$ is the supremum (4.8) of the mutual information (4.3) which is achieved on the standard entanglement, corresponding to the density operator $\omega = \oplus \omega(i,k)$ with

$$\omega(i,k) = \varkappa(i) |\sigma_i^{1/2}) (\sigma_i^{1/2} | \sigma_k^i$$

of the standard compound state (2.10) with $\widetilde{\mathcal{B}} = \mathcal{B}$, $\rho = \sigma$. Thus, $\mathsf{H}(\varsigma) = \mathsf{I}(\pi_{\mathrm{q}})$, where

$$\begin{aligned}\mathsf{I}(\pi_{\mathrm{q}}) &= \operatorname{tr}\omega(\ln\omega - \ln(\sigma\otimes I) - \ln(I\otimes\sigma)) = \mathsf{S}(\omega) - 2\mathsf{S}(\sigma)\\ &= \sum_i \varkappa(i)\ln\varkappa(i) - 2\operatorname{tr}\sigma\ln\sigma = -\sum_i \varkappa(i)(\ln\varkappa(i) + 2\operatorname{tr}_{\mathcal{H}_i}\sigma_i\ln\sigma_i).\end{aligned}$$

Here, we used that $\operatorname{tr}\omega\ln\omega = \sum_i \varkappa(i)\ln\varkappa(i)$ due to

$$\omega\ln\omega = \oplus_{i,k}\omega(i,k)\ln\omega(i,k) = \oplus_i\varkappa(i)|\sigma_i^{1/2})(\sigma_i^{1/2}|\ln\varkappa(i),$$

and that $\operatorname{tr}\sigma\ln\sigma = \sum_i \varkappa(i)(\ln\varkappa(i) - \mathsf{S}_{\mathcal{B}_i}(\varsigma_i))$ due to

$$\sigma\ln\sigma = \oplus_i\sigma(i)\ln\sigma(i) = \oplus_i\varkappa(i)\sigma_i(\ln\varkappa(i) + \ln\sigma_i)$$

for the orthogonal decomposition $\sigma = \oplus_i\varkappa(i)\sigma_i$, where $\varkappa(i) = \operatorname{tr}\sigma(i)$.

Thus, $\mathsf{H}(\varsigma) = \mathsf{H}_{\mathcal{B}|\mathcal{C}}(\varsigma) + \mathsf{H}_{\mathcal{C}}(\varsigma) = 2\mathsf{S}_{\mathcal{B}|\mathcal{C}}(\varsigma) + \mathsf{S}_{\mathcal{C}}(\varsigma) \leqslant 2\mathsf{S}(\varsigma)$, and it is bounded by

$$\begin{aligned}\mathsf{C}_{\mathcal{B}} &= \sup_{\varkappa}\sum_i \varkappa(i)(2\sup_{\varsigma_i}\mathsf{S}_{\mathcal{B}(i)}(\varsigma_i) - \ln\varkappa(i))\\ &= -\inf_{\varkappa}\sum_i \varkappa(i)(\ln\varkappa(i) - 2\ln\dim\mathcal{H}_i) = \ln\dim\mathcal{B}.\end{aligned}$$

Here, we used the fact that the supremum of von Neumann entropies $\mathsf{S}(\sigma_i)$ for the simple algebras $\mathcal{B}(i) = \mathcal{L}(\mathcal{H}_i)$, with $\dim\mathcal{B}(i) = (\dim\mathcal{H}_i)^2 < \infty$, is achieved on the tracial density operators $\sigma_i = (\dim\mathcal{H}_i)^{-1}I^i \equiv \sigma_i^{\circ}$, and the infimum of the relative entropy

$$\mathsf{R}(\varkappa : \varkappa^{\circ}) = \sum_i \varkappa(i)(\ln\varkappa(i) - \ln\varkappa^{\circ}(i)),$$

where $\varkappa^{\circ}(i) = \dim\mathcal{B}(i)/\dim\mathcal{B}$, is zero, achieved at $\varkappa = \varkappa^{\circ}$. ■

Note that, as shown in Ohya & Petz (1993) for the case of the simple algebra $\mathcal{B}$, the quantum entropy $\mathsf{H}(\varsigma)$ can also be achieved as the supremum of the von Neumann entropy $\mathsf{S}(\varrho)$ over all pure couplings given by the isometries $X : \mathcal{H} \to \mathcal{G}\otimes\mathcal{H}$, $X^{\dagger}X = I$, preserving the state ς. The latter means that the density operator ω of the corresponding compound states with the marginals $\rho = \operatorname{tr}_{\mathcal{H}}\omega$ and $\sigma = \operatorname{tr}_{\mathcal{G}}\omega$ is given as $\omega = X\sigma X^{\dagger}$.

5. Quantum channel and entropic capacities

In this section we describe a noisy quantum channel in terms of normal unital CP maps and their duals, and introduce an analogue of Shannon information for general semi-finite algebras. We consider the maximization problems for this quantity with various operational constrains on encodings, and define the entropic capacities which serve as upper bounds for the operational capacities corresponding to these constrains. The question of asymptotic equivalence of the entropic and operational capacities is not touched on here.

Let $\mathcal{H}_1$ be a Hilbert space describing a quantum input system and let $\mathcal{H}$ describe its output Hilbert space. A quantum channel is an affine operation sending each input state defined on $\mathcal{H}_1$ to an output state defined on $\mathcal{H}$, such that the mixtures of states are preserved. A deterministic quantum channel is given by a linear isometry

$U : \mathcal{H}_1 \to \mathcal{H}$ with $U^\dagger U = I^1$ (I^1 is the identify operator in $\mathcal{H}_1$), such that each input state vector $\eta_1 \in \mathcal{H}_1$, $\|\eta_1\| = 1$, is transmitted into an output state vector $\eta = U\eta_1 \in \mathcal{H}$, $\|\eta\| = 1$. The orthogonal sums $\varsigma_1 = \oplus\varsigma_1(n)$ of pure input states $\varsigma_1(B,n) = \eta_1(n)^\dagger B\eta_1(n)$ are sent into the orthogonal sums $\varsigma = \oplus\varsigma(n)$ of pure states on $\mathcal{B} = \mathcal{L}(\mathcal{H})$, corresponding to the orthogonal state vectors $\eta(n) = U\eta_1(n)$.

A noisy quantum channel sends pure input states ς_1 on an algebra $\mathcal{B}^1 \subseteq \mathcal{L}(\mathcal{H}_1)$ into mixed ones $\varsigma = \varsigma_1 \Lambda$ given by the composition with a normal completely positive unital map $\Lambda : \mathcal{B} \to \mathcal{B}^1$. We shall assume that $\mathcal{B}^1$ (as well as $\mathcal{B}$) is equipped with a normal faithful semi-finite trace ν_1 defining the pairing $\langle B, u^\dagger u\rangle_1 = \nu_1(\tilde{u}^\dagger B\tilde{u})$ of $\mathcal{B}^1$ and $\mathcal{B}^1_{\intercal} = \mathcal{B}^1_*$. Then, the input–output state transformations are described by the transposed map $\Lambda^{\intercal} : \mathcal{B}^1_{\intercal} \to \mathcal{B}_{\intercal}$:

$$\langle B, \Lambda^{\intercal}(\sigma_1)\rangle = \langle \Lambda(B), \sigma_1\rangle_1, \quad B \in \mathcal{B}, \quad \sigma_1 \in \mathcal{B}^1_{\intercal},$$

defining the output density operators $\sigma = \Lambda^{\intercal}(\sigma_1)$ for any input normal state $\varsigma_1(B) = \langle B, \sigma_1\rangle_1$. Without loss of generality, the input algebra $\mathcal{B}^1$ can be assumed to be the smallest decomposable algebra generated by the range $\Lambda(\mathcal{B})$ of the channel map Λ ($\mathcal{B}^1$ is Abelian if $\Lambda(\mathcal{B})$ consists only of commuting operators on $\mathcal{H}_1$).

The input generalized entanglements $\varpi^1 : \mathcal{A} \to \mathcal{B}^1_{\intercal}$, including encodings of the state ς_1 with the density $\sigma_1 = \varpi^1(I)$, will be defined by the couplings $\kappa^* : \mathcal{B}^1 \to \mathcal{A}_*$ as $\varpi^1 = \kappa^{\tilde{}}$. Here, $\kappa : \mathcal{A} \to \mathcal{B}^1_*$ is a normal TCP map defining the state $\varrho = \nu_1 \circ \kappa$ of a probe system $(\mathcal{A}, \mu)$ which is entangled to $(\mathcal{B}^1, \varsigma_1)$ by $\kappa^{\tilde{}}(A) = J\kappa(A^\dagger)J$, and the adjoint map κ^* is defined as usual by

$$\langle A|\kappa^*(B)\rangle_\mu = \omega_1(A^\dagger \otimes B) = \langle \kappa(A)|B\rangle_1, \quad \forall A \in \mathcal{A}, \quad B \in \mathcal{B}_1,$$

where ω_1 is the corresponding compound state on $\mathcal{A} \otimes \mathcal{B}^1$.

These (generalized) entanglements describe the quantum–quantum correspondences (q-, c-, or o-encodings) of the probe systems $(\mathcal{A}, \varrho)$ with the density operators $\rho = \kappa^{\intercal}(I^1)$, to the input $(\mathcal{B}^1, \varsigma_1)$ of the channel Λ. In particular, the most informative standard input entanglement $\varpi^1_{\mathrm{q}} : \mathcal{B}^1 \to \mathcal{B}^1_{\intercal}$ is the entanglement of the transposed input system $(\mathcal{A}^0, \varrho_0) = (\widetilde{\mathcal{B}^1}, \tilde{\varsigma}_1)$, corresponding to the TCP map $\kappa_{\mathrm{q}}(A) = J\sigma_1^{1/2}A^\dagger\sigma_1^{1/2}J$. In the case of discrete decomposable $\mathcal{A}^0 = \widetilde{\mathcal{B}^1} = \mathcal{B}^1$ with the density operator $\sigma_1 = \oplus_i\sigma_1(i)$, this extreme input q-encoding defines the following density operator

$$\omega_{\mathrm{q}} = (\mathrm{I} \otimes \Lambda^{\intercal})(\omega_{\mathrm{q}1}), \quad \omega_{\mathrm{q}1} = \oplus_i|\sigma_1(i)^{1/2})(\sigma_1(i)^{1/2}| \tag{5.1}$$

of the input–output compound state $\omega_{\mathrm{q}1}\Lambda$ on $\mathcal{A}^0 \otimes \mathcal{B} = \mathcal{B}^1 \otimes \mathcal{B}$.

The other extreme case of the generalized input entanglements, the pure c-encodings corresponding to (3.2), are less informative then the pure d-encodings $\varpi^1_{\mathrm{d}} = \kappa_{\mathrm{d}}^{\tilde{}}$ given by the decompositions $\kappa^*_{\mathrm{d}} = \sum|n\rangle\langle n|\varsigma_1(n)$ with pure states $\varsigma_1(B,n) = \eta(n)^\dagger B\eta(n)$ on $\mathcal{B}_1$. They define the density operators

$$\omega_{\mathrm{d}} = (\mathrm{I} \otimes \Lambda^{\intercal})(\omega_{\mathrm{d}1}), \quad \omega_{\mathrm{d}1} = \sum_n |n\rangle\langle n| \otimes \eta_1(n)\eta_1(n)^\dagger, \tag{5.2}$$

of the $\mathcal{B}^1 \otimes \mathcal{B}$-compound state $\omega_{\mathrm{d}1}\Lambda = \omega_{\mathrm{d}1} \circ (\mathrm{I} \otimes \Lambda)$. These are the Ohya compound states $\omega_{\mathrm{o}} = \omega_{\mathrm{o}1}\Lambda$ (Ohya 1983*a*) in the case

$$\sigma_1(n) = \eta_1^o(n)\eta_1^o(n)^\dagger, \quad \eta_1^o(n)^\dagger\eta_1^o(m) = p_1(n)\delta_n^m,$$

of orthogonality of the density operators $\sigma_1(n)$ normalized to the eigenvalues $p_1(n)$ of σ_1. The o-compound states are achieved by pure o-encodings $\varpi_{\mathrm{d}}^1 = \tilde{\kappa}_{\mathrm{o}}$ described by the couplings $\kappa_{\mathrm{o}} = \sum |n\rangle\langle n|\varsigma_1^o(n)$ with ς_1^o corresponding to η_1^o. The input–output density operator

$$\omega_{\mathrm{o}} = (\mathrm{I} \otimes \Lambda^{\intercal})\omega_{\mathrm{o}1}, \quad \omega_{\mathrm{o}1} = \sum_n |n\rangle\langle n| \otimes \eta_1^o(n)\eta_1^o(n)^{\dagger} \tag{5.3}$$

of the Ohya compound state ω_{o} is achieved by the coupling $\lambda = \kappa^* \Lambda$ of the output $(\mathcal{B}, \varsigma)$ to the extreme probe system $(\mathcal{A}^0, \varrho_0) = (\mathcal{B}^1, \varsigma_1)$ as the composition of κ^* and the channel Λ.

If $\mathrm{K} : \mathcal{A} \to \mathcal{A}^0$ is a normal completely positive unital map

$$\mathrm{K}(A) = \mathrm{tr}_{\mathcal{F}_-} \widetilde{X} A \tilde{X}^{\dagger}, \quad A \in \mathcal{A},$$

where X is a bounded operator $\mathcal{F}_- \otimes \mathcal{G}_0 \to \mathcal{G}$ with $\mathrm{tr}_{\mathcal{F}_-} X^{\dagger} X = I^0$, the compositions $\kappa = \kappa_0 \mathrm{K}$, $\pi = \Lambda^* \kappa$ describe the entanglements of the probe system $(\mathcal{A}, \varrho)$ to the channel input $(\mathcal{B}^1, \varsigma_1)$ and the output $(\mathcal{B}, \varsigma)$ via this channel, respectively. The state $\varrho = \varrho_0 \mathrm{K}$ is given by

$$\mathrm{K}^{\intercal}(\rho_0) = X(I^- \otimes \rho_0), \quad X^{\dagger} \in \mathcal{A}_*$$

for each density operator $\rho_0 \in \mathcal{A}_*^0$, where I^- is the identity operator in $\mathcal{F}_-$. The resulting entanglement $\pi = \lambda^* \mathrm{K}$ defines the compound state $\omega = \omega_{01} \circ (\mathrm{K} \otimes \Lambda)$ on $\mathcal{A} \otimes \mathcal{B}$ with

$$\omega_{01}(A^0 \otimes B^1) = \mathrm{tr}\, \tilde{A}^0 \kappa_0^*(B^1) = \mathrm{tr}\, \tilde{v}_{01}^{\dagger}(A^0 \otimes B^1)\tilde{v}_{01}$$

on $\mathcal{A}^0 \otimes \mathcal{B}^1$. Here, $v_{01} : \mathcal{G}_0 \otimes \mathcal{H}_1 \to \mathcal{F}_{01}$ is the amplitude operator uniquely defined by the input compound density operator $\omega_{01} \in \mathcal{A}_{\intercal}^0 \otimes \mathcal{B}_{\intercal}^1$ up to a unitary operator U^0 on $\mathcal{F}_{01}$. The effect of the input entanglement κ and the output channel Λ can be written in terms of the amplitude operator of the state ω as

$$v = (X \otimes Y)(I^- \otimes v_{01} \otimes I^+)U$$

up to a unitary operator U in $\mathcal{F} = \mathcal{F}_- \otimes \mathcal{F}_{01} \otimes \mathcal{F}_+$. Thus, the density operator of the input–output compound state ω is given by $\omega_{01}(\mathrm{K} \otimes \Lambda)$ with the density

$$(\mathrm{K} \otimes \Lambda)^*(\omega_{01}) = (X \otimes Y)\omega_{01}(X \otimes Y)^{\dagger}, \tag{5.4}$$

where $\omega_{01} = v_{01} v_{01}^{\dagger}$.

Let $\mathcal{K}_{\mathrm{q}}^1$ be the set of all normal TCP maps $\kappa : \mathcal{A} \to \mathcal{B}_*^1$ with any probe algebra $\mathcal{A}$ normalized as $\mathrm{tr}\, \kappa(I) = 1$, and let $\mathcal{K}_{\mathrm{q}}(\varsigma_1)$ be the subset of all $\kappa \in \mathcal{K}_{\mathrm{q}}^1$ with $\kappa(I) = \varsigma_1$. Each $\kappa \in \mathcal{K}_{\mathrm{q}}^1$ can be decomposed as $\kappa_{\mathrm{q}} \mathrm{K}$, where $\kappa_{\mathrm{q}} : \mathcal{A}^0 \to \mathcal{B}^1$ defines the standard input entanglement $\varpi_{\mathrm{q}}^1 = \tilde{\kappa}_{\mathrm{q}}$, and K is a normal unital CP map $\mathcal{A} \to \mathcal{B}^1$.

Further, let $\mathcal{K}_{\mathrm{c}}^1$ be the set of all CP-TCP maps κ described as the combinations

$$\kappa(A) = \sum_n \varrho_n(A)\sigma_1(n) \tag{5.5}$$

of the primitive maps $A \mapsto \varrho_n(A)\sigma_1(n)$, and let $\mathcal{K}_{\mathrm{d}}^1$ be the subset of the diagonalizing entanglements κ, i.e. the decompositions

$$\kappa(A) = \sum_n \langle n|A|n\rangle \sigma_1(n). \tag{5.6}$$

As in the first case, $\mathcal{K}_c(\varsigma_1)$ and $\mathcal{K}_d(\varsigma_1)$ denote the subsets corresponding to a fixed $\kappa(I) = \varsigma_1$. Each $\mathcal{K}_c(\varsigma_1)$ can be represented as the composition $\kappa = \kappa_d \mathrm{K}$, where κ_d normalized to ς_1 describes a pure d-encoding $\varpi_d^1 = \kappa_d^{\sim}$ of $(\mathcal{B}^1, \varsigma_1)$ for a proper choice of the CP map $\mathrm{K} : \mathcal{A} \to \mathcal{B}^1$.

Furthermore, let $\mathcal{K}_o^1$ (and $\mathcal{K}_o(\varsigma_1)$) be the subset of all decompositions (5.6) with orthogonal $\sigma_1(n)$ (and fixed $\sum_n \sigma_1(n) = \sigma_1$):

$$\sigma_1(m)\sigma_1(n) = 0, \quad m \neq n.$$

Each $\kappa \in \mathcal{K}_o(\varsigma_1)$ can also be represented as $\kappa = \kappa_o \mathrm{K}$, with κ_o describing the pure o-encoding $\varpi_o^1 = \kappa_o^{\sim}$ of $(\mathcal{B}^1, \varsigma_1) = (\mathcal{A}^0, \varrho_0)$.

Now, let us maximize the entangled mutual entropy for a given quantum channel Λ (and a fixed input state ς_1 on the decomposable $\mathcal{B}^1 = \bar{\mathcal{B}}^1$) by means of the above four types of entanglement κ. The mutual information (4.3) was defined in the previous section by the density operators of the corresponding compound state ω on $\mathcal{A} \otimes \mathcal{B}$, and the product-state $\varphi = \varrho \otimes \varsigma$ of the marginals ϱ, ς for ω. In each case,

$$\omega = \omega_{01}(\mathrm{K} \otimes \Lambda), \qquad \varphi = \varphi_{01}(\mathrm{K} \otimes \Lambda),$$

where K is a CP map $\mathcal{A} \to \mathcal{A}^0 = \mathcal{B}^1$, ω_{01} is one of the corresponding extreme compound states ω_{q1}, $\omega_{c1} = \omega_{d1}$, ω_{o1} on $\mathcal{B}^1 \otimes \mathcal{B}^1$ and $\varphi_{01} = \rho_0 \otimes \varsigma_1$. The density operator $\omega = (\mathrm{K} \otimes \Lambda)^{\intercal}(\omega_{01})$ is written in (5.4), and $\phi = \rho \otimes \sigma$ can be written as

$$\phi = \kappa^{\intercal}(I) \otimes \lambda^{\intercal}(I),$$

where $\lambda^{\intercal} = \Lambda^{\intercal}\pi_1^0$. This proves the following proposition.

Proposition 5.1. *The entangled mutual information achieves the following maximal values,*

$$\sup_{\kappa \in \mathcal{K}_q(\varsigma_1)} I(\kappa^* \Lambda) = I_q(\varsigma_1, \Lambda) := I(\kappa_q^* \Lambda), \tag{5.7}$$

$$I_c(\varsigma_1, \Lambda) := \sup_{\kappa \in \mathcal{K}_c(\varsigma_1)} I(\kappa^* \Lambda) = \sup_{\kappa_d} I(\kappa_d^* \Lambda) \equiv I_d(\varsigma_1, \Lambda),$$

$$\sup_{\kappa \in \mathcal{K}_o(\varsigma_1)} I(\kappa^* \Lambda) = I_o(\varsigma_1, \Lambda) := \sup_{\kappa_o} I(\kappa_o^* \Lambda), \tag{5.8}$$

where $\kappa_\cdot$ are the corresponding extremal input couplings $\mathcal{A}^0 \to \mathcal{B}_*^1$ with $\mu \circ \kappa_\cdot^* = \varsigma_1$. They are ordered as

$$I_q(\varsigma_1, \Lambda) \geqslant I_c(\varsigma_1, \Lambda) = I_d(\varsigma_1, \Lambda) \geqslant I_o(\varsigma_1, \Lambda). \tag{5.9}$$

In the following definition, the maximal information $I_c(\varsigma_1, \Lambda) = I_d(\varsigma_1, \Lambda)$ is simply denoted as $I_1(\varsigma_1, \Lambda)$.

Definition 5.2. The suprema

$$C_q(\Lambda) = \sup_{\kappa \in \mathcal{K}_q^1} I(\kappa^* \Lambda) = \sup_{\varsigma_1} I_q(\varsigma_1, \Lambda),$$

$$\sup_{\kappa \in \mathcal{K}_d^1} I(\kappa^* \Lambda) = C_1(\Lambda) := \sup_{\varsigma_1} I_1(\varsigma_1, \Lambda), \tag{5.10}$$

$$C_o(\Lambda) = \sup_{\kappa \in \mathcal{K}_o^1} I(\kappa^* \Lambda) = \sup_{\varsigma_1} I_o(\varsigma_1, \Lambda),$$

are called the q-, c- or d-, and o-capacities, respectively, for the quantum channel defined by a normal unital CP map $\Lambda : \mathcal{B} \to \mathcal{B}^1$.

Obviously, the capacities (5.10) satisfy the inequalities

$$\mathsf{C}_{\mathrm{o}}(\Lambda) \leqslant \mathsf{C}_{1}(\Lambda) \leqslant \mathsf{C}_{\mathrm{q}}(\Lambda).$$

Theorem 5.3. *Let $\Lambda(B) = U^{\dagger}BU$ be a unital CP map $\mathcal{B} \to \mathcal{B}^1$ describing a quantum deterministic channel. Then*

$$\mathsf{I}_{1}(\varsigma_1, \Lambda) = \mathsf{I}_{\mathrm{o}}(\varsigma_1, \Lambda) = \mathsf{S}(\varsigma_1), \quad \mathsf{I}_{\mathrm{q}}(\varsigma_1, \Lambda) = \mathsf{S}_{\mathrm{q}}(\varsigma_1),$$

where $\mathsf{S}_{\mathrm{q}}(\varsigma_1) = \mathsf{H}(\varsigma_1)$, and, thus, in this case

$$\mathsf{C}_{1}(\Lambda) = \mathsf{C}_{\mathrm{o}}(\Lambda) = \ln \operatorname{rank} \mathcal{B}^1, \quad \mathsf{C}_{\mathrm{q}}(\Lambda) = \ln \dim \mathcal{B}^1.$$

Proof. It was proved in the previous section for the case of the identity channel $\Lambda = \mathrm{I}$ and is thus also valid for any isomorphism $\Lambda : B \mapsto U^{\dagger}BU$ describing the state transformations $\Lambda^{\intercal} : \sigma \mapsto Y\sigma Y^{\dagger}$ by a unitary operator $U = \bar{Y}$. In the case of non-unitary Y, we can use the identity

$$\operatorname{tr} Y(\sigma_1 \otimes I^+)Y^{\dagger} \ln Y(\sigma_1 \otimes I^+)Y^{\dagger} = \operatorname{tr} S(\sigma_1 \otimes I^+) \ln S(\sigma_1 \otimes I^+),$$

where $S = Y^{\dagger}Y$. Due to this, $\mathsf{S}(\varsigma_1\Lambda) = -\operatorname{tr} S(\sigma_1 \otimes I^+) \ln S(\sigma_1 \otimes I^+)$, and

$$\mathsf{S}(\omega_{01}(\mathrm{K} \otimes \Lambda)) = -\operatorname{tr}(R \otimes S)(I^- \otimes \omega_{01} \otimes I^+) \ln(R \otimes S)(I^- \otimes \omega_{01} \otimes I^+),$$

where $R = X^{\dagger}X$. Thus, $\mathsf{S}(\varsigma_1\Lambda) = \mathsf{S}(\varsigma_1)$, $\mathsf{S}(\omega_{01}(\mathrm{K} \otimes \Lambda)) = \mathsf{S}(\omega_{01}(\mathrm{K} \otimes \mathrm{I}))$ if $Y^{\dagger}Y = I$, and

$$\begin{aligned} \mathsf{I}((\pi_1\Lambda)) &= \mathsf{S}(\varrho_0\mathrm{K}) + \mathsf{S}(\varsigma_1) - \mathsf{S}(\omega_{01}(\mathrm{K} \otimes \mathrm{I})) \\ &\leqslant \mathsf{S}(\varrho_0) + \mathsf{S}(\varsigma_1) - \mathsf{S}(\omega_{01}) = \mathsf{I}(\omega_{01}), \end{aligned}$$

for $\kappa = \kappa_0\mathrm{K}$ with any normal unital CP map $\mathrm{K} : \mathcal{A} \to \mathcal{A}^0$ and a compound state ω_{01} on $\mathcal{A}^0 \otimes \mathcal{B}^1$. The supremum (5.7), which is achieved at the standard entanglement, corresponding to $\omega_{01} = \omega_{\mathrm{q}1}$, coincides with q-entropy $\mathsf{H}(\varsigma_1)$, and the supremum (5.8), coinciding with $\mathsf{S}(\varsigma_1)$, is achieved for a pure o-entanglement, corresponding to $\omega_{01} = \omega_{\mathrm{o}1}$ given by any Schatten decomposition for σ_1. Moreover, the entropy $\mathsf{H}(\varsigma_1)$ is also achieved by any pure d-entanglement, corresponding to $\omega_{01} = \omega_{\mathrm{d}1}$ given by any extreme decomposition for σ_1 and thus is the maximal mutual information $\mathsf{I}_1(\varsigma_1, \Lambda)$ in the case of deterministic Λ. Thus, the capacity $\mathsf{C}_1(\Lambda)$ of the deterministic channel is given by the maximum $\mathsf{C}_{\mathrm{o}} = \ln \dim \mathcal{H}_1$ of the von Neumann entropy S, and the q-capacity $\mathsf{C}_{\mathrm{q}}(\Lambda)$ is equal $\mathsf{C}_{\mathcal{B}^1} = \ln \dim \mathcal{B}^1$. ∎

In the general case, d-entanglements can be more informative than o-entanglements, as can be shown by an example of a noisy quantum channel for which

$$\mathsf{I}_{1}(\varsigma_1, \Lambda) > \mathsf{I}_{\mathrm{o}}(\varsigma_1, \Lambda), \quad \mathsf{C}_{1}(\Lambda) > \mathsf{C}_{\mathrm{o}}(\Lambda).$$

The last equalities of the above theorem are related to the work on entropy by Voiculescu (1995).

The authors acknowledge the support under the JSPS Senior Fellowship Program and The Royal Society scheme for UK–Japan research collaboration.

References

Accardi, L. & Ohya, M. 2002 Compound channels, transition expectations and liftings. *J. Appl. Math. Optimiz.* (In the press.)

Araki, H. 1976 Publications Research Institute of Mathematical Sciences, Kyoto University, vol. 11, pp. 809–833.

Belavkin, V. P. 1980 *Radio Engng Electron. Phys.* **25**, 1445–1453.

Belavkin, V. P. 1994 *Found. Phys.* **24**, 685–714.

Belavkin, V. P. & Ohya, M. 1998 Quantum entanglements and entangled mutual entropy. Los Alamos Archive, quant-Ph/9812082, pp. 1–16.

Belavkin, V. P. & Ohya, M. 2000 Entanglements and compound states in quantum information theory. Los Alamos Archive, quant-Ph/0004069, pp. 1–20.

Bennett, C. H., Brassard, G., Crépeau, C., Jozsa, R., Peres, A. & Wootters, W. K. 1993 *Phys. Rev. Lett.* **70**, 1895–1899.

Bennett, C. H., Brassard, G., Popescu, S., Schumacher, B., Smolin, J. A. & Wootters, W. K. 1996 *Phys. Rev. Lett.* **76**, 722–725.

Ekert, A. 1993 *Phys. Rev. Lett.* **67**, 661–663.

Jozsa, R. & Schumacher, B. 1994 *J. Mod. Opt.* **41**, 2343–2350.

Lindblad, G. 1973 *Commun. Math. Phys.* **33**, 305–322.

Ohya, M. 1983*a* *IEEE Trans. Inform. Theory* **29**, 770–774.

Ohya, M. 1983*b* *Nuovo Cim.* **38**, 402–406.

Ohya, M. 1989 *Rep. Math. Phys.* **27** 19–47.

Ohya, M. & Petz, D. 1993 *Quantum entropy and its use*. Springer.

Schrödinger, E. 1935 *Naturwissenschaften* **23**, 807–812, 823–828, 844–849.

Schumacher, B. 1993*a* *Phys. Rev.* A **51**, 2614–2628.

Schumacher, B. 1993*b* *Phys. Rev.* A **51**, 2738–2747.

Stinespring, W. F. 1955 *Proc. Am. Math. Soc.* **6**, 211.

Uhlmann, A. 1977 *Commun. Math. Phys.* **54**, 21–32.

Umegaki, H. 1962 Kodai Mathematical Seminars Report, vol. 14, pp. 59–85.

Voiculescu, D. 1995 *Commun. Math. Phys.* **170**, 249–281.

Werner, R. F. 1989 *Lett. Math. Phys.* **17**, 359–363.

Werner, R. F. 1998 *Phys. Rev.* A **58**, 1827–1832.

JOURNAL OF MATHEMATICAL PHYSICS VOLUME 43, NUMBER 2 FEBRUARY 2002

Semiclassical properties and chaos degree for the quantum Baker's map

Kei Inoue and Masanori Ohya
Department of Information Sciences, Science University of Tokyo, Noda City, Chiba 278-8510, Japan

Igor V. Volovich[a)]
Steklov Mathematical Institute, Russian Academy of Science, Gubkin St. 8, Moscow, GSP1, 117966, Russia

(Received 27 February 2001; accepted for publication 3 October 2001)

We study the chaotic behavior and the quantum-classical correspondence for the Baker's map. Correspondence between quantum and classical expectation values is investigated and it is numerically shown that it is lost at the logarithmic timescale. The quantum chaos degree is computed and it is demonstrated that it describes the chaotic features of the model. The correspondence between classical and quantum chaos degrees is considered. © *2002 American Institute of Physics.*
[DOI: 10.1063/1.1420743]

I. INTRODUCTION

The study of chaotic behavior in classical dynamical systems dates back to Lobachevsky and Hadamard, who have studied the exponential instability property of geodesics on manifolds of negative curvature, and to Poincare, who initiated the inquiry into the stability of the solar system. One believes now that the main features of chaotic behavior in the classical dynamical systems are rather well understood (see, for example, Refs. 1 and 2). However, the status of "quantum chaos" is much less clear although significant progress has been made on this front.

Sometimes one says that an approach to quantum chaos, which attempts to generalize the classical notion of sensitivity to initial conditions, fails for two reasons: first, there is no quantum analog of the classical phase space trajectories and, second, the unitarity of linear Schrödinger equations precludes sensitivity to initial conditions in the quantum dynamics of state vector. Let us remind, however, that in fact there exists a quantum analog of the classical phase space trajectories. It is quantum evolution of expectation values of appropriate observables in suitable states. Also, let us remind that the dynamics of a classical system can be described either by the Hamilton equations or by the linear Liouville equations. In quantum theory the linear Schrödinger equation is the counterpart of the Liouville equation while the quantum counterpart of the classical Hamilton equation is the Heisenberg equation. Therefore, the study of quantum expectation values should reveal the chaotic behavior of quantum systems. In this article we demonstrate this fact for the quantum Baker's map.

If one has the classical Hamilton equations

$$dq/dt=p, \quad dp/dt=-V'(q),$$

then the corresponding quantum Heisenberg equations have the same form

$$dq_\hbar/dt=p_\hbar, \quad dp_\hbar/dt=-V'(q_\hbar),$$

where $q_\hbar$ and $p_\hbar$ are quantum canonical operators of position and momentum. For the expectation values one gets the Ehrenfest equations

[a)]Electronic mail: volovich@mi.ras.ru

Reprinted with permission from K. Inoue, M. Ohya and V. Volovich, *J. Math. Phys.* **43** (2), 734 (February 2002).
© 2002, American Institute of Physics.

$$d\langle q_h\rangle/dt=\langle p_h\rangle, \quad d\langle p_h\rangle/dt=-\langle V'(q_h)\rangle.$$

Note that the Ehrenfest equations are classical equations but for nonlinear $V'(q_h)$ they are neither Hamilton equations nor even differential equations because one can not write $\langle V'(q_h)\rangle$ as a function of $\langle q_h\rangle$ and $\langle p_h\rangle$. However, these equations are very convenient for the consideration of the semiclassical properties of quantum system. The expectation values $\langle q_h\rangle$ and $\langle p_h\rangle$ are functions of time and initial data. They also depend on the quantum states. One of important problems is to study the dependence of expectation values from the initial data. In this article we will study this problem for the quantum Baker's map.

The main objective of "quantum chaos" is to study the correspondence between classical chaotic systems and their quantum counterparts in the semiclassical limit.[3,4] The quantum-classical correspondence for dynamical systems has been studied for many years (see for example Refs. 5–10 and reference therein). A significant progress in understanding this correspondence has been achieved in the Wentzel–Kromers–Brillouin (WKB) approach when one considers the Planck constant h as a small variable parameter. Then it is well known that in the limit $h\rightarrow 0$ quantum theory is reduced to the classical one.[11] However, in physics the Planck constant is a fixed constant although it is very small. Therefore, it is important to study the relation between classical and quantum evolutions when the Planck constant is fixed. There is a conjecture[12–14,8] that a characteristic timescale τ appears in the quantal evolution of chaotic dynamical systems. For time less than τ there is a correspondence between quantum and classical expectation values, while for times greater that τ the predictions of the classical and quantum dynamics no longer coincide. The important problem is to estimate the dependence τ on the Planck constant h. Probably a universal formula expressing τ in terms of h does not exist and every model should be studied case by case. It is expected that certain quantum and classical expectation values diverge on a timescale inversely proportional to some power of h.[15] Other authors suggest that a breakdown may be anticipated on a much smaller logarithmic timescale.[16–23] The characteristic time τ associated with the hyperbolic fixed points of the classical motion is expected to be of the logarithmic form $\tau=(1/\lambda)\ln(C/h)$, where λ is the Lyapunov exponent and C is a constant which can be taken to be the classical action. Such a logarithmic timescale has been found in the numerical simulations of some dynamical models.[7] It was shown also that the discrepancy between quantum and classical evolutions is decreased by even a small coupling with the environment, which in the quantum case leads to decoherence.[7]

The chaotic behavior of the classical dynamical systems is often investigated by computing the Lyapunov exponents. An alternative quantity measuring chaos in dynamical systems, which is called the chaos degree, has been suggested in Ref. 24 in the general framework of information dynamics.[25] The chaos degree was applied to various models in Ref. 26. An advantage of the chaos degree is that it can be applied not only to classical systems but also to quantum systems as well.

In this work we study the chaotic behavior and the quantum-classical correspondence for the Baker's map.[15,27] The quantum Baker's map is a simple model invented for the theoretical study of quantum chaos. Its mathematical properties have been studied in numerical works. In particular its semiclassical properties have been considered,[16–23] quantum computing and optical realizations have been proposed,[28–30] various quantization procedures have been discussed,[18,31–33] and a symbolic dynamics representation has been given.[33]

It is well known that for the consideration of the semiclassical limit in quantum mechanics it is very useful to use coherent states. We define an analog of the coherent states for the quantum Baker's map. We study the quantum Baker's map by using the correlation functions of the special form which corresponds to the expectation values, translated in time by the unitary evolution operator and taken in the coherent states.

To explain our formalism we first discuss the classical limit for correlation functions in ordinary quantum mechanics. Correspondence between quantum and classical expectation values for the Baker's map is investigated and it is numerically shown that it is lost at the logarithmic timescale. The chaos degree for the quantum Baker's map is computed and it is demonstrated that it describes the chaotic features of the model. The dependence of the chaos degree on the Planck constant is studied and the correspondence between classical and quantum chaos degrees is established.

II. QUANTUM VERSUS CLASSICAL DYNAMICS

In this section we discuss an approach to the semiclassical limit in quantum mechanics by using the coherent states (see Ref. 6). Then in the next section an extension of this approach to the quantum Baker's map will be given. Consider the canonical system with the Hamilton function

$$H=\frac{p^2}{2}+V(x) \tag{1}$$

in the plane $(p,x)\in\mathbf{R}^2$. We assume that the canonical equations

$$\dot{x}(t)=p(t), \quad \dot{p}(t)=-V'(x(t)) \tag{2}$$

have a unique solution $(x(t),p(t))$ for times $|t|<T$ with the initial data

$$x(0)=x_0, \quad p(0)=v_0. \tag{3}$$

This is equivalent to the solution of the Newton equation

$$\ddot{x}(t)=-V'(x(t)), \tag{4}$$

with the initial data

$$x(0)=x_0, \quad \dot{x}(0)=v_0. \tag{5}$$

We denote

$$\alpha=\frac{1}{\sqrt{2}}(x_0+iv_0). \tag{6}$$

The quantum Hamilton operator has the form

$$H_h=\frac{p_h^2}{2}+V(q_h),$$

where p_h and q_h satisfy the commutation relations

$$[p_h,q_h]=-ih.$$

The Heisenberg evolution of the canonical variables is defined as

$$p_h(t)=U(t)p_hU(t)^*, \quad q_h(t)=U(t)q_hU(t)^*,$$

where

$$U(t)=\exp(-itH_h/h).$$

For the consideration of the classical limit we take the following representation,

$$p_h=-ih^{1/2}\partial/\partial x, \quad q_h=h^{1/2}x,$$

acting to functions of the variable $x\in\mathbf{R}$. We also set

$$a=\frac{1}{\sqrt{2}h^{1/2}}(q_h+ip_h)=\frac{1}{\sqrt{2}}\left(x+\frac{\partial}{\partial x}\right),\quad a^*=\frac{1}{\sqrt{2}h^{1/2}}(q_h-ip_h)=\frac{1}{\sqrt{2}}\left(x-\frac{\partial}{\partial x}\right).$$

Then,

$$[a,a^*]=1.$$

The coherent state $|\alpha\rangle$ is defined as

$$|\alpha\rangle=W(\alpha)|0\rangle, \tag{7}$$

where α is a complex number, $W(\alpha)=\exp(\alpha a^*-a\alpha^*)$ and $|0\rangle$ is the vacuum vector, $a|0\rangle=0$. The vacuum vector is the solution of the equation

$$(q_h+ip_h)|0\rangle=0. \tag{8}$$

In the x-representation one has

$$|0\rangle=\exp(-x^2/2)/\sqrt{2\pi}. \tag{9}$$

The operator $W(\alpha)$ one can write also in the form

$$W(\alpha)=Ce^{iq_hv_0/h^{1/2}}e^{-ip_hx_0/h^{1/2}}, \tag{10}$$

where $C=\exp(-v_0x_0/2h)$.

The mean value of the position operator with respect to the coherent vectors is the real valued function

$$q(t,\alpha,h)=\langle h^{-1/2}\alpha|q_h(t)|h^{-1/2}\alpha\rangle. \tag{11}$$

Now one can present the following basic formula describing the semiclassical limit

$$\lim_{h\to 0} q(t,\alpha,h)=x(t,\alpha). \tag{12}$$

Here $x(t,\alpha)$ is the solution of (4) with the initial data (5) and α is given by (6).

Let us notice that for time $t=0$ the quantum expectation value $q(t,\alpha,h)$ is equal to the classical one:

$$q(0,\alpha,h)=x(0,\alpha)=x_0 \tag{13}$$

for any h. We are going to compare the time dependence of two real functions $q(t,\alpha,h)$ and $x(t,\alpha)$; these functions are approximately equal. The important problem is to estimate for which t the large difference between them will appear. It is expected that certain quantum and classical expectation values diverge on a timescale inversely proportional to some power of h.[15] Other authors suggest that a breakdown may be anticipated on a much smaller logarithmic timescale.[16–23] One of very interesting examples[5] of classical systems with chaotic behavior is described by the Hamilton function

$$H=\frac{p_1^2}{2}+\frac{p_2^2}{2}+\lambda x_1^2x_2^2.$$

The consideration of this classical and quantum model within the described framework will be presented in another publication.

III. COHERENT STATES FOR THE QUANTUM BAKER'S MAP

The classical Baker's transformation maps the unit square $0 \leqslant q,\ p \leqslant 1$ onto itself according to

$$(q,p) \rightarrow \begin{cases} (2q, p/2), & \text{if } \ 0 \leqslant q \leqslant \frac{1}{2}, \\ (2q-1,(p+1)/2), & \text{if } \ \frac{1}{2} < q \leqslant 1. \end{cases}$$

This corresponds to compressing the unit square in the p direction and stretching it in the q direction, while preserving the area, then cutting it vertically and stacking the right part on top of the left part.

The classical Baker's map has a simple description in terms of its symbolic dynamics.[34] Each point (q,p) is represented by a symbolic string

$$\xi = \cdots \xi_{-2}\xi_{-1}\xi_0 \cdot \xi_1 \xi_2 \cdots, \tag{14}$$

where $\xi_k \in \{0,1\}$, and

$$q = \sum_{k=1}^{\infty} \xi_k 2^{-k}, \quad p = \sum_{k=0}^{\infty} \xi_{-k} 2^{-k-1}.$$

The action of the Baker's map on a symbolic string ξ is given by the shift map (Bernoulli shift) U defined by $U\xi = \xi'$, where $\xi_k' = \xi_{k+1}$. This means that, at each time step, the dot is shifted one place to the right while the entire string remains fixed. After n steps the q coordinate becomes

$$q_n = \sum_{k=1}^{\infty} \xi_{n+k} 2^{-k}. \tag{15}$$

This relation defines the classical trajectory with the initial data

$$q = q_0 = \sum_{k=1}^{\infty} \xi_k 2^{-k}. \tag{16}$$

Quantum Baker's maps are defined on the D-dimensional Hilbert space of the quantized unit square. To quantize the unit square one defines the Weyl unitary displacement operators $\hat{U}$ and $\hat{V}$ in D—dimensional Hilbert space, which produces displacements in the momentum and position directions, respectively, and the following commutation relation is obeyed,

$$\hat{U}\hat{V} = \epsilon \hat{V}\hat{U},$$

where $\epsilon = \exp(2\pi i/D)$. We choose $D = 2^N$, so that our Hilbert space will be the N qubit space $\mathbf{C}^{\otimes N}$. The constant $h = 1/D = 2^{-N}$ can be regarded as the Plank constant. The space $\mathbf{C}^2$ has a basis

$$|0\rangle = \begin{pmatrix} 1 \\ 0 \end{pmatrix}, \quad |1\rangle = \begin{pmatrix} 0 \\ 1 \end{pmatrix}.$$

The basis in $\mathbf{C}^{\otimes N}$ is

$$|\xi_1\rangle \otimes |\xi_2\rangle \otimes \cdots \otimes |\xi_N\rangle, \quad \xi_k = 0,1.$$

We write

$$\xi = \sum_{k=1}^{N} \xi_k 2^{N-k}.$$

Then $\xi=0,1,\ldots,2^N-1$ and we denote

$$|\xi\rangle=|\xi_1\xi_2\cdots\xi_N\rangle=|\xi_1\rangle\otimes|\xi_2\rangle\otimes\cdots\otimes|\xi_N\rangle.$$

We will also use for this basis notations $\{|\eta\rangle=|\eta_1\eta_2\cdots\eta_N\rangle,\ \eta_k=0,1\}$ and $\{|j\rangle=|j_1j_2\cdots j_N\rangle,j_k=0,1\}$.

The operators $\hat{U}$ and $\hat{V}$ can be written as

$$\hat{U}=e^{2\pi i\hat{q}},\quad \hat{V}=e^{2\pi i\hat{p}},$$

where the position and momentum operators $\hat{q}$ and $\hat{p}$ are operators in $\mathbf{C}^{\otimes N}$ which are defined as follows. The position operator is

$$\hat{q}=\sum_{j=0}^{2^N-1} q_j|j\rangle\langle j|=\sum_{j_1,\ldots j_N} q_j|j_N\ldots j_1\rangle\langle j_1\ldots j_N|,$$

where

$$|j\rangle=|j_1j_2\cdots j_N\rangle,\quad j_k=0,1,$$

is the basis in $\mathbf{C}^{\otimes N}$,

$$j=\sum_{k=1}^{N} j_k 2^{N-k},$$

and

$$q_j=\frac{j+\frac{1}{2}}{2^N},\quad j=0,1,\ldots,2^N-1.$$

The momentum operator is defined as

$$\hat{p}=F_N\hat{q}F_N^*,$$

where F_N is the quantum Fourier transform acting to the basis vectors as

$$F_N|j\rangle=\frac{1}{\sqrt{D}}\sum_{\xi=0}^{D-1} e^{2\pi i\xi j/D}|\xi\rangle,$$

where $D=2^N$.

The symbolic representation of quantum Baker's map T was introduced by Schack and Caves[33] and studied in Refs. 35 and 36. Let us explain the symbolic representation of the quantum Baker's map as a special case:[33] By applying a partial quantum Fourier transform

$$G_m=\overbrace{I\otimes\cdots\otimes I}^{m}\otimes F_{N-m}$$

to the position eigenstates, one obtains the following quantum Baker's map T:

$$T|.\xi_1\cdots\xi_N\rangle\equiv|\xi_1.\xi_2\cdots\xi_N\rangle,$$

where

$$T=G_{N-1}\circ G_N^{-1}$$

and

$$|\xi_1\cdots\xi_{N-m}.\xi_{N-m+1}\cdots\xi_N\rangle \equiv G_m|\xi_{N-m+1}\cdots\xi_N\xi_{N-m}\cdots\xi_1\rangle$$
$$=|\xi_{N-m+1}\rangle\otimes\cdots\otimes|\xi_N\rangle\otimes F_{N-m}|\xi_{N-m}\rangle\otimes\cdots\otimes|\xi_1\rangle.$$

The quantum Baker's map T is the unitary operator in $\mathbf{C}^{\otimes N}$ with the following matrix elements,

$$\langle\xi|T|\eta\rangle=\frac{1-i}{2}\exp\left(\frac{\pi}{2}i|\xi_1-\eta_N|\right)\prod_{k=2}^{N}\delta(\xi_k-\eta_{k-1}), \tag{17}$$

where $|\xi\rangle=|\xi_1\xi_2\cdots\xi_N\rangle$, $|\eta\rangle=|\eta_1\eta_2\cdots\eta_N\rangle$ and $\delta(x)$ is the Kronecker symbol, $\delta(0)=1$; $\delta(x)=0$, $x\neq 0$.

We define the coherent states by

$$|\alpha\rangle=Ce^{2\pi i\hat{q}v}e^{-2\pi i\hat{p}x}|\psi_0\rangle. \tag{18}$$

Here $\alpha=x+iv$, x and v are integers, C is the normalization constant and $|\psi_0\rangle$ is the vacuum vector. This definition should be compared with (10). The vacuum vector can be defined as the solution of the equation

$$(q_h+ip_h)|\psi_0\rangle=0$$

[compare with (8)]. We will use the simpler definition which in the position representation is

$$\langle q_j|\psi_0\rangle=C\exp(-q_j^2/2)$$

[compare with (9)]. Here C is a normalization constant.

IV. CHAOS DEGREE

Let us review the entropic chaos degree defined in Ref. 24. This entropic chaos degree is given by a probability distribution φ and a dynamics (channel) Λ^* sending a state to a state; $\varphi=\Sigma_k p_k\delta_k$, where δ_k is the delta measure such as

$$\delta_k(j)\equiv\begin{cases}1 & (k=j),\\ 0 & (k\neq j).\end{cases}$$

Then the entropic chaos degree is defined as

$$D(\varphi;\Lambda^*)=\sum_k p_k S(\Lambda^*\delta_k) \tag{19}$$

with the von Neumann entropy S, equivalent to the Shannon entropy because the probability distribution φ is a classical object.

A dynamics $\mathcal{F}$ of the orbit produces the above channel Λ^*, so let $\{x_n\}$ be the orbit and $\mathcal{F}$ be a map from x_n to x_{n+1}.

Take a finite partition $\{B_k\}$ of $I=[a,b]^l$ $(a,b\in\mathbf{R})\subset\mathbf{R}^l$ such as

$$I=\bigcup_k B_k \quad (B_i\cap B_j=\varnothing, i\neq j)$$

for a map $\mathcal{F}$ on I with $x_{n+1}=\mathcal{F}(x_n)$ (a difference equation). The state $\varphi^{(n)}$ of the orbit determined by the difference equation is defined by the probability distribution $(p_i^{(n)})$, that is, $\varphi^{(n)}=p^{(n)}=\Sigma_i p_i^{(n)}\delta_i$, where for an initial value $x\in I$ and the characteristic function 1_A

$$p_i^{(n)} \equiv \frac{1}{m+1} \sum_{k=n}^{m+n} 1_{B_i}(\mathcal{F}^k x).$$

When the initial value x is distributed due to a measure ν on I, the above $p_i^{(n)}$ is given as

$$p_i^{(n)} \equiv \frac{1}{m+1} \int_I \sum_{k=n}^{m+n} 1_{B_i}(\mathcal{F}^k x) d\nu.$$

In the case that $\mathcal{F}$ is a classical Baker's transformation, if the orbit is not stable and periodic, then it is shown that the $m \to \infty$ limit of $p_i^{(n)}$ exists and equals a natural invariant measure for a fixed $n \in \mathbf{N}$.[37]

The joint distribution $(p_{ij}^{(n,n+1)})$ between the time n and $n+1$ is defined by

$$p_{ij}^{(n,n+1)} \equiv \frac{1}{m+1} \sum_{k=n}^{m+n} 1_{B_i}(\mathcal{F}^k x) 1_{B_j}(\mathcal{F}^{k+1} x)$$

or

$$p_{ij}^{(n,n+1)} \equiv \frac{1}{m+1} \int_I \sum_{k=n}^{m+n} 1_{B_i}(\mathcal{F}^k x) 1_{B_j}(\mathcal{F}^{k+1} x) d\nu.$$

Then the channel Λ_n^* at n is determined by

$$\Lambda_n^* \equiv \left(\frac{p_{ij}^{(n,n+1)}}{p_i^{(n)}} \right) \Rightarrow p^{(n+1)} = \Lambda_n^* p^{(n)},$$

and the chaos degree is given by

$$D_c(p^{(n)}; \Lambda_n^*) = \sup_{\{B_k\}} \left\{ \sum_i p_i^{(n)} S(\Lambda_n^* \delta_i) = \sum_{i,j} p_{ij}^{(n,n+1)} \log \frac{p_i^{(n)}}{p_{ij}^{(n,n+1)}} ; \{B_k\} \right\}. \tag{20}$$

We can judge whether the dynamics causes a chaos or not by the value of D as

$$D > 0 \Leftrightarrow \text{chaotic}, \quad D = 0 \Leftrightarrow \text{stable}.$$

Therefore, it is enough to find a partition $\{B_k\}$ such that D is positive when the dynamics produces chaos.

This classical chaos degree was applied to several dynamical maps, such as a logistic map, a Baker's transformation and a Tinkerbel map, and it could explain their chaotic characters.[24,26] Our chaos degree has several merits compared with usual measures such as the Lyapunov exponent.

V. EXPECTATION VALUES AND CHAOS DEGREE

In this section, we show a general representation of the mean value of the position operator $\hat{q}$ for the time evolution, which is constructed by the quantum Baker's map. Then we give the algorithm to compute the chaos degree for the quantum Baker's map.

To study the time evolution and the classical limit $h \to 0$ which corresponds to $N \to \infty$ of the quantum Baker's map T, we introduce the following the mean value of the position operator $\hat{q}$ for time $n \in \mathbf{N}$ with respect to a single basis $|\xi\rangle$:

$$r_n^{(N)} = \langle \xi | T^n \hat{q} T^{-n} | \xi \rangle, \tag{21}$$

where $|\xi\rangle = |\xi_1\xi_2\cdots\xi_N\rangle$.

From (17), the following formula of the matrix elements of T^n for any $n\in\mathbf{N}$ is easily obtained:

$$\langle\xi|T^n|\zeta\rangle=\begin{cases}\left(\frac{1-i}{2}\right)^n\left(\prod_{k=1}^{N-n}\delta(\xi_{n+k}-\zeta_k)\right)\left(\prod_{l=1}^{n}A_{\xi_l\zeta_{N-n+l}}\right) & \text{if } n<N,\\ \left(\frac{1-i}{2}\right)^n\left(\prod_{k=1}^{n}A_{\xi_k\zeta_k}\right) & \text{if } n=N,\\ \left(\frac{1-i}{2}\right)^n\left(\prod_{k=1}^{p}(A^{m+1})_{\xi_k\zeta_{N-p+k}}\right)\left(\prod_{l=1}^{N-p}(A^m)_{\xi_{p+l}\zeta_l}\right) & \text{if } n=mN+p,\\ \left(\frac{1-i}{2}\right)^n\prod_{k=1}^{N}(A^m)_{\xi_k\zeta_k} & \text{if } n=mN,\end{cases} \tag{22}$$

where A is the 2×2 matrix with the element $A_{x_1x_2}=\exp((\pi/2)\,i|x_1-x_2|)$ for $x_1,x_2=0,\ 1,\ p=1,\ldots,N-1$ and $m\in\mathbf{N}$.

Using these formulas, the following theorems are obtained and their proofs are given in the Appendix.

Theorem 5.1:

$$r_n^{(N)}=\begin{cases}\sum_{k=1}^{N-n}\xi_{n+k}2^{-k}+\frac{2^n}{2^{N+1}} & \text{if } n<N,\\ \frac{1}{2} & \text{if } n=N,\\ \frac{1}{2^n}\sum_{j=0}^{2^N-1}\frac{j+\frac{1}{2}}{2^N}\prod_{k=1}^{p}|(A^{m+1})_{\xi_kj_{N-p+k}}|^2\prod_{l=1}^{N-p}|(A^m)_{\xi_{p+l}j_l}|^2 & \text{if } n=mN+p,\\ \frac{1}{2^n}\sum_{j=0}^{2^N-1}\frac{j+\frac{1}{2}}{2^N}\prod_{k=1}^{N}|(A^m)_{\xi_kj_k}|^2 & \text{if } n=mN,\end{cases} \tag{23}$$

where A is the 2×2 matrix with the element $A_{x_1x_2}=\exp((\pi/2)\,i|x_1-x_2|)$ for $x_1,x_2=0,\ 1,\ p=1,\ldots,N-1$ and $m\in\mathbf{N}$.

By diagonalizing the matrix A, we obtain the following formula of the absolute square of the matrix elements of A^n for any $n\in\mathbf{N}$.

Lemma 5.2: For any $n\in\mathbf{N}$, we have

$$|(A^n)_{kj}|^2=\begin{cases}2^n\cos^2\left(\frac{n\pi}{4}\right) & \text{if } k=j\\ 2^n\sin^2\left(\frac{n\pi}{4}\right) & \text{if } k\neq j\end{cases}$$

Combining Theorem 5.1 and Lemma 5.2, we obtain the following two theorems with respect to the mean value $r_n^{(N)}$ of the position operator.

Theorem 5.3: *For the case $n=mN+p$, $p=1,2,\ldots,N-1$ and $m\in\mathbf{N}$, we have*

$$r_n^{(N)}=\begin{cases}\displaystyle\sum_{k=1}^{N-p}\xi_{p+k}2^{-k}+\frac{2^p}{2^{N+1}} & \text{if } m=0(\text{mod}\,4),\\ \displaystyle\sum_{k=N-p+1}^{N}\eta_{k-(N-p)}2^{-k}+\frac{2^N-2^p+1}{2^{N+1}} & \text{if } m=1(\text{mod}\,4),\\ \displaystyle\sum_{k=1}^{N-p}\eta_{p+k}2^{-k}+\frac{2^p}{2^{N+1}} & \text{if } m=2(\text{mod}\,4),\\ \displaystyle\sum_{k=N-p+1}^{N}\xi_{k-(N-p)}2^{-k}+\frac{2^N-2^p+1}{2^{N+1}} & \text{if } m=3(\text{mod}\,4),\end{cases} \tag{24}$$

where $\eta_k=\xi_k+1(\text{mod}2)$, $k=1,\ldots,N$.

Theorem 5.4: *For the case* $n=mN$, $m\in\mathbf{N}$, *we have*

$$r_N^{(n)}=\begin{cases}\displaystyle\sum_{k=1}^{N}\xi_k2^{-k}+\frac{1}{2^{N+1}} & \text{if } m=0(\text{mod }4),\\ \displaystyle\frac{1}{2} & \text{if } m=1,3(\text{mod }4),\\ \displaystyle\sum_{k=1}^{N}\eta_k2^{-k}+\frac{1}{2^{N+1}} & \text{if } m=2(\text{mod }4).\end{cases} \tag{25}$$

Using formulas (23)–(25), the probability distribution $(p_i^{(n)})$ of the orbit of mean value $r_n^{(N)}$ of the position operator $\hat{q}$ for the time evolution, which is constructed by the quantum Baker's map, is given by

$$p_i^{(n)}\equiv\frac{1}{m+1}\sum_{k=n}^{m+n}1_{B_i}(r_n^{(N)})$$

for an initial value $r_0^{(N)}\in[0,1]$ and the characteristic function 1_A. The joint distribution $(p_{ij}^{(n,n+1)})$ between the time n and $n+1$ is given by

$$p_{ij}^{(n,n+1)}\equiv\frac{1}{m+1}\sum_{k=n}^{m+n}1_{B_i}(r_k^{(N)})1_{B_j}(r_{k+1}^{(N)}).$$

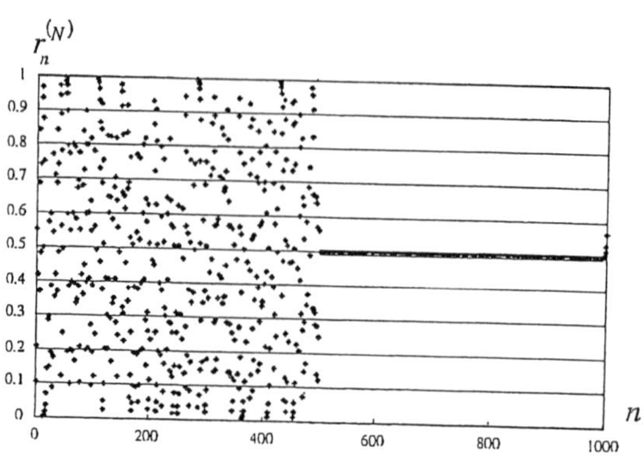

FIG. 1. The distribution of $r_N^{(n)}$ for the case $N=500$.

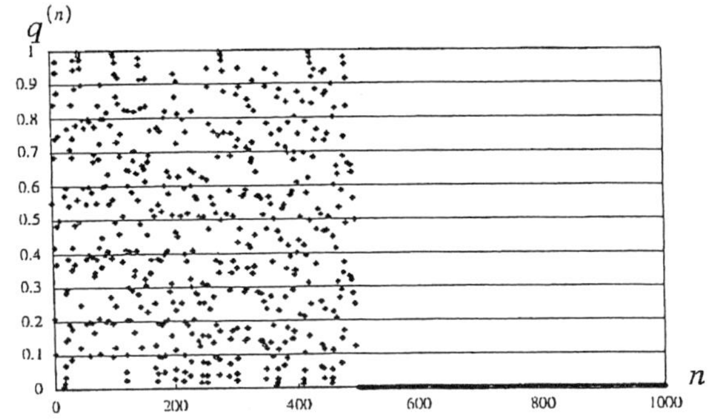

FIG. 2. The distribution of the classical value $q^{(n)}$ for the case $N=500$.

Thus the chaos degree for the quantum Baker's map is calculated by

$$D_q(p^{(n)};\Lambda_n^*)=\sum_{i,j} p_{ij}^{(n,n+1)}\log\frac{p_i^{(n)}}{p_{ij}^{(n,n+1)}}, \tag{26}$$

whose numerical value is shown in the next section.

VI. NUMERICAL SIMULATION OF THE CHAOS DEGREE AND CLASSICAL-QUANTUM CORRESPONDENCE

We compare the dynamics of the mean value $r_n^{(N)}$ of position operator $\hat{q}$ with that of the classical value q_n in the q direction. We take an initial value of the mean value as

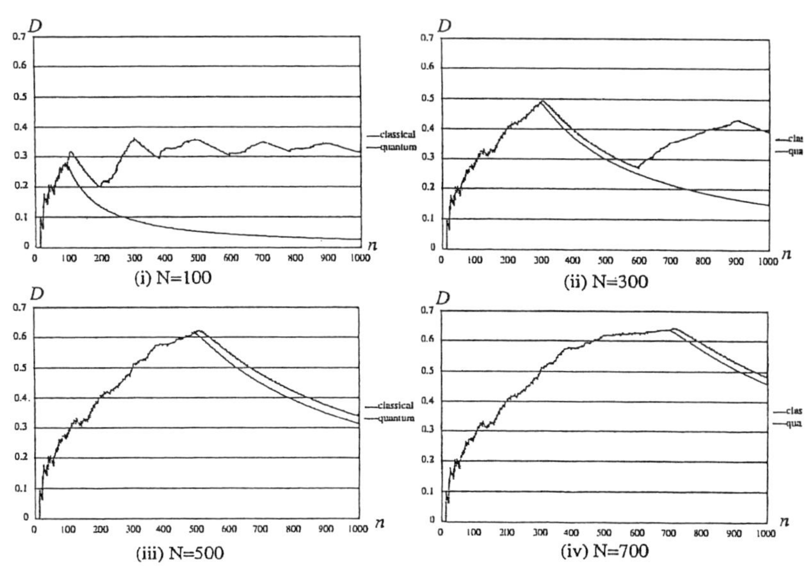

FIG. 3. The change of the chaos degree for several N's up to time $n=1000$.

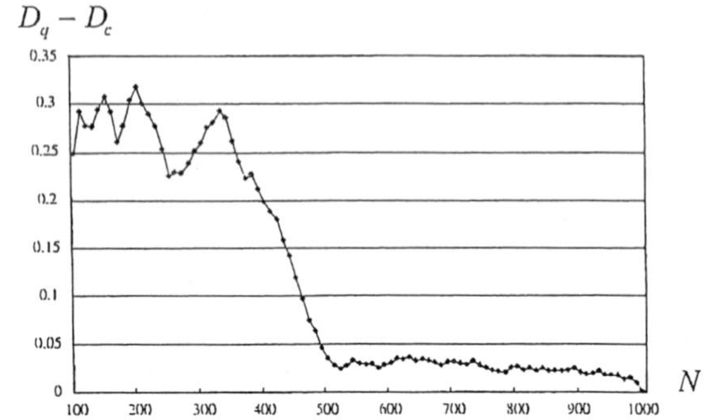

FIG. 4. The difference of the chaos degree between quantum and classical for the case $n=1000$.

$$r_0^{(N)}=\sum_{l=1}^{N} \xi_l 2^{-l}+\frac{1}{2^{N+1}}=0\cdot\xi_1\xi_2\cdots\xi_N 1,$$

where ξ_l is a pseudo-random number valued with 0 or 1. At the time zero we assume that the classical value q_0 in the q direction takes the same value as the mean value $r_0^{(N)}$ of position operator $\hat{q}$. The distribution of $r_n^{(N)}$ for the case $N=500$ is shown in Fig. 1 up to the time $n=1000$. The distribution of the classical value q_n for the case $N=500$ in the q direction is shown in Fig. 2 up to the time $n=1000$.

Figure 3 represents the change of the chaos degree for the case $N=100,300,500,700$ up to the time $n=1000$.

The correspondence between the chaos degree D_q for the quantum Baker's map and the chaos degree D_c for the classical Baker's map for some fixed N's (100,300,500,700 here) is shown for the time less than $T=\log_2(1/h)=\log_2 2^N=N$, and it is lost at the logarithtic time scale T. Here we took a finite partition $\{B_k\}$ of $I=[0,1]$ such as $B_k=[k/100,(k+1)/100))(k=0,1,\ldots,98)$ and $B_{99}=[99/100,1]$ to compute the chaos degree numerically.

The difference of the chaos degrees between the chaos degree D_q for the quantum Baker's map and the chaos degree D_c for the classical Baker's map for a fixed time n (1000, here) is displayed w.r.t. N in Fig. 4.

Thus we conclude that the dynamics of the mean value $r_n^{(N)}$ reduces the classical dynamics q_n in the q direction in the classical limit $N\to\infty(h\to 0)$.

The appearance of the logarithmic timescale have been proved rigorously in our recent paper.[38]

ACKNOWLEDGMENTS

The main part of this work was done during the visit of I.V. to the Science University of Tokyo. I.V. is grateful to JSPS for the Fellowship award. Our work was also partially supported by RFFI 99-0100866, INTAS 99-00545 and SCAT.

APPENDIX A:

Proof of Theorem 5.1: By a direct calculation, we obtain

$$
\begin{aligned}
r_n^{(N)} &= \langle \xi | T^n \hat{q} T^{-n} | \xi \rangle \\
&= \langle \xi | T^n \left(\sum_{j=0}^{2^N-1} \frac{j+\frac{1}{2}}{2^N} |j\rangle\langle j| \right) T^{-n} | \xi \rangle \\
&= \sum_{j=0}^{2^N-1} \frac{j+\frac{1}{2}}{2^N} \langle \xi | T^n | j \rangle \langle j | T^{-n} | \xi \rangle \\
&= \sum_{j=0}^{2^N-1} \frac{j+\frac{1}{2}}{2^N} \langle \xi | T^n | j \rangle \langle j | T^{*n} | \xi \rangle \\
&= \sum_{j=0}^{2^N-1} \frac{j+\frac{1}{2}}{2^N} \langle \xi | T^n | j \rangle \overline{\langle \xi | T^n | j \rangle} \\
&= \sum_{j=0}^{2^N-1} \frac{j+\frac{1}{2}}{2^N} |\langle \xi | T^n | j \rangle|^2 .
\end{aligned}
$$

Using (22), the mean value $r_n^{(N)}$ in the case $n<N$ can be expressed as

$$
\begin{aligned}
r_n^{(N)} &= \sum_{j=0}^{2^N-1} \frac{j+\frac{1}{2}}{2^N} |\langle \xi | T^n | j \rangle|^2 \\
&= \sum_{j=0}^{2^N-1} \frac{j+\frac{1}{2}}{2^N} \left| \left(\frac{1-i}{2} \right)^n \left(\prod_{k=1}^{N-n} \delta(\xi_{n+k}-j_k) \right) \left(\prod_{l=1}^{n} A_{\xi_l j_{N-n+l}} \right) \right|^2 \\
&= \sum_{j=0}^{2^N-1} \frac{j+\frac{1}{2}}{2^N} \left(\frac{1-i}{2} \right)^n \left(\prod_{k=1}^{N-n} \delta(\xi_{n+k}-j_k) \right) \left(\prod_{l=1}^{n} A_{\xi_l j_{N-n+l}} \right) \\
&\quad \times \overline{\left(\frac{1-i}{2} \right)^n \left(\prod_{k=1}^{N-n} \delta(\xi_{n+k}-j_k) \right) \left(\prod_{l-1}^{n} A_{\xi_l j_{N-n+l}} \right)} \\
&= \sum_{j=0}^{2^N-1} \frac{j+\frac{1}{2}}{2^N} \left(\frac{1-i}{2} \right)^n \left(\frac{1+i}{2} \right)^n \left(\prod_{k=1}^{N-n} \delta(\xi_{n+k}-j_k) \right) \left(\prod_{l=1}^{n} |A_{\xi_l j_{N-n+l}}|^2 \right) \\
&= \sum_{j=0}^{2^N-1} \frac{j+\frac{1}{2}}{2^N} \left(\frac{1-i}{2} \right)^n \left(\frac{1+i}{2} \right)^n \left(\prod_{k=1}^{N-n} \delta(\xi_{n+k}-j_k) \right) \\
&= \frac{1}{2^{N+n}} \sum_{j_1,\cdots j_N} \left\{ \left(\sum_{k=1}^{N} j_k 2^{N-k} \right) + \frac{1}{2} \right\} \left(\prod_{k=1}^{N-n} \delta(\xi_{n+k}-j_k) \right) \\
&= \frac{1}{2^{N+n}} \sum_{j_1,\cdots j_N} \left(\sum_{k=1}^{N} j_k 2^{N-k} \right) \left(\prod_{k=1}^{N-n} \delta(\xi_{n+k}-j_k) \right) + \frac{1}{2^{N+n+1}} \sum_{j_1,\cdots j_N} \left(\prod_{k=1}^{N-n} \delta(\xi_{n+k}-j_k) \right) \\
&= \frac{1}{2^{N+n}} \sum_{j_{N-n+1},\cdots j_N} \left(\sum_{l=1}^{N-n} \xi_{n+l} 2^{N-l} + \sum_{l=N-n+1}^{N} j_l 2^{N-l} \right) + \frac{1}{2^{N+n+1}} \left(\sum_{j_{N-n+1},\cdots j_N} 1 \right)
\end{aligned}
$$

$$
\begin{aligned}
&= \frac{1}{2^{N+n}}\left(\sum_{l=1}^{N-n} \xi_{n+l}2^{N-l}\right)\left(\sum_{j_{N-n+1},\cdots j_N} 1\right) + \frac{1}{2^{N+n}}\sum_{j_{N-n+1},\cdots j_N}\left(\sum_{l=N-n+1}^{N} j_l 2^{N-l}\right) \\
&\quad + \frac{1}{2^{N+n+1}}\left(\sum_{j_{N-n+1},\cdots j_N} 1\right) \\
&= \frac{2^n}{2^{N+n}}\left(\sum_{l=1}^{N-n} \xi_{n+l}2^{N-l}\right) + \frac{1}{2^{N+n}}\sum_{j_{N-n+1},\cdots j_N}\left(\sum_{l=N-n+1}^{N} j_l 2^{N-l}\right) + \frac{2^n}{2^{N+n+1}} \\
&= \frac{1}{2^{N}}\left(\sum_{l=1}^{N-n} \xi_{n+l}2^{N-l}\right) + \frac{1}{2^{N+n}}\sum_{j_{N-n+1},\cdots j_N}\left(\sum_{l=1}^{n} j_{N-n+l} 2^{n-l}\right) + \frac{1}{2^{N+1}} \\
&= \frac{1}{2^{N}}\left(\sum_{l=1}^{N-n} \xi_{n+l}2^{N-l}\right) + \frac{1}{2^{N+n}}\frac{1}{2}(2^n-1)2^n + \frac{1}{2^{N+1}} \\
&= \frac{1}{2^{N}}\left(\sum_{l=1}^{N-n} \xi_{n+l}2^{N-l}\right) + \frac{2^n}{2^{N+1}}.
\end{aligned}
$$

For the case $n=N$, we similarly obtain

$$
\begin{aligned}
r_n^{(N)} &= \sum_{j=0}^{2^N-1} \frac{j+\frac{1}{2}}{2^N} |\langle \xi | T^n | j\rangle|^2 \\
&= \sum_{j=0}^{2^N-1} \frac{j+\frac{1}{2}}{2^N}\left|\left(\frac{1-i}{2}\right)^N\left(\prod_{k=1}^{N} A_{\xi_k j_k}\right)\right|^2 \\
&= \sum_{j=0}^{2^N-1} \frac{j+\frac{1}{2}}{2^N}\left|\left(\frac{1-i}{2}\right)^N\right|^2 \prod_{k=1}^{N} |A_{\xi_k j_k}|^2 \\
&= \frac{1}{2^{2N}} \sum_{j=0}^{2^N-1}\left(j+\frac{1}{2}\right) \\
&= \frac{1}{2^{2N}}\frac{1}{2}(2^N-1)2^N + \frac{1}{2^{N+1}} = \frac{1}{2}.
\end{aligned}
$$

For $n=mN+p,\ p=1,2,\ldots,N-1,\ m\in\mathbf{N}$,

$$
\begin{aligned}
r_n^{(N)} &= \sum_{j=0}^{2^N-1} \frac{j+\frac{1}{2}}{2^N} |\langle \xi | T^n | j\rangle|^2 \\
&= \sum_{j=0}^{2^N-1} \frac{j+\frac{1}{2}}{2^N}\left|\left(\frac{1-i}{2}\right)^n\left(\prod_{k=1}^{p} (A^{m+1})_{\xi_k j_{N-p+k}}\right)\left(\prod_{l=1}^{N-p} (A^m)_{\xi_{p+l} j_l}\right)\right|^2 \\
&= \frac{1}{2^n}\sum_{j=0}^{2^N-1} \frac{j+\frac{1}{2}}{2^N}\prod_{k=1}^{p} |(A^{m+1})_{\xi_k j_{N-p+k}}|^2 \prod_{l=1}^{N-p} |(A^m)_{\xi_{p+l} j_l}|^2,
\end{aligned}
$$

and for $n=mN,\ m\in\mathbf{N}$,

$$\begin{aligned} r_N^{(n)} &= \sum_{j=0}^{2^N-1} \frac{j+\frac{1}{2}}{2^N} |\langle \xi | T^n | j \rangle|^2 \\ &= \sum_{j=0}^{2^N-1} \frac{j+\frac{1}{2}}{2^N} \left| \left(\frac{1-i}{2} \right)^n \prod_{k=1}^{N} (A^m)_{\xi_k j_k} \right|^2 \\ &= \frac{1}{2^n} \sum_{j=0}^{2^N-1} \frac{j+\frac{1}{2}}{2^N} \prod_{k=1}^{N} |(A^m)_{\xi_k j_k}|^2. \end{aligned}$$

Proof of Lemma 5.2: By a direct calculation, the matrix A is diagonalized as follows:

$$A = FDF^*, \tag{A1}$$

where

$$F = \frac{1}{\sqrt{2}} \begin{pmatrix} 1 & -1 \\ 1 & 1 \end{pmatrix}, \quad D = \begin{pmatrix} 1+i & 0 \\ 0 & 1-i \end{pmatrix}.$$

From (A1), we have

$$\begin{aligned} A^n &= FD^nF^* \\ &= \frac{1}{\sqrt{2}} \begin{pmatrix} 1 & -1 \\ 1 & 1 \end{pmatrix} \begin{pmatrix} (1+i)^n & 0 \\ 0 & (1-i)^n \end{pmatrix} \frac{1}{\sqrt{2}} \begin{pmatrix} 1 & 1 \\ -1 & 1 \end{pmatrix} \\ &= \frac{1}{2} \begin{pmatrix} (1+i)^n + (1-i)^n & (1+i)^n - (1-i)^n \\ (1+i)^n - (1-i)^n & (1+i)^n + (1-i)^n \end{pmatrix}. \end{aligned} \tag{A2}$$

Using (A2), it follows that for any $k=j$, $k=1,2$,

$$\begin{aligned} |(A^n)_{kj}|^2 &= \tfrac{1}{2}\{(1+i)^n + (1-i)^n\} \overline{\tfrac{1}{2}\{(1+i)^n + (1-i)^n\}} \\ &= \tfrac{1}{4}\{(1+i)^n + (1-i)^n\}\{(1-i)^n + (1+i)^n\} \\ &= \tfrac{1}{4}\{(1+i)^n + (1-i)^n\}^2 \\ &= \frac{1}{4} \left\{ \left(\sqrt{2} \frac{1+i}{\sqrt{2}} \right)^n + \left(\sqrt{2} \frac{1-i}{\sqrt{2}} \right)^n \right\}^2 \\ &= \frac{1}{4} \left\{ (\sqrt{2})^n \left(\frac{1+i}{\sqrt{2}} \right)^n + (\sqrt{2})^n \left(\frac{1-i}{\sqrt{2}} \right)^n \right\}^2 \\ &= \frac{2^n}{4} \left\{ \left(\exp\left(\frac{\pi}{4} i \right) \right)^n + \left(\exp\left(-\frac{\pi}{4} i \right) \right)^n \right\}^2 \\ &= \frac{2^n}{4} \left\{ \exp\left(\frac{n\pi}{4} i \right) + \exp\left(-\frac{n\pi}{4} i \right) \right\}^2 \\ &= \frac{2^n}{4} \left[\left\{ \cos\left(\frac{n\pi}{4} \right) + i \sin\left(\frac{n\pi}{4} \right) \right\} + \left\{ \cos\left(\frac{n\pi}{4} \right) - i \sin\left(\frac{n\pi}{4} \right) \right\} \right]^2 \\ &= \frac{2^n}{4} \left\{ 2 \cos\left(\frac{n\pi}{4} \right) \right\}^2 = 2^n \cos^2\left(\frac{n\pi}{4} \right) \end{aligned}$$

and for any $k \neq j$, $k=1,2$,

$$
\begin{aligned}
|(A^n)_{kj}|^2 &= \tfrac{1}{2i}\{(1+i)^n-(1-i)^n\}\overline{\tfrac{1}{2i}\{(1+i)^n-(1-i)^n\}} \\
&= \tfrac{1}{4i}\{(1+i)^n-(1-i)^n\}\{(1-i)^n-(1+i)^n\} \\
&= -\tfrac{1}{4}\{(1+i)^n-(1-i)^n\}^2 \\
&= -\frac{1}{4}\left\{\left(\sqrt{2}\frac{1+i}{\sqrt{2}}\right)^n-\left(\sqrt{2}\frac{1-i}{\sqrt{2}}\right)^n\right\}^2 \\
&= -\frac{1}{4}\left\{(\sqrt{2})^n\left(\frac{1+i}{\sqrt{2}}\right)^n-(\sqrt{2})^n\left(\frac{1-i}{\sqrt{2}}\right)^n\right\}^2 \\
&= -\frac{2^n}{4}\left\{\left(\exp\left(\frac{\pi}{4}i\right)\right)^n-\left(\exp\left(-\frac{\pi}{4}i\right)\right)^n\right\}^2 \\
&= -\frac{2^n}{4}\left[\left\{\cos\left(\frac{n\pi}{4}\right)+i\sin\left(\frac{n\pi}{4}\right)\right\}-\left\{\cos\left(\frac{n\pi}{4}\right)-i\sin\left(\frac{n\pi}{4}\right)\right\}\right]^2 \\
&= -\frac{2^n}{4}\left\{2i\sin\left(\frac{n\pi}{4}\right)\right\}^2 \\
&= 2^n\sin^2\left(\frac{n\pi}{4}\right).
\end{aligned}
$$

■

Proof of Theorem 5.3: For the case $n=mN+p$, $p=1,\ldots,N-1$ and $m\in\mathbf{N}$,

$$
r_n^{(N)}=\frac{1}{2^n}\sum_{j=0}^{2^N-1}\frac{j+\frac{1}{2}}{2^N}\prod_{k=1}^{p}|(A^{m+1})_{\xi_k j_{N-p+k}}|^2\prod_{l=1}^{N-p}|(A^m)_{\xi_{p+l}j_l}|^2.
$$

By a direct calculation, we obtain

$$
\begin{aligned}
r_n^{(N)} &= \frac{1}{2^n}\sum_{j=0}^{2^N-1}\frac{j+\frac{1}{2}}{2^N}\prod_{k=1}^{p}|(A^{m+1})_{\xi_k j_{N-p+k}}|^2\prod_{l=1}^{N-p}|(A^m)_{\xi_{p+l}j_l}|^2 \\
&= \frac{1}{2^n}\sum_{j=0}^{2^N-1}\frac{j+\frac{1}{2}}{2^N}\prod_{l=1}^{N-p}|(A^m)_{\xi_{p+l}j_l}|^2\prod_{k=1}^{p}|(A^{m+1})_{\xi_k j_{N-p+k}}|^2 \\
&= \frac{1}{2^{n+N}}\sum_{j=0}^{2^N-1}\left(j+\frac{1}{2}\right)\prod_{l=1}^{N-p}|(A^m)_{\xi_{p+l}j_l}|^2\prod_{k=1}^{p}|(A^{m+1})_{\xi_k j_{N-p+k}}|^2 \\
&= \frac{1}{2^{n+N}}\sum_{j_1,\cdots,j_N}\left\{\left(\sum_{k=1}^{N}j_k2^{N-k}\right)+\frac{1}{2}\right\}\prod_{l=1}^{N-p}|(A^m)_{\xi_{p+l}j_l}|^2\prod_{k=1}^{p}|(A^{m+1})_{\xi_k j_{N-p+k}}|^2 \\
&= \frac{1}{2^{n+N}}\sum_{j_1,\cdots,j_N}\left\{\left(\sum_{k=1}^{N-p}j_k2^{N-k}\right)+\left(\sum_{k=N-p+1}^{N}j_k2^{N-k}\right)+\frac{1}{2}\right\} \\
&\quad\times\prod_{l=1}^{N-p}|(A^m)_{\xi_{p+l}j_l}|^2\prod_{k=1}^{p}|(A^{m+1})_{\xi_k j_{N-p+k}}|^2 \\
&= \frac{1}{2^{(m+1)N+p}}\sum_{j_1,\cdots,j_N}\left\{\left(\sum_{k=1}^{N-p}j_k2^{N-k}\right)+\left(\sum_{k=N-p+1}^{N}j_k2^{N-k}\right)+\frac{1}{2}\right\} \\
&\quad\times\prod_{l=1}^{N-p}|(A^m)_{\xi_{p+l}j_l}|^2\prod_{k=N-p+1}^{N}|(A^{m+1})_{\xi_{k-(N-p)}j_k}|^2.
\end{aligned}
\tag{A3}
$$

(i) $m=0(\mathrm{mod}\,4)$

From the above lemma, we have

$$|(A^m)_{\xi_{p+l}j_l}|^2=\begin{cases}2^m & \text{if } j_l=\xi_{p+l},\\ 0 & \text{if } j_l\neq\xi_{p+l},\end{cases}\qquad |(A^{m+1})_{\xi_{k-(N-p)}j_k}|^2=2^m$$

for any $l=1,\cdots,N-p$ and $k=N-p+1,\cdots,N$. Using this formula the product of absolute squares can be expressed as

$$\begin{aligned}
&\prod_{l=1}^{N-p}|(A^m)_{\xi_{p+l}j_l}|^2\prod_{k=N-p+1}^{N}|(A^{m+1})_{\xi_{k-(N-p)}j_k}|^2\\
&=\begin{cases}(2^m)^{N-p}(2^m)^p & \text{if } j_l=\xi_{p+l}\quad\text{for all}\quad l=1,\cdots,N-p,\\ 0 & \text{otherwise,}\end{cases}\\
&=\begin{cases}2^{mN} & \text{if } j_l=\xi_{p+l}\quad\text{for all } l=1,\cdots,N-p,\\ 0 & \text{otherwise.}\end{cases}
\end{aligned}$$

Equation (A3) can be rewritten as

$$\begin{aligned}
r_n^{(N)}&=\frac{1}{2^{(m+1)N+p}}\sum_{j_1,\cdots,j_N}\left\{\left(\sum_{k=1}^{N-p}j_k2^{N-k}\right)+\left(\sum_{k=N-p+1}^{N}j_k2^{N-k}\right)+\frac{1}{2}\right\}\\
&\quad\times\prod_{l=1}^{N-p}|(A^m)_{\xi_{p+l}j_l}|^2\prod_{k=N-p+1}^{N}|(A^{m+1})_{\xi_{k-(N-p)}j_k}|^2\\
&=\frac{2^{mN}}{2^{(m+1)N+p}}\sum_{j_{N-p+1},\cdots,j_N}\left\{\left(\sum_{k=1}^{N-p}\xi_{p+k}2^{N-k}\right)+\left(\sum_{k=N-p+1}^{N}j_k2^{N-k}\right)+\frac{1}{2}\right\} \qquad \text{(A4)}\\
&=\frac{1}{2^{N+p}}\left(\sum_{k=1}^{N-p}\xi_{p+k}2^{N-k}\right)\left(\sum_{j_{N-p+1},\cdots,j_N}1\right)\\
&\quad+\frac{1}{2^{N+p}}\sum_{j_{N-p+1},\cdots,j_N}\left(\sum_{k=N-p+1}^{N}j_k2^{N-k}\right)+\frac{1}{2^{N+p}}\frac{1}{2}\left(\sum_{j_{N-p+1},\cdots,j_N}1\right)\\
&=\frac{1}{2^N}\sum_{k=1}^{N-p}\xi_{p+k}2^{N-k}+\frac{1}{2^{N+p}}\sum_{j_{N-p+1},\cdots,j_N}\left(\sum_{k=N-p+1}^{N}j_k2^{N-k}\right)+\frac{1}{2^{N+1}}\\
&=\frac{1}{2^N}\sum_{k=1}^{N-p}\xi_{p+k}2^{N-k}+\frac{1}{2^{N+p}}\sum_{j_{N-p+1},\cdots,j_N}\left(\sum_{k=1}^{p}j_{N-p+k}2^{p-k}\right)+\frac{1}{2^{N+1}}\\
&=\frac{1}{2^N}\sum_{k=1}^{N-p}\xi_{p+k}2^{N-k}+\frac{1}{2^{N+p}}\sum_{k=0}^{2^p-1}k+\frac{1}{2^{N+1}}\\
&=\frac{1}{2^N}\sum_{k=1}^{N-p}\xi_{p+k}2^{N-k}+\frac{1}{2^{N+p}}\frac{1}{2}(2^p-1)2^p+\frac{1}{2^{N+1}}\\
&=\sum_{k=1}^{N-p}\xi_{p+k}2^{-k}+\frac{2^p}{2^{N+1}}.
\end{aligned}$$

(ii) $m=1(\bmod 4)$

From the above lemma, we have

$$|(A^m)_{\xi_{p+l}j_l}|^2=2^{m-1}, \quad |(A^{m+1})_{\xi_{k-(N-p)}j_k}|^2=\begin{cases}2^{m+1} & \text{if } j_k\neq \xi_{k-(N-p)},\\ 0 & \text{if } j_k=\xi_{k-(N-p)},\end{cases}$$

for any $l=1,\cdots,N-p$ and $k=N-p+1,\cdots,N$. Using this formula the product of absolute squares can be expressed as

$$\prod_{l=1}^{N-p}|(A^m)_{\xi_{p+l}j_l}|^2\prod_{k=N-p+1}^{N}|(A^{m+1})_{\xi_{k-(N-p)}j_k}|$$

$$=\begin{cases}(2^{m-1})^{N-p}(2^{m+1})^p & \text{if } j_k\neq\xi_{k-(N-p)} \text{ for all } k=N-p+1,\cdots,N,\\ 0 & \text{otherwise}\end{cases}$$

$$=\begin{cases}2^{(m-1)N+2p} & \text{if } j_k\neq\xi_{k-(N-p)} \text{ for all } k=N-p+1,\cdots,N,\\ 0 & \text{otherwise.}\end{cases}$$

Let $\eta_{k-(N-p)}=\xi_{k-(N-p)}+1(\mathrm{mod}2), k=N-p+1,\ldots,N$. It follows that

$$\begin{aligned}
r_n^{(N)}&=\frac{1}{2^{(m+1)N+p}}\sum_{j_1,\cdots,j_N}\left\{\left(\sum_{k=1}^{N-p}j_k2^{N-k}\right)+\left(\sum_{k=N-p+1}^{N}j_k2^{N-k}\right)+\frac{1}{2}\right\}\\
&\quad\times\prod_{l=1}^{N-p}|(A^m)_{\xi_{p+l}j_l}|^2\prod_{k=N-p+1}^{N}|(A^{m+1})_{\xi_{k-(N-p)}j_k}|^2\\
&=\frac{2^{(m-1)N+2p}}{2^{(m+1)N+p}}\sum_{j_1,\cdots,j_{N-p}}\left\{\left(\sum_{k=1}^{N-p}j_k2^{N-k}\right)+\left(\sum_{k=N-p+1}^{N}\eta_{k-(N-p)}2^{N-k}\right)+\frac{1}{2}\right\} \qquad \text{(A5)}\\
&=\frac{1}{2^{2N-p}}\sum_{j_1,\cdots,j_{N-p}}\left\{\left(\sum_{k=1}^{N-p}j_k2^{N-k}\right)+\left(\sum_{k=N-p+1}^{N}\eta_{k-(N-p)}2^{N-k}\right)+\frac{1}{2}\right\}\\
&=\frac{1}{2^{2N-p}}\sum_{j_1,\cdots,j_{N-p}}\left(\sum_{k=1}^{N-p}j_k2^{N-k}\right)+\frac{1}{2^{2N-p}}\left(\sum_{k=N-p+1}^{N}\eta_{k-(N-p)}2^{N-k}\right)\left(\sum_{j_1,\cdots,j_{N-p}}1\right)\\
&\quad+\frac{1}{2^{2N-p}}\frac{1}{2}\left(\sum_{j_1,\cdots,j_{N-p}}1\right)\\
&=\frac{1}{2^{2N-p}}\sum_{j_1,\cdots,j_{N-p}}\left(\sum_{k=1}^{N-p}j_k2^{N-k}\right)+\frac{2^{N-p}}{2^{2N-p}}\left(\sum_{k=N-p+1}^{N}\eta_{k-(N-p)}2^{N-k}\right)+\frac{2^{N-p}}{2^{2N-p}}\frac{1}{2}\\
&=\frac{1}{2^{2N-p}}\sum_{j_1,\cdots,j_{N-p}}\left(\sum_{k=1}^{N-p}j_k2^{N-k}\right)+\sum_{k=N-p+1}^{N}\eta_{k-(N-p)}2^{-k}+\frac{1}{2^{N+1}}\\
&=\frac{2^p}{2^{2N-p}}\sum_{j_1,\cdots,j_{N-p}}\left(\sum_{k=1}^{N-p}j_k2^{N-p-k}\right)+\sum_{k=N-p+1}^{N}\eta_{k-(N-p)}2^{-k}+\frac{1}{2^{N+1}}\\
&=\frac{2^p}{2^{2N-p}}\sum_{k=0}^{2^{N-p}-1}k+\sum_{k=N-p+1}^{N}\eta_{k-(N-p)}2^{-k}+\frac{1}{2^{N+1}}\\
&=\frac{2^p}{2^{2N-p}}\frac{1}{2}(2^{N-p}-1)2^{N-p}+\sum_{k=N-p+1}^{N}\eta_{k-(N-p)}2^{-k}+\frac{1}{2^{N+1}}\\
&=\sum_{k=N-p+1}^{N}\eta_{k-(N-p)}2^{-k}+\frac{2^N-2^p+1}{2^{N+1}}.
\end{aligned}$$

(iii) $m=2(\mathrm{mod}\,4)$

From the above lemma, we have

$$|(A^m)_{\xi_{p+l}j_l}|^2=\begin{cases}2^m & \text{if } j_l\neq\xi_{p+l}\\ 0 & \text{if } j_l=j_{p+l}\end{cases},\quad |(A^{m+1})_{\xi_{k-(N-p)}j_k}|^2=2^m$$

for any $l=1,\cdots,N-p$ and $k=N-p+1,\cdots,N$. Using this formula the product of absolute squares can be expressed as

$$\prod_{l=1}^{N-p}|(A^m)_{\xi_{p+l}j_l}|^2\prod_{k=N-p+1}^{N}|(A^{m+1})_{\xi_{k-(N-p)}j_k}|$$
$$=\begin{cases}(2^m)^{N-p}(2^m)^p & \text{if } \quad j_l\neq\xi_{p+l} \text{ for all } l=1,\ldots,N-p,\\ 0 & \text{otherwise,}\end{cases}$$
$$=\begin{cases}2^{mN} & \text{if } j_l\neq\xi_{p+l} \quad \text{for all } l=1,\ldots,N-p,\\ 0 & \text{otherwise.}\end{cases}$$

Let $\eta_{p+l}=\xi_{p+l}+1(\bmod 2)$, $l=1,\ldots,N-p$. It follows that

$$r_n^{(N)}=\frac{1}{2^{(m+1)N+p}}\sum_{j_1,\cdots,j_N}\left\{\left(\sum_{k=1}^{N-p}j_k2^{N-k}\right)+\left(\sum_{k=N-p+1}^{N}j_k2^{N-k}\right)+\frac{1}{2}\right\}$$
$$\times\prod_{l=1}^{N-p}|(A^m)_{\xi_{p+l}j_l}|^2\prod_{k=N-p+1}^{N}|(A^{m+1})_{\xi_{k-(N-p)}j_k}|^2$$
$$=\frac{2^{mN}}{2^{(m+1)N+p}}\sum_{j_{N-p+1},\cdots,j_N}\left\{\left(\sum_{k=1}^{N-p}\eta_{p+k}2^{N-k}\right)+\left(\sum_{k=N-p+1}^{N}j_k2^{N-k}\right)+\frac{1}{2}\right\}.$$

Substituting η_{p+k} for ξ_{p+k} in (A4), we get

$$r_n^{(N)}=\sum_{k=1}^{N-p}\eta_{p+k}2^{-k}+\frac{2^p}{2^{N+1}}.$$

(iv) $m=3(\bmod 4)$

From the above lemma, we have

$$|(A^m)_{\xi_{p+l}j_l}|^2=2^{m-1},|(A^{m+1})_{\xi_{k-(N-p)}j_k}|^2=\begin{cases}2^{m+1} & \text{if } j_k=\xi_{k-(N-p)},\\ 0 & \text{if } j_k\neq\xi_{k-(N-p)},\end{cases}$$

for any $l=1,\ldots,N-p$ and $k=N-p+1,\ldots,N$. Using this formula the product of absolute squares can be expressed as

$$\prod_{l=1}^{N-p}|(A^m)_{\xi_{p+l}j_l}|^2\prod_{k=N-p+1}^{N}|(A^{m+1})_{\xi_{k-(N-p)}j_k}|^2$$
$$=\begin{cases}(2^{m-1})^{N-p}(2^{m+1})^p & \text{if } j_k=\xi_{k-(N-p)} \text{ for all } k=N-p+1,\ldots,N,\\ 0 & \text{otherwise,}\end{cases}$$
$$=\begin{cases}2^{(m-1)N+2p} & \text{if } j_k=\xi_{k-(N-p)} \text{ for all } k=N-p+1,\cdots,N,\\ 0 & \text{otherwise.}\end{cases}$$

Equation (A3) can be rewritten as

$$r_n^{(N)}=\frac{1}{2^{(m+1)N+p}}\sum_{j_1,\cdots,j_N}\left\{\left(\sum_{k=1}^{N-p} j_k 2^{N-k}\right)+\left(\sum_{k=N-p+1}^{N} j_k 2^{N-k}\right)+\frac{1}{2}\right\}$$

$$\times\prod_{l=1}^{N-p}|(A^m)_{\xi_{p+l}j_l}|^2\prod_{k=N-p+1}^{N}|(A^{m+1})_{\xi_{k-(N-p)}j_k}|^2$$

$$=\frac{2^{(m-1)N+2p}}{2^{(m+1)N+p}}\sum_{j_1,\cdots,j_{N-p}}\left\{\left(\sum_{k=1}^{N-p} j_k 2^{N-k}\right)+\left(\sum_{k=N-p+1}^{N}\xi_{k-(N-p)}2^{N-k}\right)+\frac{1}{2}\right\}.$$

Substituting $\xi_{k-(N-p)}$ for $\eta_{k-(N-p)}$ in (A5), we get

$$r_n^{(N)}=\sum_{k=N-p-1}^{N-p}\xi_{k-(N-p)}2^{-k}+\frac{2^N-2^p+1}{2^{N+1}}.$$

■

Proof of Theorem 5.4: For any $n=mN$, $m\in\mathbf{N}$,

$$r_n^{(N)}=\frac{1}{2^n}\sum_{j=0}^{2^N-1}\frac{j+\frac{1}{2}}{2^N}\prod_{k=1}^{N}|(A^m)_{\xi_k j_k}|^2.$$

By a direct calculation, we obtain

$$r_n^{(N)}=\frac{1}{2^n}\sum_{j=0}^{2^N-1}\frac{j+\frac{1}{2}}{2^N}\prod_{k=1}^{N}|(A^m)_{\xi_k j_k}|^2$$

$$=\frac{1}{2^{n+N}}\sum_{j=0}^{2^N-1}(j+\tfrac{1}{2})\prod_{k=1}^{N}|(A^m)_{\xi_k j_k}|^2$$

$$=\frac{1}{2^{(m+1)N}}\sum_{j_1,\cdots,j_N}\left\{\left(\sum_{k=1}^{N} j_k 2^{N-k}\right)+\frac{1}{2}\right\}\prod_{k=1}^{N}|(A^m)_{\xi_k j_k}|^2.$$

(i) $m=0 \pmod 4$

From the above lemma, we have

$$|(A^m)_{\xi_k j_k}|^2=\begin{cases}2^m & \text{if } j_k=\xi_k,\\ 0 & \text{if } j_k\neq\xi_k,\end{cases}$$

for any $k=1,\ldots,N$. Using this formula the product of absolute squares can be expressed as

$$\prod_{l=1}^{N}|(A^m)_{\xi_k j_k}|^2=\begin{cases}2^{mN} & \text{if } j_k=\xi_k \quad \text{for all } k=1,\ldots,N,\\ 0 & \text{otherwise.}\end{cases}$$

Using this formula the mean value $r_n^{(N)}$ of the position operator can be expressed as

$$r_n^{(N)}=\frac{1}{2^{(m+1)N}}\sum_{j_1,\cdots,j_N}\left\{\left(\sum_{k=1}^{N} j_k 2^{N-k}\right)+\frac{1}{2}\right\}\prod_{k=1}^{N}|(A^m)_{\xi_k j_k}|^2$$

$$=\frac{2^{mN}}{2^{(m+1)N}}\left\{\left(\sum_{k=1}^{N}\xi_k 2^{N-k}\right)+\frac{1}{2}\right\}=\sum_{k=1}^{N}\xi_k 2^{-k}+\frac{1}{2^{N+1}}.$$

(ii) $m=1,3 \pmod 4$

From the above lemma, we have

$$|(A^m)_{\xi_k j_k}|^2 = 2^{m-1}$$

for any $k=1,\ldots,N$. Note that

$$\prod_{l=1}^{N} |(A^m)_{\xi_k j_k}|^2 = 2^{(m-1)N}.$$

Using this formula the mean value $r_n^{(N)}$ of the position operator can be expressed as

$$\begin{aligned}
r_n^{(N)} &= \frac{1}{2^{(m+1)N}} \sum_{j_1,\cdots,j_N} \left\{ \left(\sum_{k=1}^{N} j_k 2^{N-k} \right) + \frac{1}{2} \right\} \prod_{k=1}^{N} |(A^m)_{\xi_k j_k}|^2 \\
&= \frac{2^{(m-1)N}}{2^{(m+1)N}} \sum_{j_1,\cdots,j_N} \left\{ \left(\sum_{k=1}^{N} j_k 2^{N-k} \right) + \frac{1}{2} \right\} \\
&= \frac{1}{2^{2N}} \cdot \left\{ \sum_{j_1,\cdots,j_N} \left(\sum_{k=1}^{N} j_k 2^{N-k} \right) + \frac{1}{2} \left(\sum_{j_1,\cdots,j_N} 1 \right) \right\} \\
&= \frac{1}{2^{2N}} \sum_{k=0}^{2^N-1} k + \frac{1}{2^{N+1}} = \frac{1}{2^{2N}} 2^{N-1}(2^N-1) + \frac{1}{2^{N+1}} = \frac{1}{2}.
\end{aligned}$$

(iii) $m = 2 (\bmod 4)$

From the above lemma, we have

$$|(A^m)_{\xi_k j_k}|^2 = \begin{cases} 2^m & \text{if } j_k \neq \xi_k, \\ 0 & \text{if } j_k = \xi_k, \end{cases}$$

for any $k=1,\ldots,N$. Using this formula the product of absolute squares can be expressed as

$$\prod_{l=1}^{N} |(A^m)_{\xi_k j_k}|^2 = \begin{cases} 2^{mN} & \text{if } j_k \neq \xi_k \text{ for all } \quad k=1,\ldots,N \\ 0 & \text{otherwise.} \end{cases}$$

Let $\eta_k = \xi_k + 1 (\bmod 2)$, $k=1,\ldots,N$. It follows that

$$\begin{aligned}
r_n^{(N)} &= \frac{1}{2^{(m+1)N}} \sum_{j_1,\cdots,j_N} \left\{ \left(\sum_{k=1}^{N} j_k 2^{N-k} \right) + \frac{1}{2} \right\} \prod_{k=1}^{N} |(A^m)_{\xi_k j_k}|^2 \\
&= \frac{2^{mN}}{2^{(m+1)N}} \left\{ \left(\sum_{k=1}^{N} \eta_k 2^{N-k} \right) + \frac{1}{2} \right\} = \sum_{k=1}^{N} \eta_k 2^{-k} + \frac{1}{2^{N+1}}.
\end{aligned}$$

■

[1] *Dynamical Systems*, edited by D. V. Anosov and V. I. Arnold (VINITI, Moscow, 1996).

[2] Ya. G. Sinai, *Introduction to Ergodic Theory* (Fasis, Moscow, 1996).

[3] M. C. Gutzwiller, *Chaos in Classical and Quantum Mechanics* (Springer, Berlin, 1990).

[4] *Quantum Chaos: Between Order and Disorder*, edited by G. Casati and B. V. Chirikov (Cambridge University Press, Cambridge, 1995).

[5] I. Y. Arefeva, P. B. Medvedev, O. A. Rytchkov, and I. V. Volovich, Chaos, Solitons Fractals **10**(2-3), 213 (1999).

[6] K. Hepp, Commun. Math. Phys. **35**, 265 (1974).

[7] W. H. Zurek, Phys. Rev. D **24**, 1516 (1981).

[8] W. H. Zurek, quant-ph/015127.

[9] G. G. Emch, H. Narnhofer, W. Thirring, and G. L. Sewell, J. Math. Phys. **35**, 5582 (1994).
[10] H. Hasegawa, Open Syst. Inf. Dyn. **4**, 359 (1997).
[11] V. P. Maslov, *Perturbation Theory and Asymptotic Methods* (MGU, Moscow, 1965).
[12] G. P. Berman and G. M. Zaslavsky, Physica A **91A**, 450 (1976).
[13] G. M. Zaslavskii, *Stochasticity of Dynamical Systems* (Nauka, Moscow, 1984).
[14] M. V. Berry, *Some Quantum to Classical Asymptotics*, Les Houches Summer School "Chaos and quantum physics," edited by M. J. Giannoni, A. Voros, and Z. Justi (North-Holland, Amsterdam, 1991).
[15] N. L. Balazs and A. Voros, Ann. Phys. **190**, 1 (1989).
[16] A. M. Ozorio de Almeida and M. Saraceno, Ann. Phys. **210**, 1 (1991).
[17] F. M. Dittes, E. Doron, and U. Smilansky, Phys. Rev. E **49**, R963 (1994).
[18] M. Saraceno and A. Voros, Physica D **79**, 206 (1994).
[19] L. Kaplan and E. J. Heller, Phys. Rev. Lett. **76**, 1453 (1996).
[20] M. G. E. da Luz and A. M. Ozorio de Almeida, Nonlinearity **8**, 43 (1995).
[21] R. Schack and C. M. Caves, Phys. Rev. Lett. **71**, 525 (1993).
[22] R. Schack and C. M. Caves, Phys. Rev. E **53**, 3257 (1996).
[23] P. W. O'Connor and S. Tomsovic, Ann. Phys. **207**, 218 (1991).
[24] M. Ohya, Int. J. Theor. Phys. **37(1)**, 495 (1998).
[25] R. S. Ingarden, A. Kossakowski, and M. Ohya, *Information Dynamics and Open Systems* (Kluwer Academic, Dordrecht, 1997).
[26] K. Inoue, M. Ohya, and K. Sato, Chaos, Solitons Fractals **11**, 1377 (2000).
[27] M. Saraceno, Ann. Phys. **199**, 37 (1990).
[28] J. H. Hannay, J. P. Keating, and A. M. Ozorio de Almeida, Nonlinearity **7**, 1327 (1994).
[29] R. Schack, Phys. Rev. A **57**, 1634 (1998).
[30] T. Brun and R. Schack, Phys. Rev. A **59**, 2649 (1999).
[31] A. Lakshminarayan and N. L. Balazs, Ann. Phys. **226**, 350 (1993).
[32] A. Lakshminarayan, Ann. Phys. **239**, 272 (1995).
[33] R. Schack and C. M. Caves, Applic. Alg. Eng. Commun. Comput. **AAECC 10**, 305 (2000).
[34] V. M. Alekseev and M. N. Yakobson, Phys. Rep. **75**, 287 (1981)
[35] A. N. Soklakov and R. Schack, quant-ph/9908040.
[36] A. N. Soklakov and R. Schack, quant-ph/0107071.
[37] E. Ott, *Chaos in Dynamical Systems* (Cambridge University Press, Cambridge, 1993).
[38] K. Inoue, M. Ohya, and I. V. Volovich, quant-ph/0108107.

Commun. Math. Phys. 225, 67 – 89 (2002)
With kind permission of Springer Science and Business Media

Communications in **Mathematical Physics**

Quantum Teleportation and Beam Splitting

Karl-Heinz Fichtner[1], **Masanori Ohya**[2]

[1] Friedrich-Schiller-Universität Jena, Fakultät für Mathematik und Informatik, Institut für Angewandte Mathematik, 07740 Jena, Germany. E-mail: fichtner@minet.uni-jena.de
[2] Department of Information Sciences, Science University of Tokyo, Noda City, Chiba 278-8510, Japan. E-mail: ohya@is.noda.sut.ac.jp

Received: 1 February 2001 / Accepted 19 July 2001

Abstract: Following the previous paper in which quantum teleportation is rigorously discussed with coherent entangled states given by beam splittings, we further discuss two types of models, the perfect teleportation model and non-perfect teleportation model, in a general scheme. Then the difference among several models, i.e., the perfect models and the non-perfect models, is studied. Our teleportation models are constructed by means of coherent states in some Fock space with counting measures, so that our model can be treated in the frame of usual optical communication.

1. Introduction

Following the previous paper [12], we further discuss non-perfect teleportation. The notion of non-perfect teleportation is introduced in [12] to construct a handy (i.e., physically more realizable) teleportation, although its mathematics becomes a little more complicated. For the completeness of the present paper, we quickly review the meaning of teleportation and some basic facts of Fock space in this section. Then we dicuss perfect teleportation in a very general (more general than that given in [12]) scheme with our previous results, and we state the main theorems obtained in [12] for non-perfect teleportation, both in Sect. 2. The main results of this paper are presented in Sect. 3, where we discuss the difference among three models, i.e., the perfect model, the non-perfect one given in [12] and that discussed in the present paper. The proofs of the main results are given in Sect. 4.

1.1. Quantum teleportation. The study of quantum teleportation was started in paper [3], whose scheme can be mathematically expressed in the following steps [11, 12]:

Step 0: A girl named Alice has an unknown quantum state ρ on (a N-dimensional) Hilbert space $\mathcal{H}_1$ and she was asked to teleport it to a boy named Bob.

Step 1: For this purpose, we need two other Hilbert spaces $\mathcal{H}_2$ and $\mathcal{H}_3$, $\mathcal{H}_2$ is attached to Alice and $\mathcal{H}_3$ is attached to Bob. Prearrange a so-called entangled state σ on $\mathcal{H}_2 \otimes \mathcal{H}_3$ having certain correlations and prepare an ensemble of the combined system in the state $\rho \otimes \sigma$ on $\mathcal{H}_1 \otimes \mathcal{H}_2 \otimes \mathcal{H}_3$.

Step 2: One then fixes a family of mutually orthogonal projections $(F_{nm})_{n,m=1}^{N}$ on the Hilbert space $\mathcal{H}_1 \otimes \mathcal{H}_2$ corresponding to an observable $F := \sum_{n,m} z_{n,m} F_{nm}$, and for a fixed pair of indices n, m, Alice performs a first kind incomplete measurement, involving only the $\mathcal{H}_1 \otimes \mathcal{H}_2$ part of the system in the state $\rho \otimes \sigma$, which filters the value z_{nm}, that is, after measurement on the given ensemble $\rho \otimes \sigma$ of identically prepared systems, only those where F shows the value z_{nm} are allowed to pass. According to the von Neumann rule, after Alice's measurement, the state becomes

$$\rho_{nm}^{(123)} := \frac{(F_{nm} \otimes \mathbf{1})\rho \otimes \sigma (F_{nm} \otimes \mathbf{1})}{\mathrm{tr}_{123}(F_{nm} \otimes \mathbf{1})\rho \otimes \sigma (F_{nm} \otimes \mathbf{1})},$$

where tr_{123} is the full trace on the Hilbert space $\mathcal{H}_1 \otimes \mathcal{H}_2 \otimes \mathcal{H}_3$.

Step 3: Bob is informed which measurement was done by Alice. This is equivalent to transmitting the information that the eigenvalue z_{nm} was detected. This information is transmitted from Alice to Bob without disturbance and by means of classical tools.

Step 4: Making only partial measurements on the third part of the system in the state $\rho_{nm}^{(123)}$ means that Bob will control a state $\Lambda_{nm}(\rho)$ on $\mathcal{H}_3$ given by the partial trace on $\mathcal{H}_1 \otimes \mathcal{H}_2$ of the state $\rho_{nm}^{(123)}$ (after Alice's measurement)

$$\begin{aligned} \Lambda_{nm}(\rho) &= \mathrm{tr}_{12}\, \rho_{nm}^{(123)} \\ &= \mathrm{tr}_{12} \frac{(F_{nm} \otimes \mathbf{1})\rho \otimes \sigma (F_{nm} \otimes \mathbf{1})}{\mathrm{tr}_{123}(F_{nm} \otimes \mathbf{1})\rho \otimes \sigma (F_{nm} \otimes \mathbf{1})}. \end{aligned}$$

Thus the whole teleportation scheme given by the family (F_{nm}) and the entangled state σ can be characterized by the family (Λ_{nm}) of channels from the set of states on $\mathcal{H}_1$ into the set of states on $\mathcal{H}_3$ and the family (p_{nm}) given by

$$p_{nm}(\rho) := \mathrm{tr}_{123}(F_{nm} \otimes \mathbf{1})\rho \otimes \sigma (F_{nm} \otimes \mathbf{1})$$

of the probabilities that Alice's measurement according to the observable F will show the value z_{nm}.

The teleportation scheme works perfectly with respect to a certain class $\mathfrak{S}$ of states ρ on $\mathcal{H}_1$ if the following conditions are fulfilled:

(E1) For each n, m there exists a unitary operator $v_{nm} : \mathcal{H}_1 \to \mathcal{H}_3$ such that

$$\Lambda_{nm}(\rho) = v_{nm}\ \rho\ v_{nm}^{*} \quad (\rho \in \mathfrak{S}),$$

(E2)

$$\sum_{nm} p_{nm}(\rho) = 1 \quad (\rho \in \mathfrak{S}).$$

(E1) means that Bob can reconstruct the original state ρ by unitary keys $\{v_{nm}\}$ provided to him.
(E2) means that Bob will succeed to find a proper key with certainty.
Such a teleportation process can be classified into two cases [1], i.e., weak teleportation and general teleportation, in which the solutions of the teleportation in each case and the conditions of the uniqueness of the unitary key were discussed. The solution of the weak teleportation is a triple $\{\sigma^{(23)}, F^{(12)}, U\}$ such that

$$\Lambda^* \rho^{(1)} = U^* \rho^{(1)} U$$

holds for any state $\rho^{(1)} \in \mathcal{S}(\mathcal{H}_1)$. Once a weak solution of a teleportation problem is given, we can construct the general solution for all n, m above [1].

In [12], we considered a teleportation model where the entangled state σ is given by the splitting of a superposition of certain coherent states, although this model doesn't work perfectly, that is, neither (E2) nor (E1) hold. In the same paper, we estimated the difference between the perfect teleportation and this non-perfect teleportation by adding a further step in the teleportation scheme:

Step 5: Bob will perform a measurement on his part of the system according to the projection

$$F_+ := \mathbf{1} - |\exp(0)\rangle\langle\exp(0)|,$$

where $|\exp(0)\rangle\langle\exp(0)|$ denotes the vacuum state (the coherent state with density 0).

Then our new teleportation channels (we denote it again by Λ_{nm}) have the form

$$\Lambda_{nm}(\rho) := \mathrm{tr}_{12} \frac{(F_{nm} \otimes F_+)\rho \otimes \sigma(F_{nm} \otimes F_+)}{\mathrm{tr}_{123}(F_{nm} \otimes F_+)\rho \otimes \sigma(F_{nm} \otimes F_+)}$$

and the corresponding probabilities are

$$p_{nm}(\rho) := \mathrm{tr}_{123}(F_{nm} \otimes F_+)\,\rho \otimes \sigma(F_{nm} \otimes F_+).$$

For this teleportation scheme, (E1) is fulfilled but (E2) is not, which we review in the next section.

1.2. Basic notions and notations. We collect some basic facts concerning the (symmetric) Fock space in a way adapted to the language of counting measures. For details we refer to [6–8, 2, 9].

Let G be an arbitrary complete separable metric space. Further, let μ be a locally finite diffuse measure on G, i.e. $\mu(B) < +\infty$ for bounded measurable subsets of G and $\mu(\{x\}) = 0$ for all singletons $x \in G$.

We denote the set of all finite counting measures on G by $M = M(G)$. Since $\varphi \in M$ can be written in the form $\varphi = \sum_{j=1}^{n} \delta_{x_j}$ for some $n = 0, 1, 2, \ldots$ and $x_j \in G$ with the Dirac measure δ_x corresponding to $x \in G$, the elements of M can be interpreted as finite (symmetric) point configurations in G. We equip M with its canonical σ–algebra $\mathfrak{W}$ (cf. [6,7]) and we consider the σ–finite measure F by setting

$$F(Y) := \mathcal{X}_Y(O) + \sum_{n\geq 1} \frac{1}{n!} \int_{G^n} \mathcal{X}_Y \left(\sum_{j=1}^{n} \delta_{x_j} \right) \mu^n(d[x_1, \ldots, x_n])(Y \in \mathfrak{W}),$$

where $\mathcal{X}_Y$ denotes the indicator function of a set Y and O represents the empty configuration, i. e., $O(G) = 0$.

Since μ was assumed to be diffuse one easily checks that F is concentrated on the set of simple configurations (i.e., without multiple points)

$$\hat{M} := \{\varphi \in M; \varphi(\{x\}) \leq 1 \text{ for all } x \in G\}.$$

$\mathcal{M} = \mathcal{M}(G) := L^2(M, \mathfrak{W}, F)$ is called the (symmetric) Fock space over G.

In [6] it was proved that $\mathcal{M}$ and the Boson Fock space $\Gamma(L^2(G))$ in the usual definition are isomorphic. For each $\Phi \in \mathcal{M}$ with $\Phi \neq 0$ we denote by $|\Phi\rangle$ the corresponding normalized vector

$$|\Phi\rangle := \frac{\Phi}{||\Phi||}.$$

Further, $|\Phi\rangle\langle\Phi|$ denotes the corresponding one–dimensional projection describing a pure state given by the normalized vector $|\Phi\rangle$. Now, for each $n \geq 1$ let $\mathcal{M}^{\otimes n}$ be the n-fold tensor product of the Hilbert space $\mathcal{M}$, which can be identified with $L^2(M^n, F^n)$.

For a given function $g : G \to \mathbb{C}$ the function exp $(g) : M \to \mathbb{C}$ defined by

$$\exp(g)(\varphi) := \begin{cases} 1 & \text{if } \varphi = 0 \\ \prod_{x\in G, \varphi(\{x\})>0} g(x) & \text{otherwise} \end{cases}$$

is called an exponential vector generated by g.

Observe that exp $(g) \in \mathcal{M}$ if and only if $g \in L^2(G)$ and one has in this case $||\exp(g)||^2 = e^{\|g\|^2}$ and $|\exp(g)\rangle = e^{-\frac{1}{2}\|g\|^2} \exp(g)$. The projection $|\exp(g)\rangle\langle\exp(g)|$ is called the coherent state corresponding to $g \in L^2(G)$. In the special case $g := 0$ we get the vacuum state

$$|\exp(0)\rangle = \mathcal{X}_{\{0\}} .$$

The linear span of the exponential vectors of $\mathcal{M}$ is dense in $\mathcal{M}$, so that bounded operators and certain unbounded operators can be characterized by their actions on exponential vectors.

The operator $D : \text{dom}(D) \to \mathcal{M}^{\otimes 2}$ given on a dense domain $\text{dom}(D) \subset \mathcal{M}$ containing the exponential vectors from $\mathcal{M}$ by

$$D\psi(\varphi_1, \varphi_2) := \psi(\varphi_1 + \varphi_2) \quad (\psi \in \text{dom}(D), \varphi_1, \varphi_2 \in M)$$

is called the compound Hida-Malliavin derivative. On exponential vectors exp (g) with $g \in L^2(G)$, one gets immediately

$$D \exp(g) = \exp(g) \otimes \exp(g). \tag{1}$$

The operator $S : \mathrm{dom}(S) \to \mathcal{M}$ given on a dense domain dom $(S) \subset \mathcal{M}^{\otimes 2}$ containing tensor products of exponential vectors by

$$S\Phi(\varphi) := \sum_{\tilde{\varphi} \leq \varphi} \Phi(\tilde{\varphi}, \varphi - \tilde{\varphi}) \quad (\Phi \in \mathrm{dom}(S),\ \varphi \in M)$$

is called the compound Skorohod integral. One gets

$$\langle D\psi, \Phi \rangle_{\mathcal{M}^{\otimes 2}} = \langle \psi, S\Phi \rangle_{\mathcal{M}} \quad (\psi \in \mathrm{dom}(D),\ \Phi \in \mathrm{dom}(S)), \tag{2}$$

$$S(\exp(g) \otimes \exp(h)) = \exp(g + h) \quad (g, h \in L^2(G)). \tag{3}$$

For more details we refer to [10].

Let T be a linear operator on $L^2(G)$ with $\|T\| \leq 1$. Then the operator $\Gamma(T)$ called the second quantization of T is the (uniquely determined) bounded operator on $\mathcal{M}$ fulfilling

$$\Gamma(T)\exp(g) = \exp(Tg) \quad (g \in L^2(G)).$$

Clearly, it holds

$$\begin{aligned} \Gamma(T_1 T_2) &= \Gamma(T_1)\Gamma(T_2), \\ \Gamma(T^*) &= \Gamma(T)^*. \end{aligned} \tag{4}$$

It follows that $\Gamma(T)$ is a unitary operator on $\mathcal{M}$ if T is a unitary operator on $L^2(G)$.

In [12] we proved.

Lemma 1.1. *Let K_1, K_2 be linear operators on $L^2(G)$ with a property*

$$K_1^* K_1 + K_2^* K_2 = \mathbf{1}\ . \tag{5}$$

Then there exists exactly one isometry v_{K_1,K_2} from $\mathcal{M}$ to $\mathcal{M}^{\otimes 2} = \mathcal{M} \otimes \mathcal{M}$ with

$$v_{K_1,K_2}\exp(g) = \exp(K_1 g) \otimes \exp(K_2 g) \quad (g \in L^2(G)). \tag{6}$$

Further it holds

$$v_{K_1,K_2} = (\Gamma(K_1) \otimes \Gamma(K_2))D \tag{7}$$

(at least on dom(D) *but one has the unique extension). The adjoint $v^*_{K_1,K_2}$ of v_{K_1,K_2} is characterized by*

$$v^*_{K_1,K_2}(\exp(h) \otimes \exp(g)) = \exp(K_1^* h + K_2^* g) \quad (g, h \in L^2(G)) \tag{8}$$

and it holds

$$v^*_{K_1,K_2} = S(\Gamma(K_1^*) \otimes \Gamma(K_2^*)). \tag{9}$$

From K_1, K_2 we get a transition expectation $\xi_{K_1K_2} : \mathcal{M} \otimes \mathcal{M} \to \mathcal{M}$, using ν_{K_1,K_2} and the lifting $\xi^*_{K_1K_2}$ may be interpreted as a certain splitting (cf. [2]). The property (5) implies

$$\|K_1 g\|^2 + \|K_2 g\|^2 = \|g\|^2 \quad (g \in L^2(G)). \tag{10}$$

Let U, V be unitary operators on $L^2(G)$. If operators K_1, K_2 satisfy (5), then the pair $\hat{K}_1 = UK_1$, $\hat{K}_2 = VK_2$ fulfill (5).

Here we explain the fundamental scheme of beam splitting [8]. We define an isometric operator $V_{\alpha,\beta}$ for coherent vectors such that

$$V_{\alpha,\beta} | \exp(g)\rangle = | \exp(\alpha g)\rangle \otimes | \exp(\beta g)\rangle$$

with $| \alpha |^2 + | \beta |^2 = 1$. This beam splitting is a useful mathematical expression for optical communication and quantum measurements [2]. As one example, take $\alpha = \beta = \frac{1}{2}$ in the above formula and let $K_1 = K_2$ be the following operator of multiplication on $L^2(G)$:

$$K_1 g = \frac{1}{\sqrt{2}} g = K_2 g \quad (g \in L^2(G)). \tag{11}$$

In this case, we put

$$\nu := \nu_{K_1,K_2},$$

then we obtain

$$\nu \exp(g) = \exp\left(\frac{1}{\sqrt{2}} g\right) \otimes \exp\left(\frac{1}{\sqrt{2}} g\right) \quad (g \in L^2(G)).$$

Another example is given by taking K_1 and K_2 as the projections to the corresponding subspaces $\mathcal{H}_1$, $\mathcal{H}_2$ of the orthogonal sum $L^2(G) = \mathcal{H}_1 \oplus \mathcal{H}_2$.

In [12] we used the first example in order to describe a teleportation model where Bob performs his experiments on the same ensemble of the systems as Alice. Further we used a special case of the second example in order to describe a teleportation model where Bob and Alice are spatially separated (cf. Sect. 5 of [12]).

2. Previous Results on Teleportation

Let us review some results obtained in [12]. We fix an ONS $\{g_1, \ldots, g_N\} \subseteq L^2(G)$, operators K_1, K_2 on $L^2(G)$ with (5), a unitary operator T on $L^2(G)$, and $d > 0$. We assume

$$TK_1 g_k = K_2 g_k \quad (k = 1, \ldots, N), \tag{12}$$

$$\langle K_1 g_k, K_1 g_j\rangle = 0 \quad (k \neq j;\ k, j = 1, \ldots, N). \tag{13}$$

Using (11) and (12) we get

$$\|K_1 g_k\|^2 = \|K_2 g_k\|^2 = \frac{1}{2}. \tag{14}$$

From (12) and (13) we get

$$\langle K_2 g_k,\ K_2 g_j\rangle = 0 \quad (k \neq j;\ k, j = 1, \dots, N). \tag{15}$$

The state of Alice asked to teleport is of the type

$$\rho = \sum_{s=1}^{N} \lambda_s |\Phi_s\rangle\langle\Phi_s|, \tag{16}$$

where

$$|\Phi_s\rangle = \sum_{j=1}^{N} c_{sj} |\exp(aK_1 g_j) - \exp(0)\rangle \quad \left(\sum_j |c_{sj}|^2 = 1; s = 1, \dots, N\right) \tag{17}$$

and $a = \sqrt{d}$. One easily checks that $(|\exp(aK_1 g_j) - \exp(0)\rangle)_{j=1}^N$ and $(|\exp aK_2 g_j) - \exp(0)\rangle)_{j=1}^N$ are ONS in $\mathcal{M}$. The set $\{\Phi_s; s = 1, \dots, N\}$ makes the N-dimensional Hilbert space $\mathcal{H}_1$ defining an input state teleported by Alice. Although we may include the vaccum state $|\exp(0)\rangle$ to define $\mathcal{H}_1$, here we take the N-dimensional Hilbert space $\mathcal{H}_1$ as above because of computational simplicity.

In order to achieve that $(|\Phi_s\rangle)_{s=1}^N$ is still an ONS in $\mathcal{M}$ we assume

$$\sum_{j=1}^{N} \bar{c}_{sj} c_{kj} = 0 \quad (j \neq k;\ j, k = 1, \dots, N). \tag{18}$$

Denote $c_s = [c_{s1,\dots,}c_{sN}] \in \mathbb{C}^N$, then $(c_s)_{s=1}^N$ is an CONS in $\mathbb{C}^N$.

Let $(b_n)_{n=1}^N$ be a sequence in $\mathbb{C}^N$,

$$b_n = [b_{n1,\dots,}b_{nN}]$$

with properties

$$|b_{nk}| = 1 \quad (n, k = 1, \dots, N), \tag{19}$$

$$\langle b_n,\ b_j\rangle = 0 \quad (n \neq j;\ n, j = 1, \dots, N). \tag{20}$$

Now, for each $m, n\ (= 1, \dots, N)$, we have unitary operators U_m, B_n on $\mathcal{M}$ given by

$$B_n |\exp(aK_1 g_j) - \exp(0)\rangle = b_{nj} |\exp(aK_1 g_j) - \exp(0)\rangle \quad (j = 1, \dots, N), \tag{21}$$

$$U_m |\exp(aK_1 g_j) - \exp(0)\rangle = |\exp(aK_1 g_{j\oplus m}) - \exp(0)\rangle \quad (j = 1, \dots, N), \tag{22}$$

where $j \oplus m := j + m (\text{mod } N)$.

2.1. A perfect teleportation. Then Alice's measurements are performed with the projection

$$F_{nm} = |\xi_{nm}\rangle\langle\xi_{nm}| \quad (n, m = 1, \dots, N) \tag{23}$$

given by

$$|\xi_{nm}\rangle = \frac{1}{\sqrt{N}} \sum_{j=1}^{N} b_{nj} |\exp(aK_1 g_j) - \exp(0)\rangle \otimes |\exp(aK_1 g_{j\oplus m}) - \exp(0)\rangle. \tag{24}$$

One easily checks that $(|\xi_{nm}\rangle)_{n,m=1}^{N}$ is an ONS in $\mathcal{M}^{\otimes 2}$. Further, the state vector $|\xi\rangle$ of the entangled state $\sigma = |\xi\rangle\langle\xi|$ is given by

$$|\xi\rangle = \frac{1}{\sqrt{N}} \sum_{k} |\exp(aK_1 g_k) - \exp(0)\rangle \otimes |\exp(aK_2 g_k) - \exp(0)\rangle. \tag{25}$$

In [12] we proved the following theorem.

Theorem 2.1. *For each $n, m = 1, \dots, N$, define a channel Λ_{nm} by*

$$\Lambda_{nm}(\rho) := \mathrm{tr}_{12} \frac{(F_{nm} \otimes 1)(\rho \otimes \sigma)(F_{nm} \otimes \mathbf{1})}{\mathrm{tr}_{123}(F_{nm} \otimes 1)(\rho \otimes \sigma)(F_{nm} \otimes \mathbf{1})} \quad (\rho \text{ normal state on } \mathcal{M}). \tag{26}$$

Then we have for all states ρ on $\mathcal{M}$ with (16) and (17),

$$\Lambda_{nm}(\rho) = \left(\Gamma(T) U_m B_n^*\right) \rho \left(\Gamma(T) U_m B_n^*\right)^*. \tag{27}$$

Remark 2.2. Using the operators B_n, U_m, $\Gamma(T)$, the projections F_{nm} are given by unitary transformations of the entangled state σ :

$$F_{nm} = \left(B_n \otimes U_m \Gamma(T^*)\right) \sigma \left(B_n \otimes U_m \Gamma(T^*)\right)^*, \tag{28}$$

or

$$|\xi_{nm}\rangle = \left(B_n \otimes U_m \Gamma(T^*)\right) |\xi\rangle.$$

If Alice performs a measurement according to the following selfadjoint operator

$$F = \sum_{n,m=1}^{N} z_{nm} F_{nm}$$

with $\{z_{nm} | n, m = 1, \dots, N\} \subseteq \mathbf{R} - \{0\}$, then she will obtain the value z_{nm} with probability $1/N^2$. The sum over all these probabilities is 1, so that the teleportation model works perfectly.

Before stating some fundamental results of [12] for the non-perfect case, we note that our perfect teleportation is obviously treated in general finite Hilbert spaces $\mathcal{H}_k$ $(k = 1, 2, 3)$ the same as the usual one [2]. Moreover, our teleportation scheme can be generalized a bit by introducing the entangled state σ_{12} on $\mathcal{H}_1 \otimes \mathcal{H}_2$ defining the projections $\{F_{nm}\}$ by the unitary operators B_n, U_m. We here discuss the perfect teleportation on general Hilbert spaces $\mathcal{H}_k$ $(k = 1, 2, 3)$. Let $\left\{\xi_j^k;\, j = 1, \cdots, N\right\}$ be CONS of the

Hilbert space $\mathcal{H}_k$ $(k = 1, 2, 3)$. Define the entangled states σ_{12} and σ_{23} on $\mathcal{H}_1 \otimes \mathcal{H}_2$ and $\mathcal{H}_2 \otimes \mathcal{H}_3$, respectively, such as

$$\sigma_{12} = |\xi_{12}\rangle\langle\xi_{12}|,\ \sigma_{23} = |\xi_{23}\rangle\langle\xi_{23}|$$

with $\xi_{12} := \frac{1}{\sqrt{N}}\sum_{j=1}^{N}\xi_j^1 \otimes \xi_j^2$ and $\xi_{23} := \frac{1}{\sqrt{N}}\sum_{j=1}^{N}\xi_j^2 \otimes \xi_j^3$. By a sequence $\left\{b_n = [b_{n1,\dots,}b_{nN}];\ n = 1,\cdots,N\right\}$ in $\mathbb{C}^N$ with the properties (19) and (20), we define the unitary operator B_n and U_m such as

$$B_n\xi_j^1 := b_{nj}\xi_j^1(n, j = 1,\cdots,N) \text{ and } U_m\xi_j^2 := \xi_{j\oplus m}^2(n, j = 1,\cdots,N)$$

with $j \oplus m := j + m \pmod N$. Then the set $\{F_{nm};\ n, m = 1,\cdots,N\}$ of the projections of Alice is given by

$$F_{nm} = (B_n \otimes U_m)\,\sigma_{12}\,(B_n \otimes U_m)^*,$$

and the teleportation channels $\left\{\Lambda_{nm}^*;\ n, m = 1,\cdots,N\right\}$ are defined as

$$\Lambda_{nm}(\rho) := \mathrm{tr}_{12}\frac{(F_{nm} \otimes 1)(\rho \otimes \sigma_{23})(F_{nm} \otimes \mathbf{1})}{\mathrm{tr}_{123}(F_{nm} \otimes 1)(\rho \otimes \sigma_{23})(F_{nm} \otimes \mathbf{1})}.$$

Finally the unitary keys $\{W_{nm};\ n, m = 1,\cdots,N\}$ of Bob are given as

$$W_{nm}\xi_j^1 = \overline{b}_{nj}\xi_{j\oplus m}^3,\quad (n, m = 1,\cdots,N),$$

by which we obtain the perfect teleportation

$$\Lambda_{nm}(\rho) = W_{nm}\rho W_{nm}^*.$$

The above perfect teleportation is unique in the sense of unitary equivalence.

2.2. A non-perfect teleportation. We will review a non-perfect teleportation model in which the probability teleporting the state from Alice to Bob is less than 1 and it depends on the density parameter d (may be the energy of the beams) of the coherent vector. There, when $d = a^2$ tends to infinity, the probability tends to 1. Thus the model can be considered as asymptotically perfect.

Take the normalized vector

$$|\eta\rangle := \frac{\gamma}{\sqrt{N}}\sum_{k=1}^{N}|\exp(ag_k)\rangle, \tag{29}$$

$$\text{with } \gamma := \left(\frac{1}{1 + (N-1)e^{-d}}\right)^{\frac{1}{2}} = \left(\frac{1}{1 + (N-1)e^{-a^2}}\right)^{\frac{1}{2}},$$

and we replace in (26) the entangled state σ by

$$\tilde{\sigma} := |\tilde{\xi}\rangle\langle\tilde{\xi}|, \tag{30}$$

$$\tilde{\xi} := \nu_{K_1,K_2}(\eta) = \frac{\gamma}{\sqrt{N}}\sum_{k=1}^{N}|\exp(aK_1g_k)\rangle \otimes |\exp(aK_2g_k)\rangle.$$

Then for each $n, m = 1, \ldots, N$, we get the channels on any normal state ρ on $\mathcal{M}$ such as

$$\tilde{\Lambda}_{nm}(\rho) := \operatorname{tr}_{12} \frac{(F_{nm} \otimes \mathbf{1})(\rho \otimes \tilde{\sigma})(F_{nm} \otimes \mathbf{1})}{\operatorname{tr}_{123}(F_{nm} \otimes \mathbf{1})(\rho \otimes \tilde{\sigma})(F_{nm} \otimes \mathbf{1})}, \tag{31}$$

$$\Theta_{nm}(\rho) := \operatorname{tr}_{12} \frac{(F_{nm} \otimes F_{+})(\rho \otimes \tilde{\sigma})(F_{nm} \otimes F_{+})}{\operatorname{tr}_{123}(F_{nm} \otimes F_{+})(\rho \otimes \tilde{\sigma})(F_{nm} \otimes F_{+})}, \tag{32}$$

where $F_{+} = \mathbf{1} - |\exp(0)\rangle\langle\exp(0)|$, i.e., F_{+} is the projection onto the space $\mathcal{M}_{+}$ of configurations having no vacuum part,

$$\mathcal{M}_{+} := \{\psi \in \mathcal{M};\ \|\,|\exp(0)\rangle\langle\exp(0)|\psi\| = 0\}.$$

One easily checks that

$$\Theta_{nm}(\rho) = \frac{F_{+}\tilde{\Lambda}_{nm}(\rho)F_{+}}{\operatorname{tr}\left(F_{+}\tilde{\Lambda}_{nm}(\rho)F_{+}\right)}, \tag{33}$$

that is, after receiving the state $\tilde{\Lambda}_{nm}(\rho)$ from Alice, Bob has to omit the vacuum.

From Theorem 2.1 it follows that for all ρ with (16) and (17),

$$\Lambda_{nm}(\rho) = \frac{F_{+}\Lambda_{nm}(\rho)F_{+}}{\operatorname{tr}\,(F_{+}\Lambda_{nm}(\rho)F_{+})}.$$

This is not true if we replace Λ_{nm} by $\tilde{\Lambda}_{nm}$, namely, in general it does not hold

$$\Theta_{nm}(\rho) = \tilde{\Lambda}_{nm}(\rho).$$

In [12] we proved the following theorem.

Theorem 2.3. *For all states ρ on $\mathcal{M}$ with (16) and (17) and each pair $n, m\ (= 1, \ldots, N)$, we have*

$$\Theta_{nm}(\rho) = \left(\Gamma(T)\,U_m B_n^*\right)\rho\left(\Gamma(T)\,U_m B_n^*\right)^* \quad \text{or} \quad \Theta_{nm}(\rho) = \Lambda_{nm}(\rho) \tag{34}$$

and

$$\sum_{n,m} p_{nm}(\rho) = \sum_{n,m} \operatorname{tr}_{123}(F_{nm} \otimes F_{+})(\rho \otimes \tilde{\sigma})(F_{nm} \otimes F_{+}) = \frac{\left(1 - e^{-\frac{d}{2}}\right)^2}{1 + (N-1)e^{-d}}. \tag{35}$$

That is, the model works only asymptotically perfectly in the sense of condition (E2). In other words, the model works perfectly for the case of high density (or energy) of the considered beams.

3. Main Results

The tools of the teleportation model considered in Sect. 2.1 are the entangled state σ and the family of projections $(F_{nm})_{n,m=1}^{N}$. In order to have a more handy model, in Sect. 2.2 we have replaced the entangled state σ by another entangled state $\tilde{\sigma}$ given by the splitting of a superposition of certain coherent states (30). In addition, we are going to replace the projectors F_{nm} by projectors $\tilde{F}_{nm}$ defined as follows:

$$\tilde{F}_{nm} := \left(B_n \otimes U_m \Gamma(T)^*\right) \tilde{\sigma} \left(B_n \otimes U_m \Gamma(T)^*\right)^* . \tag{36}$$

In order to make this definition precise we assume, in addition to (22), that it holds:

$$U_m \exp(0) = \exp(0) \quad (m = 1, \ldots , N).$$

Together with (22) that implies

$$U_m |\exp(aK_1 g_j)\rangle = |\exp(aK_1 g_{j\oplus m})\rangle \quad (m, j = 1, \ldots , N). \tag{37}$$

Formally we have the same relation between $\tilde{\sigma}$ and $\tilde{F}_{nm}$ like the relation between σ and F_{nm} (cf. Remark 2.2). Further for each pair $n, m = 1, \ldots , N$ we define channels on normal states on $\mathcal{M}$ such as

$$\tilde{\Theta}_{nm}(\rho) := \mathrm{tr}_{12} \frac{\left(\tilde{F}_{nm} \otimes F_+\right)(\rho \otimes \tilde{\sigma})\left(\tilde{F}_{nm} \otimes F_+\right)}{\tilde{p}_{nm}(\rho)}, \tag{38}$$

where

$$\tilde{p}_{nm}(\rho) := \mathrm{tr}_{123}\left(\tilde{F}_{nm} \otimes F_+\right)(\rho \otimes \tilde{\sigma})\left(\tilde{F}_{nm} \otimes F_+\right) \tag{39}$$

(cf. (33), and (34)).

In Sect. 4, we will prove the following theorem.

Theorem 3.1. *For each state ρ on $\mathcal{M}$ with (16), and (17), each pair $n, m (= 1, \ldots , N)$ and each bounded operator A on $\mathcal{M}$ it holds*

$$\left|\mathrm{tr}\left(\tilde{\Theta}_{nm}(\rho)A\right) - \mathrm{tr}\left(\Lambda_{nm}(\rho)A\right)\right| \leq \frac{2e^{-\frac{d}{2}}}{\left(1 - e^{-\frac{d}{2}}\right)}\left(N^2 + N\sqrt{N} + N\right)\|A\| , \tag{40}$$

$$\left|\tilde{p}_{nm}(\rho) - \frac{1}{N^2}\right| \leq e^{-\frac{d}{2}}\left(\frac{14}{N^2} + 2 + \frac{2}{\sqrt{N}}\right). \tag{41}$$

From Theorem 2.1 and $e^{-\frac{d}{2}} \longrightarrow 0 \; (d \to +\infty)$, Theorem 3.1 means that our modified teleportation model works asymptotically perfectly (the case of high density or energy) in the sense of conditions (E1) and (E2).

In order to obtain a deeper understanding of the whole procedure we are going to discuss another representation of the projectors $\tilde{F}_{nm}$ and of the channels $\tilde{\Theta}_{nm}$. The starting point is again the normalized vector $|\eta\rangle$ given by (29). From (14) we obtain

$$\|O_{\sqrt{2}} K_1 g_k\|^2 = \|g_k\|^2, \tag{42}$$

where O_f denotes the operator of multiplication corresponding to the number (or function) f

$$O_f\psi := f\psi \quad (\psi \in L_2(G)). \tag{43}$$

Furthermore (13) implies

$$\langle O_f K_1 g_k \, , \, O_f K_1 g_j \rangle = 0 \quad (k \neq j). \tag{44}$$

From (42), and (44) follows that we have a normalized vector $|\tilde{\eta}\rangle$ given by

$$|\tilde{\eta}\rangle := \Gamma\left(O_{\sqrt{2}} K_1\right)|\eta\rangle = \frac{\gamma}{\sqrt{N}} \sum_{k=1}^{N} |\exp\left(a\sqrt{2} K_1 g_k\right)\rangle. \tag{45}$$

Remark 3.2. In the case of the example given by (11) of Sect. 1.2, we have

$$|\tilde{\eta}\rangle = |\eta\rangle.$$

Now let V be the unitary operator on $\mathcal{M} \otimes \mathcal{M}$ characterized by

$$\begin{aligned} &V\left(\exp(f_1) \otimes \exp(f_2)\right) \\ &= \exp\left(\frac{1}{\sqrt{2}}(f_1 - f_2)\right) \otimes \exp\left(\frac{1}{\sqrt{2}}(f_1 + f_2)\right) (f_1, f_2 \in L_2(G)). \end{aligned} \tag{46}$$

One easily checks

$$\begin{aligned} &V^*\left(\exp(f_1) \otimes \exp(f_2)\right) \\ &= \exp\left(\frac{1}{\sqrt{2}}(f_1 + f_2)\right) \otimes \exp\left(\frac{1}{\sqrt{2}}(f_2 - f_1)\right) (f_1, f_2 \in L_2(G)). \end{aligned} \tag{47}$$

Remark 3.3. V describes a certain exchange procedure of particles (or energy) between two systems or beams (cf. [13]).

Now, using (12), (30), (45), and (47), (46) one gets

$$\tilde{\xi} = \nu_{K_1,K_2}(\eta) = (\mathbf{1} \otimes \Gamma(T))V^*\left(|\exp(0)\rangle \otimes |\tilde{\eta}\rangle\right), \tag{48}$$

$$\tilde{\xi} = (\mathbf{1} \otimes \Gamma(T))V\left(|\tilde{\eta}\rangle \otimes |\exp(0)\rangle\right), \tag{49}$$

and it follows

$$\begin{aligned} \tilde{\sigma} &= |\tilde{\xi}\rangle\langle\tilde{\xi}|, \\ &= (\mathbf{1} \otimes \Gamma(T))V^*\left(|\exp(0)\rangle\langle\exp(0)| \otimes |\tilde{\eta}\rangle\langle\tilde{\eta}|\right)\left((\mathbf{1} \otimes \Gamma(T))V^*\right)^*. \end{aligned} \tag{50}$$

$$\tilde{\sigma} = (\mathbf{1} \otimes \Gamma(T))V\left(|\tilde{\eta}\rangle\langle\tilde{\eta}| \otimes |\exp(0)\rangle\langle\exp(0)|\right)\left((\mathbf{1} \otimes \Gamma(T))V\right)^*. \tag{51}$$

From the definition of $\tilde{F}_{nm}$ (36) and (50) it follows

$$\tilde{F}_{nm} = (B_n \otimes U_m)\, V^* \left(|\exp(0)\rangle\langle\exp(0)| \otimes |\tilde{\eta}\rangle\langle\tilde{\eta}|\right) \left((B_n \otimes U_m)\, V^*\right)^* . \qquad (52)$$

Using (51), and (52) we obtain

$$\begin{aligned} &\left(\tilde{F}_{nm} \otimes F_+\right)(\rho \otimes \tilde{\sigma})\left(\tilde{F}_{nm} \otimes F_+\right) \quad (n, m = 1, \dots, N) \\ &= (X_{nm} \otimes \mathbf{1})\, W_{nm}\, (\rho \otimes |\tilde{\eta}\rangle\langle\tilde{\eta}| \otimes |\exp(0)\rangle\langle\exp(0)|)\, W_{nm}^* \,(X_{nm} \otimes \mathbf{1})^* , \qquad (53) \end{aligned}$$

where

$$\begin{aligned} X_{nm} &:= (B_n \otimes U_m)\, V^*, \ (n, m = 1, \cdots, N), \\ W_{nm} &:= (|\exp(0)\rangle\langle\exp(0)| \otimes |\tilde{\eta}\rangle\langle\tilde{\eta}| \otimes F_+)\,(V \otimes 1)\left(B_n^* \otimes U_m^* \otimes \Gamma(T)\right)(1 \otimes V) . \end{aligned} \qquad (54)$$

X_{nm} and consequently $X_{nm} \otimes \mathbf{1}$ are unitary operators. For that reason we get from (53),

$$\begin{aligned} &\mathrm{tr}_{123}\left(\tilde{F}_{nm} \otimes F_+\right)(\rho \otimes \tilde{\sigma})\left(\tilde{F}_{nm} \otimes F_+\right)(n, m = 1, \dots, N) \\ &\qquad = \mathrm{tr}_{123} W_{nm}\, (\rho \otimes |\tilde{\eta}\rangle\langle\tilde{\eta}| \otimes \exp(0)\rangle\langle\exp(0)|)\, W_{nm}^* \qquad (55) \end{aligned}$$

and

$$\begin{aligned} &\mathrm{tr}_{12}\left(\tilde{F}_{nm} \otimes F_+\right)(\rho \otimes \tilde{\sigma})\left(\tilde{F}_{nm} \otimes F_+\right) \\ &\qquad = \mathrm{tr}_{12} W_{nm}\, (\rho \otimes |\tilde{\eta}\rangle\langle\tilde{\eta}| \otimes |\exp(0)\rangle\langle\exp(0)|)\, W_{nm}^* \qquad (56) \end{aligned}$$

Now from (38), (39), (55) and (56) it follows

$$\tilde{p}_{nm}(\rho) = \mathrm{tr}_{123} W_{nm}\, (\rho \otimes |\tilde{\eta}\rangle\langle\tilde{\eta}| \otimes |\exp(0)\rangle\langle\exp(0)|)\, W_{nm}^*, \qquad (57)$$

$$\tilde{\Theta}_{nm}(\rho) = tr_{12} \frac{W_{nm}\, (\rho \otimes |\tilde{\eta}\rangle\langle\tilde{\eta}| \otimes |\exp(0)\rangle\langle\exp(0)|)\, W_{nm}^*}{\mathrm{tr}_{123} W_{nm}\, (\rho \otimes |\tilde{\eta}\rangle\langle\tilde{\eta}| \otimes |\exp(0)\rangle\langle\exp(0)|)\, W_{nm}^*} . \qquad (58)$$

According to (57,58) and (54), the procedure of the special teleportation model can be expressed in the following steps:

Step 0 – ***initial state***
ρ – the unknown state
Alice wants to teleport
$|\exp(0)\rangle\langle\exp(0)|$–vacuum state,
Bob's state at the beginning.

$$s_{\text{in}}(\rho) = \rho \otimes |\tilde{\eta}\rangle\langle\tilde{\eta}| \otimes |\exp(0)\rangle\langle\exp(0)|$$

Step 1 – ***Transformation according to***
that means: splitting of
the state $|\tilde{\eta}\rangle\langle\tilde{\eta}|$.

$$\mathbf{1} \otimes V$$

Step 2 – ***Transformation according to***

$$B_n^* \otimes U_m^* \otimes \Gamma(T)$$

Step 3 – ***Transformation according to***
exchange of particles (or energy)
between the first and the second
part of the system.

$$V \otimes \mathbf{1}$$

Step 4 – ***measurement according to***
checking for
– first part in the vacuum?
– in the third part is no vacuum?
– second part reconstructed?

$$|\exp(0)\rangle\langle\exp(0)| \otimes |\tilde{\eta}\rangle\langle\tilde{\eta}| \otimes F_+$$

$$\downarrow$$

Final state $s_{\text{fin}}(\rho)$
$$= \frac{W_{nm}(s_{\text{in}}(\rho))W_{nm}^*}{\text{tr}_{123} W_{nm}(s_{\text{in}}(\rho))W_{nm}^*}$$

Now from (57) we get $\tilde{\Theta}_{nm}(\rho) = \text{tr}_{12}\ s_{\text{fin}}(\rho)$. Thus Theorem 3.1 means that in the case of high density (or energy) d we have approximately (ρ with (16), and (17))

$$\text{tr}_{12}\ s_{\text{fin}}(\rho) = \left(\Gamma(T)U_m B_n^*\right) \rho \left(\Gamma(T)U_m B_n^*\right)^* .$$

The proof of Theorem 3.1 shows that we have even more, namely it holds (approximately)

$$s_{\text{fin}}(\rho) = |\exp(0)\rangle\langle\exp(0)| \otimes |\tilde{\eta}\rangle\langle\tilde{\eta}| \otimes \left(\Gamma(T)U_m B_n^*\right) \rho \left(\Gamma(T)U_m B_n^*\right)^* . \tag{59}$$

Adding in our scheme the following step:
Step 5 – ***Transformation*** $\qquad \mathbf{1} \otimes \mathbf{1} \otimes \left(\Gamma(T)U_m B_n^*\right)^*$
(that means Bob uses the key provided to him)

Then $s_{\text{fin}}(\rho)$ will change into the new final state

$$|\exp(0)\rangle\langle\exp(0)| \otimes |\tilde{\eta}\rangle\langle\tilde{\eta}| \otimes \rho.$$

Summarizing one can describe the effect of the procedure (for large d!) as follows: At the beginning Alice has (e.g., can control) a state ρ, and Bob has the vacuum state (e.g., can control nothing). After the procedure Bob has the state ρ and Alice has the vacuum. Furthermore the teleportation mechanism is ready for the next teleportation (e.g. $|\tilde{\eta}\rangle\langle\tilde{\eta}|$ is reproduced in the course of teleportation).

We have considered three different models (cf. Sects. 2.1, 2.2, 2.3). Each of them is a special case of a more general concept we are going to describe in the following:

Let H_1, H_2 be N–dimensional subspaces of $\mathcal{M}_+$ such that $\Gamma(T)$ maps H_1 onto H_2, and H_1 is invariant with respect to the unitary transformations B_n, U_m $(n, m = 1, \ldots, N)$.

Further let σ_1, σ_2 be projections of the type

$$\sigma_k = |\xi_k\rangle\langle\xi_k|, \quad \xi_k \in (H_1 \oplus \mathcal{M}_0) \otimes (H_2 \otimes \mathcal{M}_0) \quad (k = 1, 2),$$

where $\mathcal{M}_0$ is the orthogonal complement of $\mathcal{M}_+$, e.g., $\mathcal{M}_0$ is the one-dimensional subspace of $\mathcal{M}$ spanned by the vacuum vector $|\exp(0)\rangle$.

Now for each $n, m = 1, \ldots, N$ and each pair σ_1, σ_2 we define a channel $\Omega_{nm}^{\sigma_1,\sigma_2}$ from the set of all normal states ρ on H_1 into the set of all normal states on $\mathcal{M}_+$,

$$\Omega_{nm}^{\sigma_1\sigma_2}(\rho) := \mathrm{tr}_{12} \frac{\left(F_{nm}^{\sigma_1} \otimes F_+\right)(\rho \otimes \sigma_2)\left(F_{nm}^{\sigma_1} \otimes F_+\right)}{\mathrm{tr}_{123}\left(F_{nm}^{\sigma_1} \otimes F_+\right)(\rho \otimes \sigma_2)\left(F_{nm}^{\sigma_1} \otimes F_+\right)},$$

where

$$F_{nm}^{\sigma_1} := \left(B_n \otimes U_m \Gamma(T^*)\right) \sigma_1 \left(B_n \otimes U_m \Gamma(T^*)\right)^* .$$

In this paper we have considered the situation where H_1 is spanned by the ONS

$$(|\exp(aK_1 g_k) - \exp(0)\rangle)_{k=1}^N$$

and H_2 is spanned by the ONS

$$(|\exp(aK_2 g_k) - \exp(0)\rangle)_{k=1}^N .$$

Further the model discussed in Sect. 2.1 corresponds to the special case $\sigma_1 = \sigma_2 = \sigma$, e.g.

$$\Lambda_{nm} = \Omega_{nm}^{\sigma\sigma} \qquad (n, m = 1, \ldots, N)$$

(perfect in the sense of conditions (E1) and (E2)).

The model discussed in Sect. 2.2 corresponds to the special case $\sigma_1 = \sigma \neq \sigma_2 = \tilde{\sigma}$, e.g.

$$\Theta_{nm} = \Omega_{nm}^{\sigma\tilde{\sigma}}$$

(perfect in the sense of (E1), and only asymptotically perfect in the sense of (E2)).

Finally the model from this section corresponds to the special case $\sigma_1 = \sigma_2 = \tilde{\sigma}$, e.g.

$$\tilde{\Theta}_{nm} = \Omega_{nm}^{\tilde{\sigma}\tilde{\sigma}}$$

(non-perfect, neither (E2) nor (E1) hold, but asymptotically perfect in the sense of both conditions).

4. Proof of Theorem 3.1

From (14) we get

$$\|\exp\left(aK_s g_j\right) - \exp(0)\|^2 = e^{\frac{a^2}{2}} - 1 \qquad (s = 1, 2;\ j = 1, \ldots, N), \tag{60}$$

$$\|\exp\left(aK_s g_j\right)\|^2 = e^{\frac{a^2}{2}} \qquad (s = 1, 2;\ j = 1, \ldots, N). \tag{61}$$

Using (46), (60) and (61) one easily checks

$$\begin{aligned} &V\left(|\exp\left(aK_1 g_j\right) - \exp(0)\rangle \otimes |\exp\left(aK_1 g_k\right)\rangle\right) \\ &= \left(\left(e^{\frac{a^2}{2}} - 1\right)e^{\frac{a^2}{2}}\right)^{-\frac{1}{2}} \left[\exp\left(\frac{a}{\sqrt{2}}K_1\left(g_j - g_k\right)\right) \otimes \exp\left(\frac{a}{\sqrt{2}}K_1\left(g_j + g_k\right)\right)\right. \\ &\left.\quad - \exp\left(-\frac{a}{\sqrt{2}}K_1\left(g_k\right)\right) \otimes \exp\left(\frac{a}{\sqrt{2}}K_1 g_k\right)\right] \\ &(k, j = 1, \ldots, N). \end{aligned} \tag{62}$$

Lemma 4.1. *Put for $j, k = 1, \ldots, N$,*

$$\alpha_{jk} := \langle|\exp(0)\rangle \otimes |\tilde{\eta}\rangle\ ,\ V\left(|\exp\left(aK_1 g_j\right) - \exp(0)\rangle \otimes |\exp\left(aK_1 g_k\right)\rangle\right)\rangle.$$

Then it holds for $j, k = 1, \ldots, N$,

$$\alpha_{jk} = \left(\left(1 - e^{-\frac{a^2}{2}}\right)e^{-a^2}\right)^{\frac{1}{2}} \frac{\gamma}{\sqrt{N}} \qquad (k \neq j), \tag{63}$$

$$\alpha_{jj} = \left(1 - e^{-\frac{a^2}{2}}\right)^{\frac{1}{2}} \frac{\gamma}{\sqrt{N}}. \tag{64}$$

Proof. We have

$$\langle\exp(0)\ ,\ \exp(f)\rangle = 1 \qquad (f \in L_2(G))\,. \tag{65}$$

Using (62), (65), and (45) we get for $j, k = 1, \ldots, N$,

$$\begin{aligned} \alpha_{jk} &= \left(\left(e^{\frac{a^2}{2}} - 1\right)e^{\frac{a^2}{2}}\right)^{-\frac{1}{2}} \frac{\gamma}{\sqrt{N}} \sum_{s=1}^{N} \left[\left\langle|\exp(\sqrt{2}\,aK_1 g_s)\rangle\ ,\exp\left(\frac{a}{\sqrt{2}}K_1\left(g_j + g_k\right)\right)\right\rangle\right. \\ &\left. -\left\langle|\exp\left(\sqrt{2}\,aK_1 g_s\right)\rangle\ ,\ \exp\left(\frac{a}{\sqrt{2}}K_1 g_k\right)\right\rangle\right]. \end{aligned} \tag{66}$$

We have

$$\langle\exp(f_1)\ ,\ \exp(f_2)\rangle = e^{\langle f_1, f_2\rangle} \qquad (f_1, f_2 \in L_2(G))\,. \tag{67}$$

Using (13) and (67) we obtain

$$\begin{aligned} 0 = &\left\langle\exp\left(\sqrt{2}\,aK_1 g_s\right)\ ,\ \exp\left(\frac{a}{\sqrt{2}}K_1\left(g_j + g_k\right)\right)\right\rangle \qquad (s \neq j) \\ &-\left\langle\exp\left(\sqrt{2}\,aK_1 g_s\right),\ \exp\left(\frac{a}{\sqrt{2}}K_1 g_k\right)\right\rangle \end{aligned} \tag{68}$$

From (61), (66), (67), and (68) it follows

$$\alpha_{jk} = \left(\left(e^{\frac{a^2}{2}} - 1\right)e^{\frac{a^2}{2}}e^{a^2}\right)^{-\frac{1}{2}} \frac{\gamma}{\sqrt{N}} \left(e^{a^2\langle K_1 g_j\, ,\, K_1(g_j+g_k)\rangle} - e^{a^2\langle K_1 g_j\, ,\, K_1 g_k\rangle}\right). \tag{69}$$

Now (13) and (14) implies

$$\langle K_1 g_j\, ,\, K_1 g_k\rangle = \frac{1}{2}\delta_{jk}. \tag{70}$$

For that reason (63), and (64) follow from (69). □

In the following we fix a pair $n, m \in \{1, \dots, N\}$.

Remark 4.2. Without loss of generality we can assume

$$B_n = \mathbf{1}, \tag{71}$$

which we can explain as follows: Using (57)–(59), and (54) we obtain in the case (71),

$$\tilde{\Theta}_{km}(\rho) = \tilde{\Theta}_{nm}\left(B_k^* \rho B_k\right) \qquad (k = 1, \dots, N),$$
$$\tilde{p}_{km}(\rho) = \tilde{p}_{nm}\left(B_k^* \rho B_k\right) \qquad (k = 1, \dots, N).$$

On the other hand from Theorem 2.1 it follows that in the case (71) for all states ρ with (16) and (17) it holds

$$\Lambda_{km}(\rho) = \Lambda_{nm}\left(B_k^* \rho B_k\right) \qquad (k = 1, \dots, N).$$

Finally it is easy to show that $B_k^* \rho B_k$ fulfills (16), and (17) if the state ρ fulfills (16) and (17).
For those reasons Theorem 3.1 would be proved if we could prove (40), and (41) on the assumption that we have (71).

Now from (30), (49), and (37) we get

$$\left(U_m^* \otimes \Gamma(T)\right) V\left(|\tilde{\eta}\rangle \otimes |\exp(0)\rangle\right) = \frac{\gamma}{\sqrt{N}} \sum_{k=1}^{N} |\exp\left(aK_1 g_k\right)\rangle \otimes |\exp(aK_2 g_{k\oplus m})\rangle. \tag{72}$$

Lemma 4.3. *Put for* $s = 1, \dots, N$

$$\begin{aligned}\beta_s := {} & \left(\left(|\exp(0)\rangle\langle\exp(0)| \otimes |\tilde{\eta}\rangle\langle\tilde{\eta}|\right) V \otimes \mathbf{1}\right)\left(\mathbf{1} \otimes U_m^* \otimes \Gamma(T)\right) \\ & (\mathbf{1} \otimes V)|\Psi_s\rangle \otimes |\tilde{\eta}\rangle \otimes |\exp(0)\rangle\langle\exp(0)| \qquad (s = 1, \dots, N).\end{aligned} \tag{73}$$

Then it holds

$$\beta_s = \frac{\gamma^2}{N}\left(1 - e^{-\frac{a^2}{2}}\right)^{\frac{1}{2}} |\exp(0)\rangle \otimes |\tilde{\eta}\rangle \otimes \left(\left(1 - e^{-\frac{a^2}{2}}\right)\sum_{j=1}^{N} c_{sj} |\exp\left(aK_2 g_{j\oplus m}\right)\rangle \right.$$
$$\left. + e^{-\frac{a^2}{2}} \sum_{j=1}^{N} c_{sj} \sum_{k=1}^{N} |\exp\left(aK_2 g_k\right)\rangle\right). \tag{74}$$

Proof. From (17), (72), and (73) we get

$$\beta_s = \sum_{j=1}^{N} c_{sj} \frac{\gamma}{\sqrt{N}} \sum_{k=1}^{N} \left[(|\exp(0)\rangle\langle \exp(0)| \otimes |\tilde{\eta}\rangle\langle\tilde{\eta}|)\, V \left(|\exp\left(aK_1 g_j\right) - \exp(0)\rangle \right.\right.$$
$$\left.\left. \otimes |\exp\left(aK_1 g_k\right)\rangle\right)\right] \otimes |\exp\left(aK_2 g_{k\oplus m}\right)\rangle. \tag{75}$$

Further we have

$$\begin{aligned}&(|\exp(0)\rangle\langle\exp(0)| \otimes |\tilde{\eta}\rangle\langle\tilde{\eta}|)\, V \left(|\exp\left(aK_1 g_j\right) - \exp(0)\rangle \otimes |\exp\left(aK_1 g_k\right)\rangle\right) \\ &\quad = |\exp(0)\rangle \otimes |\tilde{\eta}\rangle \left\langle |\exp(0)\rangle \otimes |\tilde{\eta}\rangle\, ,\; V\left(|\exp\left(aK_1 g_j\right) - \exp(0)\rangle \otimes |\exp\left(aK_1 g_k\right)\rangle\right)\right\rangle \\ &\quad (j, k = 1, \ldots, N).\end{aligned} \tag{76}$$

Using Lemma 4.1, (75), and (76) we obtain

$$\begin{aligned}\beta_s = {}& \frac{\gamma^2}{N}\left(1 - e^{-\frac{a^2}{2}}\right)^{\frac{1}{2}} |\exp(0)\rangle \otimes |\tilde{\eta}\rangle \otimes \sum_{j=1}^{N} c_{sj} |\exp\left(aK_2 g_{j\oplus m}\right)\rangle \\ &+ \frac{\gamma^2}{N}\left(\left(1 - e^{-\frac{a^2}{2}}\right)e^{-a^2}\right)^{\frac{1}{2}} |\exp(0)\rangle \otimes |\tilde{\eta}\rangle \otimes \left(\sum_{j}\sum_{k\neq j} c_{sj} |\exp\left(aK_2 g_{k\oplus m}\right)\rangle\right).\end{aligned}$$

That implies (74). □

Now we put

$$|\Psi_0\rangle := \frac{1}{\sqrt{N}} \sum_{j=1}^{N} |\exp\left(aK_1 g_j\right) - \exp(0)\rangle. \tag{77}$$

Since

$$F_+ = \mathbf{1} - |\exp(0)\rangle\langle\exp(0)|,$$

one easily checks

$$\begin{aligned}&F_+ |\exp\left(aK_r g_k\right)\rangle = \left(1 - e^{-\frac{a^2}{2}}\right)^{\frac{1}{2}} |\exp\left(aK_r g_k\right) - \exp(0)\rangle \\ &(r = 1, 2;\; k = 1, \ldots, m).\end{aligned} \tag{78}$$

Using (77), and (78) we obtain

$$\begin{aligned}F_+\left(\sum_{k=1}^{N} |\exp\left(aK_2 g_k\right)\rangle\right) &= \left(1 - e^{-\frac{a^2}{2}}\right)^{\frac{1}{2}} \sqrt{N}\, U_m \Gamma(T) |\Psi_0\rangle \\ &= \left(1 - e^{-\frac{a^2}{2}}\right)^{\frac{1}{2}} \sqrt{N}\, \Gamma(T) U_m |\Psi_0\rangle.\end{aligned} \tag{79}$$

Using the same arguments we get

$$F_+\left(\sum_{j=1}^{N} c_{sj}|\exp\left(aK_2 g_{j\oplus m}\right)\rangle\right) = \left(1-e^{-\frac{a^2}{2}}\right)^{\frac{1}{2}} U_m\Gamma(T)|\Psi_s\rangle \quad (s=1,\dots,N) \tag{80}$$

$$= \left(1-e^{-\frac{a^2}{2}}\right)^{\frac{1}{2}} \Gamma(T)U_m|\Psi_s\rangle. \tag{81}$$

Finally we have

$$\begin{aligned}&(|\exp(0)\rangle\langle\exp(0)|\otimes|\tilde{\eta}\rangle\langle\tilde{\eta}|\otimes F_+)\,(V\otimes\mathbf{1})\\&= (\mathbf{1}\otimes\mathbf{1}\otimes F_+)\,(|\exp(0)\rangle\langle\exp(0)|\otimes|\tilde{\eta}\rangle\langle\tilde{\eta}|)\,V\otimes\mathbf{1}.\end{aligned} \tag{82}$$

Using (54), (71), (79), (80), and Lemma 4.3 one easily checks the following equality:

$$\begin{aligned}&W_{nm}\,(|\Psi_s\rangle\otimes|\tilde{\eta}\rangle\otimes|\exp(0)\rangle) \qquad (s=1,\dots,N)\\&=\frac{\gamma^2}{N}\left(1-e^{-\frac{a^2}{2}}\right)\left(|\exp(0)\rangle\otimes|\tilde{\eta}\rangle\otimes\Gamma(T)U_mB_n^*\right)\\&\otimes\left(\left(1-e^{-\frac{a^2}{2}}\right)|\Psi_s\rangle+e^{-\frac{a^2}{2}}\left(\sum_j c_{sj}\right)\sqrt{N}\,|\Psi_0\rangle\right).\end{aligned} \tag{83}$$

For that reason we have the following lemma

Lemma 4.4. *For each bounded operator A on $\mathcal{M}$ and $s=1,\dots,N$ it holds*

$$\begin{aligned}\vartheta_s(A) &:= \langle W_{nm}\,(|\Psi_s\rangle\otimes|\tilde{\eta}\rangle\otimes|\exp(0)\rangle),\ (\mathbf{1}\otimes\mathbf{1}\otimes A)W_{nm}\,(|\Psi_s\rangle\otimes|\tilde{\eta}\rangle\otimes|\exp(0)\rangle)\rangle\\&=\left(\frac{\gamma^2}{N}\right)^2\left(1-e^{-\frac{a^2}{2}}\right)^2\left[\left\langle\left(1-e^{-\frac{a^2}{2}}\right)^2\Gamma(T)U_mB_n^*|\Psi_s\rangle,\ A\Gamma(T)U_mB_N^*|\Psi_s\rangle\right\rangle\right.\\&+e^{-\frac{a^2}{2}}\left(1-e^{-\frac{a^2}{2}}\right)\left(\sum_j c_{sj}\right)\sqrt{N}\left\langle\Gamma(T)U_mB_n^*|\Psi_s\rangle,\ A\Gamma(T)U_mB_n^*|\Psi_0\rangle\right\rangle\\&+e^{-\frac{a^2}{2}}\left(1-e^{-\frac{a^2}{2}}\right)\left(\overline{\sum_j c_{sj}}\right)\sqrt{N}\left\langle\Gamma(T)U_mB_n^*|\Psi_0\rangle,\ A\Gamma(T)U_mB_n^*|\Psi_s\rangle\right\rangle\\&+e^{-a^2}|\sum_j c_{sj}|^2N\left\langle\Gamma(T)U_mB_n^*|\Psi_0\rangle,\ A\Gamma(T)U_mB_n^*|\Psi_0\rangle\right\rangle.\end{aligned}$$

Now from (16) we get

$$\begin{aligned}&\rho\otimes|\tilde{\eta}\rangle\langle\tilde{\eta}|\otimes|\exp(0)\rangle\langle\exp(0)|\\&=\sum_{s=1}^{N}\lambda_s|\Psi_s\otimes\tilde{\eta}\otimes\exp(0)\rangle\langle\Psi_s\otimes\tilde{\eta}\otimes\exp(0)|.\end{aligned} \tag{84}$$

On the other hand $(|\Psi_s \otimes \tilde{\eta} \otimes \exp(0)\rangle)_{s=1}^N$ is an ONS because $(\Psi_s)_{s=1}^N$ is an ONS. For that reason from (57,58), (84), and Lemma 4.4 with $A = \mathbf{1}$ it follows

$$\tilde{p}_{nm}(\rho) = \left(\frac{\gamma^2}{N}\right)^2 \left(1 - e^{-\frac{a^2}{2}}\right)^2 \left[\left(1 - e^{-\frac{a^2}{2}}\right)^2 + Ne^{-a^2} \sum_{s=1}^N \lambda_s | \sum_{s=1} c_{sj}|^2 \right.$$
$$\left. + \sqrt{N}\, e^{-\frac{a^2}{2}} \left(1 - e^{-\frac{a^2}{2}}\right) \sum_{s=1}^N \lambda_s \sum_{j=1}^N \left(c_{sj} \langle \Psi_s \,,\, \Psi_0\rangle + \overline{c_{sj} \langle \Psi_s \,,\, \Psi_0\rangle}\right)\right]. \tag{85}$$

As $\left(|\exp(aK_1 g_j) - \exp(0)\rangle\right)_{j=1}^N$ is an ONS we can calculate easily

$$\langle |\Psi_s\rangle, \ |\Psi_0\rangle\rangle = \frac{1}{\sqrt{N}} \sum_{k=1}^N c_{sk}.$$

For that reason from (85) follows

$$\tilde{p}_{nm}(\rho) = \left(\frac{\gamma^2}{N}\right)^2 \left(1 - e^{-\frac{a^2}{2}}\right)^2 \left[\left(1 - e^{-\frac{a^2}{2}}\right)^2 + \sum_{s=1}^N \lambda_s | \sum_{j=1}^N c_{sj}|^2 \right.$$
$$\left. \left(Ne^{-a^2} + 2\sqrt{N}\, e^{-\frac{a^2}{2}} \left(1 - e^{-\frac{a^2}{2}}\right)\right)\right]. \tag{86}$$

Further we have $\sum\limits_s \lambda_s = 1$ and

$$\left|\sum_{j=1}^N c_{sj}\right|^2 \le \sum_j \sum_k |c_{sj}|\,|c_{sk}| \le \sum_j \sum_k \left(\frac{|c_{sj}|^2}{2} + \frac{|c_{sk}|^2}{2}\right) \le N. \tag{87}$$

Using (86), (87) and the definition of γ (cf. (29)) we can estimate

$$\begin{aligned}
&\left|\tilde{p}_{nm}(\rho) - \frac{1}{N^2}\right| \\
&= \left(\frac{\gamma^2}{N}\right)^2 \left| \left(1 - e^{-\frac{a^2}{2}}\right)^2 \left[\left(1 - e^{-\frac{a^2}{2}}\right)^2 \right.\right. \\
&\quad \left.\left. + \sum_s \lambda_s \left|\sum_j c_{sj}\right|^2 \left(Ne^{-a^2} + 2\sqrt{N} e^{-\frac{a^2}{2}} \left(1 - e^{-\frac{a^2}{2}}\right)\right)\right] - \frac{1}{\gamma^4}\right| \\
&\le \frac{1}{N^2} \left| \left(1 - e^{-\frac{a^2}{2}}\right)^4 \right. \\
&\quad + \left(1 - e^{-\frac{a^2}{2}}\right)^2 \sum_s \lambda_s |\sum_j c_{sj}|^2 \left(Ne^{-a^2} + 2\sqrt{N} e^{-\frac{a^2}{2}} \left(1 - e^{-\frac{a^2}{2}}\right)\right) \\
&\quad \left. - \left(1 + (N-1)e^{-a^2}\right)^2 \right| \\
&\le \frac{1}{N^2} \left(\left| \left(1 - e^{-\frac{a^2}{2}}\right)^4 - \left(1 + (N-1)e^{-a^2}\right)^2 \right| + e^{-\frac{a^2}{2}} N \left(N + 2\sqrt{N}\right)\right) \\
&\le \frac{1}{N^2} \left(e^{-\frac{a^2}{2}} \left(14 + N^2\right) + e^{-\frac{a^2}{2}} N (N + 2\sqrt{N})\right).
\end{aligned}$$

That implies (41).

Lemma 4.5. *We use the notation $\vartheta_s(A)$ from Lemma 4.4. Then for each bounded operator A on $\mathcal{M}$ and $s = 1, \dots, N$ it holds*

$$Z_s(A) := \left| \frac{\vartheta_s(A)}{\tilde{p}_{nm}(\rho)} - \left\langle \Gamma(T) U_m B_n^* |\Psi_s\rangle \,,\; A\Gamma(T) U_m B_n^* |\Psi_s\rangle \right\rangle \right| \leq \frac{2e^{-\frac{a^2}{2}}}{\left(1 - e^{-\frac{a^2}{2}}\right)^2} \left(N^2 + N\sqrt{N} + N \right) \|A\| \,.$$

Proof. Using Lemma 4.4 and the estimation

$$\left| \left\langle \Gamma(T) U_m B_n^* |\Psi_k\rangle \,,\; A\Gamma(T) U_m B_n^* |\Psi_r\rangle \right\rangle \right| \leq \|A\| \qquad (k, r = 0, \dots, N),$$

we get

$$\begin{aligned} Z_s(A) \leq \|A\| &\left| \left(\frac{\gamma^2}{N}\right)^2 \left(1 - e^{-\frac{a^2}{2}}\right)^2 (\tilde{p}_{nm}(\rho))^{-1} - 1 \right| \\ &+ \|A\| \left(\frac{\gamma^2}{N}\right)^2 \left(1 - e^{-\frac{a^2}{2}}\right)^4 (\tilde{p}_{nm}(\rho))^{-1} \left[2e^{-\frac{a^2}{2}} \left(1 - e^{-\frac{a^2}{2}}\right) \sqrt{N} \left| \sum_j c_{sj} \right| \right. \\ &\left. + e^{-a^2} N |\sum_j c_{sj}|^2 \right] . \end{aligned}$$

Because of (86) it follows

$$\begin{aligned} Z \leq \|A\| &\left[\left| \frac{\left(1 - e^{-\frac{a^2}{2}}\right)^2}{\left(1 - e^{-\frac{a^2}{2}}\right)^2 + \sum_s \lambda_s |\sum_j c_{sj}|^2 \left(N e^{-a^2} + 2\sqrt{N} e^{-\frac{a^2}{2}} \left(1 - e^{-\frac{a^2}{2}}\right)\right)} - 1 \right| \right. \\ &\left. + \frac{2e^{-\frac{a^2}{2}} \left(1 - e^{-\frac{a^2}{2}}\right) \sqrt{N} |\sum_j c_{sj}| + e^{-a^2} N |\sum_j c_{sj}|^2}{\left(1 - e^{-\frac{a^2}{2}}\right)^2 + \sum_s \lambda_s |\sum_j c_{sj}|^2 \left(N e^{-a^2} + 2\sqrt{N} e^{-\frac{a^2}{2}} \left(1 - e^{-\frac{a^2}{2}}\right)\right)} \right] . \end{aligned}$$

Using (87) we get

$$\begin{aligned} &\left| \frac{\left(1 - e^{-\frac{a^2}{2}}\right)^2}{\left(1 - e^{-\frac{a^2}{2}}\right)^2 + \sum_s \lambda_s |\sum_j c_{sj}|^2 \left(N e^{-a^2} + 2\sqrt{N} e^{-\frac{a^2}{2}} \left(1 - e^{-\frac{a^2}{2}}\right)\right)} - 1 \right| \\ &\leq \frac{e^{-\frac{a^2}{2}}}{\left(1 - e^{-\frac{a^2}{2}}\right)^2} \left(N^2 + 2N\sqrt{N} \right) \end{aligned}$$

and

$$\frac{2e^{-\frac{a^2}{2}}\left(1-e^{-\frac{a^2}{2}}\right)\sqrt{N}|\sum_j c_{sj}| + e^{-a^2}N|\sum_j c_{sj}|^2}{\left(1-e^{-\frac{a^2}{2}}\right)^2 + \sum_s \lambda_s |\sum_j c_{sj}|^2\left(Ne^{-a^2} + 2\sqrt{N}\,e^{-\frac{a^2}{2}}\left(1-e^{-\frac{a^2}{2}}\right)\right)}$$

$$\leq \frac{e^{-\frac{a^2}{2}}}{\left(1-e^{-\frac{a^2}{2}}\right)^2}\left(2N+N^2\right)$$

That proves Lemma 4.5. □

We have the representation (84) of $\rho \otimes |\tilde{\eta}\rangle\langle\tilde{\eta}| \otimes |\exp(0)\rangle\langle\exp(0)|$ as a mixture of orthogonal projections. Thus from (56) and (57,58) we get with the notation $\vartheta_s(A)$ from Lemma 4.4,

$$\operatorname{tr}\left(\tilde{\Theta}_{nm}(\rho)A\right) = \sum_s \lambda_s \vartheta_s(A)\left(\tilde{p}_{nm}(\rho)\right)^{-1}.$$

On the other hand from Theorem 2.1 follows

$$\operatorname{tr}\left(\Lambda_{nm}(\rho)A\right) = \sum_s \lambda_s \left\langle \Gamma(T)U_m B_n^* |\Psi_s\rangle \ ,\ A\Gamma(T)U_m B_n^*|\Psi_s\rangle\right\rangle.$$

Consequently we have with notation $Z_s(A)$ from Lemma 4.5

$$|\operatorname{tr}\left(\tilde{\Theta}_{nm}(\rho)A\right) - \operatorname{tr}\left(\Lambda_{nm}(\rho)A\right)| \leq \sum_s \lambda_s Z_s(A).$$

For that reason (40) follows from Lemma 4.5, and $\sum_s \lambda_s = 1$.

That completes the proof of Theorem 3.1.

References

1. Accardi, L. and Ohya, M.: *Teleportation of general quantum states*. Voltera Center preprint, 1998
2. Accardi, L., Ohya, M.: Compound channels, transition expectations and liftings. Applied Mathematics & Optimization **39**, 33–59 (1999)
3. Bennett, C.H., Brassard, G., Crépeau, C., Jozsa, R., Peres, A. and Wootters, W.: Teleporting an unknown quantum state viaDual Classical and Einstein–Podolsky–Rosen channels. Phys. Rev. Lett. **70**, 1895–1899 (1993)
4. Bennett, C.H., Brassard, G., Popescu, S., Schumacher, B., Smolin, J.A., Wootters, W.K.: Purification of noisy entanglement and faithful teleportation via noisy channels. Phys. Rev. Lett. **76**, 722–725 (1996)
5. Ekert, A.K.: Quantum cryptography based on Bell's theorem. Phys. Rev. Lett. **67**, 661–663 (1991)
6. Fichtner, K.-H. and Freudenberg, W.: Pointprocesses and the position distrubution of infinite boson systems. J. Stat. Phys. **47**, 959–978 (1987)
7. Fichtner, K.-H. and Freudenberg, W.: Characterization of states of infinite Boson systems I. – On the construction of states. Commun. Math. Phys. **137**, 315–357 (1991)
8. Fichtner, K.-H., Freudenberg, W. and Liebscher, V.: Time evolution and invariance of Boson systems given by beam splittings. Infinite Dim. Anal. Quantum Prob. and Related Topics I, 511–533 (1998)
9. Lindsay, J.M.: Quantum and Noncausal Stochastic Calculus. Prob. Th. Rel. Fields **97**, 65–80 (1993)
10. Fichtner, K.-H. and Winkler, G.: Generalized brownian motion, point processes and stochastic calculus for random fields. Math. Nachr. **161**, 291–307 (1993)
11. Inoue, K., Ohya, M. and Suyari, H.: Characterization of quantum teleportation processes by nonlinear quantum mutual entropy. Physica D **120**, 117–124 (1998)

12. Fichtner, K.-H. and Ohya, M.: Quantum Teleportation with Entangled States given by Beam Splittings. Commun. Math. Phys. **222**, 229–247 (2001)
13. Fichtner, K.-H., Freudenberg, W. and Liebscher, V.: On Exchange Mechanisms for Bosons. Submitted to Infinite Dim. Anal., Quantum Prob. and Rel. Topics

Communicated by H. Araki

Commun. Math. Phys. 222, 229 – 247 (2001)
With kind permission of Springer Science and Business Media

Communications in **Mathematical Physics**

Quantum Teleportation with Entangled States Given by Beam Splittings

Karl-Heinz Fichtner[1], **Masanori Ohya**[2]

[1] Friedrich-Schiller-Universität Jena, Fakultät für Mathematik und Informatik, Institut für Angewandte Mathematik, 07740 Jena, Germany. E-mail: fichtner@minet.uni-jena.de
[2] Department of Information Sciences, Science University of Tokyo, Chiba 278-8510, Japan. E-mail: ohya@is.noda.sut.ac.jp

Received: 21 January 2000 / Accepted: 23 April 2001

Abstract: Quantum teleportation is rigorously demonstrated with coherent entangled states given by beam splittings. The mathematical scheme of beam splitting has been used to study quantum communication [2] and quantum stochastic [8]. We discuss the teleportation process by means of coherent states in this scheme for the following two cases: (1) Delete the vacuum part from coherent states, whose compensation provides us a perfect teleportation from Alice to Bob. (2) Use fully realistic (physical) coherent states, which gives a non-perfect teleportation but shows that it is exact when the average energy (density) of the coherent vectors goes to infinity. We show that our quantum teleportation scheme with coherent entangled state is more stable than that with the EPR pairs which was previously discussed.

It is in [3] that quantum teleportation was first studied as a part of quantum cryptography [5]. Quantum teleportation is to send a quantum state itself containing all information of a certain system from one place to another. The problem of quantum teleportation is whether there exist a physical device and a key (or a set of keys) by which a quantum state is completely transmitted and a receiver can reconstruct the state sent. It has been considered that such a teleportation would not be realistic because the usual quantum state contains information which can not be observed simultaneously. In the above paper [3], Bennett et al showed such a teleportation is possible through a device (channel) made from proper (EPR) entangled states of Bell basis. The basic idea behind their discussion is to divide the information encoded in the state into two parts, classical and quantum, and send them through different channels, a classical channel and an EPR channel. The classical channel is nothing but a simple correspondence between sender and receiver, and the EPR channel is constructed by using a certain entangled state. However the EPR channel is not so stable due to decoherence. In this paper (1) we study the quantum teleportation by means of general beam splitting processes so that it contains the EPR channel, and (2) we construct a more stable teleportation process with coherent entangled states.

The quantum teleportation scheme can be mathematically expressed in the following steps [11]:

Step 0. A girl named Alice has an unknown quantum state ρ on (a N-dimensional) Hilbert space $\mathcal{H}_1$ and she was asked to teleport it to a boy named Bob.

Step 1. For this purpose, we need two other Hilbert spaces $\mathcal{H}_2$ and $\mathcal{H}_3$, $\mathcal{H}_2$ is attached to Alice and $\mathcal{H}_3$ is attached to Bob. Prearrange a so-called entangled state σ on $\mathcal{H}_2 \otimes \mathcal{H}_3$ having certain correlations and prepare an ensemble of the combined system in the state $\rho \otimes \sigma$ on $\mathcal{H}_1 \otimes \mathcal{H}_2 \otimes \mathcal{H}_3$.

Step 2. One then fixes a family of mutually orthogonal projections $(F_{nm})_{n,m=1}^{N}$ on the Hilbert space $\mathcal{H}_1 \otimes \mathcal{H}_2$ corresponding to an observable $F := \sum\limits_{n,m} z_{n,m} F_{nm}$, and for a fixed pair of indices n, m, Alice performs a first kind of incomplete measurement, involving only the $\mathcal{H}_1 \otimes \mathcal{H}_2$ part of the system in the state $\rho \otimes \sigma$, which filters the value z_{nm}, that is, after measurement on the given ensemble $\rho \otimes \sigma$ of identically prepared systems, only those where F shows the value z_{nm} are allowed to pass. According to the von Neumann rule, after Alice's measurement, the state becomes

$$\rho_{nm}^{(123)} := \frac{(F_{nm} \otimes \mathbf{1})\rho \otimes \sigma (F_{nm} \otimes \mathbf{1})}{\mathrm{tr}_{123}(F_{nm} \otimes \mathbf{1})\rho \otimes \sigma (F_{nm} \otimes \mathbf{1})},$$

where tr_{123} is the full trace on the Hilbert space $\mathcal{H}_1 \otimes \mathcal{H}_2 \otimes \mathcal{H}_3$.

Step 3. Bob is informed which measurement was done by Alice. This is equivalent to transmitting the information that the eigenvalue z_{nm} was detected. This information is transmitted from Alice to Bob without disturbance and by means of classical tools.

Step 4. Making only partial measurements on the third part on the system in the state $\rho_{nm}^{(123)}$ means that Bob will control a state $\Lambda_{nm}(\rho)$ on $\mathcal{H}_3$ given by the partial trace on $\mathcal{H}_1 \otimes \mathcal{H}_2$ of the state $\rho_{nm}^{(123)}$ (after Alice's measurement)

$$\begin{aligned} \Lambda_{nm}(\rho) &= \mathrm{tr}_{12}\, \rho_{nm}^{(123)} \\ &= \mathrm{tr}_{12} \frac{(F_{nm} \otimes \mathbf{1})\rho \otimes \sigma (F_{nm} \otimes \mathbf{1})}{\mathrm{tr}_{123}(F_{nm} \otimes \mathbf{1})\rho \otimes \sigma (F_{nm} \otimes \mathbf{1})}. \end{aligned}$$

Thus the whole teleportation scheme given by the family (F_{nm}) and the entangled state σ can be characterized by the family (Λ_{nm}) of channels from the set of states on $\mathcal{H}_1$ into the set of states on $\mathcal{H}_3$ and the family (p_{nm}) given by

$$p_{nm}(\rho) := \mathrm{tr}_{123}(F_{nm} \otimes \mathbf{1})\rho \otimes \sigma (F_{nm} \otimes \mathbf{1})$$

of the probabilities that Alice's measurement according to the observable F will show the value z_{nm}.

The teleportation scheme works perfectly with respect to a certain class $\mathfrak{S}$ of states ρ on $\mathcal{H}_1$ if the following conditions are fulfilled:

(E1) For each n, m there exists a unitary operator $v_{nm} : \mathcal{H}_1 \to \mathcal{H}_3$ such that

$$\Lambda_{nm}(\rho) = v_{nm}\, \rho\, v_{nm}^{*} \quad (\rho \in \mathfrak{S}).$$

(E2)

$$\sum_{nm} p_{nm}(\rho) = 1 \quad (\rho \in \mathfrak{S}).$$

(E1) means that Bob can reconstruct the original state ρ by unitary keys $\{v_{nm}\}$ provided to him.
(E2) means that Bob will succeed to find a proper key with certainty.

In [3,4], the authors used an EPR spin pair to construct a teleportation model. In order to have a more convenient model, we here use coherent states to construct a model. One of the main points for such a construction is how to prepare the entangled state. The EPR entangled state used in [3] can be identified with the splitting of one particle state, so that the teleportation model of Bennett et al. can be described in terms of Fock spaces and splittings, which makes us possible to work the whole teleportation process in general beam splitting scheme. Moreover to work with beams having a fixed number of particles seems to be not realistic, especially in the case of large distance between Alice and Bob, because we have to take into account that the beams will lose particles (or energy). For that reason one should use a class of beams being insensitive to this loss of particles. That and other arguments lead to superpositions of coherent beams.

In Sect. 2 of this paper, we construct a teleportation model being perfect in the sense of conditions (E1) and (E2), where we take the Boson Fock space $\Gamma(L^2(G)) := \mathcal{H}_1 = \mathcal{H}_2 = \mathcal{H}_3$ with a certain class of states ρ on this Fock space.

In Sect. 3 we consider a teleportation model where the entangled state σ is given by the splitting of a superposition of certain coherent states. Unfortunately this model doesn't work perfectly, that is, neither (E2) nor (E1) hold. However this model is more realistic than that in Sect. 2, and we show that this model provides a nice approximation to be perfect. To estimate the difference between the perfect teleportation and non-perfect teleportation, we add a further step in the teleportation scheme:

Step 5. Bob will perform a measurement on his part of the system according to the projection

$$F_+ := \mathbf{1} - |\exp(0)><\exp(0)|,$$

where $|\exp(0)><\exp(0)|$ denotes the vacuum state (the coherent state with density 0).

Then our new teleportation channels (we denote it again by Λ_{nm}) have the form

$$\Lambda_{nm}(\rho) := \mathrm{tr}_{12} \frac{(F_{nm} \otimes F_+)\rho \otimes \sigma(F_{nm} \otimes F_+)}{\mathrm{tr}_{123}(F_{nm} \otimes F_+)\rho \otimes \sigma(F_{nm} \otimes F_+)}$$

and the corresponding probabilities are

$$p_{nm}(\rho) := \mathrm{tr}_{123}(F_{nm} \otimes F_+)\,\rho \otimes \sigma(F_{nm} \otimes F_+).$$

For this teleportation scheme, (E1) is fulfilled. Furthermore we get

$$\sum_{nm} p_{nm}(\rho) = \frac{(1 - e^{-\frac{d}{2}})^2}{1 + (N-1)e^{-d}} \quad (\rightarrow 1\ (d \rightarrow +\infty)).$$

Here N denotes the dimension of the Hilbert space and d is the expectation value of the total number of particles (or energy) of the beam, so that in the case of high density (or energy) "$d \to +\infty$" of the beam the model works perfectly.

Specializing this model we consider in Sect. 4 the teleportation of all states on a finite dimensional Hilbert space (through the space $\mathbf{R}^k$). Further specialization leads to a teleportation model where Alice and Bob are spatially separated, that is, we have to teleport the information given by the state of our finite dimensional Hilbert space from one region $X_1 \subseteq \mathbf{R}^k$ into another region $X_2 \subseteq \mathbf{R}^k$ with $X_1 \cap X_2 = \emptyset$, and Alice can only perform local measurements (inside of region X_1) as well as Bob (inside of X_2).

1. Basic Notions and Notations

First we collect some basic facts concerning the (symmetric) Fock space. We will introduce the Fock space in a way adapted to the language of counting measures. For details we refer to [6–8,2,9] and other papers cited in [8].

Let G be an arbitrary complete separable metric space. Further, let μ be a locally finite diffuse measure on G, i.e. $\mu(B) < +\infty$ for bounded measurable subsets of G and $\mu(\{x\}) = 0$ for all singletons $x \in G$. In order to describe the teleportation of states on a finite dimensional Hilbert space through the k-dimensional space $\mathbf{R}^k$, especially we are concerned with the case

$$G = \mathbf{R}^k \times \{1, \dots, N\},$$
$$\mu = l \times \#,$$

where l is the k-dimensional Lebesgue measure and # denotes the counting measure on $\{1, \dots, N\}$.

Now by $M = M(G)$ we denote the set of all finite counting measures on G. Since $\varphi \in M$ can be written in the form $\varphi = \sum_{j=1}^{n} \delta_{x_j}$ for some $n = 0, 1, 2, \dots$ and $x_j \in G$ (where δ_x denotes the Dirac measures corresponding to $x \in G$) the elements of M can be interpreted as finite (symmetric) point configurations in G. We equip M with its canonical σ-algebra $\mathfrak{W}$ (cf. [6,7]) and we consider the measure F by setting

$$F(Y) := \mathcal{X}_Y(O) + \sum_{n \geq 1} \frac{1}{n!} \int_{G^n} \mathcal{X}_Y\left(\sum_{j=1}^{n} \delta_{x_j}\right) \mu^n(d[x_1, \dots, x_n])(Y \in \mathfrak{W}).$$

Hereby, $\mathcal{X}_Y$ denotes the indicator function of a set Y and O represents the empty configuration, i. e., $O(G) = 0$. Observe that F is a σ-finite measure.

Since μ was assumed to be diffuse one easily checks that F is concentrated on the set of simple configurations (i.e., without multiple points)

$$\hat{M} := \{\varphi \in M | \varphi(\{x\}) \leq 1 \text{ for all } x \in G\}.$$

Definition 1.1. $\mathcal{M} = \mathcal{M}(G) := L^2(M, \mathfrak{W}, F)$ *is called the (symmetric) Fock space over* G.

In [6] it was proved that $\mathcal{M}$ and the Boson Fock space $\Gamma(L^2(G))$ in the usual definition are isomorphic. For each $\Phi \in \mathcal{M}$ with $\Phi \neq 0$ we denote by $|\Phi>$ the corresponding normalized vector

$$|\Phi>:=\frac{\Phi}{||\Phi||}.$$

Further, $|\Phi><\Phi|$ denotes the corresponding one–dimensional projection, describing the pure state given by the normalized vector $|\Phi>$. Now, for each $n \geq 1$ let $\mathcal{M}^{\otimes n}$ be the n-fold tensor product of the Hilbert space $\mathcal{M}$. Obviously, $\mathcal{M}^{\otimes n}$ can be identified with $L^2(M^n, F^n)$.

Definition 1.2. *For a given function $g : G \to \mathbb{C}$ the function* $\exp(g) : M \to \mathbb{C}$ *defined by*

$$\exp(g)(\varphi) := \begin{cases} 1 & \text{if } \varphi = 0 \\ \prod_{x \in G, \varphi(\{x\})>0} g(x) & \text{otherwise} \end{cases}$$

is called an exponential vector generated by g.

Observe that $\exp(g) \in \mathcal{M}$ if and only if $g \in L^2(G)$ and one has in this case $||\exp(g)||^2 = e^{||g||^2}$ and $|\exp(g)>= e^{-\frac{1}{2}||g||^2}\exp(g)$. The projection $|\exp(g)><\exp(g)|$ is called the coherent state corresponding to $g \in L^2(G)$. In the special case $g \equiv 0$ we get the vacuum state

$$|\exp(0)>= \mathcal{X}_{\{0\}}\,.$$

The linear span of the exponential vectors of $\mathcal{M}$ is dense in $\mathcal{M}$, so that bounded operators and certain unbounded operators can be characterized by their actions on exponential vectors.

Definition 1.3. *The operator D :* $\mathrm{dom}(D) \to \mathcal{M}^{\otimes 2}$ *given on a dense domain* $\mathrm{dom}(D) \subset \mathcal{M}$ *containing the exponential vectors from $\mathcal{M}$ by*

$$D\psi(\varphi_1, \varphi_2) := \psi(\varphi_1 + \varphi_2) \quad (\psi \in \mathrm{dom}(D),\ \varphi_1, \varphi_2 \in M)$$

is called a compound Hida–Malliavin derivative.

On exponential vectors $\exp(g)$ with $g \in L^2(G)$, one gets immediately

$$D\exp(g) = \exp(g) \otimes \exp(g). \tag{1}$$

Definition 1.4. *The operator S :* $\mathrm{dom}(S) \to \mathcal{M}$ *given on a dense domain* $\mathrm{dom}(S) \subset \mathcal{M}^{\otimes 2}$ *containing tensor products of exponential vectors by*

$$S\Phi(\varphi) := \sum_{\tilde{\varphi} \leq \varphi} \Phi(\tilde{\varphi}, \varphi - \tilde{\varphi}) \quad (\Phi \in \mathrm{dom}(S),\ \varphi \in M)$$

is called compound Skorohod integral.

One gets

$$\langle D\psi, \Phi\rangle_{\mathcal{M}^{\otimes 2}} = \langle \psi, S\Phi\rangle_{\mathcal{M}} \quad (\psi \in \mathrm{dom}(D),\ \Phi \in \mathrm{dom}(S)), \tag{2}$$

$$S(\exp(g) \otimes \exp(h)) = \exp(g+h) \quad (g, h \in L^2(G)). \tag{3}$$

For more details we refer to [10].

Definition 1.5. *Let T be a linear operator on $L^2(G)$ with $\|T\| \leq 1$. Then the operator $\Gamma(T)$ called second quantization of T is the (uniquely determined) bounded operator on $\mathcal{M}$ fulfilling*

$$\Gamma(T)\exp(g) = \exp(Tg) \quad (g \in L^2(G)).$$

Clearly, it holds

$$\begin{aligned} \Gamma(T_1)\Gamma(T_2) &= \Gamma(T_1T_2), \\ \Gamma(T^*) &= \Gamma(T)^*. \end{aligned} \tag{4}$$

It follows that $\Gamma(T)$ is an unitary operator on $\mathcal{M}$ if T is an unitary operator on $L^2(G)$.

Lemma 1.6. *Let K_1, K_2 be linear operators on $L^2(G)$ with property*

$$K_1^*K_1 + K_2^*K_2 = \mathbf{1}. \tag{5}$$

Then there exists exactly one isometry ν_{K_1,K_2} from $\mathcal{M}$ to $\mathcal{M}^{\otimes 2} = \mathcal{M} \otimes \mathcal{M}$ with

$$\nu_{K_1,K_2}\exp(g) = \exp(K_1 g) \otimes \exp(K_2 g) \quad (g \in L^2(G)). \tag{6}$$

Further it holds

$$\nu_{K_1,K_2} = (\Gamma(K_1) \otimes \Gamma(K_2))D \tag{7}$$

(at least on $\mathrm{dom}(D)$ *but one has the unique extension). The adjoint $\nu^*_{K_1,K_2}$ of ν_{K_1,K_2} is characterized by*

$$\nu^*_{K_1,K_2}(\exp(h) \otimes \exp(g)) = \exp(K_1^*h + K_2^*g) \quad (g, h \in L^2(G)) \tag{8}$$

and it holds

$$\nu^*_{K_1,K_2} = S(\Gamma(K_1^*) \otimes \Gamma(K_2^*)). \tag{9}$$

Remark 1.7. From K_1, K_2 we get a transition expectation $\xi_{K_1K_2} : \mathcal{M} \otimes \mathcal{M} \to \mathcal{M}$, using $\nu^*_{K_1,K_2}$ and the lifting $\xi^*_{K_1K_2}$ may be interpreted as a certain splitting (cf. [2]).

Proof of Lemma 1.6. We consider the operator

$$B := S(\Gamma(K_1^*) \otimes \Gamma(K_2^*))(\Gamma(K_1) \otimes \Gamma(K_2))D$$

on the dense domain $\mathrm{dom}(B) \subseteq \mathcal{M}$ spanned by the exponential vectors. Using (1), (3), (4) and (5) we get

$$B\exp(g) = \exp(g) \quad (g \in L^2(G)).$$

It follows that the bounded linear unique extension of B onto $\mathcal{M}$ coincides with the unity on $\mathcal{M}$,

$$B = \mathbf{1}. \tag{10}$$

On the other hand, by Eq. (7) at least on dom (D), an operator ν_{K_1,K_2} is defined. Using (2) and (4) we obtain

$$\begin{aligned}\|\nu_{K_1,K_2}\psi\|^2 &= \langle \nu_{K_1,K_2}\psi, \nu_{K_1,K_2}\psi\rangle \quad (\psi \in \mathrm{dom}\,(D))\\ &= \langle \psi, B\psi\rangle,\end{aligned}$$

which implies

$$\|\nu_{K_1,K_2}\psi\|^2 = \|\psi\|^2 \quad (\psi \in \mathrm{dom}\,(D))$$

because of (10). It follows that ν_{K_1,K_2} can be uniquely extended to a bounded operator on $\mathcal{M}$ with

$$\|\nu_{K_1,K_2}\psi\| = \|\psi\| \quad (\psi \in \mathcal{M}).$$

Now from (7) we obtain (6) using (1) and the definition of the operators of second quantization. Further, (7), (3) and (4) imply (9) and from (9) we obtain (8) using the definition of the operators of second quantization and Eq. (3). □

Here we explain the fundamental scheme of beam splitting [8]. We define an isometric operator $V_{\alpha,\beta}$ for coherent vectors such that

$$V_{\alpha,\beta}|\exp(g)\rangle = |\exp(\alpha g)\rangle \otimes |\exp(\beta g)\rangle$$

with $|\alpha|^2 + |\beta|^2 = 1$. This beam splitting is a useful mathematical expression for optical communication and quantum measurements [2].

Example 1.8 ($\alpha = \beta = 1/\sqrt{2}$ above). Let $K_1 = K_2$ be the following operator of multiplication on $L^2(G)$:

$$K_1 g = \frac{1}{\sqrt{2}} g = K_2 g \quad (g \in L^2(G)).$$

We put

$$\nu := \nu_{K_1,K_2}$$

and obtain

$$\nu \exp(g) = \exp\left(\frac{1}{\sqrt{2}} g\right) \otimes \exp\left(\frac{1}{\sqrt{2}} g\right) \quad (g \in L^2(G)).$$

Example 1.9. Let $L^2(G) = \mathcal{H}_1 \oplus \mathcal{H}_2$ be the orthogonal sum of the subspaces $\mathcal{H}_1$, $\mathcal{H}_2$. K_1 and K_2 denote the corresponding projections.

We will use Example 1.8 in order to describe a teleportation model where Bob performs his experiments on the same ensemble of the systems like Alice.

Further we will use a special case of Example 1.9 in order to describe a teleportation model where Bob and Alice are spatially separated (cf. Sect. 5).

Remark 1.10. The property (5) implies

$$\|K_1 g\|^2 + \|K_2 g\|^2 = \|g\|^2 \quad (g \in L^2(G)). \tag{11}$$

Remark 1.11. Let U, V be unitary operators on $L^2(G)$. If operators K_1, K_2 satisfy (5), then the pair $\hat{K}_1 = UK_1$, $\hat{K}_2 = VK_2$ fulfill (5).

2. A Perfect Model of Teleportation

Concerning the general idea we follow the papers [1, 11]. We fix an ONS $\{g_1, \ldots, g_N\} \subseteq L^2(G)$, operators K_1, K_2 on $L^2(G)$ with (5), a unitary operator T on $L^2(G)$, and $d > 0$. We assume

$$TK_1g_k = K_2g_k \quad (k = 1, \ldots, N), \tag{12}$$

$$\langle K_1g_k, K_1g_j\rangle = 0 \qquad (k \neq j;\ k, j = 1 \ldots, N). \tag{13}$$

Using (11) and (12) we get

$$\|K_1g_k\|^2 = \|K_2g_k\|^2 = \frac{1}{2}. \tag{14}$$

From (12) and (13) we get

$$\langle K_2g_k,\ K_2g_j\rangle = 0 \quad (k \neq j\,;\ k, j = 1, \ldots, N). \tag{15}$$

The state of Alice asked to teleport is of the type

$$\rho = \sum_{s=1}^{N} \lambda_s |\Phi_s\rangle\langle\Phi_s|, \tag{16}$$

where

$$|\Phi_s\rangle = \sum_{j=1}^{N} c_{sj} |\exp(aK_1g_j) - \exp(0)\rangle \qquad \left(\sum_j |c_{sj}|^2 = 1; s = 1, \ldots, N\right) \tag{17}$$

and $a = \sqrt{d}$. One easily checks that $(|\exp(aK_1g_j) - \exp(0)\rangle)_{j=1}^N$ and $(|\exp aK_2g_j) - \exp(0)\rangle)_{j=1}^N$ are ONS in $\mathcal{M}$. Here we took the vacuum state exp(0) off, but it is just only for computational symplicity.

In order to achieve that $(|\Phi_s\rangle)_{s=1}^N$ is still an ONS in $\mathcal{M}$ we assume

$$\sum_{j=1}^{N} \bar{c}_{sj}c_{kj} = 0 \quad (j \neq k\,;\ j, k = 1, \ldots, N)\,. \tag{18}$$

Denote $c_s = [c_{s1,\ldots,}c_{sN}] \in \mathbb{C}^N$, then $(c_s)_{s=1}^N$ is an CONS in $\mathbb{C}^N$.

Now let $(b_n)_{n=1}^N$ be a sequence in $\mathbb{C}^N$,

$$b_n = [b_{n1,\ldots,}b_{nN}]$$

with properties

$$|b_{nk}| = 1 \quad (n, k = 1, \ldots, N), \tag{19}$$

$$\langle b_n\,,\ b_j\rangle = 0 \quad (n \neq j\,;\ n, j = 1, \ldots, N). \tag{20}$$

Then Alice's measurements are performed with projection

$$F_{nm} = |\xi_{nm}\rangle\langle\xi_{nm}| \quad (n, m = 1, \ldots, N) \tag{21}$$

given by

$$|\xi_{nm}\rangle = \frac{1}{\sqrt{N}} \sum_{j=1}^{N} b_{nj} |\exp(aK_1 g_j) - \exp(0) > \otimes |\exp(aK_1 g_{j\oplus m}) - \exp(0)\rangle, \quad (22)$$

where $j \oplus m := j + m (\mathrm{mod}\, N)$.

One easily checks that $(|\xi_{nm}\rangle)_{n,m=1}^{N}$ is an ONS in $\mathcal{M}^{\otimes 2}$. Further, the state vector $|\xi\rangle$ of the entangled state $\sigma = |\xi\rangle\langle\xi|$ is given by

$$|\xi\rangle = \frac{1}{\sqrt{N}} \sum_{k} |\exp(aK_1 g_k) - \exp(0)\rangle \otimes |\exp(aK_2 g_k) - \exp(0)\rangle. \quad (23)$$

Lemma 2.1. *For each* $n, m = 1, \dots, N$ *it holds*

$$\begin{aligned}&(F_{nm} \otimes \mathbf{1})(|\Phi_s\rangle \otimes |\xi\rangle)\\ &\quad = \frac{1}{N} |\xi_{nm}\rangle \otimes \sum_{j} \bar{b}_{nj} c_{sj} |\exp(aK_2 g_{j\oplus m}) - \exp(0)\rangle \ (s = 1, \dots, N). \quad (24)\end{aligned}$$

Proof. From the fact that

$$|\gamma_j\rangle := |\exp(aK_1 g_j) - \exp(0)\rangle \quad (j = 1, \dots, N) \quad (25)$$

is an ONS, it follows

$$\langle \gamma_r \otimes \gamma_{r\oplus m}\,,\, \gamma_j \otimes \gamma_k\rangle = \begin{cases} 1 & \text{if } r = j \text{ and } k = r \oplus m \\ 0 & \text{otherwise} \end{cases}. \quad (26)$$

On the other hand, we have

$$\begin{aligned}&(F_{nm} \otimes \mathbf{1})(|\Phi_s\rangle \otimes |\xi\rangle)\\ &= \frac{1}{N} \sum_{k} \sum_{j} \sum_{r} c_{sj} \bar{b}_{ns} \langle \gamma_r \otimes \gamma_{r\oplus m}\,,\, \gamma_j \otimes \gamma_k\rangle \xi_{nm} \otimes |\exp(aK_2 g_k) - \exp(0)\rangle. \quad (27)\end{aligned}$$

Using (26) and (27), we get (24). □

Now we have

$$\begin{aligned}\rho \otimes \sigma &= \sum_{s=1}^{N} \lambda_s |\Phi_s\rangle\langle\Phi_s| \otimes |\xi\rangle\langle\xi| \\ &= \sum_{s=1}^{N} \lambda_s |\Phi_s \otimes \xi\rangle\langle\Phi_s \otimes \xi|,\end{aligned} \quad (28)$$

which implies

$$
\begin{aligned}
(F_{nm} \otimes \mathbf{1})(\rho \otimes \sigma)(F_{nm} \otimes \mathbf{1}) &= \sum_{s=1}^{N} \lambda_s (F_{nm} \otimes \mathbf{1}) |\Phi_s \otimes \xi\rangle\langle \Phi_s \otimes \xi| (F_{nm} \otimes \mathbf{1}) \\
&= \sum_{s=1}^{N} \lambda_s \|(F_{nm} \otimes \mathbf{1})(\Phi_s \otimes \xi)\|^2 \\
&\quad \times |(F_{nm} \otimes \mathbf{1})(\Phi_s \otimes \xi)\rangle\langle (F_{nm} \otimes \mathbf{1})(\Phi_s \otimes \xi)|.
\end{aligned} \tag{29}
$$

Note $|\Phi_s \otimes \xi\rangle = |\Phi_s\rangle \otimes |\xi\rangle$. From (12) it follows that

$$
\begin{aligned}
&\sum_j \bar{b}_{nj} c_{sj} |\exp(aK_2 g_{j\oplus m}) - \exp(0)\rangle \\
&\qquad = \Gamma(T) \sum_j \bar{b}_{nj} c_{sj} |\exp(aK_1 g_{j\oplus m}) - \exp(0)\rangle.
\end{aligned} \tag{30}
$$

Further, for each $m, n\ (= 1, \ldots, N)$, we have unitary operators U_m, B_n on $\mathcal{M}$ given by

$$
B_n |\exp(aK_1 g_j) - \exp(0)\rangle = b_{nj} |\exp(aK_1 g_j) - \exp(0)\rangle \quad (j = 1, \ldots, N), \tag{31}
$$

$$
U_m |\exp(aK_1 g_j) - \exp(0)\rangle = |\exp(aK_1 g_{j\oplus m}) - \exp(0)\rangle \quad (j = 1, \ldots, N). \tag{32}
$$

Therefore we get

$$
\sum_j \bar{b}_{nj} c_{sj} |\exp(aK_1 g_{j\oplus m}) - \exp(0)\rangle = U_m B_n^* (\Phi_s). \tag{33}
$$

From (30), (33) and Lemma 2.1 we obtain

$$
(F_{nm} \otimes \mathbf{1}) (|\Phi_s\rangle \otimes |\xi\rangle) = \frac{1}{N} |\xi_{nm}\rangle \otimes \left(\Gamma(T) U_m B_n^* |\Phi_s\rangle \right). \tag{34}
$$

It follows

$$
\| (F_{nm} \otimes \mathbf{1}) (|\Phi_s\rangle \otimes |\xi\rangle) \|^2 = \frac{1}{N^2}. \tag{35}
$$

Finally from (29), (34) and (35) we have

$$
(F_{nm} \otimes \mathbf{1}) (\rho \otimes \sigma) (F_{nm} \otimes \mathbf{1}) = \frac{1}{N^2} F_{nm} \otimes \left(\Gamma(T) U_m B_n^* \right) \rho \left(B_n U_m^* \Gamma(T^*) \right) \tag{36}
$$

That leads to the following solution of the teleportation problem.

Theorem 2.2. *For each $n, m = 1, \ldots, N$, define a channel Λ_{nm} by*

$$
\Lambda_{nm}(\rho) := \mathrm{tr}_{12} \frac{(F_{nm} \otimes 1) (\rho \otimes \sigma) (F_{nm} \otimes \mathbf{1})}{\mathrm{tr}_{123} (F_{nm} \otimes 1) \left(\hat{\rho} \otimes \sigma \right) (F_{nm} \otimes \mathbf{1})} \quad (\rho \text{ normal state on } \mathcal{M}). \tag{37}
$$

Then we have for all states ρ on M with (16) and (17)

$$
\Lambda_{nm}(\rho) = \left(\Gamma(T) U_m B_n^* \right) \rho \left(\Gamma(T) U_m B_n^* \right)^*. \tag{38}
$$

Remark 2.3. In case of Example 1.8 using the operators $B_n, U_m, \Gamma(T)$, the projections F_{nm} are given by unitary transformations of the entangled state σ:

$$F_{nm} = \left(B_n \otimes U_m\Gamma(T^*)\right)\sigma\left(B_n \otimes U_m\Gamma(T^*)\right)^*, \quad \text{or}$$
$$|\xi_{nm}\rangle = \left(B_n \otimes U_m\Gamma(T^*)\right)|\xi\rangle. \tag{39}$$

Remark 2.4. If Alice performs a measurement according to the following selfadjoint operator:

$$F = \sum_{n,m=1}^{N} z_{nm} F_{nm}$$

with $\{z_{nm} | n, m = 1, \ldots, N\} \subseteq \mathbf{R} - \{0\}$, then she will obtain the value z_{nm} with probability $1/N^2$. The sum over all these probabilities is 1, so that the teleportation model works perfectly.

3. A Non-Perfect Case of Teleportation

In this section we will construct a model where we also have channels with property (38). But the probability that one of these channels will work in order to teleport the state from Alice to Bob is less than 1 depending on the density parameter d (or energy of the beams, depending on the interpretation). If $d = a^2$ tends to infinity that probability tends to 1. That is, the model is asymptotically perfect in a certain sense.

We consider the normalized vector

$$|\eta\rangle := \frac{\gamma}{\sqrt{N}} \sum_{k=1}^{N} |\exp(ag_k)\rangle,$$
$$\gamma := \left(\frac{1}{1+(N-1)e^{-d}}\right)^{\frac{1}{2}} = \left(\frac{1}{1+(N-1)e^{-a^2}}\right)^{\frac{1}{2}}, \tag{40}$$

and we replace in (37) the projector σ by the projector

$$\tilde{\sigma} := |\tilde{\xi}\rangle\langle\tilde{\xi}|,$$
$$\tilde{\xi} := \nu_{K_1,K_2}(\eta) = \frac{\gamma}{\sqrt{N}} \sum_{k=1}^{N} |\exp(aK_1 g_k)\rangle \otimes |\exp(aK_2 g_k)\rangle. \tag{41}$$

Then for each $n, m = 1, \ldots, N$, we get the channels on a normal state ρ on $\mathcal{M}$ such as

$$\tilde{\Lambda}_{nm}(\rho) := \operatorname{tr}_{12} \frac{(F_{nm} \otimes \mathbf{1})(\rho \otimes \tilde{\sigma})(F_{nm} \otimes \mathbf{1})}{\operatorname{tr}_{123}(F_{nm} \otimes \mathbf{1})(\rho \otimes \tilde{\sigma})(F_{nm} \otimes \mathbf{1})}, \tag{42}$$

$$\Theta_{nm}(\rho) := \operatorname{tr}_{12} \frac{(F_{nm} \otimes F_+)(\rho \otimes \tilde{\sigma})(F_{nm} \otimes F_+)}{\operatorname{tr}_{123}(F_{nm} \otimes F_+)(\rho \otimes \tilde{\sigma})(F_{nm} \otimes F_+)}, \tag{43}$$

where $F_+ = \mathbf{1} - |\exp(0)\rangle\langle\exp(0)|$ e. g., F_+ is the projection onto the space $\mathcal{M}_+$ of configurations having no vacuum part, e. g., orthogonal to vacuum

$$\mathcal{M}_+ := \{\psi \in \mathcal{M} | \, \| |\exp(0)\rangle\langle\exp(0)|\psi\| = 0\}.$$

One easily checks that

$$\Theta_{nm}(\rho) = \frac{F_+ \tilde{\Lambda}_{nm}(\rho) F_+}{\operatorname{tr}\left(F_+ \tilde{\Lambda}_{nm}(\rho) F_+\right)}, \tag{44}$$

that is, after receiving the state $\tilde{\Lambda}_{nm}(\rho)$ from Alice, Bob has to omit the vacuum.

From Theorem 2.2 it follows that for all ρ with (16) and (17),

$$\Lambda_{nm}(\rho) = \frac{F_+ \Lambda_{nm}(\rho) F_+}{\operatorname{tr}(F_+ \Lambda_{nm}(\rho) F_+)}.$$

This is not true if we replace Λ_{nm} by $\tilde{\Lambda}_{nm}$, namely, in general it does not hold

$$\Theta_{nm}(\rho) = \tilde{\Lambda}_{nm}(\rho).$$

But we will prove that for each ρ with (16) and (17) it holds

$$\Theta_{nm}(\rho) = \Lambda_{nm}(\rho),$$

which means

$$\Theta_{nm}(\rho) = (\Gamma(T) U_m B_n^*) \rho (\Gamma(T) U_m B_n^*)^* \tag{45}$$

because of Theorem 2.2. Further we will show

$$\operatorname{tr}_{123}(F_{nm} \otimes F_+)(\rho \otimes \tilde{\sigma})(F_{nm} \otimes F_+) = \frac{\gamma^2}{N^2}\left(e^{\frac{d}{2}} - 1\right)^2 e^{-d} \tag{46}$$

and the sum over n, m $(= 1, \ldots, N)$ gives the probability

$$\frac{\left(1 - e^{-\frac{d}{2}}\right)^2}{1 + (N-1)e^{-d}} \longrightarrow 1 \ (d \longrightarrow \infty)$$

which means that the teleportation model works perfectly in the limit $d \longrightarrow \infty$, e. g., Bob will receive one of the states $\Theta_{nm}(\rho)$ given by (44). Thus we formulate the following theorem.

Theorem 3.1. *For all states ρ on $\mathcal{M}$ with* (16) *and* (17) *and each pair n, m $(= 1, \ldots, N)$, Eqs.* (44) *and* (45) *hold. Further, we have*

$$\sum_{n,m} \operatorname{tr}_{123}(F_{nm} \otimes F_+)(\rho \otimes \tilde{\sigma})(F_{nm} \otimes F_+) = \frac{\left(1 - e^{-\frac{d}{2}}\right)^2}{1 + (N-1)e^{-d}}. \tag{47}$$

In order to prove Theorem 3.1, we fix ρ with (16) and (17) and start with a lemma.

Lemma 3.2. *For each n, m, s $(= 1, \ldots, N)$, it holds*

$$(F_{nm} \otimes \mathbf{1})\left(|\Phi_s\rangle \otimes |\tilde{\xi}\rangle\right) = \frac{\gamma}{N}\left(1 - e^{-\frac{d}{2}}\right)|\xi_{nm}\rangle \otimes \left(\Gamma(T) U_m B_n^* |\Phi_s\rangle\right)$$
$$+ \frac{\gamma}{N}\left(\frac{e^{\frac{d}{2}} - 1}{e^d}\right)^{\frac{1}{2}} \langle b_n, c_s\rangle_{\mathbb{C}^N} \xi_{nm} \otimes |\exp(0)\rangle.$$

Proof. For all $k, j, r = 1, \ldots, N$, we get

$$\begin{aligned}\alpha_{k,j,r} &:= \langle |\exp(aK_1 g_r) - \exp(0)\rangle \otimes |\,|\exp(aK_1 g_{r\otimes m}) - \exp(0)\rangle, \\ &\qquad |\exp(aK_1 g_j) - \exp(0)\rangle \otimes |\exp(aK_1 g_k)\rangle\rangle \\ &= \begin{cases} \left(\dfrac{e^{\frac{a^2}{2}} - 1}{e^{\frac{a^2}{2}}}\right) & \text{if } r = j \text{ and } k = r \oplus m \\ 0 & \text{otherwise}\end{cases}\end{aligned}$$

and

$$|\exp(aK_2 g_{j\oplus m})\rangle = e^{-\frac{a^2}{2}} \left(e^{\frac{a^2}{2}} - 1\right)^{\frac{1}{2}} |\exp\left(aK_2 g_{j\oplus m}\right) - \exp(0)\rangle + e^{-\frac{a^2}{2}} |\exp(0)\rangle.$$

On the other hand, we have

$$(F_{nm} \otimes \mathbf{1})\left(|\Phi_s\rangle \otimes |\tilde{\xi}\rangle\right) = \frac{\gamma}{N} \sum_k \sum_j \sum_r c_{sj} \bar{b}_{nr} \alpha_{k,j,r} \xi_{nm} \otimes |\exp(aK_2 g_k)\rangle.$$

It follows with $a^2 = d$,

$$\begin{aligned}&(F_{nm} \otimes \mathbf{1})\left(\Phi_s \otimes \tilde{\xi}\right) \\ &\quad = \frac{\gamma}{N}\left(e^{\frac{d}{2}} - 1\right) e^{-\frac{d}{2}} \xi_{nm} \otimes \left(\sum_j c_{sj} \bar{b}_{nj} |\exp\left(aK_2 g_{j\oplus m}\right) - \exp(0)\rangle\right) \\ &\qquad + \frac{\gamma}{N}\left(e^{\frac{d}{2}} - 1\right)^{\frac{1}{2}} e^{-\frac{d}{2}} \sum_j c_{sj} \bar{b}_{nj} \xi_{nm} \otimes |\exp(0)\rangle \\ &\quad = \frac{\gamma}{N}\left(1 - e^{-\frac{d}{2}}\right) \xi_{nm} \otimes \left(\Gamma(T) U_m B_n^* \Phi_s\right) \\ &\qquad + \frac{\gamma}{N}\left(\frac{e^{\frac{d}{2}} - 1}{e^d}\right)^{\frac{1}{2}} \langle b_n, c_s\rangle_{\mathbb{C}^N} \xi_{nm} \otimes |\exp(0)\rangle. \qquad \square\end{aligned}$$

If ρ is a pure state

$$\rho = |\Phi_s\rangle\langle\Phi_s|,$$

then we obtain from Lemma 3.2,

$$\begin{aligned}&\mathrm{tr}_{123}\,(F_{nm} \otimes \mathbf{1})(\rho \otimes \tilde{\sigma})(F_{nm} \otimes \mathbf{1}) \\ &\qquad = \frac{\gamma^2}{N^2}\left(\left(1 - e^{-\frac{d}{2}}\right)^2 + \frac{e^{\frac{d}{2}} - 1}{e^d} |\langle b_n, c_s\rangle|^2\right) \\ &\qquad = \frac{1}{N^2\left(1 + (N-1)e^{-d}\right)}\left(\left(1 - e^{-\frac{d}{2}}\right)^2 + \frac{e^{\frac{d}{2}} - 1}{e^d} |\langle b_n, c_s\rangle|^2\right)\end{aligned}$$

and

$$\tilde{\Lambda}_{nm}(\rho) \not\models \left(\Gamma(T)U_m B_n^*\right) \rho \left(\Gamma(T)U_m B_n^*\right)^* .$$

Now we have

$$\Gamma(T)U_m B_n^* \Phi_s \in \mathcal{M}_+ , \ |\exp(0)\rangle \in \mathcal{M}_+^{\perp}.$$

Hence, Lemma 3.2 implies

$$(\mathbf{1} \otimes \mathbf{1} \otimes F_+)(F_{nm} \otimes \mathbf{1})\left(\Phi_s \otimes \tilde{\xi}\right) = \frac{\gamma}{N}\left(1 - e^{-\frac{d}{2}}\right) \xi_{nm} \otimes \left(\Gamma(T)U_m B_n^* \Phi_s\right)$$

that is, we have the following lemma

Lemma 3.3. *For each $n, m, s = 1, \ldots, N$, it holds*

$$(F_{nm} \otimes F_+)\left(\Phi_s \otimes \tilde{\xi}\right) = \frac{\gamma}{N}\left(1 - e^{-\frac{d}{2}}\right) \xi_{nm} \otimes \left(\Gamma(T)U_m B_n^* \Phi_s\right). \tag{48}$$

Remark 3.4. Let K_2 be a projection of the type

$$K_2 h = h\mathcal{X}_X; \ h \in L^2(G),$$

where $X \subseteq G$ is measurable. Then (48) also holds if we replace F_+ by the projection $F_{+,X}$ onto the subspace $\mathcal{M}_{+,X}$ of $\mathcal{M}$ given by

$$\mathcal{M}_{+,X} := \{\psi \in \mathcal{M} | \psi(\varphi) = 0 \text{ if } \varphi(X) = 0\}.$$

Observe that $\mathcal{M}_{+,G} = \mathcal{M}_+$.

Proof of Theorem 3.1. We have assumed that $(|\Phi_s\rangle)_{s=1}^N$ is an ONS in $\mathcal{M}$, which implies that $\left(|\xi_{nm}\rangle \otimes \left(\Gamma(T)U_m B_n^* |\Phi_s\rangle\right)\right)_{s=1}^N$ is an ONS in $\mathcal{M}^{\otimes 3}$. Hence we obtain Eqs. (45), (46) and (47) by Lemma 3.3. This proves Theorem 3.1. □

Remark 3.5. In the special case of Remark 3.4, Eqs. (45), (46) and (47) hold if we replace F_+ by $F_{+,X}$ in the definition of the channel Θ_{nm} and in (46), (47), that is, Bob will only perform "local" measurement according to the region X, about which we will discuss more details in the next sections.

4. Teleportation of States Inside $\mathbf{R}^k$

Let $\mathcal{H}$ be a finite-dimensional Hilbert space. We consider the case

$$\mathcal{H} = \mathbb{C}^N = L^2(\{1, \ldots, N\}, \#)$$

without loss of generality, where # denotes the counting measure on the set $\{1, \ldots, N\}$. We want to teleport states on $\mathcal{H}$ with the aid of the constructed channels $(\Lambda_{nm})_{n,m=1}^N$ or $(\Theta_{nm})_{n,m=1}^N$. We fix

- a CONS $(|j\rangle)_{j=1}^N$ of $\mathcal{H}$,
- $f \in L^2\left(\mathbf{R}^k\right)$, $\|f\| = 1$,
- $d = a^2 > 0$,

- $\hat{K}_1, \hat{K}_2$ linear operators on $L^2(\mathbf{R}^k)$,
- $\hat{T}$ unitary operator on $L^2(\mathbf{R}^k)$

with two properties

$$\hat{K}_1^* \hat{K}_1 f + \hat{K}_2^* \hat{K}_2 f = f, \tag{49}$$

$$\hat{T} \hat{K}_1 f = \hat{K}_2 f. \tag{50}$$

We put

$$G = \mathbf{R}^k \times \{1, \ldots, N\}, \quad \mu = l \times \#,$$

where l is the Lebesgue measure on $\mathbf{R}^k$. Then $L^2(G) = L^2(G, \mu) = L^2(\mathbf{R}^k) \otimes \mathcal{H}$. Further, put

$$g_j := f \otimes |j\rangle \quad (j = 1, \ldots, N).$$

Then $(g_j)_{j=1}^N$ is an ONS in $L^2(G)$. We consider linear operators K_1, K_2 on $L^2(G)$ with (5) and

$$K_r g_j = \left(\hat{K}_r f\right) \otimes |j\rangle \quad (j = 1, \ldots, N;\ r = 1, 2). \tag{51}$$

Remark 4.1. Equation (51) determines operators K_1, K_2 on the subspace of $\mathcal{M}$ spanned by the ONS $(g_j)_{j=1}^N$. On the orthogonal complement, one can put for instance

$$K_r \psi = \frac{1}{\sqrt{2}} \psi.$$

Then K_1, K_2 are well defined and fulfill (5) because of (49). Further, one checks that (13) and (15) hold.

Now let T be an unitary operator on $L^2(G)$ with

$$T(K_1 g_j) = \left(\hat{T} \hat{K}_2 f\right) \otimes |j\rangle.$$

From (13) one can prove the existence of T using the arguments as in Remark 4.1. Further, we get (12) from (50).

Summarizing, we obtain that $\{g_1, \ldots, g_N\}$, K_1, K_2, T fulfill all the assumptions required in Sect. 2. Thus we have the corresponding channels $\Lambda_{n,m}$, Θ_{nm} given by (37) and (43) respectively. It follows that we are able to teleport a state ρ on $\mathcal{M} = \mathcal{M}(G)$ with (16) and (17) as it was stated in Theorem 2.2 and Theorem 3.1, respectively.

In order to teleport states on $\mathcal{H}$ through the space $\mathbf{R}^k$ using the above channels, we have to consider:

First: a "lifting" $\mathcal{E}^*$ of the states $\hat{\rho}$ on $\mathcal{H}$ into the set of states on the bigger state space on $\mathcal{M}$ such that $\rho = \mathcal{E}^*(\hat{\rho})$ can be described by (16), (17), (18).

Second: a "reduction" $\mathcal{R}$ of (normal) states on $\mathcal{M}$ to states on $\mathcal{H}$ such that for all states $\hat{\rho}$ on $\mathcal{H}$ it holds

$$\mathcal{R}\left(\left(\Gamma(T) U_m B_n^*\right) \mathcal{E}^*\left(\hat{\rho}\right) \left(\Gamma(T) U_m B_n^*\right)^*\right) = V_{nm} \hat{\rho} V_{nm}^* \quad (n, m = 1, \ldots, N), \tag{52}$$

where $(V_{nm})_{n,m=1}^N$ are unitary operators on $\mathcal{H}$.

That we can obtain as follows: We have already stated in Sect. 2 that

$$\left(\left|\exp\left(aK_r(g_j)\right) - \exp(0)\right\rangle\right)_{j=1}^{N} \quad (r = 1, 2)$$

are ONS in $\mathcal{M}$. We denote by $\mathcal{M}_r$ $(r = 1, 2)$ the corresponding N- dimensional subspaces of $\mathcal{M}$. Then for each $r = 1, 2$, there exists exactly one unitary operator W_r from $\mathcal{H}$ onto $\mathcal{M}_r \subseteq \mathcal{M}$ with

$$W_r|j\rangle = \left|\exp\left(aK_r g_j\right) - \exp(0)\right\rangle \quad (j = 1, \ldots, N). \tag{53}$$

We put

$$\mathcal{E}^*\left(\hat{\rho}\right) := W_1 \hat{\rho} W_1^* \Pi_{\mathcal{M}_1} \quad \left(\hat{\rho} \text{ state on } \mathcal{H}\right), \tag{54}$$

where $\Pi_{\mathcal{M}_r}$ denotes the projection onto $\mathcal{M}_r$ $(r = 1, 2)$. Describing the state $\hat{\rho}$ on $\mathcal{H}$ by

$$\hat{\rho} = \sum_{s=1}^{N} \lambda_s |\hat{\Phi}_s\rangle\langle\hat{\Phi}_s| \tag{55}$$

with

$$|\hat{\Phi}_s\rangle = \sum_{j=1}^{N} c_{sj} |j\rangle,$$

where $\left(c_{sj}\right)_{s,j=1}^{N}$ fulfills (18), we obtain that $\rho = \mathcal{E}^*\left(\hat{\rho}\right)$ is given by (16) and (17).

Now, for each state ρ on $\mathcal{M}$ we put

$$\mathcal{R}(\rho) := \frac{W_2^* \Pi_{\mathcal{M}_2} \rho W_2}{\operatorname{tr}_{\mathcal{M}} W_2^* \Pi_{\mathcal{M}_2} \rho W_2}. \tag{56}$$

Since

$$\Pi_{\mathcal{M}_2} \Gamma(T) U_m B_n^* \mathcal{E}^*\left(\hat{\rho}\right) \left(\Gamma(T) U_m B_n^*\right)^* = \Gamma(T) U_m B_n^* \mathcal{E}^*\left(\hat{\rho}\right) \left(\Gamma(T) U_m B_n^*\right)^*,$$

we get

$$\operatorname{tr}_{\mathcal{M}} W_2^* \Pi_{\mathcal{M}_2} \Gamma(T) U_m B_n^* \mathcal{E}^*\left(\hat{\rho}\right) \left(\Gamma(T) U_m B_n^*\right)^* = 1$$

and

$$\mathcal{R}\left(\Gamma(T) U_m B_n^* \mathcal{E}^*\left(\hat{\rho}\right) \left(\Gamma(T) U_m B_n^*\right)^*\right) = W_2^* \Gamma(T) U_m B_n^* W_1 \hat{\rho} W_1^* \Pi_{\mathcal{M}_1} \left(\Gamma(T) U_m B_n^*\right)^* W_2.$$

As we have the equality

$$\Pi_{\mathcal{M}_1} \left(\Gamma(T) U_m B_n^*\right)^* W_2 = \left(\Gamma(T) U_m B_n^*\right)^* W_2,$$

which implies

$$\mathcal{R}\left(\Gamma(T) U_m B_n^* \mathcal{E}^*\left(\hat{\rho}\right) \left(\Gamma(T) U_m B_n^*\right)^*\right) = W_2^* \Gamma(T) U_m B_n^* W_1 \hat{\rho} W_1^* \left(\Gamma(T) U_m B_n^*\right)^* W_2.$$

Put

$$V_{nm} := W_2^* \Gamma(T) U_m B_n^* W_1 \quad (n, m = 1, \dots, N), \tag{57}$$

then V_{nm} $(n, m = 1, \dots, N)$ is a unitary operator on $\mathcal{H}$ and (52) holds. One easily checks

$$V_{nm}|j\rangle = \bar{b}_{nj}|j \oplus m\rangle \quad (j, m, n = 1, \dots, N).$$

Summarizing these, we have the following theorem:

Theorem 4.2. *Consider the channels on the set of states on* $\mathcal{H}$

$$\hat{\Lambda}_{nm} := \mathcal{R} \circ \Lambda_{nm} \circ \mathcal{E}^* \quad (n, m = 1, \dots, N), \tag{58}$$

$$\hat{\Theta}_{nm} := \mathcal{R} \circ \Theta_{nm} \circ \mathcal{E}^* \quad (n, m = 1, \dots, N). \tag{59}$$

where $\mathcal{R}$, $\mathcal{E}^*$, Λ_{nm}, Θ_{nm} *are given by (56), (54), (37), (43), respectively. Then for all states* $\hat{\rho}$ *on* $\mathcal{H}$, *it holds*

$$\hat{\Lambda}_{nm}\left(\hat{\rho}\right) = V_{nm}\hat{\rho}V_{nm}^* = \hat{\Theta}_{nm}\left(\hat{\rho}\right) \quad (n, m = 1, \dots, N), \tag{60}$$

where V_{nm} $(n, m = 1, \dots, N)$ *are the unitary operators on* $\mathcal{H}$ *given by (57).*

Remark 4.3. Remember that the teleportation model according to $(\Lambda_{nm})_{n,m=1}^N$ works perfectly in the sense of Remark 2.4, and the model dealing with $(\Theta_{nm})_{n,m=1}^N$ was only asymptotically perfect for large d (i.e., high density or high energy of the beams). They can transfer to $\left(\hat{\Lambda}_{n,m}\right)$, $\left(\hat{\Theta}_{nm}\right)$.

Example 4.4. We specialize

$$\hat{K}_1 h = \hat{K}_2 h = \frac{1}{\sqrt{2}} h \quad \left(h \in L^2(\mathbf{R}^k)\right), \ \hat{T} = \mathbf{1}.$$

Realizing the teleportation in this case means that Alice has to perform measurements (F_{nm}) in the whole space $\mathbf{R}^k$ and also Bob (concerning F_+).

5. Alice and Bob are Spatially Separated

We specialize the situation in Sect. 4 as follows: We fix

- $t \in \mathbf{R}^k$,
- $X_1, X_2, X_3 \subseteq \mathbf{R}^k$ are measurable decomposition of $\mathbf{R}^k$ such that $l(X_1) \neq 0$ and

$$X_2 = X_1 + t := \{x + t | x \in X_1\}.$$

Put

$$\hat{T}h(x) := h(x - t) \quad \left(x \in \mathbf{R}^k, \ h \in L^2(\mathbf{R}^k)\right),$$

$$\hat{K}_r h := h\mathcal{X}_{X_r} \quad \left(r = 1, 2, \ h \in L^2(\mathbf{R}^k)\right),$$

and assume that the function $f \in L^2(\mathbf{R}^k)$ has the properties

$$f\mathcal{X}_{X_2} = \hat{T}\left(f\mathcal{X}_{X_1}\right),\ f\mathcal{X}_{X_3} \equiv 0.$$

Then $\hat{T}$ is an unitary operator on $L^2(\mathbf{R}^k)$ and (49), (50) hold.
Using the assumption that X_1, X_2, X_3 is a measurable decomposition of $\mathbf{R}^k$ we get immediately that

$$G_s := X_s \times \{1, \ldots, N\} \quad (s = 1, 2, 3)$$

is a measurable decomposition of G. It follows that $\mathcal{M} = \mathcal{M}(G)$ is decomposed into the tensor product

$$\mathcal{M}(G) = \mathcal{M}(G_1) \otimes \mathcal{M}(G_2) \otimes \mathcal{M}(G_3)$$

[6,7,10]. According to this representation, the local algebras $\mathfrak{A}(X_s)$ corresponding to regions $X_s \subseteq \mathbf{R}^k$ $(s = 1, 2)$ are given by

$$\mathfrak{A}(X_1) := \{A \otimes \mathbf{1} \otimes \mathbf{1};\ A \text{ bounded operator on } \mathcal{M}(G_1)\},$$
$$\mathfrak{A}(X_2) := \{\mathbf{1} \otimes A \otimes \mathbf{1};\ A \text{ bounded operator on } \mathcal{M}(G_2)\}.$$

One easily checks in our special case that

$$F_{nm} \in \mathfrak{A}(X_1) \otimes \mathfrak{A}(X_1) \quad (n, m = 1, \ldots, N),$$
$$\sigma \in \mathfrak{A}(X_1) \otimes \mathfrak{A}(X_2)$$

and $\mathcal{E}^*(\hat{\rho})$ gives a state on $\mathfrak{A}(X_1)$ (the number of particles outside of G_1 is 0 with probabiliy 1). That is, Alice has to perform only local measurements inside of the region X_1 in order to realize the teleportation processes described in Sect. 4 or measure the state $\mathcal{E}^*(\hat{\rho})$. On the other hand, $\Lambda_{nm}(\mathcal{E}^*(\hat{\rho}))$ and $\Theta_{nm}(\mathcal{E}^*(\hat{\rho}))$ give local states on $\mathfrak{A}(X_2)$ such that by measuring these states Bob has to perform only local measurements inside of the region X_2. The only problem could be that according to the definition (43) of the channels Θ_{nm} Bob has to perform the measurement by F_+ which is not local. However, as we have already stated in Remark 3.5, this problem can be avoided if we replace F_+ by $F_{+.X_2} \in \mathfrak{A}(X_2)$.

Therefore we can describe the special teleportation process as follows: We have a beam being in the pure state $|\eta\rangle\langle\eta|$ (40). After splitting, one part of the beam is located in the region X_1 or will go to X_1 (cf. Remark 1.11) and the other part is located in the region X_2 or will go to X_2. Further, there is a state $\mathcal{E}^*(\hat{\rho})$ localized in the region X_1. Now Alice will perform the local measurement inside of X_1 according to $F = \sum_{n,m} z_{nm} F_{nm}$ involving the first part of the beam and the state $\mathcal{E}^*(\hat{\rho})$. This leads to a preparation of the second part of the beam located in the region X_2 which can be controlled by Bob, and the second part of the beam will show the behaviour of the state $\Lambda_{nm}(\mathcal{E}^*(\hat{\rho})) = \Theta_{nm}(\mathcal{E}^*(\hat{\rho}))$ if Alice's measurement shows the value z_{nm}. Thus we have teleported the state $\hat{\rho}$ on $\mathcal{H}$ from the region X_1 into the region X_2.

References

1. Accardi, L. and Ohya, M.: Teleportation of general quantum states. Voltera Center preprint, 1998
2. Accardi L., Ohya M.: Compound channels, transition expectations and liftings. Applied Mathematics & Optimization **39**, 33–59 (1999)
3. Bennett, C. H., Brassard, G., Crépeau, C., Jozsa, R., Peres, A. and Wootters, W.K.: Teleporting an unknown quantum state via Dual Classical and Einstein–Podolsky–Rosen channels. Phys. Rev. Lett. **70**, 1895–1899 (1993)
4. Bennett, C.H., Brassard, G., Popescu, S., Schumacher, B., Smolin, J.A., Wootters, W.K.: Purification of noisy entanglement and faithful teleportation via noisy channels. Phys. Rev. Lett. **76**, 722–725 (1996)
5. Ekert, A.K.: Quantum cryptography based on Bell's theorem. Phys. Rev. Lett. **67**, 661–663 (1991)
6. Fichtner, K.-H. and Freudenberg, W.: Point processes and the position distrubution of infinite boson systems. J. Stat. Phys. **47**, 959–978 (1987)
7. Fichtner, K.-H. and Freudenberg, W.: Characterization of states of infinite Boson systems I. – On the construction of states. Commun. Math. Phys. **137**, 315–357 (1991)
8. Fichtner, K.-H., Freudenberg, W. and Liebscher, V.: Time evolution and invariance of Boson systems given by beam splittings. Infinite Dim. Anal., Quantum Prob. and Related Topics **1**, 511–533 (1998)
9. Lindsay, J.M.: Quantum and Noncausal Stochastic Calculus. Prob. Th. Rel. Fields **97**, 65–80 (1993)
10. Fichtner, K.-H. and Winkler, G.: Generalized brownian motion, point processes and stochastic calculus for random fields. Math. Nachr. **161**, 291–307 (1993)
11. Inoue, K., Ohya, M. and Suyari, H.: Characterization of quantum teleportation processes by nonlinear quantum mutual entropy. Physica D **120**, 117–124 (1998)

Communicated by H. Araki

Open Sys. & Information Dyn. **7**: 33–39, 2000

NP Problem in Quantum Algorithm

Masanori Ohya and Natsuki Masuda

Science University of Tokyo,
Noda City, Chiba 278-8510, JAPAN

(Received: October 29, 1999)

Abstract. In complexity theory, a famous unsolved problem is whether NP is equal to P or not. In this paper, we discuss this aspect in SAT (satisfiability) problem, and it is shown that SAT can be solved in polynomial time by means of a quantum algorithm if the superposition of two orthogonal vectors $|0\rangle$ and $|1\rangle$ prepared is detected physically.

1. Introduction

Although the power of computers has highly progressed, there are several problems which still cannot be solved effectively, namely, in polynomial time. Among such problems, NP problems and NP-complete problems are fundamental. It is known [5] that all NP complete problems are equivalent and the essential question is whether there exists an algorithm solving an NP-complete problem in polynomial time.

After pioneering papers of Feynman [4] and Deutsch [1], several important works have been done on quantum algorithms by Deutsch and Josa [2], Shor [7], Ekert and Jozsa [3] and many others [11]. Computation in a quantum computer is performed on a tensor product Hilbert space, and its fundamental point is to use quantum coherence of states. All mathematical features of quantum computers and computations are summarized in [11].

In this paper, we discuss the quantum algorithm of the SAT problem and we point out that this problem, hence all other NP problems, can be solved in polynomial time by a quantum computer if the superposition of two orthogonal vectors $|0\rangle$ and $|1\rangle$ is physically detected.

2. The SAT Problem

Let $X \equiv \{x_1, \ldots, x_n\}$ be a set. Then x_k and its negation $\overline{x}_k$, $k = 1, 2, \ldots, n$, are called literals and the set of all such literals is denoted by $X' = \{x_1, \overline{x}_1, \ldots, x_n, \overline{x}_n\}$. The set of all subsets of X' is denoted by $\mathcal{F}(X')$ and an element $C \in \mathcal{F}(X')$ is called a clause. We consider a truth assignment to all variables x_k. If we can assign the truth value to at least one element of C, then C is called satisfiable. When C is satisfiable, the truth value $t(C)$ of C is regarded as true, otherwise, that of C is false. We denote the truth values as "true → 1, false → 0". Then

$$C \text{ is satisfiable iff } t(C) = 1.$$

Let $L = \{0, 1\}$ be a Boolean lattice with usual join $\vee$ and meet $\wedge$, and let $t(x)$ be the truth value of a literal x in X. Then the truth value of a clause C is written as $t(C) \equiv \vee_{x \in C} t(x)$. Moreover the set $\mathcal{C}$ of all clauses C_j, $j = 1, 2, \ldots, m$, is called satisfiable iff the meet of all truth values of C_j is 1; $t(\mathcal{C}) \equiv \wedge_{j=1}^{m} t(C_j) = 1$. Thus the SAT problem is written as follows:

DEFINITION 1 (SAT Problem).
Given a set $X \equiv \{x_1, \ldots, x_n\}$ and a set $\mathcal{C} = \{C_1, C_2, \ldots, C_m\}$ of clauses, determine whether $\mathcal{C}$ is satisfiable or not.

That is, this problem is to ask whether there exsits a truth assignment to make $\mathcal{C}$ satisfiable.

It is known [5] in usual algorithmic complexity theory that it takes polynomial time to check the satisfiability when a specific truth assignment is given, but we cannot determine the satisfiability in polynomial time when an assignment is not specified.

3. Quantum Algorithm of SAT

Let 0 and 1 of the Boolean lattice L be denoted by the vectors

$$|0\rangle \equiv \begin{pmatrix} 1 \\ 0 \end{pmatrix} \quad \text{and} \quad |1\rangle \equiv \begin{pmatrix} 0 \\ 1 \end{pmatrix}$$

in the Hilbert space $\mathbb{C}^2$, respectively. That is, the vector $|0\rangle$ corresponds to false and the vector $|1\rangle$ to truth.

As we explained in the previous section, an element $x \in X$ can be denoted by 0 or 1, and so by $|0\rangle$ or $|1\rangle$. In order to describe a clause C with length at most n by a quantum state, we need the n-tuple tensor product Hilbert space $\mathcal{H} \equiv \otimes_1^n \mathbb{C}^2$. For instance, in the case of $n = 2$, given $C = \{x_1, x_2\}$ with an assignment $x_1 = 0$ and $x_2 = 1$, then the corresponding quantum state vector is $|0\rangle \otimes |1\rangle$, so that the quantum state vector describing C is generally written as $|C\rangle = |x_1\rangle \otimes |x_2\rangle \in \mathcal{H}$ with $x_k = 0$ or 1, $k = 1, 2$.

The quantum computation is performed by a unitary gate constructed from several fundamental gates such as Not gate, Controlled-Not gate, Controlled-Controlled Not gate [3, 11]. Once $X \equiv \{x_1, \ldots, x_n\}$ and $\mathcal{C} = \{C_1, C_2, \ldots, C_m\}$ are given, the SAT is to find the vector $|f(\mathcal{C})\rangle \equiv \wedge_{j=1}^{m} \vee_{x \in C_j} t(x)$, where $t(x)$ is $|0\rangle$ or $|1\rangle$ when $x = 0$ or 1, respectively, and $t(x) \wedge t(y) \equiv t(x \wedge y)$, $t(x) \vee t(y) \equiv t(x \vee y)$.

We consider a quantum algorithm for the SAT problem. Since we have n variables x_k, $k = 1, \ldots n$, and a quantum computation produces some dust bits, the assignments of the n variables and the dusts are represented by n qubits and l qubits in the Hilbert space $\otimes_1^n \mathbb{C}^2 \otimes_1^l \mathbb{C}^2$. Moreover the resulting state vector $|f(C)\rangle$ should be added, so that the total Hilbert space is

$$\mathcal{H} \equiv \otimes_1^n \mathbb{C}^2 \otimes_1^l \mathbb{C}^2 \otimes \mathbb{C}^2 .$$

Let us start the quantum computation of SAT problem from an initial vector $|v_0\rangle \equiv \otimes_1^n |0\rangle \otimes_1^l |0\rangle \otimes |0\rangle$ when C contains n Boolean variables $x_1, \ldots x_n$. We apply the discrete Fourier transform denoted by

$$U_F \equiv \otimes_1^n \frac{1}{\sqrt{2}} \begin{pmatrix} 1 & 1 \\ 1 & -1 \end{pmatrix}$$

to the part of the Boolean variables of the vector $|v_0\rangle$, then the resulting state vector becomes

$$|v\rangle \equiv U_F \otimes_1^l \mathbb{1} |v_0\rangle = \frac{1}{\sqrt{2^n}} \otimes_1^n (|0\rangle + |1\rangle) \otimes_1^l |0\rangle \otimes |0\rangle ,$$

where $\mathbb{1}$ is the identity matrix in $\mathbb{C}^2$. This vector can be written as

$$|v\rangle = \frac{1}{\sqrt{2^n}} \sum_{x_1,\ldots,x_n=0}^{1} \otimes_{j=1}^n |x_j\rangle \otimes_1^l |0\rangle \otimes |0\rangle .$$

Now, we use the quantum computer to check the satisfiability, which will be done by a unitary operator U_f properly constructed by unitary gates. Then after the computation by U_f, the vector $|v\rangle$ goes to

$$\begin{aligned} |v_f\rangle \equiv U_f |v\rangle &= \frac{1}{\sqrt{2^n}} \sum_{x_1,\ldots,x_n=0}^{1} U_f \otimes_{j=1}^n |x_j\rangle \otimes_1^l |0\rangle \otimes |0\rangle \\ &= \frac{1}{\sqrt{2^n}} \sum_{x_1,\ldots,x_n=0}^{1} \otimes_{j=1}^n |x_j\rangle \otimes_{i=1}^l |y_i\rangle \otimes |f(x_1, \ldots, x_n)\rangle, \end{aligned}$$

where $f(x_1, \ldots, x_n) \equiv f(C)$ because C contains $x_1, \ldots x_n$, and $|y_i\rangle$ are the dust bits produced by the computation. As we will explain in an example below, the unitary operator U_f is concretely constructed.

The final step to check the satisfiability of C is to apply the projection $E \equiv \otimes_1^{n+l} \mathbb{1} \otimes |1\rangle\langle 1|$ to the state $|v_f\rangle$, mathematically equivalent , to compute the value $\langle v_f|E|v_f\rangle$. If the vector $E|v_f\rangle$ exists or the value $\langle v_f|E|v_f\rangle$ is not 0, then we conclude that C is satisfiable. The value of $\langle v_f|E|v_f\rangle$ corresponds to that of a random algorithm as we will see in an example of the next section and it may or may not be obtained in polynomial time. Let us consider an operator V_θ given by

$$V_\theta \equiv \otimes_1^{n+l} (A|0\rangle\langle 0| + B|1\rangle\langle 1|) \otimes e^{i\theta f(C)} \mathbb{1} ,$$

and apply it to the vector $|v_f\rangle$, where

$$A \equiv \frac{1}{\sqrt{2}}\begin{pmatrix} 1 & 1 \\ 1 & -1 \end{pmatrix} \quad \text{and} \quad B \equiv \frac{1}{\sqrt{2}}\begin{pmatrix} 1 & 1 \\ -1 & 1 \end{pmatrix}$$

and θ is a certain constant describing the phase of the vector $|f(C)\rangle$. The resulting vector is the superposition of two vectors with some constants α, β, such as

$$V_\theta |v_f\rangle = \frac{1}{\sqrt{2^{n+l}}} \otimes_1^{n+l} (|0\rangle + |1\rangle) \otimes (\alpha|0\rangle + \beta e^{i\theta}|1\rangle),$$

one of which is polarized with θ and the other one is non-polarized. The existence of the mixture of two vectors $|0\rangle$ and $e^{i\theta}|1\rangle$ is the starting point of quantum computation which implies the satisfiability.

4. An Example of Computation

Let us explain the quantum computation for SAT in the case $X = \{x_1, x_2, x_3\}$ and $C = \{\{x_1\}, \{x_2, x_3\}, \{x_1, \overline{x}_3\}, \{\overline{x}_1, \overline{x}_2, x_3\}\}$. The resulting state $|f(x_1, x_2, x_3)\rangle$ is written as

$$|f(x_1, x_2, x_3)\rangle = |x_1\rangle \wedge (|x_2\rangle \vee |x_3\rangle) \wedge (|x_1\rangle \vee |\overline{x}_3\rangle) \wedge (|\overline{x}_1\rangle \vee |\overline{x}_2\rangle \vee |x_3\rangle).$$

In the quantum computation, it is not necessary to substitute all values of x_j, $j = 1, 2, 3$ as in the classical computation, we only have to use a unitary operator U_f for the computation of $|f(x_1, x_2, x_3)\rangle$. This unitary operator U_f is constructed as follows: Let U_{NOT}, U_{CN} and U_{CCN} be Not gate on $\mathbb{C}^2$, Controlled-Not gate on $\mathbb{C}^2 \otimes \mathbb{C}^2$ and Controlled-Controlled-Not gate on $\mathbb{C}^2 \otimes \mathbb{C}^2 \otimes \mathbb{C}^2$, respectively, which are given by

$$\begin{aligned} U_{NOT} &= |0\rangle\langle 1| + |1\rangle\langle 0| \\ U_{CN} &= |0\rangle\langle 0| \otimes \mathbb{1} + |1\rangle\langle 1| \otimes (|0\rangle\langle 1| + |1\rangle\langle 0|) \\ U_{CCN} &= |0\rangle\langle 0| \otimes |0\rangle\langle 0| \otimes \mathbb{1} + |0\rangle\langle 0| \otimes |1\rangle\langle 1| \otimes \mathbb{1} + |1\rangle\langle 1| \otimes |0\rangle\langle 0| \otimes \mathbb{1} \\ &\quad + |1\rangle\langle 1| \otimes |1\rangle\langle 1| (|0\rangle\langle 0| \otimes |1\rangle\langle 0|). \end{aligned}$$

Then the unitary operator U_f is determined by the combination of the above three unitaries as

$$U_f \equiv U_{36} U_{35} \ldots U_2 U_1,$$

where, for instance,

$$\begin{aligned} U_1 &\equiv |0\rangle\langle 0| \otimes_1^{23} \mathbb{1} + |1\rangle\langle 1| \otimes_1^2 \mathbb{1} \otimes (|0\rangle\langle 1| + |1\rangle\langle 0|) \otimes_1^{20} \mathbb{1} \\ U_2 &\equiv \mathbb{1} \otimes |0\rangle\langle 0| \otimes_1^{22} \mathbb{1} + \mathbb{1} \otimes |1\rangle\langle 1| \otimes_1^2 \mathbb{1} \otimes (|0\rangle\langle 1| + |1\rangle\langle 0|) \otimes_1^{19} \mathbb{1} \\ U_3 &\equiv \otimes_1^2 \mathbb{1} \otimes |0\rangle\langle 0| \otimes_1^{21} \mathbb{1} + \otimes_1^2 \mathbb{1} \otimes |1\rangle\langle 1| \otimes_1^2 \mathbb{1} \otimes (|0\rangle\langle 1| + |1\rangle\langle 0|) \otimes_1^{18} \mathbb{1} \end{aligned}$$

and other $U_4, \ldots, U_{36}$ are similarly constructed (see the computation diagram 1). In this case, we need 20 dust bits (the number of the dust bits needed in a general case is counted in the next section), so that U_f operates on the Hilbert space $\otimes_1^{24}\mathbb{C}^2$.

Starting from the initial vector $|v_0\rangle \equiv \otimes_1^3|0\rangle \otimes_1^{20}|0\rangle \otimes |0\rangle$, the final vector is

$$
\begin{aligned}
|v_f\rangle &= U_{36}\ldots U_1 U_F|v_0\rangle = U_f U_F|v_0\rangle \\
&= \frac{1}{\sqrt{2^3}}\Big(|0,0,0,0,1,1,0,1,0,1,0,0,0,1,1,0,0,0,1,1,1,0,1;0\rangle \\
&\quad + |1,0,0,1,1,1,0,0,0,1,1,0,0,1,1,1,0,0,1,1,1,0,1;0\rangle \\
&\quad + |0,1,0,0,0,1,1,1,0,1,0,1,0,1,1,0,1,0,1,1,1,0,1;0\rangle \\
&\quad + |0,0,1,0,1,0,1,1,1,0,0,0,0,0,1,0,1,0,0,1,0,0,0;0\rangle \\
&\quad + |1,1,0,1,0,1,1,0,0,1,1,1,1,1,0,1,1,1,1,1,0,1,0;0\rangle \\
&\quad + |1,0,1,1,1,0,1,0,1,1,1,0,0,0,1,1,1,1,1,1,1,1,1;1\rangle \\
&\quad + |0,1,1,0,0,0,1,1,1,0,0,1,0,0,1,0,1,0,0,1,0,0,0;0\rangle \\
&\quad + |1,1,1,1,0,0,1,0,1,1,1,1,1,0,1,1,1,1,1,1,1,1,1;1\rangle\Big),
\end{aligned}
$$

where we used the notation

$$
|x_1, x_2, x_3, y_1, \ldots, y_{20}; f(x_1,x_2,x_3)\rangle \equiv \otimes_{j=1}^3|x_j\rangle \otimes_{i=1}^{20}|y_i\rangle \otimes |f(x_1,x_2,x_3)\rangle
$$

Applying the operator V_θ of the previous section to the vector $|v_f\rangle$, we obtain

$$
\frac{1}{\sqrt{2^{23}}} \otimes_1^{23} (|0\rangle + |1\rangle) \otimes \Big(\frac{3}{\sqrt{10}}|0\rangle + \frac{1}{\sqrt{10}}e^{i\theta}|1\rangle\Big),
$$

which is the superposition of two vectors $|0\rangle$ and the polarized vector $|f(\mathcal{C})\rangle = |1\rangle$. Moreover, when we measure the operator $E \equiv \otimes_1^{23}\mathbb{1} \otimes |1\rangle\langle 1|$ in the state $V_\theta|v_f\rangle$, we obtain

$$
\langle v_f V_\theta | E | V_\theta v_f\rangle = \frac{1}{10}.
$$

This concludes the satisfiability of $\mathcal{C}$.

5. Complexity of the Quantum Algorithm for SAT

Here we discuss the number of steps for the quantum algorithm solving the SAT problem. The size of input with n variables x_k and their negations $\overline{x}_k$, $k = 1, \ldots, n$, and also $\mathcal{C} = \{C_1, \ldots, C_m\}$ is $N_1 = \log n + 2mn$ like for the classical algorithm because $|C_k|$, the number of elements in C_k, is at most n, so that it is the polynomial order ($O(mn)$). The number of dust bits to compute $f(\mathcal{C})$ is related to that of the operations and substitutions of AND and OR, so that the maximum number (complexity) N_2 of the dust bits needed is the same as that in the classical case,

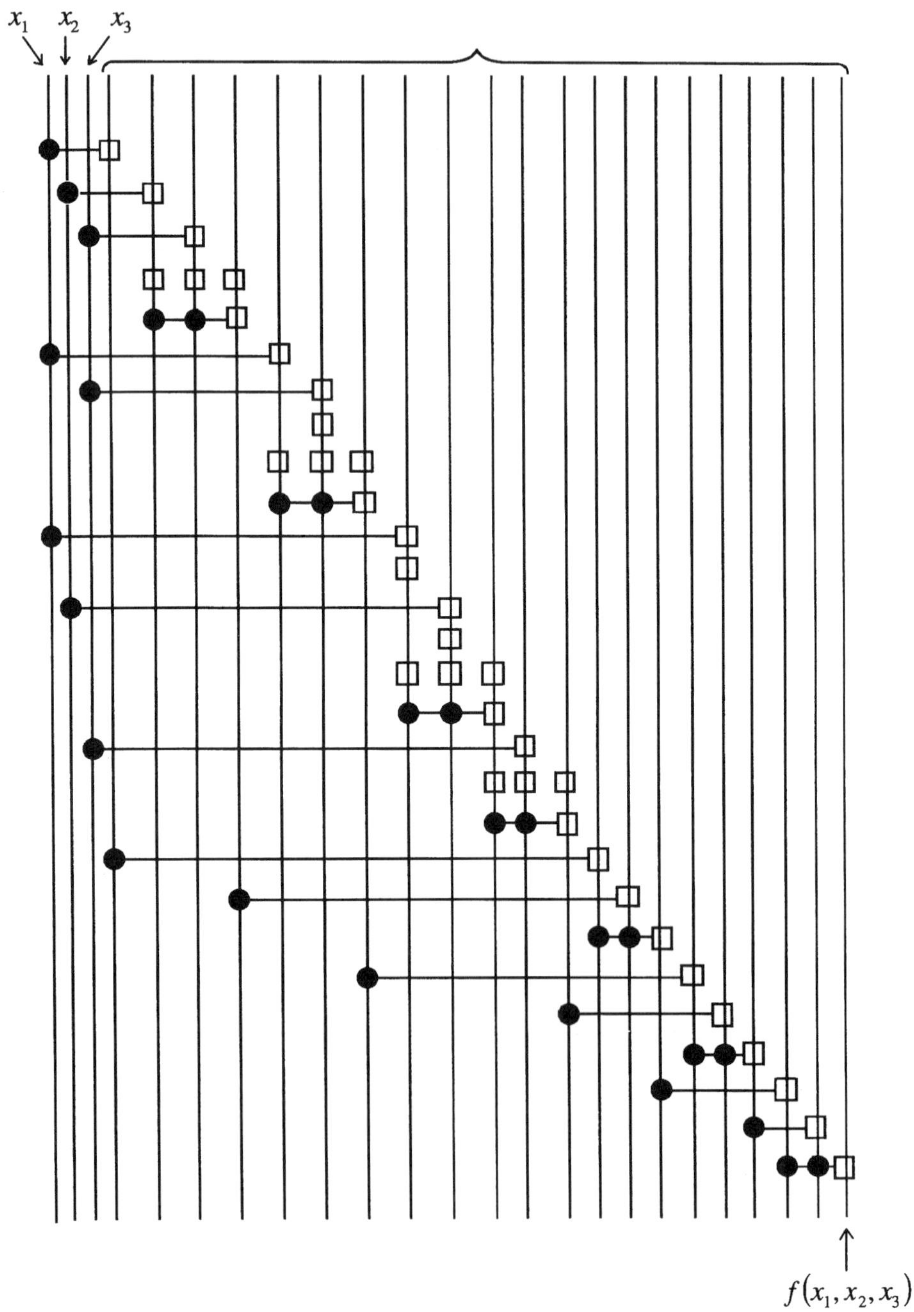

Diagram 1.

namely, $N_2 =$ (the numbers of AND and OR operations)$-$ (the steps to take the negation) $= (5mn - 1) - mn = 4mn - 1$.

The number of steps N_3 needed to obtain the vector $|f(\mathcal{C})\rangle$ is counted as follows: First take 1 step for the discrete Fourier transform to get the entangled vector, next we need $3mn$ steps for truth assignment and substitution to n variables. Secondly to compute the logical sum in each clause and take their logical product, the complexity corresponding to the logical sum, whose gate is made of four unitaries, is $4m(2n-1)$ and that corresponding to the logical product is $m-1$. Thus $N_3 = 1 + 3mn + 4m(2n-1) + m - 1 = 11mn - 3m$.

Finally to check the satisfiability of $\mathcal{C}$, we have to look at the value of $|f(\mathcal{C})\rangle$ registered at the last position of the tensor product state, such as the last position of physical registers (e.g., spins) lined up. This can be easily done by applying the operator V_θ to $|v_f\rangle$, and the resulting vector is a superposition of two vectors $|0\rangle$ and $e^{i\theta}|1\rangle$. We can obtain in polynomial time (at most $n+l+1$ steps) the vector $|V_\theta v_f\rangle$ and the value of $\langle V_\theta v_f, EV_\theta v_f\rangle| = |\beta|^2$ for the projection E if needed. The existence of the above superposition is the starting point of quantum computation, so that it should be physically detected being different from both $|0\rangle$ and $|1\rangle$, which implies the satisfiability. Thus, the quantum algorithm of the SAT problem is of polynomial order.

In this paper, we assumed the physical detection of the vector $\alpha|0\rangle + \beta|1\rangle$ to prove the SAT, whose assumption can be omitted in [8].

Acknowledgment

One of the authors (MO) appreciate Professor A. Ekert for his critical reading and valuable comments.

Bibliography

1. D. Deutsch, Proc. Roy. Soc. London series A **400**, 97 (1985).
2. D. Deutsch and R. Jozsa, Proc. Roy. Soc. London series A, **439**, 553 (1992).
3. A. Ekert and R. Jozsa, Rev. Mod. Phys. **68**, 733 (1996).
4. R. Feynman, Optics News **11**, 11 (1985).
5. M. Garey and D. Johnson, *Computers and Intractability — a guide to the theory of NP-completeness*, Freeman, 1979.
6. M. Ohya, *Mathematical Foundation of Quantum Computer*, *Mathematical Foundation of Quantum Information*, Maruzen Publ. Company, 1998.
7. P. W. Shor, Algorithm for quantum computation: Discrete logarithm and factoring, in: *Proceedings of the 35th Annual IEEE Symposium on Foundation of Computer Science*, 124 (1994).
8. M. Ohya and I. V. Volvich, "Quantum computing, NP-complete problems and chaotic dynamics", quant-ph/9912100.

PERGAMON

Chaos, Solitons and Fractals 11 (2000) 1377–1385

CHAOS
SOLITONS & FRACTALS

www.elsevier.nl/locate/chaos

Application of chaos degree to some dynamical systems

Kei Inoue [a], Masanori Ohya [a,*], Keiko Sato [b]

[a] *Department of Information Sciences, Science University of Tokyo, Noda City, Chiba 278-8510, Japan*
[b] *Department of Control and Computer Engineering, Numazu College of Technology, Numazu City, Shizuoka 410-8501, Japan*

Communicated by Prof. Y.H. Ichikawa

Accepted 15 March 1999

Abstract

Chaos degree defined through two complexities in information dynamics is applied to some deterministic dynamical models. It is shown that this degree well describes the chaotic feature of the models.

1. Introduction

There exist several approaches in the study of chaotic behavior of dynamical systems using the concepts such as entropy, complexity, chaos, fractal, stochasticity [1,3–6,13]. In 1991, one of the authors proposed information dynamics (ID for short) [10,11] to treat such chaotic behavior of systems altogether. ID is a synthesis of the dynamics of state change and the complexity of systems, and it is applied to several different fields such as quantum physics, fractal theory, quantum information and genetics [7].

A quantity measuring chaos in dynamical systems was defined by means of two complexities in ID, and it is called chaos degree. In particular, among several chaos degrees, an entropic chaos degree was introduced in [8,12], and is applied to logistic map to study its chaotic behavior. This chaos degree has several merits compared with usual measures of chaos such as Lyapunov exponent.

In Section 2, we review the complexity in ID and the chaos degree (CD for short). In Section 3, we remind the entropic chaos degree and the Lyapunov exponent (LE for short). In Section 4, the algorithm computing the entropic chaos degree is shown. In Section 5, we compute CD and LE for Bernoulli shift, Baker's transformation and Tinkerbell's map, then we discuss the merits of the entropic chaos degree.

2. Complexity of information dynamics and chaos degree

Information dynamics provides a frame to study the state change and the complexity associated with a dynamical system. We briefly explain the concept of the complexity of ID in a bit simplified version (see [11]).

Let $(\mathcal{A}, \mathfrak{S}, \alpha(G))$ be an input (or initial) system and $(\overline{\mathcal{A}}, \overline{\mathfrak{S}}, \overline{\alpha}(\overline{G}))$ be an output (or final) system. Here $\mathcal{A}$ is the set of all objects to be observed and $\mathfrak{S}$ is the set of all means for measurement of $\mathcal{A}$, $\alpha(G)$ is a certain evolution of system. Often we have $\mathcal{A} = \overline{\mathcal{A}}$, $\mathfrak{S} = \overline{\mathfrak{S}}$, $\alpha = \overline{\alpha}$.

* Corresponding author. Tel.: +81-471241501 ext. 3358; fax: +81-471241532.

For instance, when $\mathscr{A}$ is the set $M(\Omega)$ of all measurable functions on a measurable space $(\Omega, \mathscr{F})$ and $\mathfrak{S}(\mathscr{A})$ is the set $P(\Omega)$ of all probability measures on Ω, we have the usual probability theory, by which the classical dynamical system is described. When $\mathscr{A}$ is the set $B(\mathscr{H})$ of all bounded linear operators on a Hilbert space $\mathscr{H}$ and $\mathfrak{S}(\mathscr{A})$ is the set $\mathfrak{S}(\mathscr{H})$ of all density operators on $\mathscr{H}$, we have a quantum dynamical system.

Once an input and output system are set, the situation of the input system is described by a state, an element of $\mathfrak{S}$, and the change of the state is expressed by a mapping from $\mathfrak{S}$ to $\overline{\mathfrak{S}}$, called a channel. The concept of channel is fundamental both in physics and mathematics [7]. Moreover, there exist two complexities in ID, which are axiomatically given as follows:

Let $(\mathscr{A}_t, \mathfrak{S}_t, \alpha^t(G^t))$ be the total system of $(\mathscr{A}, \mathfrak{S}, \alpha)$ and $(\overline{\mathscr{A}}, \overline{\mathfrak{S}}, \overline{\alpha})$, and let $C(\varphi)$ be the complexity of a state φ and $T(\varphi; \Lambda^*)$ be the transmitted complexity associated with the state change $\varphi \to \Lambda^*\varphi$. These complexities C and T are the quantities satisfying the following conditions:

1. For any $\varphi \in \mathfrak{S}$,
$$C(\varphi) \geqslant 0, \quad T(\varphi; \Lambda^*) \geqslant 0.$$
2. For any orthogonal bijection $j : ex\mathfrak{S} \to ex\mathfrak{S}$ (the set of all extreme points in $\mathfrak{S}$),
$$C(j(\varphi)) = C(\varphi),$$
$$T(j(\varphi); \Lambda^*) = T(\varphi; \Lambda^*).$$
3. For $\Phi \equiv \varphi \otimes \psi \in \mathfrak{S}_t$,
$$C(\Phi) = C(\varphi) + C(\psi).$$
4. For any state φ and a channel Λ^*,
$$0 \leqslant T(\varphi; \Lambda^*) \leqslant C(\varphi).$$
5. For the identity map "id" from $\mathfrak{S}$ to $\mathfrak{S}$.
$$T(\varphi; id) = C(\varphi).$$

When a state φ changes to the state $\Lambda^*\varphi$, a *chaos degree* (CD) [11] w.r.t. φ and Λ^* is given by
$$D(\varphi; \Lambda^*) = C(\Lambda^*\varphi) - T(\varphi; \Lambda^*).$$

Using the above CD, we observe chaos of a dynamical system as
$$CD > 0 \Longleftrightarrow \text{chaotic},$$
$$CD = 0 \Longleftrightarrow \text{stable}.$$

3. Entropic chaos degree and Lyapunov exponent

Chaos degree in ID was applied to a smooth map on **R** and it is shown that this degree enables to describe the chaotic aspects of a logistic map as well [8,12]. Here we briefly review the chaos degree defined through classical entropies. For an input state described by a probability distribution $p = (p_i)$ and the joint distribution $r = (r_{i,j})$ between p and the output state $\bar{p} = \Lambda^* p = (\bar{p}_i)$ through a channel Λ^*, the Shannon entropy $S(p) = -\sum_i p_i \log p_i$ and the mutual entropy $I(p; \Lambda^*) = \sum_{i,j} r_{i,j} \log r_{i,j}/p_i\bar{p}_j$ satisfy all conditions of the complexities C and T, then the entropic chaos degree is defined by
$$D(p; \Lambda^*) \equiv C(\Lambda^* p) - T(p; \Lambda^*)$$
$$\equiv S(\bar{p}) - I(p; \Lambda^*),$$

This entropic chaos degree is nothing but the conditional entropy of aposteriori state w.r.t. the channel Λ^*. The characteristic point of the entropic chaos degree is easy to get the probability distribution of the orbit for a deterministic dynamics, which is discussed in the next section.

Lyapunov exponent (LE) is used to study chaotic behavior of a deterministic dynamics. The Lyapunov exponent $\lambda(f)$ for a smooth map f on $\mathbf{R}$ is defined by [2]

$$\lambda(f) = \lim_{n\to\infty} \lambda_n(f), \quad \lambda_n(f) = \frac{1}{n}\sum_{k=1}^{n} \log\left|\frac{\mathrm{d}f}{\mathrm{d}x}\left(x^{(k)}\right)\right|,$$

where $x^{(n)} = f(x^{(n-1)})$ for any $n \in \mathbf{N}$.

For a smooth map $f = (f_1, \ldots, f_m)$ on $\mathbf{R}^m$, the vector version of LE is defined as follows: Let x^0 be an initial point of $\mathbf{R}^m$ and $x^{(n)} = f(x^{(n-1)})$ for any $n \in \mathbf{N}$. After n times iterations of f to x^0, the Jacobi matrix $J_n(x^0)$ of $x^{(n)} = (x_1^{(n)}, \ldots, x_m^{(n)})$ w.r.t. $x^{(0)}$ is

$$J_n\left(x^{(0)}\right) = Df^n\left(x^{(0)}\right) = \begin{bmatrix} \left(\partial f_1^{(n)}/\partial x_1\right)\left(x^{(0)}\right) & \cdots & \left(\partial f_1^{(n)}/\partial x_m\right)\left(x^{(0)}\right) \\ \vdots & \vdots & \vdots \\ \left(\partial f_m^{(n)}/\partial x_1\right)\left(x^{(0)}\right) & \cdots & \left(\partial f_m^{(n)}/\partial x_m\right)\left(x^{(0)}\right) \end{bmatrix}.$$

Then the Lyapunov exponent $\lambda(f)$ of $x^{(0)}$ is defined by

$$\lambda(f) = \log \bar{\mu}_1, \; \bar{\mu}_k = \lim_{n\to\infty} \left(\mu_k^n\right)^{1/n} \quad (k = 1, 2, \ldots, m).$$

Here μ_k^n is the kth largest square root of the eigenvalues of the matrix $J_n(x^{(0)})J_n(x^{(0)})^{\mathrm{T}}$.

An orbit of the dynamical system described by f is said to be chaotic when the exponent $\lambda(f)$ is positive, and to be stable when the exponent is negative. The positive exponent means that the orbit is very sensitive to the initial value, so that it describes a chaotic behavior. Lyapunov exponent is difficult to compute for some models (e.g., Tinkerbell map) and its negative value is not clearly explained.

4. Algorithm for computation of entropic chaos degree

It is proved [9] that if a piecewise monotone mapping f from $[a,b]^m$ to $[a,b]^m$ has non-positive Schwarzian derivatives and does not have a stable and periodic orbit, then there exists an ergodic measure μ on the Borel set $\mathfrak{B}$ of $[a,b]^m$, absolutely continuous w.r.t. the Lebesgue measure.

Take a finite partition $\{A_k\}$ of $I = [a,b]^m$ such as

$$I = \bigcup_k A_k \quad \left(A_i \bigcap A_j = \emptyset, \; i \neq j\right).$$

Let $|S|$ be the number of the elements in a set S. Suppose that n is a sufficiently large natural number and m is a fixed natural number. Let $p^{(n)} \equiv (p_i^{(n)})$ be the probability distribution of the orbit up to nth step, that is, how many $x^{(k)}$ $(k = m+1, \ldots, m+n)$ are in A_i

$$p_i^{(n)} \equiv \frac{\left|\left\{k \in \mathbf{N}; x^{(k)} \in A_i, m < k \leqslant m+n\right\}\right|}{n}.$$

It is shown that the $n \to \infty$ limit of $p_i^{(n)}$ exists and equals to $\mu(A_i)$. The channel Λ^* is a map given by $p^{(n+1)} = \Lambda^* p^{(n)}$. Further, the joint distribution $r^{(n,n+1)} = (r_{i,j}^{(n,n+1)})$ for a sufficient large n is approximated as

$$r_{i,j}^{(n,n+1)} = \frac{\left|\left\{k \in \mathbf{N}; \left(x^{(k)}, f\left(x^{(k)}\right)\right) \in A_i \times A_j, m < k \leqslant m+n\right\}\right|}{n}.$$

Then the entropic chaos degree is computed as

$$\begin{aligned} D\left(p^{(n)}; \Lambda^*\right) &= S\left(p^{(n+1)}\right) - I\left(p^{(n)}; \Lambda^*\right) \\ &= -\sum_i p_i^{(n+1)} \log p_i^{(n+1)} - \sum_{i,j} r_{i,j}^{(n,n+1)} \log \frac{r_{i,j}^{(n,n+1)}}{p_i^{(n)} p_j^{(n+1)}}. \end{aligned}$$

5. Entropic chaos degree for some deterministic dynamical models

In this section, we study the chaotic behavior of several well-known deterministic maps by the entropic chaos degree.

5.1. Bernoulli shift

Let f be a map from $[0,1]$ to itself such that

$$f(x_n) = \begin{cases} 2ax^{(n)} & (0 \leqslant x^{(n)} \leqslant 0.5), \\ a(2x^{(n)} - 1) & (0.5 < x^{(n)} \leqslant 1), \end{cases} \tag{5.1}$$

where $x^{(n)} \in [0,1]$ and $0 \leqslant a \leqslant 1$.

Let us compute the Lyapunov exponent and the entropic chaos degree (ECD for short) for the above Bernoulli shift f.

The orbit of Eq. (5.1) is shown in Fig. 1.

The Lyapunov exponent $\lambda_n(f)$ is $\log 2a$ for the Bernoulli shift (Fig. 2).

On the other hand, the entropic chaos degree of the Bernoulli shift is shown in Fig. 3.

Here we took 740 different a's between 0 and 1 with

$$A_i = \left[\frac{i}{2000}, \frac{i+1}{2000}\right] \quad (i = 0, \ldots, 1999),$$
$$n = 100\,000.$$

5.2. Baker's transformation

We apply the chaos degree to a smooth map on R^2. Let us compute the Lyapunov exponent and the ECD for the following Baker's transformation f_a:

$$f_a(x^{(n)}) = f_a\left(x_1^{(n)}, x_2^{(n)}\right) = \begin{cases} \left(2ax_1^{(n)}\ \frac{1}{2}ax_2^{(n)}\right) & \left(0 \leqslant x_1^{(n)} \leqslant 0.5\right), \\ \left(a\left(2x_1^{(n)} - 1\right), \frac{1}{2}a\left(x_2^{(n)} + 1\right)\right) & \left(0.5 < x_1^{(n)} \leqslant 1\right), \end{cases}$$

where $(x_1^{(n)}, x_2^{(n)}) \in [0,1] \times [0,1]$ and $0 \leqslant a \leqslant 1$.

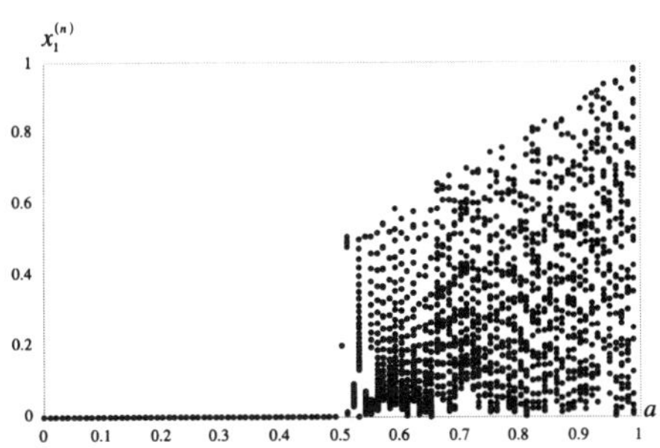

1

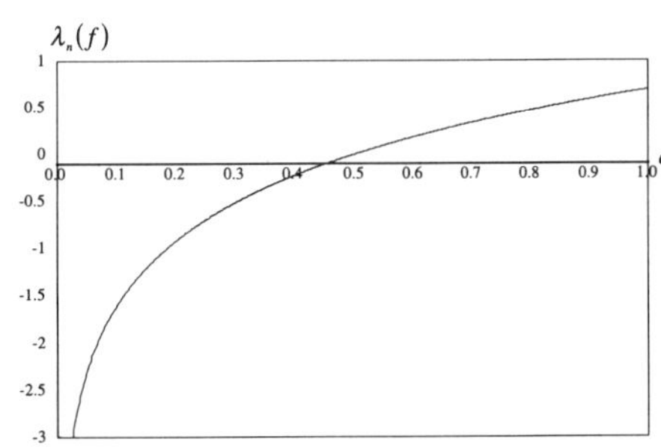

2

Fig. 1. The bifurcation diagram for Bernoulli shift.

Fig. 2. Lyapunov exponent of Bernoulli shift.

K. Inoue et al. / Chaos, Solitons and Fractals 11 (2000) 1377–1385

The orbit for each a is shown in Figs. 4–9.

These figures show that the larger a is, the more complicated the orbit is. The maximum Lyapunov exponent $\lambda_n^1(f)$ is $\log 2a$ for Baker's transformation (Fig. 10).

On the other hand, the ECD of Baker's transformation is shown in Fig. 11.

Here we took 740 different a's between 0 and 1 with

$$A_{i,j} = \left[\frac{i}{100}, \frac{i+1}{100}\right] \times \left[\frac{j}{100}, \frac{j+1}{100}\right] \quad (i,j = 0,\ldots,99),$$

$$n = 100\,000.$$

5.3. Tinkerbell map

Let us compute the CD for the following two type of Tinkerbell maps f_a and f_b on $I = [-1.2, 0.4] \times [-0.7, 0.3]$.

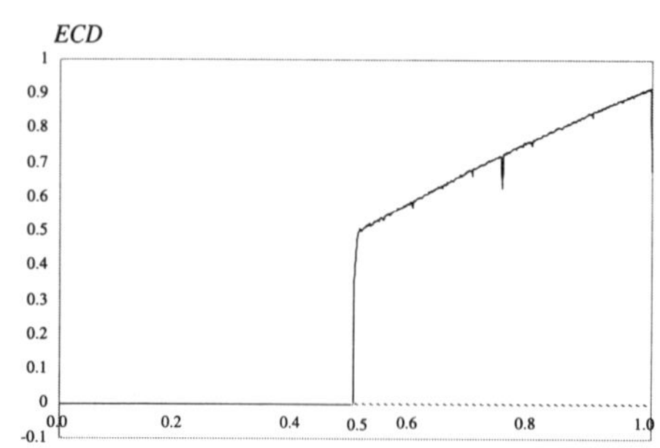

3

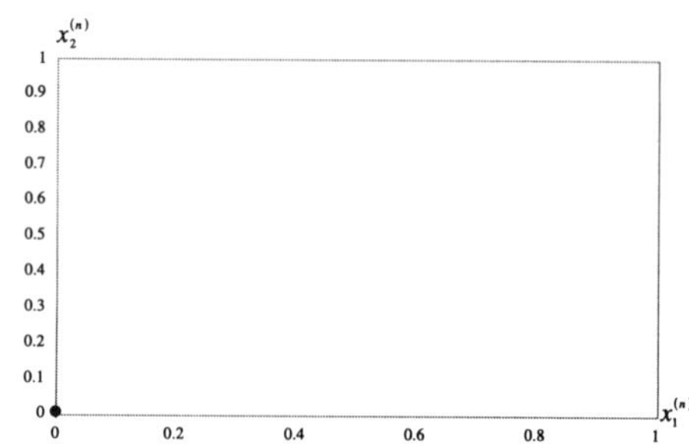

4

Fig. 3. ECD of Bernoulli shift.

Fig. 4. Orbit of f_a w.r.t. $0 \leqslant a < 0.5$.

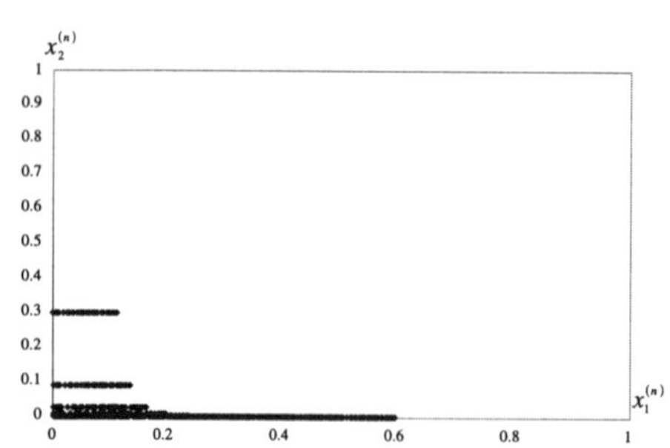

5

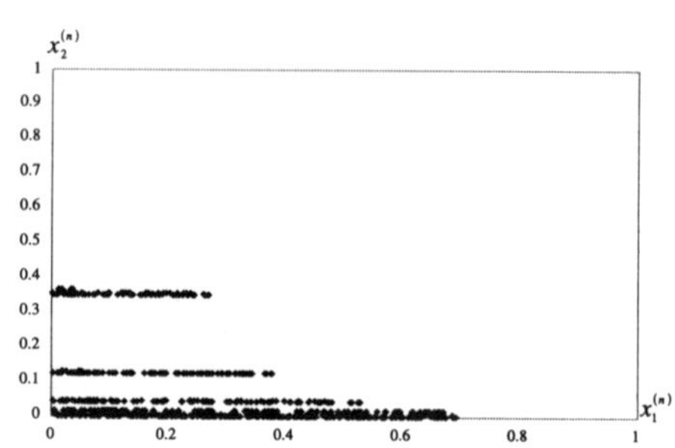

6

Fig. 5. Orbit of f_a w.r.t. $a = 0.6$.

Fig. 6. Orbit of f_a w.r.t. $a = 0.7$.

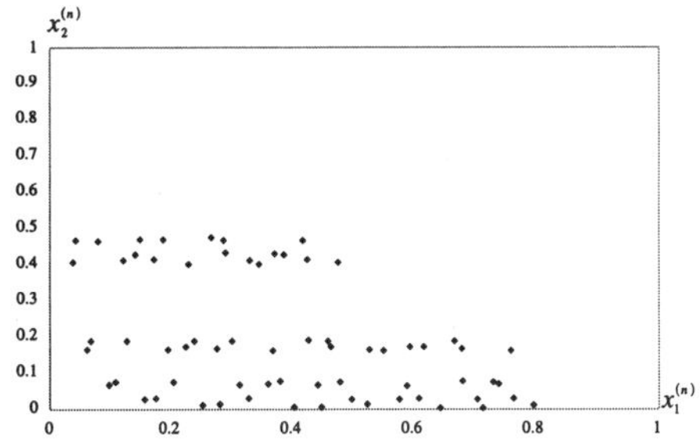

7

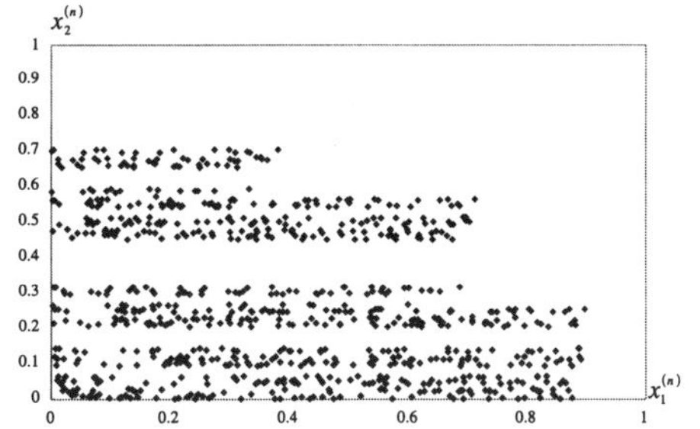

8

Fig. 7. Orbit of f_a w.r.t. $a = 0.8$.
Fig. 8. Orbit of f_a w.r.t. $a = 0.9$.

$$\begin{aligned} f_a(x^{(n)}) &= f_a\left(x_1^{(n)}, x_2^{(n)}\right) \\ &= \left(\left(x_1^{(n)}\right)^2 - \left(x_2^{(n)}\right)^2 + a x_1^{(n)} + c_2 x_2^{(n)},\ 2x_1^{(n)}x_2^{(n)} + c_3 x_1^{(n)} + c_4 x_2^{(n)}\right), \\ f_b(x^{(n)}) &= f_b\left(x_1^{(n)}, x_2^{(n)}\right) \\ &= \left(\left(x_1^{(n)}\right)^2 - \left(x_2^{(n)}\right)^2 + c_1 x_1^{(n)} + c_2 x_2^{(n)},\ 2x_1^{(n)}x_2^{(n)} + b x_1^{(n)} + c_4 x_2^{(n)}\right), \end{aligned}$$

where $(x_1^{(n)}, x_2^{(n)}) \in I$, $-0.4 \leqslant a \leqslant 0.9$, $1.9 \leqslant b \leqslant 2.9$, $(c_1, c_2, c_3, c_4) = (-0.3, -0.6, 2.0, 0.5)$, and $(x_1^{(0)}, x_2^{(0)}) = (0.1, 0.1)$.

Let us plot the points $(x_1^{(n)}, x_2^{(n)})$ for 3000 different n's between 1001 and 4000.

In stable domain, the number of points $(x_1^{(n)}, x_2^{(n)})$ is finite because the point $(x_1^{(n)}, x_2^{(n)})$ periodically appears in time n. Figs. 12 and 13 are examples of the orbits of f_a and f_b in a stable domain.

On the other hand, the point $(x_1^{(n)}, x_2^{(n)})$ takes random value in time n in chaotic domain. Figs. 14 and 15 are examples of the orbits of f_a and f_b in a chaotic domain.

The ECD of Tinkerbell map f_a and f_b are shown in Figs. 16 and 17.

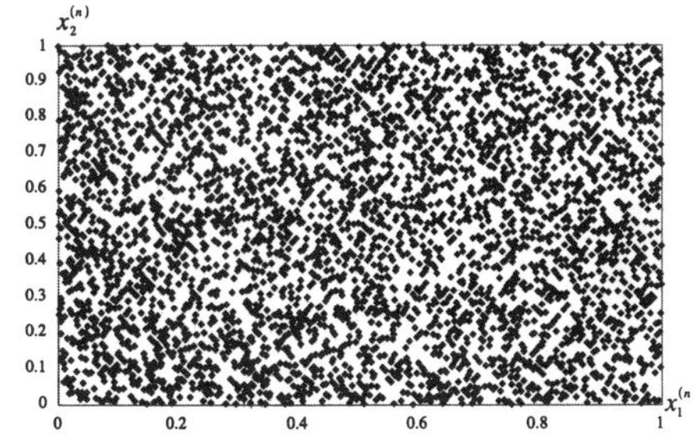

9

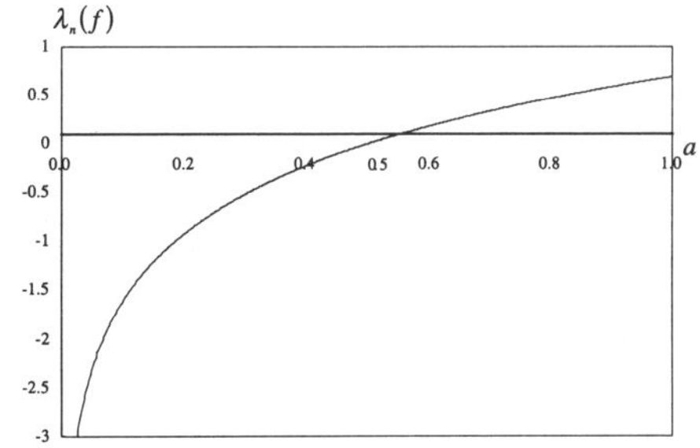

10

Fig. 9. Orbit of f_a w.r.t. $a = 1.0$.
Fig. 10. Lyapunov exponent of Baker's transformation.

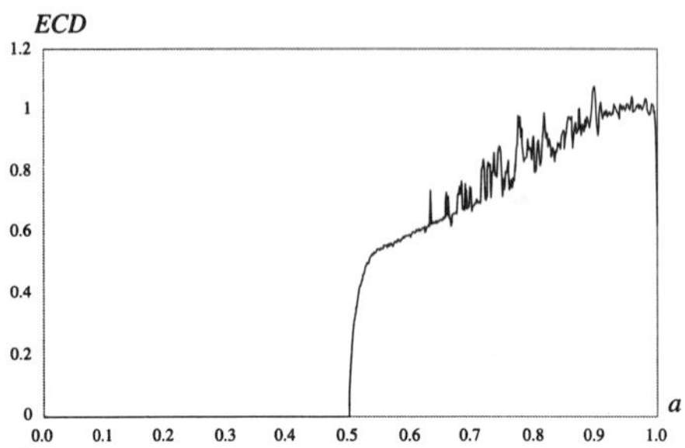

11

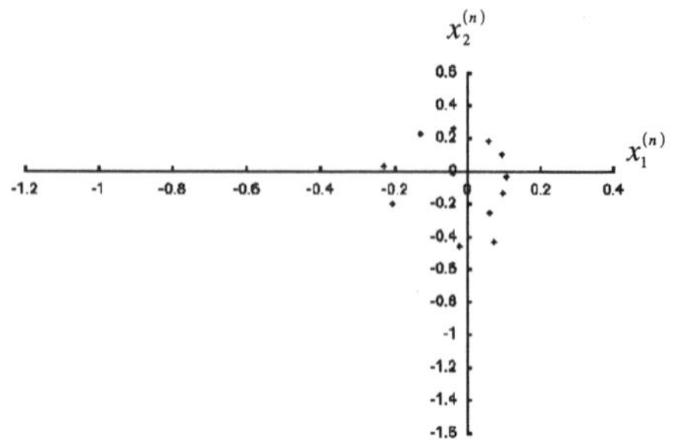

12

Fig. 11. ECD of Baker's transformation.
Fig. 12. Orbit of f_a w.r.t. $a = 0.243$.

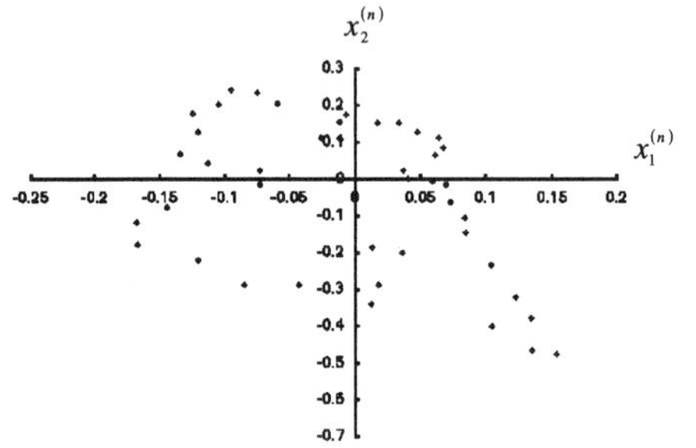

13

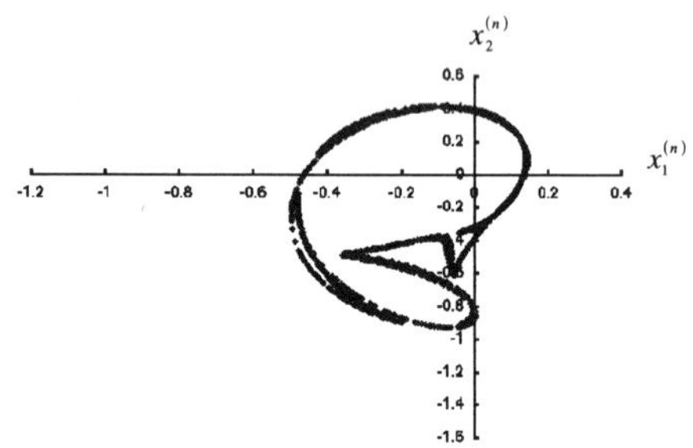

14

Fig. 13. Orbit of f_b w.r.t. $b = 2.65$.
Fig. 14. Orbit of f_a w.r.t. $a = 0.670$.

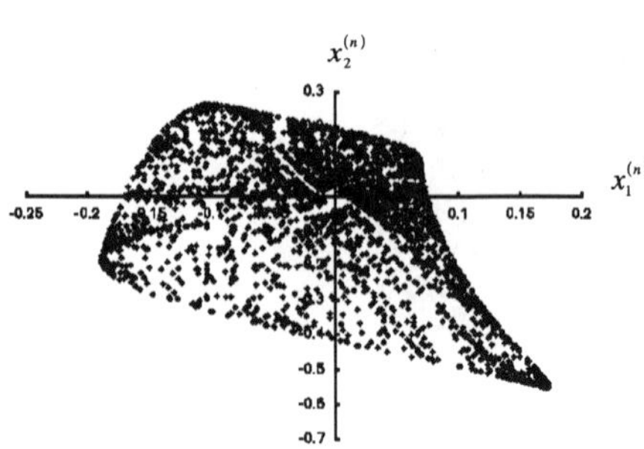

Fig. 15. Orbit of f_b w.r.t. $b = 2.8$.

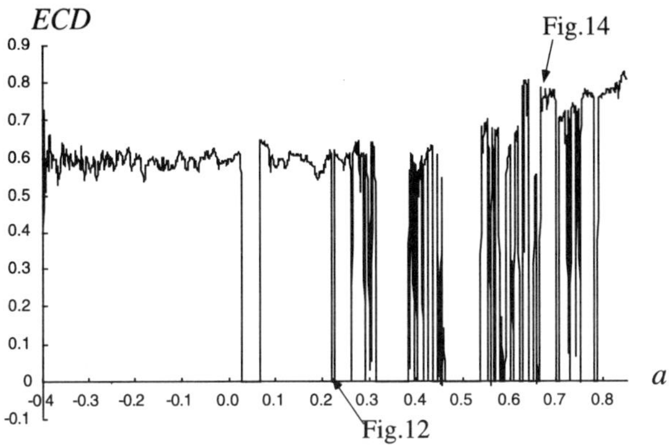

Fig. 16. ECD for Tinkerbell map f_a.

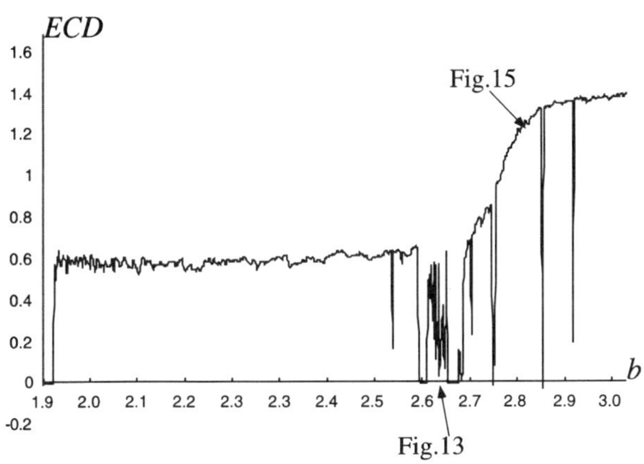

Fig. 17. ECD for Tinkerbell map f_b.

Here we took 740 different a's between -1.2 and 0.9 and 740 different b's between 1.9 and 2.9 with

$$A_{i,j} = \left[\frac{i}{100}, \frac{i+1}{100}\right] \times \left[\frac{j}{100}, \frac{j+1}{100}\right]$$
$$(i = -120, -119, \ldots, -1, 0, 1, \ldots, 38, 39)$$
$$(j = -70, -69, \ldots, -1, 0, 1, \ldots, 28, 29)$$
$$n = 100\,000.$$

6. Conclusion

From our results, when the orbits for Bernoulli shift, Baker's transformation or Tinkerbell map are chaotic, both Lyapunov exponent and chaos degree are positive. However, our chaos degree can resolve some inconvenient properties of the Lyapunov exponent in the following senses:

1. Lyapunov exponent takes negative and sometimes $-\infty$. For instance, although the exponent of Bernoulli shift and Baker's transformation cannot be defined for $a = 0$, the ECD is always positive and defined for any $a \geqslant 0$.
2. It is difficult to compute the Lyapunov exponent for the Tinkerbell maps f_a and f_b because it is difficult to compute f_a^n and f_b^n for large n. On the other hand, the ECD of f_a and f_b are easily computed.
3. Generally, the algorithm for CD is much easier than that for Lyapunov exponent.

References

[1] Akashi S. The asymptotic behavior of ε-entropy of a compact positive operator. J Math Anal Appl 1990;153:250.

[2] Alligood KT, Sauer TD, Yorke JA. Chaos – an introduction to dynamical systems. Textbooks in mathematical sciences. Berlin: Springer, 1996.

[3] Alicki R. Quantum geometry of noncommutative Bernoulli shifts. Banach center publications. Mathematics Subject Classification 46L87, 1991.

[4] Bennatti F. Deterministic chaos in infinite quantum systems. Berlin: Springer, 1993.

[5] Devaney RL. An introduction to chaotic dynamical systems, New York: Benjamin, 1986.

[6] Hasegawa H. Dynamical formulation of quantum level statistics. Open Systems and Information Dynamics 1997;4:350.

[7] Ingarden RS, Kossakowski A, Ohya M. Information dynamics and open systems. Dordrecht: Kluwer Academic Publishers, 1997.

[8] Kosaka M, Ohya M. A study of chaotic phenomena by information dynamics IEICE 1997; J80-A (12):2138 [in Japanese].

[9] Misiurewicz M. Absolutely continuous measutres for certain maps of interval. Publ Math IHES 1981;53:17.

[10] Ohya M. Information dynamics and its applications to optical communication processes. Lecture Notes in Physics 1991;378:81.

[11] Ohya M. Complexity and fractal dimensions for quantum states. Open Systems and Information Dynamics 1997;4:141.

[12] Ohya M. Complexities and their applications to characterization of chaos. Int J Theoret Phys 1998; 37(1): 495.

[13] Toda M. Crisis in chaotic scattering of a highly excited van der Waals complex. Phys Rev Lett 1995;74(14):2970.

Open Sys. & Information Dyn. **6**: 69–78, 1999 69

Fundamentals of Quantum Mutual Entropy and Capacity

Masanori Ohya

Department of Information Sciences
Science University of Tokyo
Noda City, Chiba 278-8510, Japan

(Received: March 3, 1998)

Abstract. The rigorous formulation of the quantum mutual entropy is reviewed. The capacities for various channels, such as quantum channel, classical-quantum channel and classical-quantum-classical channel are discussed exhaustively and some misuses of these capacities are indicated.

1. Introduction

The study of mutual entropy (information) and capacity in classical system was extensively done after Shannon by several authors like Kolmogorov [13] and Gelfand [8]. In quantum systems, there have been several definitions of the mutual entropy for classical input and quantum output [6, 9, 10, 15]. In 1983, the author defined [22] the fully quantum mechanical mutual entropy by means of the relative entropy of Umegaki [36], and he extended it [24] to general quantum systems by the relative entropy of Araki [4] and Uhlmann [35]. When introducing the quantum mutual entropy, the author did not indicate that it contained other definitions of the mutual entropy including the classical one, so that there still exists a misunderstanding concerning the use of the mutual entropy (information) to compute the capacity of quantum channels. In this note, we point out that our quantum mutual entropy generalizes the other and where the misuse occurs.

2. Mutual Entropy

The quantum mutual entropy was introduced in [22] for a quantum input and quantum output, namely, for a purely quantum channel, and it was later formulated for a general quantum system described in C*-algebraic terminology [24]. We review here the mutual entropy in usual quantum systems described Hilbert space language.

Let $\mathcal{H}$ be the input Hilbert space, $B(\mathcal{H})$ be the set of all bounded linear operators on $\mathcal{H}$, and $\mathcal{S}(\mathcal{H})$ be the set of all density operators on $\mathcal{H}$. The output space is described by another Hilbert space $\widetilde{\mathcal{H}}$, but often $\mathcal{H} = \widetilde{\mathcal{H}}$. A channel from the input system to the output system is a mapping Λ^* from $\mathcal{S}(\mathcal{H})$ to $\mathcal{S}(\widetilde{\mathcal{H}})$ [21]. A channel

Λ^* is said to be completely positive if the dual map Λ satisfies the following condition: $\sum_{k,j=1}^{n} A_k^* \Lambda(B_k^* B_j) A_j \geq 0$ for any $n \in \mathbb{N}$ and any $A_j \in B(\mathcal{H})$, $B_j \in B(\widetilde{\mathcal{H}})$. This condition is not strong at all because almost all physical transformations satisfy it [11, 24].

An input state $\rho \in \mathcal{S}(\mathcal{H})$ is sent to the output system through the channel Λ^*, so that the output state is written as $\widetilde{\rho} \equiv \Lambda^* \rho$. Then it is important to ask how much information of ρ is correctly sent to the output state $\Lambda^* \rho$. This amount of information transmitted from input to output is expressed by the mutual entropy.

In order to define the quantum mutual entropy, we first mention the entropy of a quantum state introduced by von Neumann [20]. For a state ρ, there exists a unique spectral decomposition

$$\rho = \sum_k \lambda_k P_k \,, \tag{2.1}$$

where λ_k is an eigenvalue of ρ and P_k is the associated projection for each λ_k. The projection P_k is not one-dimensional when λ_k is degenerated, so that the spectral decomposition can be further decomposed into one-dimensional projections. Such a decomposition is called a Schatten decomposition [34], namely,

$$\rho = \sum_k \lambda_k E_k \,, \tag{2.2}$$

where E_k is the one-dimensional projection associated with λ_k and the degenerated eigenvalue λ_k is repeated $\dim P_k$ times; for instance, if the eigenvalue λ_1 has the degeneracy 3, then $\lambda_1 = \lambda_2 = \lambda_3 < \lambda_4$. This Schatten decomposition is not unique unless every eigenvalue is non-degenerated. Then the entropy (von Neumann entropy) $S(\rho)$ of a state ρ is defined by

$$S(\rho) = -\mathrm{tr}\, \rho \log \rho \,, \tag{2.3}$$

which equals to the Shannon entropy of the probability distribution $\{\lambda_k\}$:

$$S(\rho) = -\sum_k \lambda_k \log \lambda_k \,. \tag{2.4}$$

The quantum mutual entropy was introduced on the basis of the above von Neumann entropy for purely quantum communication processes. The mutual entropy depends on input state ρ and channel Λ^*, so it is denoted by $I(\rho; \Lambda^*)$. It should satisfy the following conditions:

(1) The quantum mutual entropy is well-matched to the von Neumann entropy. Furthermore, if the channel is trivial, i.e., $\Lambda^* =$ identity map, then the mutual entropy equals the von Neumann entropy: $I(\rho; \mathrm{id}) = S(\rho)$.

(2) For a classical system, the quantum mutual entropy reduces to the classical one.

(3) Shannon's fundamental inequality, [33], $0 \le I(\rho; \Lambda^*) \le S(\rho)$ is satisfied.

Before turning to the quantum mutual entropy, we briefly review the classical mutual entropy [7]. Let $(\Omega, \mathcal{F})$, $(\tilde{\Omega}, \tilde{\mathcal{F}})$ be an input and output measurable spaces, respectively, and $P(\Omega)$, $P(\tilde{\Omega})$ are the corresponding sets of all probability measures (states) on Ω and $\tilde{\Omega}$, respectively. A channel Λ^* is a mapping from $P(\Omega)$ to $P(\tilde{\Omega})$ and its dual Λ is a map from the set $B(\Omega)$ of all Baire measurable functions on Ω to $B(\tilde{\Omega})$. For an input state $\mu \in P(\Omega)$, the output state $\tilde{\mu} = \Lambda^*\mu$ and the joint state (probability measure) Φ is given by

$$\Phi(Q \times \tilde{Q}) = \int_{\tilde{Q}} \Lambda(1\!\!1_Q)d\mu, \qquad Q \in \mathcal{F},\, \tilde{Q} \in \tilde{\mathcal{F}}, \tag{2.5}$$

where $1\!\!1_Q$ is the characteristic function on Ω:

$$1\!\!1_Q(\omega) = \begin{cases} 1 & \omega \in Q \\ 0 & \omega \notin Q \end{cases}.$$

The classical entropy, relative entropy and mutual entropy are defined as follows:

$$S(\mu) = \sup\Big\{-\sum_{k=1}^{n} \mu(A_k)\log\mu(A_k)\,;\ \{A_k\} \in \mathcal{P}(\Omega)\Big\}, \tag{2.6}$$

$$S(\mu,\nu) = \sup\Big\{\sum_{k=1}^{n} \mu(A_k)\log\frac{\mu(A_k)}{\nu(A_k)}\,;\ \{A_k\} \in \mathcal{P}(\Omega)\Big\}, \tag{2.7}$$

$$I(\mu;\Lambda^*) = S(\Phi, \mu \otimes \Lambda^*\mu), \tag{2.8}$$

where $\mathcal{P}(\Omega)$ is the set of all finite partitions on Ω, that is, $\{A_k\} \in \mathcal{P}(\Omega)$ iff $A_k \in \mathcal{F}$ with $A_k \cap A_j = \emptyset$ $(k \neq j)$ and $\cup_{k=1}^{n} A_k = \Omega$.

The quantum mutual entropy is defined as follows: In order to define the quantum mutual entropy, we need the quantum relative entropy and the joint state (it is called "compound state" in the sequel) describing the correlation between an input state ρ and the output state $\Lambda^*\rho$ through the channel Λ^*. A finite partition of Ω in classical case corresponds to an orthogonal decomposition $\{E_k\}$ of the identity operator I of $\mathcal{H}$ in quantum case because the set of all orthogonal projections is considered to be the event system in a quantum system. It is known [29] that the following equality holds

$$\sup\Big\{-\sum_{k} \operatorname{tr}\rho E_k \log \operatorname{tr}\rho E_k\,;\ \{E_k\}\Big\} = -\operatorname{tr}\rho\log\rho,$$

and the supremum is attained when $\{E_k\}$ is a Schatten decomposition of ρ. Therefore, the Schatten decomposition is used to define the compound state and the quantum mutual entropy.

The compound state σ_E (corresponding to joint state in CS) of ρ and $\Lambda^*\rho$ was introduced in [22, 23]; it is given by

$$\sigma_E = \sum_k \lambda_k E_k \otimes \Lambda^* E_k , \tag{2.9}$$

where E stands for a Schatten decomposition $\{E_k\}$ of ρ, so that the compound state depends on how we decompose the state ρ into basic states (elementary events) or, in other words, on how we see the input state.

The relative entropy for two states ρ and σ is defined by Umegaki [36] and Lindblad [16], which is written as

$$S(\rho, \sigma) = \begin{cases} \operatorname{tr} \rho(\log \rho - \log \sigma) & \text{when } \overline{\operatorname{ran} \rho} \subset \overline{\operatorname{ran} \sigma}) \\ \infty & \text{otherwise} \end{cases} \tag{2.10}$$

Then we can define the mutual entropy by means of the compound state and the relative entropy [22], that is,

$$I(\rho; \Lambda^*) = \sup\{S(\sigma_E, \rho \otimes \Lambda^*\rho) \, ; \, E = \{E_k\}\} , \tag{2.11}$$

where the supremum is taken over all Schatten decompositions because this decomposition is not unique in general. Some computations reduce it to the following form:

$$I(\rho; \Lambda^*) = \sup\Big\{\sum_k \lambda_k S(\Lambda^* E_k, \Lambda^*\rho) \, ; \, E = \{E_k\}\Big\} . \tag{2.12}$$

This mutual entropy satisfies all conditions (1)–(3) mentioned above.

When the input system is classical, an input state ρ is given by a probability distribution or a probability measure, in either case the Schatten decomposition of ρ is unique, namely, for the case of probability distribution, $\rho = \{\lambda_k\}$,

$$\rho = \sum_k \lambda_k \delta_k , \tag{2.13}$$

where δ_k is the delta measure, that is,

$$\delta_k(j) = \delta_{k,j} = \begin{cases} 0 & k \neq j \\ 1 & k = j , \end{cases} \qquad \forall j . \tag{2.14}$$

Therefore, for any channel Λ^*, the mutual entropy becomes

$$I(\rho; \Lambda^*) = \sum_k \lambda_k S(\Lambda^* \delta_k, \Lambda^*\rho) , \tag{2.15}$$

which equals the following usual expression of Shannon when the minus is well-defined:

$$I(\rho; \Lambda^*) = S(\Lambda^*\rho) - \sum_k \lambda_k S(\Lambda^* \delta_k) . \tag{2.16}$$

The above equality has been taken as the definition of the mutual entropy for a classical-quantum channel [5, 6, 9, 10, 15].

Note that the definition (2.12) of the mutual entropy is written as

$$I(\rho;\Lambda^*) = \sup\Big\{\sum_k \lambda_k S(\Lambda^*\rho_k,\Lambda^*\rho)\,;\ \rho=\sum_k \lambda_k\rho_k \in F_o(\rho)\Big\},$$

where $F_o(\rho)$ is the set of all orthogonal finite decompositions of ρ. Here ρ_k is orthogonal to ρ_j (denoted by $\rho_k \perp \rho_j$) meaning that the range of ρ_k is orthogonal to that of ρ_j. The above equality is easily proved as follows. Put

$$I_f(\rho;\Lambda^*) = \sup\Big\{\sum_k \lambda_k S(\Lambda^*\rho_k,\Lambda^*\rho)\,;\ \rho=\sum_k \lambda_k\rho_k \in F_o(\rho)\Big\}.$$

The inequality $I(\rho;\Lambda^*) \le I_f(\rho;\Lambda^*)$ is obvious. Let us prove the converse. Each ρ_k in an orthogonal decomposition of ρ is further decomposed into one-dimensional projections, $\rho_k = \sum_j \mu_j^{(k)} E_j^{(k)}$, a Schatten decomposition of ρ_k. From the following equalities for the relative entropy [4, 29]: (1) $S(a\rho, b\sigma) = aS(\rho,\sigma) - a\log(b/a)$, for any positive numbers a, b (2) $\rho_1 \perp \rho_2 \Longrightarrow S(\rho_1+\rho_2,\sigma) = S(\rho_1,\sigma)+S(\rho_2,\sigma)$, we have

$$\begin{aligned}\sum_k \lambda_k S(\Lambda^*\rho_k,\Lambda^*\rho) &= \sum_{k,j}\lambda_k\mu_j^{(k)} S(\Lambda^*E_j^{(k)},\Lambda^*\rho) + \sum_{k,j}\lambda_k\mu_j^{(k)}\log\mu_j^{(k)} \\ &\le \sum_{k,j}\lambda_k\mu_j^{(k)} S(\Lambda^*E_j^{(k)},\Lambda^*\rho),\end{aligned}$$

which implies the converse inequality $I(\rho;\Lambda^*) \ge I_f(\rho;\Lambda^*)$ because $\sum_{k,j}\lambda_k\mu_j^{(k)}E_j^{(k)}$ is a Schatten decomposition of ρ. Thus $I(\rho;\Lambda^*) = I_f(\rho;\Lambda^*)$.

More general formulation of the mutual entropy for general quantum systems was done [24, 11] for C*-dynamical systems by using Araki's or Uhlmann's relative entropy [4, 35, 29]. This general mutual entropy contains all other cases including measure-theoretic definition of Gelfand and Yaglom [8].

The mutual entropy is not only a measure of information transmission but it also describes the change of state, so that this quantity can be applied in several fields [1, 2, 3, 17, 19, 24, 25, 28, 32].

3. Communication Processes

In this section, we discuss communication processes [7, 10, 29]. Let $\{a_1, a_2, \ldots, a_n\}$ be an alphabet and Ω be the infinite direct product of A: $\Omega = A^{\mathbb{Z}} \equiv \prod_{-\infty}^{\infty} A$ called a message space. In order to send an information represented by an element of this message space to a receiver, we often need to transform the message into a form proper for a communication channel. This transformation of message is called coding. Precisely, coding is a measurable one-to-one map ξ from Ω to a proper space X. For instance, we have the following codings: (1) When a message is expressed in binary symbols 0 and 1, such coding is a map from Ω to $\{0,1\}^{\mathbb{N}}$. (2) A message expressed by a $\{0,1\}$ sequence in (1) is represented by an electric signal. (3) Instead of an electric signal, we use optical signal. Coding is a combination of several maps like the ones above. One of the main targets of the coding theory is to find the most efficient coding and also decoding for information transmission.

Let $(\Omega, \mathcal{F}_\Omega, P(\Omega))$ be an input probability space and X be the coded input space. This space X may be a classical object or a quantum object. For instance, X is a Hilbert space $\mathcal{H}$ of a quantum system, then the coded input system is described by $(B(\mathcal{H}), \mathcal{S}(\mathcal{H}))$ of Sec. 2.

An output system is similarly described as the input system: The coded output space is denoted by $\widetilde{X}$ and the decoded output space is $\widetilde{\Omega}$, possibly over a different alphabet. A transmission (map) from X to $\widetilde{X}$ is described by a channel reflecting all properties of a physical device, which is denoted by γ here. With a decoding $\widetilde{\xi}$, the whole information transmission process is written as

$$\Omega \xrightarrow{\xi} X \xrightarrow{\gamma} \widetilde{X} \xrightarrow{\widetilde{\xi}} \widetilde{\Omega} . \tag{3.1}$$

That is, a message $\omega \in \Omega$ is coded to $\xi(\omega)$ and it is sent to the output system through the channel γ, then the output coded message becomes $\gamma \circ \xi(\omega)$ and it is decoded to $\widetilde{\xi} \circ \gamma \circ \xi(\omega)$ at the receiver.

This transmission process is mathematically set as follows: M messages are sent to the receiver, the k-th message $\omega^{(k)}$ occuring with probability λ_k. Then the occurrence probability of each message in the sequence $(\omega^{(1)}, \omega^{(2)}, \ldots, \omega^{(M)})$ of M messages is denoted by $\rho = \{\lambda_k\}$, which is a state in a classical system. If ξ is a classical coding, then $\xi(\omega)$ is a classical object such as an electric pulse. If ξ is a quantum coding, then $\xi(\omega)$ is a quantum object (state) such as a coherent state. Here we consider such a quantum coding, that is, $\xi(\omega^{(k)})$ is a quantum state, and we denote $\xi(\omega^{(k)})$ by σ_k. Thus the coded state for the sequence $(\omega^{(1)}, \omega^{(2)}, \ldots, \omega^{(M)})$ is written as

$$\sigma = \sum_k \lambda_k \sigma_k . \tag{3.2}$$

This state is transmitted through a channel γ. This channel is expressed by a completely positive mapping Γ^*, in the sense of Sec. 1, from the state space of X to that of $\widetilde{X}$, hence the output coded quantum state $\widetilde{\sigma}$ is $\Gamma^*\sigma$. Since the information transmission process can be understood as a process of state (probability) change, when Ω and $\widetilde{\Omega}$ are classical and X and $\widetilde{X}$ are quantum, the process (3.1) is written as

$$P(\Omega) \xrightarrow{\Xi^*} \mathcal{S}(\mathcal{H}) \xrightarrow{\Gamma^*} \mathcal{S}(\widetilde{\mathcal{H}}) \xrightarrow{\widetilde{\Xi}^*} P(\widetilde{\Omega}) , \tag{3.3}$$

where Ξ^* (resp. $\widetilde{\Xi}^*$) is the channel corresponding to the coding ξ (resp. $\widetilde{\xi}$) and $\mathcal{S}(\mathcal{H})$ (resp. $\mathcal{S}(\widetilde{\mathcal{H}})$) is the set of all density operators (states) on $\mathcal{H}$ (resp. $\widetilde{\mathcal{H}}$).

We have to study the objects in the above transmission process (3.1) or (3.3) carefully. Namely, we have to make clear which object we are going to study. For instance, if we want to know the information capacity of a quantum channel γ $(= \Gamma^*)$, then we have to take X describing a quantum system, i.e. a Hilbert space, and we need to start the study from a quantum state in the quantum space X, not from a classical state associated to a message. If we like to know the capacity of the whole process including the coding and decoding, which means the capacity of the channel $\tilde{\xi} \circ \gamma \circ \xi = \tilde{\Xi}^* \circ \Gamma^* \circ \Xi^*$, then we have to start from a classical state. In any case, when we are concerned with the capacity of a channel, we have to take the supremum of the mutual entropy $I(\rho; \Lambda^*)$ over quantum or classical states ρ in a proper set determined by what we want to study. We explain this more precisely in the next section.

4. Channel Capacity

We discuss two types of channel capacity in communication processes, namely, the capacity of a quantum channel Γ^* and that of a classical (classical-quantum-classical) channel $\tilde{\Xi}^* \circ \Gamma^* \circ \Xi^*$.

4.1. Capacity of quantum channel

The capacity of a quantum channel is the ability of the channel itself to transmit information, so that it does not depend on how a message (treated as a classical object) is coding; we have to begin with an arbitrary quantum state and find the supremum of the mutual entropy. One often makes a mistake at this point. For example, one begins with the coding of a message, and computes the supremum of the mutual entropy, identifying the supremum with the capacity of the quantum channel, which is not correct. Even when the coding is quantum and the coded message is sent to a receiver through a quantum channel, if one starts from a classical state, then the capacity is not the capacity of the quantum channel itself. In his case, the usual Shannon theory is applied because one can easily compute the conditional distribution in a usual (classical) way. Such a supremum is the capacity of a classical-quantum-classical channel, and it is in the second category discussed below.

The capacity of a quantum channel Γ^* is defined as follows: Let $\mathcal{S}_0 \subset \mathcal{S}(\mathcal{H})$ be the set of all states prepared for expression of information. Then the capacity of the channel Γ^* with respect to $\mathcal{S}_0$ is defined by

$$C^{\mathcal{S}_0}(\Gamma^*) = \sup\{I(\rho; \Gamma^*)\,;\ \rho \in \mathcal{S}_0\}. \tag{4.1}$$

Here $I(\rho; \Gamma^*)$ is the mutual entropy given in (2.11) or (2.12) with $\Lambda^* = \Gamma^*$. When $\mathcal{S}_0 = \mathcal{S}(\mathcal{H})$, $C^{\mathcal{S}(\mathcal{H})}(\Gamma^*)$ is denoted by $C(\Gamma^*)$ for simplicity. In [30, 18], we also considered the pseudo-quantum capacity $C_p(\Gamma^*)$ defined by (4.1) with the pseudo-mutual entropy $I_p(\rho;\Gamma^*)$ where the supremum is taken over all finite decompositions instead of all orthogonal pure decompositions:

$$I_p(\rho;\Gamma^*) = \sup\left\{\sum_k \lambda_k S(\Gamma^*\rho_k, \Gamma^*\rho)\,;\ \rho = \sum_k \lambda_k\rho_k\,,\ \text{finite decomposition}\right\}. \tag{4.2}$$

However, the pseudo-mutual entropy is not well-matched with the conditions explained in Sec. 2, and it is difficult to compute it numerically [31]. From the monotonicity of the mutual entropy [29], we have

$$0 \le C^{\mathcal{S}_0}(\Gamma^*) \le C_p^{\mathcal{S}_0}(\Gamma^*) \le \sup\{S(\rho)\,;\ \rho\in\mathcal{S}_0\}\,.$$

4.2. Capacity of classical-quantum-classical channel

The capacity of C-Q-C channel $\widetilde{\Xi}^*\circ\Gamma^*\circ\Xi^*$ is the capacity of the information transmission process starting from the coding of messages, therefore it can be considered as the capacity including a coding (and a decoding). As is discussed in Sec. 3, the input state ρ is the probability distribution $\{\lambda_k\}$ of messages, and its Schatten decomposition is unique as (2.13), so the mutual entropy is written by (2.15):

$$I(\rho;\widetilde{\Xi}^*\circ\Gamma^*\circ\Xi^*) = \sum_k \lambda_k S(\widetilde{\Xi}^*\circ\Gamma^*\circ\Xi^*\delta_k\,,\ \widetilde{\Xi}^*\circ\Gamma^*\circ\Xi^*\rho)\,. \tag{4.3}$$

If the coding Ξ^* is a quantum coding, then $\Xi^*\delta_k$ is expressed by a quantum state. Let us denote the coded quantum state by σ_k and put $\sigma = \Xi^*\rho = \sum_k \lambda_k\sigma_k$. Then the above mutual entropy is written as

$$I(\rho;\widetilde{\Xi}^*\circ\Gamma^*\circ\Xi^*) = \sum_k \lambda_k S(\widetilde{\Xi}^*\circ\Gamma^*\sigma_k\,,\ \widetilde{\Xi}^*\circ\Gamma^*\sigma)\,. \tag{4.4}$$

This is the expression of the mutual entropy of the whole information transmission process starting from a coding of classical messages. Hence the capacity of C-Q-C channel is

$$C^{P_0}(\widetilde{\Xi}^*\circ\Gamma^*\circ\Xi^*) = \sup\left\{I(\rho;\widetilde{\Xi}^*\circ\Gamma^*\circ\Xi^*)\,;\ \rho\in P_0\right\}, \tag{4.5}$$

where $P_0 \subset P(\Omega)$ is the set of all probability distributions prepared for input (*a-priori)* states (distributions or probability measures). Moreover, the capacity for coding free channel is found by taking the supremum of the mutual entropy (4.4) over all probability distributions and all codings Ξ^*:

$$C_c^{P_0}(\widetilde{\Xi}^*\circ\Gamma^*) = \sup\left\{I(\rho;\widetilde{\Xi}^*\circ\Gamma^*\circ\Xi^*)\,;\ \rho\in P_0, \Xi^*\right\}. \tag{4.6}$$

The last capacity is for both coding and decoding free channels and it is given by

$$C_{cd}^{P_0}(\Gamma^*) = \sup\left\{I(\rho; \widetilde{\Xi}^* \circ \Gamma^* \circ \Xi^*)\,;\ \rho \in P_0, \Xi^*, \widetilde{\Xi}^*\right\}. \tag{4.7}$$

These capacities $C_c^{P_0}$, $C_{cd}^{P_0}$ do not measure the ability of the quantum channel Γ^* itself, but measure the ability of Γ^* through the coding and decoding.

Note that when $\sum_k \lambda_k S(\Gamma^*\sigma_k)$ is finite, then (4.4) becomes

$$I(\rho; \widetilde{\Xi}^* \circ \Gamma^* \circ \Xi^*) = S(\widetilde{\Xi}^* \circ \Gamma^* \sigma) - \sum_k \lambda_k S(\widetilde{\Xi}^* \circ \Gamma^* \sigma_k)\,. \tag{4.8}$$

Further, if ρ is a probability measure having a density function $f(\lambda)$ and each λ corresponds to a quantum coded state $\sigma(\lambda)$, then $\sigma = \int f(\lambda)\sigma(\lambda)d\lambda$ and

$$I(\rho; \widetilde{\Xi}^* \circ \Gamma^* \circ \Xi^*) = S(\widetilde{\Xi}^* \circ \Gamma^* \sigma) - \int f(\lambda) S(\widetilde{\Xi}^* \circ \Gamma^* \sigma(\lambda))\, d\lambda\,, \tag{4.9}$$

which is less than

$$S(\Gamma^*\sigma) - \int f(\lambda) S(\Gamma^*\sigma(\lambda))\, d\lambda\,,$$

because of the monotonicity of the quantum mutual entropy. The above bound is called the Holevo bound, and it is computed in several cases [30, 37].

The above three capacities C^{P_0}, $C_c^{P_0}$, $C_{cd}^{P_0}$ satisfy the following inequalities

$$0 \le C^{P_0}(\widetilde{\Xi}^* \circ \Gamma^* \circ \Xi^*) \le C_c^{P_0}(\widetilde{\Xi}^* \circ \Gamma^*) \le C_{cd}^{P_0}(\Gamma^*) \le \sup\{S(\rho)\,;\ \rho \in P_o\}\,,$$

where $S(\rho)$ is not the von Neumann entropy but the Shannon one: $-\sum \lambda_k \log \lambda_k$.

The capacities (4.1), (4.6), (4.7) and (4.8) are in general different. Some misunderstanding occurs due to forgetting which channel is considered. That is, we have to make clear whether the capacity of a quantum channel itself or that of a classical-quantum(-classical) channel or that of a coding free, is considered.

Bibliography

1. L. Accardi, M. Ohya, and H. Suyari, Open Sys. Information Dyn. **2**, 337 (1994).
2. S. Akashi, J. Math. Anal. Appl. **153**, 250 (1990).
3. R. Alicki, Open Sys. Information Dyn. **4**, 53 (1997).
4. H. Araki, Publ. RIMS Kyoto Univ. **11**, 809 (1976).
5. V. P. Belavkin, *Proc. of VIII-th Conference on Coding Theory and Information Transmission*, Moscow-Kuibishev, 1982, p. 15.
6. V. P. Belavkin and P. L. Stratonovich, Radio Eng. Electron. Phys. **18**, 1839 (1973).
7. P. Billingsley, *Ergodic Theory and Information*, Wiley, NY.,1965.
8. I. M. Gelfand and A. M. Yaglom, Amer. Math. Soc. Transl. **12**, 199 (1959).
9. A. S. Holevo, Problemy Peredachi Informacii **9**, 3 (1973), in Russian.
10. R. S. Ingarden, Rep. Math. Phys. **10**, 43 (1976).
11. R. S. Ingarden, A. Kossakowski and M. Ohya, *Information Dynamics and Open Systems*, Kluwer Academic Publisher, 1997.

12. K. Inoue, M. Ohya, and H. Suyari, Physica D **120**, 117 (1998).
13. A. N. Kolmogorov, Amer. Math. Soc. Translation, Ser.2 **33**, 291 (1963).
14. S. Kullback and R. Leibler, Ann. Math. Stat. **22**, 79 (1951).
15. L. B. Levitin, Springer Lect. Note in Phys. **378**, 101 (1991).
16. G. Lindblad, Commun. Math. Phys., **33**, 111 (1973).
17. T. Matsuoka and M. Ohya, Rep. Math. Phys. **36**, 365 (1995).
18. N. Muraki, M. Ohya, and D. Petz, Open Sys. Information Dyn. **1**, 43 (1992).
19. N. Muraki and M. Ohya, Lett. Math. Phys. **36**, 327 (1996).
20. J. von Neumann, *Die Mathematischen Grundlagen der Quantenmechanik*, Springer-Berlin, 1932.
21. M. Ohya, J. Math. Anal. Appl. **84**, 318 (1981).
22. M. Ohya, IEEE Trans. Information Theory, **29**, 770 (1983).
23. M. Ohya, L. Nuovo Cimento **38**, 402 (1983).
24. M. Ohya, Rep. Math. Phys. **27**, 19 (1989).
25. M. Ohya, Quantum Probability and Related Topics **6**, World Scientific, Singapore, 1991.
26. M. Ohya, Quantum Communications and Measurement, p. 309, 1995.
27. M. Ohya, Open Sys. Information Dyn. **4**, 141 (1997).
28. M. Ohya, Int. J. Theor. Phys. **37**, 495 (1998).
29. M. Ohya and D. Petz, *Quantum Entropy and Its Use*, Springer, 1993.
30. M. Ohya, D. Petz, and N. Watanabe, Probability and Mathematical Statistics, **17**, 179 (1997).
31. M. Ohya, D. Petz, and N. Watanabe, Int. J. Theor. Phys. **38**, 507 (1998).
32. M. Ohya and N. Watanabe, Physica D **120**, 206 (1998).
33. C. E. Shannon, Bell System Tech. J. **27**, 379 (1948).
34. R. Schatten, *Norm Ideals of Completely Continuous Operators*, Springer-Verlag, 1970.
35. A. Uhlmann, Commun. Math. Phys. **54**, 21 (1977).
36. H. Umegaki, Kodai Math. Sem. Rep. **14**, 59 (1962).
37. H. P. Yuen and M. Ozawa, Phys. Rev. Lett. **70**, 363 (1993).

Appl Math Optim 39:33–59 (1999)
With kind permission of Springer Science and Business Media

Applied Mathematics & Optimization

Compound Channels, Transition Expectations, and Liftings

L. Accardi[1] and M. Ohya[2]

[1]Dipartimento di Matematica, Centro Matematico V. Volterra,
Universitá di Roma II, Rome, Italy

[2]Department of Information Sciences,
Science University of Tokyo,
Noda City, Chiba 278, Japan

Abstract. In Section 1 we introduce the notion of lifting as a generalization of the notion of compound state introduced in [21] and [22] and we show that this notion allows a unified approach to the problems of quantum measurement and of signal transmission through quantum channels. The dual of a linear lifting is a transition expectation in the sense of [3] and we characterize those transition expectations which arise from compound states in the sense of [22].

In Section 2 we characterize those liftings whose range is contained in the closed convex hull of product states and we prove that the corresponding quantum Markov chains [2] are uniquely determined by a classical generalization of both the quantum random walks of [4] and the locally diagonalizable states considered in [3].

In Section 4, as a first application of the above results, we prove that the attenuation (beam splitting) process for optical communication treated in [21] can be described in a simpler and more general way in terms of liftings and of transition expectations. The error probabilty of information transmission in the attenuation process is rederived from our new description. We also obtain some new results concerning the explicit computation of error probabilities in the squeezing case.

Key Words. Compound state, Transition expectation, Lifting, Channel, Quantum probability, Quantum Markov chain, Beam splitting, Optical communication.

AMS Classification. 81Q99, 94A40, 60J27.

Introduction

The following situation is very common both in classical and quantum physics: one considers two systems, denoted respectively 1, 2, and their algebras of observables, $\mathcal{A}_1, \mathcal{A}_2$. One usually assumes that the interaction between the two systems is switched on at a sharp time t_0 before which the two systems are considered to be independent. During the interaction the two systems merge into a larger system denoted $(1, 2)$ whose algebra of observables $\mathcal{A}$ contains both $\mathcal{A}_1$ and $\mathcal{A}_2$, in the sense that there are embeddings

$$j_1\colon \mathcal{A}_1 \to \mathcal{A}; \qquad j_2\colon \mathcal{A}_2 \to \mathcal{A} \tag{0.1}$$

and that any physical information on the state of system 1 or of system 2 after the interaction can be obtained by choosing a state φ on $\mathcal{A}$, i.e., a state of the composite system $(1, 2)$, and considering its restriction on the algebra $j_1(\mathcal{A}_1)$ (resp. $j_2(\mathcal{A}_2)$). In most applications one chooses

$$\mathcal{A} = \mathcal{A}_1 \otimes \mathcal{A}_2; \qquad \varphi = \varphi_1 \otimes \varphi_2, \tag{0.2}$$

$$j_1(a_1) = a_1 \otimes 1_2; \qquad j_2(a_2) = 1_1 \otimes a_2, \qquad a_1 \in \mathcal{A}_1, \quad a_2 \in \mathcal{A}_2, \tag{0.3}$$

i.e., the algebra of the compound system is described by a tensor product. In this paper we confine our analysis of such a situation.

The choice of the state φ depends on the initial states of the two systems and on the interaction between them. In connection with this situation one studies several problems depending on the interpretation of systems 1 and 2. For example:

(i) The state φ_2 of system 2, **after** the interaction, is known (e.g., an output signal, a pointer in a measurement apparatus) and one wants to know the state φ_1 of system 1 **before** the interaction (e.g., an input signal, the state of a microsystem which has interacted with the apparatus).

(ii) As in (i), exchanging the roles of 1 and 2, from the mathematical point of view, this exchange is trivial, but we want to underline that our approach avoids the separation of a *macroworld*, described by classical physics, from a *microworld*, described by quantum physics.

(iii) The initial state of the composite system $(1, 2)$ is known and one wants to know the state of system 1 (system 2).

(iv) The state φ_1 of system 1, **before** the interaction (e.g., the preparation of a microsystem) and the form of the interaction, is known and one wants to know the state of system 1 **after** the interaction.

In all these cases the goal is to construct a map from the state space of a system to the state space of another system. In the literature on quantum information and communication systems, such a map is called a **channel** [20]. An important class of channels are those from the state space of an algebra $\mathcal{A}_1$ into the state space of the algebra $\mathcal{A}_1 \otimes \mathcal{A}_2$. These channels are called **liftings**; more generally, a lifting should be through as a channel from a subsystem to a compound system. An important example of liftings are the duals of transition expectations.

Recall (see Definition 1.3 below) that if $\mathcal{A}_1, \mathcal{A}_2$ are C*-algebras, a **transition expectation** from $\mathcal{A}_1 \otimes \mathcal{A}_2$ to $\mathcal{A}_1$ is a completely positive linear map $\mathcal{E}\colon \mathcal{A}_1 \otimes \mathcal{A}_2 \to \mathcal{A}_1$ satisfying (1.6).

Transition expectations play a crucial role in the construction of quantum Markov chains and they arise naturally within the framework of measurement theory in the following way: the composite system $(1, 2)$ undergoes an evolution u_t: $\mathcal{A} \to \mathcal{A}$ $(t \in R)$, which is a one-parameter group of $*$-automorphisms of $\mathcal{A}$. This means that the state φ of $(1, 2)$ evolves according to the law

$$\varphi_t := \varphi \circ u_t \tag{0.4}$$

and the state φ_1 of the system 1 evolves according to the **reduced evolution**:

$$\varphi_{1,t}(a_1) := \varphi_1(E_2 \circ u_t \circ j_1(a_1)); \qquad a_1 \in \mathcal{A}_1, \tag{0.5}$$

where j_1 is given by (0.3) and E_2: $\mathcal{A} = \mathcal{A}_1 \otimes \mathcal{A}_2 \to \mathcal{A}_1$ is the Umegaki conditional expectation characterized by

$$E_2(a_1 \otimes a_2) = a_1\varphi_2(a_2); \qquad a_1 \in \mathcal{A}_1, \quad a_2 \in \mathcal{A}_2. \tag{0.6}$$

We fix a time T representing the moment when the experiment ends (ideally $T = +\infty$) and consider the linear map $\mathcal{E}_T$: $\mathcal{A}_1 \otimes \mathcal{A}_2 \to \mathcal{A}_1$ characterized by

$$\mathcal{E}_T(a_1 \otimes a_2) = E_2(u_T(a_1 \otimes a_2)); \qquad a_1 \in \mathcal{A}_1, \quad a_2 \in \mathcal{A}_2. \tag{0.7}$$

Then $\mathcal{E}_T$ is a transition expectation. If one is ready to accept that the evolution u_T does not take place inside the algbra $\mathcal{A}_1 \otimes \mathcal{A}_2$ but is a representation of $\mathcal{A}_1 \otimes \mathcal{A}_2$ into another algebra (usually much larger), then in some cases and in a certain technical sense, (0.7) represents the most general class of transition expectations (see Theorem 1.4 below).

An *instrument* in the sense of the operational approach to quantum measurment is obtained by taking the restriction of a transition expectation $\mathcal{E}$ to a subalgebra $\mathcal{C}_1 \otimes \mathcal{A}_2$ of $\mathcal{A}_1 \otimes \mathcal{A}_2$ where $\mathcal{C}_1$ is a σ-finite abelian von Neumann sup-algebra of $\mathcal{A}_1$. In this case it is known that if $\mathcal{C}_1$ is σ-finite, then there exists a probability space $(\Omega, \mathcal{F}, P)$ such that $\mathcal{C}_1$ is isomorphic to $L^\infty(\Omega, \mathcal{F}, P)$ and the points $\omega \in \Omega$ are interpreted as macroscopic parameters of the apparatus. If $\mathcal{A}_1 = \mathcal{C}_1$, i.e., if $\mathcal{A}_1$ is an abelian von Neumann algebra, the isomorphism (see [28])

$$\mathcal{A} = \mathcal{A}_1 \otimes \mathcal{A}_2 = L^\infty(\Omega, \mathcal{F}, P) \otimes \mathcal{A}_2 \cong L^\infty(\Omega, \mathcal{F}, P; \mathcal{A}_2) \tag{0.8}$$

implies that the elements of $\mathcal{A}$, i.e., the observables of the composite system $(1, 2)$ can be interpreted as functions $(\Omega, \mathcal{F}, P) \to \mathcal{A}_2$, i.e., as **operator-valued** random variables. Thus interpreting $(\Omega, \mathcal{F}, P)$ as the sample space of a classical stochastic process, the operational scheme becomes equivalent to the theory of **operator-valued classical processes** (see [7]).

From this point of view an instrument in the operational sense is an object which is only *half-quantum*. The physical motivations for this choice go back to some ideas of Ludwig, according to which the measurement apparatus is usually a macroscopic body so that classical probability should be sufficient for its description. Several authors have introduced variations and modifications of Ludwig's ideas, however, since all the examples of physical interest of instruments in an operational sense produced up to now are the restrictions of liftings, we feel that the latter notion plays a more natural and fundamental role.

In conclusion we show the theory of lifting includes the so-called *operational approach.*

1. Channels, Liftings, and Transition Expectations

For a C*-algebra $\mathcal{A}$, we denote by $\mathcal{S}(\mathcal{A})$ the convex set of its states. In this paper all C*- and W*-algebras are realized on some separable Hilbert spaces and, unless explicitly stated, the tensor products are those induced by the tensor products of the corresponding Hilbert spaces. If $\mathcal{A}$ is a von Neumann algebra, $\mathcal{S}(\mathcal{A})$ denotes the set of its normal states and $\mathcal{S}(\mathcal{A})_{\text{extr}}$ the set of extremal states. Both $\mathcal{S}(\mathcal{A})$ and $\mathcal{S}(\mathcal{A})_{\text{extr}}$ are measurable spaces with their Borel structure and the set of probability measures on $\mathcal{S}(\mathcal{A})$ (resp. $\mathcal{S}(\mathcal{A})_{\text{extr}}$) is denoted by $\text{Prob}\,\mathcal{S}(\mathcal{A})$ (resp. $\text{Prob}\,\mathcal{S}(\mathcal{A})_{\text{extr}}$). If $\mathcal{A}$ and $\mathcal{B}$ are C*-algebras, a **channel** from $\mathcal{A}$ to $\mathcal{B}$ is a map $\Lambda^*\colon\ \mathcal{S}(\mathcal{A}) \to \mathcal{S}(\mathcal{B})$. If Λ^* is affine we speak of **a linear channel**. If Λ^* is $w*$-continuous and linear, then it can be extended by linearity to a linear map (still denoted Λ^*) from $\mathcal{A}^*$ to $\mathcal{B}^*$. Its dual $\Lambda\colon\ \mathcal{B} \to \mathcal{A}$ is a positive map. If it is completely positive, we call it a **Markovian operator**. Such channels have been studied with some applications by several authors (see [20], [24], [25], and references quoted therein). Certain quantum channels are naturally associated to classical Markovian kernels on the measurable space $\mathcal{S}(\mathcal{B}) \times \mathcal{S}(\mathcal{A})$. In fact, given such a Markovian kernel, i.e., a measurable map

$$p\colon\ \omega \in \mathcal{S}(\mathcal{A}) \to p(\,\cdot \mid \omega) \in \text{Prob}(\mathcal{S}(\mathcal{B})),$$

one can define a channel in the following way: for any state $\varphi \in \mathcal{S}(\mathcal{A})$ one fixes a convex decomposition

$$\varphi = \int_{\mathcal{S}(\mathcal{A})} \omega\, d\mu_\varphi(\omega)$$

and defines the channel $\Lambda^*\colon\ \mathcal{S}(\mathcal{A}) \to \mathcal{S}(\mathcal{B})$ through the identity

$$\Lambda^*\varphi := \int_{\mathcal{S}(\mathcal{A})} d\mu_\varphi(\omega) \int_{\mathcal{S}(\mathcal{B})} \omega' p(d\omega' \mid \omega).$$

The channel Λ^* is usually nonlinear since the map $\varphi \to \mu_\varphi$ is affine if and only if $\mathcal{S}(\mathcal{A})$ is a simplex [8, Theorem (4.1.15) and Corollary (4.1.17)] and this is the case if and only if $\mathcal{A}$ is Abelian [8, Example (4.1.6)]. On the other hand, given a linear channel Λ^* one might try to associate to it a Markovian kernel $p(\,\cdot\,|\omega)$ on the measurable space $\mathcal{S}(\mathcal{B}) \times \mathcal{S}(\mathcal{A})$, by fixing, for each $\omega \in \mathcal{S}(\mathcal{A})$, a convex decomposition

$$\Lambda^*\omega := \int_{\mathcal{S}(\mathcal{B})} \omega' p(d\omega' \mid \omega).$$

The possibility of choosing such a decomposition to assure the measurability of the map

$$p\colon\ \omega \in \mathcal{S}(\mathcal{A}) \to p(\,\cdot \mid \omega) \in \text{Prob}(\mathcal{S}(\mathcal{B}))$$

as well as study of the support of these measures give rise to some subtle measure-theoretic problems which will be discussed elsewhere. In many examples, however, these Markovian kernels can be explicitly constructed and, at least on a subset of the states, have good support and measurability properties.

Definition 1.1. Let $\mathcal{A}_1$, $\mathcal{A}_2$ be C*-algebras and let $\mathcal{A}_1 \otimes \mathcal{A}_2$ be a fixed C*-tensor product of $\mathcal{A}_1$ and $\mathcal{A}_2$. A **lifting** from $\mathcal{A}_1$ to $\mathcal{A}_1 \otimes \mathcal{A}_2$ is a $w*$-continuous map

$$\mathcal{E}^*\colon\ \mathcal{S}(\mathcal{A}_1) \to \mathcal{S}(\mathcal{A}_1 \otimes \mathcal{A}_2). \tag{1.1}$$

If $\mathcal{E}^*$ is affine and its dual is a completely positive map, we call it a linear lifting; if it maps pure states into pure states, we call it **pure**.

Remark. Also in the nonlinear case some kind of *complete positivity requirement* should be included in the definition of lifting. However, the theory of nonlinear completely positive maps is still in its infancy but some is true for a satisfactory *dualization* of it. Therefore we leave the general question open for further developments and we limit ourselves to presenting some examples of nonlinear liftings which are of some interest for the applications.

To every lifting from $\mathcal{A}_1$ to $\mathcal{A}_1 \otimes \mathcal{A}_2$ we can associate two channels: one from $\mathcal{A}_1$ to $\mathcal{A}_1$, defined by

$$\Lambda_1^*\rho_1(a_1) := (\mathcal{E}^*\rho_1)(a_1 \otimes 1); \qquad \forall a_1 \in \mathcal{A}_1, \tag{1.2}$$

and another from $\mathcal{A}_1$ to $\mathcal{A}_2$, defined by

$$\Lambda_2^*\rho_1(a_2) := (\mathcal{E}^*\rho_1)(1 \otimes a_2); \qquad \forall a_2 \in \mathcal{A}_2. \tag{1.3}$$

In general, a state $\varphi \in \mathcal{S}(\mathcal{A}_1 \otimes \mathcal{A}_2)$ such that

$$\varphi|_{\mathcal{A}_1 \otimes 1} = \rho_1; \qquad \varphi|_{1 \otimes \mathcal{A}_2} = \rho_2 \tag{1.4}$$

has been called [21], [24] a **compound state** of the states $\rho_1 \in \mathcal{S}(\mathcal{A}_1)$ and $\rho_2 \in \mathcal{S}(\mathcal{A}_2)$. In classical probability theory, the term **coupling** between ρ_1 and ρ_2 is also used [13].

The following problem is important in several applications: Given a state $\rho_1 \in \mathcal{S}(\mathcal{A}_1)$ and a channel Λ^*: $\mathcal{S}(\mathcal{A}_1) \to \mathcal{S}(\mathcal{A}_2)$, find a standard lifting $\mathcal{E}^*$: $\mathcal{S}(\mathcal{A}_1) \to \mathcal{S}(\mathcal{A}_1 \otimes \mathcal{A}_2)$ such that $\mathcal{E}^*\rho_1$ is a compound state of ρ_1 and $\Lambda^*\rho_1$. Several particular solutions of this problem have been proposed in [9]–[11] and [21]–[24], however, an explicit description of all the possible solutions to this problem is still missing.

Definition 1.2. A lifting from $\mathcal{A}_1$ to $\mathcal{A}_1 \otimes \mathcal{A}_2$ is called **nondemolition** for a state $\rho_1 \in \mathcal{S}(\mathcal{A}_1)$ if ρ_1 is invariant for Λ_1^*, i.e., if, for all $a_1 \in \mathcal{A}_1$,

$$(\mathcal{E}^*\rho_1)(a_1 \otimes 1) = \rho_1(a_1). \tag{1.5}$$

The idea of this definition being that the interaction with system 2 does not alter the state of system 1.

Definition 1.3. Let $\mathcal{A}_1$, $\mathcal{A}_2$ be C*-algebras and let $\mathcal{A}_1 \otimes \mathcal{A}_2$ be a fixed C*-tensor product of $\mathcal{A}_1$ and $\mathcal{A}_2$. A **transition expectation** from $\mathcal{A}_1 \otimes \mathcal{A}_2$ to $\mathcal{A}_1$ is a completely positive linear map $\mathcal{E}$: $\mathcal{A}_1 \otimes \mathcal{A}_2 \to \mathcal{A}_1$ satisfying

$$\mathcal{E}(1_{\mathcal{A}_1} \otimes 1_{\mathcal{A}_2}) = 1_{\mathcal{A}_1}. \tag{1.6}$$

Remark. The notion of *nondemolition lifting* discussed here is essentially (i.e., up to minor technicalities) included in the more abstract notion of **state extension** introduced by Cecchini and Petz [11] (see also [10]).

The two interpretations of the notion of standard lifting, which are used in this paper, are the following:

(1) **The measurement process.** $\mathcal{A}_1$ (resp. $\mathcal{A}_2$) is interpreted as the algebra of observables of a system (resp. a measurement apparatus) and $\mathcal{E}^*$ describes the interaction between system and apparatus as well as the preparation of the apparatus. If $\rho_1 \in \mathcal{S}(\mathcal{A}_1)$ is the preparation of the system, i.e., its state before the interaction with the apparatus, then $\Lambda_1^*\rho_1 \in \mathcal{S}(\mathcal{A}_1)$ (resp. $\Lambda_2^*\rho_1 \in \mathcal{S}(\mathcal{A}_2)$) is the state of the system (resp. of the apparatus) after the measurement.

(2) **The signal transmission process.** An input signal is transmitted and received by an apparatus which produces an output signal. Here $\mathcal{A}_1$ (resp. $\mathcal{A}_2$) is interpreted as the algebra of observables of the input (resp. output) signal and $\mathcal{E}^*$ describes the interaction between the input signal and the receiver as well as the preparation of the receiver. If $\rho_1 \in \mathcal{S}(\mathcal{A}_1)$ is the input signal, then the state $\Lambda_2^*\rho_1 \in \mathcal{S}(\mathcal{A}_2)$, defined by (1.3), is the state of the (observed) output signal.

    An important lifting related to this signal transmission is one due to a quantum communication process discussed below (Examples 1a and 4).

In several important applications the state ρ_1 of the system before the interaction (preparation, input signal) is not known and one would like to know this state knowing only $\Lambda_2^*\rho_1 \in \mathcal{S}(\mathcal{A}_2)$, i.e., the state of the apparatus after the interaction (output signal). From a mathematical point of view this problem is not well posed, since usually the map Λ_2^* is not invertible. The best one can do in such cases is to acquire a control on the description of those input states which have the same image under Λ_2^* and then choose among them according to some statistical criterion. Another widely applied procedure is to postulate, on the basis of some experimental information, that the input state belongs to an a priori given restricted class of states and to choose among these ones by some statistical criterion.

In the following we describe several examples of liftings which appear frequently in the applications.

Example 1 (Isometric Liftings). Let $V\colon \mathcal{H}_1 \to \mathcal{H}_1 \otimes \mathcal{H}_2$ be an isometry

$$V^*V = 1_{\mathcal{H}_1}. \tag{1.7}$$

Then the map

$$\mathcal{E}\colon x \in \mathcal{B}(\mathcal{H}_1) \otimes \mathcal{B}(\mathcal{H}_2) \to V^*xV \in \mathcal{B}(\mathcal{H}_1) \tag{1.8}$$

is a transition expectation, and the associated lifting maps a density matrix $w_1 \in T(\mathcal{H}_1)$ into $\mathcal{E}^*w_1 = Vw_1V^*$. Liftings of this type are called isometric. Every isometric lifting is a pure lifting. Isometric liftings have turned out to play a relevant role in some mathematical models for superconductivity [14].

Example 1a (The Attenuation (or Beam Splitting) Lifting). It is the particular isometric lifting characterized by the properties

$$\mathcal{H}_1 = \mathcal{H}_2 =: \Gamma(\mathbf{C}) = \text{the Fock space over } \mathbf{C}.$$

$$V\colon \Gamma(\mathbf{C}) \to \Gamma(\mathbf{C}) \otimes \Gamma(\mathbf{C}) \tag{1.9}$$

is characterized by the expression

$$V|\theta\rangle = |\alpha\theta\rangle \otimes |\beta\theta\rangle, \tag{1.10}$$

where $|\theta\rangle$ is the normalized coherent vector parametrized by $\theta \in \mathbf{C}$ and $\alpha, \beta \in \mathbf{C}$ are such that

$$|\alpha|^2 + |\beta|^2 = 1. \tag{1.11}$$

Notice that this liftings maps coherent states into products of coherent states. So it maps the simplex of the so-called **classical states** (i.e., the convex combinations of coherent vectors) into itself. Restricted to these states it is of convex product type in the sense of Definition 2.1 below, but it is not of convex product type on the set of all states.

Denoting, for $\theta \in \mathbf{C}$, ω_θ as the state on $\mathcal{B}(\Gamma(\mathbf{C}))$ defined by

$$\omega_\theta(b) = \langle\theta, b\theta\rangle; \qquad b \in \mathcal{B}(\Gamma(\mathbf{C})), \tag{1.12}$$

we see that, for any $b \in \mathcal{B}(\Gamma(\mathbf{C}))$,

$$(\mathcal{E}^*\omega_\theta)(b \otimes 1) = \omega_{\alpha\theta}(b), \tag{1.13}$$

hence this lifting is not nondemolition.

Interpretation. $\Gamma(\mathbf{C})$ is the space of a one-mode EM field (signal). V represents the interaction, of the signal with an apparatus (e.g., a receiver or a semitransparent mirror). In $\Gamma(\mathbf{C}) \otimes \Gamma(\mathbf{C})$ the second factor is the space of the apparatus.

Equation (1.10) means that, by the effect of the interaction, a coherent signal (beam) $|\theta\rangle$ splits into two signals (beams) still coherent, but of lower intensity. Because of (1.11), the total intensity (energy) is preserved by the transformation.

Example 1b (Superposition Beam Splitting). The only difference with Example 1a is the form of V, which in this case is

$$V|\theta\rangle = \frac{1}{\sqrt{2}}(|\alpha\theta\rangle \otimes |\beta\theta\rangle - i|\beta\theta\rangle \otimes |\alpha\theta\rangle). \tag{1.14}$$

One easily checks that V extends linearly to an isometry of the form (1.9).

This isometric lifting is not of convex product type in the sense of Definition 2.1 of the next section, neither it is a nondemolition lifting.

Example 2 (Compound Lifting). Let $\Lambda^*\colon \mathcal{S}(\mathcal{A}_1) \to \mathcal{S}(\mathcal{A}_2)$ be a channel. For any $\rho_1 \in \mathcal{S}(\mathcal{A}_2)$ in the closed convex hull of the external states, fix a decomposition of ρ_1 as a convex combination of extremal states in $\mathcal{S}(\mathcal{A}_1)$:

$$\rho_1 = \int_{\mathcal{S}(\mathcal{A}_1)} \omega_1 p(d\omega_1 \mid \rho_1), \tag{1.15}$$

where $p(\cdot \mid \rho_1)$ is a Borel measure on $\mathcal{S}(\mathcal{A}_1)$ with support in the extremal states, and define

$$\mathcal{E}^*\rho_1 := \int_{\mathcal{S}(\mathcal{A}_1)} \omega_1 \otimes \Lambda^*\omega_1 p(d\omega_1 \mid \rho_1). \tag{1.16}$$

Then $\mathcal{E}^*$: $\mathcal{S}(\mathcal{A}_1) \to \mathcal{S}(\mathcal{A}_2 \otimes \mathcal{A}_2)$ is a lifting, nonlinear even if Λ^* is linear, and non-demolition for ρ_1. In Section 2 we shall see that the most general lifting, mapping $\mathcal{S}(\mathcal{A}_1)$ into the closed convex hull of the external product states on $\mathcal{A}_1 \otimes \mathcal{A}_2$, is essentially of this type. Here "essentially" means that, in order to recover the most general case, we shall weaken, from the original definition of compound state in [22], the condition that $p(d\omega_1 \mid \rho_1)$ is concentrated on the extremal states.

Therefore once a channel is given, a lifting of convex product type can be constructed by (1.16), and the converse is also true due to (1.3):

channel $\leftrightarrow$ **lifting**.

For example, the von Neumann quantum measurement process is written, in our terminology, as follows: Having measured an observable $A = \sum_n a_n P_n$ (spectral decomposition with discrete spectrum and $\sum_n P_n = I$) in a state ρ, the state after this measurement will be

$$\Lambda^*\rho = \sum_n P_n \rho P_n$$

and a lifting $\mathcal{E}^*$, of convex product type, associated to this channel Λ^* and to a fixed decomposition of ρ as $\rho = \sum_n \lambda_n \rho_n$ ($\rho_n \in \mathcal{S}(\mathcal{A}_1)$) is given by

$$\mathcal{E}^*\rho = \sum_n \lambda_n \rho_n \otimes \Lambda^*\rho_n. \tag{1.17}$$

A more sophisticated example of lifting of this type is a reduction of a state associated with an open system dynamics. Namely, if a system Σ_1 described by a Hilbert space $\mathcal{H}$ interacts with an external system Σ_2 described by another Hilbert space $\mathcal{K}$ and the initial states of Σ_1 and Σ_2 are ρ and σ, respectively, then the combined state θ_t of Σ_1 and Σ_2 at time t after the interaction between the two systems is given by

$$\theta_t = U_t^*(\rho \otimes \sigma)U_t,$$

where $U_t = \exp(itH)$ with the total Hamiltonian H of Σ_1 and Σ_2. A channel is obtained by taking the partial trace with respect to $\mathcal{K}$, i.e.,

$$\rho \to \Lambda_t^*\rho = \mathrm{tr}_{\mathcal{K}}\, \theta_t.$$

A lifting associated to this channel is given by (1.17) with Λ_t^* above.

Example 3 (Canonical Form of a Lifting). Let $\mathcal{A}_1 = \mathcal{B}(\mathcal{H}_1)$, $\mathcal{A}_2 = \mathcal{B}(\mathcal{H}_2)$. The most general linear lifting $\mathcal{E}^*$: $\mathcal{S}(\mathcal{A}_1) \to \mathcal{S}(\mathcal{A}_1 \otimes \mathcal{A}_2)$ has the form

$$w_1 \in T(\mathcal{H}_1) \to \sum K_i(w_1 \otimes 1)K_i^* \in T(\mathcal{H}_1 \otimes \mathcal{H}_2) \tag{1.18}$$

for some $K_i \in \mathcal{B}(\mathcal{H}_1 \otimes \mathcal{H}_2)$ such that

$$\sum K_i^* K_i = 1. \tag{1.19}$$

This is a simple consequence of Krein's lemma.

Example 4 (The Lifting for a Quantum Communication Channel). Let $\mathcal{H}_1$, $\mathcal{H}_2$, $\mathcal{K}_1$, $\mathcal{K}_2$ be Hilbert spaces. Denote by α the amplification

$$\alpha\colon b_2 \in \mathcal{B}(\mathcal{H}_2) \to \alpha(b_2) = b_2 \otimes 1_{\mathcal{K}_2} \in \mathcal{B}(\mathcal{H}_2 \otimes \mathcal{K}_2). \tag{1.20}$$

Let

$$\gamma\colon \mathcal{B}(\mathcal{H}_2 \otimes \mathcal{K}_2) \to \mathcal{B}(\mathcal{H}_1 \otimes \mathcal{K}_1)$$

be a completely positive identity preserving map and, for $\sigma_1 \in \mathcal{S}(\mathcal{B}(\mathcal{K}_1))$, denote by $\bar{\sigma}_1^{(2)}$ the conditional expectation

$$\bar{\sigma}_1^{(2)}\colon a_1 \otimes b_1 \in \mathcal{B}(\mathcal{H}_1) \otimes \mathcal{B}(\mathcal{K}_1) \to a_1\sigma_1(b_1) \in \mathcal{B}(\mathcal{H}_1) \cong \mathcal{B}(\mathcal{H}_1) \otimes 1_{\mathcal{K}_1}. \tag{1.21}$$

Then the lifting and the channel describing quantum communication processes are defined by

$$\mathcal{E}^* = \gamma^* \circ \bar{\sigma}_1^{(2)*}, \tag{1.22}$$

$$\Lambda^*\rho = \alpha^* \circ \mathcal{E}^*(\rho) = \mathrm{tr}_{\mathcal{K}_2}\, \gamma^*(\rho \otimes \sigma_1); \qquad \rho \in \mathcal{S}(\mathcal{B}(\mathcal{H}_1)), \tag{1.23}$$

where ρ and σ_1 correspond to an input state and a noise state, respectively (see [24]).

Moreover, the following remark, extending an unpublished result of A. Frigerio, shows that the above model of the quantum communication process is universal among the transition expectations, provided one chooses the space of the representation large enough.

Theorem 1.4. *Let $\mathcal{B} = \mathcal{B}(\mathcal{H})$ for a separable, infinite-dimensional, Hilbert space $\mathcal{H}$ and let $\mathcal{E}\colon \mathcal{B} \otimes \mathcal{B} \to \mathcal{B}$ be a normal transition expectation. Then there exist a normal state φ on the* W*-*algebra* $(\otimes\mathcal{B})^3 \otimes \mathbf{M}_2 =: \mathcal{C}$ ($\mathbf{M}_2$ *is the algebra of* 2×2 *complex matrices), and a unitary element U of*

$$\mathcal{A} := \mathcal{C} \otimes \mathcal{B} \cong (\otimes\mathcal{B})^4 \otimes \mathbf{M}_2 \tag{1.24}$$

such that, denoting $\pi\colon \mathcal{B} \otimes \mathcal{B} \to \mathcal{A}$ as the normal representation (amplification)

$$\pi(x) = \begin{pmatrix} 1_{\mathcal{B}^2} \otimes x & 0 \\ 0 & 1_{\mathcal{B}^2} \otimes x \end{pmatrix} \tag{1.25}$$

and $\bar{\varphi}^{(1)} := \varphi \otimes \iota_{\mathcal{B}} : \mathcal{A} \to \mathcal{B} \cong 1_{\mathcal{C}} \otimes \mathcal{B}$ as the conditional expectation ($\iota_{\mathcal{B}}$ is the identical map on $\mathcal{B}$)

$$(\varphi \otimes \iota_{\mathcal{B}})(c \otimes b) = \varphi(c)1 \otimes b, \qquad c \in \mathcal{C}, \quad b \in \mathcal{B}, \tag{1.26}$$

one has, for all $x \in \mathcal{B} \otimes \mathcal{B}$,

$$\mathcal{E}(x) = (\varphi \otimes \iota_{\mathcal{B}})(U^*\pi(x)U). \tag{1.27}$$

Proof. From Kraus' lemma [18] we know that, since $\mathcal{E}$ is normal, there exist a countable (since $\mathcal{H}$ is separable family) $a_i \in \mathcal{B} \otimes \mathcal{B}$ $(i \in N)$ such that, identifying $\mathcal{B}$ with $\mathcal{B} \otimes 1$, one has

$$\mathcal{E}(x) = \sum_{i \in N} a_i^* x a_i; \qquad x \in \mathcal{B} \otimes \mathcal{B}.$$

If $\mathcal{H}$ is infinite-dimensional in $\mathcal{B} \otimes \mathcal{B}$ there exist isometries u_i $(i \in N)$ such that, for each $i, j, u_i^* u_j = \delta_{ij}$. Thus, defining

$$v := \sum_{i=1}^{d} u_i \otimes a_i \in (\otimes \mathcal{B})^4, \tag{1.28}$$

one finds

$$v^*(1_{\mathcal{B}^2} \otimes x)v = 1_{\mathcal{B}^2} \otimes \mathcal{E}(x), \qquad x \in \mathcal{B} \otimes \mathcal{B}. \tag{1.29}$$

In particular, since $v^* v = 1_{\mathcal{B}^4}$, v is a partial isometry with initial projection as the identity. Denote $e = vv^*$ as its final projection and define the unitary operator

$$U = \begin{pmatrix} v & 1-e \\ 0 & v^* \end{pmatrix} : (\otimes H)^4 \otimes \mathbf{C}^2 \cong (\otimes H)^4 \oplus (\otimes H)^4 \to (\otimes H)^4 \otimes \mathbf{C}^2. \tag{1.30}$$

Then for any vector $\eta \in H$ and for any fixed unit vector $\Phi \in (\otimes H)^3$ one finds

$$\langle \eta, \mathcal{E}(x)\eta \rangle = \langle \Phi \otimes \eta, (1_{\mathcal{B}^3} \otimes \mathcal{E}(x))\Phi \otimes \eta \rangle = \langle \Phi \otimes \eta, (v^*(1_{\mathcal{B}^2} \otimes x)v)\Phi \otimes \eta \rangle. \tag{1.31}$$

Using the notation

$$\xi = \Phi \otimes \eta \in (\otimes H)^4 \tag{1.32}$$

the right-hand side of (1.31) can be rewritten as

$$\begin{aligned} \langle v\xi, (1_{\mathcal{B}^2} \otimes x)v\xi \rangle &= \left\langle \begin{pmatrix} v\xi \\ 0 \end{pmatrix}, \begin{pmatrix} 1_{\mathcal{B}^2} \otimes x & 0 \\ 0 & 1_{\mathcal{B}^2} \otimes x \end{pmatrix} \begin{pmatrix} v\xi \\ 0 \end{pmatrix} \right\rangle \\ &= \left\langle \begin{pmatrix} \xi \\ 0 \end{pmatrix}, U^* \pi(x) U \begin{pmatrix} \xi \\ 0 \end{pmatrix} \right\rangle \end{aligned} \tag{1.33}$$

and, using Kraus' representation and the identifications (1.30) and (1.32),

$$\begin{pmatrix} \xi \\ 0 \end{pmatrix} = \begin{pmatrix} \Phi \otimes \eta \\ 0 \end{pmatrix} \cong \begin{pmatrix} \Phi \\ 0 \end{pmatrix} \otimes \eta.$$

The right-hand side of (1.33) then becomes equal to

$$\left\langle \begin{pmatrix} \Phi \\ 0 \end{pmatrix} \otimes \eta, U^* \pi(x) U \begin{pmatrix} \Phi \\ 0 \end{pmatrix} \otimes \eta \right\rangle. \tag{1.34}$$

Thus, defining the state φ on $(\mathcal{B}^3) \otimes M(2, \mathbf{C})$ by

$$\varphi(y) = \left\langle \begin{pmatrix} \Phi \\ 0 \end{pmatrix}, y \begin{pmatrix} \Phi \\ 0 \end{pmatrix} \right\rangle; \qquad y \in (\mathcal{B})^3 \otimes \mathbf{C}^2, \tag{1.35}$$

from (1.31), (1.33), (1.34), and (1.35) we obtain that, for each $\eta \in H$,

$$\langle \eta, \mathcal{E}(x)\eta \rangle = \langle \eta, (\varphi \otimes \iota_{\mathcal{B}}) \left(U^*\pi(x)U \right) \eta \rangle,$$

and this is equivalent to (1.27). □

Remark. If $\mathcal{H}$ is finite-dimensional, then the identity (1.27) still holds, but on a different algebra and with a slightly less explicit form of the representation π. The argument which can be applied also to the infinite-dimensional case, if $\mathcal{E}$ is faithful, is the following: by Stinespring's theorem, $\mathcal{E}$ has the form

$$\mathcal{E}(x) = V^*{}_0\pi_0(x)V_0; \qquad x \in \mathcal{B} \otimes \mathcal{B}, \tag{1.36}$$

where $\mathcal{K}_0$ is a (separable) Hilbert space and π_0 is a faithful normal (since $\mathcal{E}$ is) representation of $\mathcal{B} \otimes \mathcal{B}$ into $\mathcal{B}(\mathcal{K}_0)$. Being faithful and normal, π_0 must be (isomorphic to) a multiple of the identity representation (see Section 3 of [18]). Therefore we may assume that $\mathcal{K}_0$ has the form $\mathcal{K}_1 \otimes \mathcal{H}$ and

$$\pi_0(1 \otimes b) = 1_{\mathcal{K}_1} \otimes b; \qquad b \in \mathcal{B}. \tag{1.37}$$

From (1.37) it follows that the restriction of π_0 on $\mathcal{B} \otimes 1$ has the form

$$\pi_0(b \otimes 1_{\mathcal{B}}) = \pi_1(b) \otimes 1_{\mathcal{H}}; \qquad b \in \mathcal{B}, \tag{1.38}$$

for some representation π_1 of $\mathcal{B}$ into $\mathcal{B}(\mathcal{K}_1)$. Define now a contraction $W^*\colon \mathcal{K}_0 \to \mathcal{K}_0 = \mathcal{K}_1 \otimes \mathcal{H}$, by

$$W^*g = u \otimes V_0^*g; \qquad g \in \mathcal{K}_0, \tag{1.39}$$

where $u \in \mathcal{K}_1$ is a unit vector fixed arbitrarily. Denoting W as the adjoint of W^*, one easily verifies that

$$W(f \otimes h) = \langle u, f \rangle V_0 h, \tag{1.40}$$

so W is a partial isometry with initial and final projections given respectively by

$$W^*W(f \otimes h) = \langle u, f \rangle u \otimes h =: P(f \otimes h), \tag{1.41}$$

$$WW^*(g) = V_0V_0^*g =: Q(g). \tag{1.42}$$

Denote $\mathcal{K}_2 := \mathcal{K}_0 \otimes \mathbf{C}^2 \cong \mathcal{K}_0 \oplus \mathcal{K}_0$;

$$U := \begin{pmatrix} W & 1-Q \\ 1-P & W^* \end{pmatrix}, \tag{1.43}$$

the Riesy–Neagy unitary dialation of the partial isometry W, and using the identifications

$$\mathcal{K}_2 := \mathcal{K}_0 \otimes \mathbf{C}^2 = \mathcal{K}_1 \otimes \mathcal{H} \otimes \mathbf{C}^2 = \mathcal{K}_1 \otimes \mathbf{C}^2 \otimes \mathcal{H} = (\mathcal{K}_1 \oplus \mathcal{K}_1) \otimes \mathcal{H}$$

denote, for $b \in \mathcal{B}$,

$$v := \begin{pmatrix} u \\ 0 \end{pmatrix} \in \mathcal{K}_1 \oplus \mathcal{K}_1; \quad \pi(b) := \begin{pmatrix} \pi_1(b) & 0 \\ 0 & \pi_1(b) \end{pmatrix} \in \mathcal{B}(\mathcal{K}_1 \oplus \mathcal{K}_1), \tag{1.44}$$

$$\varphi(\cdot) := \langle v, (\cdot)v \rangle.$$

Then π is a representation of $\mathcal{B}$ in $\mathcal{B}(\mathcal{K}_1 \oplus \mathcal{K}_1)$ and φ is a state on $\mathcal{K}_1 \oplus \mathcal{K}_1$ such that, for any $b, b' \in \mathcal{B}$ and $\xi, \eta \in \mathcal{H}$, one has, using (1.40), (1.43), and the fact that any vector of the form $u \otimes \xi$ is in the range of P,

$$\begin{aligned} &\langle \xi, \varphi \otimes i_{\mathcal{B}}(U^*(\pi(b) \otimes b')U)\eta \rangle \\ &\quad = \langle v \otimes \xi, U^*(\pi(b) \otimes b')Uv \otimes \eta \rangle \\ &\quad = \left\langle \begin{pmatrix} V_0\xi \\ 0 \end{pmatrix}, \begin{pmatrix} \pi_1(b) \otimes b' & 0 \\ 0 & \pi_1(b) \otimes b' \end{pmatrix} \begin{pmatrix} V_0\eta \\ 0 \end{pmatrix} \right\rangle \\ &\quad = \langle \xi, V^*{}_0(\pi_1(b) \otimes b')V_0\eta \rangle = \langle \xi, \mathcal{E}(b \otimes b')\eta \rangle \end{aligned}$$

and this proves (1.27) with π given by (1.44).

Remark. The relevance of the above result for quantum probability is due to the fact that it allows us to prove the essential equivalence of different notions of quantum Markov chains.

2. Convex Combinations of Product States

One of the main differences between classical and quantum probability is that while all the measures on a product space are in the closed convex hull (for the weak topology) of product measures, it is not true that all the states on the tensor product $\mathcal{A}_1 \otimes \mathcal{A}_2$ of two general C*-algebras are the limits (in some topology) of convex combinations of product states.

In particular, the image under a general lifting $\mathcal{E}^*$ of a state φ will usually not be a convex combination of product states.

However, the class of liftings with this property is particulary interesting because we expect that in this class some features of quantum probability will mix with some features of classical probability. This class is defined as follows:

Definition 2.1. Let $\mathcal{A}_1$ and $\mathcal{A}_2$ be W*-algebras. A lifting $\mathcal{E}^*$: $\mathcal{S}(\mathcal{A}_1) \to \mathcal{S}(\mathcal{A}_1 \otimes \mathcal{A}_2)$ is called **of convex product type**, or a **convex product lifting**, if any state $\omega \in \mathcal{S}(\mathcal{A}_1)$ is mapped by $\mathcal{E}^*$ into a convex combination of product states on $\mathcal{A}_1 \otimes \mathcal{A}_2$. If this property holds only for any state ω in a subset $\mathcal{F} \subseteq \mathcal{S}(\mathcal{A}_1)$, then $\mathcal{E}^*$ is called a convex product lifting with respect to the family $\mathcal{F}$.

For any von Neumann algebra $\mathcal{A}$, the set $\mathcal{S}(\mathcal{A})$ of all its states has a natural structure of measurable space with its Borel σ-algebra. In the sequel, any probability measure on $\mathcal{S}(\mathcal{A})$ will be meant with respect to this σ-algebra.

Definition 2.2. A **convex decomposition** of $\varphi \in \mathcal{S}(\mathcal{A})$ is a probability measure μ on $\mathcal{S}(\mathcal{A})$ satisfying

$$\varphi = \int_{\mathcal{S}(\mathcal{A})} \omega \, d\mu(\omega). \tag{2.1}$$

If μ is pseudosupported, in the sense of [8], in the set of extremal states of $\mathcal{S}(\mathcal{A})$, we speak of an **extremal convex decomposition** of φ.

Proposition 2.1. *To every lifting of convex product type $\mathcal{E}^*\colon \mathcal{S}(\mathcal{A}_1) \to \mathcal{S}(\mathcal{A}_1 \otimes \mathcal{A}_2)$, one can associate a pair*

$$\{p_\rho(d\omega_1),\ p_\rho(d\omega_2 \mid \omega_1)\} \tag{2.2}$$

with the following properties:

(i) $p_\rho(d\omega_1)$ *is a probability measure on* $\mathcal{S}(\mathcal{A}_1)$.
(ii) $p_\rho(d\omega_2 \mid \omega_1)$ *is a Markovian kernel from* $\mathcal{S}(\mathcal{A}_1)$ *to* $\mathcal{S}(\mathcal{A}_2)$.

Conversely every pair (2.2) *satisfying* (i) *and* (ii) *above determines, via* (2.4) *and* (2.5), *a unique convex product lifting.*

Proof. For $\mathcal{E}^*$ as in Definition 2.1, we fix a state $\rho_1 \in \mathcal{S}(\mathcal{A}_1)$ and also a decomposition of $\mathcal{E}^*\rho_1$ as a convex combination of product states

$$\mathcal{E}^*\rho_1 = \int_{\mathcal{S}(\mathcal{A}_1)\times\mathcal{S}(\mathcal{A}_2)} \omega_1 \otimes \omega_2 \, dp(\omega_1, \omega_2 \mid \rho_1). \tag{2.3}$$

Denoting $p_{\rho_1}(d\omega_2 \mid \omega_1)$ as the conditional probability of $p(\,\cdot \mid \rho_1)$ on the σ-algebra of the first factor and $dp_{\rho_1}(\omega_1)$ as the marginal of $p(\,\cdot \mid \rho_1)$ on the first factor, we obtain

$$\begin{aligned}\mathcal{E}^*\rho_1 &= \int_{\mathcal{S}(\mathcal{A}_1)} \int_{\mathcal{S}(\mathcal{A}_2)} \omega_1 \otimes \omega_2 \, dp_{\rho_1}(\omega_1) p_{\rho_1}(d\omega_2|\omega_1) \\ &= \int_{\mathcal{S}(\mathcal{A}_1)} \omega_1 \otimes \Lambda^*_\rho \omega_1 \, dp_{\rho_1}(\omega_1),\end{aligned} \tag{2.4}$$

$$\Lambda^*_{\rho_1}\omega_1 := \int_{\mathcal{S}(\mathcal{A}_2)} \omega_2 p_{\rho_1}(d\omega_2 \mid \omega_1). \tag{2.5}$$

Thus any lifting $\mathcal{E}^*\colon \mathcal{S}(\mathcal{A}_1) \to \mathcal{S}(\mathcal{A}_1 \otimes \mathcal{A}_2)$, of convex product type, has the form (2.4), where p_{ρ_1} is a probability measure on $\mathcal{S}(\mathcal{A}_1)$ and the map $\Lambda^*_{\rho_1}\colon \mathcal{S}(\mathcal{A}_1) \to \mathcal{S}(\mathcal{A}_2)$ is given by (2.5). Notice that $\Lambda^*_{\rho_1}$ is a channel in the sense of Section 1 and is usually nonlinear both in ω_1 and ρ_1.

Conversely, given p_{ρ_1} and $\Lambda^*_{\rho_1}$ as above, if we define $\mathcal{E}^*$ by (2.3), then clearly $\mathcal{E}^*$ is a convex product lifting from $\mathcal{S}(\mathcal{A}_1)$ to $\mathcal{S}(\mathcal{A}_1 \otimes \mathcal{A}_2)$. Finally, it is clear that the map

$$(\omega_1, S_2) \in \mathcal{S}(\mathcal{A}_1) \times \mathrm{Borel}(\mathcal{S}(\mathcal{A}_2)) \to p_{\rho_1}(S_2|\omega_1) \in [0, 1]$$

is a classical Markovian kernel on the Borel space $\mathcal{S}(\mathcal{A}_1) \times \mathcal{S}(\mathcal{A}_2)$. □

Remark. If in (2.3) one conditions on the σ-algebra of the second factor rather than on the first one, the resulting lifting is

$$\mathcal{E}^*\rho_1 = \int_{\mathcal{S}(\mathcal{A}_1)}\int_{\mathcal{S}(\mathcal{A}_2)} \omega_1 \otimes \omega_2 \, dq_{\rho_1}(\omega_2)\, dq_{\rho_1}(d\omega_1 \mid \omega_2),$$

where $dq_{\rho_1}(\omega_2)$ is a probability measure on $\mathcal{S}(\mathcal{A}_2)$ and $dq_{\rho_1}(d\omega_1 \mid \omega_2)$ is a Markovian kernel from $\mathcal{S}(\mathcal{A}_2)$ to $\mathcal{S}(\mathcal{A}_1)$.

We now consider **the relation between the liftings of convex product type and Markov chains**.

The dual of a linear lifting is a transition expectation, therefore to any linear lifting one can associate a quantum Markov chain [2] in a standard way. If the lifting is of convex product type, then we can take advantage of this special structure to extend the construction of quantum Markov chains to the case of a not necessarily linear lifting $\mathcal{E}^*$. In what follows we describe this procedure.

If $\mathcal{E}^*\colon \mathcal{S}(\mathcal{A}_2) \to \mathcal{S}(\mathcal{A}_1 \otimes \mathcal{A}_2)$ is a lifting of convex product type, then it has the form

$$\mathcal{E}^*\rho_2 = \int_{\mathcal{S}(\mathcal{A}_1)}\int_{\mathcal{S}(\mathcal{A}_2)} \omega^1 \otimes \omega^2 p(d\omega^1, d\omega^2 \mid \rho_2). \tag{2.6}$$

Notice that $p(d\omega^1, d\omega^2 \mid \rho_2)$ can be considered as a Markovian kernel on the space

$$\mathcal{S}_{12} := \mathcal{S}(\mathcal{A}_1) \times \mathcal{S}(\mathcal{A}_2),$$

which is constant on the first conditioning, i.e.,

$$\begin{aligned} &p(d\omega^1, d\omega^2 \mid \rho_1, \rho_2) = p(d\omega^1, d\omega^2 \mid \rho_2); \\ &\quad \omega_1, \rho_1, \in \mathcal{S}(\mathcal{A}_1), \quad \omega_2, \rho_2 \in \mathcal{S}(\mathcal{A}_2). \end{aligned} \tag{2.7}$$

Clearly, (2.6) is a state on $\mathcal{A}_1 \otimes \mathcal{A}_2$. If we apply $\mathcal{E}^*$ to ω^2 in (2.6), we obtain the following state on $(\mathcal{A}_1 \otimes \mathcal{A}_2) \otimes \mathcal{A}_2$:

$$\begin{aligned} &\int_{\mathcal{S}_{12}} p(d\omega_1^1, d\omega_1^2, \mid \rho_2)\omega_1^1 \otimes \mathcal{E}^*\omega_1^2 \\ &\quad = \int_{\mathcal{S}_{12}}\int_{\mathcal{S}_{12}} p(d\omega_1^1, d\omega_1^2 \mid \rho_2) p(d\omega_2^1, d\omega_2^2 \mid \omega_1^2)\omega_1^1 \otimes \omega_2^1 \otimes \omega_2^2, \end{aligned}$$

where

$$\omega_i^1 \in \mathcal{S}(\mathcal{A}_1) \quad \text{and} \quad \omega_i^2 \in \mathcal{S}(\mathcal{A}_2).$$

Applying again $\mathcal{E}^*$ to ω_2^2 we find

$$\int_{S_{12}}\int_{S_{12}}\int_{S_{12}} p(d\omega_1^1, d\omega_1^2 \mid \rho_2) p(d\omega_2^1, d\omega_2^2 \mid \omega_1^2) p(d\omega_3^1, d\omega_3^2 \mid \omega_2^2)\omega_1^1 \otimes \omega_2^1 \otimes \omega_3^1 \otimes \omega_3^2.$$

At the nth iteration we obtain the state $\mathcal{E}_{n]}^*\rho_2$ on $(\otimes \mathcal{A}_1)^n \otimes \mathcal{A}_2$, defined by

$$\mathcal{E}_{n]}^*\rho_2 := \int_{S_{12}^n} \left(\otimes_{i=1}^n \omega_i^1\right) \otimes \omega_n^2 \cdot \Pi_{i=2}^n p(d\omega_i^1, d\omega_i^2 | \omega_{i-1}^2) p(d\omega_1^1, d\omega_1^2, |\rho_2). \tag{2.8}$$

This suggests introducing the classical Markov process

$$\xi_n := (\xi_n^1, \xi_n^2) : (\Omega, \mathcal{F}, P) \to \mathcal{S}(\mathcal{A}_1) \times \mathcal{S}(\mathcal{A}_2) = S_{12} \tag{2.9}$$

with the transition function given by (2.7) and initial distribution $p(\cdot|\rho_2)$. This transition probability has a nice interpretation in terms of signal and noise: if system 1 represents the noise and system 2 the signal, then condition (2.7) means that the joint distribution at time $(n+1)$ depends on the signal at time n, but not on the noise at time n: a natural assumption if we think of the noises at different times as generated by independent causes. Now let $\mathcal{A} := \otimes_{\mathbf{N}} \mathcal{A}_1$. The identification

$$a_1 \otimes \cdots \otimes a_n \cong a_1 \otimes a_2 \otimes \cdots \otimes a_n \otimes 1 \otimes 1 \otimes \cdots$$

induces a natural identification of $(\otimes \mathcal{A}_1)^n$ with a subalgebra $\mathcal{A}_{[2,n]}$ of $\mathcal{A} = \otimes_{\mathbf{N}} \mathcal{A}_1$ (the product of the first n-factors).

In particular, if $\rho_2 \in \mathcal{S}(\mathcal{A}_2)$ is a state on $\mathcal{A}_2$, the restriction of $\mathcal{E}_{n]}^*\rho_2$ on $(\otimes \mathcal{A}_1)^n$ is a state on $(\otimes^n \mathcal{A}_1)$ and, with the above identification, we can consider it a state $\rho_{[1,n]}$ on $\mathcal{A}$. Following from above, in particular (2.8), we obtain

Proposition 2.2. *For any $\rho_2 \in \mathcal{S}(\mathcal{A}_2)$ the limit*

$$\lim_{n\to\infty} \rho_{[1,n]} =: \varphi \tag{2.10}$$

exists pointwise weakly on $\mathcal{A}$. Moreover, if E_ξ denotes the mean with respect to the process $\{\xi_n\}$, defined by (2.9), *then one has*

$$\varphi = E_\xi \left(\otimes_{n\in\mathbf{N}} \xi_n^1\right). \tag{2.11}$$

3. Centralizer Liftings

In this section we introduce an interesting class of nonlinear liftings generalizing the construction of [22]. It is shown that the Cecchini–Petz notion of **state extension** [11], introduced after [22] and for totally independent reasons, is a generalization of our construction hence, a fortiori, of the one in [22].

Recall that a linear map E from a C*-algebra $\mathcal{A}$ to a C*-algebra $\mathcal{B}$ is called **anti-completely positive** if the map $\bar{E} : \mathcal{A} \to \mathcal{B}$, defined by

$$\bar{E}(a) := E(a^*); \qquad a \in \mathcal{A}, \tag{3.1}$$

is completely positive antilinear, i.e., for any natural integer n, any $a_1, \ldots, a_n \in \mathcal{A}$, and any $b_1, \ldots, b_n \in \mathcal{B}$, one has

$$\sum_{jk} b_j^* E(a_k^* a_j) b_k = \sum_{jk} b_j^* \bar{E}(a_j^* a_k) b_k \geq 0.$$

Proposition 3.1. *Let $\mathcal{A}_1, \mathcal{A}_2$ be W*-algebras, let $\mathcal{A}_1 \otimes^{(o)} \mathcal{A}_2$ denote their algebraic tensor product. For $\rho \in \mathcal{S}(\mathcal{A}_1)$ let $\mathcal{A}_1^\rho$ denote the centralizer of ρ and let $E\colon \mathcal{A}_2 \to \mathcal{A}_1^\rho$ be any anticompletely positive identity preserving linear map. Then there exists a unique state φ_ρ on $\mathcal{A}_1 \otimes^{(o)} \mathcal{A}_2$ such that*

$$\varphi_\rho(a_1 \otimes a_2) := \rho(a_1 E(a_2)); \qquad a_1 \in \mathcal{A}_1, \quad a_2 \in \mathcal{A}_2.$$

Proof. Let n be a natural integer and let $b_1, \ldots, b_n \in \mathcal{A}_1$ and $a_1, \ldots, a_n \in \mathcal{A}_2$. By assumption the $\mathcal{A}_1^\rho$-valued $n \times n$ matrix $B = (B_{kj})$ defined by

$$B_{kj} := E(a_j^* a_k) = \bar{E}(a_k^* a_j)$$

is of positive type, hence it has the form $B = M^* M$ for some $\mathcal{A}_1^\rho$-valued $n \times n$ matrix $M = (M_{kj})$. One has therefore

$$\begin{aligned} \varphi_\rho\left(\left|\sum_j a_j \otimes b_j\right|^2\right) &= \sum_{jk} \rho(b_j^* b_k E(a_j^* a_k)) \\ &= \sum_{jkh} \rho(b_j^* b_k M_{hk}^* M_{hj}) \\ &= \sum_{jkh} \rho(M_{hj} b_j^* b_k M_{hk}^*) \\ &= \sum_h \rho\left(\left[\sum_j b_j M_{hj}^*\right]^* \cdot \left[\sum_k b_k M_{hk}^*\right]\right) \geq 0. \end{aligned}$$
□

Remark. Clearly,

$$|\varphi_\rho(a_1 \otimes a_2)| \leq \|a_1\| \cdot \|a_2\|; \qquad a_1 \in \mathcal{A}_1, \quad a_2 \in \mathcal{A}_2,$$

hence φ_ρ is continuous for the greatest cross norm on $\mathcal{A}_1 \otimes^{(o)} \mathcal{A}_2$. Cecchini and Petz [12] have proved that it is also continuous for the smallest C*-norm [28]. (This is clear if the centralizer of ρ, i.e., $\mathcal{A}_1^\rho$, is abelian because in that case all the C*-norms on $\mathcal{A}_1 \otimes^{(o)} \mathcal{A}_2$ coincide with the minimal C*-norm [28, Proposition 1.22.5]. Moreover it is easy to check that the operator E, defined by (3.1) is an example of Cecchini's λ-operator [9]. In this case in fact the Tomita involution J_1 acts as the identity on the cyclic space of $\mathcal{A}_1^\rho$, the centralizer of $\mathcal{A}_1$, therefore the identity (3.1) is precisely the defining relation of the λ-operator.

If $\mathcal{A}_1^\rho$ is a discrete abelian algebra generated by a partition (e_j) of the identity, then any positive map E, from $\mathcal{A}_2$ to $\mathcal{A}_1^\rho$, is also completely and anticompletely positive and has the form

$$E(a_2) = \sum_j \varphi_j(a_2)e_j; \qquad a_2 \in \mathcal{A}_2, \tag{3.2}$$

with $\varphi_j \in \mathcal{S}(\mathcal{A}_2)$. In this case it is immediate to verify that

$$\varphi_\rho = \sum_j \rho_j \otimes \varphi_j, \tag{3.3}$$

where φ_j is given by (3.2) and

$$\rho_j := \rho(e_j(\,\cdot\,)e_j).$$

In general, whenever the state φ_ρ, defined by (3.3), is continuous, the map $\rho \mapsto \varphi_\rho$ defines a lifting $\mathcal{E}^*$ in the sense of Definition 1.1. This lifting is in general nonlinear since the map E in (3.1) may depend on ρ.

For example, if $\mathcal{A}_1$ is the algebra of all operators on some Hilbert space and ρ has the form $\rho = \operatorname{tr}(w\,\cdot\,)$ for some density matrix w with spectral decomposition given by

$$w = \sum_j p_j e_j, \tag{3.4}$$

then the centralizer $\mathcal{A}_1^\rho$ is the closed linear span of the (e_j) and if the φ_j are chosen to be of the form

$$\varphi_j = \Lambda^* \left(\frac{\rho_j}{\rho_j(1)} \right) \tag{3.5}$$

for some channel $\Lambda^*\colon \mathcal{S}(\mathcal{A}_1) \to \mathcal{S}(\mathcal{A}_2)$, then (3.3) becomes of the same form as (1.17) giving an example of nonlinear compound lifting.

4. Error Probability for Optical Communication

An optical communication process studied by several authors (see [21] for a mathematical analysis), the so-called **attenuation process**, can be described by the isometric lifting described in Example 1a. This description is simpler than the previous ones and allows quicker computations. This statement is illustrated with the computation of several error probabilities related to this model.

Before introducing these computations, we briefly review some basic facts about the notions of quantum coding and of error probability in quantum control communication processes along the lines of [24].

Suppose that, by some procedure, we encode an information representing it by a sequence of letters $c^{(1)}, \ldots, c^{(n)}, \ldots$, where $c^{(k)}$ is an element in a set C of symbols called the alphabet.

A **quantum code** is a map which associates to each symbol (or sequence of symbols) in C a quantum state, representing an optical signal. Sometimes one uses a state as two codes: one for input and one for output.

In what follows we only consider a two-symbol alphabet:

$$\mathcal{C} = \{0, 1\}. \tag{4.1}$$

One example of quantum code $\Xi = \{\xi_0, \xi_1\}$, where ξ_i is the quantum state corresponding to the symbol $i_i \in \mathcal{C}$, is obtained by choosing ξ_0 as the vacuum state and ξ_1 as another state such as a coherent or a squeezed state of a one-mode field.

Two states (quantum codes) $\xi_0^{(1)}$ and $\xi_1^{(1)}$ in the input system are transmitted to the output system through a channel Λ^*. We here assume a Z-type signal transmission, namely, that the input signal "0", represented by the state $\xi_0^{(1)}$, is error free in the sense that it always goes to the output signal "0" represented by $\xi_0^{(2)}$, while the input signal "1," represented by the state $\xi_1^{(1)}$, is not error free in the sense that its output can give rise to both states $\xi_0^{(2)}$ or $\xi_1^{(2)}$ with different probabilities.

The **error probability** q_e is then the probability that the input signal "1" is recognized as the output signal "0," so that it is given by

$$q_e = \text{tr}\, \Lambda^*(\xi_1^{(1)})\xi_0^{(2)}. \tag{4.2}$$

In the case of the quantum attenuation process, this error probability is written by using the attenuation operator V given in Example 1a with the construction (Example 4) of quantum lifting:

$$q_e = \text{tr}_{\mathcal{H}_2}\, \text{tr}_{\mathcal{K}_2}(V\xi_1^{(1)}V^*)\xi_0^{(2)}. \tag{4.3}$$

There are two main ways, called **pulse modulation**, to code the symbols of the alphabet $\mathcal{C}$. We briefly explain them for completeness. A pulse is an optical signal, represented by a nonvacuum state of the EM field; its energy is here called the **height** of the pulse. To a single symbol of the alphabet $\mathcal{C}$ one associates one or more pulses. Time is discretized and each time interval between t_k and t_{k+1} has length τ. Each time interval corresponds to a single symbol of the alphabet.

(1) **PCM (Pulse Code Modulation)** To the kth symbol a_k of the input sequence, one associates N pulses starting at a time t_k. The ordered set of these pulses is denoted x_k. For instance, for the alphabet $\{a_0, a_1\}$, for $N = 5$ and choosing the elementary pulses to be the vacuum (i.e., no pulse) denoted 0, and another fixed pulse, e.g., a coherent state, denoted 1, the code x_k corresponding to a_k is determined by $x_0 = (1, 0, 1, 0, 0)$, $x_1 = (0, 1, 1, 0, 1)$, and so on. For this modulation, we need N slots (sites) in one time interval (e.g., between t_k and t_{k+1}) to represent fully all M signals; $2^{N-1} < M \leq 2^N$.

(2) **PPM (Pulse Position Modulation)** In this case there is only one nonvacuum pulse in each time interval of length τ. The code x_k expressing a signal a_k is determined by the position of the nonvacuum pulse, so that we need M slots (sites) in each time interval in order to express M signals. For instance, in the same notations as above, $x_0 = (0, 0, 0, 1, 0)$, $x_2 = (0, 1, 0, 0, 0)$.

Given (4.2), the error probability of PCM with the t_0-tuple error correcting the following (4.3) and (4.4), respectively:

$$P_e^{\mathrm{PCM}} = \sum_{j=t_0+1}^{\nu} {}_\nu C_j q_e^j (1-q_e)^{\nu-j}, \tag{4.4a}$$

$$P_e^{\mathrm{PPM}} = q_e, \tag{4.4b}$$

where ${}_\nu C_j = \nu!/\{(\nu-j)!j!\}$.

The most general case for the computation of q_e is one where both $\xi_1^{(1)}$ and $\xi_0^{(2)}$ are squeezed states, but in usual optical communication it is often enough to take a coherent or squeezed state as $\xi_1^{(1)}$ and the vacuum state as $\xi_0^{(2)}$. Hence we first calculate the error probability q_e for the latter two cases and compare them with the results previously obtained in [24]. Secondly, we show the computation for the most general case, $\xi_1^{(1)}$ and $\xi_0^{(2)}$ squeezed, for mathematical interest and generality, although this somehow does not fit the assumption of our Z-type transmission.

Case I: $\xi_1^{(1)} = |\theta\rangle\langle\theta| =$ *coherent state and* $= |0\rangle\langle 0|$. The error probability (4.2) becomes

$$\begin{aligned} q_e &= \mathrm{tr}_{\mathcal{H}_2}(\mathrm{tr}_{\mathcal{K}_2}\, V^*|\theta\rangle\langle\theta|V)|0\rangle\langle 0| \\ &= \mathrm{tr}_{\mathcal{H}_2}(\mathrm{tr}_{\mathcal{K}_2}|\alpha\theta\rangle\langle\alpha\theta| \otimes |\beta\theta\rangle\langle\beta\theta|)|0\rangle\langle 0| \\ &= \mathrm{tr}_{\mathcal{H}_2}|\alpha\theta\rangle\langle\alpha\theta||0\rangle\langle 0| \\ &= |\langle 0,\ \alpha\theta\rangle|^2 \\ &= \exp(-|\alpha\theta|^2), \end{aligned}$$

which is equal to the usual result (see [16] and [19]), but our new derivation is much simpler than the old one.

Case II: $\xi_1^{(1)} =$ *squeezed state and* $\xi_0^{(2)} = |0\rangle\langle 0|$. A squeezed state can be expressed by a unitary operator $U(z)$ ($z \in \mathbf{C}$), given in the Appendix, such that

$$\xi_1^{(1)} = U(z)|\gamma\rangle\langle\gamma|U(z)^*,$$

where $|\gamma\rangle$ is a certain coherent state. Therefore the error probability q_e is

$$\begin{aligned} q_e &= \mathrm{tr}_{\mathcal{H}_2}(\mathrm{tr}_{\mathcal{K}_2}\, VU(z)|\gamma\rangle\langle\gamma|U(z)^*V^*)|0\rangle\langle 0| \\ &= \mathrm{tr}\, V^*(|0\rangle\langle 0| \otimes I)VU(z)|\gamma\rangle\langle\gamma|U(z)^* \\ &= \langle U(z)\gamma,\ V^*(|0\rangle\langle 0| \otimes I)VU(z)\gamma\rangle. \end{aligned}$$

To carry this calculation, we have to know the effect of V^* on $\mathcal{H}_2 \otimes \mathcal{K}_2$.

$$\begin{aligned} &\langle\gamma,\ (V^*|\gamma'\rangle \otimes |\gamma''\rangle)\rangle \\ &\quad= \langle V\gamma, (|\gamma'\rangle \otimes |\gamma''\rangle)\rangle \\ &\quad= \langle((\langle\alpha\gamma| \otimes \langle\beta\gamma|),\ (|\gamma'\rangle \otimes |\gamma''\rangle)\rangle \\ &\quad= \langle\alpha\gamma,\ \gamma'\rangle\langle\beta\gamma,\ \gamma''\rangle \\ &\quad= \exp\{\tfrac{1}{2}(-|\alpha\gamma|^2 - |\gamma'|^2 + 2\bar{\alpha}\bar{\gamma}\gamma')\}\exp\{\tfrac{1}{2}(-|\gamma''|^2 - |\beta\gamma|^2 + 2\bar{\beta}\bar{\gamma}\gamma'')\} \\ &\quad= \exp\{-\tfrac{1}{2}(|\beta\gamma'|^2 + |\alpha\gamma''|^2)\}\langle\gamma,\ \bar{\alpha}\gamma' + \bar{\beta}\gamma''\rangle\exp(\mathrm{Re}(\bar{\alpha}\beta\gamma'\bar{\gamma}'')), \end{aligned}$$

which implies

$$V^*|\gamma'\rangle \otimes |\gamma''\rangle = \exp\{-\tfrac{1}{2}(|\beta\gamma'|^2 + |\alpha\gamma''|^2)\} \times \exp(\mathrm{Re}(\bar{\alpha}\beta\gamma'\gamma''))|\bar{\alpha}\gamma' + \bar{\beta}\gamma''\rangle.$$

Therefore q_e is

$$\begin{aligned} q_e &= \frac{1}{\pi}\int d^2w \langle U(z)\gamma,\ V^*(|0\rangle\langle 0| \otimes I)Vw\rangle\langle w,\ U(z)\gamma\rangle \\ &= \frac{1}{\pi}\int d^2w \exp\{\tfrac{1}{2}(|\alpha|^2|\beta|^2|w|^2)\} \\ &\quad \times\langle U(z)\gamma,\ |\beta|^2 w\rangle\langle 0,\ \alpha w\rangle\langle w,\ U(z)\gamma\rangle. \end{aligned}$$

This can be computed by expression (A.27) given in the Appendix and a Gaussian type integration:

$$\begin{aligned} \langle w,\ U(z)\gamma\rangle &= \left\langle \exp\left(-\frac{i}{2}\varphi\right) w,\ U(r)\exp\left(-\frac{i}{2}\varphi\right)\gamma\right\rangle \\ &= \exp\{-\tfrac{1}{2}(|w|^2 + |\gamma|^2)\}(\cosh r)^{-1/2} \\ &\quad \times \exp\{\bar{w}\theta(\cosh r)^{-1} + \tanh r\{\tfrac{1}{2}(\exp(-i\varphi)\theta^2 - \exp(i\varphi)\bar{w}^2)\}\}, \end{aligned}$$

$$\begin{aligned} &\frac{1}{\pi}\int d^2w \exp\{-|w|^2 + aw + b\bar{w} + cw^2 + d\bar{w}^2\} \\ &\quad = \frac{1}{\sqrt{1-4cd}}\exp\left\{\frac{a^2d + ab + b^2c}{1-4cd}\right\}. \end{aligned}$$

The result is

$$\begin{aligned} q_e &= \frac{(\cosh r)^{-1}\exp\{\tfrac{1}{2}(\gamma^2 + \bar{\gamma}^2)(\tanh r) - |\gamma|^2\}}{\sqrt{1-(1-|\alpha|^2)^2(\tanh r)^2}} \\ &\quad \times \exp\left(\frac{-\tfrac{1}{2}(1-|\alpha|^2)^2(\cosh r)^{-2}(\tanh r)(\gamma^2+\bar{\gamma}^2) + (1-|\alpha|^2)|\gamma|^2(\cosh r)^{-2}}{1-(1-|\alpha|^2)^2(\tanh r)^2}\right) \\ &= \frac{1}{\sqrt{(\cosh r)^2 - (1-|\alpha|^2)^2(\sinh r)^2}} \\ &\quad \times \exp\left\{\left(\frac{1-|\alpha|^2}{(\cosh r)^2 - (1-|\alpha|^2)^2(\sinh r)^2} - 1\right)|\gamma|^2\right. \\ &\quad \left. + \left(1 - \frac{(1-|\alpha|^2)^2}{(\cosh r)^2 - (1-|\alpha|^2)^2(\sinh r)^2}\right)(\tfrac{1}{2}(\gamma^2+\bar{\gamma}^2)(\tanh r))\right\}, \end{aligned}$$

which is same as the result obtained in [24]:

$$q_e = \sqrt{\tau}\exp\left[\{(1-\eta)\tau - 1\}|\gamma|^2 + [1-(1-\eta)^2\tau]\left\{\frac{\bar{\mu}\gamma^2}{2\lambda} + \frac{\mu\bar{\gamma}^2}{2\bar{\lambda}}\right\}\right],$$

where $\eta = |\alpha|^2$, $\tau = \{|\lambda|^2 - (1-\eta)^2|\mu|^2\}^{-1}$ with $\lambda = \exp(i\varphi)\cosh r$, $\mu = \sinh r$.

Case III: $\xi_1^{(1)}$ = *squeezed state* $U(p)|\gamma\rangle\langle\gamma|U(p)^*$ *and* $\xi_0^{(2)}$ = *squeezed state* $U(q)|\sigma\rangle\langle\sigma|U(q)^*$. By a similar way as in case II,

$$\begin{aligned}
q_e &= \mathrm{tr}_{\mathcal{H}_2}(\mathrm{tr}_{\mathcal{K}_2}\, VU(p)|\gamma\rangle\langle\gamma|U(p)^*V^*)U(q)|\sigma\rangle\langle\sigma|U(q)^* \\
&= \langle U(p)\gamma,\ V^*(|U(q)\sigma\rangle\langle U(q)\sigma| \otimes I)VU(p)\gamma\rangle \\
&= \frac{1}{\pi^2}\iint d^2w\, d^2z \langle U(p)\gamma, w\rangle\langle w, V^*(|U(q)\sigma\rangle\langle U(q)\sigma| \otimes I)Vz\rangle\langle z, U(p)\gamma\rangle \\
&= \frac{1}{\pi^2}\iint d^2w\, d^2z \langle U(p)\gamma, w\rangle\langle \alpha w, U(q)\sigma\rangle\langle \beta w, \beta z\rangle\langle U(q)\sigma, \alpha z\rangle\langle z, U(p)\gamma\rangle.
\end{aligned}$$

Applying the above formula presented in case II, we can compute this error probability q_e as

$$\begin{aligned}
q_e &= \frac{1}{\pi^2}\iint \langle U(p)\gamma, w\rangle\langle \alpha w, U(q)\sigma\rangle\langle \beta w, \beta z\rangle\langle U(q)\sigma, \alpha z\rangle\langle z, U(p)\gamma\rangle d^2w\, d^2z \\
&= \frac{1}{(\cosh q)(\cosh p)} \\
&\quad \times \frac{1}{\sqrt{1-(\alpha^2+\overline{\alpha}^2)(\tanh q)(\tanh p) + |\alpha|^4(\tanh q)^2(\tanh p)^2 - |\beta|^4(\tanh p)^2}} \\
&\quad \times \exp\Bigg[-\left(|\alpha\gamma|^2 + |\sigma|^2 - \frac{(\alpha\gamma)^2 + \overline{(\alpha\gamma)}^2 - \sigma^2 - \overline{\sigma}^2}{2}(\tanh q)\right) \\
&\qquad - \frac{(\overline{\sigma}\alpha)^2 + (\sigma\overline{\alpha})^2 + (|\alpha|^4 - 2|\alpha|^2)(\gamma^2+\overline{\gamma}^2)}{2}(\tanh p) \\
&\qquad + ((\alpha^2|\alpha|^2 + \alpha^2 + \overline{\alpha}^2)|\gamma|^2 + (2\alpha^2 + \overline{\alpha}^2)|\sigma|^2)(\tanh q)(\tanh p) \\
&\qquad + ((|\alpha|^4 - |\alpha|^2)|\gamma|^2 + (2|\alpha|^4 - 3|\alpha|^2 + 1)|\sigma|^2)(\tanh p)^2 \\
&\qquad + \tfrac{1}{2}((\alpha^4 + |\alpha|^4)\gamma^2 + 2|\alpha|^4\overline{\gamma}^2 - 2\alpha^2\sigma^2 - (\alpha^2+\overline{\alpha}^2)\overline{\sigma}^2) \\
&\qquad \times (\tanh q)^2(\tanh p) \\
&\qquad + \tfrac{1}{2}((\alpha^2|\alpha|^4 - 2\alpha^2|\alpha|^2 - \overline{\alpha}^2)\gamma^2 + (\alpha^2|\alpha|^4 - 2\alpha^2|\alpha|^2 - \alpha^2)\overline{\gamma}^2 \\
&\qquad\quad + (|\alpha|^4 + 2|\alpha|^2 - 1)\sigma^2 + (\alpha^4 + 2|\alpha|^2 - 1)\overline{\sigma}^2)(\tanh q)(\tanh p)^2 \\
&\qquad - ((\alpha^4 + 2|\alpha|^4)|\gamma|^2 + (\alpha^4 + |\alpha|^4 + |\alpha|^2)|\sigma|^2)(\tanh q)^2(\tanh p)^2 \\
&\qquad + ((\alpha^2|\alpha|^2 - \alpha^2|\alpha|^4)|\gamma|^2 + (\alpha^2|\alpha|^2 - \alpha^2)|\sigma|^2)(\tanh q)(\tanh p)^3 \\
&\qquad + \tfrac{1}{2}(\alpha^4\sigma^2 + |\alpha|^4\overline{\sigma}^2 - \alpha^2|\alpha|^4(\gamma^2+\overline{\gamma}^2))(\tanh q)^3(\tanh p)^2 \\
&\qquad + \tfrac{1}{2}(|\alpha|^4\gamma^2 + \alpha^4\overline{\gamma}^2 + \alpha^2(1-2|\alpha|^2)(\sigma^2+\overline{\sigma}^2))(\tanh q)^2(\tanh p)^3 \\
&\qquad + \alpha^2|\alpha|^2(|\alpha|^2+1)(|\gamma|^2+|\sigma|^2)(\tanh q)^3(\tanh p)^3 \Bigg] \\
&\quad \times \exp\Bigg[(\overline{\sigma}\alpha\gamma + \sigma\overline{\alpha}\overline{\gamma} + (|\alpha|^2-1)(\overline{\sigma}\alpha\overline{\gamma} + \sigma\overline{\alpha}\gamma)(\tanh p) \\
&\qquad -(\overline{\sigma}\alpha(\alpha^2+\overline{\alpha}^2)\gamma + 2\sigma\overline{\alpha}\alpha^2\overline{\gamma})(\tanh q)(\tanh p) \\
&\qquad +\alpha^2(1-|\alpha|^2)(\overline{\sigma}\alpha\overline{\gamma} + \sigma\overline{\alpha}\gamma)(\tanh q)(\tanh p)^2 \\
&\qquad +(\overline{\sigma}\alpha|\alpha|^4\gamma + \sigma\overline{\alpha}\alpha^4\overline{\gamma})(\tanh q)^2(\tanh p)^2)\frac{1}{(\cosh q)(\cosh p)}\Bigg].
\end{aligned}$$

Appendix. Squeezed States

In this section we recall the definition and the basic properties of a class of states which in the past have been the object of several studies in the field of quantum optics [17], [29], [30]. We mainly follow [30].

Let **C** denote the set of all complex numbers. A **Fock representation of the Canonical Commutation Relations (CCR)** over **C** is a triple

$$\{\mathcal{H}, W, \Phi\},$$

where $\mathcal{H}$ is a Hilbert space and $W\colon z \in \mathbf{C} \mapsto W(z) \in U_n(\mathcal{H})$ is a map from **C** to the unitary operators on $\mathcal{H}$ such that $W(0) = \text{id}$ and

$$W(u)W(v) = \exp\{i \text{ Im } \bar{u}v\}W(u+v); \qquad u, v \in \mathbf{C}, \tag{A.1}$$

and $\Phi \in \mathcal{H}$ is a unit vector, called the Fock vacuum, satisfying

$$\langle \Phi, W(z)\Phi \rangle = \exp\{-\tfrac{1}{2}|z|^2\}; \qquad z \in \mathbf{C}. \tag{A.2}$$

It is moreover assumed that the weak closure of the complex vector space generated by the $\{W(z)\colon z \in \mathbf{C}\}$ coincides with the algebra of all bounded operators on $\mathcal{H}$. This property is called **irreducibility**. Clearly, any two Fock representations are canonically isomorphic. The Stone–von Neumann theorem asserts that if $\{W(z)\colon z \in \mathbf{C}\}$ is any irreducible family of unitary operators on a Hilbert space $\mathcal{H}$ satisfying (A.1), then it is isomorphic to the Fock representation. In particular, for any such a family, there will exist a (necessarily unique) vector Φ satisfying (A.2), i.e., a Fock vacuum for this family. A corollary of the Stone–von Neumann theorem is the following: let $T\colon \mathbf{C} \to \mathbf{C}$ be any real linear transformation such that

$$\text{Im}(Tu)^{-}(Tv) = \text{Im } \bar{u}v; \qquad \forall u, v \in \mathbf{C}, \tag{A.3}$$

where the overbar denotes the complex conjugate, and define

$$W_T(z) = W(Tz); \qquad z \in \mathbf{C}. \tag{A.4}$$

Then the set $\{W_T(z)\colon z \in \mathbf{C}\}$ (because any T satisfying (A.3) must be invertible) is irreducible. Moreover, it satisfies (A.1) because of (A.3). Hence by the Stone–von Neumann theorem, there exists a vector $\Phi_T \in \mathcal{H}$ and a unitary operator $U_T\colon \mathcal{H} \to \mathcal{H}$, characterized by the property

$$U_T W(z)\Phi = W_T(z)\Phi_T; \qquad z \in \mathbf{C}. \tag{A.5}$$

The vector $\Phi_T = U_T\Phi$ (i.e., the vacuum for the W_T-representation) is called a **squeezed vector** for the W-representation. The most general operator T, satisfying (A.3), is given by the following:

Proposition A.1. *Let $V\colon \mathbf{C} \to \mathbf{R}^2$ be the isomorphism of real linear spaces characterized by*

$$V(1) = \begin{pmatrix} 1 \\ 0 \end{pmatrix}; \qquad V(i) = \begin{pmatrix} 0 \\ 1 \end{pmatrix}. \tag{A.6}$$

Then a real 2×2 *matrix* T *induces on* $\mathbf{C}$ *a transformation satisfying* (A.3) *if and only if* $\det T = 1$.

Proof. A direct calculation: The identity (A.2) implies that for each $z \in \mathbf{C}$ the one-parameter unitary group $\{W(tz)\}$ $(t \in \mathbf{R})$ is strongly continuous, hence

$$W(tz) = \exp\{itB(z)\} \tag{A.7}$$

for some self-adjoint operator $B(z)$. Moreover, the map $z \in \mathbf{C} \mapsto B(z)$ is real linear. The operators

$$\tfrac{1}{2}B(1) = q; \qquad \tfrac{1}{2}B(i) = p \tag{A.8}$$

are called momentum and position operators, respectively. Condition (A.1) implies that

$$[B(u), B(v)] = 2i \operatorname{Im} \bar{u}v \tag{A.9}$$

so that, in particular

$$[q, p] = \frac{i}{2}.$$

Finally, denoting

$$a = p - iq; \qquad a^* = p + iq \tag{A.10}$$

one has

$$[a, a^*] = 1,$$

$$iB(z) = za^* - \bar{z}a; \qquad z \in \mathbf{C}. \tag{A.11}$$

The vectors

$$|\theta\rangle = W(\theta)\Phi; \qquad \theta \in \mathbf{C}, \tag{A.12}$$

are called **coherent vectors**. Now let $T : \mathbf{C} \to \mathbf{C}$ be a real linear map satisfying (A.3) and let W_T, U_T, Φ_T be characterized respectively by (A.4) and (A.5). Then one has, for $z \in \mathbf{C}$,

$$W_T(z) = \exp(iB_T(z)) = \exp(za_T^* - \bar{z}a_T). \tag{A.13}$$

On the other hand, by definition (A.4) of W_T, one also has

$$\begin{aligned} W_T(z) &= W(Tz) = \exp((Tz)a^* - (Tz)^- a) \\ &= \exp\left\{(a^*, -a)\begin{pmatrix} Tz \\ (Tz)^- \end{pmatrix}\right\}. \quad \square \end{aligned} \tag{A.14}$$

Proposition A.2. *In the above notations, if $T: \mathbf{C} \to \mathbf{C}$ is represented, in the identification of Proposition* A.1, *by the matrix*

$$\begin{pmatrix} a & b \\ c & d \end{pmatrix} \in \mathrm{SL}(2; \mathbf{R}), \tag{A.15}$$

then, for each $z \in \mathbf{C}$,

$$\begin{pmatrix} Tz \\ (Tz)^- \end{pmatrix} = \begin{pmatrix} \bar{c} & -s \\ -\bar{s} & c \end{pmatrix} \begin{pmatrix} z \\ \bar{z} \end{pmatrix} \tag{A.16}$$

with

$$\bar{c} = \tfrac{1}{2}([a+d] + i[c-b]); \qquad -s = \tfrac{1}{2}([a-d] + i[c+b]). \tag{A.17}$$

Remark. Notice that any c, s given by (A.17) satisfy

$$|c|^2 - |s|^2 = \det \begin{pmatrix} a & b \\ c & d \end{pmatrix} = 1. \tag{A.18}$$

Conversely, given $c, s \in \mathbf{C}$ such that $|c|^2 - |s|^2 = 1$, we can define a, b, c, d by (A.17) and the resulting matrix is in SL(2; **R**).

Proof. Denote by $\mathbf{C}^2(\mathbf{R})$ the real vector space

$$\mathbf{C}^2(\mathbf{R}) = \left\{ \lambda \begin{pmatrix} z \\ \bar{z} \end{pmatrix}; \ z \in \mathbf{C}, \lambda \in \mathbf{R} \right\},$$

and by V_0: $\mathbf{R}^2 \to \mathbf{C}^2(\mathbf{R})$ the isomorphism of real vector spaces characterized by

$$V_0 \begin{pmatrix} 1 \\ 0 \end{pmatrix} = \begin{pmatrix} 1 \\ 1 \end{pmatrix}; \qquad V_0 \begin{pmatrix} 0 \\ 1 \end{pmatrix} = \begin{pmatrix} i \\ -i \end{pmatrix}.$$

Then, if V: $\mathbf{C} \to \mathbf{R}^2$ is the isomorphism of Proposition A.1 and $z = x + iy$, one has

$$V_0 V T z = V_0 \begin{pmatrix} a & b \\ c & d \end{pmatrix} \begin{pmatrix} x \\ y \end{pmatrix} = \begin{pmatrix} (ax+by) + i(cx+dy) \\ (ax+by) - i(cx+dy) \end{pmatrix}.$$

Expressing x, y in terms of $z, \bar{z}$, one finds (A.16), (A.17).

Putting together Proposition (A.2) and the identity (A.14), we obtain

$$\begin{aligned} W_T(z) &= \exp\left\{ (a^*, -a) \begin{pmatrix} \bar{c} & -s \\ -\bar{s} & c \end{pmatrix} \begin{pmatrix} z \\ \bar{z} \end{pmatrix} \right\} \\ &= \exp\{ z(a^*\bar{c} + a\bar{s}) - \bar{z}(ac + a^*s) \}. \end{aligned} \tag{A.19}$$

Comparing (A.19) with (A.13) we finally find

$$a_T = ac + a^*s \tag{A.20}$$

or equivalently

$$\begin{pmatrix} a_T \\ a_T^* \end{pmatrix} = \begin{pmatrix} c & s \\ \bar{s} & \bar{c} \end{pmatrix} \begin{pmatrix} a \\ a^* \end{pmatrix} \Leftrightarrow \tag{A.21}$$

However, from (A.5), (A.7), and (A.11), it follows that the operator U_T is characterized by the property

$$U_T a_T U_T^* = a$$

or, in view of (A.20), by

$$U_T^* a U_T = ca + sa^*. \tag{A.22}$$

Our goal is to find the operator U_T satisfying (A.22) for given c and s satisfying (A.18). To this goal first notice that, in view of (A.18), there exist real numbers r, η, φ such that

$$c = \exp\{i\eta\}\cosh r = \exp\{i\eta\}c_r; \quad s = \exp\{-i\varphi\}\sinh r = \exp\{-i\varphi\}s_r. \tag{A.23}$$

Moreover, due to the identities

$$\begin{aligned} &\exp\{xa^*a\}a\exp\{-xa^*a\} = \exp\{-x\}a; \\ &\exp\{xa^*a\}a^*\exp\{-xa^*a\} = \exp\{x\}a^*; \qquad x \in \mathbf{C}, \end{aligned} \tag{A.24}$$

by replacing the representation $W(z)$ with the equivalent representation $W(\exp\{i\eta\}z)$, we can always suppose that c, in (A.22), is real, i.e., $\eta = 0$ in (A.23). □

Proposition A.3. *Let c, s be given by* (A.23) *with $\eta = 0; r > 0$. Then the operator U_T, characterized by* (A.22), *is given by*

$$U_T = \exp\{\tfrac{1}{2}(za^2 - \bar{z}a^{*2})\} \equiv U(z), \tag{A.25}$$

$$z = r\exp\{i\varphi\}. \tag{A.26}$$

Proof. Denote $D_z = \frac{1}{2}(za^2 - \bar{z}a^{*2})$ and define

$$f(t) = \exp\{tD_z\}a\exp\{-tD_z\}.$$

Then, due to the easily verified commutation relations

$$[D_z, a] = \bar{z}a^*; \qquad [D_z, a^*] = za,$$

one deduces the equation

$$\frac{d}{dt}\begin{pmatrix} f(t) \\ f^*(t) \end{pmatrix} = \begin{pmatrix} 0 & \bar{z} \\ z & 0 \end{pmatrix}\begin{pmatrix} f(t) \\ f^*(t) \end{pmatrix}$$

with initial condition

$$\begin{pmatrix} f(0) \\ f^*(0) \end{pmatrix} = \begin{pmatrix} a \\ a^* \end{pmatrix},$$

whence

$$\begin{pmatrix} f(t) \\ f^*(t) \end{pmatrix} = \exp\left(t \begin{pmatrix} 0 & \bar{z} \\ z & 0 \end{pmatrix} \begin{pmatrix} a \\ a^* \end{pmatrix} \right)$$
$$= \begin{pmatrix} \cosh tr & \exp\{-i\varphi\} \sinh tr \\ \exp\{i\varphi\} \sinh tr & \cosh tr \end{pmatrix} \begin{pmatrix} a \\ a^* \end{pmatrix}.$$

For $t = 1$, using (A.23) and the assumption $\eta = 0$, one finds

$$\exp\{D_z\} a \exp\{-D_z\} = ca + sa^*,$$

so that $U_T = U(z) \equiv \exp\{D_z\}$. □

Remark. Let $z = r\exp\{i\varphi\}$ and denote by V_t the one-parameter unitary group $V_t = \exp\{ita^*a\}$. Then one easily checks, using (A.24), that

$$\exp\{D_z\} = V_{-\varphi/2} \exp\{D_r\} V_{\varphi/2}. \tag{A.27}$$

So we can limit ourselves to studying the operator D_z in the case of real z. In several applications it is useful to know the matrix elements of the operator $\exp\{D_r\} = U(r)$ with respect to the coherent states in the W-representation.

Proposition A.4. *In the notation* (A.12), (A.23), (A.25) *one has*

$$\langle \alpha, U(r)\beta \rangle = \langle 0, U(r)\beta \rangle \exp\left\{ \frac{\bar{\alpha}\beta}{c_r} - \bar{\alpha}^2 \frac{s_r}{2c_r} \right\}, \tag{A.28}$$

$$\langle 0, U(r)\beta \rangle = c_r^{-1/2} \exp\left(\frac{\beta^2 s_r}{c_r} \right). \tag{A.29}$$

Proof. Denote $f(\bar{\alpha}) = \langle \alpha, U(r)\beta \rangle$. Then

$$\frac{d}{d\bar{\alpha}} f(\bar{\alpha}) = \frac{\beta}{c_r} \left(1 - \frac{\bar{\alpha} c_r}{\beta} \right) f(\bar{\alpha}); \qquad f(0) = \langle 0, U(r)\beta \rangle.$$

Solving this equation, we find (A.27). Now put $f(r) = \langle 0, U(r)\beta \rangle$. Then $f(0) = 1$ and

$$\frac{d}{dr} f(r) = \frac{1}{2} \left(\frac{\beta^2}{c_r^2} - \frac{s_r}{c_r} \right) f(r).$$

The solution of this equation is

$$f(r) = \exp \int_0^r \frac{1}{2} \left(\frac{\beta^2}{c_\tau^2} - \frac{s_\tau}{c_\tau} \right) d\tau. \tag{A.30}$$

Taking into account (A.23), the integral in (A.30) is easily evaluated and leads to (A.29). □

References

1. Accardi, L.; Quantum Kalman filtering. Contribution to the memorial volume for the 60th birthday of R.E. Kalman.
2. Accardi, L.; Noncommutative Markov chains with preassigned evolution: an application to the quantum theory measurement, Advances in Mathematics, 29 (1978), 226–243.
3. Accardi, L.; An outline of quantum probability. Unpublished manuscript (1990).
4. Accardi, L., Watson, G.S.; Quantum random walks and coherent quantum chains. Quantum Probability and Applications, IV. LNM 1396, Springer-Verlag, Berlin (1987), pp. 73–88.
5. Alicki, R., Frigerio, A.; Quantum Poisson noise and linear quantum Boltzmann equation. Preprint (March 1989).
6. Bach, A.; The simplex structure of the classical states of the quantum harmonic oscillator, Communications in Mathematical Physics, 107 (1986), 553–560.
7. Barchielli, A.; Stochastic processes and continual measurements in quantum mechanics. Preprint.
8. Bratteli, O., Robinson, D.V.; Operator algebras and quantum statistical mechanics, I and II. Springer-Verlag, Berlin (1981).
9. Cecchini, C.; Stochastic couplings for von Neumann algebras. Quantum Probability and Applications, III. LNM 1303, Springer-Verlag, Berlin (1988), pp. 1–5.
10. Cecchini, C., Petz, D.; Classes of conditional expectations over von Neumann algebras, Journal of Functional Analysis 91 (1990).
11. Cecchini, C., Petz, D.; State extention and a Radon-Nikodym theorem for conditional expectations on von Neumann algebras, Pacific Journal of Mathematics, 138 (1989), 9–24.
12. Cecchini, C., Kümmerer, B.; Stochastic transitions on preduals of von Neumann algebras. Quantum Probability and Applications, V. LNM 1442, Springer-Verlag, Berlin (1990), pp. 126–130.
13. Chen Mu-Fa, Selected topics in probability theory. Preprint, Volterra N.94, Universita di Roma, (1992).
14. Fannes, M., Nachtergaele, B., Werner, R.F.; Construction and study of exact ground states for a class of quantum antiferromagnets. Preprint (1989).
15. Haagerup, U., A new proof of the equivalence of injectivity and hyperfiniteness for factors on a separable Hilbert space, Journal of Functional Analysis 62 (1985), 160–201.
16. Hirota, O., Yamazaki, K., Nakagawa, M., Ohya, M.; Properties of error correcting code using photon pulse, Transaction of IECE Japan, E69 (1986), 917–919.
17. Hollenhorst, J.N.; Quantum limits on resonant-mass gravitational-radiation detectors, Physical Review, D19 (1979), 1669–1679.
18. Kraus, K.; States, Effects and Operations. LNP 190 Springer-Verlag, Berlin (1983).
19. McEliece, R.J.; Practical code for photon communication, IEEE Transaction of Information Theory, 27 (1981), 393–398.
20. Ohya, M.; Quantum ergodic channels in operator algebras, Journal of Mathematical Analysis and Applications, 84 (1981), 318–327.
21. Ohya, M.; On compound state and mutual information in quantum information theory, IEEE Transactions of Information Theory, 29 (1983), 770–774.
22. Ohya, M.; Note on quantum probability, Lettere al Nuovo Cimento, 38 (1983), 402–404.
23. Ohya, M.; State change and entropies in quantum dynamical systems. Accardi, L., von Waldenfels, W. (eds.). LNM 1136, Springer-Verlag, Berlin (1985), pp. 397–408.
24. Ohya, M.; Some aspects of quantum information theory and their applications to irreversible processes, Reports on Mathematical Physics, 27 (1989), 19–47.
25. Petz, D.; Characterization of sufficient observation channels. Preprint.
26. Petz, D.; Sufficient subalgebras and the relative entropy of states of a von Neumann algebra. Communications in Mathematical Physics, 105 (1986), 123–131.
27. Rondoni, L.; Nonlinear Boltzman maps in classical and quantum probability. Quantum Probability and Related Topics, VIII. World Scientific, Singapore (1993).
28. Sakai, S.; C^*-Algebras and W^*-Algebras. Springer-Verlag, Berlin (1971).
29. Stoler, D.; Equivalence class of minimum uncertainty packets, Physical Review, D1 (1970), 3217–3219 and D4 (1971), 1925.
30. Yuen, H.P.; Two-photon coherent states of the radiation field, Physical Review, A13 (1976), 2226–2243.

Accepted 17 *February* 1997

Infinite Dimensional Analysis, Quantum Probability and Related Topics
Vol. 2, No. 2 (1999) 267–282

QUANTUM DYNAMICAL ENTROPY FOR COMPLETELY POSITIVE MAP

A. KOSSAKOWSKI

Institute of Physics, Nicholas Copernicus University, Grudziadzka 5, 87-100 Torun, Poland

M. OHYA and N. WATANABE*

Department of Information Sciences, Science University of Toko, Noda City, Chiba 278-8510, Japan

Received 5 November 1998

A dynamical entropy for not only shift but also completely positive (CP) map is defined by generalizing the AOW entropy[1] defined through quantum Markov chain and AF entropy defined by a finite operational partition. Our dynamical entropy is numerically computed for several models.

0. Introduction

Quantum dynamical entropy was first introduced by Emch[2] and Connes–Störmer[3] around 1975. In 1987, Connes, Narnhoffer and Thirring[4] defined a dynamical entropy (CNT entropy) in C^*-dynamical systems. Park computed the CNT entropy for several models.[5] In 1994, Alicki and Fannes[6] defined a quantum dynamical entropy (AF entropy) by means of a finite operational partition of unity. In 1994, Hudetz[7] discussed the dynamical entropy in terms of topological entropy. In 1995, Ohya[8] introduced a quantum dynamical entropy and a quantum dynamical mutual entropy based on the C^*-mixing entropy[9] and the complexity in information dynamics.[10] In 1995, Voiculescu[11] introduced a dynamical approximation entropy for C^*- and W^*-algebra automorphisms based on a general approximation approach. In 1997, Accardi, Ohya and Watanabe[1] defined a quantum dynamical entropy (AOW entropy) through quantum Markov chain. The difference among some definitions of the dynamical entropy were discussed in Refs. 12–14. Some computations of the dynamical entropy are done in several papers such as Refs. 15 and 16.

We introduce a dynamical entropy for a completely positive (CP) map in Sec. 1. The Stinespring type expression of transition expectation associated to CP maps is

*E-mail: watanabe@is.noda.sut.ac.jp

discussed in Sec. 2. It is shown that the dynamical entropy of Sec. 1 is a generalization of both AF and AOW dynamical entropies in Sec. 3. Several computations of our dynamical entropy giving a finite value are carried for several models in Sec. 4.

1. Quantum Dynamical Entropy for CP Map

Let $\mathbf{B}(\mathcal{K})$ (resp. $\mathbf{B}(\mathcal{H})$) be the set of all bounded linear operators on separable Hilbert space $\mathcal{K}$ (resp. $\mathcal{H}$). We denote the set of all density operators on $\mathcal{K}$ (resp. $\mathcal{H}$) by $\mathfrak{S}(\mathcal{K})$ (resp. $\mathfrak{S}(\mathcal{H})$). Let

$$\Gamma : \mathbf{B}(\mathcal{K}) \otimes \mathbf{B}(\mathcal{H}) \to \mathbf{B}(\mathcal{K}) \otimes \mathbf{B}(\mathcal{H}) \tag{1.1}$$

be a normal, unital CP linear map, i.e. Γ satisfies

$$\tilde{B}_\alpha \uparrow \tilde{B} \Longrightarrow \Gamma(\tilde{B}_\alpha) \uparrow \Gamma(\tilde{B}),$$

$$\Gamma(I_\mathcal{K} \otimes I_\mathcal{H}) = I_\mathcal{K} \otimes I_\mathcal{H} \qquad (I_\mathcal{H}(\text{resp. } I_\mathcal{K}) \text{ is the unity in } \mathcal{H}(\text{resp. } \mathcal{K}))$$

for any increasing net $\{\tilde{B}_\alpha\} \subset \mathbf{B}(\mathcal{K}) \otimes \mathbf{B}(\mathcal{H})$ converging to $\tilde{B} \in \mathbf{B}(\mathcal{K}) \otimes \mathbf{B}(\mathcal{H})$ and

$$\sum_{i,j=1}^{n} \tilde{B}_j^* \Gamma(\tilde{A}_j^* \tilde{A}_i) \tilde{B}_i \geq 0$$

hold for any $n \in \mathbb{N}$ and any $\tilde{A}_i, \tilde{B}_j \in \mathbf{B}(\mathcal{K}) \otimes \mathbf{B}(\mathcal{H})$. For a normal state ω in $\mathbf{B}(\mathcal{K})$, there exists a density operator $\tilde{\omega} \in \mathfrak{S}(\mathcal{K})$ associated to ω (i.e. $\omega(A) = \operatorname{tr} \tilde{\omega} A$, $\forall\, A \in \mathbf{B}(\mathcal{K})$). Then a map

$$E^{\Gamma,\omega} : \mathbf{B}(\mathcal{K}) \otimes \mathbf{B}(\mathcal{H}) \to \mathbf{B}(\mathcal{H}) \tag{1.2}$$

defined as

$$E^{\Gamma,\omega}(\tilde{A}) = \omega(\Gamma(\tilde{A})) = \operatorname{tr}_\mathcal{K} \tilde{\omega} \Gamma(\tilde{A}), \qquad \forall\, \tilde{A} \in \mathbf{B}(\mathcal{K}) \otimes \mathbf{B}(\mathcal{H}) \tag{1.3}$$

is a transition expectation in the sense of Ref. 14 (i.e. $E^{\Gamma,\omega}$ is a linear unital CP map from $\mathbf{B}(\mathcal{K}) \otimes \mathbf{B}(\mathcal{H})$ to $\mathbf{B}(\mathcal{H})$), whose dual is a map

$$E^{*\Gamma,\omega}(\rho) : \mathfrak{S}(\mathcal{H}) \to \mathfrak{S}(\mathcal{K} \otimes \mathcal{H}) \tag{1.4}$$

given by

$$E^{*\Gamma,\omega}(\rho) = \Gamma^*(\tilde{\omega} \otimes \rho). \tag{1.5}$$

The dual map $E^{*\Gamma,\omega}$ is a lifting in the sense of Ref. 17; i.e. it is a continuous map from $\mathfrak{S}(\mathcal{H})$ to $\mathfrak{S}(\mathcal{K} \otimes \mathcal{H})$.

For a normal, unital CP map $\Lambda : \mathbf{B}(\mathcal{H}) \to \mathbf{B}(\mathcal{H})$, $\mathrm{id} \otimes \Lambda : \mathbf{B}(\mathcal{K}) \otimes \mathbf{B}(\mathcal{H}) \to \mathbf{B}(\mathcal{K}) \otimes \mathbf{B}(\mathcal{H})$ is a normal, unital CP map, where id is the identity map on $\mathbf{B}(\mathcal{K})$. Then one defines the transition expectation

$$E_\Lambda^{\Gamma,\omega}(\tilde{A}) = \omega((\mathrm{id} \otimes \Lambda)\Gamma(\tilde{A})), \qquad \forall\, \tilde{A} \in \mathbf{B}(\mathcal{K}) \otimes \mathbf{B}(\mathcal{H}) \tag{1.6}$$

and the lifting

$$E_\Lambda^{*\Gamma,\omega}(\rho) = \Gamma^*(\tilde{\omega} \otimes \Lambda^*(\rho)), \qquad \forall\, \rho \in \mathfrak{S}(\mathcal{H}). \tag{1.7}$$

The above Λ^* has been called a quantum channel[18] from $\mathfrak{S}(\mathcal{H})$ to $\mathfrak{S}(\mathcal{H})$, in which ρ is regarded as an input signal state and $\tilde{\omega}$ is as a noise state.

The equality

$$\begin{aligned} &\mathrm{tr}_{(\bigotimes_1^n \mathcal{K})\otimes\mathcal{H}} \Phi_{\Lambda,n}^{*\Gamma,\omega}(\rho)(A_1 \otimes \cdots \otimes A_n \otimes B) \\ &\quad \equiv \mathrm{tr}_{\mathcal{H}}\ \rho(E_\Lambda^{\Gamma,\omega}(A_1 \otimes E_\Lambda^{\Gamma,\omega}(A_2 \otimes \cdots \otimes A_{n-1} \otimes E_\Lambda^{\Gamma,\omega}(A_n \otimes B)\cdots))) \end{aligned} \tag{1.8}$$

for all $A_1, A_2, \ldots, A_n \in \mathbf{B}(\mathcal{K}), B \in \mathbf{B}(\mathcal{H})$ and any $\rho \in \mathfrak{S}(\mathcal{H})$ defines

(a) a lifting

$$\Phi_{\Lambda,n}^{*\Gamma,\omega} : \mathfrak{S}(\mathcal{H}) \to \mathfrak{S}\left(\left(\bigotimes_1^n \mathcal{K}\right) \otimes \mathcal{H}\right) \tag{1.9}$$

and

(b) marginal states

$$\rho_{\Lambda,n}^{\Gamma,\omega} \equiv \mathrm{tr}_{\mathcal{H}}\ \Phi_{\Lambda,n}^{*\Gamma,\omega}(\rho) \in \mathfrak{S}\left(\bigotimes_1^n \mathcal{K}\right), \tag{1.10}$$

$$\bar{\rho}_{\Lambda,n}^{\Gamma,\omega} \equiv \mathrm{tr}_{\bigotimes_1^n \mathcal{K}} \Phi_{\Lambda,n}^{*\Gamma,\omega}(\rho) \in \mathfrak{S}(\mathcal{H}). \tag{1.11}$$

Here, the state

$$\Phi_{\Lambda,n}^{*\Gamma,\omega}(\rho) \in \mathfrak{S}\left(\left(\bigotimes_1^n \mathcal{K}\right) \otimes \mathcal{H}\right) \tag{1.12}$$

is a compound state for $\bar{\rho}_{\Lambda,n}^{\Gamma,\omega}$ and $\rho_{\Lambda,n}^{\Gamma,\omega}$ in the sense of Ref. 19. Note that generally $\bar{\rho}_{\Lambda,n}^{\Gamma,\omega}$ is not equal to ρ.

Definition 1.1. The quantum dynamical entropy with respect to Λ, ρ, Γ and ω is defined by

$$\tilde{S}(\Lambda; \rho, \Gamma, \omega) \equiv \limsup_{n\to\infty} \frac{1}{n} S(\rho_{\Lambda,n}^{\Gamma,\omega}), \tag{1.13}$$

where $S(\cdot)$ is von Neumann entropy[20]; i.e. $S(\sigma) \equiv -\mathrm{tr}\,\sigma \log \sigma$, $\sigma \in \mathfrak{S}(\bigotimes_1^n \mathcal{K})$. The dynamical entropy with respect to Λ and ρ is defined as

$$\tilde{S}(\Lambda; \rho) \equiv \sup\{\tilde{S}(\Lambda; \rho, \Gamma, \omega); \Gamma, \omega\}.$$

2. Transition Expectation

In this section, we discuss the transition expectation $E_\Lambda^{\Gamma,\omega}$ associated to the CP maps Γ, Λ and a state ω.

For a complete orthonormal system (CONS) $\{e_i\}$ in $\mathcal{K}$, put $E_{ij} = |e_i\rangle\langle e_j|$. There exist operators $v_\alpha \in \mathbf{B}(\mathcal{H})$ and complex numbers $\lambda_{kl\alpha,mn\beta} \in \mathbb{C}$ such that the unital CP map Γ of (1.1) can be written in the form (p. 145 of Ref. 16)

$$\Gamma(\tilde{A}) = \sum_{k,l,m,n} \sum_{\alpha,\beta} \lambda_{kl\alpha,mn\beta}(E_{kl}^* \otimes v_\alpha^*)\tilde{A}(E_{mn} \otimes v_\beta), \qquad \tilde{A} \in \mathbf{B}(\mathcal{K}) \otimes \mathbf{B}(\mathcal{H}). \tag{2.1}$$

Let $\tilde{A} \in \mathbf{B}(\mathcal{K}) \otimes \mathbf{B}(\mathcal{H})$ be

$$\tilde{A} = \sum_{i,j} E_{ij} \otimes A_{ij}, \qquad A_{ij} \in \mathbf{B}(\mathcal{H}). \tag{2.2}$$

From (2.1) and (2.2), we have

$$\begin{aligned}
\Gamma(\tilde{A}) &= \sum_{k,l,m,n} \sum_{\alpha,\beta} \lambda_{kl\alpha,mn\beta} (E_{kl}^* \otimes v_\alpha^*) \left(\sum_{i,j} E_{ij} \otimes A_{ij} \right) (E_{mn} \otimes v_\beta) \\
&= \sum_{k,l,m,n} \sum_{\alpha,\beta} \sum_{i,j} \lambda_{kl\alpha,mn\beta} (E_{kl}^* \otimes v_\alpha^*)(E_{ij} \otimes A_{ij})(E_{mn} \otimes v_\beta) \\
&= \sum_{k,l,m,n} \sum_{\alpha,\beta} \sum_{i,j} \lambda_{kl\alpha,mn\beta} (E_{kl}^* E_{ij} E_{mn\beta} \otimes v_\alpha^* A_{ij} v) \\
&= \sum_{k,l,m,n} \sum_{\alpha,\beta} \sum_{i,j} \lambda_{kl\alpha,mn\beta} (|e_l\rangle\langle e_k, e_i\rangle\langle e_j, e_m\rangle\langle e_n| \otimes v_\alpha^* A_{ij} v_\beta) \\
&= \sum_{k,l,m,n} \sum_{\alpha,\beta} \lambda_{kl\alpha,mn\beta} (E_{l\,n} \otimes v_\alpha^* A_{km} v_\beta) .
\end{aligned} \tag{2.3}$$

The equality

$$\Gamma(I_{\mathcal{K}} \otimes I_{\mathcal{H}}) = I_{\mathcal{K}} \otimes I_{\mathcal{H}} \tag{2.4}$$

implies

$$\sum_k \sum_{\alpha,\beta} \lambda_{kl\alpha,kn\beta} v_\alpha^* v_\beta = \delta_{l\,n} I_{\mathcal{H}} . \tag{2.5}$$

As the matrix $\sigma \equiv (\lambda_{kl\alpha,mn\beta})$ is positive definite, it can always be diagonalized as

$$\sigma = \sum_q \tau_q |\xi^{(q)}\rangle\langle\xi^{(q)}| = \left(\sum_q \tau_q \tilde{\xi}_{kl\alpha}^{*(q)} \tilde{\xi}_{mn\beta}^{(q)} \right),$$

where the positive number τ_q is the eigenvalue of σ and the complex number $\tilde{\xi}_{kl\alpha}^{(q)}$ is the component of the orthogonal eigenvectors $|\xi^{(q)}\rangle$ associated to τ_q. From the positivity of $\tilde{\omega}$, we have

$$\begin{aligned}
\omega(E_{l\,n}) &= \mathrm{tr}(\tilde{\omega}(E_{l\,n})) = \mathrm{tr}(\tilde{\omega}|e_l\rangle\langle e_n|) = \langle e_n, \tilde{\omega} e_l\rangle \\
&= \sum_p \langle \sqrt{\tilde{\omega}} e_n, f_p\rangle\langle f_p, \sqrt{\tilde{\omega}} e_l\rangle ,
\end{aligned}$$

where $\{f_p\}$ is a CONS of $\mathcal{K}$. From the above equality, the transition expectation $E^{\Gamma,\omega}$ is expressed as (2.6) below.

$$\begin{aligned}
&E^{\Gamma,\omega}\left(\sum_{i,j} E_{ij}\otimes A_{ij}\right)\\
&\quad=\omega\left(\Gamma\left(\sum_{i,j} E_{ij}\otimes A_{ij}\right)\right)\\
&\quad=\sum_{k,l,m,n}\sum_{\alpha,\beta}\lambda_{kl\alpha,mn\beta}\,\omega(E_{l\,n})v_\alpha^* A_{km}v_\beta\\
&\quad=\sum_{k,l,m,n}\sum_{\alpha,\beta}\left(\sum_q \tau_q \tilde{\xi}^{*(q)}_{kl\alpha}\tilde{\xi}^{(q)}_{mn\beta}\right)\sum_p\langle\sqrt{\tilde{\omega}}e_n, f_p\rangle\langle f_p,\sqrt{\tilde{\omega}}e_l\rangle v_\alpha^* A_{km}v_\beta\\
&\quad=\sum_{k,m}\left(\sum_p\sum_q\tau_q\left(\sum_{l,\alpha}\tilde{\xi}^{*(q)}_{kl\alpha}\overline{\langle\sqrt{\tilde{\omega}}e_l, f_p\rangle}v_\alpha^*\right)A_{km}\left(\sum_{n,\beta}\tilde{\xi}^{(q)}_{mn\beta}\langle\sqrt{\tilde{\omega}}e_n, f_p\rangle v_\beta\right)\right)\\
&\quad=\sum_{k,m}\left(\sum_p\sum_q\left(\sqrt{\tau_q}\sum_{l,\alpha}\tilde{\xi}^{*(q)}_{kl\alpha}\overline{\langle\sqrt{\tilde{\omega}}e_l, f_p\rangle}v_\alpha^*\right)\right.\\
&\qquad\left.\times A_{km}\left(\sqrt{\tau_q}\sum_{n,\beta}\tilde{\xi}^{(q)}_{mn\beta}\langle\sqrt{\tilde{\omega}}e_n, f_p\rangle v_\beta\right)\right).
\end{aligned}\tag{2.6}$$

Putting $u_{pqm}=\sqrt{\tau_q}\sum_{n,\beta}\tilde{\xi}^{(q)}_{mn\beta}\langle\sqrt{\tilde{\omega}}e_n, f_p\rangle v_\beta$,

$$E^{\Gamma,\omega}\left(\sum_{i,j} E_{ij}\otimes A_{ij}\right)=\sum_{k,m}\sum_p\sum_q u_{pqk}^* A_{km}u_{pqm},\tag{2.7}$$

so that we can rewrite Γ,ω by u, and the dual of $E^{\Gamma,\omega}(=E^u)$ is written as

$$E^{*u}(\rho)=\sum_{k,m}\sum_p\sum_q E_{km}\otimes u_{pqk}\rho u_{pqm}^*\,.\tag{2.8}$$

Since $\Gamma(I_{\mathcal{K}}\otimes I_{\mathcal{H}})=I_{\mathcal{K}}\otimes I_{\mathcal{H}}$ holds, one obtains

$$\sum_{p,q,k} u_{pqk}^* u_{pqk}=I_{\mathcal{H}}\,.\tag{2.9}$$

For a channel $\Lambda^*:\mathfrak{S}(\mathcal{H})\to\mathfrak{S}(\mathcal{H})$, the transition expectation (1.6) and the lifting (1.7) are expressed as follows:

$$E_\Lambda^u\left(\sum_{i,j} E_{ij}\otimes A_{ij}\right)=\sum_{k,m}\sum_p\sum_q\Lambda(u_{pqk}^* A_{km}u_{pqm})\tag{2.10}$$

and

$$E_{\Lambda}^{*u}(\rho)=\sum_{k,m}\sum_{p}\sum_{q}E_{km}\otimes u_{pqk}\Lambda^{*}(\rho)u_{pqm}^{*}\,. \tag{2.11}$$

3. Dynamical Entropy

In this section, we generalize both the AOW entropy and the AF entropy. Then we compare the generalized AF entropy with the generalized AOW entropy.

Let θ be an automorphism of $\mathbf{B}(\mathcal{H})$, ρ be a density operator on $\mathcal{H}$ and E_{θ}^{u} be the transition expectation on $\mathbf{B}(\mathcal{K})\otimes\mathbf{B}(\mathcal{H})$ with $\Lambda=\theta$ defined in Sec. 2.

One introduces a transition expectation E_{θ}^{u} from $\mathbf{B}(\mathcal{K})\otimes\mathbf{B}(\mathcal{H})$ to $\mathbf{B}(\mathcal{H})$ such as

$$\begin{aligned} E_{\theta}^{u}\left(\sum_{i,j}E_{ij}\otimes A_{ij}\right) &\equiv \sum_{k,m,p,q}\theta(u_{pqk}^{*}A_{km}u_{pqm}) \\ &= \sum_{k,m,p,q}\theta(u_{pqk}^{*})\theta(A_{km})\theta(u_{pqm})\,. \end{aligned} \tag{3.1}$$

The quantum Markov state $\{\rho_{\theta,n}^{u}\}$ on $\bigotimes_{1}^{n}\mathbf{B}(\mathcal{K})$ is defined through this transition expectation E_{θ}^{u} by

$$\begin{aligned} &\mathrm{tr}_{\bigotimes_{1}^{n}\mathcal{K}}[\rho_{\theta,n}^{u}(A_{1}\otimes\cdots\otimes A_{n})] \\ &\quad\equiv \mathrm{tr}_{\mathcal{H}}[\rho E_{\theta}^{u}(A_{1}\otimes E_{\theta}^{u}(A_{2}\otimes\cdots\otimes A_{n-1}\otimes E_{\theta}^{u}(A_{n}\otimes I)\cdots))] \end{aligned} \tag{3.2}$$

for all $A_{1},\ldots,A_{n}\in\mathbf{B}(\mathcal{K})$ and any $\rho\in\mathfrak{S}(\mathcal{H})$.

Let us consider another transition expectation $e_{\theta^{m}}^{u}$ such that

$$e_{\theta^{m}}^{u}\left(\sum_{i,j}E_{ij}\otimes A_{ij}\right)\equiv\sum_{k,l,p,q}\theta^{m}(u_{pqk}^{*})A_{kl}\theta^{m}(u_{pql})\,. \tag{3.3}$$

One can define the quantum Markov state $\{\tilde{\rho}_{\theta,n}^{u}\}$ in terms of e_{θ}^{u}

$$\begin{aligned} &\mathrm{tr}_{\bigotimes_{1}^{n}\mathcal{K}}[\tilde{\rho}_{\theta,n}^{u}(A_{1}\otimes\cdots\otimes A_{n})] \\ &\quad\equiv \mathrm{tr}_{\mathcal{H}}[\rho e_{\theta}^{u}(A_{1}\otimes e_{\theta^{2}}^{u}(A_{2}\otimes\cdots\otimes A_{n-1}\otimes e_{\theta^{n}}^{u}(A_{n}\otimes I)\cdots))] \end{aligned} \tag{3.4}$$

for all $A_{1},\ldots,A_{n}\in\mathbf{B}(\mathcal{K})$ and any $\rho\in\mathfrak{S}(\mathcal{H})$. Then we have the following theorem.

Theorem 3.1. $\rho_{\theta,n}^{u}=\tilde{\rho}_{\theta,n}^{u}$.

Proof. It is easily seen that

$$e_{\theta^{i}}^{u}(A\otimes B)=\theta^{i}e^{u}(A\otimes\theta^{-i}(B))=E_{\theta^{i}}^{u}(A\otimes\theta^{-i}(B)) \tag{3.5}$$

for any $A\in\mathbf{B}(\mathcal{K})$ and any $B\in\mathbf{B}(\mathcal{H})$, where $e^{u}=e_{\theta=\text{identity}}^{u}$. From the equality (3.5), we have

$$e_{\theta}^{u}(A_{1}\otimes I)=\theta e^{u}(A_{1}\otimes I)=E_{\theta}^{u}(A_{1}\otimes I) \tag{3.6}$$

and

$$\begin{aligned} e^u_\theta(A_1 \otimes e^u_{\theta^2}(A_2 \otimes I)) &= e^u_\theta(A_1 \otimes \theta^2 e^u(A_2 \otimes I)) \\ &= \theta e^u(A_1 \otimes \theta^{-1} \circ \theta^2 e^u(A_2 \otimes I)) \\ &= \theta e^u(A_1 \otimes \theta e^u(A_2 \otimes I)) \\ &= E^u_\theta(A_1 \otimes E^u_\theta(A_2 \otimes I)) \,. \end{aligned} \tag{3.7}$$

Generally, we observe

$$\begin{aligned} &e^u_\theta(A_1 \otimes e^u_{\theta^2}(A_2 \otimes \cdots \otimes e^u_{\theta^n}(A_n \otimes I))) \\ &\quad = E^u_\theta(A_1 \otimes E^u_\theta(A_2 \otimes \cdots \otimes E^u_\theta(A_n \otimes I))) \end{aligned}$$

for any $n \geq 1$. Hence $\rho^u_{\theta,n} = \tilde{\rho}^u_{\theta,n}$ holds. □

Let $\mathcal{B}_0$ be a subalgebra of $\mathbf{B}(\mathcal{K})$. Taking the restriction of a transition expectation $E : \mathbf{B}(\mathcal{K}) \otimes \mathbf{B}(\mathcal{H}) \to \mathbf{B}(\mathcal{H})$, to $\mathcal{B}_0 \otimes \mathbf{B}(\mathcal{H})$, i.e. $E_0 = E|_{\mathcal{B}_0 \otimes \mathbf{B}(\mathcal{H})}$, E_0 is the transition expectation from $\mathcal{B}_0 \otimes \mathbf{B}(\mathcal{H})$ to $\mathbf{B}(\mathcal{H})$. The QMC (quantum Markov chain) defines the state $\rho^{u(0)}_{\theta,n}$ on $\bigotimes_1^n \mathcal{B}_0$ through (3.4), which is

$$\rho^{u(0)}_{\theta,n} = \rho^u_{\theta,n}\big|_{\bigotimes_1^n \mathcal{B}_0} \,. \tag{3.8}$$

The subalgebra $\mathcal{B}_0$ of $\mathbf{B}(\mathcal{K})$ can be constructed as follows: Let $P_1, \ldots, P_m$ be projection operators on mutually orthogonal subspaces of $\mathcal{K}$ such that $\sum_{i=1}^m P_i = I_{\mathcal{K}}$. Putting $\mathcal{K}_i = P_i\mathcal{K}$, the subalgebra $\mathcal{B}_0$ is generated by

$$\sum_{i=1}^m P_i A P_i, \qquad A \in \mathbf{B}(\mathcal{K}) \,. \tag{3.9}$$

One observes that in the case of $n = 1$

$$\rho^{u(0)}_{\theta,1} \equiv \rho^u_{\theta,1}|_{\mathcal{B}_0} = \sum_{i=1}^m P_i \rho^u_{\theta,1} P_i \tag{3.10}$$

and one has for any $n \in \mathbb{N}$

$$\rho^{u(0)}_{\theta,n} = \sum_{i_1,\ldots,i_n} (P_{i_1} \otimes \cdots \otimes P_{i_n}) \rho^u_{\theta,n} (P_{i_1} \otimes \cdots \otimes P_{i_n}) \tag{3.11}$$

from which the following theorem is proved (cf., see Ref. 16).

Theorem 3.2.

$$S(\rho^u_{\theta,n}) \leqslant S(\rho^{u(0)}_{\theta,n}) \,.$$

Proof. Let $\{\tilde{P}_{i_1,\ldots,i_n}\}$ be PVM (projection valued measure) on $\otimes_1^n \mathcal{K}$. If $\mathcal{E}_{\tilde{P}}(\tilde{A})$ is a map defined by

$$\mathcal{E}_{\tilde{P}}(\tilde{A}) \equiv \sum_{i_1,\ldots,i_n} \tilde{P}_{i_1,\ldots,i_n} \tilde{A} \tilde{P}_{i_1,\ldots,i_n}, \qquad \forall\, \tilde{A} \in \bigotimes_1^n \mathbf{B}(\mathcal{K}) \,,$$

then $\mathcal{E}_{\tilde{P}}$ satisfies (a) $\mathcal{E}_{\tilde{P}}(I_{\otimes_1^n \mathcal{K}}) = I \in \otimes_1^n \mathcal{B}_0$, (b) $\mathcal{E}_{\tilde{P}} \circ \mathcal{E}_{\tilde{P}}(\tilde{A}) = \mathcal{E}_{\tilde{P}}(\tilde{A})$, $\forall\ \tilde{A} \in \bigotimes_1^n \mathbf{B}(\mathcal{K})$, (c) $\operatorname{tr} \mathcal{E}_{\tilde{P}}(\tilde{A})\tilde{B} = \operatorname{tr} \tilde{A}\tilde{B}$, $\forall\ \tilde{A} \in \bigotimes_1^n \mathbf{B}(\mathcal{K})$, $\forall\ \tilde{B} \in \bigotimes_1^n \mathcal{B}_0$. Therefore $\mathcal{E}_{\tilde{P}}$ is a conditional expectation from $\bigotimes_1^n \mathbf{B}(\mathcal{K})$ to $\bigotimes_1^n \mathcal{B}_0$, so that

$$
\begin{aligned}
S(\rho_{\theta,n}^{u(0)}) &= S(\mathcal{E}_{\tilde{P}}(\rho_{\theta,n}^u)) = \operatorname{tr} \eta(\mathcal{E}_{\tilde{P}}(\rho_{\theta,n}^u)) \\
&\geq \operatorname{tr} \mathcal{E}_{\tilde{P}}(\eta(\rho_{\theta,n}^u)) = \operatorname{tr} \eta(\rho_{\theta,n}^u) = S(\rho_{\theta,n}^u)\,, \qquad (3.12)
\end{aligned}
$$

where $\eta(t) \equiv -t \log t$ $(t \geq 0)$. We used the concavity:

$$\eta(\mathcal{E}_{\tilde{P}}(\rho_{\theta,n}^u)) \geq \mathcal{E}_{\tilde{P}}(\eta(\rho_{\theta,n}^u))\,.$$

□

Taking into account the construction of subalgebra $\mathcal{B}_0$ of $\mathbf{B}(\mathcal{K})$, one can construct a transition expectation in the case that $\mathbf{B}(\mathcal{K})$ is a finite subalgebra of $\mathbf{B}(\mathcal{H})$.

Let $\mathbf{B}(\mathcal{K})$ be the $d \times d$ matrix algebra M_d ($d \leq \dim \mathcal{H}$) in $\mathbf{B}(\mathcal{H})$ and $E_{ij} = |e_i\rangle\langle e_j|$ with normalized vectors $e_i \in \mathcal{H}(i = 1, 2, \ldots, d)$. Let $\gamma_1, \ldots, \gamma_d \in \mathbf{B}(\mathcal{H})$ be a finite operational partition of unity, i.e. $\sum_{i=1}^d \gamma_i^* \gamma_i = I$, then a transition expectation

$$E^\gamma : M_d \otimes \mathbf{B}(\mathcal{H}) \to \mathbf{B}(\mathcal{H}) \qquad (3.13)$$

is defined by

$$E^\gamma \left(\sum_{i,j=1}^d E_{ij} \otimes A_{ij} \right) \equiv \sum_{i,j=1}^d \gamma_i^* A_{ij} \gamma_j\,. \qquad (3.14)$$

Remark that the above type complete positive map E^γ is also discussed in Ref. 21.

Let M_d^0 be a subalgebra of M_d consisting of diagonal elements of M_d. Since an element of M_d^0 has the form $\sum_{i=1}^d b_i E_{ii} (b_i \in \mathbb{C})$, one can see that the restriction $E^{\gamma(0)}$ of E^γ to M_d^0 is defined as

$$E^{\gamma(0)} \left(\sum_{i,j=1}^d E_{ij} \otimes A_{ij} \right) \equiv \sum_{i=1}^d \gamma_i^* A_{ii} \gamma_i\,. \qquad (3.15)$$

When $\Lambda : \mathbf{B}(\mathcal{H}) \to \mathbf{B}(\mathcal{H})$ is a normal unital CP map, the transition expectations E_Λ^γ and $E_\Lambda^{\gamma(0)}$ are defined by

$$E_\Lambda^\gamma \left(\sum_{i,j=1}^d E_{ij} \otimes A_{ij} \right) \equiv \sum_{i,j=1}^d \Lambda(\gamma_i^* A_{ij} \gamma_j)\,,$$

$$E_\Lambda^{\gamma(0)} \left(\sum_{i,j=1}^d E_{ij} \otimes A_{ij} \right) \equiv \sum_{i=1}^d \Lambda(\gamma_i^* A_{ii} \gamma_i)\,.$$

Then one obtains the quantum Markov states $\{\rho_{\Lambda,n}^\gamma\}$ and $\{\rho_{\Lambda,n}^{\gamma(0)}\}$

$$
\begin{aligned}
\rho_{\Lambda,n}^\gamma = &\sum_{i_1,\ldots,i_n=1}^d \sum_{j_1,\ldots,j_n=1}^d \operatorname{tr}_{\mathcal{H}} \rho \Lambda(W_{j_1 i_1}(\Lambda(W_{j_2 i_2}(\cdots \Lambda(W_{j_n i_n}(I_{\mathcal{H}})))))) \\
&\times E_{i_1 j_1} \otimes \cdots \otimes E_{i_n j_n} \qquad (3.16)
\end{aligned}
$$

and

$$\rho_{\Lambda,n}^{\gamma(0)} = \sum_{i_1,\ldots,i_n=1}^{d} \mathrm{tr}_{\mathcal{H}}\, \rho\Lambda(W_{i_1i_1}(\Lambda(W_{i_2i_2}(\cdots\Lambda(W_{i_ni_n}(I_{\mathcal{H}}))))))E_{i_1i_1}\otimes\cdots\otimes E_{i_ni_n}$$

$$= \sum_{i_1,\ldots,i_n=1}^{d} p_{i_1,\ldots,i_n} E_{i_1i_1}\otimes\cdots\otimes E_{i_ni_n}\,, \tag{3.17}$$

where we put

$$W_{ij}(A) \equiv \gamma_i^* A\gamma_j, \qquad A\in \mathbf{B}(\mathcal{H})\,, \tag{3.18}$$

$$W_{ij}^*(\rho) \equiv \gamma_j\rho\gamma_i^*, \qquad \rho\in \mathfrak{S}(\mathcal{H})\,, \tag{3.19}$$

$$p_{i_1,\ldots,i_n} \equiv \mathrm{tr}_{\mathcal{H}}\, \rho\Lambda(W_{i_1i_1}(\Lambda(W_{i_2i_2}(\cdots\Lambda(W_{i_ni_n}(I_{\mathcal{H}}))))))$$

$$= \mathrm{tr}_{\mathcal{H}}\, W_{i_ni_n}^*(\Lambda^*\cdots\Lambda^*(W_{i_2i_2}^*(\Lambda^*(W_{i_1i_1}^*(\Lambda^*(\rho))))))\,. \tag{3.20}$$

The above $\rho_{\Lambda,n}^{\gamma}$, $\rho_{\Lambda,n}^{\gamma(0)}$ become the special cases of $\rho_{\Lambda,n}^{\Gamma,\omega}$ defined in Sec. 2 by taking Γ and ω in (2.10). Therefore the dynamical entropy (1.13) becomes

$$\tilde{S}(\Lambda;\rho,\{\gamma_i\}) \equiv \limsup_{n\to\infty} \frac{1}{n} S(\rho_{\Lambda,n}^{\gamma})\,, \tag{3.21}$$

$$\tilde{S}^{(0)}(\Lambda;\rho,\{\gamma_i\}) \equiv \limsup_{n\to\infty} \frac{1}{n} S(\rho_{\Lambda,n}^{\gamma(0)})\,. \tag{3.22}$$

The dynamical entropies of Λ with respect to a finite-dimensional subalgebra $\mathcal{B}\subset \mathbf{B}(\mathcal{H})$ and the transition expectations E_Λ^{γ} and $E_\Lambda^{\gamma(0)}$ are given by

$$\tilde{S}_{\mathcal{B}}(\Lambda;\rho) \equiv \sup\{\tilde{S}(\Lambda;\rho,\{\gamma_i\}), \{\gamma_i\}\subset \mathcal{B}\}\,, \tag{3.23}$$

$$\tilde{S}_{\mathcal{B}}^{(0)}(\Lambda;\rho) \equiv \sup\{\tilde{S}^{(0)}(\Lambda;\rho,\{\gamma_i\}), \{\gamma_i\}\subset \mathcal{B}\}\,. \tag{3.24}$$

We call (3.23) and (3.24) a generalized AF entropy and a generalized AOW entropy, respectively. When $\{\gamma_i\}$ is PVM (projection valued measure) and Λ is an automorphism θ, $\tilde{S}_{\mathcal{B}}^{(0)}(\theta;\rho)$ is equal to the AOW entropy.[1] When $\{\gamma_i^*\gamma_i\}$ is POV (positive operater valued measure) and $\Lambda=\theta$, $\tilde{S}_{\mathcal{B}}(\theta;\rho)$ is equal to the AF entropy.[6]

From Theorem 3.2, one obtains an inequality

Theorem 3.3.

$$\tilde{S}_{\mathcal{B}}(\Lambda;\rho) \leqslant \tilde{S}_{\mathcal{B}}^{(0)}(\Lambda;\rho)\,. \tag{3.25}$$

That is, the generalized AOW entropy is greater than the generalized AF entropy. Moreover the dynamical entropy $\tilde{S}_{\mathcal{B}}(\Lambda;\rho)$ is rather difficult to compute because there exist off-diagonal parts in (3.16), so we mainly consider the dynamical entropy $\tilde{S}_{\mathcal{B}}^{(0)}(\Lambda;\rho)$ in the next section.

Here, we note that the dynamical entropy defined in terms of $\rho_{\theta,n}^{u}$ on $\bigotimes_1^n \mathbf{B}(\mathcal{K})$ is related to that of flows by Emch,[2] which was defined in terms of the conditional expectation, provided $\mathbf{B}(\mathcal{K})$ is a subalgebra of $\mathbf{B}(\mathcal{H})$.

4. Some Models

We numerically compute the generalized AOW entropy for several models.

Let $\gamma_i = \gamma_i^*$ be projection operators on one-dimensional mutually orthogonal subspaces of $\mathcal{H}$ such that $\sum_i \gamma_i = I_{\mathcal{H}}$ holds. For unitary operators U_i and V on $\mathcal{H}$, $\widetilde{\gamma}_i \equiv U_i \gamma_i V$ satisfies

$$\sum_i \widetilde{\gamma}_i^* \widetilde{\gamma}_i = \sum_i V^* \gamma_i^* U_i^* U_i \gamma_i V = I_{\mathcal{H}} . \tag{4.1}$$

Let us consider a transition expectation

$$E^{\gamma(0)} : \mathbf{A}(\mathcal{H}) \otimes \mathbf{B}(\mathcal{H}) \to \mathbf{B}(\mathcal{H}) \tag{4.2}$$

defined by

$$E^{\gamma(0)} \left(\sum_{i,j=1} E_{ij} \otimes A_{ij} \right) \equiv \sum_i \tilde{\gamma}_i A_{ii} \tilde{\gamma}_i, \qquad A_{ii} \in \mathbf{B}(\mathcal{H}) , \tag{4.3}$$

where $\mathbf{A}(\mathcal{H})$ is an Abelian subalgebla of $\mathbf{B}(\mathcal{H})$ generated by $\sum_i b_i E_{ii} (b_i \in \mathbb{C})$. Let $\Lambda : \mathbf{B}(\mathcal{H}) \to \mathbf{B}(\mathcal{H})$ be a normal unital CP map given by

$$\Lambda(A) \equiv \sum_{k,l,m,n} \lambda_{kl,mn} E_{kl}^* A E_{mn} \tag{4.4}$$

satisfying

$$\sum_k \lambda_{kl,kn} = \delta_{l\,n} \qquad (\Leftrightarrow \Lambda(I_{\mathcal{H}}) = I_{\mathcal{H}}) \quad \text{and} \quad \lambda_{kl,kn} = \sum_j \lambda_{kj,kj} \lambda_{jl,jn} . \tag{4.5}$$

We will compute the dynamical entropy of Λ based on $E^{\gamma(0)}$ above.

Theorem 4.1. *When $\gamma_i = E_{ii}$, the quantum dynamical entropy with respect to Λ, ρ and $\{\gamma_i = E_{ii}\}$ is*

$$\tilde{S}^{(0)}(\Lambda; \rho, \{\gamma_i = E_{ii}\}) = -\sum_{i,j} \sum_{l,k} \lambda_{ij,ij} \lambda_{jl,jk} \operatorname{tr}_{\mathcal{H}}(\rho E_{lk}) \log \lambda_{ij,ij} , \tag{4.6}$$

where $\{\lambda_{ij,ij}\}$ are the coefficients of (4.4) *associated to the CP map Λ.*

Proof. From (3.17), it follows that one obtains

$$\rho_{\Lambda,n}^{\gamma(0)} = \sum_{j_1,\ldots,j_n=1} p_{j_1,\ldots,j_n} E_{j_1 j_1} \otimes \cdots \otimes E_{j_n j_n} , \tag{4.7}$$

where

$$\begin{aligned} p_{j_1,\ldots,j_n} &\equiv \operatorname{tr}_{\mathcal{H}}(\rho \Lambda(W_{j_1 j_1}(\Lambda(W_{j_2 j_2}(\cdots \Lambda(W_{j_n j_n}(I_{\mathcal{H}})))))))\\ &= \operatorname{tr}_{\mathcal{H}}(\rho \Lambda(W_{j_1 j_1}(\Lambda(W_{j_2 j_2}(\cdots W_{j_{n-1} j_{n-1}}(\Lambda(E_{j_n j_n})))))))\\ &= \operatorname{tr}_{\mathcal{H}} \left(\rho \Lambda \left(W_{j_1 j_1} \left(\Lambda \left(W_{j_2 j_2} \left(\cdots W_{j_{n-1} j_{n-1}} \left(\sum_{l,k} \lambda_{j_n l, j_n k} E_{lk} \right) \right) \right) \right) \right) \right) \end{aligned}$$

$$= \lambda_{j_n j_{n-1}, j_n j_{n-1}} \mathrm{tr}_{\mathcal{H}}(\rho\Lambda(W_{j_1 j_1}(\Lambda(W_{j_2 j_2}(\cdots\Lambda(E_{j_{n-1} j_{n-1}}))))))$$

$$= \lambda_{j_n j_{n-1}, j_n j_{n-1}} \cdots \lambda_{j_2 j_1, j_2 j_1} \mathrm{tr}_{\mathcal{H}}(\rho\Lambda(E_{j_1 j_1}))$$

$$= \prod_{k=2}^{n} \lambda_{j_k j_{k-1}, j_k j_{k-1}} \sum_{l,k} \lambda_{j_1 l, j_1 k} \mathrm{tr}_{\mathcal{H}}(\rho E_{lk}) . \tag{4.8}$$

From the completely positivity of Λ and $\Lambda(I_{\mathcal{H}}) = I_{\mathcal{H}}$, it follows that

$$\lambda_{kl,kl} \geqslant 0 \tag{4.9}$$

and

$$\sum_k \lambda_{kl,kl} = 1 , \tag{4.10}$$

which implies $P_n = \{p_{j_1,\ldots,j_n}\}$ has the Markov property. Therefore the quantum dynamical entropy with respect to Λ, ρ and $\{\gamma_i = E_{ii}\}$ is computed as

$$\tilde{S}^{(0)}(\Lambda; \rho, \{\gamma_i = E_{ii}\}) = -\sum_{i,j}\sum_{l,n} \lambda_{ij,ij}\lambda_{jl,jn} \mathrm{tr}_{\mathcal{H}}(\rho E_{ln}) \log \lambda_{ij,ij} .$$

□

Theorem 4.2. *When $\gamma_i = U^* E_{ii} U (= |f_i\rangle\langle f_i| = F_{ii})$ for a unitary operator U on $\mathcal{H}$, the quantum dynamical entropy with respect to Λ, ρ and $\{\gamma_i = F_{ii}\}$ is*

$$\tilde{S}^{(0)}(\Lambda; \rho, \{\gamma_i = F_{ii}\}) = -\sum_{i,j}\sum_{l,k} \lambda^{(F)}_{ij,ij}\lambda_{jl,jk} \ \mathrm{tr}_{\mathcal{H}}(\rho U^* E_{lk} U) \log \lambda^{(F)}_{ij,ij} , \tag{4.11}$$

where $\{\lambda_{ij,ij}\}$ are the coefficients of (4.4) associated to the CP map Λ and $\lambda^{(F)}_{ij,ij} = \sum_{p,q,r,s} \lambda_{pq,rs} \ \mathrm{tr}_{\mathcal{H}} \, E_{qp} U^* E_{ii} U E_{rs} U^* E_{jj} U$.

Proof. Let $F_{ij} \equiv U^* E_{ij} U$. Then

$$|f_i\rangle = U^*|e_i\rangle, \qquad F_{ij} = |f_i\rangle\langle f_j|, \qquad e_i, f_i \in \mathcal{H}$$

and one has

$$\begin{aligned} E_{kl}x &= UF_{kl}U^*x = U|f_k\rangle\langle f_l|U^*x = Uf_k\langle f_l, U^*x\rangle = Uf_k\langle Uf_l, x\rangle \\ &= \sum_{m,n} |f_m\rangle\langle f_m, Uf_k\rangle\langle Uf_l, f_n\rangle\langle f_n, x\rangle \\ &= \sum_{m,n} \langle f_m, Uf_k\rangle\langle Uf_l, f_n\rangle|f_m\rangle\langle f_n|x \\ &= \sum_{m,n} \langle f_m, Uf_k\rangle\langle Uf_l, f_n\rangle F_{mn}x, \qquad \forall\, x \in \mathcal{H} , \end{aligned} \tag{4.12}$$

which implies

$$E_{kl} = \sum_{m,n} u_{kl,mn}F_{mn}, \qquad u_{kl,mn} \equiv \overline{\langle Uf_k, f_m\rangle}\langle Uf_l, f_n\rangle . \tag{4.13}$$

Inserting (4.13) into (4.4), one obtains

$$\begin{aligned}\Lambda(A) &= \sum_{p,q,r,s} \lambda_{pq,rs} E_{pq}^* A E_{rs} \\ &= \sum_{p,q,r,s} \lambda_{pq,rs} \left(\sum_{k,l} \bar{u}_{pq,kl} F_{kl}^*\right) A \left(\sum_{m,n} u_{rs,mn} F_{mn}\right) \\ &= \sum_{k,l,m,n} \left(\sum_{p,q,r,s} \lambda_{pq,rs} \bar{u}_{pq,kl} u_{rs,mn}\right) F_{kl}^* A F_{mn} \\ &= \sum_{k,l,m,n} \lambda_{kl,mn}^{(F)} F_{kl}^* A F_{mn}\,, \end{aligned} \tag{4.14}$$

where

$$\begin{aligned}\lambda_{kl,mn}^{(F)} &\equiv \sum_{p,q,r,s} \lambda_{pq,rs} \bar{u}_{pq,kl} u_{rs,mn} \\ &= \sum_{p,q,r,s} \lambda_{pq,rs} \operatorname{tr}_{\mathcal{H}} E_{qp} U^* E_{ii} U E_{rs} U^* E_{jj} U\,. \end{aligned} \tag{4.15}$$

Proceeding as before, it is shown that

$$\begin{aligned} p_{j_1,\dots,j_n}^{(F)} &= \lambda_{j_n j_{n-1};j_n j_{n-1}}^{(F)} \cdots \lambda_{j_2 j_1;j_2 j_1}^{(F)} \sum_{l,k} \lambda_{j_1 l, j_1 k} \operatorname{tr}_{\mathcal{H}}(\rho F_{l\,k}) \\ &= \lambda_{j_n j_{n-1};j_n j_{n-1}}^{(F)} \cdots \lambda_{j_2 j_1;j_2 j_1}^{(F)} \sum_{l,n} \lambda_{j_1 l, j_1 k} \operatorname{tr}_{\mathcal{H}}(\rho U^* E_{lk} U) \end{aligned} \tag{4.16}$$

because of $\gamma_i = F_{ii}$. Since $P_n^{(F)} = \{p_{j_1,\dots,j_n}^{(F)}\}$ has the Markov property, we obtain the quantum dynamical entropy with respect to Λ, ρ and $\{\gamma_i = F_{ii}\}$ as

$$\tilde{S}^{(0)}(\Lambda; \rho, \{\gamma_i = F_{ii}\}) = -\sum_{i,j} \sum_{l,k} \lambda_{ij,ij}^{(F)} \lambda_{jl,jk} \operatorname{tr}_{\mathcal{H}}(\rho U^* E_{lk} U) \log \lambda_{ij,ij}^{(F)}\,.$$

□

Theorem 4.3. *Take $\Lambda_U(A) \equiv U^* A U$ for a unitary operator U on $\mathcal{H}$. When U has a simple point spectral $\{e^{i\varphi_k}\}$ and its eigenvector f_k, the dynamical entropy with respect to Λ_U, ρ and $\{\gamma_k = |f_k\rangle\langle f_k|\}$ is*

$$\tilde{S}^{(0)}(\Lambda_U; \rho, \{\gamma_k = |f_k\rangle\langle f_k|\}) = 0\,. \tag{4.17}$$

Proof. For a CONS $e = \{e_m\}$ of $\mathcal{H}$, U is written by

$$U = \sum_{m,n} \lambda_{mn}^{(e)} E_{mn}\,,$$

where

$$\lambda_{mn}^{(e)} \equiv \langle e_m, U e_n\rangle\,, \tag{4.18}$$

$$E_{mn} \equiv |e_m\rangle\langle e_n|\,. \tag{4.19}$$

From the definition of Λ_U, one obtains

$$\Lambda_U(A) = \sum_{k,l,m,n} \lambda^{(e)}_{kl,mn} E_{lk} A E_{mn} \,,$$

where $\lambda^{(e)}_{kl,mn} \equiv \bar{\lambda}^{(e)}_{kl} \lambda^{(e)}_{mn}$. Therefore, one has

$$p^{(e)}_{j_1,\dots,j_n} \equiv \lambda^{(e)}_{j_nj_{n-1},j_nj_{n-1}} \cdots \lambda^{(e)}_{j_2j_1,j_2j_1} \sum_{l,k} \lambda^{(e)}_{j_1l,j_1k} \, \mathrm{tr}_{\mathcal{H}}(\rho E_{lk}) \,,$$

$$\lambda^{(e)}_{mn,mn} \equiv |\langle e_m, Ue_n\rangle|^2 \,.$$

$P_n = \{p_{j_1,\dots,j_n}\}$ has Markov property. Moreover, for the eigenvectors $\{f_k\}$ of the simple point spectral $\{e^{i\varphi_k}\}$ of U such that

$$Uf_k = e^{i\varphi_k} f_k \,, \tag{4.20}$$

since

$$\lambda^{(f)}_{mn,mn} = |\langle f_m, Uf_n\rangle|^2 = \delta_{mn} \,, \tag{4.21}$$

$$p^{(f)}_{j_1} \equiv \sum_{l,k} \lambda^{(f)}_{j_1l,j_1k} \, \mathrm{tr}_{\mathcal{H}}(\rho |f_l\rangle\langle f_k|) = \langle f_{j_1},\ \rho\, f_{j_1}\rangle \,, \tag{4.22}$$

the dynamical entropy with respect to Λ_U, ρ and $\{\gamma_k = |f_k\rangle\langle f_k|\}$ is

$$\tilde{S}^{(0)}(\Lambda_U; \rho, \{\gamma_k = |f_k\rangle\langle f_k|\}) = \lim_{n\to\infty} \frac{1}{n} \left(- \sum_{j_1=1}^{d} \langle f_{j_1},\ \rho\, f_{j_1}\rangle \log\langle f_{j_1},\ \rho\, f_{j_1}\rangle \right) = 0 \,.$$ □

We remark here that for another choice of base $\{g_i\} \subset \mathcal{H}$, one has

$$\tilde{S}^{(0)}(\Lambda_U; \rho, \{\gamma_k = |g_k\rangle\langle g_k|\}) > 0 \,.$$

Now we study the dynamical entropy for a quantum communication process, in particular, the attenuation process. That is, Λ^* is the attenuation channel[22] defined as follows: Let $\mathcal{H} = \mathcal{L}^2(\mathbb{R})$, $|\theta\rangle$ be a coherent state vector in $\mathcal{H}$ and $\gamma \equiv \{\gamma_j\}_{j=1}^2$, where $\gamma_j = |x_j\rangle\langle x_j|$ and

$$|x_j\rangle = a_j|0\rangle + b_j|\theta\rangle \,,$$

$$|a_j|^2 = \frac{\tau_j^2}{\tau_j^2 + 2\exp(-\frac{1}{2}|\theta|^2)\tau_j + 1} \,,$$

$$|b_j|^2 = \frac{1}{\tau_j^2 + 2\exp(-\frac{1}{2}|\theta|^2)\tau_j + 1} \,,$$

$$a_j\bar{b}_j = \bar{a}_j b_j = \frac{\tau_j}{\tau_j^2 + 2\exp(-\frac{1}{2}|\theta|^2)\tau_j + 1} \,,$$

$$\tau_j = \frac{-(1-2\lambda) - (-1)^j \sqrt{1 - 4\lambda(1-\lambda)(1-\exp(-|\theta|^2))}}{2(1-\lambda)\exp(-\frac{1}{2}|\theta|^2)} \,.$$

The attenuation channel Λ^* with a transmission rate η is defined by

$$\Lambda^*(|\theta\rangle\langle\theta|) \equiv |\sqrt{\eta}\theta\rangle\langle\sqrt{\eta}\theta| \,.$$

Theorem 4.4. *When ρ is given by $\rho = \lambda|0\rangle\langle 0| + (1-\lambda)|\theta\rangle\langle\theta|$ and Λ^* is the attenuation channel with a transmission rate η satisfying $\sum_j p_{k,j}p_j = p_k$, the quantum dynamical entropy with respect to Λ, ρ and $\{\gamma_j\}$ is obtained by*

$$\tilde{S}^{(0)}(\Lambda;\rho,\{\gamma_j\}) = -\sum_{j,k} p_{k,j}p_j \log p_{k,j}\,, \tag{4.23}$$

where $p_j = \lambda|\langle 0, x_j\rangle|^2 + (1-\lambda)|\langle\sqrt{\eta}\theta, x_j\rangle|^2$ and

$$\begin{aligned}
p_{k,j} &= \upsilon_j^+|\langle x_k, x_j^+\rangle|^2 + (1-\upsilon_j^+)|\langle x_k, x_j^-\rangle|^2\,,\\
|x_j^+\rangle &= a_j^+|0\rangle + b_j^+|\sqrt{\eta}\theta\rangle, |x_j^-\rangle = a_j^-|0\rangle + b_j^-|\sqrt{\eta}\theta\rangle\,,\\
a_j^+ &= \varepsilon_j^+ a_j, a_j^- = \varepsilon_j^- a_j, b_j^+ = \varepsilon_j^+ b_j, b_j^- = \varepsilon_j^- b_j\,,\\
\varepsilon_j^+ &= \sqrt{\frac{\tau_j^2+2\exp(-\frac{1}{2}|\theta|^2)\tau_j+1}{\tau_j^2+2\exp(-\frac{1}{2}\eta|\theta|^2)\tau_j+1}}\,,\\
\varepsilon_j^- &= \sqrt{\frac{\tau_j^2+2\exp(-\frac{1}{2}|\theta|^2)\tau_j+1}{\tau_j^2-2\exp(-\frac{1}{2}\eta|\theta|^2)\tau_j+1}}\,,\\
\upsilon_j^+ &= \frac{1}{2}\left(1+\exp\left(-\frac{1}{2}(1-\eta)|\theta|^2\right)\right)\frac{1}{(\varepsilon_j^+)^2}\,.
\end{aligned}$$

Proof. The formula (3.17) can be rewritten in the form ($n \geqslant 3$).

$$\rho_{\Lambda,n}^{\gamma(0)} = \sum_{j_1,\ldots,j_n=1}^{2} p_{j_1,\ldots,j_n}\bigotimes_{k=1}^{n}|x_{j_k}\rangle\langle x_{j_k}| \tag{4.24}$$

where

$$p_{j_1,\ldots,j_n} \equiv \mathrm{tr}_{\mathcal{H}}\, W_{j_nj_n}^*(\Lambda^*(\cdots\Lambda^*(W_{j_2j_2}^*(\Lambda^*(W_{j_1j_1}^*(\Lambda^*(\rho)))))\cdots))\,, \tag{4.25}$$

$$\Lambda^*(\rho) = \lambda|0\rangle\langle 0| + (1-\lambda)|\sqrt{\eta}\theta\rangle\langle\sqrt{\eta}\theta|\,,$$

$$W_{jj}^*(\Lambda^*(\rho)) \equiv \gamma_j^*\Lambda^*(\rho)\gamma_j = (\lambda|\langle 0, x_j\rangle|^2 + (1-\lambda)|\langle\sqrt{\eta}\theta, x_j\rangle|^2)|x_j\rangle\langle x_j|\,. \tag{4.26}$$

From Refs. 23 and 24, we have

$$\Lambda^*(|x_j\rangle\langle x_j|) = \upsilon_j^+|x_j^+\rangle\langle x_j^+| + (1-\upsilon_j^+)|x_j^-\rangle\langle x_j^-|\,. \tag{4.27}$$

Thus one has

$$\begin{aligned}
p_{j_1,\ldots,j_n} &= \prod_{k=2}^{n}(\upsilon_{j_{k-1}}^+|\langle x_{j_k}, x_{j_{k-1}}^+\rangle|^2 + (1-\upsilon_{j_{k-1}}^+)|\langle x_{j_k}, x_{j_{k-1}}^-\rangle|^2)\\
&\quad\times(\lambda|\langle 0, x_{j_1}\rangle|^2 + (1-\lambda)|\langle\sqrt{\eta}\theta, x_{j_1}\rangle|^2)\\
&= \prod_{k=2}^{n} p_{j_k,j_{k-1}}p_{j_1}\,.
\end{aligned}$$

When $\sum_j p_{k,j} p_j = p_k$ is hold, we obtain the dynamical entropy with respect to Λ, ρ and $\{\gamma_j\}$ such as

$$\tilde{S}^{(0)}(\Lambda; \rho, \{\gamma_j\}) = -\sum_{j,k} p_{k,j} p_j \log p_{k,j} \,. \qquad \square$$

Acknowledgment

We thank Prof. Petz for his useful comments to our present work. We also thank Prof. Accardi for his encouragement.

References

1. L. Accardi, M. Ohya and N. Watanabe, *Dynamical entropy through quantum Markov chain, Open System Infor. Dynamics* **4** (1997) 71–87.
2. G. G. Emch, *Positivity of the K-entropy on non-Abelian K-flows, Z. Wahrscheinlichkeitstheory Gebiete* **29** (1974) 241.
3. A. Connes and E. Störmer, *Entropy for automorphisms of II_1 von Neumann algebras, Acta Math.* **134** (1975) 289–306.
4. A. Connes, H. Narnhoffer and W. Thirring, *Dynamical entropy of C^*-algebras and von Neumann algebras, Commun. Math. Phys.* **112** (1987) 691–719.
5. Y. M. Park, *Dynamical entropy of generalized quantum Markov chains, Lett. Math. Phys.* **32** (1994) 63–74.
6. R. Alicki and M. Fannes, *Defining quantum dynamical entropy, Lett. Math. Phys.* **32** (1994) 75–82.
7. T. Hudetz, *Topological entropy for appropriately approximated C^*-algebras, J. Math. Phys.* **35** (1994) 4303–4333.
8. M. Ohya, *State change, complexity and fractal in quantum systems, Quantum Commun. Measurement* **2** (1995) 309–320.
9. M. Ohya, *Some aspects of quantum information theory and their applications to irreversible processes, Rep. Math. Phys.* **27** (1989) 19–47.
10. R. S. Ingarden, A. Kossakowski and M. Ohya, *Information Dynamics and Open Systems* (Kluwer, 1997).
11. D. Voiculescu, *Dynamical approximation entropies and topological entropy in operator algebras, Commun. Math. Phys.* **170** (1995) 249–281.
12. F. Benatti, **Deterministic Chaos in Infinite Quantum Systems** (Springer, 1993).
13. N. Muraki and M. Ohya, *Entropy functionals of Kolmogorov Sinai type and their limit theorems, Lett. Math. Phys.* **36** (1996) 327–335.
14. L. Accardi, M. Ohya and N. Watanabe, *Note on quantum dynamical entropies, Rep. Math. Phys.* **38** (1996) 457–469.
15. M. Choda, *Entropy for extensions of Bernoulli shifts, Ergodic Theory Dynamic Systems* **16** (1996) 1197–1206.
16. M. Ohya and D. Petz, **Quantum Entropy and Its Use** (Springer, 1993).
17. L. Accardi and M. Ohya, *Compound channels, transition expectations and liftings*, to appear in *J. Multivariate Anal.*
18. M. Ohya, *Quantum ergodic channels in operator algebras, J. Math. Anal. Appl.* **84** (1981) 318–328.
19. M. Ohya, *Note on quantum probability, Lett. Nuovo Cimento* **38** (1983) 402–404.
20. J. von Neumann, **Die Mathematischen Grundlagen der Quantenmechanik** (Springer-Verlag, 1932).

21. P. Tuyls, *Comparing quantum dynamical entropies, Banach Centre Publication* **43** (1998) 411–420.
22. M. Ohya, *On compound state and mutual information in quantum information theory, IEEE Trans. Information Theory* **29** (1983) 770–774.
23. M. Ohya, D. Petz and N. Watanabe, *Numerical computation of quantum capacity, Internat. J. Theor. Phys.* **37** (1998) 507–510.
24. M. Ohya, D. Petz and N. Watanabe, *Capacity of a noisy quantum channel,* SUT preprint.

International Journal of Theoretical Physics, Vol. 37, No. 1, 1998

Complexities and Their Applications to Characterization of Chaos[2]

Masanori Ohya[1]

Received July 4, 1997

The concept of complexity in Information Dynamics is discussed. The chaos degree defined by the complexities is applied to examine chaotic behavior of logistic map.

1. INTRODUCTION

There are several tools to describe chaotic aspects of natural or nonnatural phenomena such as entropy. The concept of complexity is one such tool. In 1991 the author proposed Information Dynamics (ID, for short) to synthesize the dynamics of state change and the complexity of a system. In this paper, I briefly review the concept of ID and discuss some applications of the entropic complexities in ID to the characterization of chaos.

2. INFORMATION DYNAMICS

Information Dynamics is an attempt to provide a new view for the study of chaotic behavior of systems (Ohya, 1995).

Let $(\mathcal{A}, \mathfrak{S}, \alpha(G))$ be an input (or initial) system and $(\overline{\mathcal{A}}, \overline{\mathfrak{S}}, \overline{\alpha}(\overline{G}))$ be an output (or final) system. Here $\mathcal{A}$ is the set of all objects to be observed and $\mathfrak{S}$ is the set of all means of getting the observed value, $\alpha(G)$ is a certain evolution of the system. Often we have $\mathcal{A} = \overline{\mathcal{A}}$, $\mathfrak{S} = \overline{\mathfrak{S}}$, $\alpha = \overline{\alpha}$.

[1] Department of Information Sciences, Science University of Tokyo, Noda City, Chiba 278, Japan.

[2] Dedicated to Professor Gérard G. Emch on his 60th birthday.

Therefore we claim

Giving a mathematical structure to input and output triples

≡ Having a theory

For instance, when $\mathcal{A}$ is the set $M(\Omega)$ of all measurable functions on a measurable space $(\Omega, \mathcal{F})$ and $\mathfrak{S}(\mathcal{A})$ is the set $P(\Omega)$ of all probability measures on Ω, we have usual probability theory, by which the classical dynamical system is described. When $\mathcal{A} = B(\mathcal{H})$, the set of all bounded linear operators on a Hilbert space $\mathcal{H}$, and $\mathfrak{S}(\mathcal{A}) = \mathfrak{S}(\mathcal{H})$, the set of all density operators on $\mathcal{H}$, we have a quantum dynamical system.

The dynamics of state change is described by a channel Λ^*: $\mathfrak{S} \to \overline{\mathfrak{S}}$ (sometimes $\mathfrak{S} \to \mathfrak{S}$). The fundamental point of ID is that there exist two complexities in ID itself.

Let $(\mathcal{A}_i, \mathfrak{S}_i, \alpha^t(G_t))$ be the total system of $(\mathcal{A}, \mathfrak{S}, \alpha)$ and $(\overline{\mathcal{A}}, \overline{\mathfrak{S}}, \overline{\alpha})$, and $\mathcal{S}$ be a subset of $\mathfrak{S}$ from which we measure the observables and we call this subset a reference system [e.g., $\mathcal{S} = I(\alpha)$, the set of all invariant elements of α].

$G^{\mathcal{S}}(\varphi)$ is the complexity of a state φ measured from $\mathcal{S}$ and $T^{\mathcal{S}}(\varphi; \Lambda^*)$ is the transmitted complexity associated with the state change $\varphi \to \Lambda^*\varphi$, which satisfy the following properties:

(i) For any $\varphi \in \mathcal{S} \subset \mathfrak{C}$,

$$C^{\mathcal{S}}(\varphi) \geq 0, \qquad T^{\mathcal{S}}(\varphi; \Lambda^*) \geq 0$$

(ii) For any orthogonal bijection $\bar{j} : ex\,\mathcal{S} \to ex\,\mathcal{S}$ (the set of all extreme points in $\mathcal{S}$),

$$C^{j(\mathcal{S})}(j(\varphi)) = C^{\mathcal{S}}(\varphi)$$

$$T^{j(\mathcal{S})}(j(\varphi); \Lambda^*) \doteq T^{\mathcal{S}}(\varphi; \Lambda^*)$$

(iii) For $\Phi \equiv \varphi \otimes \psi \in \mathcal{S}_t \subset \mathfrak{S}_t$,

$$C^{\mathcal{S}_t}(\Phi) = C^{\mathcal{S}}(\varphi) + C^{\overline{\mathcal{S}}}(\psi)$$

(iv) For any state φ and a channel Λ^*,

$$0 \leq T^{\mathcal{S}}(\varphi; \Lambda^*) \leq C^{\mathcal{S}}(\varphi)$$

(v) For the identity map *id* from $\mathfrak{S}$ to GS,

$$T^{\mathcal{S}}(\varphi; id) = C^{\mathcal{S}}(\varphi)$$

Instead of (iii), when "(iii′) $\Phi \in \mathcal{S}_t \subset \mathfrak{S}_t$, put $\varphi \equiv \Phi \restriction_{\mathcal{A}}$ (i.e., the restriction of Φ to $\mathcal{A}$), $\psi \equiv \Phi \restriction_{\overline{\mathcal{A}}}$, $C^{\mathcal{S}_t}(\Phi) \leq C^{\mathcal{S}}(\varphi) + C^{\overline{\mathcal{S}}}(\psi)$" is satisfied,

C and T are called a pair of strong complexity. Therefore ID can be considered as follows.

Definition 1. Information Dynamics (ID) is defined by

$$(\mathcal{A}, \mathfrak{S}, \alpha(G); \overline{\mathcal{A}}, \overline{\mathfrak{S}}, \overline{\alpha}(\overline{G}); \Lambda^*; C^{\mathcal{S}}(\varphi), T^{\mathcal{S}}, (\varphi; \Lambda^*))$$

and some relations R among them.

Thus, in the framework of ID, we have to:
(i) Determine mathematically

$$\mathcal{A}, \mathfrak{S}, \alpha(G); \overline{\mathcal{A}}, \overline{\mathfrak{S}}, \overline{\alpha}(\overline{G})$$

(ii) Choose Λ^* and R.
(iii) Define $C^S(\varphi)$, T^S $(\varphi; \Lambda^*)$.

Information Dynamics can be applied to the study of chaos in the following ways:

(a) ψ is more chaotic than φ as seen from the reference system $\mathcal{S}$ if $C^{\mathcal{S}}(\psi) \geq C^{\mathcal{S}}(\varphi)$.
(b) When φ changes to $\Lambda^*\varphi$, a *degree of chaos* associated to this state change is given by

$$D^{\mathcal{S}}(\varphi; \Lambda^*) = C^{\overline{\mathcal{S}}}(\Lambda^*\varphi) - T^{\mathcal{S}}(\varphi; \Lambda^*)$$

In ID, several different topics can be treated from a common standpoint (Matsuoka and Ohya, 1995; Ohya, 1991a, n.d.-a, c; Ohya and Watanabe, 1993). Although there exist several complexities (Ohya, 1997), one of the most fundamental pairs of C and T in quantum system is the von Neumann entropy and the mutual entropy, whose C and T are modified to formulate the entropic complexities such as ϵ-entropy (ϵ-entropic complexity) (Ohya, 1989, 1991b, 1995) and Kolmogorov–Sinai type dynamical entropy (entropic complexity) (Accardi *et al.*, 1996; Muraki and Ohya, 1996).

In this paper, we discuss some applications of entropic complexities to the study of chaos.

3. CHANNEL

The concept of channel or channeling transformation is fundamental in ID and it is a convenient mathematical tool to treat several physical dynamics in a unified way (Ohya, 1981).

In classical systems, an input (or initial) system is described by the set of all random variables $\mathcal{A} = M(\Omega)$ and its state space $\mathfrak{S} = P(\Omega)$, and an output (or final) system by M $(\overline{\Omega})$ and P $(\overline{\Omega})$.

A quantum system is described on a Hilbert space $\mathcal{H}$. That is, an input $\mathcal{A}$ is the set $B(\mathcal{H})$ of all bounded linear operators on $\mathcal{H}$, and $\mathfrak{S}$ is the set $T(\mathcal{H})$ of all density operators on $\mathcal{H}$. An output system is $\overline{\mathcal{A}} = B(\overline{\mathcal{H}})$ and $\overline{\mathfrak{S}} = T(\overline{\mathcal{H}})$. A more general quantum system is described by a C*-algebra and its space, but this general frame is not used in this paper.

In any case, a channel is a mapping from $\mathfrak{S}(P(\Omega))$ or $T(\mathcal{H})$, resp.) to $\overline{\mathfrak{S}}$ ($P(\overline{\Omega})$ or $T(\overline{\mathcal{H}})$, *resp.*). Almost all physical transformations are described by this mapping.

Definition 2. Let Λ^* be a channel from $\mathfrak{S}$ to $\overline{\mathfrak{S}}$.

(1) Λ^* is linear if $\Lambda^*(\lambda\varphi + (1-\lambda)\psi) = \lambda\Lambda^*\varphi + (1-\lambda)\Lambda^*\psi$ holds for all $\varphi, \psi \in \mathfrak{S}$ and any $\lambda \in [0, 1]$.

(2) Λ^* is completely positive (C. P.) if Λ^* is linear and its dual $\Lambda: \overline{\mathcal{A}} \to \mathcal{A}$ satisfies

$$\sum_{i,j=1}^{n} A_i^*\Lambda(\overline{A}_i^*\overline{A}_j)A_j \geq 0$$

for any $n \in \mathbf{N}$ and any $\{\overline{\mathcal{A}}_i\} \subset \overline{\mathcal{A}}$, $\{A_i\} \subset \mathcal{A}$.

Most channels appearing in physical processes are C.P. channels. We here list a few examples of such channels (Ohya, 1989). Take a density operator ρ as an input (initial) state.

(1) *Time evolution*: Let $\{U_t\,;\, t \in \mathbf{R}^+\}$ be one-parameter group or semigroup on $\mathcal{H}$. We have

$$\rho \to \Lambda_t^*\rho = U_t\rho U_t^*$$

(2) *Quantum measurement*: When a measuring apparatus is described by a positive operator-valued measure $\{Q_n\}$ and the measurement is carried out in a state ρ, the state ρ changes to a state $\Lambda^*\rho$ by this measurement such that

$$\rho \to \Lambda^*\rho = \sum_n Q_n^{1/2}\rho Q_n^{1/2}$$

(3) *Reduction*: If a system Σ_1 interacts with an external system Σ_2 described by another Hilbert space $\mathcal{K}$ and the initial states of Σ_1 and Σ_2 are ρ and σ, respectively, then the combined state θ_t of Σ_1 and Σ_2 at time t after the interaction between two systems is given by

$$\theta_t \equiv U_t(\rho \otimes \sigma)U_t^*$$

where $U_t = \exp(-itH)$ with the total Hamiltonian H of Σ_1 and Σ_2. A channel is obtained by taking the partial trace w.r.t. $\mathcal{K}$ such as

$$\rho \to \Lambda_t^*\rho \equiv \mathrm{tr}_{\mathcal{K}}\theta_t$$

4. QUANTUM ENTROPY AS COMPLEXITY

The concept of entropy was introduced and developed to study the following topics: irreversible behavior, symmetry breaking, amount of information transmission, chaotic properties of states, etc. Here we review quantum entropies as an example of our complexities C and T.

A state in quantum systems is described by a density operator on a Hilbert space $\mathcal{H}$. The entropy of a state ρ was introduced by von Neumann (1932; Ohya and Petz, 1993) as

$$S(\rho) = -\mathrm{tr}\, \rho \log \rho$$

If $\rho = \Sigma_k p_k E_k$ is the Schatten decomposition (i.e., p_k is the eigenvalue of ρ and E_k is the one-dimensional projection associated with p_k, this decomposition is not unique unless every eigenvalue is nondegenerate of ρ, then

$$S(\rho) = -\sum_k p_k \log p_k$$

because $\{p_k\}$ is a probability distribution. Therefore the von Neumann entropy contains the Shannon entropy as a special case.

For two states $\rho, \sigma \in \mathfrak{S}(\mathcal{H})$, the relative entropy (Umegaki, 1962) is defined by

$$S(\rho, \sigma) = \begin{cases} \mathrm{tr}\, \rho(\log \rho - \log \sigma), & \rho << \sigma \\ +\infty, & \text{otherwise} \end{cases}$$

where $\rho << \sigma$ means that $\mathrm{tr}\, \sigma A = 0 \Rightarrow \mathrm{tr}\, \rho A = 0$ for any $A \geq 0$.

Let $\Lambda^*: \mathfrak{S}(\mathcal{H}) \rightarrow \overline{\mathfrak{S}}(\overline{\mathcal{H}})$ be a channel and define the compound state by

$$\theta_E = \sum_k p_k E_k \otimes \Lambda^* E_k$$

which expresses the correlation between the initial state ρ and the final state $\Lambda^*\rho$ (Ohya, 1983a, b). The mutual entropy (Ohya, 1983a) for a state $\rho \in \mathfrak{S}(\mathcal{H})$ and a channel Λ^* is given by

$$\begin{aligned} I(\rho; \Lambda^*) &= \sup\{S(\theta_E, \rho \otimes \Lambda^*\rho); E = \{E_k\}\} \\ &= \sup\left\{\sum_k p_k S(\Lambda^* E_k, \Lambda^*\rho); E = \{E_k\}\right\} \end{aligned}$$

where the supremum is taken over all Schatten decompositions. The above entropy and mutual entropy become a pair of our two complexities according to the following facts:

(1) The fundamental inequality of Shannon type (Shannon 1948; Ohya, 1983a)

$$0 \le I(\rho; \Lambda^*) \le \min\{S(\rho), S(\Lambda^*\rho)\}$$

because of $S(\Lambda^*E_k, \Lambda^*\rho) = S(\Lambda^*\rho) - \Sigma_k p_k S(\Lambda^* E_k) \le S(\Lambda^*\rho)$ and the monotonicity (Uhlmann, 1977; Ohya and Petz, 1993) of the relative entropy: $S(\Lambda^*E_k, \Lambda^*\rho) \le S(E_k, \rho)$.

(2) $I(\rho; id) = S(\rho)$, which is proved as follows:

$$I(\rho; \mathrm{id}) = \sup\left\{\sum_k p_k S(E_k, \rho); E = \{E_k\}\right\}$$

$$= \sup\left\{S(\rho) - \sum_k p_k S(E_k) E = \{E_k\}\right\} = S(\rho)$$

because of $S(E_k) = 0$.

In Shannon's communication theory in classical systems, ρ is a probability distribution $p = (p_k)$ and Λ^* is a transition probability (t_{ij}), so that the Schatten decomposition of ρ is unique and the compound state of ρ and its output $\bar{\rho}$ $[\equiv \bar{p} = (\bar{p}_i)]$ is the joint distribution $r = (r_{ij})$ with $r_{ij} \equiv t_{ij}p_j$. Then the above complexities C and T become the Shannon entropy and mutual entropy, respectively,

$$S(p) = -\sum_k p_k \log p_k$$

$$I(p; \Lambda^*) = \sum_{i,j} r_{ij} \log \frac{r_{ij}}{p_j \bar{p}_i}$$

We can construct several other types of entropic complexities and they are used to define the quantum dynamical entropy (Muraki and Ohya, 1996; Ohya, n.d.-a; Accardi *et al.*, 1996), which is one of fundamental tools to describe chaotic aspects of a dynamical system (Billingsley, 1965; Benatti, 1993; Connes *et al.*, 1987; Connes and Stormer, 1975; Emch, 1975). For instance, one pair of the complexities is

$$T(\rho; \Lambda^*) = \sup\left\{\sum_k p_k S(\Lambda^*\rho_k, \Lambda^*\rho); \rho = \sum_k p_k\rho_k\right\}, \qquad C(\rho) = T(\rho; id)$$

where $\rho = \Sigma_k p_k \rho_k$ is a finite decomposition of ρ and the supremum is taken over all such finite decompositions.

The mutual entropy given above contains other definitions of the mutual information (Holevo, 1973; Ingarden, 1976; Levitin, 1991). Moreover it is not only a fundamental quantity to study quantum communication processes such as the capacity of a quantum channel (Ohya, n.d.-b; Ohya *et al.*, 1997), but also can be used to study irreversible processes (Ohya, 1989).

5. CHAOS DEGREE

We apply the chaos degree in ID to a deterministic dynamical system and discuss its usefulness. The degree of chaos for a state ρ (density operator or probability distribution) and a channel Λ^* is defined as

$$\begin{aligned} D(\rho; \Lambda^*) &\equiv C(\Lambda^*\rho) - T(\rho; \Lambda^*) \\ &\equiv S(\Lambda^*\rho) - I(\rho; \Lambda^*) \end{aligned}$$

We shall see how this degree works to describe the chaotic aspects of a logistic map. The logistic map is given by the following equation:

$$x_{n+1} = f_a(x_n) = ax_n(1 - x_n), \qquad x_n \in [0, 1], \qquad 0 \le a \le 4 \tag{5.1}$$

The solution of this equation bifurcates as shown in Fig. 1.

The Lyapunov exponent of this map has been calculated by Shaw (Shaw, 1981) (Fig. 2). The Lyapunov exponent λ is defined as

$$\lambda_n = \frac{1}{n} \sum_{k=1}^{n} \log \left| \frac{df_a}{dx}(x_k) \right|, \qquad \lambda = \lim_{n\to\infty} \lambda_n$$

Figure 2 is the result of computing λ_n for 1000 a's from 3.0 to 4.0 with

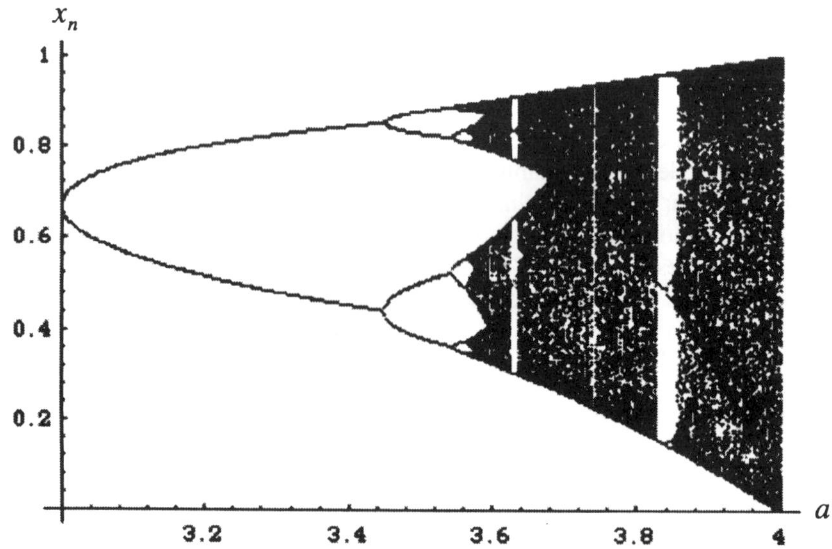

Fig. 1. Bifurcation diagram of logistic map.

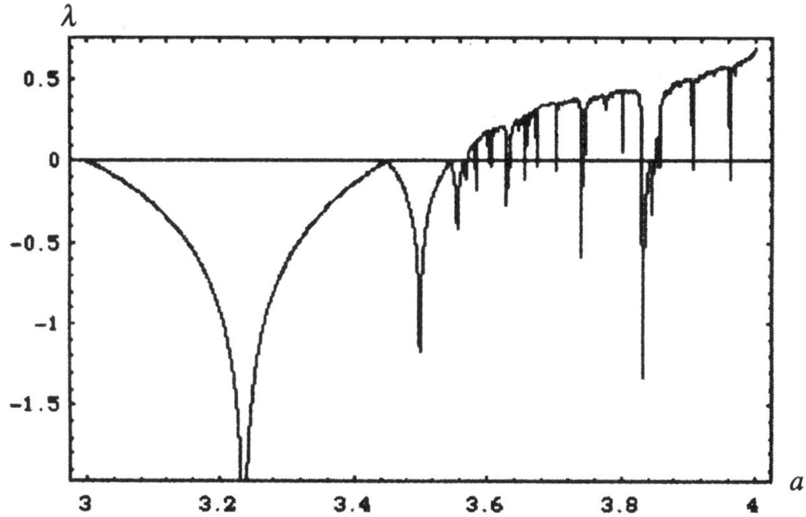

Fig. 2. Lyapunov exponent of logistic map.

$n = 100{,}000$ steps. A positive exponent means that the trajectory is very sensitive to the initial value and is called chaotic; a negative exponent means that the trajectory is stable.

If a logistic map f_a does not have a stable and periodic trajectory, then there exists an ergodic probability measure μ on the Borel set of [0,1], absolutely continuous with respect to the Lebesgue measure (Misiurewicz, 1981).

Take a finite partition $\{A_k\}$ of $I = [0,1]$ such as

$$I = \bigcup_{k=1}^{m} A_k \qquad (A_i \cap A_j = \emptyset,\ i \neq j)$$

Let $|Q|$ be the number of the elements in a set Q and $p^{(n)} \equiv (p_k^{(n)})$ be the probability distribution of the trajectory up the nth step, that is, how many x_j ($j = 1, \ldots, n-1$) are in A_k:

$$p_k^{(n)} = \frac{|\{j;\ x_j \in A_k,\ 1 \leq j \leq n\}|}{n}, \qquad n \geq 1$$

This probability distribution is obviously from the difference equation of (5.1), hence it depends on the initial value x_1 and f_a. The channel Λ^* is a map given by $p^{(n+1)} = \Lambda^* p^{(n)}$. It can be shown (Misiurewicz, 1981) that the $n \to \infty$ limit of $p_k^{(n)}$ exists and is equal to $\mu(A_k)$. Further, the joint distribution $r^{(n,n+1)} = (r_{ij}^{(n,n+1)})$ for a sufficiently large n is approximated as

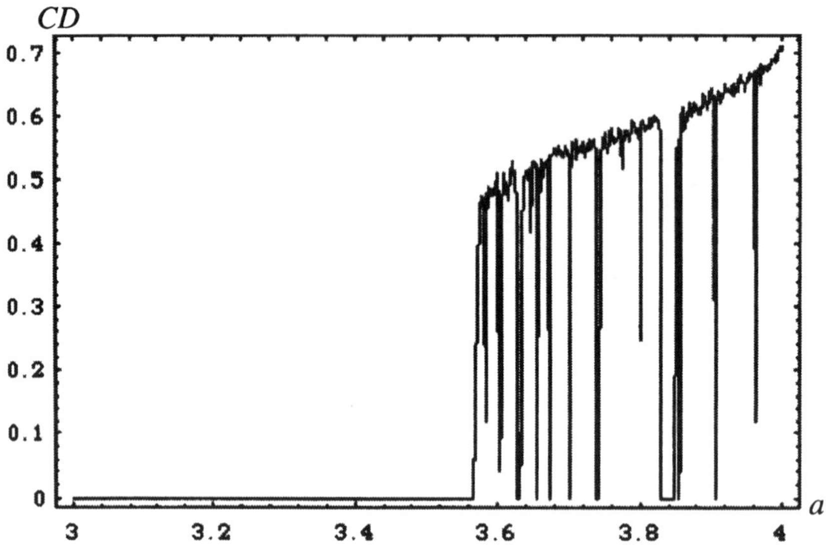

Fig. 3. Chaos degree.

$$r_{ij}^{(n,n+1)} = \frac{|\{k;\ (x_k, x_{k+1}) \in A_i \times A_j,\ 1 \le k \le n\}|}{n}$$

Then the chaos degree (CD) is

$$\begin{aligned} D(p^{(n)}; \Lambda^*) &\equiv C(\Lambda^* p^{(n)}) - T(p^{(n)}; \Lambda^*) \\ &= S(\Lambda^* p^{(n)}) - I(p^{(n)}; \Lambda^*) \\ &= S(p^{(n+1)}) - I(p^{(n)}; \Lambda^*) \end{aligned}$$

For a computer simulation, we take 1000 a's from 3.0 to 4.0 and

$$A_i = \left[\frac{i}{2000}, \frac{i+1}{2000}\right], \qquad i = 0, \ldots, 1999$$

$$n = 100{,}000$$

The choice of these quantities does not alter the results so much if we take large n.

The result is shown in Fig. 3, from which we conclude that our chaos degree describes the chaotic aspects of the logistic map, namely,

$$D > 0 \Leftrightarrow \text{chaotic}$$

$$D = 0 \Leftrightarrow \text{non-chaotic}$$

Although the Lyapunov exponent becomes negative, sometimes-∞, our degree is always nonnegative, so that it might be useful to make the chaotic domain clear-cut. More rigorous study of the chaos degree and its use for other dynamical channels including quantum systems is now in progress.

REFERENCES

Accardi, L., Ohya, M., and Watanabe, N. (1996). Note on quantum dynamical entropy, *Reports on Mathematical Physics,* **38**, 457–469.

Benatti, F. (1993). *Deterministic Chaos in Infinite Quantum Systems*, Springer-Verlag, Berlin.

Billingsley, P. (1965). *Ergodic Theory and Information*, Wiley, New York.

Connes, A., and Størmer, E. (1975). Entropy for automorphisms of II_1 von Neumann algebras, *Acta Mathematica,* **134**, 289–306.

Connes, A., Narnhofer, H., and Thirring, W. (1987). Dynamical entropy of C*-algebras and von Neumann algebras, *Communications in Mathematical Physics,* **112**, 691–719.

Emch, G. G. (1974). Positivity of the K-entropy on non-abelian K-flows, *Zeitschrift für Wahrscheinlichkeits theorie Verwandte Gebiete,* **29**, 241–252.

Holevo, A. S. (1973). Some estimates for the amount of information transmittable by a quantum communication channel, *Problemy Peredachi Informacii,* **9**, 3–11 [in Russian].

Ingarden, R. S. (1976). Quantum information theory, *Reports on Mathematical Physics,* **10**, 43–73.

Ingarden, R. S., Kossakowski, A., and Ohya, M. (1997). *Information Dynamics and Open System*, Kluwer Academic Publishers.

Levitin, L. B. (1991). Physical information theory for 30 years: basic concepts and results, in *Springer Lecture Notes in Physics*, Vol. **378**, pp. 101–110.

Matsuoka, T., and Ohya, M. (1995). Fractal dimensions of states and its application to Ising model *Reports on Mathematical Physics,* **36**, 365–379.

Misiurewicz, M. (1981). Absolutely continuous measures for certain maps of interval, *Publications Mathematiques IHES,* **53**, 17–51.

Muraki, N., and Ohya, M. (1996). Entropy functionals of Kolmogorov Sinai type and their limit theorems, *Letters in Mathematical Physics,* **36**, 327–335.

Ohya, M. (1981). Quantum ergodic channels in operator algebras, *Journal of Mathematical Analysis and Applications,* **84**, 318–327.

Ohya, M. (1983a). On compound state and mutual information in quantum information theory, *IEEE Transaction of Information Theory,* **29**, 770–774.

Ohya, M. (1983b). Note on quantum probability, *Lettere al Nuovo Cimento,* **38**, 402–406.

Ohya, M. (1989). Some aspects of quantum information theory and their applications to irreversible processes, *Reports on Mathematical Physics,* **27**, 19–47.

Ohya, M. (1991a). Information dynamics and its application to optical communication processes, *Springer Lecture Notes in Physics*, Vol. **378**, pp. 81–92.

Ohya, M. (1991b). Fractal dimension of states, *Quantum Probability and Related Topics,* **6**, 359–369.

Ohya, M. (1995). State change, complexity and fractal in quantum systems, *Quantum Communications and Measurement,* **2**, 309–320.

Ohya, M. (n.d.-a). Foundation of entropy, complexity and fractal in quantum systems, in *International Congress of Probability toward 2000*, to appear.

Ohya, M. (n.d.-b). Fundamentals of quantum mutual entropy and capacity, submitted.

Ohya, M. (n.d.-c). Applications of information dynamics to genome sequences, SUT preprint.

Ohya, M. (1997). Complexity and fractal dimensions for quantum states, *Open Systems and Information Dynamics,* **4**, 141–157.

Ohya, M., and Petz, D. (1993). *Quantum Entropy and Its Use*, Springer-Verlag, Berlin.

Ohya M., and Watanabe N. (1993). Information dynamics and its application to Gaussian communication process, *Maximum Entropy and Bayesian Method,* **12**, 195–203.

Ohya, M., Petz, D., and Watanabe, N. (1997). On capacities of quantum channels, *Probability and Mathematical Statistics,* **17**, 179–196.

Shannon, C. E. (1948). Mathematical theory of communication, *Bell System Technical Journal,* **27**, 379–423.

Shaw, R. (1981). Strange attractors, chaotic behavior and information flow, *Zeitschrift für Naturforschung,* **36a**, 80–112.

Uhlmann, A. (1977). Relative entropy and the Wigner–Yanase–Dyson–Lieb concavity in interpolation theory, *Communication in Mathematical Physics,* **54**, 21–32.

Umegaki, H. (1962). Conditional expectations in an operator algebra IV (entropy and information), *Kodai Mathematical Seminar Reports,* **14**, 59–85.

Von Neumann, J. (1932). Mathematischen Grundlagen *Quantenmechanik*, Springer, Berlin.

Amino Acids (1998) 14:343–352
With kind permission of Springer Science and Business Media

Amino
Acids

Analysis of HIV by entropy evolution rate

K. Sato[1], **S. Miyazaki**[2] and **M. Ohya**[1]

[1] Department of Information Sciences, Science University of Tokyo, Noda City, Chiba, Japan
[2] Center for Information Biology, National Institute of Genetics, Mishima City, Shizuoka, Japan

Accepted October 29, 1997

Summary. We analyze the variation of HIV after infection by means of an information measure, called the entropy evolution rate. In our analysis, we use a part of the external glycoprotein gp120 including the V3 region observed from six patients.
Then we could make the following two aspects clear;
(1) the relation between the change of the entropy evolution rate and the appearance of symptoms of disease, and
(2) the relation between the change of the entropy evolution rate and that of the CD4 count of the patients.

Keywords: Amino acids – HIV – Entropy evolution rate

1. Introduction

The main purpose of this study is to find a new criterion grasping the processes of the change of CD4 count and the immunity of patients from the gene level after HIV infection.

In Section 2, we summarize the data of HIV genes of the six patients used in this paper. The entropy evolution rate and the method of how to use it for our analysis are discussed in this section. In Section 3, we present our results by the graphs. We discuss our results and the usefulness of our method in Section 4.

2. Material and methods

We consider two aligned amino acid (resp. base) sequences $\mathcal{A}$ and $\mathcal{B}$, which are composed of 20 (resp. 4) kinds of amino acids (resp. bases) and the gap *. The complete event system $(\mathcal{A},p)$ of $\mathcal{A}$ is determined by the occurrence probability p_i of each amino acid (resp. base) a_i and the gap * $(0 \le i \le 20)$ (resp. $0 \le i \le 4$) with $a_0 = *$;

$$\begin{pmatrix} \mathcal{A} \\ p \end{pmatrix} = \begin{pmatrix} *, & a_1, \ldots, & a_{20} \\ p_0, & p_1, \ldots, & p_{20} \end{pmatrix} \quad \left(\text{resp.} \begin{pmatrix} *, & a_1, \ldots, & a_4 \\ p_0, & p_1, \ldots, & p_4 \end{pmatrix} \right)$$

In the same way, the complete event system $(\mathcal{B},q)$ of $\mathcal{B}$ is

$$\begin{pmatrix} \mathcal{B} \\ q \end{pmatrix} = \begin{pmatrix} *, & a_1, \ldots, & a_{20} \\ q_0, & q_1, \ldots, & q_{20} \end{pmatrix} \quad \left(\text{resp.} \begin{pmatrix} *, & a_1, \ldots, & a_4 \\ q_0, & q_1, \ldots, & q_4 \end{pmatrix} \right)$$

We can construct the compound event system $(\mathcal{A} \times \mathcal{B}, r)$ for two sequences $\mathcal{A}$ and $\mathcal{B}$.

$$\begin{pmatrix} \mathcal{A} \times \mathcal{B} \\ r \end{pmatrix} = \begin{pmatrix} (**), & (*a_1), \ldots, & (a_{20}a_{20}) \\ r_{00}, & r_{01}, \ldots, & r_{2020} \end{pmatrix}$$

$$\left(\text{resp.} \begin{pmatrix} (**), & (*a_1), \ldots, & (a_4a_4) \\ r_{00}, & r_{01}, \ldots, & r_{44} \end{pmatrix} \right),$$

where r_{ij} represents the joint probability of the event i of $\mathcal{A}$ and the event j of $\mathcal{B}$.

These event systems define various entropies, among which the following two are important:

(1) Shannon entropy

$$S(\mathcal{A}) = -\sum_i p_i \log p_i,$$

which expresses the amount of information carried by $(\mathcal{A},p)$.

(2) The mutual entropy

$$I(\mathcal{A},\mathcal{B}) = \sum_{i,j} r_{i,j} \log \frac{r_{i,j}}{p_i q_j},$$

which expresses the amount of information transmitted from $\mathcal{A}$(resp. $\mathcal{B}$) to $\mathcal{B}$(resp. $\mathcal{A}$).

Using the above information measures, a measure indicating the difference between two amino acid sequences was introduced in (Ohya, 1989). This measure is called the entropy evolution rate and defined as follows: Put

$$r(\mathcal{B}/\mathcal{A}) = \frac{I(\mathcal{A},\mathcal{B})}{S(\mathcal{A})},$$

which is the rate how much information is transmitted from $\mathcal{A}$ to $\mathcal{B}$, and it is symmetrized as

$$r(\mathcal{A},\mathcal{B}) = \frac{1}{2}\{r(\mathcal{A}/\mathcal{B}) + r(\mathcal{B}/\mathcal{A})\}$$

The entropy evolution rate $\varrho(\mathcal{A},\mathcal{B})$ is defined by

$$\varrho(\mathcal{A},\mathcal{B}) = 1 - r(\mathcal{A},\mathcal{B})$$

In this paper, we use this entropy evolution rate to examine the variation of HIV sequences of six patients. The entropy evolution rate takes the value in [0, 1]; $\varrho(\mathcal{A},\mathcal{B}) = 0$ if $\mathcal{A}$ and $\mathcal{B}$ are completely same and $\varrho(\mathcal{A},\mathcal{B}) = 1$ if they are completely different. Therefore the variation of HIV becomes larger, the entropy evolution rate is getting larger.

Data used in our analysis are the base sequences of HIV for the six patients reported in (Wolfs et al., 1991; Holmes et al., 1992; McNearney et al., 1992). We obtained the data from the International Nucleotide Sequence Database (DDBJ/EMBL/GenBank). Here, the six patients are designated as patient A to patient F. The facts reported for the six patients are summarized in Table 1.

Table 1. Data used in our analysis

Designation in our analysis	Patient A	Patient B	Patient C	Patient D	Patient E	Patient F
Designation in our analysis	Patient 1	Patient 495	Patient 82	s 1	s 2	s 4
Presumed trans mission mode	homosexual contact	homosexual contact	a single batch of factor VIII	no information	no information	no information
Clinical status	p24antigenemia (1988)	AIDS (1989) p24antigenemia	asymptomatic	no information	no information	no information
CD4 counts Antiviral therapy	decreasing None	decreasing AZT (1989)	decreasing None	fluctuating None	decreasing None	decreasing None
Term	1985– (about 5 years period)	1985– (about 5 years period)	1984–1991 (7 years period)	1985.11–89.5 (4.5 years period)	1985.5–87.10 (2.5 years period)	1985.1–89.6 (4.5 years period)
Length	183–276nt	183–276nt	234nt	332–335nt	332–335nt	332–335nt
Tissue	serum	serum	plasma	peripheral blood leucocyte	peripheral blood leucocyte	peripheral blood leucocyte
Molecular type	RNA	RNA	RNA	DNA	DNA	DNA

We use the sequence, a part of the gp120 region including the third variable (V3) region whose mutation rate is particularly high in HIV (Watson et al., 1993; de Jong et al., 1992). The V3 region is composed of disulfide bounds of cysteine residues located in the amino acids 296 and 330 of the gp120, as shown in Fig. 1. Although it has been called the principal neutralization domain (PND) as the antibody for the segment is able to block the HIV infection, the antibody for a specific virus is gradually losing its effect because of the variation (mutation) of the virus.

It is reported (Wolfs et al., 1991) that patient B was diagnosed as having AIDS in about 5 years after the primary infection and that patient A remained healthy although the p24-antigenemia reappeared at 3 years after his infection. The p24-antigenemia in blood reflects the amount of the virus, it is reappeared when the patient has AIDS, so that it is used as a value measuring the condition of the patient.

The CD4 count has been used as a measure to know having AIDS by several researcher and medical doctors. The number of CD4 for patient D fluctuates, and that for other patients gradually decreases. This CD4 count represents the number of immunocyte destroyed by HIV. The immunocyte for healthy people is around from 800 to 1,000 per $1\mu\ell$-blood. When the CD4 count of a patient decreases and it becomes less than 200, various infections are considered to appear. Therefore, according to the diagnosis standard of CDC (Centers for Disease Control), when the CD4 become less than 200, the patient is recognized to have AIDS. The CD4 count is reported only for patient D, E, and F. For patient D, it changes as 470, 826, 273, and 515 from the primary stage (called 1 year) up to the fourth stage (called 4 year). For the patient E and F, it changes as 1225, 756, 368 and 943, 575, 187, respectively.

The number of the sequence data observed from the six patients are listed in Table 2.

We used the base sequences having the same length of bases for each patient. For example, in the primary stage of patient A, we used 6 data out of 8 data because the length of six data is 276 and that of other two is 183. Moreover, in order to carry out our analysis, first we translate the base sequences of HIV collected from the six patients into the amino acid sequences, and secondly, we directly used the base sequences. Our analysis is done in the following two cases (I) and (II).

(I) In order to compare the genome sequences of HIV in successive years, the entropy evolution rate is computed for the sequences obtained at one year with respect to those obtained at the next year (we call it the entropy evolution rate for each year), and we examine the variation (mutation) rate with the mean of the entropy evolution rates for each year and their standard deviation for each year.

(II) In order to check the variation of HIV from the primary stage, we compute the entropy evolution rate for the sequences of each year w.r.t. the primary year (we call it the entropy evolution rate for the primary year). Similarly as the case (I), we examine the variation by means of the entropy evolution rates for the primary year and their standard deviations.

As an example, we explain how to compute the entropy evolution rate and others mentioned above in (I) and (II) for the patient A. From Table 2, the number of genome sequences for the patient A are as follows: n = 6 (year 0), n = 7 (year 1), n = 7 (year 2), n = 5 (year 3), n = 6 (year 4), n = 6 (year 5). For the case (I), we compute every entropy evolution rate for the aligned sequences in successive years, for instance, $\varrho(A_i^3, A_j^4)$ ($i = 1$,

IVIRSDNITDNAKTIIVQLKEAVQIN CTRPNNNTRKSIHIGPGKAFYATGEIIGDIRQAHC NLSRVDWEDTLKQIAEKLREQFRNKTIVFNQ

IVIRSDNITDNSKTIIVQLKEAVQIN CTRPNNNTRKSIHIGPGKAFYATGEIIGDIRQAHC NLSRVDWEDTLKQIAEKLREQFRNKTIVFNQ

IVVRSDNITDNAKTIIVQLKKAVQIN CIRPNNNTRKSIHIGPGKAFYATGETIGDIRQAHC NLSGGDWENTLKQIAEKLREQFRNKTIVFNQ

Fig. 1. 3 sequence data collected from patient A in primary stage of infection. It's V3 region is underlined

Table 2. The number of sequences used in this paper

Patient A	year 0	year 1	year 2	year 3	year 4	year 5
Number of data collected from GenBank	8	7	9	9	9	8
Number of data used	6	7	7	5	6	6
Patient B	year 0	year 1	year 2	year 3	year 4	year 5
Number of data collected from GenBank	11	6	6	6	7	8
Number of data used	7	3	4	4	4	4
Patient C	year 0	year 3	year 4	year 5	year 6	year 7
Number of data collected from GenBank	1	15	11	23	15	13
Number of data used	1	15	11	23	15	13
Patient D	year 0	year 2	year 3	year 4		
Number of data collected from GenBank	5	2	4	3		
Number of data used	5	2	4	3		
Patient E	year 0	year 2	year 2.5			
Number of data collected from GenBank	5	5	6			
Number of data used	5	5	6			
Patient F	year 0	year 4	year 4.5			
Number of data collected from GenBank	5	6	6			
Number of data used	5	6	6			

$\ldots, 5.\ j = 1, \ldots, 6$) for the sequence A_i^3 of the third year and the sequence A_j^4 of the fourth year. Then we compute their mean value given by

$$\bar{\varrho}\left(A^3, A^4\right) \equiv \frac{\sum_{i=1}^{5}\sum_{j=1}^{6} \varrho\left(A_i^3, A_j^4\right)}{30},$$

which enables us to examine the variation of HIV. In the same way, we compute $\bar{\varrho}(A^0,A^1)$, $\bar{\varrho}(A^1,A^2)$, $\bar{\varrho}(A^2,A^3)$, $\bar{\varrho}(A^3,A^4)$, $\bar{\varrho}(A^4,A^5)$. The standard deviation of the entropy evolution rate for year 3 and year 4 is defined as follows:

$$\sqrt{\frac{\sum_{i=1}^{5}\sum_{j=1}^{6}\left\{\varrho\left(A_i^3, A_j^4\right) - \bar{\varrho}\left(A^3, A^4\right)\right\}^2}{30}}$$

For the case (II), we compute the mean entropy evolution rates for every sequence of each year with respect to that of the primary year. For instance, the mean entropy evolution rate for the fifth year w.r.t. the primary year is given by

$$\bar{\varrho}\left(A^0, A^5\right) = \frac{\sum_{i=1}^{6}\sum_{j=1}^{6} \varrho\left(A_i^0, A_j^5\right)}{36}.$$

We similarly compute $\bar{\varrho}(A^0,A^1)$, $\bar{\varrho}(A^0,A^2)$, $\bar{\varrho}(A^0,A^3)$, $\bar{\varrho}(A^0,A^4)$, $\bar{\varrho}(A^0,A^5)$, and their standard deviations.

All six patients are examined with these quantities, and our results are shown in the next section.

Here we note that we should align the sequences to compute the entropy evolution rate, and the alignment is done by the method in (Ohya and Uesaka, 1970; Neeleman and Wunsch, 1970).

3. Results

The following figure (Fig. 2) is the results of the mean entropy evolution rates for each year and the standard deviations obtained from the amino acid

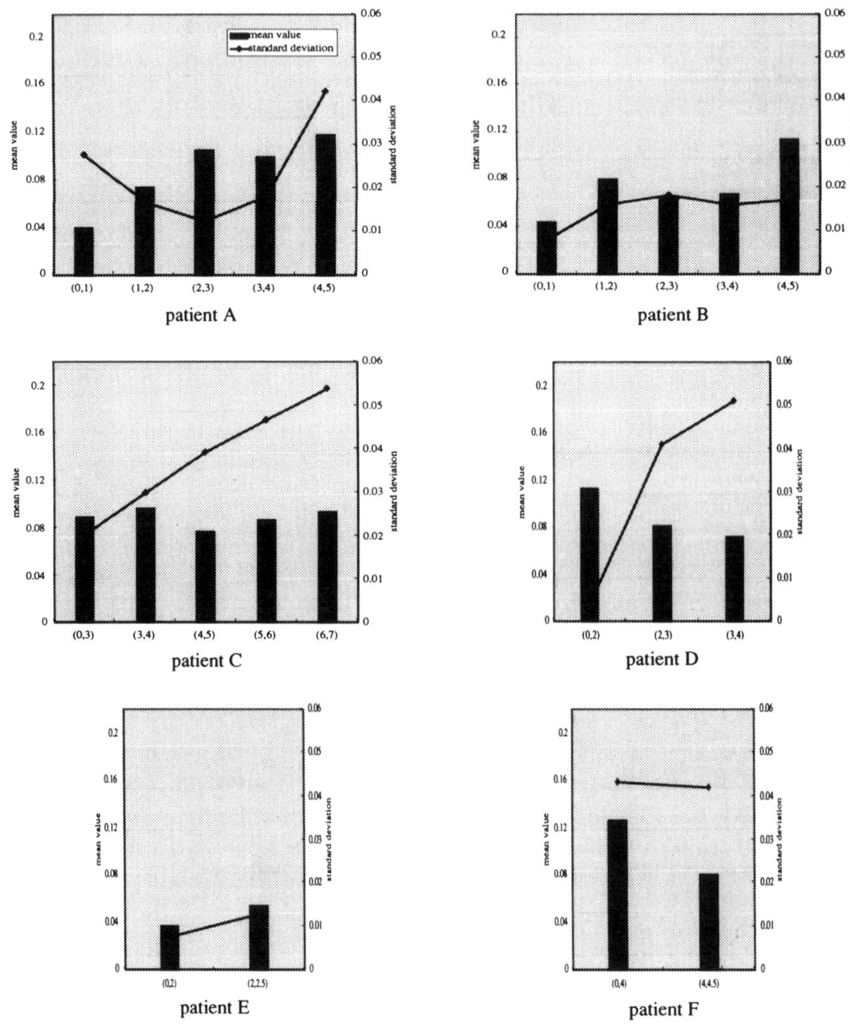

Fig. 2. Mean entropy evolution rate (bars) and standard deviation (lines) for each year

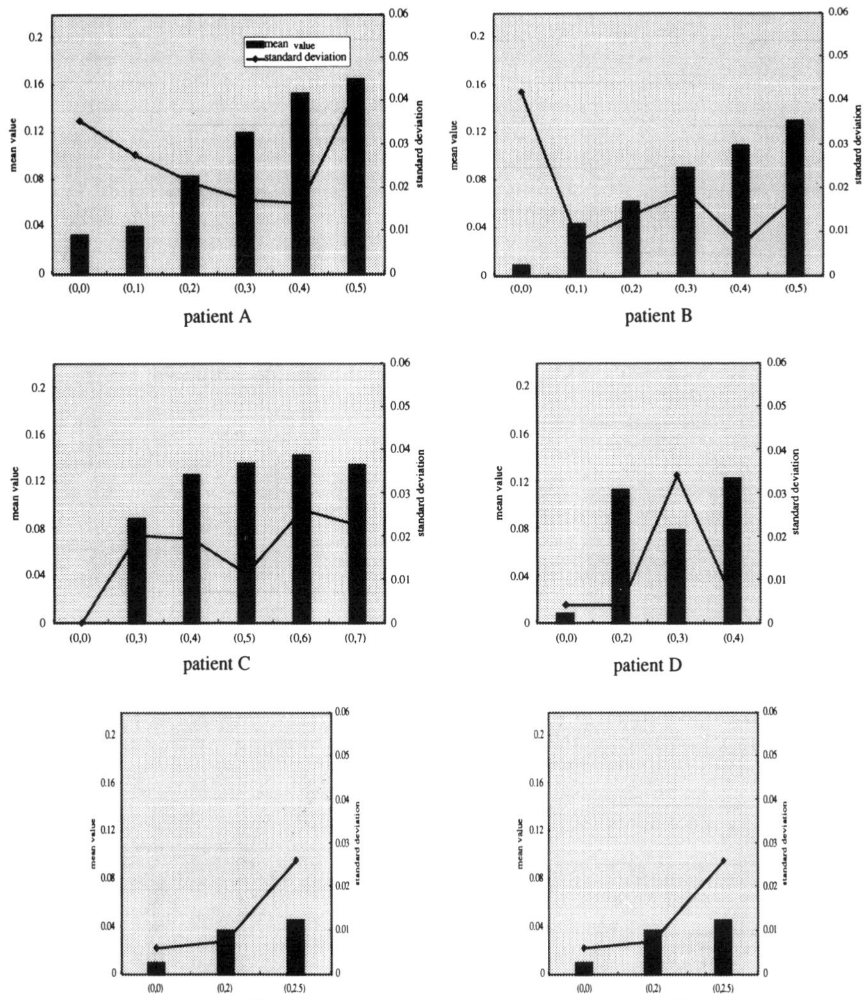

Fig. 3. Mean entropy evolution rate (bars) and standard deviation (lines) measured from primary year

sequences. Here $(i, i + 1)$ denotes the $(i + 1)$-th year w.r.t. i-th year and the mean value means the mean entropy evolution rate.

Fig. 3 shows the results of the mean entropy evolution rates for the primary year and their standard deviations, so that $(0, i)$ denotes the i-th year w.r.t. the primary year.

The following figures (Fig. 4, Fig. 5) are the results obtained from the base sequences.

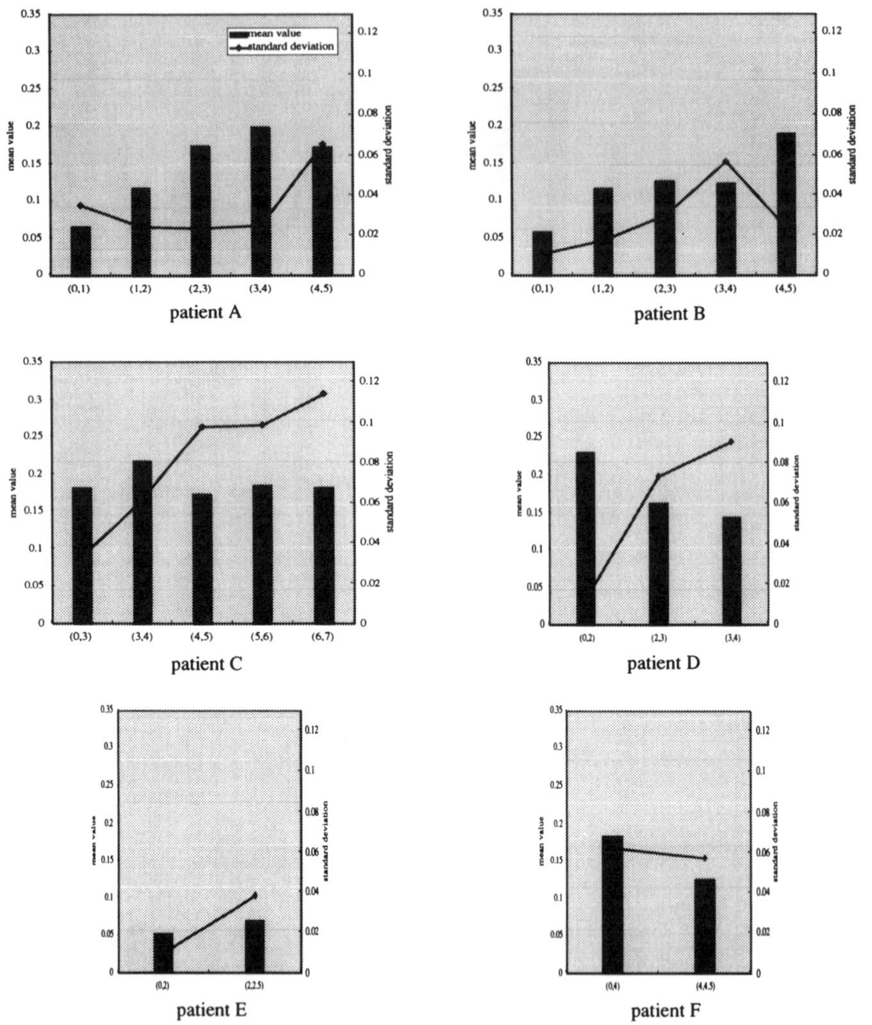

Fig. 4. Mean entropy evolution rate (bars) and standard deviation (lines) for each year

4. Discussion

Patient B was diagnosed as having AIDS at about 5 years after the primary infection. According to the result of the mean entropy evolution rate (m-EER for short) for each year, the change of the m-EER for the patient B is met the second extreme increase at that time. The change of the m-EER (Fig. 2) for patient B is considered as a fundamental pattern of the outbreak of AIDS. Based on this pattern, we may say the following conclusions for other patients. Patient A will be diagnosed as having AIDS in a few years, because the second extreme increase seems starting. Similarly, patient C will have an attack of AIDS soon, because the second moderate increase is occurred.

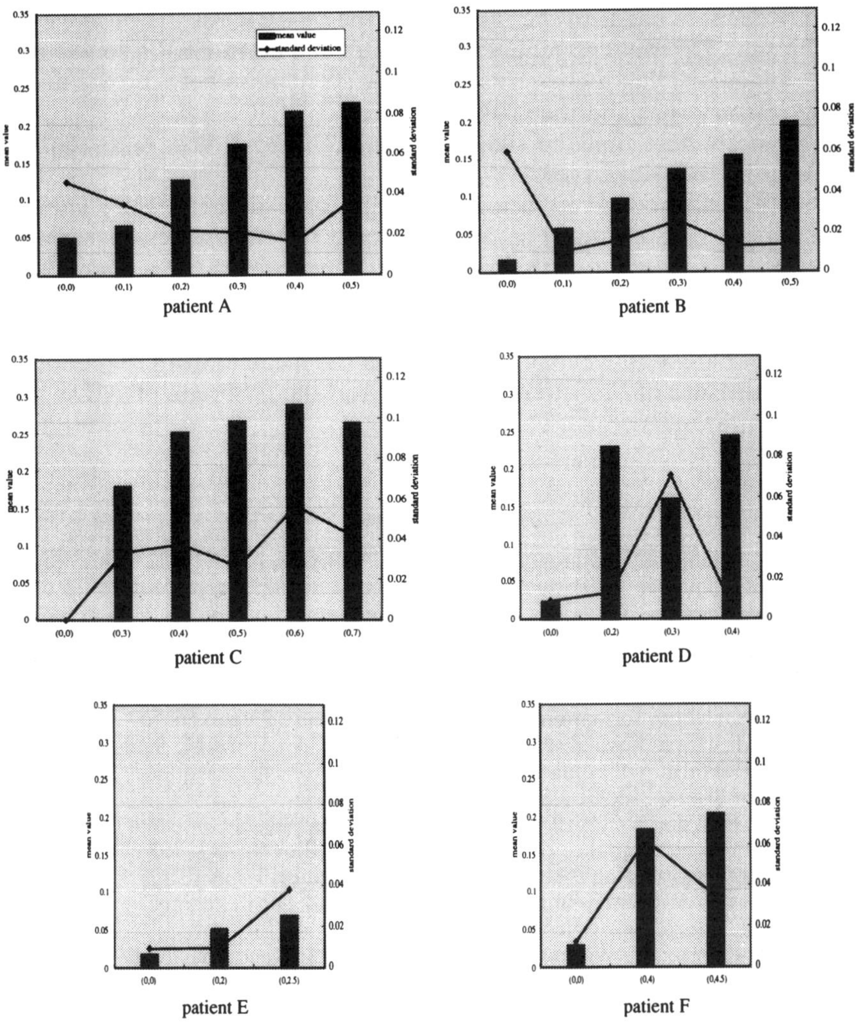

Fig. 5. Mean entropy evolution rate (bars) and standard deviation (lines) measured from primary year

Patient D, patient E and patient F have few number of data, so that we merely say a few comments. Patient D may possibly increase here after. Patient E may be far from the outbreak of AIDS. Since the data of patient F are lacked a first few years, it is very difficult to judge the variation of his HIV.

According to the result of the mean entropy evolution rate measured from the primary stage, the m-EERs of the patients except the patient D increase as shown in Fig. 3. This consequence agrees with a report that the CD4 counts of the patients except patient D decrease. That is, the gradual decrease of the CD4 count is equal to the increase of the m-EER for the primary year. Further, the CD4 count of patient D fluctuates and his m-EER also fluctuates. This result means that there exists a positive correlation between the m-EER for the primary year and the CD4.

We merely note that the standard deviation shows how many different HIV exist in each stage (year).

The results obtained by using the whole base sequences are shown in Fig. 4 for the case (I) and Fig. 5 for the case (II). These results show that patient B, like the case of the amino acid sequences, have a characteristic variation and other patients are also similar to the case of the amino acid sequences.

From our analysis, we may conclude that the mean entropy evolution rate can be a measure of the variation of HIV and the outbreak of AIDS as the CD4 count.

References

Holmes EC, Zhang LQ, Simmonds P, Ludlam CA, Brown AJL (1992) Convergent and divergent sequence evolution in the surface envelope glycoprotein of human immunodeficiency virus type 1 within a single infected patient. Evolution 89: 4835–4839

de Jong JJ, Goudsmit J, Keulen W, Klaver B, Krone W, Tersmette M, de Ronde A (1992) Human immunodeficiency virus type 1 clones chimeric for the envelope V3 domain differ in syncytium formation and replication capacity. J Virol 66: 757–765

McNearney T, Hornickova Z, Markham R, Birdwell A, Arens M, Saah A, Ratner L (1992) Relationship of human immunodeficiency virus type 1 sequence heterogeneity to stage of disease. Medical Sciences 89: 10247–10251

Needleman SB, Wunsch CD (1970) A general method applicable to search for similarities in the amino acid sequence of two proteins. J Mol Biol 48: 443–453

Ohya M (1989) Information theoretical treatment of genes. Trans IEICE E 725: 556–560

Ohya M, Uesaka Y (1992) Amino acid sequences and DP matching: new method for alignment. Information Sciences 63: 139–151

Watson JD, Gilman M, Witkowski J, Zoller M (1993) Recombinant DNA, 2nd edn. Freeman and Company

Wolfs TW, Zwart G, Bakker M, Valk M, Kuiken C, Goudsmit J (1991) Naturally occurring mutations within HIV-1 V3 genomic RNA lead to antigenic variation dependent on a single amino acid substitution. Virology 185: 195–205

Authors' address: Dr. Keiko Sato and Prof. Dr. M. Ohya, Department of Information Sciences, Science University of Tokyo, Noda City, Chiba 278, Japan.

Received May 29, 1997

PROBABILITY
AND
MATHEMATICAL STATISTICS
Vol. 17, Fasc. 1 (1997), pp. 179–196

ON CAPACITIES OF QUANTUM CHANNELS

BY

MASANORI OHYA, DÉNES PETZ* AND NOBORU WATANABE (TOKYO)

Abstract. Capacities of quantum mechanical channels are defined in terms of mutual information quantities. Geometry of the relative entropy is used to express capacity as a divergence radius. The symmetric quantum spin 1/2 channel and the attenuation channel of Boson fields are discussed as examples.

1. Introduction. A discrete communication system – as modeled by Shannon – is capable of transmitting successively symbols of a finite input alphabet $\{x_1, x_2, \ldots, x_m\}$. In the stochastic approach to the communication model it is assumed that the input symbols show up with certain probability. Let p_{ji} be the probability that a symbol x_i is sent over the channel and the output symbol y_j appears at the destination. The joint distribution p_{ji} yields marginal distributions $(p_1, p_2, \ldots, p_m)$ and $(q_1, q_2, \ldots, q_k)$ on the set of input symbols and output symbols, respectively. Shannon introduced the mutual information

$$I = \sum_{i,j} p_{ji} \log \frac{p_{ji}}{p_i q_j} \tag{1.1}$$

in order to measure the amount of information going through the channel.

The interest in quantum communication channels arose in the late 1960'. The scheme of a quantum communication system is not different from a classical one, however, zero point fluctuation (noise) cannot be avoided in quantum systems. Important recent devices for communication are based on optical fiber which is a quantum object. Hence we may assume that the actual signal transmission is over a quantum mechanical medium which is described in the usual Hilbert space formalism of quantum theory. Coding, actual signal transmission and decoding (or measurement) are the main components of the communication chain. The splitting of the communication chain into these three parts C–T–M is somewhat arbitrary. The parts can be investigated individually and their capacity can be defined by means of mutual information.

---

* On leave from the Technical University Budapest and from the Mathematical Institute of the Hungarian Academy of Sciences. The author was partially supported by JSPS and by OTKA T016924.

Our approach to capacity is based on quantum mutual information which is defined in terms of relative entropy (or informational divergence). Therefore, relative entropy is the basic tool in the paper. The capacity is not compared with performance bounds of classical coding (as in [5] and [13]) because we are mainly interested in the purely quantum part of the channel. Section 2 contains some generalities of quantum communication channels, mutual information and relative entropy. Kholevo's bound is also discussed and we show that it is rarely achievable. In Section 3 our capacities are introduced and the quantum mechanical counterpart of Csiszár's information geometry is used to realize the pure quantum capacity as the divergence radius of the range. The toy example of symmetric quantum spin 1/2 channel is used to demonstrate our ideas. We verify that the pure quantum capacity of this channeling transformation is the same as the performance bound from coding (see [5]). Section 4 treats an infinite-dimensional example, the attenuation channel of Boson fields. The channeling transformation is treated in the abstract Weyl algebra setting as well as in Fock representation. It is proved that the capacity of the attenuation channel is infinite, however, the transmission of arbitrarily much information requires infinite energy.

2. Generalities of quantum mechanical channels. To each input symbol x_i there corresponds a signal state φ_i of the quantum communication system, φ_i functions as the codeword of x_i. The signal states φ_i are mostly pure but they can be non-orthogonal. However, we do not make any assumption on them at this level of generality. The channel state is a convex combination

$$\varphi = \sum_i p_i \varphi_i \tag{2.1}$$

whose coefficients are the corresponding probabilities, p_i is the probability that the letter x_i should be transmitted over the channel.

In the mathematical sense the *quantum channeling transformation* Λ^* is an affine transformation of the state space of the input quantum system into the state space of the output quantum system. (The notation Λ^* is used here because very often Λ^* is the dual mapping of a linear transformation of observables.) At the output some sort of detection scheme retrieves the transmitted information. To each output symbol y_j there corresponds a non-negative observable A_j, that is a self-adjoint operator A_j on the Hilbert space $\mathscr{H}$, such that $\sum_j A_j = I$. (Some people speak about *effects*, or call (A_j) a *generalized measurement*). In terms of the quantum states the transition probabilities are $(\Lambda^*\varphi_i)(A_j)$ and the probability that x_i was sent and y_j is read is

$$p_{ji} = p_i(\Lambda^*\varphi_i)(A_j). \tag{2.2}$$

On the basis of these joint probability distributions the classical mutual information (1.1) is given. Kholevo's theorem provides a fundamental bound for the mutual information in terms of the quantum von Neumann entropy. Before stating

Kholevo's result, we review the simplest entropy quantities used in quantum information theory, for details see [10].

The relative entropy of two states is defined (following Umegaki [14], Lindblad [8] and Araki [1]) as

$$S(\varphi_1, \varphi_2) = \operatorname{Tr} D_1 (\log D_1 - \log D_2), \tag{2.3}$$

where D_1 and D_2 are the corresponding density operators. (This formula extends the Kullback–Leibler information measure.) We shall use a kind of algebraic language and view the states as linear functionals on (operator) algebras. The basic property of relative entropy is its monotonicity under channeling transformation. More precisely, if $\alpha\colon \mathscr{A} \to \mathscr{B}$ is a unitial (completely) positive mapping between the algebras $\mathscr{A}$ and $\mathscr{B}$, that is, the dual α^* is a channeling transformation from the state space of $\mathscr{B}$ into that of $\mathscr{A}$, then

$$S(\varphi_1 \circ \alpha, \varphi_2 \circ \alpha) \leqslant S(\varphi_1, \varphi_2). \tag{2.4}$$

Relative entropy (or *information gain*) is the fundamental information quantity, many other information quantities are expressed by it. For example, the von Neumann entropy is

$$S(\varphi) = -\operatorname{Tr}(D \log D) = \sup\{\sum_j \lambda_j S(\varphi_j, \varphi)\colon \sum_j \lambda_j \varphi_j = \varphi,\ \lambda_j \geqslant 0\}. \tag{2.5}$$

Let $\alpha\colon \mathscr{A} \to \mathscr{B}$ be positive unitial mapping and φ be a state of $\mathscr{B}$. So φ is an initial state of the channel α^*. The *quantum mutual entropy* is defined after [9] as

$$I(\varphi; \alpha) = \sup\{\sum_j \lambda_j S(\varphi_j \circ \alpha, \varphi \circ \alpha)\colon \sum_j \lambda_j \varphi_j = \varphi\}, \tag{2.6}$$

where the least upper bound is over all orthogonal extremal decompositions. (One checks easily that this formula reduces to (1.1) when $\mathscr{A} = \boldsymbol{C}^k$ and $\mathscr{B} = \boldsymbol{C}^m$ since in this case the orthogonal extremal decomposition is unique and $(p_1, p_2, \ldots, p_m) = \sum_i p_i \delta_{i\cdot}$.)

THEOREM 2.1. *With the above notation the inequality*

$$I = \sum_{i,j} p_{ji} \log \frac{p_{ji}}{p_i q_j} \leqslant S(\Lambda^*(\varphi)) - \sum_i p_i S(\Lambda^*(\varphi_i))$$

holds true.

Kholevo [6] proved this inequality in 1973 when the concept of quantum relative entropy was not well understood yet. Kholevo's upper bound is

$$S(\Lambda^* \varphi) - \sum_i p_i S(\Lambda^* \varphi_i) = \sum_i p_i S(\Lambda^* \varphi_i, \Lambda^* \varphi). \tag{2.7}$$

Let $\boldsymbol{C}^m$, $B(\mathscr{H})$, $B(\mathscr{K})$ and $\boldsymbol{C}^k$ be operator algebras and consider the mappings

$$\alpha\colon \boldsymbol{C}^k \to B(\mathscr{K}), \quad (c_1, c_2, \ldots, c_k) \mapsto \sum_j c_j A_j,$$

$$\Lambda\colon B(\mathcal{H}) \to B(\mathcal{K}),$$

$$\beta\colon B(\mathcal{H}) \to \mathbf{C}^m, \qquad H \mapsto (\varphi_1(H), \varphi_2(H), \dots, \varphi_m(H)).$$

So the duals of the positive unitial mappings α, Λ and β correspond to the measurement, quantum state transmission and coding procedures, respectively. In this terminology the upper bound is the mutual information $I(\mu, \beta\circ\Lambda)$ of the quantum channel α with input quantum input state μ (where $\mu((c_1, c_2, \dots, c_m)) = \sum_i p_i c_i$) and the classical mutual information I becomes the quantum mutual information $I(\mu, \beta\circ\Lambda\circ\alpha)$ of a composite quantum channel. So Kholevo's theorem is read as

$$I(\mu, \beta\circ\Lambda\circ\alpha) \leqslant I(\mu, \beta\circ\Lambda) \tag{2.8}$$

which follows from the monotonicity of the quantum mutual information. For the details see [10], in particular pp. 139–140. It is noteworthy that these ideas work in the continuous case as well as it was observed also in [15].

Yuen and Ozawa [15] propose to call Theorem 1 the *fundamental theorem of quantum communication*. The theorem bounds the performance of the detecting scheme. We see that in most cases the bound cannot be achieved. Namely, the bound may be achieved in the only case when the output states $\Lambda^*(\varphi_i)$ have commuting densities.

PROPOSITION 2.2. *If the states $\Lambda^*(\varphi_i)$, $1 \leqslant i \leqslant m$, do not commute, then*

$$I = \sum_{i,j} p_{ji} \log \frac{p_{ji}}{p_i q_j} < S(\Lambda^*(\varphi)) - \sum_i p_i S(\Lambda^*(\varphi_i))$$

is a strict inequality.

In the terminology of Chapter 8 of [10] the equality in Kholevo's theorem means that the measurement channel α is sufficient for the states $\Lambda^*(\varphi_i)$, $1 \leqslant i \leqslant m$, and the sufficiency has several characterizations, for example, the existence of states ω_j of the output quantum system such that

$$\sum_j \varphi_i(A_j)\,\omega_j = \varphi_i \qquad \text{for every } i.$$

In particular, if the bound is achieved, the states $\Lambda^*(\varphi_i)$ have to commute. Suppose that the state $\Lambda^*(\varphi_i)$ has a density D_i and let $D = \sum_i p_i D_i$ (which is the density of $\Lambda^*(\varphi)$). Then the generalized measurement $A_i = D_i D^{-1}$ achieves Kholevo's bound. (The operator $D_i D^{-1}$ is well defined even if D is not invertible, because the kernel of D is contained in that of D_i.) The technicalities of the detailed proof depend very much on the level of generality. For finite dimension one has to investigate the equality case in the Jensen inequality and this was carried out by Kholevo [6]. We consider now the infinite-dimensional case but under the restrictive assumption of faithfulness. Since Λ^* does not play any role, we skip it from the notation. The proof of the next theorem uses the idea of the paper [12] and the proof presented here is a bit sketchy. (The interested reader may consult [12] for the more detailed justification of the steps.)

THEOREM 2.3. *Assume that* $\varphi = \sum_i p_i \varphi_i$ *and there is a sequence* (α_n) *of generalized measurements such that*

$$\sum_i p_i S\big(\alpha_n^*(\varphi_i), \alpha_n^*(\varphi)\big) \to \sum_i p_i S(\varphi_i, \varphi).$$

If the limit is finite and the states φ_i *are faithful, then the family* (φ_i) *must commute.*

Proof. Since $S\big(\alpha_n^*(\varphi_i), \alpha_n^*(\varphi)\big) \leqslant S(\varphi_i, \varphi)$, the assumption implies that

$$S\big(\alpha_n^*(\varphi_i), \alpha_n^*(\varphi)\big) \to S(\varphi_i, \varphi) \quad \text{for every } i.$$

Let D_i, D be the statistical operators of φ_i, φ, respectively. In the sequel we shall use the relative modular operator technique and we work on the Hilbert space $\mathscr{K}$ of Hilbert–Schmidt operators. There exists a positive operator Δ_i such that $\{AD_i^{1/2}\colon A \in B(\mathscr{H})\}$ is a core for $\Delta_i^{1/2}$, and

$$\|\Delta_i^{1/2} A D_i^{1/2}\|^2 = \varphi(AA^*) \quad (A \in B(\mathscr{H})).$$

In fact, the relative modular operator Δ_i is the extension of the linear operator $AD_i^{1/2} \mapsto D^{1/2}A$ defined on a dense linear subspace of $\mathscr{K}$. In terms of the relative modular operator, we have

$$S(\varphi_i, \varphi) = -\langle D_i^{1/2}, (\log \Delta) D_i^{1/2}\rangle = \int_0^\infty \langle D_i^{1/2}, (\Delta_i + t)^{-1} D_i^{1/2}\rangle - (1+t)^{-1}\, dt. \tag{2.9}$$

Similarly,

$$S\big(\alpha^*(\varphi_i), \alpha^*(\varphi)\big) = \int_0^\infty \langle d_{n,i}^{1/2}, (\delta_{n,i} + t)^{-1} d_{n,i}^{1/2}\rangle - (1+t)^{-1}\, dt,$$

where the probability vector $d_{n,i}^{1/2}$ (d_n) corresponds to the state $\alpha_n^* \varphi_i$ $(\alpha_n^* \varphi)$ and

$$\delta_{n,i} = d_n / d_{n,i}.$$

(Note that $\delta_{n,i}$ is the relative modular operator of $\alpha_n^* \varphi$ with respect to $\alpha_n^*(\varphi_i)$ but, due to the simple situation coming from finite dimension and commutativity, we may just regard it as a vector.) Since

$$\langle d_{n,i}^{1/2}, (\delta_{n,i} + t)^{-1} d_{n,i}^{1/2}\rangle \leqslant \langle D_i^{1/2}, (\Delta_i + t)^{-1} D_i^{1/2}\rangle \quad \text{for every } t > 0 \text{ and } i,$$

our assumption implies that

$$\langle d_{n,i}^{1/2}, (\delta_{n,i} + t)^{-1} d_{n,i}^{1/2}\rangle \to \langle D_i^{1/2}, (\Delta_i + t)^{-1} D_i^{1/2}\rangle \quad \text{as } n \to \infty.$$

From this we infer that

$$\alpha_n\big((t + \delta_{n,i})^{-1} d_{n,i}^{1/2}\big) \to (t + \Delta_i)^{-1} D_i^{1/2}. \tag{2.10}$$

Now we consider the function $F_i(z) = \Delta_i^z D_i^{1/2}$. We know that $D_i^{1/2}$ is in the domain of $\Delta_i^{1/2}$. The range of $\Delta_i^{1/2}$ contains the operators $D^{1/2}A$ for a bounded A. Since $\|D^{-1/2} D_i^{1/2}\| \leqslant p_i^{-1/2}$, we may choose $A = D^{-1/2} D_i^{1/2}$, and infer that $D_i^{1/2}$ is in the domain of $\Delta_i^{-1/2}$. As a consequence, the function $F_i(z)$ is analytic on the strip $\{z \in C\colon -1/2 < \mathrm{Re} z < 1/2\}$. We should not gather so much with $f_{n,i}(z) = \delta_{n,i}^z d_{n,i}^{1/2}$, because it is analytic on the whole complex plain.

Our next aim is to show that

$$V_{n,i}\big(f_{n,i}(z)\big) \to F_i(z) \tag{2.11}$$

if $-1/2 < \mathrm{Re} z < 1/2$ for the contraction $V_{n,i}$ defined by

$$V_{n,i}(a_n d_{n,i}^{1/2}) = \alpha_n(a_n) D_i^{1/2}.$$

Since we have an analytic function at our disposal, it suffices to prove (2.11) for $0 < s < 1/2$ in place of z. For $0 < s < 1/2$ we obtain

$$\begin{aligned} V_{n,i}\big(f_{n,i}(s)\big) &= \frac{\sin s\pi}{\pi} \int_0^\infty t^{s-1} V_{n,i}\big(\delta_{n,i}(t+\delta_{n,i})^{-1} d_{n,i}^{1/2}\big)\, dt \\ &= \frac{\sin s\pi}{\pi} \int_0^\infty t^{s-1} \Delta_i (t+\Delta_i)^{-1} D_i^{1/2}\, dt = F_i(s). \end{aligned}$$

So we may consider in (2.11) a pure imaginary $z = it$:

$$V_{n,i}\big(F_n(it)\big) = V_{n,i}(\delta_{n,i}^{it} d_{n,i}^{1/2}) \to \Delta_i^{it} D_i^{1/2} = D^{it} D_i^{(-it+1)/2}$$

and

$$\alpha_n(\delta_{n,i}^{it}) D_i^{1/2} \to (D^{it} D_i^{-it}) D_i^{1/2}$$

in the Hilbert–Schmidt norm. The strong operator convergence follows from the Hilbert–Schmidt norm convergence and we arrive at

$$\alpha_n(\delta_{n,i}^{it}) \to (D^{it} D_i^{-it}) \text{ (strongly).} \tag{2.12}$$

In particular, $D^{it} D_i^{-it}$ is a unitary group for fixed i, and D and D_i must commute. ∎

3. Capacity of channels. Let $\mathscr{H}$ and $\mathscr{K}$ be the input and output Hilbert spaces of a quantum communication system. The *channeling transformation* $\Lambda^*\colon \Sigma(\mathscr{H}) \to \Sigma(\mathscr{K})$ sends density operators acting on $\mathscr{H}$ into those acting on $\mathscr{K}$. A *pseudo-quantum code* is a probability distribution on $\Sigma(\mathscr{H})$ with finite support. So $\{(p_i), (\varphi_i)\}$ is a pseudo-quantum code if (p_i) is a probability vector and φ_i are states of $B(\mathscr{H})$. The quantum states φ_i are sent over the quantum mechanical media, for example, optical fiber, and yield the output quantum states $\Lambda^*\varphi_i$. The performance of coding and transmission is measured by the *mutual information*

$$I\big((p_i), (\varphi_i), \Lambda^*\big) = \sum_i p_i S(\Lambda^*\varphi_i, \Lambda^*\varphi). \tag{3.1}$$

Taking the supremum over certain classes of pseudo-quantum codes, we obtain various capacities of the channel. Here we consider one subclass of pseudo-quantum codes. A *quantum code* is defined by the additional requirement that $\{\varphi_i\}$ is a set of pairwise orthogonal pure states. Correspondingly, we arrive at two alternative concepts of capacity:

$$C_{\mathrm{pq}}(\Lambda^*) = \sup\{I((p_i), (\varphi_i), \Lambda): ((p_i), (\varphi_i)) \text{ is a pseudo-quantum code}\} \tag{3.2}$$

and

$$C_{\mathrm{q}}(\Lambda^*) = \sup\{I((p_i), (\varphi_i), \Lambda^*): ((p_i), (\varphi_i)) \text{ is a quantum code}\}. \tag{3.3}$$

We can write $C_{\mathrm{q}}(\Lambda^*)$ in a slightly different form by using the notation (2.6):

$$C_{\mathrm{q}}(\Lambda^*) = \sup\{I(\varphi, \Lambda^*): \varphi \text{ is an input state}\}. \tag{3.4}$$

The capacity C_{q} may be viewed as the characteristic of the purely quantum mechanical signal transmission.

It follows from the definition that

$$C_{\mathrm{q}}(\Lambda^*) \leqslant C_{\mathrm{pq}}(\Lambda^*)$$

holds for every channel.

EXAMPLE 3.1. Let Λ^* be a channel on the 2×2 density matrices such that

$$\Lambda^*: \begin{pmatrix} a & b \\ \bar{b} & c \end{pmatrix} \mapsto \begin{pmatrix} a & 0 \\ 0 & c \end{pmatrix}.$$

Consider the input density matrix

$$D_\lambda = \frac{1}{2}\begin{pmatrix} 1 & 1-2\lambda \\ 1-2\lambda & 1 \end{pmatrix} \qquad (0 < \lambda < 1).$$

For $\lambda \neq 1/2$ the orthogonal extremal decomposition is unique; in fact,

$$D_\lambda = \frac{\lambda}{2}\begin{pmatrix} 1 & -1 \\ -1 & 1 \end{pmatrix} + \frac{1-\lambda}{2}\begin{pmatrix} 1 & 1 \\ 1 & 1 \end{pmatrix}$$

and we have

$$I(D_\lambda, \Lambda^*) = 0 \qquad \text{for } \lambda \neq 1/2.$$

However, $I(D_{1/2}, \Lambda^*) = \log 2$. Since $C_{\mathrm{q}}(\Lambda^*) \leqslant C_{\mathrm{pq}}(\Lambda^*) \leqslant \log 2$, we conclude that $C_{\mathrm{q}}(\Lambda^*) = C_{\mathrm{pq}}(\Lambda^*) = \log 2$.

The example shows that the quantity $I(\varphi, \Lambda^*)$ may be discontinuous at φ when φ has some degeneracy in the spectrum.

In order to estimate the quantum mutual information, we introduce the concept of *divergence center*. Let $\{\omega_i: i \in I\}$ be a family of states and $R > 0$. We say that the state ω is a *divergence center for* $\{\omega_i: i \in I\}$ *with radius* $\leqslant R$ if

$$S(\omega_i, \omega) \leqslant R \quad \text{for every } i \in I.$$

In the following discussion about the geometry of relative entropy (or divergence as it is called in information theory) the ideas of [3] can be recognized very well.

LEMMA 3.2. *Let* $((p_i), (\varphi_i))$ *be a pseudo-quantum code for the channel* Λ^* *and* ω *be a divergence center with radius* $\leqslant R$ *for* $\{\Lambda^*\varphi_i\}$. *Then*

$$I((p_i), (\varphi_i), \Lambda^*) \leqslant R.$$

Proof. We assume that the states $\Lambda^*\varphi_i$, $\Lambda^*\varphi = \sum_i p_i \Lambda^*\varphi_i$ and ω have finite entropy and their densities are denoted by D_i, D and D', respectively. We have

$$-S(\Lambda^*\varphi_i) - \operatorname{Tr} D_i \log D' \leqslant R,$$

and hence

$$\begin{aligned} (3.5) \qquad \sum_i p_i S(\Lambda^*\varphi_i, \Lambda^*\varphi) &= -\sum_i p_i S(\Lambda^*\varphi_i) - \operatorname{Tr} D \log D \\ &\leqslant R - \operatorname{Tr} D(\log D - \log D') = R - S(\Lambda^*\varphi, \omega). \end{aligned}$$

The extra assumption we made holds always in finite dimension. When the entropies are not finite but the relative entropies are so, one has to use more sophisticated methods for the proof. It is quite clear that inequality (3.5) is close to equality if $S(\Lambda^*\varphi_i, \omega)$ is about R and $\sum_i p_i \Lambda^*\varphi_i$ is about ω. ∎

Let $\{\omega_i\colon i \in I\}$ be a family of states. We say that the state ω is an *exact divergence center with radius R* if

$$R = \inf_{\varphi} \sup_{i} \{S(\omega_i, \varphi)\}$$

and ω is a minimizer for the right-hand side. (When R is finite, then there exists a minimizer, because $\varphi \mapsto \sup\{S(\omega_i, \varphi)\colon i \in I\}$ is lower semicontinuous with compact level sets; cf. Proposition 5.27 in [10].)

LEMMA 3.3. *Let* ψ_0, ψ_1 *and* ω *be states of* $B(\mathscr{K})$ *such that the Hilbert space* $\mathscr{K}$ *is finite dimensional and set*

$$\psi_\lambda = (1-\lambda)\psi_0 + \lambda\psi_1 \qquad (0 \leqslant \lambda \leqslant 1).$$

If $S(\psi_0, \omega)$ *and* $S(\psi_1, \omega)$ *are finite and*

$$S(\psi_\lambda, \omega) \geqslant S(\psi_1, \omega) \qquad (0 \leqslant \lambda \leqslant 1),$$

then

$$S(\psi_1, \omega) + S(\psi_0, \psi_1) \leqslant S(\psi_0, \omega).$$

Proof. Let the densities of ψ_λ and ω be D_λ and D, respectively. Due to the assumption $S(\psi_\lambda, \omega) < +\infty$, the kernel of D is smaller than that of D_λ. The function $f(\lambda) = S(\varphi_\lambda, \omega)$ is convex on $[0, 1]$ and $f(\lambda) \geqslant f(1)$ (cf. Proposition 3.1 in [10]). It follows that $f'(1) \leqslant 0$. Hence we have

$$f'(1) = \mathrm{Tr}(D_1 - D_0)(I + \log D_1) - \mathrm{Tr}(D_1 - D_0)\log D$$
$$= S(\psi_1, \omega) - S(\psi_0, \omega) + S(\psi_0, \psi_1) \leqslant 0.$$

This is the inequality we had to obtain.

We note that in the differentiation of the function $f(\lambda)$ the well-known formula

$$\frac{\partial}{\partial t}\mathrm{Tr}\, F(A + tB)|_{t=0} = \mathrm{Tr}\big(F'(A)B\big)$$

can be used. ∎

LEMMA 3.4. *Let $\{\omega_i\colon i \in I\}$ be a finite set of states of $B(\mathscr{K})$ such that the Hilbert space $\mathscr{K}$ is finite dimensional. Then the exact divergence center is unique and it is in the convex hull on the states ω_i.*

Proof. Let K be the (closed) convex hull of the states $\omega_1, \omega_2, \ldots, \omega_n$ and let ω be an arbitrary state such that $S(\omega_i, \omega) < +\infty$. There is a unique state $\omega \in \mathscr{K}$ such that $S(\omega', \omega)$ is minimal (where ω' runs over K), see Theorem 5.25 in [10]. Then

$$S\big(\lambda\omega_i + (1-\lambda)\omega', \omega\big) \geqslant S(\omega', \omega) \quad \text{for every } 0 \leqslant \lambda \leqslant 1 \text{ and } 1 \leqslant i \leqslant n.$$

It follows from the previous lemma that

$$S(\omega_i, \omega) \geqslant S(\omega_i, \omega').$$

Hence the divergence center of ω_i's must be in K. The uniqueness of the exact divergence center follows from the fact that the relative entropy functional is strictly convex in the second variable. ∎

THEOREM 3.5. *Let $\Lambda^*\colon \Sigma(\mathscr{H}) \to \Sigma(\mathscr{K})$ be a channel with finite-dimensional $\mathscr{K}$. Then the capacity $C_{\mathrm{pq}}(\Lambda^*)$ is the divergence radius of the range of Λ^*.*

Proof. Let $((p_i), (\varphi_i))$ be a pseudo-quantum code. Then $I\big((p_i), (\varphi_i), \Lambda^*\big)$ is at most the divergence radius of $\{\Lambda^*\varphi_i\}$ (according to Lemma 3.2), which is obviously majorized by the divergence radius of the range of Λ^*. Therefore, the capacity does not exceed the divergence radius of the range.

To prove the converse inequality we assume that the exact divergence radius of $\Lambda^*\big(\Sigma(\mathscr{H})\big)$ is larger than $t \in \mathbf{R}$. Then we can find $\varphi_1, \varphi_2, \ldots, \varphi_n \in \Sigma(\mathscr{H})$ such that the exact divergence radius R of $\Lambda^*(\varphi_1), \ldots, \Lambda^*(\varphi_n)$ is larger than t. Lemma 3.4 states that the divergence center ω of $\Lambda^*(\varphi_1), \ldots, \Lambda^*(\varphi_n)$ lies in their convex hull K. By possible reordering of the states φ_i we can achieve that

$$S(\Lambda^*(\varphi_i), \omega)\begin{cases} = R & \text{if } 1 \leqslant i \leqslant k, \\ < R & \text{if } i < k \leqslant n. \end{cases}$$

Let K' be the convex hull of $\Lambda^*(\varphi_1), \ldots, \Lambda^*(\varphi_n)$. We claim that $\omega \in K'$; we choose $\omega' \in K'$ such that $S(\omega', \omega)$ is minimal (ω' is running over K'). Then

$$S(\Lambda^*\varphi_i, \varepsilon\omega' + (1-\varepsilon)\omega) < R$$

for every $1 \leqslant i \leqslant k$ and $0 < \varepsilon < 1$, due to Lemma 3.3. However,

$$S(\Lambda^*\varphi_i, \varepsilon\omega' + (1-\varepsilon)\omega) < R$$

for $k \leqslant i \leqslant n$ and for a small ε by a continuity argument. In this way, we conclude that there exists a probability distribution $(p_1, p_2, \ldots, p_k)$ such that

$$\sum_{i=1}^{k} p_i \Lambda^*\varphi_i = \omega, \qquad S(\Lambda^*\varphi_i, \omega) = R.$$

Consider now the pseudo-quantum code $((p_i), (\varphi_i))$ such that

$$\sum_{i=1}^{k} p_i S\big(\Lambda^*\varphi_i, \Lambda^*\big(\sum_{j=1}^{k} p_j\varphi_j\big)\big) = \sum_{i=1}^{k} p_i S(\Lambda^*\varphi_i, \omega) = R.$$

So we have found a pseudo-quantum code which has quantum mutual information larger than t. The channel capacity must exceed the entropy radius of the range. ∎

Up to now our discussion has concerned the capacities of coding and transmission, which are bounds for the performance of quantum coding and quantum transmission. After a measurement is performed, the quantum channel becomes classical and Shannon's theory applied. The *total capacity* (or *classical capacity*) of a quantum channel Λ^* is

$$C_{\text{cl}}(\Lambda^*) = \sup\{I((p_i), (\varphi_i), \gamma^* \circ \Lambda^*)\}, \tag{3.6}$$

where the supremum is taken over both all pseudo-quantum codes (p_i), (φ_i) and all measurements γ^*. Due to the monotonicity of the mutual information we have

$$C_{\text{cl}}(\Lambda^*) \leqslant C_{\text{pq}}(\Lambda^*). \tag{3.7}$$

EXAMPLE 3.6. Consider the Stokes parametrization of 2×2 density matrices:

$$D_x = \tfrac{1}{2}(I + x_1\sigma_1 + x_2\sigma_2 + x_3\sigma_3),$$

where σ_1, σ_2, σ_3 are the Pauli matrices and $(x_1, x_2, x_3) \in \boldsymbol{R}^3$ with $x_1^2 + x_2^2 + x_3^2 \leqslant 1$. For a positive semidefinite (3×3)-matrix A the application Γ^*: $D_x \mapsto D_{Ax}$ gives a channeling transformation when $\|A\| \leqslant 1$. This channel was introduced in [5] under the name of *symmetric binary quantum channel.* We want to compute the capacities of Γ^*. Since a unitary conjugation does not obviously change capacity, we may assume that A is diagonal with eigenvalues $1 \geqslant \lambda_1 \geqslant \lambda_2 \geqslant \lambda_3 \geqslant 0$. The range of Γ^* is visualized as an ellipsoid with (Euclidean) diameter $2\lambda_1$. It is not difficult to see that the trace state τ is the exact divergence center of the segment connected the states $(I \pm \lambda_1 \sigma_1)/2$, and hence τ must be the divergence center of the whole range. The divergence radius is

$$S\left(\frac{1}{2}\begin{pmatrix}1 & 0\\ 0 & 0\end{pmatrix}+\frac{\lambda}{2}\begin{pmatrix}1 & 0\\ 0 & -1\end{pmatrix}, \tau\right)$$

$$= \log 2 - S\left(\frac{1}{2}\begin{pmatrix}1+\lambda & 0\\ 0 & 1-\lambda\end{pmatrix}\right) = \log 2 - \eta\left((1+\lambda)/2\right) - \eta\left((1-\lambda)/2\right).$$

This gives the capacity $C_{pq}(\Gamma^*)$ according to Theorem 3.5. Inequality (3.7) states that the capacity $C_q(\Gamma^*)$ cannot exceed this value. On the other hand,

$$I(\tau, \Gamma^*) = \log 2 - \eta\left((1+\lambda)/2\right) - \eta\left((1-\lambda)/2\right),$$

and we have $C_{pq}(\Gamma^*) = C_q(\Gamma^*)$. ∎

Shannon's communication theory is largely of asymptotic character, the message length N is supposed to be very large. So we consider the N-fold tensor product of the input and output Hilbert spaces $\mathscr{H}$ and $\mathscr{K}$,

$$\mathscr{H}_N = \bigotimes_{i=1}^{N} \mathscr{H}, \qquad \mathscr{K}_N = \bigotimes_{i=1}^{N} \mathscr{K}.$$

Note that

$$B(\mathscr{H}_N) = \bigotimes_{i=1}^{N} B(\mathscr{H}), \qquad B(\mathscr{K}_N) = \bigotimes_{i=1}^{N} B(\mathscr{K}).$$

The *(multi-) channeling transformation* is a mapping

$$\Lambda_N^*\colon\ \Sigma(\mathscr{H}_N) \to \Sigma(\mathscr{K}_N).$$

The main example is the *memoryless channel*, which is the tensor product of the same *single site channels*:

$$\Lambda_N^* = \Gamma^* \otimes \ldots \otimes \Gamma^* \ \ (N\text{-fold}).$$

The sequences $C_{pq}(\Lambda_N^*)$ and $C_q(\Lambda_N^*)$ of capacities are defined as above for a single channel.

For a memoryless channel the sequences $C_{pq}(\Lambda_N^*)$ and $C_q(\Lambda_N^*)$ are superadditive. Indeed, if $((p_i), (\varphi_i))$ and $((q_j), (\psi_j))$ are (pseudo-) quantum codes of order N and M, then $((p_i, q_j), (\varphi_i \otimes \psi_j))$ is a (pseudo-) quantum code of order $N+M$ and

$$I\left((p_i, q_j), (\varphi_i \otimes \psi_j), \Lambda_{N+M}^*\right) = I\left((p_i), (\varphi_i), \Lambda_N^*\right) + I\left((q_j), (\psi_j), \Lambda_N^*\right) \tag{3.8}$$

follows from the additivity of relative entropy under taking tensor product.

One can check that if the initial codes are (pseudo-) quantum, then the product code is (pseudo-) quantum as well. After taking the supremum, the additivity (3.8) yields the superadditivity of the sequences $C_{pq}(\Lambda_N^*)$ and $C_q(\Lambda_N^*)$. So the following limits exist and they are well known to coincide with the suprema:

$$(3.9)\quad C_{pq}^{\infty} = \lim \frac{1}{N} C_{pq}(\Lambda_N^*), \qquad C_q^{\infty} = \lim \frac{1}{N} C_q(\Lambda_N^*), \qquad C_{cl}^{\infty} = \lim \frac{1}{N} C_{cl}(\Lambda_N^*).$$

(For multiple channels with some memory effect, one may take the limsup in (3.9) to get a good concept of capacity per single use.) We have

$$(3.10)\qquad C_{cl}^{\infty}(\Lambda_N^*) \leqslant C_{pq}^{\infty}(\Lambda_N^*) \leqslant C_q^{\infty}(\Lambda_N^*)$$

for the capacities per single use.

Example 3.7. In the case of the memoryless symmetric binary channel we have

$$C_{pq}^{\infty}(\Gamma^*) = C_q^{\infty}(\Gamma^*) = \log 2 - \eta((1+\lambda)/2) - \eta((1-\lambda)/2),$$

that is the capacity of the single channel coincides with the capacity per single use for the multiple channel.

The proof consists in checking that the trace state remains the divergence center of certain states in the range. Since $\tau = (\varphi_1 + \varphi_2)/2$ for certain output states φ_1, φ_2 such that $S(\varphi_i, \tau)$ is the capacity, we have

$$\tau \otimes \ldots \otimes \tau = 2^{-n} \sum \varphi_{i(1)} \otimes \ldots \otimes \varphi_{i(n)}.$$

Due to symmetry, the trace state is the divergence center, the exact divergence radius is n times $S(\varphi_i, \tau)$ according to the additivity of the relative entropy. This implies that the entropy C_{pq}^{∞} equals the single site one. The argument for C_q^{∞} is similar to the single site case.

The work [5] deals with $C_{cl}(\Lambda^*)$ in detail, and, among other things, a coding theorem relates C_{cl}^{∞} to the code rate of a sequence of pseudo-quantum codes and measurements with asymptotically vanishing average error probability. The picture looks rather similar to Shannon's coding theorem. (Note that in [5] our capacity C_{pq} was called *pseudo-capacity* because the authors were interested in the classical capacity.)

The relations among C_{pq}, C_q and C_{cl} form an important problem, worthy of study. For a noiseless channel, $C_{cl} = \log n$ was obtained in [5], where n is the dimension of the output Hilbert space (actually identical to the input one). Since the trace state is the exact divergence center of all density matrices, we have $C_{pq} = \log n$ and also $C_q = \log n$. We expect that $C_{pq} < C_{cl}$ for "truly quantum mechanical channels" but $C_{cl}^{\infty} = C_{pq}^{\infty} = C_q^{\infty}$ must hold for a large class of memoryless channels.

In the case of the binary symmetric channel, all the three capacities coincide as computed in Example 3.7 and in [5].

4. The attenuation channel. First we discuss the attenuation channel in the context of the Weyl algebra in a representation free way. It will turn out somewhat later that what we are describing is identical to the attenuation channel defined in terms of the bosonic Fock space in [9].

Let σ be a non-degenerate symplectic form on a linear space $\mathscr{H}$. Typically, $\mathscr{H}$ is a complex Hilbert space and $\sigma(f, g) = -\operatorname{Im}\langle f, g\rangle$. The Weyl algebra $\mathrm{CCR}(\mathscr{H})$ is generated by unitaries $\{W(f): f \in \mathscr{H}\}$ satisfying the Weyl form of the canonical commutation relation:

$$W(f)\,W(g) = e^{\mathrm{i}\sigma(f,g)}\,W(f+g) \qquad (f, g \in \mathscr{H}). \tag{4.1}$$

Since the linear hull of the unitaries $W(f)$ is dense in $\mathrm{CCR}(\mathscr{H})$, any state is determined uniquely by its values taken on the Weyl unitaries. The most important state of the Weyl algebra is the Fock state which is given as

$$\varphi(W(f)) = \exp\{-\|f\|^2/2\} \qquad (f \in \mathscr{H}). \tag{4.2}$$

The GNS Hilbert space corresponding to the Fock state is called the *(bosonic) Fock space* $\Gamma(\mathscr{H})$ and the cyclic vector Φ is said to be a *vacuum*. The states

$$\varphi_f(\cdot) = \varphi(W(f)^* \cdot W(f)) \tag{4.3}$$

are called *coherent states* and they are induced by the coherent vectors

$$\pi_{\mathrm{F}}(W(f))\Phi = \Phi_f$$

in the Fock representation π_{F}. We have

$$\begin{aligned} \langle \Phi_f, \Phi_g\rangle &= \varphi(W(f)^*\,W(g)) = \exp\{-\tfrac{1}{2}\|g-f\|^2\}\exp\{-\mathrm{i}\sigma(f, g)\} \\ &= \exp\{-\tfrac{1}{2}(\|f\|^2+\|g\|^2)+\langle f, g\rangle\}, \end{aligned} \tag{4.4}$$

and

$$\begin{aligned} \varphi_f(W(g)) &= \exp\{-\tfrac{1}{2}\|g\|^2 - 2\mathrm{i}\sigma(f, g)\} \\ &= \exp\{-\tfrac{1}{2}\|g\|^2 + 2\mathrm{i}\operatorname{Im}\langle f, g\rangle\} \qquad (f, g \in \mathscr{H}). \end{aligned} \tag{4.5}$$

The field operators are obtained as the generators of the unitary groups $t \mapsto \pi_{\mathrm{F}}(W(tf))$ in the Fock representation. In other words, $B(f)$ is an unbounded self-adjoint operator on $\Gamma(\mathscr{H})$ such that

$$B(f) = -\mathrm{i}\frac{d}{dt}\pi_{\mathrm{F}}(W(tf))|_{t=0}$$

with an appropriate domain. The creation and annihilation operators are defined as

$$a^*(f) = \tfrac{1}{2}(B(\mathrm{i}f) - \mathrm{i}B(f)), \qquad a(f) = \tfrac{1}{2}(B(\mathrm{i}f) + \mathrm{i}B(f)).$$

The positive self-adjoint operator $N(f) = a^*(f)\,a(f)$ has spectrum $\boldsymbol{Z}^+$ and it is called the *particle number operator* (for the "f-mode").

Let T be a symplectic transformation of $\mathcal{H}$ to $\mathcal{H} \oplus \mathcal{K}$, i.e., $\sigma(f, g) = \sigma(Tf, Tg)$. Then there is a homomorphism

$$\alpha_T\colon \mathrm{CCR}(\mathcal{H}) \to \mathrm{CCR}(\mathcal{H} \oplus \mathcal{K})$$

such that

$$\alpha_T(W(f)) = W(Tf). \tag{4.6}$$

We may regard the Weyl algebra $\mathrm{CCR}(\mathcal{H} \oplus \mathcal{K})$ as $\mathrm{CCR}(\mathcal{H}) \otimes \mathrm{CCR}(\mathcal{K})$ and, given a state ψ on $\mathrm{CCR}(\mathcal{K})$, a channeling transformation arises as

$$(\Lambda^*\omega)(A) = (\omega \otimes \psi)(\alpha_T(A)), \tag{4.7}$$

where the input state ω is an arbitrary state of $\mathrm{CCR}(\mathcal{H})$ and $A \in \mathrm{CCR}(\mathcal{H})$. (In the language of optical communication, ψ is called a *noise state.*) To see a concrete example discussed in [9], we choose $\mathcal{H} = \mathcal{K}$, $\psi = \varphi$ and

$$S(\xi) = a\xi \oplus b\xi. \tag{4.8}$$

If $|a|^2 + |b|^2 = 1$ holds for the numbers a and b, this S is an isometry and a symplectic transformation, and we arrive at the channeling transformation

$$(\Lambda^*\omega)\, W(g) = \omega(W(ag)) \exp\{-\tfrac{1}{2}\|bg\|^2\} \qquad (g \in \mathcal{H}). \tag{4.9}$$

In order to have an alternative description of Λ^* in terms of density operators acting on $\Gamma(\mathcal{H})$ we introduce the linear operator $V\colon \Gamma(\mathcal{H}) \to \Gamma(\mathcal{H}) \otimes \Gamma(\mathcal{H})$ defined by

$$V\pi_F(A)\Phi = \pi_F(\alpha_T(A))\Phi \otimes \Phi.$$

We have

$$V\pi_F(W(f))\Phi = (\pi_F(W(af)) \otimes \pi_F(W(bf)))\,\Phi \otimes \Phi,$$

and hence

$$V\Phi_f = \Phi_{af} \otimes \Phi_{bf}. \tag{4.10}$$

LEMMA 4.1. *Let ω be a state of $\mathrm{CCR}(\mathcal{H})$ which has density D in the Fock representation. Then the output state $\Lambda^*\omega$ of the attenuation channel has density* $\mathrm{Tr}_2\, VDV^*$ *in the Fock representation.*

Proof. Since we work only in the Fock representation, we skip π_F in the formulas. First we show that

$$V^*(W(f) \otimes I)\, V = W(af) \exp\{-\tfrac{1}{2}\|bf\|^2\} \tag{4.11}$$

for every $f \in \mathcal{H}$. (This can be done by computing the quadratic form of both operators on coherent vectors.) Now we proceed as follows:

$$\mathrm{Tr}(\mathrm{Tr}_2 VDV^*)W(f) = \mathrm{Tr}\, VDV^*(W(f)\otimes I) = \mathrm{Tr}\, DV^*(W(f)\otimes I)V$$
$$= \mathrm{Tr}\, DW(af)\exp\{-\tfrac{1}{2}\|bf\|^2\} = \omega(W(af))\exp\{-\tfrac{1}{2}\|bf\|^2\},$$

which is nothing else but $(\Lambda^*\omega)(W(f))$ due to (4.9). ∎

The lemma states that Λ^* is really the same (attenuation) channel discussed in [9] or [10], p. 305.

We note that Λ is a so-called quasi-free completely positive mapping of $\mathrm{CCR}(\mathscr{H})$ given as

$$\Lambda(W(f)) = W(af)\exp\{-\tfrac{1}{2}\|bf\|^2\} \tag{4.12}$$

(cf. [4] or Chapter 8 of [11]).

PROPOSITION 4.2. *If ψ is a regular state of* $\mathrm{CCR}(\mathscr{H})$, *that is $t\mapsto\psi(W(tf))$ is a continuous function on $\mathbf{R}$ for every $f\in\mathscr{H}$, then $(\Lambda^*)^n(\psi)\to\varphi$ pointwise. (φ denotes the Fock state.)*

Proof. It is enough to look at the formula

$$((\Lambda^*)^n\psi)(W(f)) = \varphi(W(a^nf))\exp\{-\tfrac{1}{2}\sum_{k=0}^{n-1}\|a^kbf\|^2\}$$

and the statement is concluded. ∎

It is worth noting that the singular state

$$\tau(W(f)) = \begin{cases} 0 & \text{if } f\neq 0,\\ 1 & \text{if } f=0\end{cases}$$

is an invariant state of $\mathrm{CCR}(\mathscr{H})$. On the other hand, the proposition applies to states with density operator in the Fock representation. Therefore, we have

COROLLARY 4.3. *Λ^* regarded as a channel of $B(\Gamma(\mathscr{H}))$ has a unique invariant state, the Fock state, and correspondingly Λ is ergodic.*

Λ is not only ergodic but it is *completely dissipative* in the sense that

$$\Lambda(A^*A) = \Lambda(A^*)\Lambda(A) \tag{4.13}$$

may happen only in the trivial case when A is a multiple of the identity. The authors are grateful to M. Fannes and A. Verbeure for this information (private communication). In fact,

$$\Lambda = (\mathrm{id}\otimes\omega)\circ\alpha_S, \tag{4.14}$$

where α_S is given by (4.6) and (4.8), and $\omega(W(f)) = \exp\{-\|bf\|^2\}$ is a quasi-free state. Here $\mathrm{id}\otimes\omega$ is just a conditional expectation which leaves invariant a separating product state.

LEMMA 4.4. *Let Λ^* be the attenuation channel. Then*

$$\sup I((p_i), (\varphi_i), \Lambda^*) = \log n$$

when the supremum is taken over all pseudo-quantum codes $((p_i)_{i=1}^n, (\varphi_{f(i)})_{i=1}^n)$ *applying n coherent states.*

Proof. We know that $\Lambda^*\varphi_f = \varphi_{af}$, so the output $\{\Lambda^*\varphi_{f(1)}, \ldots, \Lambda^*\varphi_{f(n)}\}$ consists of n pure states. The corresponding vectors of $\Gamma(\mathscr{H})$ span a Hilbert space of dimension $k \leqslant n$. Since the trace state on that Hilbert space is a divergence center with radius $\leqslant \log k \leqslant \log n$, $\log n$ is always a bound for the mutual information according to Lemma 3.2.

In order to show that the bound $\log n$ is really achieved we choose the vectors $f(k)$ such that

$$f(k) = \lambda k f \quad (1 \leqslant k \leqslant n),$$

where $f \in \mathscr{H}$ is a fixed non-zero vector. Then in the limit $\lambda \to \infty$ the states $\varphi_{f(k)}$ become orthogonal, since

$$|\langle \Phi_{\lambda k f} \mid \Phi_{\lambda m f} \rangle|^2 = \exp\{-\lambda^2 (k-m)^2 \|f\|^2/2\} \to 0 \tag{4.15}$$

whenever $k \neq m$. In the limit $\lambda \to \infty$ the trace state (of a subspace) becomes the exact divergence center and we have

$$\lim_{\lambda \to \infty} I((1/n), (\varphi_{f(i)}), \Lambda^*) = \log n.$$

This proves the lemma. ∎

The next theorem follows directly from the previous lemma.

THEOREM 4.5. *The capacity* C_{pq} *of the attenuation channel is infinite.*

Some remarks are in order. Since the argument of the proof of Lemma 4.4 works for any quasi-free channel, we can conclude $C_{\mathrm{pq}} = \infty$ also in that more general case. Another remark concerns the classical capacity C_{cl}. Since the states $\varphi_{f(n)}$ used in the proof of Lemma 4.4 commute in the limit $\lambda \to \infty$, the total capacity C_{cl} is infinite as well. $C_{\mathrm{cl}} = \infty$ follows also from the proof of the next theorem.

THEOREM 4.6. *The capacity* C_{q} *of the attenuation channel is infinite.*

Proof. We follow the strategy of the proof of the previous theorem, but we use the number states in place of the coherent ones. The attenuation channel sends the number state $|n\rangle\langle n|$ into the binomial mixture of the number states

$$|0\rangle\langle 0| = \Phi, |1\rangle\langle 1|, \ldots, |n\rangle\langle n|.$$

Hence the commuting family of convex combination of number states is invariant under the attenuation channel, and the channel restricted to those states is classical with obviously infinite capacity. Since C_{q} (as well as C_{cl}) cannot have a smaller value, the claim follows. ∎

Let us make some comments on the previous results. The theorems mean that arbitrarily large amount of information can go through the attenuation channel, however the theorems do not say anything about the price for it. The expectation value of the number of particles needed in the pseudo-quantum code of Lemma 4.4 tends to infinity. Indeed,

$$\sum_i \frac{1}{n} \varphi_{f(i)}(N) = \frac{1}{n} \sum_{i=1}^{n} \|f(i)\|^2 = \lambda(n+1)(2n+1)\|f\|^2/6,$$

which increases rapidly with n (here N denotes the number operator). Hence the good question is to ask for the capacity of the attenuation channel when some energy constraint is posed:

$$C(E_0) = \sup\{I((p_i), (\varphi_i), \Lambda^*): \sum_i p_i \varphi_i(N) \leqslant E_0\}. \tag{4.16}$$

(To be more precise, we have posed a bound on the average energy, different constraints are also possible, cf. [2].) Since $\Lambda(N) = a^2 N$ for the number operator N, we have

$$C(E_0) = \sup\{\sum_i p_i S(\varphi_i, \sum_j p_j \varphi_j): \sum_i p_i \varphi_i(N) \leqslant a^{-2} E_0\}. \tag{4.17}$$

The solution of this problem is the same as that of

$$\sup\{S(\psi): \psi(N) = a^2 E_0\}$$

and the well-known maximizer of this problem is a so-called *Gibbs state*. Therefore, we have

$$C(E_0) \leqslant a^2 E_0 + \log(a^2 E_0 + 1). \tag{4.18}$$

This value can be realized as a classical capacity if the number states can be output states of the attenuation channel.

REFERENCES

[1] H. Araki, *Relative entropy for states of von Neumann algebras*, Publ. Res. Inst. Math. Sci., Kyoto Univ., 11 (1976), pp. 809–833.

[2] C. M. Caves and P. D. Drummond, *Quantum limits on bosonic communication rates*, Rev. Modern Phys. 66 (1994), pp. 481–537.

[3] I. Csiszár, *I-divergence geometry of probability distributions and minimization problems*, Ann. Probab. 3 (1975), pp. 146–158.

[4] B. Demoen, P. Vanheuverzwijn and A. Verbeure, *Completely positive maps on the CCR-algebra*, Lett. Math. Phys. 2 (1977), pp. 161–166.

[5] A. Fujiwara and H. Nagaoka, *Capacity of memoryless quantum communication channels*, Math. Eng. Tech. Rep. 94–22, University of Tokyo, 1994.

[6] A. S. Kholevo, *Some estimates for the amount of information transmittable by a quantum communication channel*, Problemy Peredachi Informatsii 9 (1973), pp. 3–11.

[7] — *Capacity of a quantum communication channel*, Problems Inform. Transmission 15 (1979), pp. 247–253.

[8] G. Lindblad, *Completely positive maps and entropy inequalities*, Comm. Math. Phys. 40 (1975), pp. 147–151.

[9] M. Ohya, *On compound state and mutual information in quantum information theory*, IEEE Trans. Inform. Theory 29, pp. 770–777.

[10] – and D. Petz, *Quantum Entropy and Its Use*, Springer, 1993.

[11] D. Petz, *The Algebra of the Canonical Commutation Relation*, Leuven University Press, 1990.

[12] – *Discrimination between states of a quantum system by observations*, J. Funct. Anal. 120 (1994), pp. 82–97.

[13] B. Schumacher, *Quantum coding*, Phys. Rev. A (1995), pp. 2738–2747.

[14] H. Umegaki, *Conditional expectations in an operator algebra IV (entropy and information)*, Kodai Math. Sem. Rep. 14 (1962), pp. 59–85.

[15] H. P. Yuen and M. Ozawa, *Ultimate information carrying limit of quantum systems*, Phys. Rev. Lett. 70 (1993), pp. 363–366.

Department of Information Sciences
Science University of Tokyo
Noda City, Chiba 278, Japan

Received on 9.5.1996

Open Sys. & Information Dyn. 4: 141–157, 1997 141

Complexity, Fractal Dimension for Quantum States

Masanori Ohya

Department of Information Sciences
Science University of Tokyo
Noda City, Chiba 278, Japan

(Received November 28, 1996)

Abstract. The complexities in information dynamics are reviewed and their examples are given. The fractal dimensions of a quantum state are discussed from a general point of view of complexity. It is shown trough a model that the fractal dimensions of a state provide measures for order structure of chaotic systems.

Introduction

There exists several approaches in the study of chaotic behavior of systems using of concepts such as entropy, complexity, chaos, fractality, stochasticity. In 1991, the author proposed Information Dynamics (ID) to find a common frame to treat such chaotic behavior of systems altogether. That is, ID is an attempt to synthesize the dynamics of state changes and the complexity of systems [32]. Since then, the author and his coworkers attempted to refine this concept and apply it to several topics. In particular, the fractal dimensions of states are defined [33] not only for geometrical sets but also for general states, so that we can examine whether a complicated (chaotic) object obeys a certain rule (i.e., the fractal structure like self-similarity) or not by means of the fractal dimensions of states. This means that this and other complexities might provide measures for some order structures of chaotic systems.

In Section 1, we briefly review ID and an axiomatic approach to the complexity. In Section 2, various examples of the state changes (channels) are presented, some of which are new expressions of physical processes. In Section 3, some examples of the complexities are discussed. In Section 4, fractal dimensions of states are discussed on the basis of two complexities. In Section 5, the use of the fractal dimensions is discussed to characterize chaotic aspects of physical phenomena.

1. Information Dynamics

Information dynamics (ID) is a synthesis of the dynamics of the state change and the complexity of states [15, 32, 36]. It is an attempt to provide a new view for the study of chaotic behavior of systems. We briefly review what ID is.

Let $(\mathcal{A}, \mathfrak{S}, \alpha(G))$ be an input (or initial) system and $(\overline{\mathcal{A}}, \overline{\mathfrak{S}}, \overline{\alpha}(\overline{G}))$ be an output (or final) system. Here $\mathcal{A}$ is the set of some objects to be observed and $\mathfrak{S}$ is the set of some means to get the observed value, $\alpha(G)$ describes certain evolution of the system. We often have $\mathcal{A} = \overline{\mathcal{A}}$, $\mathfrak{S} = \overline{\mathfrak{S}}$, $\alpha = \overline{\alpha}$. Therefore, we claim

[Giving a mathematical structure to input and output triples
$\equiv$ Having a theory].

The dynamics of the state change is described by a channel, which will be explained in the next section, Λ^*: $\mathfrak{S} \to \overline{\mathfrak{S}}$ (sometimes $\mathfrak{S} \to \mathfrak{S}$). The fundamental point of ID is that ID contains two complexities in itself. Let $(\mathcal{A}_t, \mathfrak{S}_t, \alpha^t(G^t))$ be the total system of $(\mathcal{A}, \mathfrak{S}, \alpha)$ and $(\overline{\mathcal{A}}, \overline{\mathfrak{S}}, \overline{\alpha})$, and $\mathcal{S}$ be a subset of $\mathfrak{S}$ in which we are measuring observables (e.g., $\mathcal{S}$ is the set of all KMS or stationary states in a C*-system). Two complexities are denoted by C and T. C is the complexity of a state φ measured from the reference system $\mathcal{S}$, in which we actually observe the objects in $\mathcal{A}$, and T is the transmitted complexity associated with the state change $\varphi \to \Lambda^*\varphi$, both of which should satisfy the following properties:

Axiom of complexity

(i) For any $\varphi \in \mathcal{S} \subset \mathfrak{S}$,

$$C^{\mathcal{S}}(\varphi) \geq 0, \qquad T^{\mathcal{S}}(\varphi; \Lambda^*) \geq 0$$

(ii) For any orthogonal bijection j: ex$\mathfrak{S} \to$ ex$\mathfrak{S}$, where ex$\mathfrak{S}$ is the set of all extremal points of $\mathfrak{S}$,

$$C^{j(\mathcal{S})}(j(\varphi)) = C^{\mathcal{S}}(\varphi),$$
$$T^{j(\mathcal{S})}(j(\varphi); \Lambda^*) = T^{\mathcal{S}}(\varphi; \Lambda^*).$$

(iii) For $\Phi \equiv \varphi \otimes \psi \in \mathcal{S}_t \subset \mathfrak{S}_t$,

$$C^{\mathcal{S}_t}(\Phi) = C^{\mathcal{S}}(\varphi) + C^{\overline{\mathcal{S}}}(\psi).$$

(iv) $$0 \leq T^{\mathcal{S}}(\varphi; \Lambda^*) \leq C^{\mathcal{S}}(\varphi).$$

(v) $T^{\mathcal{S}}(\varphi; id) = C^{\mathcal{S}}(\varphi)$, where "$id$" is the identity map from $\mathfrak{S}$ to $\mathfrak{S}$.

If instead of (iii), the following is satisfied
(iii') $\Phi \in \mathcal{S}_t \subset \mathfrak{S}_t$, put $\varphi \equiv \Phi \mid \mathcal{A}$, $\psi \equiv \Phi \mid \overline{\mathcal{A}}$ (i.e., the restriction of Φ to $\overline{\mathcal{A}}$), $C^{\mathcal{S}_t}(\Phi) \leq C^{\mathcal{S}}(\varphi) + C^{\overline{\mathcal{S}}}(\psi)$, C and T is called a pair of strong complexity. Therefore, ID is defined as follows:

DEFINITION 1.1. Information Dynamics is described by

$$\left(\mathcal{A}, \mathfrak{S}, \alpha(G); \overline{\mathcal{A}}, \overline{\mathfrak{S}}, \overline{\alpha}(\overline{G}); \Lambda^*; C^{\mathcal{S}}(\varphi), T^{\mathcal{S}}(\varphi; \Lambda^*)\right)$$

and some relations R among them.

Therefore, in the framework of ID, we have to

(i) mathematically determine

$$\mathcal{A}, \mathfrak{S}, \alpha(G); \overline{\mathcal{A}}, \overline{\mathfrak{S}}, \overline{\alpha}(\overline{G}),$$

(ii) choose Λ^* and R, and

(iii) define $C^{\mathcal{S}}(\varphi)$, $T^{\mathcal{S}}(\varphi; \Lambda^*)$.

Information Dynamics can be applied to the study of chaos in the following meaning

(a) ψ is more chaotic than φ seen from the reference system $\mathcal{S}$ if $C^{\mathcal{S}}(\psi) \geq C^{\mathcal{S}}(\varphi)$.

(b) When φ changes to $\Lambda^*\varphi$, the degree of chaos associated to this state change is given by

$$D^{\mathcal{S}}(\varphi; \Lambda^*) = C^{\overline{\mathcal{S}}}(\Lambda^*\varphi) - T^{\mathcal{S}}(\varphi; \Lambda^*).$$

This degree of chaos plays similar rôle as several other expressions of chaos in classical dynamical systems (CDS) such as Lyapunov's number or topological entropy.

In ID several different topics can be treated on common grounds so that we can find a new clue bridging several different fields [3, 16, 21, 22, 32, 33, 38, 39].

2. State Changes

ID contains the dynamics of the state change as its part. A state change is mathematically described by a channeling transformation (it is called "channel" [26]) or a bit restricted notion of "lifting" [2]. In this section, we discuss the notions of channel and lifting, and we show that several dynamics encountered in physics can be expressed by these notions.

Before defining the above notions, we set the notation used throughout this paper. Let $(\Omega, \mathcal{F})$ be a measurable space and $P(\Omega)$ be the set of all probability measures, $M(\Omega)$ be the set of all random variables on $(\Omega, \mathcal{F})$. The usual quantum system is described on a Hilbert space denoted by $\mathcal{H}$ and $B(\mathcal{H})$ is the set of all bounded linear operators on $\mathcal{H}$, and $\mathfrak{S}(\mathcal{H})$ is the set of all density operators (normal states) on $B(\mathcal{H})$. A general quantum system is described in a C^*-algebraic or von Neumann algebraic framework by a C^*-algebra or von Neumann algebra $\mathcal{A}$, and the set $\mathfrak{S}(\mathcal{A})$ of all states on $\mathcal{A}$.

The descriptions of classical dynamical systems (CDS), quantum dynamical systems (QDS) and general quantum dynamical systems (GQDS) are given in Table 2.1.

	CDS	QDS	GQDS
observable	real r.v. in $M(\Omega)$	Hermitian operator A on $\mathcal{H}$ (self adjoint operator in $B(\mathcal{H})$)	self-adjoint A in C*-algebra $\mathcal{A}$
state	probability measure $\mu \in P(\Omega)$	density operator ρ on $\mathcal{H}$	p.l.fnal $\varphi \in \mathfrak{S}$ with $\varphi(I) = 1$
expectation	$\int_\Omega f d\omega$	$\text{tr}\rho A$	$\varphi(A)$

Tab. 2.1
Descriptions of CDS, QDS and GQDS

The input and output triple $(\mathcal{A}, \mathfrak{S}, \alpha(G))$ and $(\overline{\mathcal{A}}, \overline{\mathfrak{S}}, \overline{\alpha}(\overline{G}))$ are the above sets, that is, $\mathcal{A}$ is $M(\Omega)$ or $B(\mathcal{H})$ or $\mathcal{A}$ (C^*-algebra), and $\mathfrak{S}$ corresponds to the state space in each case, and $\alpha(G)$ is an inner evolution of $\mathcal{A}$ with a parameter group G (or semigroup) and so is the output system.

A channel is a mapping from $\mathfrak{S}(\mathcal{A})$ to $\overline{\mathfrak{S}}(\overline{\mathcal{A}})$. Almost all physical transformations are described by this mapping. We first give the mathematical definition of various types of channels.

DEFINITION 2.2. Let $(\mathcal{A}, \mathfrak{S}(\mathcal{A}), \alpha)$ be an input system and $(\overline{\mathcal{A}}, \overline{\mathfrak{S}}(\overline{\mathcal{A}}), \overline{\alpha})$ be an output system. Take any $\varphi, \psi \in \mathfrak{S}(\mathcal{A})$. For $\Lambda^*: \mathfrak{S}(\mathcal{A}) \to \overline{\mathfrak{S}}(\overline{\mathcal{A}})$ we have:

(1) Λ^* is linear if $\Lambda^*(\lambda\varphi + (1-\lambda)\psi) = \lambda\Lambda^*\varphi + (1-\lambda)\Lambda^*\psi$ holds for any $\lambda \in [0,1]$.

(2) Λ^* is completely positive (C.P.) if Λ^* is linear and its dual map $\Lambda: \overline{\mathcal{A}} \to \mathcal{A}$ satisfies

$$\sum_{i,j=1}^{n} A_i^* \Lambda(\overline{A}_i^* \overline{A}_j) A_j \geq 0$$

for any $n \in \mathbb{N}$ and any $\overline{A}_i \in \overline{\mathcal{A}}$, $A_i \in \mathcal{A}$.

(3) Λ^* is of Schwarz type if $\Lambda(\overline{A}^*) = \Lambda(\overline{A})^*$ and $\Lambda(\overline{A})^*\Lambda(\overline{A}) \leq \Lambda(\overline{A}^*\overline{A})$.

(4) Λ^* is stationary if $\Lambda \circ \alpha_t = \overline{\alpha}_t \circ \Lambda$ for any $t \in \mathbb{R}$.

(5) Λ^* is ergodic if Λ^* is stationary and $\Lambda^*(\text{ex}I(\alpha)) \subset \text{ex}I(\overline{\alpha})$, where $I(\alpha)$ (resp. $I(\overline{\alpha})$) is the set of all α (resp. $\overline{\alpha}$) invariant states in $\mathfrak{S}$ (resp. $\overline{\mathfrak{S}}$).

(6) Λ^* is orthogonal if for any two orthogonal states $\varphi_1, \varphi_2 \in \mathfrak{S}(\mathcal{A})$ (denoted by $\varphi_1 \perp \varphi_2$) one has $\Lambda^*\varphi_1 \perp \Lambda^*\varphi_2$.

(7) Λ^* is deterministic if Λ^* is orthogonal and bijective.

(8) For a subset S of $\mathfrak{S}(\mathcal{A})$, Λ^* is chaotic for $\mathcal{S}$ if $\Lambda^*\varphi_1 = \Lambda^*\varphi_2$ for any $\varphi_1, \varphi_2 \in \mathcal{S}$.

(9) Λ^* is chaotic if Λ^* is chaotic for $\mathfrak{S}(\mathcal{A})$.

When we take $\overline{\mathcal{A}} = \mathcal{A} \otimes \mathcal{B}$, $\mathcal{B}$ is another algebra, the channel is called a "lifting". This special channel is useful to arrange several processes.

DEFINITION 2.3. (1) A continuous map $\mathcal{E}^*: \mathfrak{S}(\mathcal{A}) \to \mathfrak{S}(\mathcal{A} \otimes \mathcal{B})$ is called a lifting.

(2) A lifting $\mathcal{E}^*$ is nondemolition for a state $\varphi \in \mathfrak{S}(A)$ if $\mathcal{E}^*\varphi(A \otimes I) = \varphi(A)$ for any $A \in \mathcal{A}$.

Lifting is not necessary linear. An important example of nonlinear lifting is a compound state, which will be discussed in Section 3. A linear lifting is the dual map of a transition expectation of Accardi [1] from $\mathcal{A} \otimes \mathcal{B}$ to $\mathcal{A}$.

We show several examples of channels and liftings which appear in physics and quantum communication [15, 30].

2.1. Unitary evolution

For any density operator $\rho \in \mathfrak{S}(\mathcal{H})$

$$\rho \to \Lambda_t^* \rho = AdU_t(\rho) \equiv U_t \rho U_t^*, \quad t \in \mathbb{R}, \quad U_t = \exp(itH).$$

2.2. Semigroup evolution

$\rho \to \Lambda_t^* \rho = V_t \rho V_t^*, t \in \mathbb{R}^+$, where $V_t\,(t \in \mathbb{R}^+)$ is a one-parameter semigroup on $\mathcal{H}$.

2.3. Quantum measurement

When we measure an observable $A = \sum_n a_n P_n$ (spectral decomposition with $\sum_n P_n = I$) in a state ρ, then ρ changes to a state $\Lambda^*\rho$ by this measurement, such as

$$\rho \to \Lambda^* \rho = \sum_n P_n \rho P_n .$$

2.4. Reduction (Open System Dynamics)

If a system Σ_1 interacts with an external system Σ_2 described by another Hilbert space $\mathcal{K}$ and the initial states of Σ_1 and Σ_2 are ρ and σ, respectively, then the combined state θ_t of Σ_1 and Σ_2 at time t after the interaction between two systems is given by

$$\mathcal{E}_t^* \rho = \theta_t \equiv U_t(\rho \otimes \sigma) U_t^* ,$$

where $U_t = \exp(itH)$ with the total Hamiltonian H of Σ_1 and Σ_2. A channel is obtained by taking the partial trace w.r.t. $\mathcal{K}$, such as

$$\rho \rightarrow \Lambda_t^* \rho \equiv \mathrm{tr}_{\mathcal{K}} \mathcal{E}_t^* \rho .$$

If Σ_1 is an observed system and Σ_2 is a measuring apparatus, then $\mathcal{E}_t^*$ exhibits the interaction between Σ_1 and Σ_2. Namely, when an initial state of Σ_1 is ρ then $\Lambda_t^* \rho$ is the final state after the interaction between Σ_1 and Σ_2 and $\Lambda'^*_t \rho \equiv \mathrm{tr}_{\mathcal{H}} \mathcal{E}_t^*(\rho)$ is the final state of Σ_2. Therefore, a measuring process can be described by a lifting.

2.5. Optical communication processes

Quantum communication process is described by the following scheme.

$$\begin{array}{ccc} & \sigma \in \mathfrak{S}(\mathcal{K}) & \\ & \downarrow & \\ \mathfrak{S}(\mathcal{H}) \ni \rho & \longrightarrow & \overline{\rho} = \Lambda^* \rho \in \mathfrak{S}(\mathcal{H}) \\ & \downarrow & \\ & \text{Loss} & \end{array}$$

$$\begin{array}{ccc} \mathfrak{S}(\mathcal{H}) & \xrightarrow{\Lambda^*} & \mathfrak{S}(\mathcal{H}) \\ \gamma^* \downarrow & & \uparrow a^* \\ \mathfrak{S}(\mathcal{H} \otimes \mathcal{K}) & \underset{\pi^*}{\longrightarrow} & \mathfrak{S}(\mathcal{H} \otimes \mathcal{K}) \end{array}$$

The above maps γ^*, a^* are given as

$$\begin{aligned} \gamma^* \rho &= \rho \otimes \sigma , & \rho &\in \mathfrak{S}(\mathcal{H}) , \\ a^* \xi &= \mathrm{tr}_{\mathcal{K}} \xi , & \xi &\in \mathfrak{S}(\mathcal{H} \otimes \mathcal{K}) , \end{aligned}$$

where σ is a noise coming from the outside of the system. The map π^* is a certain channel determined by physical properties of the combined system. Hence the lifting and the channel for the above process are given as

$$\begin{aligned} \mathcal{E}^* \rho &\equiv \pi^* (\rho \otimes \sigma) , \\ \Lambda^* \rho &\equiv (a^* \circ \pi^* \circ \gamma^*)(\rho) = \mathrm{tr}_{\mathcal{K}} \pi^* (\rho \otimes \sigma) . \end{aligned}$$

2.6. Beam Splitting [2]

V is an isometry from $\mathcal{H}$ to $\mathcal{H} \otimes \overline{\mathcal{H}}$ and a lifting defined by

$$\mathcal{E}^* \rho = V \rho V^* , \qquad \rho \in \mathfrak{S}(\mathcal{H}) ,$$

is called an isometric lifting. One of this type is the beam splitting (attenuation) process, where the isometry $V_{\alpha\beta}$ ($\alpha, \beta \in \mathbb{C}$ with $|\alpha|^2 + |\beta|^2 = 1$) is defined on $\mathcal{H}$ (the usual Fock space) as

$$V_{\alpha\beta} |\theta\rangle = |\alpha\theta\rangle \otimes |\beta\theta\rangle$$

for a coherent state vector $|\theta\rangle \in \mathcal{H}$. Then the lifting associated with $V_{\alpha\beta}$ is

$$\mathcal{E}^*_{\alpha\beta} \rho = V_{\alpha\beta} \rho V^*_{\alpha\beta} , \qquad \rho \in \mathfrak{S}(\mathcal{H}) .$$

This beam splitting lifting was used to construct a new quantum Markov processes [14]. The attenuation process, a special channel of the type (2.5), is written by

$$\Lambda^*_{\alpha\beta} \rho = \mathrm{tr}_{\overline{\mathcal{H}}} \mathcal{E}^*_{\alpha\beta} \rho .$$

2.7. Amplifier process

In quantum optics, a linear amplifier is usually expressed by means of annihilation operators a and b on $\mathcal{H}$ and $\mathcal{K}$, respectively:

$$c = \sqrt{G} a \otimes I + \sqrt{G-1} I \otimes b^\dagger ,$$

where G (≥ 1) is a constant and c satisfies CCR (i.e., $[c, c^\dagger] = I$) on $\mathcal{H} \otimes \mathcal{K}$. This expression is not convenient to compute several measures of information like entropy. The lifting expression of the amplifier is as follows [35]:

Let $c = \mu a \otimes I + \nu I \otimes b^\dagger$ with $|\mu|^2 - |\nu|^2 = 1$ and $|\gamma\rangle$ be the eigenvector of c: $c|\gamma\rangle = \gamma|\gamma\rangle$. For two coherent vectors $|\theta\rangle$ on $\mathcal{H}$ and $|\theta'\rangle$ on $\mathcal{K}$, $|\gamma\rangle$ can be written by the squeezing expression: $|\gamma\rangle = |\theta \otimes \theta'; \mu, \nu\rangle$ and the lifting is defined by an isometry

$$V_{\theta'} |\theta\rangle = |\theta \otimes \theta'; \mu, \nu\rangle$$

such that

$$\mathcal{E}^*_{\theta'} \rho = V_{\theta'} \rho V^*_{\theta'} , \qquad \rho \in \mathfrak{S}(\mathcal{H}) .$$

The channel of the amplifier is

$$\Lambda^*_{\theta'} \rho = \mathrm{tr}_{\mathcal{K}} \mathcal{E}^*_{\theta'} \rho .$$

3. Examples of Complexities

In this section, we give several examples of complexities C and T related mainly to information.

1. The first examples of C and T are the entropy S and the mutual entropy I, respectively. Both classical and quantum S and I satisfied the conditions of the complexities. Here we only discuss the quantum case. For a density operator ρ in a Hilbert space and a channel Λ^*, the entropy $S(\rho)$ and the mutual entropy $I(\rho;\Lambda^*)$ are defined in [28] as
$$\begin{aligned} S(\rho) &= -\mathrm{tr}\rho\log\rho\,, \\ I(\rho;\Lambda^*) &= \sup\Big\{\sum_k \lambda_k S(\Lambda^* E_k, \Lambda^*\rho);\ \{E_k\}\Big\}, \end{aligned}$$
where the supremum is taken over all Schatten decompositions $\{E_k\}$ of ρ; $\rho = \sum_k \lambda_k E_k$. From fundamental properties [34] of the entropy $S(\rho)$, it satisfies (i) $S(\rho) \geq 0$, (ii) $S(j(\rho)) = S(\rho)$ for an orthogonal bijection j, that is, it is a map from a set of orthogonal pure states to another set of orthogonal pure states, (iii) $S(\rho_1 \otimes \rho_2) = S(\rho_1) + S(\rho_2)$, so that $S(\rho)$ is a complexity C of ID.
The mutual entropy $I(\rho;\Lambda^*)$ satisfies the conditions (i), (ii), (iv) from the fundamental inequality of mutual entropy [28]:
$$0 \leq I(\rho;\Lambda^*) \leq \min\{S(\rho), S(\Lambda^*\rho)\}\,.$$
Further, for the identity channel $\Lambda^* = id$,
$$\begin{aligned} I(\rho;id) &= \sup\Big\{\sum_k \lambda_k S(E_k;\rho);\ E_k\}\Big\} \\ &= \sup\Big\{\sum_k \lambda_k \mathrm{tr}E_k(\log E_k - \log\rho);\ E_k\}\Big\} \\ &= -\mathrm{tr}\rho\log\rho \end{aligned}$$
because of $S(E_k) = 0$, hence it satisfies the condition (v). Thus S and I become a pair of the complexity. Moreover, S satisfies the condition of the strong complexity (subadditivity).

2. Fuzzy entropy has been defined by several authors like Zadeh [43], DeLuca and Termini [11] and Ebanks [13]. Here we take Ebanks's fuzzy entropy and we show that we can use it to construct the complexity C. Let X (this is $\mathcal{A}$ of ID) be a countable set $\{x_1, \ldots, x_n\}$ and f_A be a membership function from X to $[0,1]$ associated with a subset $A \subset X$. If $f_A = 1_A$, then A is a usual set, which is called a sharp set, and if $f_A \neq 1_A$, then A is called a fuzzy set. Therefore, the correspondence between a fuzzy set and a membership function is one to one. Take a membership function f and let us denote $f_i = f(x_i)$ for each $x_i \in X$. Then Ebanks's fuzzy entropy $S(f)$ for a membership function f is defined by
$$S(f) = -\sum_{i=1}^{n} f_i^{\nu} \log f_i\,, \qquad (\nu = \log_2 e)\,.$$

When f is sharp, that is, $f_i = 0$ or 1 for any $x_i \in X$, $S(f) = 0$. When $f_i = \frac{1}{2}$ for any i, $S(f)$ attains the maximum value. Moreover, any two membership functions (or equivalently fuzzy sets) f and f' have the following order $\prec$:

$$f \prec f' \equiv \begin{cases} f(x) \geq f'(x) & \left(\text{when } f'(x) \geq \frac{1}{2}\right) \\ f(x) \leq f'(x) & \left(\text{when } f'(x) \leq \frac{1}{2}\right). \end{cases}$$

Then $f \prec f' \Leftrightarrow \left|f_i - \frac{1}{2}\right| \geq \left|f_i' - \frac{1}{2}\right|$, which implies

$$S(f) \leq S(f').$$

This fuzzy entropy $S(f)$ defines the complexity $C(f)$

$$C(f) = S\left(\frac{f}{n}\right).$$

The positivity of $C(f)$ is proved as follows: from Klein's inequality, $\log \frac{1}{x} \geq 1 - x$ for any $x > 0$, we have

$$\begin{aligned} C(f) &= -\sum_{i=1}^{n} \left(\frac{f_i}{n}\right)^{\nu} \log \frac{f_i}{n} \geq -\sum_{i=1}^{n} \left(\frac{f_i}{n}\right)^{\nu} \left(1 - \frac{n}{f_i}\right) \\ &= \sum_{i=1}^{n} \left(\frac{f_i}{n}\right)^{\nu-1} \left(1 - \frac{f_i}{n}\right) \geq 0. \end{aligned}$$

The invariance under a permutation π of indices i of x_i (i.e, $i \to \pi(i)$), comes directly from the invariance of S under π.

This $C(f)$ satisfies not only the additivity but also the subadditivity. Let Y be another set $\{y_1, \ldots, y_m\}$ and g be a membership function from Y to $[0,1]$. Moreover, let h be a membership function on $X \times Y$ to $[0,1]$ satisfying

$$\sum_{j=1}^{m} h(x_i, y_j) = f(x_i), \qquad \sum_{i=1}^{n} h(x_i, y_j) = g(y_j).$$

What we have to show is the inequality

$$C(h) \leq C(f) + C(g).$$

Without loss of generality, we assume $n \geq m \geq 2$. Put

$$\eta(t) = -t^{\nu} \log t, \qquad (\nu = \log_2 e).$$

$\eta(t)$ is monotone increasing in $0 \leq t \leq \frac{1}{2}$, so that we have

$$\eta\left(\frac{f_i}{nm}\right) \geq \eta\left(\frac{h_{ij}}{nm}\right), \qquad (n \geq m \geq 2),$$

because of $h_{ij} \equiv h(x_i, y_j) \leq f_i \equiv f(x_i)$, and hence $0 \leq h_{ij}/nm \leq f_i/nm \leq \frac{1}{2}$ for any i, j.

Thus we have

$$m\eta\left(\frac{f_i}{nm}\right) \geq \sum_{j=1}^{m} \eta\left(\frac{h_{ij}}{nm}\right).$$

Now

$$\begin{aligned}
\eta\left(\frac{f_i}{n}\right) - \frac{1}{2}m\eta\left(\frac{f_i}{nm}\right) &= \\
&= \left(\frac{f_i}{n}\right)^{\nu}\left\{\left(\frac{1}{2m^{\nu-1}} - 1\right)\log f_i + \left(1 - \frac{1}{2m^{\nu-1}}\right)\log n - \frac{1}{2m^{\nu-1}}\log m\right\} \\
&\geq \left(\frac{f_i}{n}\right)^{\nu}\left\{\left(\frac{1}{2m^{\nu-1}} - 1\right)\log f_i + \left(1 - \frac{1}{m^{\nu-1}}\right)\log n\right\}
\end{aligned}$$

which is positive since $n \geq m \geq 2$. Hence

$$\eta\left(\frac{f_i}{n}\right) \geq \frac{1}{2}\sum_{j=1}^{m} \eta\left(\frac{h_{ij}}{nm}\right)$$

which implies

$$C(f) \geq \frac{1}{2}C(h).$$

Similarly we can prove

$$C(g) \geq \frac{1}{2}C(h).$$

Therefore, we have the subadditivity

$$C(h) \leq C(f) + C(g).$$

3. Kolmogorov [17] and Chaitin [8] discussed the complexity of sequences. Consider e.g. the following two sequences a and b composed of 0-s and 1-s:

$$\begin{aligned}
a &: 010101010101, \\
b &: 011010000110.
\end{aligned}$$

In both a and b, the occurrence probabilities $p(0)$ and $p(1)$ are the same, $p(0) = p(1) = 1/2$. However, the sequence b seems to be more complicated than a. It suffices to know the first two letters to guess the whole a, but one may need the whole sequence of letters to know b. In general, we consider a computer (an automaton) transforming binary input sequences into output ones. Formalizing the description of such a machine it is possible to introduce the notion of the *minimum programme* as the "simplest" algorithm which produces a given output sequence. The amount of information contained in the specification of the minimum programme (measured e.g. in "bits") is called the complexity of a given (output) sequence.

Let $\mathcal{A}$ be the set of all finite sequences over an alphabet, say $\{0,1\}$, and $\bar{\mathcal{A}}$ be another set. Further, let f be a *partial function* from $\mathcal{A}$ to $\bar{\mathcal{A}}$ (i.e., f is not necessarily defined on the whole $\mathcal{A}$). The triple $(\mathcal{A}, \bar{\mathcal{A}}, f)$ can be regarded as a language describing certain objects. For an element $a \in \mathcal{A}$, the length of a is denoted by $\ell(a)$. The minimum length of $a \in \mathcal{A}$ describing $\bar{a} \in \bar{\mathcal{A}}$ (that is $f(a) = \bar{a}$) is called the *complexity of description*. If there is no $a \in \mathcal{A}$ such that $f(a) = \bar{a}$, then we set the complexity of $\bar{a}$ to ∞.

When both $\mathcal{A}$ and $\bar{\mathcal{A}}$ are sets of binary sequences, we consider partial *computable* functions $f: \mathcal{A} \to \bar{\mathcal{A}}$ (i.e., such that there exists a programme which, given an input $a \in \mathcal{A}$, terminates (halts) with the output $f(a) \in \bar{\mathcal{A}}$ whenever $a \in$ Domain (f), but it need not halt at all for inputs $a \notin$ Domain (f)). The *complexity* $H_f(\bar{a})$ determined by $(\mathcal{A}, \bar{\mathcal{A}}, f)$ is defined as

$$H_f(\bar{a}) = \begin{cases} \min \{\ell(a); a \in \mathcal{A}, f(a) = \bar{a}\} & (\text{when } \exists a \in \mathcal{A} \text{ s.t. } f(a) = \bar{a}) \\ \infty & (\text{otherwise}) . \end{cases}$$

For $a \in \mathcal{A}$ let $0^k 1a$ denote the sequence obtained from a by appending k symbols 0 and a single 1 on the left, i.e. $0 \ldots 01a$. Then it is shown [8] that there exist $k \in \mathbb{N}$ and a computable partial function f_U such that for any f, $f_U(0^k 1a) = f(a)$. This f_U is called the universal partial function. With this function certain *universal computer* U is associated. In particular, U is capable of computing $f(a)$ for $a \in$ Domain (f). Some important consequences of the above are: (1) there exists a constant ε such that $H_{f_U}(\bar{a}) \leq H_f(\bar{a}) + \varepsilon$ for any f, and (2) there exists a constant ε' for two universal partial functions $f_U, f_{U'}$ satisfying $|H_{f_U}(\bar{a}) - H_{f_{U'}}(\bar{a})| \leq \varepsilon'$.

The above facts imply that H_{f_U} gives the minimum value for H_f if we neglect the constant ε. Kolmogorov and Chaitin introduced the following complexity

$$H(\bar{a}) = H_{f_U}(\bar{a})$$

which does not depend on a choice of f_U because of the statement (2). Moreover, Chaitin introduced the mutual entropy type complexity in the same framework as above. This complexity and mutual entropy type complexity can be associated with our complexities C and T, respectively.

4. Generalizing the entropy S and the mutual entropy I, we can construct complexities of entropy type:

   Let $(\mathcal{A}, \mathfrak{S}(\mathcal{A}), \alpha(G))$, $(\overline{\mathcal{A}}, \overline{\mathfrak{S}}(\overline{\mathcal{A}}), \overline{\alpha}(\overline{G}))$ be C^* systems as before. Let $\mathcal{S}$ be a weak *-compact convex subset of $\mathfrak{S}(\mathcal{A})$ and $M_\varphi(\mathcal{S})$ be the set of all maximal measures μ on $\mathcal{S}$ with the fixed barycenter φ [9]

$$\varphi = \int_{\mathcal{S}} \omega d\mu .$$

Moreover, let $F_\varphi(\mathcal{S})$ be the set of all measures of finite support with the fixed barycenter φ. The following three pairs C and T satisfy all conditions of the complexities:

$$\begin{aligned} T^{\mathcal{S}}(\varphi;\Lambda^*) &\equiv \sup\left\{\int_{\mathcal{S}} S(\Lambda^*\omega,\Lambda^*\varphi)d\mu;\ \mu\in M_\varphi(\mathcal{S})\right\} \\ C_T^{\mathcal{S}}(\varphi) &\equiv T^{\mathcal{S}}(\varphi;id) \\ I^{\mathcal{S}}(\varphi;\Lambda^*) &\equiv \sup\left\{S\left(\int_{\mathcal{S}}\omega\otimes\Lambda^*\omega d\mu,\varphi\otimes\Lambda^*\varphi\right);\ \mu\in M_\varphi(\mathcal{S})\right\} \\ C_I^{\mathcal{S}}(\varphi) &\equiv I^{\mathcal{S}}(\varphi;id) \\ J^{\mathcal{S}}(\varphi;\Lambda^*) &\equiv \sup\left\{\int_{\mathcal{S}} S(\Lambda^*\omega,\Lambda^*\varphi)d\mu;\ \mu\in F_\varphi(\mathcal{S})\right\} \\ C_J^{\mathcal{S}}(\varphi) &\equiv J^{\mathcal{S}}(\varphi;id)\,. \end{aligned}$$

Here, the state $\int_{\mathcal{S}}\omega\otimes\Lambda^*\omega d\mu$ is the compound state exhibiting the correlation between the initial state and the final state $\Lambda^*\varphi$. This compound state was introduced in [27] as a quantum generalization of the joint probability measure in CDS because it does not exist in QDS [42].

Moreover, it is a nondemolition nonlinear lifting from $\mathfrak{S}(\mathcal{A})$ to $\mathfrak{S}(\mathcal{A}\otimes\overline{\mathcal{A}})$.

These complexities with the mixing $\mathcal{S}$-entropy $S^{\mathcal{S}}(\varphi)$ and the CNT (Connes-Narnhofer-Thirring) entropy $H_\varphi(\mathcal{A})$ satisfy the relations of Theorem 3.1 [24, 36]. Before stating a theorem, we review the definition of the $\mathcal{S}$-entropy and the CNT entropy: For a state $\varphi\in\mathcal{S}\subset\mathfrak{S}(\mathcal{A})$, put

$$D_\varphi(\mathcal{S})\equiv$$
$$\left\{\mu\in M_\varphi(\mathcal{S});\exists\{\mu_k\}\subset\mathbb{R}^+\text{and}\{\varphi_k\}\subset\text{ex}\mathcal{S}\text{ s.t. }\sum_k\mu_k=1,\mu=\sum_k\mu_k\delta(\varphi_k)\right\},$$

where $\delta(\varphi)$ is the delta measure concentrated on $\{\varphi\}$, and put

$$H(\mu) = -\sum_k \mu_k\log\mu_k$$

for a measure $\mu\in D_\varphi(\mathcal{S})$. Then the $\mathcal{S}$-entropy of a state $\varphi\in\mathcal{S}$ is defined in [29, 30] as

$$S^{\mathcal{S}}(\varphi) = \begin{cases}\inf\{H(\mu);\mu\in D_\varphi(\mathcal{S})\} & \text{if } D_\varphi(\mathcal{S})\neq\emptyset \\ +\infty & \text{if } D_\varphi(\mathcal{S})=\emptyset,\end{cases}$$

Connes, Narnhofer and Thirring introduced the entropy of a subalgebra $\mathfrak{M}$ of $\mathcal{A}$ [10]. The CNT-entropy $H_\varphi(\mathfrak{M})$ is defined as follows:

For a state φ and a subalgebra $\mathfrak{M}$

$$H_{\varphi}(\mathfrak{M}) \equiv$$

$$\sup\left\{\sum_j \mu_j S\left(\varphi_j\left|_{\mathfrak{M}}, \varphi\right|_{\mathfrak{M}}\right); \varphi = \sum_j \mu_j \varphi_j (\text{finite decomposition of } \varphi)\right\},$$

where $S(\cdot\,,\,\cdot)$ is the relative entropy for C^*-algebra according to the definition of Araki [4] or Uhlmann [41] and $\varphi|_{\mathfrak{M}}$ is the restriction of φ to $\mathfrak{M}$.

THEOREM 3.1. *(1)* $0 \le I^{\mathcal{S}}(\varphi;\Lambda^*) \le T^{\mathcal{S}}(\varphi;\Lambda^*) \le J^{\mathcal{S}}(\varphi;\Lambda^*)$.

(2) $C_I^{\mathfrak{S}}(\varphi) = C_T^{\mathfrak{S}}(\varphi) = C_J^{\mathfrak{S}}(\varphi) = S^{\mathfrak{S}}(\varphi) = H_{\varphi}(\mathcal{A})$.

(3) When $\mathcal{A} = \overline{\mathcal{A}} = B(\mathcal{H})$, *for any density operator* ρ

$$0 \le I^{\mathcal{S}}(\rho;\Lambda^*) = T^{\mathcal{S}}(\rho;\Lambda^*) \le J^{\mathcal{S}}(\rho;\Lambda^*)\,.$$

These complexities provide us with a new definition of the quantum dynamical entropy [36], which can be used to characterize quantum communication processes [37].

It is possible to construct other complexities, not of the entropy type, in several fields like genetics [38], economics and computer sciences [39].

4. Fractal Dimension of States

Usual fractal theory treats mostly geometrical sets. It is desirable to extend the fractal theory so as to be applicable to some other objects. For this purpose, we introduce the fractal dimension of general states. First, we recall two fractal dimensions of geometrical sets [20].

4.1. Scaling dimension

We observe a complex set F built from a fundamental pattern. If the number of the patterns observed is $N(1)$ when the scale is very rough, say 1, and the number is $N(r)$ when the scale is r, then we call the dimension defined through

$$d_s(F) = \frac{\log(N(r)/N(1))}{\log(1/r)}$$

the scaling dimension of the set F.

4.2. Capacity dimension

Let us cover a set F in the n-dimensional Euclidean space $\mathbb{R}^n$ by copies of a certain convex set with the diameter ε. If the smallest number of the convex sets needed to cover the set F is $N(\varepsilon)$, then we call the dimension given by

$$d_c(F) = \lim_{\varepsilon\to 0} \frac{\log N(\varepsilon)}{\log(1/\varepsilon)}$$

the capacity dimension (or the ε-entropy dimension) of the set F.

These two fractal dimensions become equal for almost all sets for which they can be computed.

The ε-entropy was extensively studied by Kolmogorov [18] and his ε-entropy was defined for a probability measure, which gives us an idea to define the ε-entropy for a general state.

Kolmogorov introduced the notion of ε-entropy in probability space $(\Omega, \mathcal{F}, \mu)$. His formulation is as follows: for two random variables $f, g \in M(\Omega)$, the mutual entropy $I(f,g)$ is defined by the joint probability measure $\mu_{f,g}$ and the direct product measure $\mu_f \otimes \mu_g$ such that

$$I(f,g) = S(\mu_{f,g}, \mu_f \otimes \mu_g),$$

where $S(\cdot,\cdot)$ is the relative entropy [19]. The ε-entropy for a random variable f is given by

$$S_{\mathrm{K}}(f,\varepsilon) \equiv \inf\{I(f,g);\ g \in M_d(f,\varepsilon)\},$$

where $M_d(f,\varepsilon) \equiv \{g \in M(\Omega);\ \|f-g\| \le \varepsilon\}$ with $\|f-g\| \equiv \sqrt{\int_\Omega |f-g|^2 d\mu_{fg}}$.

For a general probability measure μ on $(\Omega, \mathcal{F})$, the Kolmogorov ε-entropy $S_{\mathrm{K}}(\mu;\varepsilon)$ is given by

$$S_{\mathrm{K}}(\mu;\varepsilon) = \inf\{S(\mu_{co}, \mu\otimes\bar{\mu});\ \bar{\mu} \in P_0(\Omega)\},$$

where μ_{co} is the joint (compound) probability measure of μ and $\bar{\mu}$ and $P_0(\Omega)$ is the set of all probability measures $\bar{\mu}$ satisfying $\|\mu - \bar{\mu}\| \le \varepsilon$.

We introduced the ε-entropy of a general quantum state φ and the fractal dimensions of states in [31, 33, 36]. Let $\mathcal{C}$ be a set of some physically interesting channels and define two sets:

$$\begin{aligned} \mathcal{C}_1(\Lambda^*;\varphi) &= \{\Gamma^* \in \mathcal{C}; \Gamma^*\varphi = \Lambda^*\varphi\}, \\ \mathcal{C}_2(\varphi;\varepsilon) &= \{\Gamma^* \in \mathcal{C};\ \|\varphi - \Gamma^*\varphi\| \le \varepsilon\}. \end{aligned}$$

Then the ε-entropy of a state of φ w.r.t. $\mathcal{S}$ is defined by means of the transmitted complexity $T^{\mathcal{S}}(\varphi;\Lambda^*)$ as

$$S^{\mathcal{S}}_{\mathcal{C},0}(\varphi;\varepsilon) = \inf\left\{J^{\mathcal{S}}_{\mathcal{C}}(\varphi;\Lambda^*);\ \Lambda^* \in \mathcal{C}_2(\varphi;\varepsilon)\right\},$$

where

$$J_{\mathcal{C}}^{\mathcal{S}}(\varphi;\Lambda^*) = \sup\left\{T^{\mathcal{S}}(\varphi;\Gamma^*);\ \Gamma^* \in \mathcal{C}_1(\Lambda^*;\varphi)\right\}.$$

When $\mathcal{S} = \mathfrak{S}$ and $\mathcal{C}$ is the set of all channels on $\mathfrak{S}$, the above ε-entropy is denoted by $S_{\mathrm{O}}(\varphi;\varepsilon)$ for simplicity.

The capacity dimension of a state φ w.r.t. $\mathcal{S}$ and $\mathcal{C}$ is defined by

$$d_{\mathcal{C},\mathrm{O}}^{\mathcal{S}}(\varphi) \equiv \lim_{\varepsilon\to 0} d_{\mathcal{C},\mathrm{O}}^{\mathcal{S}}(\varphi;\varepsilon),$$

where

$$d_{\mathcal{C},\mathrm{O}}^{\mathcal{S}}(\varphi;\varepsilon) = \frac{S_{\mathcal{C},\mathrm{O}}^{\mathcal{S}}(\varphi;\varepsilon)}{\log(1/\varepsilon)}.$$

The above $d_{\mathcal{C}}^{\mathcal{S}}(\varphi;\varepsilon)$ is called the capacity dimension of ε-order. The information dimension of a state φ for $\mathcal{S}$ of ε order is defined by

$$d_{I,\mathcal{C},\mathrm{O}}^{\mathcal{S}}(\varphi;\varepsilon) = \frac{S_{\mathcal{C},\mathrm{O}}^{\mathcal{S}}(\varphi;\varepsilon)}{S^{\mathcal{S}}(\varphi)}.$$

The above new ε-entropy and fractal dimensions of states provide measures to classify chaotic phenomena in both classical and quantum systems. We will discuss some applications of such uses in the next section.

5. Some Applications of Fractal Dimensions of States

These ε-entropy and fractal dimensions are applied to several physical phenomena and mathematical objects. For instance, we can classify the shapes of river basins or seas on the moon [22], as well as consider symmetry breaking in Ising systems [21]. We state some results concerning the classification of Gaussian measures [15, 16]. For a random variable $f = (f_1, \ldots, f_n)$ from Ω to $\mathbb{R}^n$ and the measure μ_f associated with f, the random variable norm $\|\cdot\|_{\mathrm{R.V}}$ of μ_f is defined by

$$\|\mu_f\|_{\mathrm{R.V}} = \sqrt{\frac{1}{n}\sum_{i=1}^{n}\|f_i\|^2}, \qquad \|f_i\| = \sqrt{\int_\Omega |f_i|^2 d\mu}.$$

THEOREM 5.2. *If the distance of two states is defined through the above random variable norm on $\Omega = \mathbb{R}^n$ and the transmitted complexity T is the mutual entropy in CDS, then*

(1)

$$S_{\mathrm{O}}(\mu_f;\varepsilon) = S_{\mathrm{K}}(f;\varepsilon) = \frac{1}{2}\sum_{i=1}^{n}\log\max\left(\frac{\lambda_i}{\theta^2},1\right),$$

where $\lambda_1, \ldots, \lambda_n$ *are the eigenvalues of the covariance operator* R *for* μ_f *and* θ^2 *is the constant uniquely determined by the equation*

$$\sum_{i=1}^{n} \min(\lambda_i, \theta^2) = \varepsilon^2 .$$

(2) $d_{\mathrm{O}}(\mu_f) = d_{\mathrm{K}}(\mu_f) = n.$

According to this theorem, the Kolmogorov ε-entropy coincides with our ε-entropy. The difference between S_{O} and S_{K} comes from the norm taken for measures. For instance, when we take the norm of measures by the total variation, namely, $\|\nu\|_{TV} = |\nu|(\Omega)$ for any finite measure ν. We show the difference for the case $\Omega = \mathbb{R}$. If an input state μ is described by the mean 0 and the covariance σ^2 and a noise of a channel is exhibited by one-dimensional Gaussian measure $\mu_0 = [0, \sigma_0^2] \in P(\Omega)$, then the output state $\Lambda^*\mu$ is given by $[0, \bar{\alpha}^2\sigma^2 + \sigma_0^2]$ with some constant $\bar{\alpha}$. Since the channel Λ^* depends on $\bar{\alpha}$ and σ_0^2, we denote $\Lambda^* = \Lambda^*_{(\bar{\alpha},\sigma_0^2)}$.

THEOREM 5.3. *Let* $\Lambda^*_{\delta(\bar{\alpha},\sigma_0^2)}$ *be a channel satisfying* $\|\mu - \Lambda^*_{\delta(\bar{\alpha},\sigma_0^2)}\mu\|_{TV} = \delta$ *for any* $\delta \in M(\varepsilon)$. *If a Gaussian channel* $\Lambda^*_{\delta(\bar{\alpha},\sigma_0^2)}$ *satisfies the condition* $\bar{\alpha}^2 \le \dfrac{C_\delta - \delta}{\sigma^2}$, *then we have*

(1) $$S_{C,\mathrm{O}}(\mu;\varepsilon) = \frac{1}{2}\log\frac{1}{\varepsilon} + \frac{1}{2}\log\frac{\sigma^2}{\left(1 + \frac{\sqrt{2\pi}}{4}(\varepsilon + o(\varepsilon))\right)^2} > S_{\mathrm{K}}(\mu\,\varepsilon) = 0,$$

(2) $$d_{C,\mathrm{O}}(\mu) = \frac{1}{2},$$

where $o(\varepsilon)$ *is the order of* ε : $\lim_{\varepsilon\to 0} o(\varepsilon) = 0$ *and* $M(\delta) = \{\delta \in \mathbb{R}; 0 \le \delta \le \varepsilon\}$.

This theorem tells us the difference between the Kolmogorov ε-entropy and our ε-entropy. It concludes that our fractal dimension shows the fractal structure of Gaussian measures. Moreover, our fractal dimensions enable to characterize chaos in the following sense: when a system is random, our fractal dimension is infinite, but when it has some order like self-similar structure, it will yield a finite value.

Bibliography

1. L. Accardi, *Noncommutative Markov Chains*, International School of Mathematical Physics, Camerino, pp. 268–295, 1974.
2. L. Accardi and M. Ohya, Compound Channels, Transition Expectations and Liftings, to appear in J. Multivariate Analysis.
3. L. Accardi, M. Ohya and N. Watanabe, Rep. Math. Phys. **38**, 1 (1996).
4. H. Araki, Publ. RIMS, Kyoto Univ. **11**, 809 (1976).

5. F. Benatti, *Deterministic Chaos in Infinite Quantum Systems*, Trieste Notes in Physics, Springer-Verlag, 1993.
6. P. Billingsley, *Ergodic Theory and Information*, Wiley, New York, 1965.
7. O. Bratteli and D.W. Robinson, *Operator Algebras and Quantum Statistical Mechanics II*, Springer, New York, Berlin, Heidelberg, 1981.
8. G.J. Chaitin, *Algorithmic Information Theory*, Cambridge Uni. Press, 1987.
9. G. Choquet, *Lecture Analysis I, II, III, Bengamin*, New York, 1969.
10. A. Connes, H. Narnhofer, and W. Thirring, Commun. Math. Phys. **112**, 691 (1987).
11. A. DeLuca and S. Termini, Inform. Control. **20**, 301 (1972).
12. G.G. Emch, Z. Wahrscheinlichkeitstheorie Verw. Gebiete **29**, 241 (1974).
13. B.R. Ebanks, J. Math. Anal. Appl. **94**, 24 (1983).
14. K.H. Fichtner, W. Freudenberg, and V. Liebscher, *Beam Splitting and Time Evolutions of Boson Systems*, preprint.
15. R.S. Ingarden, A. Kossakowski, and M. Ohya, *Open Systems and Information Dynamics*, to be published in Kluwer.
16. K. Inoue, T. Matsuoka, and M. Ohya, *New Approach to ε-Entropy and Its Comparison with Kolmogolov's ε-entropy*, SUT preprint.
17. A.N. Kolmogorov, Dokl. Akad. Nauk SSSR **119**, 861 (1958).
18. A.N. Kolmogorov, Amer. Math. Soc. Translation, Ser. 2 **33**, 291 (1963).
19. S. Kullback and R. Leibler, Ann. Math. Stat. **22**, 79 (1951).
20. B.B. Mandelbrot, *The Fractal Geometry of Nature*, W.H. Freemann and Company, San Francisco, 1982.
21. T. Matsuoka and M. Ohya, Rep. Math. Phys. **36**, 27 (1995).
22. T. Matsuoka and M. Ohya, *Fractal Dimension of States and Its Application to Shape Analysis Problem*, SUT preprint.
23. N. Muraki, M. Ohya, and D. Petz, Open Sys. Information Dyn. **1**, 43 (1992).
24. N. Muraki and M. Ohya, Lett. Math. Phys. **36**, 327 (1996).
25. J. von Neumann, *Die Mathematischen Grundlagen der Quantenmechanik*, Springer, Berlin, 1932.
26. M. Ohya, J. Math. Anal. Appl. **84**, 318 (1981).
27. M. Ohya, L. Nuovo Cimento **38**, 402 (1983).
28. M. Ohya, IEEE Trans. Information Theory **29**, 770 (1983).
29. M. Ohya, J. Math. Anal. Appl. **100**, 222 (1984).
30. M. Ohya, Rep. Math. Phys. **27**, 19 (1989).
31. M. Ohya, Proc. Symp. Appl. Func. Anal. **11**, 45 (1989).
32. M. Ohya, Lecture Notes in Physics **378**, Springer, 81 (1991).
33. M. Ohya, Quantum Probability and Related Topics **6**, 359 (1991).
34. M. Ohya and D. Petz, *Quantum Entropy and Its Use*, Springer-Verlag, 1993.
35. M. Ohya and H. Suyari, Rep. Math. Phys. **36**, 403 (1995).
36. M. Ohya, Quantum Communications and Measurement **2**, Plenum, 309 (1995).
37. N. Ohya and N. Watanabe, Note on Irreversible Dynamics and Quantum Information, to appear in the Alberto Frigerio Conference Proceedings.
38. M. Ohya, *Analysis of Geneome Sequences by Complexity*, SUT preprint.
39. N. Ohya and N. Watanabe, *On Mathematical Treatment of Fredkin-Toffoli-Milburn Gate*, SUT preprint.
40. C.E. Shannon, Bell System Tech. J. **27**, 379 (1948).
41. A. Uhlmann, Commun. Math. Phys. **54**, 21 (1977).
42. K. Urbanik, Stud. Math. **21**, 119 (1961).
43. L.A. Zadeh, J. Math. Anal. Appl. **23**, 421 (1968).

Studia Scientiarum Mathematicarum Hungarica 31 (1996), 423–430

NOTES ON QUANTUM ENTROPY

M. OHYA and D. PETZ

The present paper consists of two parts. In the first one it will be proved that the von Neumann entropy governs the size of rather sure projections in the course of independent trials. The second part is devoted to the extension of the von Neumann entropy to states of arbitrary unital C^*-algebras.

By a finite quantum system we mean an algebra of matrices which is stable under taking adjoint. (In other words, a finite quantum system is a finite dimensional C^*-algebra.) If $\mathcal{A}$ is such an algebra then there is a linear functional Tr which takes the value 1 at each minimal projection. It is "tracial" in the sense that

$$\operatorname{Tr} ab = \operatorname{Tr} ba \qquad (a, b \in \mathcal{A}).$$

Every functional ω on $\mathcal{A}$ is determined by a density operator $D_\omega \in \mathcal{A}$ in the form

$$\omega(a) = \operatorname{Tr} D_\omega a \qquad (a \in \mathcal{A}).$$

The entropy $S(\omega)$ of a functional ω is defined by means of its density operator as

$$S(\omega) = \operatorname{Tr} \eta(D_\omega).$$

This notion was introduced by von Neumann in 1927 and we term it von Neumann's entropy or shortly entropy (cf. [8]).

It is understood in probability theory that the notion of (Shannon or measure theoretic) entropy has successful applications in a variety of subjects because it determines the asymptotic behaviour of certain probabilities in the course of independent trials (see, for example [1] or [11]). Now we will discuss this phenomenon for finite quantum systems.

Let $\mathcal{A}$ be a finite quantum system with a faithful state ω. The n-fold algebraic tensor product $\mathcal{A}_n = \mathcal{A} \otimes \ldots \otimes \mathcal{A}$ is again a finite quantum system and the product functional $\omega_n = \omega \otimes \ldots \otimes \omega$ is a state of $\mathcal{A}_n$. Using the obvious identifications the inclusion $(\mathcal{A}_n, \omega_n) \subset (\mathcal{A}_m, \omega_m)$ holds for $n \leqq m$ and we set

$$(\mathcal{A}_\infty, \omega_\infty) = \cup\{(\mathcal{A}_n, \omega_n) : n \in \mathbb{N}\}.$$

1991 *Mathematics Subject Classification.* Primary 46L30; Secondary 82B10, 94A17.

Key words and phrases. Density matrix, state space, von Neumann entropy, relative entropy.

On the $*$-algebra $\mathcal{A}_\infty$ the right shift endomorphism γ is defined for $a_1 \otimes a_2 \otimes \otimes \ldots \otimes a_n \in \mathcal{A}_n$ as

$$\gamma(a_1 \otimes a_2 \otimes \ldots \otimes a_n) = I \otimes a_1 \otimes a_2 \otimes \ldots \otimes a_n \in \mathcal{A}_{n+1}$$

and ω_∞ is invariant under γ. Now perform the GNS-constructions with the state ω_∞ and arrive at the triplet $(\pi, \mathcal{H}, \Omega)$. We identify $\mathcal{A}_\infty$ through its faithful representation π with a subalgebra of the generated von Neumann algebra $\mathcal{M} = \pi(\mathcal{A}_\infty)'' \subset \mathcal{B}(\mathcal{H})$. The normal state

$$\omega(a) = \langle \Omega, a\Omega \rangle \qquad (a \in \mathcal{M})$$

is an extension of ω_∞ and the endomorphism γ extends to $\mathcal{M}$ such that the relation $\omega \circ \gamma = \omega$ is preserved. (For the sake of simpler notation we do not use a new letter for the extension.)

The following result may be called the weak law of large numbers (for independent finite quantum systems). Since it is well known its proof will be omitted.

PROPOSITION 1. *In the above described situation the following statements hold.*

(i) *If $a \in \mathcal{M}$ and $\gamma(a) = a$ then $a \in \mathbf{C}I$.*

(ii) *For every $a \in \mathcal{M}$ the sequence $s_n(a) = n^{-1}(a + \gamma(a) + \ldots + \gamma^{n-1}(a))$ converges to $\omega(a)I$ in the strong operator topology.*

(iii) *If $a \in \mathcal{M}^{sa}$ and $J \subset \mathbf{R}$ is closed interval such that $\omega(a) \notin J$ and p_n is the spectral projection of $s_n(a)$ corresponding to the interval J then $p_n \to 0$ in the strong operator topology.*

Let us fix a positive number $\varepsilon < 1$. For a while we say that a projection $Q_n \in \mathcal{A}_n$ is rather sure if $\omega_n(Q_n) \geqq 1 - \varepsilon$. On the other hand, the size of Q_n, the cardinality of a maximal pairwise orthogonal family of projections contained in Q_n, is given by $\mathrm{Tr}_n Q_n$. (The subscript n in Tr_n indicates that the algebraic trace functional on $\mathcal{A}_n$ is meant here.) The theorem below says that the von Neumann's entropy of ω governs asymptotically the size of rather sure projections: A rather sure projection in $\mathcal{A}_n$ contains at least $\exp(nS(\omega))$ pairwise orthogonal minimal projection.

THEOREM 2. *Under the above conditions and with the above notation the limit relation*

$$\lim_{n\to\infty} \frac{1}{n} \inf \{\log \mathrm{Tr}_n Q_n : Q_n \in \mathcal{A}_n \text{ is a projection, } \omega_n(Q_n) \geqq 1 - \varepsilon\} = S(\omega)$$

holds.

PROOF. If D_n denotes the density of ω_n then one can see easily that

$$-\log D_n = \sum_{i=0}^{n-1} \gamma^i(-\log D_1)$$

where γ stands for the right shift. The sequence $(\gamma^i(-\log D_1))$ behaves as independent identically distributed random variables with respect to the state ω_∞. More precisely, the previous proposition applies for $a = -\log D_1$ and tells

$$\frac{1}{n}\log D_n \to S(\omega)I$$

strongly. Let $P(n,\delta)$ be the spectral projection of the selfadjoint operator $-n^{-1}\log D_n$ corresponding to the interval $(S(\omega)-\delta, S(\omega)+\delta)$. According to (iii) of Proposition 1 one has

$$P(n,\delta) \to I \tag{1}$$

strongly for every $\delta > 0$. In particular,

$$\omega(P(n,\delta)) = \langle P(n,\delta)\Omega, \Omega\rangle \to 1$$

as $n \to \infty$ and $P(n,\delta)$ is a rather sure projection if n is large enough. It follows from the definition of $P(n,\delta)$ that

$$D_n P(n,\delta)\exp(nS(\omega)-n\delta) \leqq P(n,\delta) \leqq D_n \exp(nS(\omega)+n\delta) \tag{2}$$

which gives

$$\frac{1}{n}\log \mathrm{Tr}_n P(n,\delta) \leqq S(\omega)+\delta.$$

Since $\delta > 0$ was arbitrary we establish

$$\limsup_{n\to\infty} \frac{1}{n}\inf\{\log \mathrm{Tr}_n Q_n : Q_n\} \leqq S(\omega). \tag{3}$$

To prove that $S(\omega)$ is actually the limit we shall argue by contradiction. Assume that there exist a sequence $n(1) < n(2) < \ldots$ of integers, a number $t > 0$ and projections $Q(n(k)) \in \mathcal{A}_{n(k)}$ $(k = 1, 2, \ldots)$ such that

(i) $\omega_\infty(Q(n(k)) \geqq 1-\varepsilon$,
(ii) $\log \mathrm{Tr}_{n(k)} Q(n(k)) \leqq n(k)(S(\omega)-t)$.

The bounded sequence $(Q(n(k)))_k$ has a weak limit point in the von Neumann algebra $\mathcal{M}$, say $T \in \mathcal{M}$. Instead of selecting a subsequence we suppose that $Q(n(k)) \to T$ weakly. It is straightforward to show that from (1) the weak limit

$$Q(n(k))P(n(k),\delta) \to T$$

follows. Consequently,

$$\liminf_{k\to\infty} \omega_\infty(Q(n(k))P(n(k),\delta)) \geqq \omega(T) \geqq 1-\varepsilon. \tag{4}$$

Using the first part of (2) we estimate

$$\begin{aligned}\mathrm{Tr}\, Q(n(k)) &\geqq \mathrm{Tr}\, Q(n(k))P(n(k),\delta)\\ &\geqq \mathrm{Tr}\, D_{n(k)} Q(n(k))P(n(k),\delta)\exp(nS(\omega)-n\delta)\\ &= \exp(nS(\omega)-n\delta)\omega_\infty(Q(n(k))P(n(k),\delta))\end{aligned}$$

and

$$\liminf_{k\to\infty} \frac{1}{n(k)} \log \mathrm{Tr}_{n(k)} Q(n(k))$$
$$\geqq S(\omega) - \delta + \lim_{k\to\infty} \frac{1}{n(k)} \log \omega_\infty (Q(n(k))P(n(k),\delta)).$$

The limit term on the right-hand side vanishes due to (4) and we arrive at a contradiction with (ii) if $0 < \delta < t$. This proves the theorem.

Opposite to the commutative case the state space of a quantum system is not a Choquet simplex in the sense that states admit several extremal decompositions. For example, for $\mathcal{A} = M_2(\mathbf{C})$ the general form of a density matrix is

$$D = \frac{1}{2}\begin{pmatrix} 1+a & b+ic \\ b-ic & 1-a \end{pmatrix} \tag{5}$$

where a, b, c are real numbers and $a^2 + b^2 + c^2 \leqq 1$. Thanks to the affine correspondence $D \leftrightarrow (a,b,c)$ we can visualize the state space as a ball (of radius 1) and surface points correspond to pure states.

Let φ be a state of a finite quantum system and $\varphi = \sum_i \lambda_i \psi_i$ be an extremal decomposition (that is, every ψ_i is pure). Approaching from information theory one might think that the entropy of φ is $-\sum_i \lambda_i \log \lambda_i$. This, however, would not be satisfactory because the λ_i's are not in general the probabilities of mutually exclusive events. In fact,

$$S(\varphi) \leqq -\sum_i \lambda_i \log \lambda_i \tag{6}$$

and the equality holds if and only if the extremal decomposition $\sum \lambda_i \psi_i$ is orthogonal. This was obtained in [4] a long time ago and here it will be deduced by means of the relative entropy. The inequality (6) is interpreted as follows. In the sense of information content, the most economical extremal decomposition is the orthogonal one, which is implemented by the density matrix.

The entropy of ω with respect to φ is defined by

$$S(\omega,\varphi) = \begin{cases} \mathrm{Tr}\, D_\omega (\log D_\omega - \log D_\varphi) & \text{if } \operatorname{supp} D_\varphi \geqq \operatorname{supp} D_\omega \\ +\infty & \text{otherwise.} \end{cases} \tag{7}$$

Here $\operatorname{supp} D_\psi$ denotes the smallest projection p such that $\psi(p) = \psi(I)$. In particular, $S(\omega,\varphi)$ is always finite if the density of φ has strictly positive eigenvalues. (Such a φ is called faithful.) When D_φ commutes with D_ω and their eigenvalue lists are $(\lambda_1, \lambda_2, \ldots, \lambda_n)$ and $(\kappa_1, \ldots, \kappa_n)$, respectively, then $S(.,.)$ reduces to the classical expression due to Kullback and Leibler.

Although we mostly speak of the relative entropy of states it is convenient to allow ω and φ in the definition of $S(\omega,\varphi)$ to be arbitrary positive functionals.

The relative entropy may be defined for linear functionals of an arbitrary C^*-algebra. Now we do not give the details of the rather technical chain of definitions going through von Neumann algebras and normal functionals to an arbitrary C^*-algebra. We just mention a possible extension of (7), the so-called Kosaki's formula ([1], [5]).

$$S(\omega,\varphi)=\sup\sup\left\{\omega(I)\log n-\int_{1/n}^{\infty}[\omega(y(t)^*y(t))+t^{-1}\varphi(x(t)x(t)^*]t^{-1}dt\right\}$$

where the first sup is taken over all natural numbers n, the second one is over all step functions $x\colon(1/n,\infty)\to N$ with finite range and $y(t)=I-x(t)$.

The relative entropy of positive functionals of a C^*-algebra shares the following properties (see [1], [5] and [9]).

(i) $(\omega,\varphi)\mapsto S(\omega,\varphi)$ is convex and weakly lower semicontinuous.

(ii) $\|\varphi-\omega\|^2\leqq 2S(\omega,\varphi)$ if $\varphi(I)=\omega(I)=1$.

(iii) $S(\omega,\varphi_1)\geqq S(\omega,\varphi_2)$ if $\varphi_1\leqq\varphi_2$.

(iv) For a unital Schwarz map $\alpha\colon\mathcal{A}_0\mapsto\mathcal{A}_1$ the relation $S(\omega\circ\alpha,\varphi\circ\alpha)\leqq$ $\leqq S(\omega,\varphi)$ holds.

(v) For $\omega=\sum_i\omega_i$ we have $S(\omega,\varphi)+\sum_{i=1}^n S(\omega_i,\omega)=\sum_{i=1}^n S(\omega_i,\varphi)$.

Below the properties (ii) and (v) will not be required.

PROPOSITION 3. *Let φ be a state of a finite quantum system $\mathcal{A}$. If φ is a convex combination $\sum_j\mu_j\varphi_j$ of states then*

$$S(\varphi)\geqq\sum_j\mu_j S(\varphi,\varphi_j).$$

When all the φ_j's are pure then the equality sign applies.

PROOF. It suffices to prove the equality because the inequality follows by convexity of the relative entropy. By simple computation we have

$$\sum_j\mu_j S(\varphi,\varphi_j)=\sum_j\mu_j\operatorname{Tr}\eta(D_{\varphi_j})+\operatorname{Tr}\eta(D_\varphi)$$

and the first term on the right-hand side vanishes when all the D_{φ_j}'s are projections.

Now let $\varphi=\sum\lambda_i\psi_i$ be an extremal decomposition. Combining Proposition 3 with the monotonicity (iii) of the relative entropy in the first variable we infer (6) as follows.

$$S(\varphi)=\sum_i\lambda_i S(\varphi,\psi_i)\leqq\sum_i\lambda_i S(\lambda_i\psi_i,\psi_i)=-\sum_i\lambda_i\log\lambda_i.$$

We note that Proposition 3 and its consequence (6) remain valid if $\mathcal{A}$ is a von Neumann algebra which is the direct sum of type I factors and φ is an arbitrary normal state. (In this case the functional Tr in the proof should be understood as the faithful normal semifinite trace which takes the value 1 at each minimal projection.)

Proposition 3 allowed the following definition of the entropy of states of arbitrary C^*-algebras in [7].

$$S(\varphi) = \sup\Big\{ \sum_i \lambda_i S(\varphi, \varphi_i) : \sum_i \lambda_i \varphi_i = \varphi \Big\}. \tag{8}$$

Here the supremum is taken over all decompositions of φ into finite (or equivalently countable) convex combinations of other states. Apparently the background uniform distribution provided by the trace functional in finite quantum systems is not present in this definition. Some properties of $S(\varphi)$ are immediate from those of the relative entropy. The quantity $S(\varphi)$ is nonnegative and vanishes when and only when, φ is a pure state. Moreover, the entropy is lower semicontinuous because it is the supremum of lower semicontinuous relative entropy functionals (see (i) above). The invariance of $S(\varphi)$ under automorphisms is obvious as well.

PROPOSITION 4. *Let $\mathcal{A}_0 \subset \mathcal{A}$ be C^*-algebras and assume that there exists a conditional expectation of $\mathcal{A}$ onto $\mathcal{A}_0$ preserving a given state φ of $\mathcal{A}$. Then*

$$S(\varphi) \geqq S(\varphi \mid \mathcal{A}_0). \tag{9}$$

PROOF. Indeed, if $\sum \lambda_i \psi_i = \varphi \mid \mathcal{A}_0$ for some states ψ_i of $\mathcal{A}_0$ then $\sum \lambda_i \psi_i \circ E$ is a decomposition of φ where E was written for the conditional expectation in the statement. The rest follows from $S(\varphi \mid \mathcal{A}_0, \psi_i) \leqq S(\varphi, \psi_i \circ E)$.

The existence of the conditional expectation preserving the given state is an essential hypothesis in Proposition 4. The monotonicity property (9) does not hold in general. The simplest counterexample is due to the fact that in the quantum case a pure state of the algebra can yield a restriction which is a mixed state on the subalgebra. The next observation is obvious.

PROPOSITION 5. *Let $\mathcal{A} = \mathcal{A}_1 \oplus \mathcal{A}_2$ be C^*-algebras and $\varphi = \lambda\varphi_1 \oplus (1-\lambda)\varphi_2$ a state of $\mathcal{A}$ $(0 < \lambda < 1)$. Then*

$$S(\varphi) = \lambda S(\varphi_1) + (1-\lambda) S(\varphi_2).$$

Now let φ be a normal state on a von Neumann algebra $\mathcal{M}$. Then a decomposition $\varphi = \sum \lambda_i \varphi_i$ is necessarily built from normal states φ_i if $\lambda_i \neq 0$. Hence, if we wish, in the definition (8) we may restrict ourselves to normal states φ_i. When p is the support projection of φ then $S(\psi, \varphi) = S(\psi \mid p\mathcal{M}p, \varphi \mid p\mathcal{M}p)$ whenever $\psi(p) = 1$ for the state ψ. Consequently

$$S(\varphi) = S(\varphi \mid p\mathcal{M}p). \tag{10}$$

The following result is due ot Hiai ([3]) and its proof uses the structure theory of von Neumann factors.

THEOREM 6. *Let φ be normal state of a von Neumann algebra $\mathcal{M}$ and let p be the support projection of φ. If $p\mathcal{M}p$ is a countable direct sum of type I factors then $S(\varphi) = \mathrm{Tr}\,\eta(D_\varphi)$ where D_φ is the density of φ with respect to the canonical semifinite normal trace* Tr *on $p\mathcal{M}p$. Otherwise, $S(\varphi) = \infty$.*

Now we turn to entropy of states of C^*-algebras. Let ψ be a state of a C^*-algebra $\mathcal{A}$. We write $\overline{\psi}$ for the vector state induced by the cyclic vector Ψ on the von Neumann algebra $\pi_\psi(\mathcal{A})''$ when (H_ψ, Ψ, π_ψ) is the GNS-triple corresponding to ψ.

LEMMA 7. *With the above notation we have $S(\psi) = S(\overline{\psi})$.*

PROOF. The key to the proof is the fact that a finite relative entropy may be computed in the GNS representation of the reference state. This yields readily that $S(\psi) \geqq S(\overline{\psi})$. On the other hand, if $\psi = \sum_i \lambda_i \psi_i$ is a convex decomposition in the state space of $\mathcal{A}$ then one can find (normal) states $\overline{\psi}_i$ of $\pi_\psi(\mathcal{A})''$ such that $\psi_i = \overline{\psi}_i \circ \pi_\psi$ and $\overline{\psi} = \sum_i \lambda_i \overline{\psi}_i$. Since $S(\psi, \psi_i) = S(\overline{\psi}, \overline{\psi}_i)$ the converse inequality $S(\psi) \leqq S(\overline{\psi})$ follows.

THEOREM 8. *Let ψ be a state of a C^*-algebra $\mathcal{A}$. Then*

$$S(\psi) = \inf \Big\{ \sum_i \eta(\lambda_i) \Big\} \tag{11}$$

where the infimum is taken over all possible decompositions $\psi = \sum_i \lambda_i \psi_i$ into pure states. If ψ is not a countable convex combination of pure states then $S(\psi) = \infty$.

PROOF. Lemma 7 allows us to reduce the theorem to the von Neumann algebra version. We have to consider two cases. If ψ is not a countable convex combination of pure states, then the von Neumann algebra $p\pi_\psi(\mathcal{A})''p$ is not a countable direct sum of type I factors. (p is the support projection of $\overline{\psi}$.) Theorem 6 tells us that $S(\overline{\psi}) = \infty$ in this case. If ψ is a countable convex combination of pure states then $p\pi_\psi(\mathcal{A})''p$ is a countable direct sum of type I factors and we may refer to Theorem 6 again.

The expression (11) appeared as the definition of entropy in [6]. It is noteworthy that in [9] this formula was generalized to define the entropy of a state with respect to a compact convex subset.

Theorem 6 and Lemma 7 with the additivity of the von Neumann entropy under tensor product gives that

$$S(\varphi \otimes \psi) = S(\varphi) + S(\psi) \tag{12}$$

if φ and ψ are arbitrary states of C^*-algebras. It is quite remarkable that this property does not follow readily from the definition (8).

REFERENCES

[1] ARAKI, H., Recent progress on entropy and relative entropy, *VIIIth Int. Congr. on Math. Phys.* (Marseille, 1986), World Sci. Publishing, Singapore, 1987, 354–365. *MR* **88j**:81004

[2] CSISZÁR, I. and KÖRNER, J., *Information Theory. Coding theorems for discrete memoryless systems*, Probability and Mathematical Statistics, Academic Press, New York–London, 1981. *MR* **84e**:94007

[3] HIAI, F., Minimum index for subfactors and entropy II, *J. Math. Soc. Japan* **43** (1991), 347–379.

[4] JAYNES, E. T., Information theory and statistical mechanics. II, *Phys. Rev.* (2) **108** (1957), 171–190. *MR* **20** #2898

[5] KOSAKI, H., Relative entropy of states: a variational expression, *J. Operator Theory* **16** (1986), 335–348. *MR* **87j**:46110

[6] MANUCEAU, J., NAUDTS, J. and VERBEURE, A., Entropy of normal states, *Comm. Math. Phys.* **27** (1972), 327–338.

[7] NARNHOFER, H. and THIRRING, W., From relative entropy to entropy, *Fizika* **17** (1985), 257–265.

[8] NEUMANN, J. VON, *Mathematische Grundlagen der Quantenmechanik*, Die Grundlehren der mathematischen Wissenschaften in Einzeldarstellungen, Bd. 38, Springer, Berlin, 1932. *Jb.Fortschritte Math.* **58**, 929–930. Dover Publications, New York, N. Y., 1943. *MR* **5**–165

[9] OHYA, M., Entropy transmission in C^*-dynamical systems, *J. Math. Anal. Appl.* **100** (1984), 222–235.

[10] PETZ, D., Properties of quantum entropy, *Quantum probability and applications*, II (Heidelberg, 1984), ed. by L. Accardi and W. von Waldenfels, Lecture Notes in Math., No. 1136, Springer, Berlin–New York, 1985, 428–441. *MR* **87i**:46144

[11] RÉNYI, A., On the foundations of information theory, *Rev. Inst. Internat. Statist.* **33** (1965), 1–14. *MR* **31** #5712

[12] SHANNON, C. E., A mathematical theory of communication, *Bell System Tech. J.* **27** (1948), 379–423. *MR* **10**–133

(Received April 12, 1992)

DEPARTMENT OF INFORMATION SCIENCES
SCIENCE UNIVERSITY OF TOKYO
NODA CITY
CHIBA 278
JAPAN

MTA MATEMATIKAI KUTATÓINTÉZETE
POSTAFIÓK 127
H–1364 BUDAPEST
HUNGARY

Vol. 38 (1996) *REPORTS ON MATHEMATICAL PHYSICS* No. 3

NOTE ON QUANTUM DYNAMICAL ENTROPIES

LUIGI ACCARDI

Centro Matematico Vito Volterra, Università di Roma II, via di Tor Vergata, Roma, Italy

MASANORI OHYA and NOBORU WATANABE

Department of Information Sciences, Science University of Tokyo, Noda, Chiba 278, Japan

(*Received September 12, 1996*)

Classical dynamical entropy is an important tool to analyse the efficiency of information transmission in communication processes. Quantum dynamical entropy was first studied by Connes, Størmer and Emch. Since then, there have been many attempts to formulate or compute the dynamical entropy for some models. Here we review four formulations due to (a) Connes, Narnhofer and Thirring, (b) Ohya, (c) Accardi, Ohya and Watanabe, (d) Alicki and Fannes. We consider mutual relations between these formulations and we show some concrete computations for a model.

Introduction

The classical dynamical (or Kolmogorov–Sinai) entropy $S(T)$ [13, 26] for a measure preserving transformation T was defined on a message space through finite partitions of the measurable space. The classical coding theorems of Shannon are important tools to analyse communication processes which have been formulated by the mean dynamical entropy and the mean dynamical mutual entropy. The mean dynamical entropy represents the amount of information per one letter of a signal sequence sent from an input source, and the mean dynamical mutual entropy does the amount of information per one letter of the signal received in an output system.

The quantum dynamical entropy (QDE) has been studied by Connes and Størmer [10], Emch [11], Connes, Narnhofer and Thirring [9], Alicki and Fannes [5], and others [7, 22]. Their dynamical entropies were defined in the observable spaces.

Recently, the quantum dynamical entropy and the quantum dynamical mutual entropy were studied by one of the present authors [14, 23]. They are formulated in the state spaces through the complexity of Information Dynamics [21, 23]. Furthermore, another formulation of the dynamical entropy through the quantum Markov chain (QMC) was done in [4].

In Section 1 we briefly review the formulation by Connes, Narnhofer and Thirring (CNT). In Section 2 we explain the formulation by complexity [23, 24]. In Section 3 we review the formulation by QMC, and in Section 4 we briefly explain the formulation by Alicki and Fannes (AF). Mutual relations among these four formulations are discussed in Section 4, and in Section 6 we compute the mean dynamical entropies in quantum communication processes.

1. The CNT formulation

Let $\mathcal{A}$ be a unital C^*-algebra, θ be an automorphism of $\mathcal{A}$, and φ be a stationary state over $\mathcal{A}$ with respect to θ; $\varphi \circ \theta = \varphi$. Let $\mathcal{B}$ be a finite dimensional C^*-subalgebra of $\mathcal{A}$.

The CNT entropy [9] for a subalgebra $\mathcal{B}$ is given by

$$H_\varphi(\mathcal{B}) = \sup\left\{\sum_k \lambda_k S(\omega_k|\mathcal{B}, \varphi|\mathcal{B}); \quad \varphi = \sum_k \lambda_k \omega_k \text{ finite decomposition of } \varphi\right\},$$

where $\varphi|\mathcal{B}$ is the restriction of the state φ to $\mathcal{B}$ and $S(\cdot, \cdot)$ is the relative entropy for C^*-algebra [6, 27, 28].

The CNT dynamical entropy with respect to θ and $\mathcal{B}$ is given by

$$\tilde{H}_\varphi(\theta, \mathcal{B}) = \limsup_{N\to\infty} \frac{1}{N} H_\varphi(\mathcal{B} \vee \theta\mathcal{B} \vee \cdots \vee \theta^{N-1}\mathcal{B}),$$

and the dynamical entropy for θ is defined by

$$\tilde{H}_\varphi(\theta) = \sup_{\mathcal{B}} \tilde{H}_\varphi(\theta, \mathcal{B}).$$

2. Formulation by complexity

In this section, we first review the concepts of channel and complexity, which are the key concepts of ID (Information Dynamics) introduced by Ohya [21, 23].

Let $(\mathcal{A}, \Sigma(\mathcal{A}), \alpha(G))$, $(\overline{\mathcal{A}}, \overline{\Sigma}(\overline{\mathcal{A}}), \overline{\alpha}(\overline{G}))$ be an input (initial) and an output (final) C^*-systems, respectively, where $\mathcal{A}$ (resp. $\overline{\mathcal{A}}$) is a unital C^*-algebra, $\Sigma(\mathcal{A})$ (resp. $\overline{\Sigma}(\overline{\mathcal{A}})$) is the set of all states on $\mathcal{A}$ (resp. $\overline{\mathcal{A}}$) and $\alpha(G)$ (resp. $\overline{\alpha}(\overline{G})$) is the group of automorphisms of $\mathcal{A}$ (resp. $\overline{\mathcal{A}}$) indexed by a group G (resp. $\overline{G}$).

The channel is a map Λ^* from $\Sigma(\mathcal{A})$ to $\overline{\Sigma}(\bar{\mathcal{A}})$. If the dual map Λ from $\bar{\mathcal{A}}$ to $\mathcal{A}$ of Λ^* satisfies the complete positivity, the channel Λ^* is called a complete positive channel (CP channel).

For a weak * compact convex subset $\mathcal{S}$ of Σ, there exists a maximum measure μ with the barycenter φ such that

$$\varphi = \int_{\mathcal{S}} \omega \, d\mu.$$

The compound state, introduced in [17, 18], which exhibits the correlation between the initial φ and final $\Lambda^*\varphi$ states is given by

$$\mathcal{E}_\mu^* \varphi = \int_{\mathcal{S}} \omega \otimes \Lambda^* \omega \, d\mu.$$

In the sequel, we use a CP channel Λ^* and the compound state to formulate the dynamical entropy.

There are two types of complexities in ID. One of them is the complexity $C_T^{\mathcal{S}}(\varphi)$ of the system itself, the other is the transmitted complexity $T^{\mathcal{S}}(\varphi\,;\Lambda^*)$ from an initial system to a final system. These complexities should satisfy the following conditions:

(i) For any $\varphi \in \mathcal{S} \subset \Sigma$,

$$C^{\mathcal{S}}(\varphi) \geq 0\,, \qquad T^{\mathcal{S}}(\varphi\,;\Lambda^*) \geq 0\,.$$

(ii) If there exists an orthogonal bijection $j : \mathrm{ex}\Sigma \to \mathrm{ex}\Sigma$; the set of all extremal points in Σ, then

$$\begin{aligned} C^{j(\mathcal{S})}(j(\varphi)) &= C^{\mathcal{S}}(\varphi)\,, \\ T^{j(\mathcal{S})}(j(\varphi)\,;\Lambda^*) &= T^{\mathcal{S}}(\varphi\,;\Lambda^*)\,. \end{aligned}$$

(iii) For a state $\Psi = \varphi \otimes \psi \in \mathcal{S}_t$, with $\varphi \in \mathcal{S}$, $\psi \in \overline{\mathcal{S}}$,

$$C^{\mathcal{S}_t}(\Psi) = C^{\mathcal{S}}(\varphi) + C^{\overline{\mathcal{S}}}(\psi)\,.$$

(iv) $0 \leq T^{\mathcal{S}}(\varphi\,;\Lambda^*) \leq C^{\mathcal{S}}(\varphi)\,.$

(v) $T^{\mathcal{S}}(\varphi\,;\mathrm{id}) = C^{\mathcal{S}}(\varphi)\,.$

Next we review the definitions of the three types of entropic complexities introduced in [23].

Let $(\mathcal{A},\ \Sigma(\mathcal{A}),\ \alpha(G))$, $\left(\overline{\mathcal{A}},\ \overline{\Sigma}(\overline{\mathcal{A}}),\ \overline{\alpha}\left(\overline{G}\right)\right)$ and $\mathcal{S}$ be as above. Let $M_\varphi(\mathcal{S})$ be the set of all maximal measures μ on $\mathcal{S}$ with the fixed barycenter φ, and let $F_\varphi(\mathcal{S})$ be the set of all measures having finite support with the fixed barycenter φ. Then the three pairs of complexities are

$$\begin{aligned} T^{\mathcal{S}}(\varphi\,;\Lambda^*) &\equiv \sup\left\{\int_{\mathcal{S}} S(\Lambda^*\omega, \Lambda^*\varphi)\,d\mu\,; \quad \mu \in M_\varphi(\mathcal{S})\right\}, \\ C_T^{\mathcal{S}}(\varphi) &\equiv T^{\mathcal{S}}(\varphi\,;\mathrm{id})\,, \\ I^{\mathcal{S}}(\varphi\,;\Lambda^*) &\equiv \sup\left\{S\left(\int_{\mathcal{S}} \omega \otimes \Lambda^*\omega\,d\mu\,,\, \varphi \otimes \Lambda^*\varphi\right); \quad \mu \in M_\varphi(\mathcal{S})\right\}, \\ C_I^{\mathcal{S}}(\varphi) &\equiv I^{\mathcal{S}}(\varphi\,;\mathrm{id})\,, \\ J^{\mathcal{S}}(\varphi\,;\Lambda^*) &\equiv \sup\left\{\int_{\mathcal{S}} S(\Lambda^*\omega, \Lambda^*\varphi)\,d\mu_f\,; \quad \mu_f \in F_\varphi(\mathcal{S})\right\}, \\ C_J^{\mathcal{S}}(\varphi) &\equiv J^{\mathcal{S}}(\varphi\,;\mathrm{id})\,. \end{aligned}$$

Based on the above complexities, we explain the quantum dynamical complexity (QDC) [17].

Let θ (resp. $\bar{\theta}$) be a stationary automorphism of $\mathcal{A}$ (resp. $\bar{\mathcal{A}}$); $\varphi \circ \theta = \varphi$, and Λ be a covariant CP map (i.e., $\Lambda \circ \theta = \bar{\theta} \circ \Lambda$) from $\bar{\mathcal{A}}$ to $\mathcal{A}$. $\mathcal{B}_k$ (resp. $\bar{\mathcal{B}}_k$) is a finite subalgebra of $\mathcal{A}$ (resp. $\bar{\mathcal{A}}$). Moreover, let α_k (resp. $\bar{\alpha}_k$) be a CP unital map from $\mathcal{B}_k$ (resp. $\bar{\mathcal{B}}_k$) to $\mathcal{A}$ (resp. $\bar{\mathcal{A}}$) and α^M and $\bar{\alpha}_\Lambda^N$ are given by

$$\begin{aligned} \alpha^M &= (\alpha_1, \alpha_2, \ldots, \alpha_M), \\ \bar{\alpha}_\Lambda^N &= (\Lambda \circ \bar{\alpha}_1, \Lambda \circ \bar{\alpha}_2, \ldots, \Lambda \circ \bar{\alpha}_N). \end{aligned}$$

Two compound states for α^M and $\bar{\alpha}_\Lambda^N$, with respect to $\mu \in M_\varphi(\mathcal{S})$, are defined as

$$\begin{aligned} \Phi_\mu^{\mathcal{S}}(\alpha^M) &= \int_{\mathcal{S}} \bigotimes_{m=1}^{M} \alpha_m^* \omega \, d\mu, \\ \Phi_\mu^{\mathcal{S}}(\alpha^M \cup \bar{\alpha}_\Lambda^N) &= \int_{\mathcal{S}} \bigotimes_{m=1}^{M} \alpha_m^* \omega \bigotimes_{n=1}^{N} \bar{\alpha}_n^* \Lambda^* \omega \, d\mu. \end{aligned}$$

Using the above compound states, the three transmitted complexities [23] are defined by

$$T_\varphi^{\mathcal{S}}(\alpha^M, \bar{\alpha}_\Lambda^N) \equiv \sup\left\{ \int_{\mathcal{S}} S\Big(\bigotimes_{m=1}^{M} \alpha_m^* \omega \bigotimes_{n=1}^{N} \bar{\alpha}_n^* \Lambda^* \omega,\ \Phi_\mu^{\mathcal{S}}(\alpha^M) \otimes \Phi_\mu^{\mathcal{S}}(\bar{\alpha}_\Lambda^N)\Big) d\mu;\quad \mu \in M_\varphi(\mathcal{S}) \right\},$$

$$I_\varphi^{\mathcal{S}}(\alpha^M, \bar{\alpha}_\Lambda^N) \equiv \sup\left\{ S\left(\Phi_\mu^{\mathcal{S}}(\alpha^M \cup \bar{\alpha}_\Lambda^N),\ \Phi_\mu^{\mathcal{S}}(\alpha^M) \otimes \Phi_\mu^{\mathcal{S}}(\bar{\alpha}_\Lambda^N)\right);\quad \mu \in M_\varphi(\mathcal{S}) \right\},$$

$$J_\varphi^{\mathcal{S}}(\alpha^M, \bar{\alpha}_\Lambda^N) \equiv \sup\left\{ \int_{\mathcal{S}} S\Big(\bigotimes_{m=1}^{M} \alpha_m^* \omega \bigotimes_{n=1}^{N} \bar{\alpha}_n^* \Lambda^* \omega,\ \Phi_\mu^{\mathcal{S}}(\alpha^M) \otimes \Phi_\mu^{\mathcal{S}}(\bar{\alpha}_\Lambda^N)\Big) d\mu_f;\ \mu_f \in F_\varphi(\mathcal{S}) \right\}.$$

When $\mathcal{B}_k = \bar{\mathcal{B}}_k = \mathcal{B}$, $\mathcal{A} = \bar{\mathcal{A}}$, $\theta = \bar{\theta}$, $\alpha_k = \theta^{k-1} \circ \alpha = \bar{\alpha}_k$, where α is a unital CP map from $\mathcal{A}_0$ to $\mathcal{A}$, the mean transmitted complexities are

$$\begin{aligned} \tilde{T}_\varphi^{\mathcal{S}}(\theta, \alpha, \Lambda^*) &\equiv \limsup_{N \to \infty} \frac{1}{N} T_\varphi^{\mathcal{S}}(\alpha^N, \bar{\alpha}_\Lambda^N), \\ \tilde{T}_\varphi^{\mathcal{S}}(\theta, \Lambda^*) &\equiv \sup_\alpha \tilde{T}_\varphi^{\mathcal{S}}(\theta, \alpha, \Lambda^*), \end{aligned}$$

and the same for $\tilde{I}_\varphi^{\mathcal{S}}$ and $\tilde{J}_\varphi^{\mathcal{S}}$. These quantities have properties similar to those of the CNT entropy [14, 23].

3. Formulation by QMC

A construction of dynamical entropy is due to the quantum Markov chain [4].

Let $\mathcal{A}$ be a von Neumann algebra acting on a Hilbert space $\mathcal{H}$ and let φ be a state on $\mathcal{A}$ and $\mathcal{A}_0 = M_d$ ($d \times d$ matrix algebra). Take the transition expectation $\mathcal{E}_\gamma : \mathcal{A}_0 \otimes \mathcal{A} \to \mathcal{A}$ of Accardi [1, 2] such that

$$\mathcal{E}_\gamma(\tilde{A}) = \sum_i \gamma_i A_{ii} \gamma_i,$$

where $\tilde{A} = \sum_{i,j} e_{ij} \otimes A_{ij} \in \mathcal{A}_0 \otimes \mathcal{A}$ and $\gamma = \{\gamma_j\}$ is a finite partition of unity $I \in \mathcal{A}$. Quantum Markov chain is defined by $\psi \equiv \{\varphi, \mathcal{E}_{\gamma,\theta}\} \in \Sigma(\bigotimes_1^\infty \mathcal{A}_0)$ such that

$$\psi(j_1(A_1) \cdots j_n(A_n)) \equiv \varphi\left(\mathcal{E}_{\gamma,\theta}(A_1 \otimes \mathcal{E}_{\gamma,\theta}(A_2 \otimes \cdots \otimes A_{n-1} \mathcal{E}_{\gamma,\theta}(A_n \otimes I) \cdots))\right),$$

where $\mathcal{E}_{\gamma,\theta} = \theta \circ \mathcal{E}_\gamma$, $\theta \in \mathrm{Aut}\,(\mathcal{A})$, and j_k is an embedding of $\mathcal{A}_0$ into $\overset{\infty}{\underset{1}{\otimes}} \mathcal{A}_0$ such that $j_k(A) = I \otimes \cdots \otimes I \otimes \underset{k\text{-th}}{A} \otimes I \cdots$.

Suppose that for φ there exists a unique density operator ρ such that $\varphi(A) = \mathrm{Tr}\,\rho A$ for any $A \in \mathcal{A}$. Let us define a state ψ_n on $\overset{n}{\underset{1}{\otimes}} \mathcal{A}_0$ expressed as

$$\psi_n(A_1 \otimes \cdots \otimes A_n) = \psi(j_1(A_1) \cdots j_n(A_n)) .$$

The density operator ξ_n for ψ_n is given by

$$\xi_n \equiv \sum_{i_1} \cdots \sum_{i_n} \mathrm{Tr}_{\mathcal{A}}(\theta^n(\gamma_{i_n}) \cdots \gamma_{i_1} \rho \gamma_{i_1} \cdots \theta^n(\gamma_{i_n})))e_{i_1 i_1} \otimes \cdots \otimes e_{i_n i_n} .$$

Put

$$P_{i_n \cdots i_1} = \mathrm{Tr}_{\mathcal{A}}(\theta^n(\gamma_{i_n}) \cdots \gamma_{i_1} \rho \gamma_{i_1} \cdots \theta^n(\gamma_{i_n}))) .$$

The dynamical entropy through QMC is defined by

$$\tilde{S}_\varphi(\theta;\gamma) \equiv \limsup_{n\to\infty} \frac{1}{n}\Big(-\mathrm{Tr}\,\xi_n \log \xi_n\Big) = \limsup_{n\to\infty} \frac{1}{n}\Big(-\sum_{i_1,\ldots,i_n} P_{i_n \ldots i_1} \log P_{i_n \cdots i_1}\Big).$$

If $P_{i_n \cdots i_1}$ satisfies the Markov property, then the above equality is written by

$$\tilde{S}_\varphi(\theta;\gamma) = -\sum_{i_1,i_2} P(i_2|i_1)P(i_1) \log P(i_2|i_1) .$$

The dynamical entropy through QMC with respect to θ and a von Neumann subalgebra $\mathcal{B}$ of $\mathcal{A}$ is given by

$$\tilde{S}_\varphi(\theta;\mathcal{B}) \equiv \sup\Big\{\tilde{S}_\varphi(\theta;\gamma)\,; \quad \gamma \subset \mathcal{B}\Big\} .$$

4. Formulation by AF

Let $\mathcal{A}$ be a C^*-algebra, θ be an automorphism on $\mathcal{A}$ and φ be a stationary state with respect to θ and $\mathcal{B}$ be a unital *-subalgebra of $\mathcal{A}$. A set $\gamma = \{\gamma_1, \gamma_2, \ldots, \gamma_k\}$ of elements of $\mathcal{B}$ is called a finite operational partition of unity of size k if γ satisfies the following condition:

$$\sum_{i=1}^{k} \gamma_i^* \gamma_i = I . \tag{4.1}$$

The operation $\circ$ is defined by

$$\gamma \circ \xi \equiv \{\gamma_i \xi_j\,; \quad i = 1,2,\ldots,k, \quad j = 1,2,\ldots,l\}$$

for any partitions $\gamma = \{\gamma_1, \gamma_2, \ldots, \gamma_k\}$ and $\xi = \{\xi_1, \xi_2, \ldots, \xi_l\}$. For any partition γ of size k, a $k \times k$ density matrix $\rho[\gamma] = (\rho[\gamma]_{i,j})$ is given by

$$\rho[\gamma]_{i,j} = \varphi(\gamma_j^* \gamma_i) .$$

Then the dynamical entropy $\tilde{H}_\varphi(\theta, \mathcal{B}, \gamma)$ with respect to the partition γ and shift θ is defined by von Neumann entropy $S(\,\cdot\,)$;

$$\tilde{H}_\varphi(\theta, \mathcal{B}, \gamma) = \limsup_{n\to\infty} \frac{1}{n} S(\rho[\theta^{n-1}(\gamma) \circ \cdots \circ \theta(\gamma) \circ \gamma]) . \tag{4.2}$$

The dynamical entropy $\tilde{H}_\varphi(\theta, \mathcal{B})$ is given by taking the supremum over operational partition of unity in $\mathcal{B}$ as

$$\tilde{H}_\varphi(\theta, \mathcal{B}) = \sup\Big\{\tilde{H}_\varphi(\theta, \mathcal{B}, \gamma); \gamma \subset \mathcal{B}\Big\}. \tag{4.3}$$

5. Relations among the four formulations

In this section we discuss relations among the above four formulations. The $\mathcal{S}$-mixing entropy in GQS introduced in [20] is

$$S^{\mathcal{S}}(\varphi) = \inf\{H(\mu)\,; \quad \mu \in M_\varphi(\mathcal{S})\},$$

where $H(\mu)$ is given by

$$H(\mu) = \sup\Big\{-\sum_{A_k\in\tilde{A}} \mu(A_k)\log\mu(A_k) : \quad \tilde{A} \in P(\mathcal{S})\Big\},$$

and $P(\mathcal{S})$ is the set of all finite partitions of $\mathcal{S}$.

The following theorem [14, 23] shows the relation between the formulation by CNT and that by complexity.

THEOREM 5.1. *Under the above settings, we have the following relations:*
(1) $0 \le I^{\mathcal{S}}(\varphi\,; \Lambda^*) \le T^{\mathcal{S}}(\varphi\,; \Lambda^*) \le J^{\mathcal{S}}(\varphi\,; \Lambda^*)$,
(2) $C_I^{\Sigma}(\varphi) = C_T^{\Sigma}(\varphi) = C_J^{\Sigma}(\varphi) = S^{\Sigma}(\varphi) = H_\varphi(\mathcal{A})$,
(3) $\mathcal{A} = \bar{\mathcal{A}} = B(\mathcal{H})$, *for any density operator* ρ, *and*

$$0 \le I^{\mathcal{S}}(\rho\,; \Lambda^*) = T^{\mathcal{S}}(\rho; \Lambda^*) \le J^{\mathcal{S}}(\rho\,; \Lambda^*) .$$

Since there exists a model showing that $S^{I(\alpha)}(\varphi) \ge H_\varphi(\mathcal{A}_\alpha)$, $S^{\mathcal{S}}(\varphi)$ distinguishes states more sharply than $H_\varphi(\mathcal{A})$, where $\mathcal{A}_\alpha = \{A \in \mathcal{A};\ \alpha(A) = A\}$.

Furthermore, we have the following results [24].

(1) When $\mathcal{A}_n, \mathcal{A}$ are abelian C^*-algebras and α_k is an embedding map, then

$$\begin{aligned} T^{\Sigma}(\mu; \alpha^M) &= S_\mu^{\text{classical}}\Big(\bigvee_{m=1}^{M} \tilde{A}_m\Big), \\ I^{\Sigma}(\mu; \alpha^M, \bar{\alpha}^N) &= I_\mu^{\text{classical}}\Big(\bigvee_{m=1}^{M} \tilde{A}_m, \bigvee_{n=1}^{N} \tilde{B}_n\Big) \end{aligned}$$

are satisfied for any finite partitions $\tilde{A}_m, \tilde{B}_n$ on the probability space ($\Omega = \text{spec}(\mathcal{A})$, $\mathcal{F}$, μ).

(2) When Λ is the restriction of $\mathcal{A}$ to a subalgebra $\mathcal{M}$ of $\mathcal{A}$; $\Lambda = |\mathcal{M}$,

$$H_\varphi(\mathcal{M}) = J^\Sigma(\varphi; |\mathcal{M}) = J_\varphi^\Sigma(\mathrm{id}; |\mathcal{M}).$$

Moreover, when

$$\mathcal{N} \subset \mathcal{A}_0, \qquad \mathcal{A} = \bigotimes^{\mathbb{N}} \mathcal{A}_0, \qquad \theta \in \mathrm{Aut}(\mathcal{A});$$

$$\alpha^N \equiv (\alpha, \theta \circ \alpha, \ldots; \theta^{N-1} \circ \alpha);$$

$$\alpha = \bar{\alpha}; \qquad \mathcal{A}_0 \to \mathcal{A} \quad \text{an embedding};$$

$$\mathcal{N}_N \equiv \bigotimes_1^N \mathcal{N},$$

we have

$$\tilde{H}_\varphi(\theta; \mathcal{N}) = \tilde{J}_\varphi^\Sigma(\theta; \mathcal{N}) = \limsup_{N\to\infty} \frac{1}{N} J_\varphi^\Sigma(\alpha^N; |\mathcal{N}_N).$$

We show the relation between the formulation by complexity and that by QMC. Under the same settings in Section 3, we define a map $\mathcal{E}^*_{(n,\gamma)}$ from $\Sigma(\mathcal{H})$, the set of all density operators in $\mathcal{H}$, to $\Sigma((\overset{n}{\underset{1}{\otimes}} \mathbb{C}^d) \otimes \mathcal{H})$ by

$$\begin{aligned} \mathcal{E}^*_{(n,\gamma)}(\rho) \;&=\; \sum_{i_1} \cdots \sum_{i_{n-1}} \sum_{i_n} e_{i_1 i_1} \otimes \cdots \otimes e_{i_{n-1} i_{n-1}} \otimes e_{i_n i_n} \otimes \\ &\otimes \theta^{n-1}(\gamma_{i_n}) \theta^{(n-2)}(\gamma_{i_{n-1}}) \cdots \gamma_{i_1} \rho \gamma_{i_1} \cdots \theta^{(n-2)}(\gamma_{i_{n-1}}) \theta^{n-1}(\gamma_{i_n}), \end{aligned}$$

for any density operator $\rho \in \Sigma(\mathcal{H})$. Let us take a map $E^*_{(n)}$ from $\Sigma((\overset{n}{\underset{1}{\otimes}} \mathbb{C}^d) \otimes \mathcal{H})$ to $\Sigma(\overset{n}{\underset{1}{\otimes}} \mathbb{C}^d)$ such that

$$E^*_{(n)}(\sigma) = \mathrm{Tr}_{\mathcal{H}} \sigma, \qquad \forall \sigma \in \Sigma\Big(\Big(\bigotimes_1^n \mathbb{C}^d\Big) \otimes \mathcal{H}\Big).$$

Then a map $\Gamma^*_{(n,\gamma)}$ from $\Sigma(\mathcal{H})$ to $\Sigma(\overset{n}{\underset{1}{\otimes}} \mathbb{C}^d)$ is given by

$$\Gamma^*_{(n)}(\rho) \equiv E^*_{(n)}\left(\mathcal{E}^*_{(n,\gamma)}(\rho)\right), \quad \forall \rho \in \Sigma(\mathcal{H}),$$

so that $\Gamma^*_{(n,\gamma)}(\rho) = \xi_n$ and

$$\tilde{S}_\varphi(\theta; \gamma) = \limsup_{n\to\infty} \frac{1}{n} S(\Gamma^*_{(n,\gamma)}(\rho)).$$

From Theorem 5.1, we have $C_I^\Sigma(\Gamma^*_{(n,\gamma)}(\rho)) = S(\Gamma^*_{(n,\gamma)}(\rho))$. Hence

$$\tilde{S}_\varphi(\theta; \gamma) = \tilde{C}_I^\Sigma(\Gamma^*_{(\gamma)}(\rho))\left(\equiv \limsup_{n\to\infty} \frac{1}{n} C_I^\Sigma(\Gamma^*_{(n,\gamma)}(\rho))\right).$$

Now we briefly show the relation between the formulation by complexity and that by AF.

Under the same settings in Section 4, for any partitions $\gamma = \{\gamma_1, \cdots, \gamma_k\}$ of size k, the $k \times k$ density matrix $\rho[\gamma] = (\rho[\gamma]_{i,j})$ is defined by

$$\rho[\gamma]_{i,j} = \varphi(\gamma_j^* \gamma_i)$$

which is acting on the k-dimensional Hilbert space $\mathcal{H}_k$. We define a map $\Xi^*_{(m,\gamma)}$ from $\Sigma(\mathcal{H}_k)$ to $\Sigma(\mathcal{H}_{k^m})$ by

$$\Xi^*_{(m,\gamma)}(\rho[\gamma]) = \rho[\theta^{m-1}(\gamma) \circ \cdots \circ \theta(\gamma) \circ \gamma]$$

for any partitions $\gamma = \{\gamma_1, \cdots, \gamma_k\}$ of size k and any density matrices $\rho[\gamma] \in \Sigma(\mathcal{H}_k)$. The dynamical entropy by AF is given by

$$\tilde{H}_\varphi(\theta, \mathcal{B}, \gamma) = \limsup_{m \to \infty} \frac{1}{m} S(\Xi^*_{(m,\gamma)}(\rho[\gamma])) .$$

Since $C_I^\Sigma(\Xi^*_{(m,\gamma)}(\rho[\gamma])) = S(\Xi^*_{(m,\gamma)}(\rho[\gamma]))$, we have

$$\tilde{H}_\varphi(\theta, \mathcal{B}, \gamma) = \tilde{C}_I^\Sigma(\Xi^*_{(\gamma)}(\rho[\gamma])).$$

In any case, the formulation by the entropic complexities contains other formulations. Moreover it opens other possibilities to classify dynamical systems more fine [5].

6. Computation of quantum dynamical complexity for a model

Let $X = \{a_1, \ldots, a_M\}$ be an alphabet used to construct the input signals and let $\mathcal{S} = \{E_1, \ldots, E_M\}$ be a set of one-dimensional projections on a Hilbert space $\mathcal{H}_0$ satisfying (1) $E_n \perp E_m \quad (n \neq m)$ and (2) E_n corresponds to the character a_n.

By Σ_0 we denote the set of density operators generated by $\mathcal{S}$

$$\Sigma_0 \equiv \left\{ \sigma = \sum_{n=1}^{M} \lambda_n E_n ; \quad \sum_{n=1}^{M} \lambda_n = 1 , \quad \lambda \geq 0 \right\}.$$

Suppose that the input quantum state is an element of Σ_0. To send the information effectively, the state is first transmitted through a quantum modulator; the transmitted state is called the quantum modulated state. Let $\gamma^*_{(\mathcal{M})}$ be a map from Σ_0 to $\Sigma_0^{(\mathcal{M})}$ such that $\gamma_{(\mathcal{M})}$ is a completely positive unital map from $\mathcal{A}$ to $\mathcal{A}$ induced by the modulator $(\mathcal{M})$. For any $E_n \in \mathcal{S}$, the modulated state $E_n^{(\mathcal{M})}$ is given by

$$E_n^{(\mathcal{M})} = \gamma^*_{(\mathcal{M})}(E_n) .$$

By $\Sigma_0^{(\mathcal{M})}$ we denote the set of modulated states

$$\Sigma_0^{(\mathcal{M})} \equiv \left\{ \sigma^{(\mathcal{M})} = \sum_{n=1}^{M} \lambda_n E_n^{(\mathcal{M})} ; \quad \sum_{n=1}^{M} \lambda_n = 1 , \quad \lambda \geq 0 \right\}.$$

In this paper, we consider the modulated states constructed by the photon number states.

(1) For any $E_n \in \Sigma_0$, the modulated state $E_n^{(PAM)}$ for PAM (Pulse Amplitude Modulator) is defined by

$$E_n^{(PAM)} = \gamma_{(PAM)}^*(E_n) = |n\rangle\langle n|\,,$$

where $|n\rangle\langle n|$ is the n photon number state on $\mathcal{H}$.

(2) For any $E_n \in \Sigma_0$, the modulated state $E_n^{(PPM)}$ for PPM (Pulse Position Modulator) is defined by

$$\begin{aligned} E_n^{(PPM)} &= \gamma_{(PPM)}^*(E_n) \\ &= \underbrace{E_0^{(PAM)} \otimes \cdots \otimes E_0^{(PAM)} \otimes \overbrace{E_d^{(PAM)}}^{n\text{-th}} \otimes E_0^{(PAM)} \otimes \cdots \otimes E_0^{(PAM)}}_{M}, \end{aligned}$$

where $E_0^{(PAM)}$ is the vacuum state and $E_d^{(PAM)} = |d\rangle\langle d|$ (d is fixed).

The transmission efficiency using the MER (mutual entropy — entropy ratio) [29] is calculated for some modulators.

Now we compute the mean dynamical mutual entropy for PAM and PPM expressed by the photon number state (as above).

Let $B(\mathcal{H}_0)$ (resp. $B(\overline{\mathcal{H}}_0)$) be the set of all bounded linear operators on a Hilbert space $\mathcal{H}_0$ (resp. $\overline{\mathcal{H}}_0$), and let $\mathcal{B}_0$ (resp. $\bar{\mathcal{B}}_0$) be a finite subset in $B(\mathcal{H}_0)$ (resp. $B(\overline{\mathcal{H}}_0)$). Let $\mathcal{A}$ (resp. $\bar{\mathcal{A}}$) be an infinite tensor product space of $B(\mathcal{H}_0)$ (resp. $B(\overline{\mathcal{H}}_0)$) denoted by

$$\mathcal{A} \equiv \bigotimes_{i=-\infty}^{\infty} B(\mathcal{H}_0)\,, \qquad \bar{\mathcal{A}} \equiv \bigotimes_{i=-\infty}^{\infty} B(\overline{\mathcal{H}}_0)\,.$$

Moreover, let θ (resp. $\bar{\theta}$) be a shift transformations on $\mathcal{A}$ (resp. $\bar{\mathcal{A}}$) defined by

$$\begin{aligned} \theta\Big(\bigotimes_{i=-\infty}^{\infty} A_i\Big) &= \bigotimes_{i'=-\infty}^{\infty} A_{i'} \quad (i' = i-1) \quad \text{for any} \quad \bigotimes_{i=-\infty}^{\infty} A_i \in \mathcal{A}\,, \\ \bar{\theta}\Big(\bigotimes_{j=-\infty}^{\infty} \bar{A}_j\Big) &= \bigotimes_{j'=-\infty}^{\infty} \bar{A}_{j'} \quad (j' = j-1) \quad \text{for any} \quad \bigotimes_{j=-\infty}^{\infty} \bar{A}_j \in \bar{\mathcal{A}}\,. \end{aligned}$$

Let α (resp. $\bar{\alpha}$) be the embedding map from $\mathcal{B}_0$ into $\mathcal{A}$ (resp. $\bar{\mathcal{B}}_0$ to $\bar{\mathcal{A}}$) given by

$$\begin{aligned} \alpha(A) &= \cdots I \otimes I \otimes A \otimes I \otimes \cdots \in \mathcal{A}\,, \qquad \text{for any} \quad A \in \mathcal{B}_0\,, \\ \bar{\alpha}(\bar{A}) &= \cdots I \otimes I \otimes \bar{A} \otimes I \otimes \cdots \in \bar{\mathcal{A}}\,, \qquad \text{for any} \quad \bar{A} \in \bar{\mathcal{B}}_0\,. \end{aligned}$$

The set of all density operators on $\mathcal{H}_0$ (resp. $\overline{\mathcal{H}}_0$) we denote by Σ_0 (resp. $\overline{\Sigma}_0$), and let Σ (resp. $\overline{\Sigma}$) be the set of all density operators on $\mathcal{A}$ (resp. $\overline{\rho}$ on $\bar{\mathcal{A}}$).

The maps $\alpha^N_{(\mathcal{M})}$ and $\bar{\alpha}^N_{\tilde{\Lambda}(\mathcal{M})}$ are given by

$$\begin{aligned}\alpha^N_{(\mathcal{M})} &\equiv (\alpha\circ\tilde{\gamma}_{(\mathcal{M})},\theta\circ\alpha\circ\tilde{\gamma}_{(\mathcal{M})},\ldots,\theta^{N-1}\circ\alpha\circ\tilde{\gamma}_{(\mathcal{M})}),\\ \bar{\alpha}^N_{\tilde{\Lambda}(\mathcal{M})} &\equiv (\tilde{\gamma}_{(\mathcal{M})}\circ\tilde{\Lambda}\circ\bar{\alpha},\tilde{\gamma}_{(\mathcal{M})}\circ\tilde{\Lambda}\circ\bar{\theta}\circ\bar{\alpha},\ldots,\tilde{\gamma}_{(\mathcal{M})}\circ\tilde{\Lambda}\circ\bar{\theta}^{N-1}\circ\bar{\alpha}),\end{aligned}$$

where we took a special channel and modulator such that $\tilde{\Lambda}\equiv\bigotimes_{i=-\infty}^{\infty}\Lambda$ and $\tilde{\gamma}_{(\mathcal{M})}\equiv\bigotimes_{i=-\infty}^{\infty}\gamma_{(\mathcal{M})}$.

(I) PAM. Let us take a stationary initial state $\rho\in\Sigma$

$$\rho=\sum_m\mu_m\bigotimes_{i=-\infty}^{\infty}\rho_m^{(i)}\qquad\Big(\sum_m\mu_m=1,\quad\mu_m\neq\mu_{m'},\quad(m\neq m')\Big)\tag{6.1}$$

with the unique Schatten decomposition $\rho_m^{(i)}=\sum_{n_i=1}^M\lambda_{n_i}^{(m)}E_{n_i}\in\Sigma_0$ $(\sum_{n_i=1}^M\lambda_{n_i}^{(m)}=1,\ \lambda_{n_i}^{(m)}\neq\lambda_{k_i}^{(m)},\ (n_i\neq k_i))$ of $\rho_m^{(i)}$. Then we have

$$\begin{aligned}\Phi_E(\alpha^N_{(PAM)}) &= \sum_{n_0=1}^M\cdots\sum_{n_{N-1}=1}^M\Big(\sum_m\mu_m\prod_{k=0}^{N-1}\lambda_{n_k}^{(m)}\Big)\Big(\bigotimes_{i=0}^{N-1}E_{n_i}^{(PAM)}\Big),\qquad(6.2)\\ \Phi_E(\bar{\alpha}^N_{\tilde{\Lambda}(PAM)}) &= \sum_{n_0=1}^M\cdots\sum_{n_{N-1}=1}^M\Big(\sum_m\mu_m\prod_{k=0}^{N-1}\lambda_{n_k}^{(m)}\Big)\Big(\bigotimes_{i=0}^{N-1}\Lambda^*E_{n_i}^{(PAM)}\Big).\qquad(6.3)\end{aligned}$$

When Λ^* is an attenuation channel, we obtain [17]

$$\Lambda^*E_{n_i}^{(PAM)}=\sum_{j_i=0}^{n_i}|C_{j_i}^{n_i}|^2F_{j_i}^{(PAM)},$$

where $F_{j_i}^{(PAM)}=|j_i\rangle\langle j_i|$ is the j_i-photon number state in the output space $\overline{\Sigma}_0$ and

$$|C_{j_i}^{n_i}|^2=\frac{n_i!}{j_i!(n_i-j_i)!}\eta^{j_i}(1-\eta)^{(n_i-j_i)},$$

where η is the transmission rate of the channel [17, 19]. The compound states related to the channel $\tilde{\Lambda}^*$ become

$$\begin{aligned}&\Phi_E(\alpha^N_{(PAM)}\cup\bar{\alpha}^N_{\tilde{\Lambda}(PAM)})\\ &\quad=\sum_{n_0=1}^M\cdots\sum_{n_{N-1}=1}^M\Big(\sum_m\mu_m\prod_{k=0}^{N-1}\lambda_{n_k}^{(m)}\Big)\sum_{j_0=0}^{n_0}\cdots\sum_{j_{N-1}=0}^{n_{N-1}}\Big(\prod_{k'=0}^{N-1}|C_{j_{k'}}^{n_{k'}}|^2\Big)\times\\ &\qquad\times\Big(\bigotimes_{i=0}^{N-1}E_{n_i}^{(PAM)}\Big)\Big(\bigotimes_{i'=0}^{N-1}F_{j_{i'}}^{(PAM)}\Big),\end{aligned}\tag{6.4}$$

$$\Phi_E(\alpha^N_{(PAM)}) \otimes \Phi_E(\bar{\alpha}^N_{\bar{A}(PAM)})$$
$$= \sum_{n_0=1}^{M} \cdots \sum_{n_{N-1}=1}^{M} \Big(\sum_m \mu_m \prod_{k=0}^{N-1} \lambda^{(m)}_{n_k}\Big) \sum_{n'_0=1}^{M} \cdots \sum_{n'_{N-1}=1}^{M} \Big(\sum_{m'} \prod_{k'=0}^{N-1} \lambda^{(m')}_{n'_{k'}}\Big) \times$$
$$\times \sum_{j'_0=0}^{n'_0} \cdots \sum_{j'_{N-1}=0}^{n'_{N-1}} \Big(\prod_{k"=0}^{N-1} |C^{n'_{k"}}_{j'_{k"}}|^2\Big)\Big(\bigotimes_{i=0}^{N-1} E^{(PAM)}_{n_i}\Big)\Big(\bigotimes_{i'=0}^{N-1} F^{(PAM)}_{j'_{i'}}\Big). \tag{6.5}$$

For an initial state ρ in (6.1), we have

$$\mathrm{I}_E(\rho; \alpha^N_{(PAM)}, \bar{\alpha}^N_{\bar{A}(PAM)})$$
$$= \sum_{j_0=0}^{M} \cdots \sum_{j_{N-1}=0}^{M} \sum_{n_0=J_0}^{M} \cdots \sum_{n_{N-1}=J_{N-1}}^{M} \Big(\sum_m \mu_m \prod_{k=0}^{N-1} \lambda^{(m)}_{n_k}\Big)\Big(\prod_{k'=0}^{N-1} |C^{n_{k'}}_{j_{k'}}|^2\Big) \times$$
$$\times \log \frac{\prod_{k=0}^{N-1} |C^{n_k}_{j_k}|^2}{\sum_{n'_0=J_0}^{M} \cdots \sum_{n'_{N-1}=J_{N-1}}^{M} \Big(\sum_{m'} \mu_{m'} \prod_{k'=0}^{N-1} \lambda^{(m')}_{n'_{k'}}\Big)\Big(\prod_{k"=0}^{N-1} |C^{n'_{k"}}_{j_{k"}}|^2\Big)}.$$

Having the above equality, we obtain the following theorem.

THEOREM 6.1. (1) *For an initial state* ρ *in* (6.1), *we have the lower bound of* $\tilde{C}_\rho(\theta, \alpha_{(PAM)})$ *such that*

$$\tilde{C}_\rho(\theta, \alpha_{(PAM)}) \geq \sum_m \mu_m S(\rho^{(0)}_m).$$

(2) *Let* $\mu_0 = 1, \mu_k = 0$ $(\forall k \geq 1)$ *and* $\lambda^{(0)}_n = \lambda_n$ *in* (6.1). *When* $\mathcal{A} = \bar{\mathcal{A}}$, $\theta = \bar{\theta}$ *and* $\alpha = \bar{\alpha}$, *we obtain the following equalities*:

$$\tilde{C}_\rho(\theta, \alpha_{(PAM)}) = -\sum_{n=1}^{M} \lambda_n \log \lambda_n,$$
$$\tilde{I}^{\Sigma}_\rho(\theta, \alpha_{(PAM)}, \Lambda^*) = \sum_{j=0}^{M} \sum_{n=J}^{M} \lambda_n |C^n_j|^2 \log \frac{|C^n_j|^2}{\sum_{n'=J}^{M} \lambda_{n'} |C^{n'}_j|^2}.$$

(II) PPM. For an initial state ρ in (6.1), we obtain the following compound states:

$$\Phi_E(\alpha^N_{(PPM)}) = \sum_{n_0=1}^{M} \cdots \sum_{n_{N-1}=1}^{M} \Big(\sum_m \mu_m \prod_{k=0}^{N-1} \lambda^{(m)}_{n_k}\Big)\Big(\bigotimes_{i=0}^{N-1}\Big(\bigotimes_{j=0}^{N-1} E^{(PAM)}_{d\delta_{j,n_i}}\Big)\Big). \tag{6.6}$$

$$\Phi_E(\bar{\alpha}^N_{\bar{A}(PPM)}) = \sum_{n_0=1}^{M} \cdots \sum_{n_{N-1}=1}^{M} \Big(\sum_m \mu_m \prod_{k=0}^{N-1} \lambda^{(m)}_{n_k}\Big) \times$$
$$\times \sum_{\ell_0=0}^{d\delta_{1,n_0}} \cdots \sum_{\ell_{N-1}=0}^{d\delta_{M,n_{N-1}}} |C^{d\delta_{1,n_0}}_{\ell_0}|^2 \cdots |C^{d\delta_{M,n_{N-1}}}_{\ell_{N-1}}|^2 \Big(\bigotimes_{i=0}^{N-1}\Big(\bigotimes_{j=1}^{M} F^{(PAM)}_{\ell_{iM+j}}\Big)\Big). \tag{6.7}$$

$$\Phi_E(\alpha^N_{(PPM)} \cup \bar{\alpha}^N_{\bar{\Lambda}(PPM)})$$
$$= \sum_{n_0=1}^{M} \cdots \sum_{n_{N-1}=1}^{M} \Big(\sum_m \mu_m \prod_{k=0}^{N-1} \lambda^{(m)}_{n_k}\Big) \sum_{\ell_1=0}^{d\delta_{1,n_0}} \cdots \sum_{\ell_{N-1}=0}^{d\delta_{M,n_{N-1}}} |C^{d\delta_{1,n_0}}_{\ell_1}|^2 \cdots |C^{d\delta_{M,n_{N-1}}}_{\ell_{N-1}}|^2 \times$$
$$\times \Big(\bigotimes_{i=0}^{N-1}\Big(\bigotimes_{j=1}^{M} E^{(PAM)}_{d\delta_{j,n_i}}\Big)\Big)\Big(\bigotimes_{i'=0}^{N-1}\Big(\bigotimes_{j'=0}^{M} F^{(PAM)}_{\ell_{i'M+j'}}\Big)\Big). \tag{6.8}$$

$$\Phi_E(\alpha^N_{(PPM)}) \otimes \Phi_E(\bar{\alpha}^N_{\bar{\Lambda}(PPM)})$$
$$= \sum_{n_0=1}^{M} \cdots \sum_{n_{N-1}=1}^{M} \sum_{n'_0=1}^{M} \cdots \sum_{n'_{N-1}=1}^{M} \Big(\sum_m \mu_m \prod_{k=0}^{N-1} \lambda^{(m)}_{n_k}\Big) \times$$
$$\times \sum_{\ell_1=0}^{d\delta_{1,n_0}} \cdots \sum_{\ell_{N-1}=0}^{d\delta_{M,n_{N-1}}} |C^{d\delta_{1,n_0}}_{\ell_1}|^2 \cdots |C^{d\delta_{M,n_{N-1}}}_{\ell_{N-1}}|^2 \sum_{n'_0=1}^{M} \cdots \sum_{n'_{N-1}=1}^{M} \Big(\sum_{m'} \mu_{m'} \prod_{k'=0}^{N-1} \lambda_{n'_{k'}}\Big) \times$$
$$\times \Big(\bigotimes_{i=0}^{N-1}\Big(\bigotimes_{j=1}^{M} E^{(PAM)}_{d\delta_{j,n_i}}\Big)\Big)\Big(\bigotimes_{i'=0}^{N-1}\Big(\bigotimes_{j'=1}^{M} F^{(PAM)}_{\ell_{i'M+j'}}\Big)\Big). \tag{6.9}$$

For an initial state ρ in (6.1), we have

$$\mathrm{I}_E(\rho; \alpha^N_{(PPM)}, \bar{\alpha}^N_{\bar{\Lambda}(PPM)})$$
$$= -\sum_{n_0=1}^{M} \cdots \sum_{n_{N-1}=1}^{M} \Big(\sum_m \mu_m \prod_{k=0}^{N-1} \lambda^{(m)}_{n_k}\Big) \sum_{p=1}^{N} \sum_{\{q_1,\cdots,q_p\}\subset\{1,2,\cdots,N\}} \times$$
$$\times \sum_{\ell_1=1}^{d} \cdots \sum_{\ell_p=1}^{d} |C^d_{\ell_1}|^2 \cdots |C^d_{\ell_p}|^2 (1-\eta)^{N-p}\eta^p \log\Big(\sum_{m'} \mu_{m'} \prod_{k'=0}^{p} \lambda^{(m')}_{n_{q_{k'}}}\Big).$$

THEOREM 6.2. (1) *For an initial state ρ in (6.1), we have the lower bound of $\tilde{C}_\rho(\theta, \alpha_{(PPM)})$ such that*

$$\tilde{C}_\rho(\theta, \alpha_{(PPM)}) \geq \sum_m \mu_m S(\rho^{(0)}_m).$$

(2) *Let $\mu_0 = 1, \mu_k = 0$ $(\forall k \geq 1)$ and $\lambda^{(0)}_n = \lambda_n$ in (6.1). When $\mathcal{A} = \bar{\mathcal{A}}$, $\theta = \bar{\theta}$ and $\alpha = \bar{\alpha}$, we obtain the following equalities:*

$$\tilde{C}_\rho(\theta, \alpha_{(PPM)}) = -\sum_{n=1}^{M} \lambda_n \log \lambda_n,$$
$$\tilde{I}^{\Sigma}_\rho(\theta, \alpha_{(PPM)}, \Lambda^*) = (1-(1-\eta)^d)\tilde{C}_\rho(\theta, \alpha_{(PPM)}).$$

(3) *If $\mathcal{A} = \bar{\mathcal{A}}$, $\theta = \bar{\theta}$, $\alpha = \bar{\alpha}$ and $d \geq N$, then*

$$\tilde{I}^{\Sigma}_\rho(\theta, \alpha_{(PPM)}, \Lambda^*) \geq \tilde{I}^{\Sigma}_\rho(\theta, \alpha_{(PAM)}, \Lambda^*).$$

REFERENCES

[1] L. Accardi: *Noncommutative Markov chains*, Internatinal School of Mathematical Physics, Camerino (1974), 268.
[2] L. Accardi, A. Frigerio and J. Lewis: *Quantum stochastic processes*, Publications of the Research institute for Mathematical Sciences Kyoto University, **18** (1982), 97.
[3] L. Accardi and M. Ohya: Compound channels, transition expectations and liftings, to appear in *J. Multivariate Analysis.*
[4] L. Accardi, M. Ohya and N. Watanabe: Dynamical entropy through quantum Markov chain, to appear in *Open System and Information Dynamics.*
[5] R. Alicki and M. Fannes: *Lett. Math. Phys.* **32** (1994), 75.
[6] H. Araki: *Relative entropy for states of von Neumann algebras, Publications of the Research institute for Mathematical Sciences Kyoto University* **11** (1976), 809.
[7] F. Benatti: *Deterministic Chaos in Infinite Quantum Systems*, Springer, Berlin 1993.
[8] L. Bilingsley: *Ergodic Theory and Information*, Wiley, New York 1965.
[9] A. Connes, H. Narnhoffer and W. Thirring: *Commun. Math. Phys.* **112** (1987), 691.
[10] A. Connes and E. Størmer: *Acta Math.* **134** (1975), 289.
[11] G. G. Emch: *Z. Wahrscheinlichkeitstheorie verw. Gebiete* **29** (1974), 241.
[12] L. Feinstein: *Foundations of Information Theory*, MacGrow-Hill, 1965.
[13] A. N. Kolmogorov: *Dokl. Akad. Nauk SSSR* **119** (1958), 861.
[14] N. Muraki and M. Ohya: *Lett. Math. Phys.* **36** (1996), 327.
[15] J. von Neumann: *Die Mathematischen Grundlagen der Quantenmechanik*, Springer, Berlin 1932.
[16] M. Ohya: *J. Math. Anal. Appl.* **84** (1981), 318.
[17] M. Ohya: *IEEE Trans. Information Theory* **29** (1983), 770.
[18] M. Ohya: *L. Nuovo Cimento* **38** (1983), 402.
[19] M. Ohya: *J. Math. Anal. Appl.* **100** (1984), 222.
[20] M. Ohya: *Rep. Math. Phys.* **27** (1989), 19.
[21] M. Ohya: *Information dynamics and its application to optical communication processes*, Springer Lecture Notes in Physics, **378**, p. 81, Springer, Berlin 1991.
[22] M. Ohya and D. Petz: *Quantum Entropy and its Use*, Springer, Berlin 1993.
[23] M. Ohya: *Quantum Communications and Measurement* **2** (1995), 309.
[24] M. Ohya: Foundation of entropy, complexity and fractal in quantum systems, to appear in International Congress of Probability Towards 2000, 1996.
[25] M. Ohya and N. Watanabe: Note on irreversible dynamics and quantum information, to appear in the *Alberto Frigerio Conference Proceedings.*
[26] J. G. Sinai: *Dokl. Akad. Nauk SSSR* **124** (1959), 768.
[27] A. Uhlmann: *Commun. Math. Phys.* **54** (1977), 21.
[28] H. Umegaki: *Kodai Math. Sem. Rep.* **14** (1962), 59.
[29] N. Watanabe: *Quantum Probability and Related Topics* **6** (1991), 489.

Letters in Mathematical Physics **36**: 327–335, 1996.

Entropy Functionals of Kolmogorov–Sinai Type and Their Limit Theorems

NAOHUMI MURAKI[1] and MASANORI OHYA[2]
[1]*Department of Applied Science, Yamaguchi University, Tokiwadai 2557, Ube City, Yamaguchi 755, Japan*
[2]*Department of Information Sciences, Science University of Tokyo, Noda City, Chiba 278, Japan*

(Received: 14 March 1995)

Abstract. Two functionals $\tilde{S}$ and $\tilde{I}$ are introduced for C^*-dynamical systems with invariant states and stationary channels. It is shown that the Kolmogorov–Sinai-type theorems hold for these functionals $\tilde{S}$ and $\tilde{I}$. Our functionals $\tilde{S}$ and $\tilde{I}$ are set within the framework of quantum information theory and generalize a quantum KS entropy by CNT and the mutual entropy by Ohya.

Mathematics Subject Classification (1991). 94A17.

Key words: Kolmogorov–Sinai theorem, entropy functionals, quantum information theory.

1. Introduction

Quantum information theory was formulated in the operator algebraic setting [6]. In general quantum information theory, a stationary information source is described by a C^*-dynamical system $(\mathscr{A}, \theta_{\mathscr{A}}, \phi)$, where $\mathscr{A}$ is a unital C^*-algebra, $\theta_{\mathscr{A}}$ is an automorphism of $\mathscr{A}$, and ϕ is an invariant state over $\mathscr{A}$ with respect to $\theta_{\mathscr{A}}$, that is, $\phi \circ \theta_{\mathscr{A}} = \phi$. A stationary information channel from an input C^*-dynamical system $(\mathscr{A}, \theta_{\mathscr{A}})$ to an output C^*-dynamical system $(\mathscr{B}, \theta_{\mathscr{B}})$ is here the transpose $\Lambda^*: \mathscr{A}^* \to \mathscr{B}^*$ of a completely positive unital (c.p.u.) map $\Lambda: \mathscr{B} \to \mathscr{A}$ being covariant with $\theta_{\mathscr{A}}$ and $\theta_{\mathscr{B}}$, i.e., $\Lambda \circ \theta_{\mathscr{B}} = \theta_{\mathscr{A}} \circ \Lambda$. Such Λ^* is called a stationary channel from an input system $(\mathscr{A}, \theta_{\mathscr{A}})$ to an output system $(\mathscr{B}, \theta_{\mathscr{B}})$, and maps an input state ϕ over $\mathscr{A}$ to the output state $\Lambda^*\phi$ over $\mathscr{B}$. In [7, 8], Ohya introduced the concept of the compound state of an input state ϕ and the output state $\Lambda^*\phi$, and, using them, he defined the mutual entropy $I(\phi; \Lambda^*)$ which expresses the amount of information transmitted from the input state ϕ to the output state $\Lambda^*\phi$ through the channel Λ^*. In this Letter, in order to formulate the entropy of KS Type [4] in a new direction, we extend the construction of mutual entropy in [6, 7] to the situation involving automorphisms $\theta_{\mathscr{A}}, \theta_{\mathscr{B}}$ of C^*-algebras $\mathscr{A}, \mathscr{B}$, certain evolutions of systems. That is, we introduce two functionals $\tilde{S}(\phi)$ and $\tilde{I}(\phi; \Lambda^*)$ for an invariant state ϕ over $\mathscr{A}$ and a stationary channel Λ^* from $(\mathscr{A}, \theta_{\mathscr{A}})$ to $(\mathscr{B}, \theta_{\mathscr{B}})$. The functional $\tilde{S}$ (resp. $\tilde{I}$) is interpreted as a kind of quantum analogue of entropy rate (resp. mutual entropy rate) in information theory.

The contents of this Letter are as follows. In Section 2, using the concept of the compound state (or compound lifting [1]), we define the functionals $S_\mu(\alpha^M)$ and

$I_\mu(\alpha^M, \beta^N)$ for a pair of finite sequences

$$\alpha^M = (\alpha_1, \alpha_2, \ldots, \alpha_M), \qquad \beta^N = (\beta_1, \beta_2, \ldots, \beta_N)$$

of completely positive unital maps $\alpha_m: A_m \to \mathcal{A}$, $\beta_n: B_n \to \mathcal{A}$ from finite-dimensional unital C^*-algebras A_m $(m=1, \ldots, M)$, B_n $(n=1, \ldots, N)$ to $\mathcal{A}$, and for a decomposition measure μ of ϕ. In Section 3, we introduce the functionals $\tilde{S}(\phi)$ and $\tilde{I}(\phi; \Lambda^*)$ based on the functionals $S_\mu(\alpha^M)$ and $I_\mu(\alpha^M, \beta^N)$, and we show that the Kolmogorov–Sinai-type theorems hold for the functionals $\tilde{S}(\phi)$ and $\tilde{I}(\phi; \Lambda^*)$.

2. Functionals S and I for Completely Positive Unital Maps

Let $\mathcal{A}$ be a unital C^*-algebra with a state ϕ. In this section, we introduce functionals $S_\mu(\alpha^M)$, $S_\phi(\alpha^M)$, $I_\mu(\alpha^M, \beta^N)$, and $I_\phi(\alpha^M, \beta^N)$, for a pair of finite sequences of

$$\alpha^M = (\alpha_1, \alpha_2, \ldots, \alpha_M), \qquad \beta^N = (\beta_1, \beta_2, \ldots, \beta_N)$$

of completely positive unital maps $\alpha_m: A_m \to \mathcal{A}$, $\beta_n: B_n \to \mathcal{A}$ from finite-dimensional unital C^*-algebras

$$A_m (m=1, \ldots, M), \qquad B_n (n=1, \ldots, N)$$

to $\mathcal{A}$. At the end of this section, these functionals S and I will be interpreted as some kinds of 'joint entropy' and 'joint mutual entropy', respectively.

Let $M(\phi)$ be the set of all regular Borel probability measures μ on the state space $S(\mathcal{A})$ of $\mathcal{A}$ such that μ is maximal in the Choquet ordering and μ represents ϕ: $\int_{S(\mathcal{A})} \omega \, d\mu(\omega) = \phi$. We refer to such measures as extremal decomposition measure for ϕ. For any state ϕ over $\mathcal{A}$, such measures exist from Choquet's theorem. For a given finite sequence of completely positive unital maps α_m from finite-dimensional unital C^*-algebras A_m $(m=1, \ldots, M)$ to $\mathcal{A}$ and a given extremal decomposition measure μ of ϕ, we define on the tensor product algebra $\bigotimes_{m=1}^M A_m$ a state $\Phi_\mu(\alpha^M)$ by

$$\Phi_\mu(\alpha^M) = \int_{S(\mathcal{A})} \bigotimes_{m=1}^M \alpha_m^* \omega \, d\mu(\omega). \tag{2.1}$$

The state $\Phi_\mu(\alpha^M)$ is a *compound state* of $\alpha_1^*\phi, \alpha_2^*\phi, \ldots, \alpha_M^*\phi$ in the sense of [7, 8], so that

$$\Phi_\mu(\alpha^M)(x \otimes I) = \alpha_m^* \phi(x),$$

is satisfied for any $x \in A_m$, $I \in \otimes_{m' \neq m} A_{m'}$. Furthermore, $\Phi_\mu(\alpha^M \cup \beta^N)$ is a compound state of $\Phi_\mu(\alpha^M)$ and $\Phi_\mu(\beta^N)$ with $\alpha^M \cup \beta^N \equiv (\alpha_1, \alpha_2, \ldots, \alpha_M, \beta_1, \beta_2, \ldots, \beta_N)$ constructed as (2.1). This compound state satisfies the following marginal conditions

$$\Phi_\mu(\alpha^M \cup \beta^N)(x \otimes I) = \Phi_\mu(\alpha^M)(x),$$

for $x \in \otimes_{m=1}^M A_m$, $I \in \otimes_{n=1}^N B_n$, and

$$\Phi_\mu(\alpha^M \cup \beta^N)(I \otimes y) = \Phi_\mu(\beta^N)(y),$$

for $I \in \bigotimes_{m=1}^{M} A_m$, $y \in \bigotimes_{n=1}^{N} B_n$.

For any pair (α^M, β^N) of finite sequences

$$\alpha^M = (\alpha_1, \ldots, \alpha_M) \quad \text{and} \quad \beta^N = (\beta_1, \ldots, \beta_N)$$

of completely positive unital maps (c.p.u. maps for short) such that $\alpha_m: A_m \to \mathscr{A}$, $\beta_n: B_n \to \mathscr{A}$ and for any extremal decomposition measure μ of ϕ, we define two functionals; the entropy functional S_μ and the mutual entropy functional I_μ, by

$$S_\mu(\alpha^M) = \int_{S(\mathscr{A})} S\left(\bigotimes_{m=1}^{M} \alpha_m^* \omega \,|\, \Phi_\mu(\alpha^M)\right) d\mu(\omega),$$

$$I_\mu(\alpha^M, \beta^N) = S(\Phi_\mu(\alpha^M \cup \beta^N) \,|\, \Phi_\mu(\alpha^M) \otimes \Phi_\mu(\beta^N)),$$

respectively, where $S(\cdot|\cdot)$ stands for the relative entropy [2, 11]

$$S(\omega_1 \,|\, \omega_2) = \mathrm{Tr}\{\rho_{\omega_1}(\log \rho_{\omega_1} - \log \rho_{\omega_2})\}$$

where ρ_{ω_1} (resp. ρ_{ω_2}) is the density operator of a state ω_1 (resp. ω_2) over some finite-dimensional unital C^*-algebra.

The following is a fundamental relation of functionals S_μ and I_μ.

PROPOSITION 2.1.

$$S_\mu(\alpha^M) + S_\mu(\beta^N) = S_\mu(\alpha^M \cup \beta^N) + I_\mu(\alpha^M, \beta^N).$$

Proof. Let ρ_ω (resp. σ_ω) be the density operator in $\otimes_{m=1}^{M} A_m$ (resp. $\otimes_{n=1}^{N} B_n$) corresponding to a state $\otimes_{m=1}^{M} \alpha_m^* \omega$ (resp. $\otimes_{n=1}^{N} \beta_n^* \omega$) in the fixed trace Tr, then

$$\begin{aligned}
&S_\mu(\alpha^M \cup \beta^N) + I_\mu(\alpha^M, \beta^N) \\
&= \int S\left(\rho_\omega \otimes \sigma_\omega \,\middle|\, \int \rho_\omega \otimes \sigma_\omega \, d\mu(\omega)\right) d\mu(\omega) + \\
&\quad + S\left(\left(\int \rho_\omega \otimes \sigma_\omega \, d\mu(\omega)\right) \middle| \left(\int \rho_\omega \, d\mu(\omega)\right) \otimes \left(\int \sigma_\omega \, d\mu(\omega)\right)\right) \\
&= \int \mathrm{Tr}\left\{(\rho_\omega \otimes \sigma_\omega) \log(\rho_\omega \otimes \sigma_\omega) - (\rho_\omega \otimes \sigma_\omega) \log\left(\int \rho_\omega \otimes \sigma_\omega \, d\mu(\omega)\right)\right\} d\mu(\omega) \\
&\quad + \mathrm{Tr}\left\{\left(\int \rho_\omega \otimes \sigma_\omega \, d\mu(\omega)\right) \log\left(\int \rho_\omega \otimes \sigma_\omega \, d\mu(\omega)\right)\right. \\
&\quad \left. - \left(\int \rho_\omega \otimes \sigma_\omega \, d\mu(\omega)\right) \log\left(\int \rho_\omega \, d\mu(\omega) \otimes \int \sigma_\omega \, d\mu(\omega)\right)\right\} \\
&= \int \mathrm{Tr}\left\{\rho_\omega \log \rho_\omega - \rho_\omega \log \int \rho_\omega \, d\mu(\omega)\right\} d\mu(\omega) \\
&\quad + \int \mathrm{Tr}\left\{\sigma_\omega \log \sigma_\omega - \sigma_\omega \log \int \sigma_\omega \, d\mu(\omega)\right\} d\mu(\omega)
\end{aligned}$$

$$= \int S\left(\rho_\omega \middle| \int \rho_\omega \, \mathrm{d}\mu(\omega)\right) \mathrm{d}\mu(\omega) + \int S\left(\sigma_\omega \middle| \int \sigma_\omega \, \mathrm{d}\mu(\omega)\right) \mathrm{d}\mu(\omega)$$

$$= S_\mu(\alpha^M) + S_\mu(\beta^N). \qquad \square$$

Note that this relation between S_μ and I_μ is analogous to the classical relation

$$H_P(\tilde{A}) + H_P(\tilde{B}) = H_P(\tilde{A} \vee \tilde{B}) + I_P(\tilde{A}, \tilde{B})$$

of the entropy H and the mutual entropy I for measurable finite partitions

$$\tilde{A} = \{X_1, \ldots, X_{N_1}\}, \qquad \tilde{B} = \{Y_1, \ldots, Y_{N_2}\}$$

of a probability space $(\Omega, \mathscr{F}, P)$ (see [2, 9]). If $\mathscr{A}$,

$$A_m (m = 1, \ldots, M), \qquad B_n (n = 1, \ldots, N)$$

are Abelian C^*-algebras and each $\alpha_m: A_m \to \mathscr{A}$ (resp. $\beta_n: B_n \to \mathscr{A}$) equals to embedding j_{A_m} (resp. j_{B_n}), of A_m (resp. B_n) into $\mathscr{A}$, then there exists the unique extremal decomposition measure μ for each state ϕ over A, and our functional S_μ coincides with the classical entropy

$$S_\mu(\alpha^M) = H_\mu(\vee_{m=1}^{M} \tilde{A}_m),$$

and I_μ coincides with the classical mutual entropy,

$$I_\mu(\alpha^M, \beta^N) = I_\mu(\vee_{m=1}^{M} \tilde{A}_m, \vee_{n=1}^{N} \tilde{B}_n),$$

where $\tilde{A}_m$ (resp. $\tilde{B}_n$) is a finite partition of the compact Hausdorff space $\Omega = \mathrm{Spec}(\mathscr{A})$ of all pure states over $\mathscr{A}$.

For a given pair of finite sequences of c.p.u. maps

$$\alpha^M = (\alpha_1, \ldots, \alpha_M), \qquad \beta^N = (\beta_1, \ldots, \beta_N),$$

we define the functional S_ϕ (resp. I_ϕ) by taking the supremum of $S_\mu(\alpha^M)$ (resp. $I_\mu(\alpha^M, \beta^N)$) for all possible extremal decompositions μ's of ϕ:

$$S_\phi(\alpha^M) = \sup_{\mu \in M_+^1(\phi)} S_\mu(\alpha^M),$$

$$I_\phi(\alpha^M, \beta^N) = \sup_{\mu \in M_+^1(\phi)} I_\mu(\alpha^M, \beta^N).$$

Recall that the original mutual entropy $I(\phi; \Lambda^*)$ in [6] was defined by

$$I(\phi; \Lambda^*) = \sup_{\mu \in M_+^1(\phi)} I_\mu(\phi; \Lambda^*),$$

where $I_\mu(\phi; \Lambda^*) = S(\Phi_\mu \,|\, \phi \otimes \Lambda^* \phi)$ with $\Phi_\mu = \int_{S(\mathscr{A})} \omega \otimes \Lambda^* \omega \, \mathrm{d}\mu(\omega)$, and $S(\cdot|\cdot)$ denotes Uhlmann's relative entropy [10] for infinite-dimensional C^*-algebras. If we drop the assumption in the definition of $I_\phi(\alpha^M, \beta^N)$ that A's and B's are finite-dimensional, and we take Uhlmann's relative entropy to set $S(\cdot|\cdot)$, then the corresponding functional $I_\phi(\alpha^M, \beta^N)$ can be formally viewed as a natural generalization of Ohya's mutual entropy $I(\phi;\Lambda^*)$ to the cases involving many c.p.u. maps. In fact, one clearly has

$$I(\phi;\Lambda^*)=I_\phi(\mathrm{id}_{\mathscr{A}},\Lambda),$$

where $\mathrm{id}_{\mathscr{A}}$ stands for the identity map of $\mathscr{A}$.

Using the well-known properties of relative entropy such as joint convexity, monotonicity, scaling property, and others [5, 9], the following proposition is easily proved due to the finite-dimensionality of $\{A_m\}$ and $\{B_n\}$.

PROPOSITION 2.2. *Let* $\gamma_m\colon A'_m\to A_m$ (resp. $\delta_n\colon B'_n\to B_n$) *be c.p.u. maps from a finite-dimensional* C^**-algebra* A'_m *to* A_m $(m=1,\ldots,M)$ (resp. B'_n *to* B_n $(n=1,\ldots,N)$), *and let* $\theta\colon\mathscr{A}\to\mathscr{A}$ *is a c.p.u. map with* $\phi\circ\theta=\phi$. *Then*

(1) *Fundamental inequality:*

$$0\leqslant I_\phi(\alpha^M,\beta^N)\leqslant\min\{S_\phi(\alpha^M),S_\phi(\beta^N)\}.$$

(2) *For finite sequences of c.p.u. maps*

$$\alpha^{M'}=(\alpha_1,\alpha_2,\ldots,\alpha_L)\quad\text{and}\quad\alpha^{M''}=(\alpha_{L+1},\alpha_{L+2},\ldots,\alpha_M)$$

with $L<M$ *and* $M=M'\cup M''$,

$$\max\{S_\phi(\alpha^{M'}),S_\phi(\alpha^{M''})\}\leqslant S_\phi(\alpha^M)\leqslant S_\phi(\alpha^{M'})+S_\phi(\alpha^{M''}).$$

(3) *For finite sequences of c.p.u. maps*

$$\alpha^M\circ\gamma^M\equiv(\alpha_1\circ\gamma_1,\alpha_2\circ\gamma_2,\ldots,\alpha_M\circ\gamma_M)\quad\text{and}$$
$$\beta^N\circ\delta^N\equiv(\beta_1\circ\delta_1,\beta_2\circ\delta_2,\ldots,\beta_N\circ\delta_N),$$
$$S_\phi(\alpha^M\circ\gamma^M)\leqslant S_\phi(\alpha^M),$$
$$I_\phi(\alpha^M\circ\gamma^M,\beta^N\circ\delta^N)\leqslant I_\phi(\alpha^M,\beta^N).$$

Equality holds if the γ_m *and* δ_n *are conditional expectations for all* m *and* n, *associated to* $\omega\circ\alpha_m$ *and* $\omega\circ\beta_n$, *respectively.*

(4) *For finite sequences of c.p.u. maps*

$$\theta\circ\alpha^M\equiv(\theta\circ\alpha_1,\theta\circ\alpha_2,\ldots,\theta\circ\alpha_M)\quad\text{and}\quad\theta\circ\beta^N\equiv(\theta\circ\beta_1,\theta\circ\beta_2,\ldots,\theta\circ\beta_N),$$
$$S_\phi(\theta\circ\alpha^M)\leqslant S_\phi(\alpha^M),$$
$$I_\phi(\theta\circ\alpha^M,\theta\circ\beta^N)\leqslant I_\phi(\alpha^M,\beta^N).$$

3. Functionals $\tilde{S}$ and $\tilde{I}$ for Invariant States and Stationary Channels

In this section, we introduce two functionals $\tilde{S}(\phi)$ and $\tilde{I}(\phi,\Lambda^*)$ for a stationary channel Λ^* with an invariant input state ϕ, and we prove the Kolmogorov–Sinai-type theorems for these functionals.

Let $\mathscr{A}$ (resp. $\mathscr{B}$) be a unital C^*-algebra with a fixed automorphism $\theta_{\mathscr{A}}$ (resp. $\theta_{\mathscr{B}}$), Λ be a covariant c.p.u. map from $\mathscr{B}$ to $\mathscr{A}$, and ϕ be an invariant state over $\mathscr{A}$, that is, $\phi\circ\theta_{\mathscr{A}}=\phi$. For each c.p.u. map $\alpha\colon A\to\mathscr{A}$ (resp. $\beta\colon B\to\mathscr{B}$) from a finite-dimensional unital C^*-algebra A (resp. B) to $\mathscr{A}$ (resp. $\mathscr{B}$), put

$$\tilde{S}_\phi(\alpha)=\limsup_{n\to\infty}\frac{1}{n}S_\phi(\alpha^n)$$

$$\tilde{I}_{\phi,\Lambda^*}(\alpha,\beta)=\limsup_{n\to\infty}\frac{1}{n}I_\phi(\alpha^n,\Lambda\circ\beta^n_\Lambda),$$

where

$$\alpha^n=(\alpha,\theta_{\mathscr{A}}\circ\alpha,\ldots,\theta^{n-1}_{\mathscr{A}}\circ\alpha)\quad\text{and}\quad\beta^n_\Lambda=(\Lambda\circ\beta,\Lambda\circ\theta_{\mathscr{B}}\circ\beta,\ldots,\Lambda\circ\theta^{n-1}_{\mathscr{B}}\circ\beta).$$

Note that indeed $\tilde{S}_\phi(\alpha)$ (resp. $\tilde{I}_{\phi,\Lambda^*}(\alpha,\beta)$) is the limit of $S_\phi(\alpha^n)/n$ (resp. $1/n\,I_\phi(\alpha^n,\Lambda\circ\beta^n_\Lambda)$) from the subadditivity (3) in Proposition 2.2. Then we define the functionals $\tilde{S}(\phi)$ and $\tilde{I}(\phi,\Lambda^*)$ by taking the supremum for all possible A's, α's, B's, and β's:

$$\tilde{S}(\phi)=\sup_\alpha\tilde{S}_\phi(\alpha),$$

$$\tilde{I}(\phi,\Lambda^*)=\sup_{\alpha,\beta}\tilde{I}_{\phi,\Lambda^*}(\alpha,\beta).$$

Note that the above functionals $\tilde{S}(\phi)$ and $\tilde{I}(\phi,\Lambda^*)$ depend on $\theta_{\mathscr{A}}$ and $(\theta_{\mathscr{A}},\theta_{\mathscr{B}})$, respectively, but we abbreviate them for simplicity here.

The fundamental inequality in information theory certainly holds for our $\tilde{S}$ and $\tilde{I}$ from Proposition 2.2:

PROPOSITION 3.1.

$$0\leqslant\tilde{I}(\phi,\Lambda^*)\leqslant\min\{\tilde{S}(\phi),\tilde{S}(\Lambda^*\phi)\}.$$

In ergodic theory and information theory, the Kolmogorov–Sinai theorem for the entropy of automorphisms is very important because it makes possible to compute the mean entropy for concrete examples. In [4], Connes–Narnhofer–Thirring proved the non-Abelian Kolmogorov–Sinai theorem for their dynamical entropy. For the case of our functionals $\tilde{S}(\phi)$ and $\tilde{I}(\phi,\Lambda^*)$, the Kolmogorov–Sinai-type theorem still holds (Theorems 3.5, 3.6). The proof of this Kolmogorov–Sinai-type theorem for functionals $\tilde{S}$ and $\tilde{I}$ is similarly as that in [3], because our functionals obey similar estimates to that of the Connes–Narnhofer–Thirring joint entropy. The points we stress are that our functionals $\tilde{S}(\phi)$ and $\tilde{I}(\phi,\Lambda^*)$ are constructed from the functionals $S_\mu(\alpha^M)$ and $I_\mu(\alpha^M,\beta^N)$ which are related in information theory and our functionals are obtained by using a channel transformation so that ours contains the dynamical entropy in some cases.

For the proof of the Kolmogorov–Sinai-type theorem for functionals $\tilde{S}$ and $\tilde{I}$, we need a Lemma from [4]

LEMMA 3.2 [4]. *Let $\mathscr{A},\mathscr{B}$ be unital C^*-algebras with B finite-dimensional and $d=\dim B$. Let $\phi=\int\omega\,\mathrm{d}\mu(\omega)$ be any decomposition measure of a state ϕ over $\mathscr{A}$. Then* *for any c.p.u. maps* $\alpha, \bar{\alpha}: B \to \mathscr{A}$ *with* $\|\alpha - \bar{\alpha}\| \leqslant \varepsilon$ *one has*

$$\left| \int_{S(\mathscr{A})} S(\alpha^* \omega \mid \alpha^* \phi)\, \mathrm{d}\mu(\omega) - \int_{S(\mathscr{A})} S(\bar{\alpha}^* \omega \mid \bar{\alpha}^* \phi)\, \mathrm{d}\mu(\omega) \right| \leqslant$$

$$\leqslant 6\varepsilon \left(\frac{1}{2} + \log\left(1 + \frac{d}{\varepsilon}\right)\right).$$

Using this, we can show the continuity of the functionals $S_\phi(\alpha^M)$ and $I_\phi(\alpha^M, \beta^N)$ with respect to α's and β's.

PROPOSITION 3.3. *Let* $\mathscr{A}$ *be a unital* C^**-algebra with a state* ϕ *and* A_m, $m=1, \ldots, M$ *be finite-dimensional unital* C^**-algebras,* $\alpha_m, \bar{\alpha}_m$ *be c.p.u. maps from* A_m to $\mathscr{A}$. *Let* d *be the* max *of the dimensions of the* A_m*'s and* $\max_m \|\alpha_m - \bar{\alpha}_m\| \leqslant \varepsilon$, *then*

$$|S_\phi(\alpha^M) - S_\phi(\bar{\alpha}^M)| \leqslant 6M\varepsilon \left(\frac{1}{2} + \log\left(1 + \frac{d}{\varepsilon}\right)\right).$$

PROPOSITION 3.4. *Let* $\mathscr{A}$ *be a unital* C^**-algebra with a state* ϕ, *and* A_m, $m=1, \ldots, M$, *and* B_n, $n=1, \ldots, N$, *be finite-dimensional* C^**-algebras,* $\alpha_m, \bar{\alpha}_m$ *be c.p.u. maps from* A_m *to* $\mathscr{A}$, *and* $\beta_n, \bar{\beta}_n$ *be c.p.u. maps from* B_n *to* $\mathscr{A}$. *Let* d *be the* max *of the dimensions of the* A_m*'s and* B_n*'s, and*

$$\max\{\|\alpha_m - \bar{\alpha}_m\|, \|\beta_n - \bar{\beta}_n\| \mid m=1, \ldots, M; n=1, \ldots, N\} \leqslant \varepsilon,$$

then

$$|I_\phi(\alpha^M, \beta^N) - I_\phi(\bar{\alpha}^M, \bar{\beta}^N)| \leqslant 12(M+N)\varepsilon \left(\frac{1}{2} + \log\left(1 + \frac{d}{\varepsilon}\right)\right).$$

Under the conditions of the above propositions with an invariant state ϕ, we get the following Kolmogorov–Sinai-type theorems (Theorems 3.5 and 3.6).

THEOREM 3.5. *Let* α_m *be a sequence of c.p.u. maps* $\alpha_m: A_m \to \mathscr{A}$ *such that for suitable c.p.u. maps* $\alpha'_m: \mathscr{A} \to A_m$, *one has* $\alpha_m \circ \alpha'_m \to \mathrm{id}_{\mathscr{A}}$ *in the pointwise topology. Then*

$$\tilde{S}(\phi) = \lim_{m \to \infty} \tilde{S}_\phi(\alpha_m).$$

THEOREM 3.6. *Let* α_m *and* β_m *be sequences of c.p.u. maps* $\alpha_m: A_m \to \mathscr{A}$ *and* $\beta_m: B_m \to \mathscr{B}$ *such that for suitable c.p.u. maps* $\alpha'_m: \mathscr{A} \to A_m$ *and* $\beta'_m: \mathscr{B} \to B_m$, *one has* $\alpha_m \circ \alpha'_m \to \mathrm{id}_{\mathscr{A}}$ *and* $\beta_m \circ \beta'_m \to \mathrm{id}_{\mathscr{B}}$ *in the pointwise topology. Then*

$$\tilde{I}(\phi, \Lambda^*) = \lim_{m \to \infty} \tilde{I}_{\phi, \Lambda^*}(\alpha_m, \beta_m).$$

Proof. Put $\gamma_m = \alpha_m \circ \alpha'_m \circ \gamma$ for any c.p.u. map $\gamma: A \to \mathscr{A}$ with a finite C^*-algebra A. Then $(\gamma_m(x))$ converges to $\gamma(x)$ for each $x \in A$ and, hence,

$$\lim_{m \to \infty} \|\alpha_m - \alpha\| = 0,$$

since γ_m and γ are bounded linear maps from a finite-dimensional C^*-algebra A. In the same way,

$$\lim_{m\to\infty} \|\delta_m - \delta\| = 0, \qquad \delta_m = \beta_m \circ \beta'_m \circ \delta$$

for any c.p.u. $\delta: B \to \mathscr{B}$ with a finite C^*-algebra B. From Proposition 3.4, we have

$$\left| \frac{I_\phi((\theta^n_{\mathscr{A}} \circ \gamma_m)^{N-1}_{n=0}, (\theta^n_{\mathscr{B}} \circ \Lambda \circ \delta_m)^{N-1}_{n=0})}{N} - \frac{I_\phi((\theta^n_{\mathscr{A}} \circ \gamma)^{N-1}_{n=0}), (\theta^n_{\mathscr{B}} \circ \Lambda \circ \delta)^{N-1}_{n=0})}{N} \right|$$

$$\leqslant 24N\varepsilon\left(\frac{1}{2} + \log\left(1 + \frac{d}{\varepsilon}\right)\right)$$

for $\|\gamma_m - \gamma\| \leqslant \varepsilon$ and $\|\delta_m - \delta\| \leqslant \varepsilon$ and, hence,

$$|\tilde{I}_{\phi,\Lambda^*}(\gamma_m, \delta_m) - \tilde{I}_{\phi,\Lambda^*}(\gamma, \delta)| \leqslant 24N\varepsilon\left(\frac{1}{2} + \log\left(1 + \frac{d}{\varepsilon}\right)\right),$$

and

$$\lim_{m\to\infty} \tilde{I}_{\phi,\Lambda^*}(\gamma_m, \delta_m) = \tilde{I}_{\phi,\Lambda^*}(\gamma, \delta).$$

From the monotonicity of the functional I (Proposition 2.3 (3)), we have

$$\tilde{I}_{\phi,\Lambda^*}(\gamma_m, \delta_m) \leqslant \tilde{I}_{\phi,\Lambda^*}(\alpha_m, \beta_m)$$

and, hence,

$$\tilde{I}_{\phi,\Lambda^*}(\gamma, \delta) \leqslant \liminf_{m\to\infty} \tilde{I}_{\phi,\Lambda^*}(\alpha_m, \beta_m) \leqslant \tilde{I}(\phi, \Lambda^*).$$

Since γ and δ are arbitrary completely positive unital maps from A to $\mathscr{A}$, we have

$$\lim_{m\to\infty} \tilde{I}_{\phi,\Lambda^*}(\alpha_m, \beta_m) = \tilde{I}(\phi, \Lambda^*). \qquad \square$$

The assumption of the above two theorems are always satisfied if we take C^*-algebras to be nuclear. If we apply these theorems for AF algebras (the norm closure of an increasing union $\cup A_n$ of sequences of finite-dimensional subalgebras) we can get the followings.

COROLLARY 3.7. *Let* $\mathscr{A} = \overline{\cup A_n}$ *be an AF algebra with an automorphism* $\theta_{\mathscr{A}}$, *then for any invariant state* ϕ *over* $\mathscr{A}$,

$$\tilde{S}(\phi, \theta) = \lim_{n\to\infty} \tilde{S}_{\phi,\theta}(A_n).$$

COROLLARY 3.8. *Let* $\mathscr{A} = \overline{\cup A_n}$ *(*resp. $\mathscr{B} = \overline{\cup B_m}$*) be an AF algebra with an automorphism* $\theta_{\mathscr{A}}$ *(*resp. $\theta_{\mathscr{B}}$*). Then for each covariant c.p.u. map* $\Lambda: \mathscr{B} \to \mathscr{A}$ *and an invariant state* ϕ *over* $\mathscr{A}$,

$$\tilde{I}(\phi, \Lambda^*(\theta_{\mathscr{A}}, \theta_{\mathscr{B}})) = \lim_{n\to\infty} \tilde{I}_{\phi,\Lambda^*}(A_m, B_n).$$

In the above corollaries, the subalgebra $A_n \subset \mathscr{A}$ stands for the inclusion map $j_{A_n}: A_n \to \mathscr{A}$ for every n.

Some extensions and verification of two functionals $\tilde{S}(\phi)$ and $\tilde{I}(\phi, \Lambda^*)$ are discussed in [12], in order to study communication processes and complexities of various physical systems.

References

1. Accardi, L. and Ohya, M., Compound channels, transition expectations and liftings, to appear in *J. Multivariate Anal.*
2. Araki, H., Relative entropy of states of von Neumann algebras, *Publ. RIMS, Kyoto Univ.* **11**, 809–833 (1976).
3. Billingsley, P., *Ergodic Theory and Information*, Wiley, New York, 1965.
4. Connes, A., Narnhofer, H., and Thirring, W., Dynamical entropy of C^* algebras and von Neumann algebras, *Comm. Math. Phys.* **112**, 691–719 (1987).
5. Muraki, N., Ohya, M., and Petz, D., Entropies of general quantum states, *Open Systems Inform. Dynam.* **1**(1), 43–56 (1992).
6. Ohya, M., Some aspects of quantum information theory and their applications to irreversible processes, *Rep. Math. Phys.* **27**, 19–47 (1989).
7. Ohya, M., On compound state and mutual information in quantum information theory, *IEEE Trans. Inform. Theory* **29**, 770–774 (1983).
8. Ohya, M., Note on quantum probability, *Lett. Nuovo Cim.* **38**, 402–406 (1983).
9. Ohya, M. and Petz, D., *Quantum Entropy and its Use*, Springer, New York, 1993.
10. Uhlmann, A., Relative entropy and the Wigner–Yanase–Dyson–Lieb concavity in an interpolation theory, *Comm. Math. Phys.* **54**, 21–32 (1977).
11. Umegaki, H., Conditional expectation in an operator algebra IV (entropy and information), *Kôdai Math. Sem. Rep.* **14**, 59–85 (1962).
12. Ohya, M., State change, complexity and fractal in quantum systems, to appear in *Quantum Communications and Measurement*, Plenum, New York.

Information Dynamics and Its Applications to Optical Communication Processes

Masanori OHYA

Department of Information Sciences
Science University of Tokyo
Noda City, Chiba 278, Japan

Abstract

Various physical or nonphysical systems can be described by states, so that the dynamics of a system is described by the state change. One of essential characters of a state is expressed by its complexity. Complexity such as entropy is a key concept in Information theory. We call the study of the state change together with such complexities "Information dynamics", which is a kind of synthesis of dynamics of state change and information theory. Here we explain what information dynamics is and indicate how it can be used in optical communication. Some of concrete applications of information dynamics are discussed in the papers [9,10] of this volume.

§1. What is Information Dynamics ?

Let an input dynamical system and an output dynamical system be described by $(\mathcal{A}, \mathfrak{S}, \alpha)$ and $(\overline{\mathcal{A}}, \overline{\mathfrak{S}}, \overline{\alpha})$, respectively. Here $\mathcal{A}$ is a set of all objects to be observed and $\mathfrak{S}$ is a set of all means getting the observed value for each element A in $\mathcal{A}$, and α describes an inner evolution of the input system. We call $\mathfrak{S}$ a "state space" here. Same for the output system $(\overline{\mathcal{A}}, \overline{\mathfrak{S}}, \overline{\alpha})$. Thus we may say

[Giving a mathematical structure to input and output triples $\equiv$ Having a theory].

A map providing a bridge between two systems is called a "channel" if it sends a state of the input system to that of the output system ; $\Lambda^* : \mathfrak{S} \rightarrow \overline{\mathfrak{S}}$. There exist several channels, whose properties specify the character of two systems.

Mathematical structure of almost all systems can be expressed by the following charts.

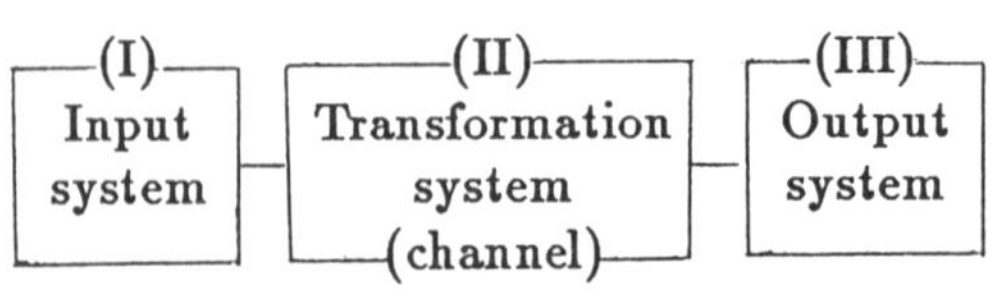

(chart 1)

Lecture Notes in Physics, Vol. 378
C. Bendjaballah, O. Hirota, S. Reynaud (Eds.),
Quantum Aspects of Optical Communications

82

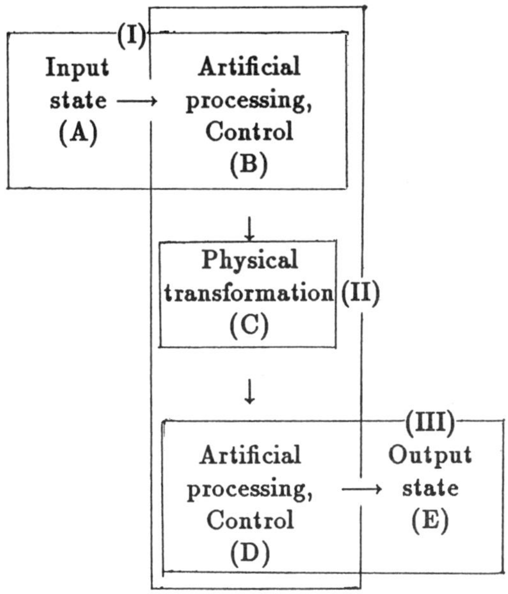

(Chart 2)

In the above chart 2, (I) (= (A) + (B)) corresponds to an input system, (II) (= (B) + (C) + (D)) does to a transformation (channel) system (III) (= (D) + (E)) to an output system. More complex systems are constructed from this first structure.

A "naked" state (A) is artificially processed or controlled to a "dressed" state, and it is suffered to change by a physical (natural) transformation, and it is again artificially processed and controlled. The fundamental part of the process for this state change is obviously "A$\rightarrow$ C$\rightarrow$ E".

Let us give some examples of an input system (I), a transformation system (channel) (II) and an output system (III).

(1) Causal System : (A) $x \in R^n$; (B) $\dot{x} = f(x)$; (C) $x(t) = \Phi_t(x)$, where Φ_t is an evolution (semi)group generated by f.

Remark: If $x(t)$ can not be obtained directly, then examines the properties of f and find some proper approximation to obtain Φ_t.
$\Longrightarrow$ We may have a chaotic system.

(2) Signal Transmission : (A) causal signal $x(t)$; (B) coding causal signal $x(t)$ to $\{x_n\}$ by e.g., sampling theorem with cut off ; (C) some transformation $y_n = f(x_n)$ for any n ; (III) interpolating or decoding $\{y_n\}$ to $y(t)$.

(3) Discrete systems : (A) input probability distribution $p = \{p_1, \ldots, p_n\}$ of events $X = \{x_1, \ldots, x_n\}$; (C) transition probability $(p(i|j))$; (E) output probability distribution $q = (q_i); q_i = \sum_j p(i|j)p_j$.

(4) Continuous systems : (A) probability measure μ on measurable space $(\Omega, \mathfrak{F})$; (C) Markov kernel $\lambda : \Omega \times \overline{\mathfrak{F}} \rightarrow [0,1]$ s.t. $\lambda(x, \cdot) \in P(\overline{\Omega})$ and $\lambda(\cdot, A) \in M(\Omega)$; (E) $\overline{\mu} \equiv \int_\Omega \lambda(x, \cdot)d\mu$.

83

(5) Quantum system 1 : (A) $x \in \mathcal{H}$ (Hilbert space) ; (C) unitary operator U or isometric operator V ; (E) $y = Ux$ or $Vx \in \mathcal{H}$.

(6) Quantum system 2 : (A) density operator $\rho \in T(\mathcal{H})_{+,1}$; (C) $\Lambda^* \equiv \mathrm{Ad}U$ or $\mathrm{Ad}V$. (E) $\overline{\rho} = \Lambda^* \rho$.

(7) C*-system containing all above: (A) $(\mathcal{A}, \mathfrak{S}, \alpha(G))$ C*-triple; (C) $\Lambda^* : \mathfrak{S}(\mathcal{A}) \to \mathfrak{S}(\overline{\mathcal{A}})$ a dual map of a completely positive map $\Lambda : \overline{\mathcal{A}} \to \mathcal{A}$, where $(\overline{\mathcal{A}}, \overline{\mathfrak{S}}, \alpha(\overline{G}))$ C*-triple; (E) $\overline{\varphi} = \Lambda^* \varphi$.

Once input and output systems are mathematically fixed and a transformation rule (channel) is given, we next consider some complexities of the state associated with the systems, which are a corner stone of information dynamics. The first complexity is one for a state itself : For a state φ, the complexity seen from a reference system $\mathcal{S}$, a certain subset of $\mathfrak{S}$, is dented by $C^{\mathcal{S}}(\varphi)$. The second complexity is determined by both input and output states φ, $\overline{\varphi}$ or an input state φ and a channel Λ^*, so that it is denoted by $T^{\mathcal{S}}(\varphi; \overline{\varphi})$ or $T^{\mathcal{S}}(\varphi; \Lambda^*)$, which is called a transmitted complexity from φ to $\overline{\varphi}$ or $\Lambda^* \varphi$. Typical examples of these complexities are entropy and mutual entropy, which play essential roles in several fields.

These complexities should satisfy the following properties: Let $(\mathcal{A}_1, \mathfrak{S}_1, \alpha^1(G))$, $(\mathcal{A}_2, \mathfrak{S}_2, \alpha^2(G))$ be two systems and $(\mathcal{A}, \mathfrak{S}, \alpha(G))$ be the compound system such that $\mathcal{A} = \mathcal{A}_1 \otimes \mathcal{A}_2$, and Λ^* be a channel from $\mathfrak{S}_1$ to $\mathfrak{S}_2$.

(i) For any state $\varphi \in \mathcal{S} \subset \mathfrak{S}_1$, $C^{\mathcal{S}}(\varphi) \geq 0$ and $T^{\mathcal{S}}(\varphi; \Lambda^*) \geq 0$.

(ii) For any bijection j from $\mathfrak{S}_1$ to $\mathfrak{S}_1$,

$$C^{j(\mathcal{S})}(j(\varphi)) = C^{\mathcal{S}}(\varphi)$$
$$T^{j(\mathcal{S})}(j(\varphi); \Lambda^*) = T^{\mathcal{S}}(\varphi; \Lambda^*).$$

(iii) For any $\Phi \in \mathcal{S} \subset \mathfrak{S}$, put $\varphi \equiv \Phi \restriction \mathcal{A}_1 \in \mathcal{S}_1 \equiv \mathcal{S} \restriction \mathcal{A}_1$ and $\psi \equiv \Phi \restriction \mathcal{A}_2 \in \mathcal{S}_2 \equiv \mathcal{S} \restriction \mathcal{A}_2$

$$C^{\mathcal{S}}(\Phi) \leq C_1^{\mathcal{S}}(\varphi) + C_2^{\mathcal{S}}(\psi).$$

Moreover

(iv) $0 \leq T^{\mathcal{S}}(\varphi; \Lambda^*) \leq C^{\mathcal{S}}(\varphi)$.

(v) $T^{\mathcal{S}}(\varphi; id) = C^{\mathcal{S}}(\varphi)$, where "id" is an identity channel from $\mathfrak{S}_1$ to $\mathfrak{S}_1$; $id(\varphi) = \varphi$, $\forall \varphi \in \mathfrak{S}_1$.

Definition 1.1: *Information dynamics is a dynamics described by a set*

$$\{\mathcal{A}, \mathfrak{S}, \mathcal{S}, \alpha ; \overline{\mathcal{A}}, \overline{\mathfrak{S}}, \overline{\mathcal{S}}, \overline{\alpha} ; \Gamma ; C^{\mathcal{S}}(\varphi), T^{\mathcal{S}}(\varphi; \Gamma) ; R\},$$

where R is a certain relation among above quantities.

Therefore, for a system of interest, we have to

(1) mathematically determine $\mathcal{A}, \mathfrak{S}, \mathcal{S}, \alpha$; $\overline{\mathcal{A}}, \overline{\mathfrak{S}}, \overline{\mathcal{S}}, \overline{\alpha}$

(2) choose Γ and R ;

(3) fix $C^{\mathcal{S}}(\varphi)$ and $T^{\mathcal{S}}(\varphi; \Gamma)$.

By setting the above (1) ~ (3) in general quantum systems (GQS), we can apply this general frames to several topics such as optical communication[1,5,6],fractal theory [8], molecular genetics [7]. We here mainly discuss an application of information dynamics to optical communication processes.

§2. Channel and Lifting

In order to consider an application of information dynamics to optical communication processes, we have to mathematically set all quantities needed. Let both input system $(\mathcal{A}, \mathfrak{S}, \alpha)$ and output system $(\overline{\mathcal{A}}, \overline{\mathfrak{S}}, \overline{\alpha})$ be described by certain C*-algebraic triples. Very often $\mathcal{A} = \overline{\mathcal{A}}$, $\mathfrak{S} = \overline{\mathfrak{S}}$, $\alpha = \overline{\alpha}$.

A map $\Lambda^* : \mathfrak{S} \to \overline{\mathfrak{S}}$ is called a channel. In particular, Λ^* is called completely positive channel if the dual map $\Lambda : \overline{\mathcal{A}} \to \mathcal{A}$ satisfies

$$\sum_{i,j} B_i^* \Lambda(A_i^* A_j) B_j \geq 0 \quad \text{for } \forall B_j \in \mathcal{A} \text{ and } \forall A_i \in \overline{\mathcal{A}}.$$

A certain class of channels may be a lifting $\mathcal{E}^*$ from $\mathcal{A}$ to $\mathcal{A} \otimes \overline{\mathcal{A}}$ [1]; that is, a continuous map such as

$$\mathcal{E}^* : \mathfrak{S}(\mathcal{A}) \to \mathfrak{S}(\mathcal{A} \otimes \overline{\mathcal{A}}).$$

A lifting $\mathcal{E}^*$ is said to be nondemolition for a state φ if

$$(\mathcal{E}^*\varphi)(A \otimes I) = \varphi(A) \text{ for } \forall A \in \mathcal{A}.$$

Given a lifting $\mathcal{E}^*$, we can construct channels Λ^* and $\overline{\Lambda^*}$ such that

$$\Lambda^* : \mathfrak{S} \to \overline{\mathfrak{S}}$$

by $\Lambda^*\varphi(\overline{A}) = (\mathcal{E}^*\varphi)(I \otimes \overline{A}), \forall \overline{A} \in \overline{\mathcal{A}}$, and

$$\overline{\Lambda}^* : \mathfrak{S} \to \mathfrak{S}$$

by $\overline{\Lambda}^*\varphi(\mathcal{A}) = (\mathcal{E}^*\varphi)(A \otimes I), \forall \overline{A} \in \mathcal{A}$. Conversely, given a channel $\Lambda^* : \mathfrak{S} \to \overline{\mathfrak{S}}$ and $\varphi \in \mathcal{S} \subset \mathfrak{S}$ and a decomposition of φ ; $\varphi = \int_{\mathcal{S}} \omega d\mu$

$$\mathcal{E}^*\varphi = \int_{\mathcal{S}} \omega \otimes \Lambda^*\omega d\mu \quad \text{(compound state [3])}$$

is a nondemolition lifting. Hence we have

[Channels ⇔ Liftings]

< Examples of channel and lifting >

Here we give some examples of channel and we construct a quantum communication channel. Let $\rho = \sum \lambda_n \rho_n$ be a certain decomposition of a density operator when $\mathcal{A} = \mathrm{B}(\mathcal{H})$ and $\mathfrak{S} = \mathrm{T}(\mathcal{H})_{+,1}$ on a Hilbert space $\mathcal{H}$.

<1> Unitary evolution :

$$\rho \to \Lambda_t^*\rho = \mathrm{Ad}U_t(\rho) \equiv U_t^*\rho U_t, \ t \in R \ \Rightarrow \mathcal{E}^*\rho = \sum \lambda_n \rho_n \otimes \Lambda^*\rho_n$$

where U_t is a unitary operator $U_t = \exp(\mathrm{it}H)$.

<u><2> Semigroup evolution :</u>

$$\rho \to \Lambda_t^* \rho = \mathrm{Ad}V_t(\rho) \equiv V_t^* \rho V_t, \ t \in R^+ \ \Rightarrow \mathcal{E}^* \rho = \sum \lambda_n \rho_n \otimes \Lambda^* \rho_n$$

where $\{V_t \ ; \ \mathrm{t} \in R^+\}$ is a one parameter semigroup on $\mathcal{H}$.

<u><3> Measurement :</u> Measure $\mathrm{A} = \sum_n a_n P_n$ (spectral decomposition of A) in a state ρ,

$$\rho \to \Lambda^* \rho = \sum_n P_n \rho P_n \Rightarrow \mathcal{E}^* \rho = \sum \lambda_n \rho_n \otimes \Lambda^* \rho_n.$$

<u><4> Reduction :</u> When two systems described by Hilbert spaces $\mathcal{H}$ and $\mathcal{K}$ interacts and we look at the state change of the first system.

$$\begin{array}{ccc} \mathfrak{S}(\mathcal{H}) & \xrightarrow{\text{interaction}} & \mathfrak{S}(\mathcal{K}). \\ \Cup & & \Cup \\ \rho & & \sigma \end{array}$$

Then

$$\mathcal{E}_t^* \rho \equiv \theta_t = U_t^* (\rho \otimes \sigma) U_t,$$
$$\rho \to \Lambda_t^* \rho = tr_{\mathcal{K}} \theta_t.$$

More generally for <3> and <4>, $\mathcal{A}$ (resp. $\overline{\mathcal{A}}$) is interpreted as the algebra of observables of a system (resp. a measuring apparatus) and $\mathcal{E}^*$ describes an interaction between the system and the apparatus as well as the preparation of the apparatus. If $\varphi \in \mathcal{S}(\mathcal{A})$ is the preparation of the system, i.e., the state before the interaction with the apparatus, then $\overline{\Lambda}^* \varphi \in \mathcal{S}(\mathcal{A})$ (resp. $\Lambda^* \varphi \in \mathcal{S}(\overline{\mathcal{A}})$) is the state of the system (resp. of the apparatus) after the measurement.

<u><5> Isometric lifting :</u> For an isometry V ($V^*V = I_{\mathcal{H}}$) from $\mathcal{H}$ to $\mathcal{H} \otimes \mathcal{K}$, a lifting defined by

$$\mathcal{E}^* \rho = V \rho V^*, \quad \forall \rho \in \mathfrak{S}(\mathcal{H}_1)$$

is called an isometric lifting [1].

<u><6> Quantum Communication Channel :</u>

Let us construct a quantum (communication) channel.

$$\begin{array}{ccc} (Noise)\ \mathfrak{S}(\mathcal{K}_1) & & \\ \downarrow & & \\ \mathfrak{S}(\mathcal{H}_1) \ni \rho & \longrightarrow & \bar{\rho} = \Lambda^* \rho \in \mathfrak{S}(\mathcal{H}_2) \\ & & \downarrow \\ & & (Loss)\ \mathfrak{S}(\mathcal{K}_2) \end{array}$$

Let $\nu \in \mathfrak{S}(\mathcal{K}_1)$ be a state representing the noise and a, π, γ be the following maps: (1) $a : B(\mathcal{H}_2) \rightarrow B(\mathcal{H}_2 \otimes \mathcal{K}_2)$ given by $a(A) = A \otimes I$ for any $A \in B(\mathcal{H}_2)$, (2) $\pi : B(\mathcal{H}_2 \otimes \mathcal{K}_2) \rightarrow B(\mathcal{H}_1 \otimes \mathcal{K}_1)$ completely positive with $\pi(I) = I$, (3) $\gamma : B(\mathcal{H}_1 \otimes \mathcal{K}_1) \rightarrow B(\mathcal{H}_1)$ by $\gamma(Q) = tr_{\mathcal{K}_1} \nu Q$ for any $Q \in B\,(\mathcal{H}_1 \otimes \mathcal{K}_1)$.

$$\Lambda = \gamma \circ \pi \circ a.$$

Then

$$\Lambda^* = a^* \circ \pi^* \circ \gamma^*$$

or equivalently,

$$\Lambda^* \rho = tr_{\mathcal{K}_2} \pi^*(\rho \otimes \nu), \ \forall \rho \in \mathfrak{S}(\mathcal{H}_1).$$

We call this channel "quantum communication channel".

When $\mathcal{H}_1 = \mathcal{H}_2 = \mathcal{H}$,

$$\mathcal{E}^* : \rho \in \mathfrak{S}(\mathcal{H}) \rightarrow \pi^*(\rho \otimes \nu) \in \mathfrak{S}(\mathcal{H} \otimes \mathcal{K})$$

is a lifting, and

$$\Lambda^* \rho = tr_{\mathcal{K}} \mathcal{E}^* \rho.$$

In <5> and <6>, an input signal is transmitted and received by an apparatus which produces an output signal. Here $\mathcal{A}$ (resp. $\overline{\mathcal{A}}$) is interpreted as the algebra of observables of the input (resp. output) signal and $\mathcal{E}^*$ describes an interaction between the input signal and the receiver as well as the surroundings of the receiver. If $\rho_1 \in \mathcal{S}(\mathcal{A})$ is the input signal, then the state $\Lambda^* \varphi \in \mathcal{S}(\overline{\mathcal{A}})$ is the state of the (observed) output signal.

§3. Entropies describing complexities in GQS

Next we discuss two types of entropy for general quantum states as the complexities of information dynamics which are needed for optical communication.

Let $(\mathcal{A}, \mathfrak{S}, \alpha(R))$ be a C*-dynamical system and $\mathcal{S}$ be a weak* compact and convex subset of $\mathfrak{S}$, ex$\mathcal{S}$ be the set of all extreme points of $\mathcal{S}$.

Every state $\varphi \in \mathcal{S}$ has a maximal measure μ pseudosupported on ex $\mathcal{S}$ such that

$$\varphi = \int_{\mathcal{S}} \omega d\mu.$$

The measure μ giving the above decomposition is not unique unless $\mathcal{S}$ is a Choquet simplex, so that we denote the set of all such measures by $M_\varphi(\mathcal{S})$. For the probability measure μ, define

$$H(\mu) = \sup\{- \sum_{A_k \in \tilde{A}} \mu(A_k) \log \mu(A_k); \ \tilde{A} \in P(\mathcal{S})\},$$

where $P(\mathcal{S})$ is the set of all finite partitions of $\mathcal{S}$. Then the entropy of a state $\varphi \in \mathcal{S}$ w.r.t. $\mathcal{S}$ is defined by

$$S^{\mathcal{S}}(\varphi) = \inf\{H(\mu); \ \mu \in M_\varphi(\mathcal{S})\}.$$

This entropy does depend on the set $\mathcal{S}$ chosen, and we call it "$\mathcal{S}$ - entropy". Even when $S(\varphi) = +\infty$, $S^{\mathcal{S}}(\varphi) < +\infty$ for some $\mathcal{S}$, which is a remarkable property of $\mathcal{S}$ - entropy [4].

A compound state (lifting) $\Phi_\mu^{\mathcal{S}}(\equiv \mathcal{E}_\mu^* \varphi)$ of φ and $(\bar{\varphi} \equiv)\Lambda^* \varphi$ with respect to $\mathcal{S}$ and μ was introduced as

$$\Phi_\mu^{\mathcal{S}} = \int_{\mathcal{S}} \omega \otimes \Lambda^* \omega d\mu.$$

The mutual entropy w.r.t. an initial state $\varphi \in \mathcal{S}$, the decomposition measure μ and a channel Λ^* is defined by [2,6]

$$I_\mu^{\mathcal{S}}(\varphi; \Lambda^*) = S(\Phi_\mu^{\mathcal{S}} | \Phi_0),$$

where $S(\cdot|\cdot)$ is the relative entropy for two states. In some cases, this mutual entropy can be written as

$$I_\mu^{\mathcal{S}}(\varphi; \Lambda^*) = \int_\Omega S(\Lambda^* \omega | \Lambda^* \varphi) d\mu.$$

The mutual entropy w.r.t. an initial state $\varphi \in \mathcal{S}$ and a channel Λ^* is then defined by

$$I^{\mathcal{S}}(\varphi; \Lambda^*) = \lim_{\epsilon \to 0} \sup \{ I_\mu^{\mathcal{S}}(\varphi; \Lambda^*); \mu \in F_\varphi(\mathcal{S}; \varepsilon) \},$$

where $F_\varphi(\mathcal{S})$ is the subset of the set $M_\varphi(\mathcal{S})$ such that $F_\varphi(\mathcal{S}; \varepsilon) = \{ \mu \in M_\varphi(\mathcal{S}); S^{\mathcal{S}}(\varphi) \leq H(\mu) < S^{\mathcal{S}}(\varphi) + \varepsilon < +\infty \}$ or $F_\varphi(\mathcal{S}; \varepsilon) = M_\varphi(\mathcal{S})$ when $S^{\mathcal{S}}(\varphi) = \infty$. Note that the mutual entropy should be used when the decomposition measure is fixed. In the sequel we use the simple notations $S(\varphi), \Phi_\mu, I_\mu(\varphi; \Lambda^*)$ and $I(\varphi; \Lambda^*)$ when $\mathcal{S} = \mathfrak{S}$.

Let us write the mutual entropy in usual quantum system, namely, when $\mathcal{A}$ is the full algebra $B(\mathcal{H})$ and any normal state φ is described by a density operator ρ such as $\varphi(\cdot) = tr\rho\,\cdot$. Then our entropy $S(\varphi)$ is shown to be equal to that of von Neumann: $S(\varphi) = S(\rho) = -tr\rho \log \rho$. Every Schatten decomposition $\rho = \sum_n \lambda_n E_n$, $E_n = |x_n >< x_n|$ (i.e., λ_n is the eigenvalue of ρ and x_n is its associated eigenvector) provides every orthogonal measure in $M_\varphi(\mathfrak{S})$ defining the entropy $S(\varphi)$. Since the Schatten decomposition of ρ is not unique unless every eigenvalue λ_n is nondegenerate, the compound state Φ is expressed as

$$\Phi_E(Q) = tr\sigma_E Q, \quad Q \in \mathcal{A} \otimes \overline{\mathcal{A}},$$

with

$$\sigma_E = \sum_n \lambda_n E_n \otimes \Lambda^* E_n,$$

where E represents a Schatten decomposition $\{E_n\}$. Then the mutual entropy for φ and the channel Λ^* is given by

$$I(\rho; \Lambda^*)(\equiv I(\varphi; \Lambda^*)) = \sup \{ I_E(\rho; \Lambda^*); \ E = \{E_n\} \ of \ \rho \},$$

with

$$I_E(\rho; \Lambda^*) = S(\sigma_E | \sigma_0) = tr\sigma_E(\log \sigma_E - \log \sigma_0),$$

where $\sigma_0 = \rho \otimes \Lambda^* \rho$. This form of the mutual entropy was introduced in [2] to study optical communication processes.

Fundamental properties of $S^{\mathcal{S}}(\varphi)$ and $I^{\mathcal{S}}(\varphi;\Lambda^*)$:

Theorem 3.1: *When* $\mathcal{A} = B(\mathcal{H})$ *and* $\alpha_t = Ad(U_t)$ *with a unitary operator* U_t, *for any state* φ *given by* $\varphi(\cdot) = tr\rho\,\cdot$ *with a density operator* ρ, *we have the followings:*
(1) $S(\varphi) = -tr\rho \log \rho$.
(2) *If* φ *is an* α*-invariant faithful state and every eigenvalue of* ρ *is nondegenerate, then* $S^I(\varphi) = S(\varphi)$.
(3) *If* $\varphi \in K(\alpha)$, *then* $S^K(\varphi) = 0$.

Two states φ_1 and φ_2 are said to be orthogonal each other (denoted by $\varphi_1 \perp \varphi_2$) if their supports $s(\varphi_1)$ and $s(\varphi_2)$ are orthogonal, where the support $s(\varphi)$ of φ means the smallest projection E satisfying φ(I-E) = 0. The measure $\mu \in M_\varphi(\mathcal{S})$ is said to be orthogonal if $(\int_Q \omega d\mu)\perp(\int_{\Omega/Q} \omega d\mu)$ is satisfied for every Borel set Q in $\mathcal{S}$. A channel is called normal if it sends a normal state to a normal state.

Theorem 3.2: *For a normal C.P. channel* Λ^* *and a normal state* φ, *if a measure* μ *has a discrete support and is orthogonal, then*

$$I_\mu(\varphi;\Lambda^*) = \int_{\mathfrak{S}} S(\Lambda^*\omega|\Lambda^*\varphi)d\mu < S(\varphi) + \varepsilon.$$

Theorem 3.3: *For a density operator* ρ

$$0 \leq I(\rho,\Lambda^*) \leq \min\{S(\rho), S(\Lambda^*\rho)\}.$$

In the notatins of Section 1, two complexities are

$$C^{\mathcal{S}}(\varphi) = S^{\mathcal{S}}(\varphi)$$
$$T^{\mathcal{S}}(\varphi;\Lambda^*) = I^{\mathcal{S}}(\varphi;\Lambda^*).$$

Without seriously taking the original meaning of the entropy S (i.e., forgetting the correspondence between each element of φ and that of $\Lambda^*\varphi$), and as far as the complexity is concerned, $C^{\mathcal{S}}(\varphi)$ and $T^{\mathcal{S}}(\varphi;\Lambda^*)$ can be given as follows : For any $\varphi \in \mathcal{S}$ and any decomposition of φ

$$\varphi = \int_{\mathcal{S}} \omega d\mu,$$

the complexities are

$$T^{\mathcal{S}}(\varphi;\Lambda^*) \equiv \int_{\mathfrak{S}} S(\Lambda^*\omega|\Lambda^*\varphi)d\mu$$
$$C^{\mathcal{S}}(\varphi) \equiv T^{\mathcal{S}}(\varphi; id).$$

Watanabe [10] discusses the efficiency of a modulation M by

$$r(M) = \frac{\mathrm{I}(\rho_M;\Lambda^*)}{\mathrm{S}(\rho_M)},$$

where ρ_M is a modulated input state.

Remark: The channel capacity is given by

$$\mathrm{C}^{\mathcal{S}}(\Lambda^*) = \sup\{\mathrm{I}^{\mathcal{S}}(\varphi;\Lambda^*)\,;\,\varphi\in\mathcal{S}\},$$

which is also useful to study the efficiency of channel for communication of information.

§4. Applications of Auantum Channel to Optical Communication Processes

As applications of a mathematical expression of quantum channel given in <6> of §2, we here discuss an attenuation optical communication process and derivation of error probability.

4.1 Attenuation process

4.1.1 Conventional expression:

A quantum system composed of photons is described by the Hamiltonian $H = a^*a+1/2$, where a^* and a are creation and annihilation operators of photon, respectively. The Schrödinger equation $Hx(q) = Ex(q)$ is easily solved, whose eigenvalue is $E_n = \mathrm{n}+1/2$ ($\mathrm{n}\geq 0$) and the eigenvector $x_n(\mathrm{q})$ for E_n is $(1/(\pi^{1/2}\mathrm{n}!)^{1/2})H_n(\sqrt{2q})\exp(-q^2/2)$, where $H_n(q)$ is the n-th Hermite function. The Hilbert space of this system is a certain closed linear span of linear combinations of $x_n(q)$ (n=0,1,2 ...).

The model is considered as follows: When n_1 photons are transmitted from the input system, m_1 photons from the noise system add to the signal. Then m_2 photons are lost to the loss system through the channel, and n_2 photons are detected in the output system. In this model, the Hilbert spaces are denoted by $\mathcal{H}_1$, $\mathcal{H}_2$, $\mathcal{K}_1$, $\mathcal{K}_2$ and their coordinates represented wave functions are respectively denoted by

$$x^{(1)}_{n_1}(q_1) = \{2^{n_1}\pi(n_1!)\}^{-\frac{1}{2}}\exp\{\frac{-q_1^2}{2}\}H_{n_1}(q_1),$$

$$y^{(1)}_{m_1}(t_1) = \{2^{m_1}\pi(m_1!)\}^{-\frac{1}{2}}\exp\{\frac{-t_1^2}{2}\}H_{m_1}(t_1),$$

$$x^{(2)}_{n_2}(q_2) = \{2^{n_2}\pi(n_2!)\}^{-\frac{1}{2}}\exp\{\frac{-q_2^2}{2}\}H_{n_2}(q_2),$$

$$y^{(2)}_{m_2}(t_2) = \{2^{m_2}\pi(m_2!)\}^{-\frac{1}{2}}\exp\{\frac{-t_2^2}{2}\}H_{m_2}(t_2).$$

According to the conservation of energy ($n_1+m_1 = n_2+m_2$), we take a following linear transformation among the coordinates q_1, t_1, q_2, t_2 of the input, noise, output and loss systems, respectively:

$$\begin{cases} q_2 = \alpha q_1 + \beta t_1, \\ t_2 = -\beta q_1 + \alpha t_1. \end{cases} \qquad (\alpha^2 + \beta^2 = 1)$$

For simplicity, we put $m_1 = 0$. We define a mappings $U : \mathcal{H}_1 \otimes \mathcal{K}_1 \to \mathcal{H}_2 \otimes \mathcal{K}_2$ and π^* by

$$\begin{aligned} &U(x_n^{(1)} \otimes y_0^{(1)})(q_2, t_2) \\ &= x_n^{(2)} \otimes y_0^{(2)}(\alpha q_2 - \beta t_2, \beta q_2 + \alpha t_2) \\ &= \sum_{j=0}^{n} \sqrt{\frac{n!}{(n-j)!j!}} \alpha^j (-\beta)^{n-j} x_j^{(2)} \otimes y_{n-j}^{(2)}(q_2, t_2). \\ &\pi^* = U(\cdot)U^*. \end{aligned}$$

Thus from the expression Λ^* of quantum channel of <6> in §2, the attenuation channel Λ^* is given by

$$\Lambda^* \rho = tr_{\mathcal{K}_2} U^* (\rho \otimes \nu) U, \tag{4.1}$$

with the noise $\nu = |y_0^{(1)} >< y_0^{(1)}| \in \mathfrak{S}(\mathcal{K}_1)$ due to the "zero point fluctuation" of electromagnetic field ($y_0^{(1)}$ is a vacuum state vector in $\mathcal{K}_1$). Note that we may take $\mathcal{H} = \mathcal{H}_2 = \mathcal{K}_2$.

4.1.2 A simple expression [1]:

The above attenuation process (4.1) can be written by a little simple way.

Let $\mathcal{H} = \mathcal{K} = \Gamma(\mathrm{C})$ (Fock space) and let

$$|\theta> = \exp\Big(-\frac{|\theta|^2}{2}\Big) \sum_n \frac{\theta^n}{\sqrt{n!}} |n>, \ \theta \in C$$

denote the coherent vector, where $|n>$ is the number state. Define a mapping V from $\Gamma(\mathcal{H})$ to $\Gamma(\mathcal{H}) \otimes \Gamma(\mathcal{K})$ as

$$V|\theta> = |\alpha\theta> \otimes |\beta\theta>$$

with $\alpha, \beta \in \mathrm{C}$, $|\alpha|^2 + |\beta|^2 = 1$.

V represents the interaction of the signal with an apparatus or a receiver and it means that by the effect of the interaction a coherent signal (beam) $|\theta>$ splits into two signals (beams) still coherent but of lower intensity although the total intensity (energy) is preserved by the transformation.

Now, let us show the equivalence of the above operator V and the operator U in the conventional expression.

$$\begin{aligned} &V|\theta> = |\alpha\theta> \otimes |\beta\theta> \\ &= \exp(-\frac{|\alpha\theta|^2}{2}) \sum_n \frac{(\alpha\theta)^n}{\sqrt{n!}} |n> \exp(-\frac{|\beta\theta|^2}{2}) \sum_m \frac{(\beta\theta)^m}{\sqrt{m!}} |m> \\ &= \exp(-\frac{|\theta|^2}{2}) \sum_n \sum_m \frac{(\alpha\theta)^n (\beta\theta)^m}{\sqrt{n!m!}} |n> \otimes |m> \\ &= \exp(-\frac{|\theta|^2}{2}) \{\sum_{N=0}^{\infty} \frac{\theta^N}{\sqrt{N!}}\} \{\sum_{n=0}^{\infty} \alpha^n \beta^{N-n} \sqrt{\frac{N!}{n!(N-n)!}} |n> \otimes |N-n>\}, \end{aligned}$$

which implies, for any nonnegative integer N,

$$V|N>=\sum_{n=0}^{N}\alpha^{n}\beta^{N-n}\sqrt{\frac{N!}{n!(N-n)!}}|n>\otimes|N-n>.$$

Thus U equals to V by replacing β with $-\beta$.

Therefore the attenuation channel can be written as

$$\Lambda^{*}\rho=tr_{\mathcal{K}}V\rho V^{*}.$$

4.2 Error probability

Let ξ_i be the quantum code corresponding to a symbol $c_i \in \mathcal{C}$: For simplicity, take

$$\mathcal{C}=\{0,1\}\Longleftrightarrow\Xi=\{\xi_0,\xi_1\}.$$

One expression of quantum code is due to a state of photon; for instance, ξ_0 is the vacuum state and ξ_1 is another state such as a coherent or a squeesed state.

Two states (quantum mechanical codes) $\xi_0^{(1)}$ and $\xi_1^{(1)}$ in the input system are transmitted to the output system through a channel Λ^*. Consider a Z-type signal transmission, namely, the signal "0" represented by the state $\xi_0^{(1)}$ goes always to "0" represented by $\xi_0^{(2)}$ and the signal "1" represented by the state $\xi_1^{(1)}$ goes to $\xi_0^{(2)}$ or other states. Then the error probability q_e comes from that the signal "1" is recognized as the signal "0", so that it is given by

$$\begin{aligned}q_e&=tr\Lambda^{*}(\xi_1^{(1)})\xi_0^{(2)}\\&=tr_{\mathcal{H}_2}(tr_{\mathcal{K}_2}\pi^{*}(\xi_1^{(1)}\otimes\nu))\xi_0^{(2)}\end{aligned}$$

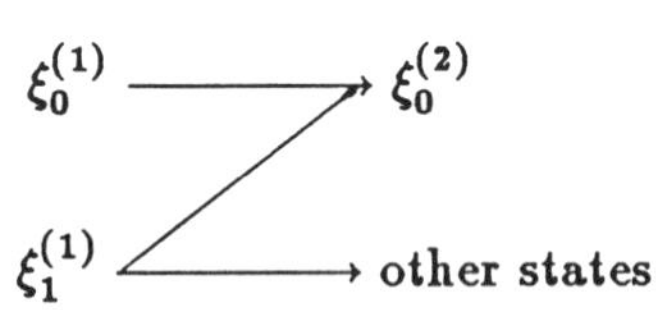

Based on this error probability q_e, the error probability of PCM modulation with the t_0-tuple error correcting code with the weight N and that for PPM modulation are given by

$$\begin{aligned}P_e^{PCM}&=\sum_{j=t_0+1}^{N}{}_{N}C_j q_e^j(1-q_e)^{N-j},\\P_e^{PPM}&=q_e,\end{aligned}$$

where ${}_NC_j=N!/\{(N-j)!j!\}$.

Concrete computation and physical discussion of error probability for some optical processes are given in the paper [9] of this volume, by which we obtain an interesting observation. A certain input squeezed state gives us a better error probability than the input coherent state; that is, the error probability very much depends on the way of squeezing the coherent state.

References: I here simply give references of mine and see [2] and [6] for a complete reference.

[1] L. Accardi and M. Ohya : "Compoud channels, transition expectations and liftings", preprint.

[2] M. Ohya : "On compound state and mutual information in quantum information theory", IEEE.Trans.Inf.Theory, 29, pp.770-774 (1983).

[3] _______ : "Note on quantum probability", L. Nuovo Cimento, 38,pp.402-404 (1983).

[4] _______ : "Entropy transmission in C*-dynamical systems", J. Math. Anal. Appl., 100, pp.222-235 (1984).

[5] M. Ohya, H. Yoshimi and O. Hirota : "Rigorous derivation of error probability in quantum control communication processes", IEICE of Japan, J71-B, No.4, 533-539 (1988).

[6] M. Ohya : "Some aspects of quantum information theory and their applications to irreversible processes", Rep. on Math. Phys., 27, pp.19-47 (1989).

[7] _______ : "Information theoretical treatment of genes", Trans. IEICE, E70, No.5, pp.556-560 (1989).

[8] _______ : "Fractal dimensions of states", to appear in Quantum Probability and Applications (edited by L. Accardi and W. von Waldenfels), KLUWER Publishing Company.

[9] M. Ohya and H. Suyari : "Rigorous derivation of error probability in coherent optical communication", in this volume.

[10] N. Watanabe : "Efficiency of optical modulations with coherent state", in this volume.

556

THE TRANSACTIONS OF THE IEICE, VOL. E **72**, NO. **5** MAY **1989**

LETTER <Special Issue on Information Theory and Its Applications>

Information Theoretical Treatments of Genes

Masanori OHYA†, *Member*

SUMMARY Some concepts in information theory are tried to apply to the study of genes. The mutual entropy is used to define a measure indicating the similarity between two genetic sequences. The alignment of sequences is briefly discussed. Some phylogenetic trees are written by using the entropy measure. According to this results, usefulness of information theory is discussed in the study of genes such as molecular evolution.

1. Introduction

Recently it becomes possible to study the biological evolution from genes, more precisely, from information carried by DNA. For rather long time, the evolution has been studied through the forms of several species existing now and of fossils found in stratums. When the evolutionary process is discussed, the form of organism is considered to be an important clue, but there are several different interpretations of the form. Thus nowadays, DNA or amino acid sequences are often used to establish more objective interpretation of biological evolution. A genetic information preserved on DNA is regarded as a message made by a sequence of four bases (adenine (A), guanine (G), thymine (T) and cytosine (C)). This information is transmitted to RNA and is used to make twenty amino acids and proteins. If we regard biological replication or multiplication as a communication of the message carried by DNA, then information theory is applicable in analysis of base sequences (DNA) and amino acid sequences (proteins).

This letter proceeds as follows: In the first place, we briefly mention how to compare genetic sequences. Namely, we explain the idea of the alignment of sequences for this purpose. Secondly, we construct complete event systems for genetic sequences. Thirdly, a measure indicating the similarity of two sequences is formulated by using the mutual entropy (information). Then the UPG method writing phylogenetic trees is briefly reviewed. Finally, we write phylogenetic trees for biological evolution by using the genetic matrix obtained through our measure of similarity, and we compare our results with those derived by other kinds of genetic matrices.

Manuscript received December 5, 1988.
Manuscript revised February 20, 1989.
† The author is with the Faculty of Science and Technology, Science University of Tokyo, Noda-shi, 278 Japan.

2. DNA Sequences and Amino Acid Sequences

We briefly review fundamental facts of DNA and animo acids for self-consistency of this letter.

Self-replication and multiplication of a living system are caused by a gene having the information of the form and functions of the system. In 1944, Avery, MacLeod and McCarty found that a gene is a part of DNA itself. In 1953, by the X-ray structure analysis, Watson and Crick showed that the shape of DNA is a double helix. The main chain of DNA consists of a regular sequence of alternating deoxyribose-phosphate. Each of four bases, (A), (G), (T), (C), joins with a part of deoxyribose. The bases on two main chains are joined according to the constraint that (A) only bonds to (T) and (C) only bonds to (G), which is called the Watson-Crick base-pairing rule. The character of a protein is determined by three of four bases. The mechanism for the production of a protein from DNA is the following: The information stored in the base sequence of a main chain of DNA is copied by messenger-RNA (mRNA). The four bases in DNA are transformed to the following four bases on mRNA; (A), (G), (C), (U) (uracil corresponding to thymine (T) in DNA), respectively. The amino acid sequence of RNA is translated into a protein in ribosome. A triplet of mRNA bases, called a codon, specifies one of twenty amino acids. A protein is synthesized by a combination of several amino acids starting from the codon named *f*-methionine to a stop codon.

Here we list the names and symbols of twenty amino acids for the later use: alanine (A), cysteine (C), aspartic acid (D), glutamic acid (E), phenylalanine (F), glutamine (G), histidine (H), isoleucine (I), lysine (K), leucine (L), methionine (M), asparagine (N), proline (P), glutamine (Q), arginine (R), serine (S), threonine (T), valine (V), tryptophan (W) and tyrosine (Y).

3. Alignment of Genetic Sequences

Let $\mathscr{A}$ and $\mathscr{B}$ be amino acid sequences two organisms specifying an identical protein. These sequences are generally considered to be close each other, but there might exist some difference betweeen them because of the biological evolution. For instance, suppose that these sequences are given as

$\mathcal{A}$: A C D A C D E

$\mathcal{B}$: A E D E A C D,

where each alphabet A, C, D, ⋯ represents each amino acid existing in the sequences. The above two amino acid sequences look not so close each other, whose difference might come from the fact that some amino acids in $\mathcal{A}$ (resp. $\mathcal{B}$) change, delete or insert in $\mathcal{B}$ (resp. $\mathcal{A}$) during the course of the biological evolution. Therefore we have to align two sequences by taking account of this fact. When an amino acid in $\mathcal{A}$ (resp. $\mathcal{B}$) is considered to be lacking in $\mathcal{B}$ (resp. $\mathcal{A}$), we insert the gap (dummy) "$*$" in the corresponding place of $\mathcal{B}$ (resp. $\mathcal{A}$), and when an amino acid, say A, in $\mathcal{A}$ is considered to change to another amino acid, say B, in $\mathcal{B}$, we correspond A with B. This arrangement makes us possible to take the matching (alignment) of two amino acid sequences, whose result will be

$\mathcal{A}$: A C D * A C D E

$\mathcal{B}$: A E D E A C D *.

After this alignment, we can see the similarity of two sequences $\mathcal{A}$ and $\mathcal{B}$ as expected. The alignment can been done by using a computer on the basis of mathematical formulation of the distance between two sequences[(1)-(3)]. The fundamental points of such a mathematical formulation are to define the distance $d(X, Y)$ of two amino acids $X \in \mathcal{A}$, $Y \in \mathcal{B}$ and to minimize the total distance between $\mathcal{A}$ and $\mathcal{B}$ such that $d_{total} = \Sigma\{d(X, Y);\ X \in \mathcal{A}, Y \in \mathcal{B}\}$.

For instance, the alignment of the amino acid sequences of Hemoglobin α for human and carp becomes

```
Human : V L S P A D K T N V K A A W G K V G A
Carp  : S L S D K D K A A V K I A W A K I S P

        H A G E Y G A E A L E R M F L S F P T
        K A D D I G A E A L G R M L T V Y P Q

        T K T Y F P H * F D L S H G S A Q V K
        T K T Y F A H W A D L S P G S G P V K

        G H G K K V * A D A L T N A V A H V D
        * H G K K V I M G A V G D A V S K I D

        D M P N A L S A L S D L H A H K L R V
        D L V G G L A S L S E L H A S K L R V

        D P V N F K L L S H C L L V T L A A H
        D P A N F K I L A N H I V V G I M F Y

        L P A E F T P A V H A S L D K F L A S
        L P G D F P P E V H M S V D K F F Q N

        V S T V L T S K Y R
        L A L A L S E K Y R
```

4. Event Systems and Entropy Ratio

For a set $\mathcal{A} = \{A_1, A_2, \cdots, A_n\}$ and the occurrence probability $p = \{p_1, p_2, \cdots, p_n\}$ of each event A_k (i. e. $p_k = p(A_k)$, $\Sigma_k p_k = 1$), a pair $(\mathcal{A}, p)$ is called a complete event system. For two complete event systems $(\mathcal{A}, p)$ and $(\mathcal{B}, q)$ the compound event system is denoted by $(\mathcal{A}\mathcal{B}, r)$ where $\mathcal{A}\mathcal{B} = \{(A, B;\ A \in \mathcal{A},\ B \in \mathcal{B}\}$ and $r_{ij} = p(A_i, B_j)$ with $\Sigma_i r_{ij} = q_j$, $\Sigma_j r_{ij} = p_i$.

Shannon introduced several information measures to study the communication of information[(9)]. One of them is the information (entropy) carried by a system $(\mathcal{A}, p)$, which is given by

$$S(\mathcal{A}) = -\Sigma_i p_i \log p_i$$

The most fundamental information measure in Shannon's communication theory is the mutual entropy expressing the amount of information correctly transmitted from an input system $(\mathcal{A}, p)$ to an output system $(\mathcal{B}, q)$, which is described as

$$I(\mathcal{A}, \mathcal{B}) = \sum_{i,j} r_{ij} \log \frac{r_{ij}}{p_i q_j}$$

In order to apply these two information measures to the study of amino acid (or DNA) sequences, we have to set the complete event systems of amino acid (DNA) sequences. For an amino acid (DNA) sequence $\mathcal{A}$, the complete event system associated to $\mathcal{A}$ is nothing but $(\mathcal{A}, p)$ with the occurrence probability $p_k = p(A_k)$ for each event A_k in $\mathcal{A}$, where A_k represents each amino acid (base).

When we consider two amino acid sequences $\mathcal{A}$ and $\mathcal{B}$, it is not so easy to set a proper compound event system of two sequences. However, once we know the correspondence between $\mathcal{A}$ and $\mathcal{B}$, we can construct the compound event system $(\mathcal{A}\mathcal{B}, r)$, so that the information transmitted from $(\mathcal{A}, p)$ to $(\mathcal{B}, q)$ can be calculated. Remark that the event system $(\mathcal{A}, p)$ here contains "$*$" as an event. Indeed

$$\begin{pmatrix} \mathcal{A} \\ p \end{pmatrix} = \begin{pmatrix} *, & A, & C, & \cdots, & Y \\ p_0, & p_1, & p_2, & \cdots, & p_{20} \end{pmatrix}$$

$$\begin{pmatrix} \mathcal{B} \\ q \end{pmatrix} = \begin{pmatrix} *, & A, & C, & \cdots, & Y \\ q_0, & q_1, & q_2, & \cdots, & q_{20} \end{pmatrix}$$

$$\begin{pmatrix} \mathcal{A}\mathcal{B} \\ r \end{pmatrix} = \begin{pmatrix} (*, *), & (*, A), & (*, C), & \cdots, & (Y, Y) \\ r_{0,0}, & r_{0,1}, & r_{0,2}, & \cdots, & r_{20,20} \end{pmatrix}$$

By constructing these event systems, a measure indicating the difference between two amino acid sequences can be introduced. This measure is called the entropy ratio[(4)] defined by

$$r(\mathcal{B}|\mathcal{A}) = \frac{I(\mathcal{A}, \mathcal{B})}{S(\mathcal{A})}$$

This quantity is regarded as the ratio of the information transmitted from $\mathcal{A}$ into $\mathcal{B}$ to the information carried by $\mathcal{A}$. By symmetrizing the entropy ratio, we here introduce a more suitable measure

$$r(\mathcal{A},\mathcal{B})=\frac{1}{2}\{r(\mathcal{B}|\mathcal{A})+r(\mathcal{A}|\mathcal{B})\},$$

which may be called the symmetrized entropy ratio or the evolution entropy rate.

5. A Method Constructing Phylogenetic Trees

In this section, we briefly review the UPG (unweighted pair group clustering) method writing a phylogenetic tree. This method is initiated by P. H. A. Sneath and R. R. Sokai and by Nei[5] it is now understood as a way to divide organisms into several groups. The pair having the smallest difference makes the first group. Then we try to find a next group (pair or triple) giving the second smallest difference calculated for any pair out of organisms and the first group. Moreover we consider the difference between two groups, that is, the averaged difference of all pairs of organisms in two groups. We repeat this procedure and make a final relation among all organisms. Let us show this procedure by an example.

Let the difference between an organism $\mathcal{A}$ and an organism $\mathcal{B}$ be denoted by $\rho(\mathcal{A},\mathcal{B})$, which is an element of the properly defined genetic (distance) matrix given in advance. The averaged difference between an organism $\mathcal{A}$ and a group $(\mathcal{B},\mathcal{C})$ is denoted by $\rho(\mathcal{A},(\mathcal{B},\mathcal{C}))$, and it is computed by

$$\rho(\mathcal{A},(\mathcal{B},\mathcal{C}))=\frac{\rho(\mathcal{A},\mathcal{B})+\rho(\mathcal{A},\mathcal{C})}{2}$$

The averaged difference between a group $(\mathcal{A},(\mathcal{B},\mathcal{C}))$ and a group $(\mathcal{D},\mathcal{E})$ is computed as

$$\rho((\mathcal{A},(\mathcal{B},\mathcal{C})),(\mathcal{D},\mathcal{E}))=$$

$$\frac{\rho(\mathcal{A},\mathcal{D})+\rho(\mathcal{A},\mathcal{E})+\rho(\mathcal{B},\mathcal{D})+\rho(\mathcal{B},\mathcal{E})+\rho(\mathcal{C},\mathcal{D})+\rho(\mathcal{C},\mathcal{E}))}{6}$$

Now, suppose that the genetic distance matrix is given as in Table 1. Then an organism $\mathcal{A}$ and an organism $\mathcal{B}$ are first combined together because the difference between $\mathcal{A}$ and $\mathcal{B}$ is the smallest. Secondly, we compute the differeces for a group $(\mathcal{A},\mathcal{B})$ and one of three organisms $\mathcal{C},\mathcal{D},\mathcal{E}$:

$$\rho((\mathcal{A},\mathcal{B}),\mathcal{C})=\frac{\rho(\mathcal{A},\mathcal{C})+\rho(\mathcal{B},\mathcal{C})}{2}=\frac{5+6}{2}=5.5,$$

$$\rho((\cdot\,,\mathcal{B}),\mathcal{D})=\frac{\rho(\mathcal{A},\mathcal{D})+\rho(\mathcal{B},\mathcal{D})}{2}=\frac{7+8}{2}=7.5,$$

$$\rho((\mathcal{A},\mathcal{B}),\mathcal{E})=\frac{\rho(\mathcal{A},\mathcal{E})+\rho(\mathcal{B},\mathcal{E})}{2}=\frac{8+8}{2}=8,$$

$(\rho(\mathcal{C},\mathcal{D})=9$, $\rho(\mathcal{C},\mathcal{E})=10$ and $\rho(\mathcal{D},\mathcal{E})=7$.)

Table 1 Genetic distance matrix.

	$\mathcal{A}$	$\mathcal{B}$	$\mathcal{C}$	$\mathcal{D}$	$\mathcal{E}$
$\mathcal{A}$	0	3	5	7	8
$\mathcal{B}$	3	0	6	8	8
$\mathcal{C}$	5	6	0	9	10
$\mathcal{D}$	7	8	9	0	7
$\mathcal{E}$	8	8	10	7	0

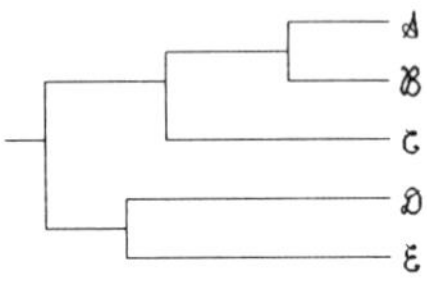

Fig. 1 Example of phylogenetic tree.

Since the pair having the smallest difference forms a group, the group $(\mathcal{A},\mathcal{B})$ and the organism $\mathcal{C}$ are combined.

We next compute the following three differences:

$$\rho(((\mathcal{A},\mathcal{B}),\mathcal{C}),\mathcal{D})=\frac{\rho(\mathcal{A},\mathcal{D})+\rho(\mathcal{B},\mathcal{D})+\rho(\mathcal{C},\mathcal{D})}{3}$$

$$=\frac{7+8+9}{3}=8,$$

$$\rho(((\mathcal{A},\mathcal{B}),\mathcal{C},\mathcal{E})=\frac{\rho(\mathcal{A},\mathcal{E})+\rho(\mathcal{B},\mathcal{E})+\rho(\mathcal{C},\mathcal{E})}{3}$$

$$=\frac{8+8+10}{3}=8.3,$$

$(\rho(\mathcal{D},\mathcal{E})=7)$.

Accordingly, we have a group $(\mathcal{D},\mathcal{E})$. Finally,

$$\rho(((\mathcal{A},\mathcal{B}),\mathcal{C}),(\mathcal{D},\mathcal{E}))=8.3.$$

On the basis of the above results, we can write a phylogenetic tree of these organisms as Fig. 1.

In the next section, we write phylogenetic trees by using this UPG method with the genetic matrix constructed from our symmetrized entropy ratio $r(\mathcal{A},\mathcal{B})$.

6. Phylogenetic Tree by Hemoglobin α

In this section, we write phylogenetic trees for hemoglobin α. We here consider the following species: Monodelphia (human, horse), Marsupialia (gray kangaroo), Monotremata (platypus), Aves (ducks, greylag goose), Crocodilia (alligator, nile crocodile), viper, bullfrog tadpole, Osteinthyes (carp, goldfish), port jackson shark. All data are taken from[6]. Let the degree of difference between two organisms $\mathcal{A}$ and $\mathcal{B}$ be given as

$$\rho_1(\mathcal{A},\mathcal{B})=1-r(\mathcal{A},\mathcal{B})$$

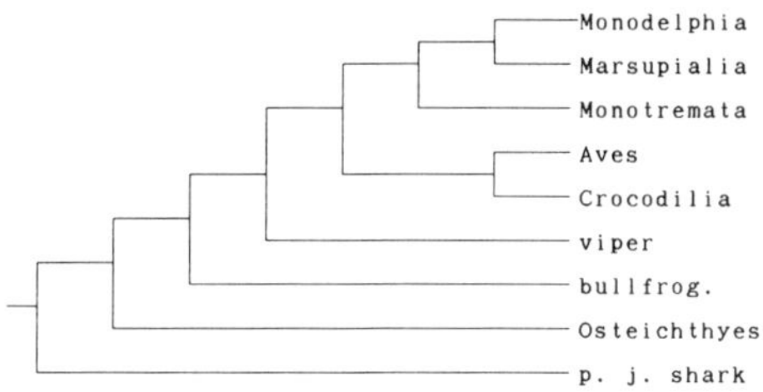

Fig. 2 Tree constructed by entropy ratio.

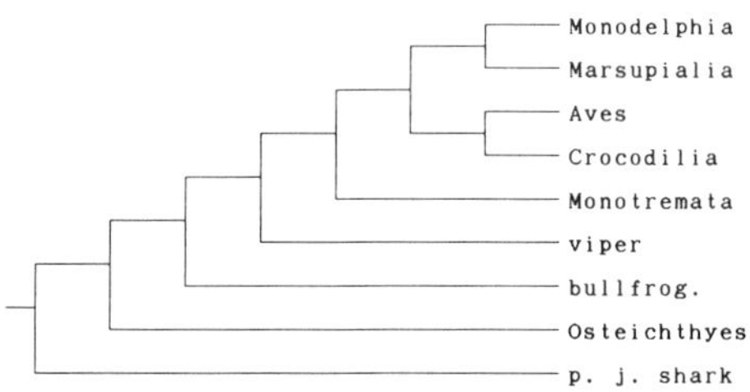

Fig. 3 Tree constructed by substitution rate.

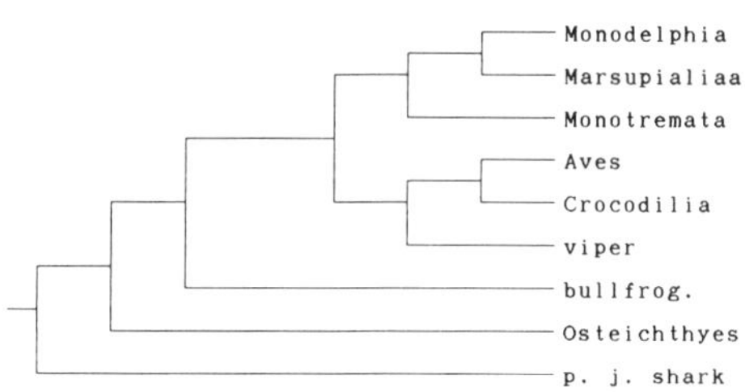

Fig. 4 Tree constructed by fossils.

or

$$\rho_2(\mathscr{A}, \mathscr{B}) = \frac{n}{N},$$

where n is the number of replacing amino acids between the sequences $\mathscr{A}$ and $\mathscr{B}$ and N is the number of amino acids in $\mathscr{A}$ or $\mathscr{B}$ both after the alignment. Then we can make the genetic matrix ρ_k from $\rho_k(\mathscr{A}, \mathscr{B})(k=1, 2)$ such that $\rho_k = (\rho_k(\mathscr{A}, \mathscr{B}))$, by which we can construct the phylogenetic tree for the above species. The difference ρ_1 is new, but the difference ρ_2 (or its modification) is one often used in several occasions.

The phylogenetic trees written by ρ_1 and ρ_2 are shown in Fig. 2 and Fig. 3, respectively.

The Fig. 4 is a result estimated from the fossils of species. At first glance, Fig. 2 is closer to Fig. 4 than Fig. 3.

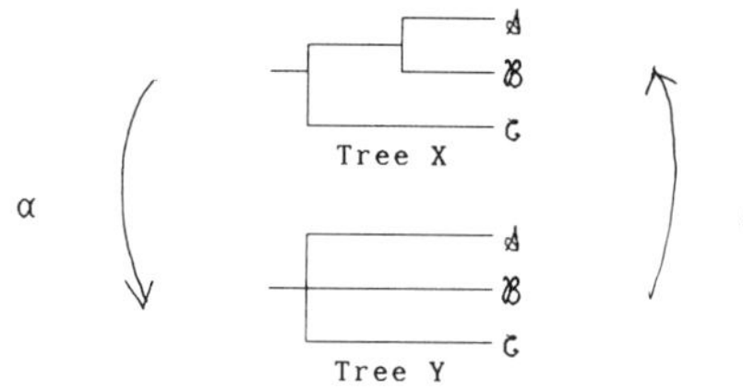

Fig. 5 RF operations.

Table 2 RF-distance between trees.

	Fig.1	Fig.2	Fig.3
Fig.1		2	2
Fig.2	2		4
Fig.3	2	4	

For more scientific judgement among the resulted phylogenetic trees, Robinson and Foulds considered[7] a certain operation which expresses the movement of a branch between two phylogenetic trees. For two phylogenetic trees, say X and Y, if we can overlap X with Y by moving n branches in X, then the difference between X and Y is said to be n. There are two such operations α and β : α is the operation adding a branch in a tree and β is that eliminating a branch, as shown in Fig.5 for an example.

The difference of three phylogenetic trees Fig. 2, Fig. 3, and Fig. 4 are shown in the Table 2. Therefore if we believe the phylogenetic tree written by the data of fossils and if the UPG method is a plausible way to write the phylogenetic tree, then the genetic matrix ρ_1 constructed by the entropy ratio will be better than the genetic matrix ρ_2. Even so, we have a little difference between Fig. 2 and Fig. 4, so that we might need to refine both UPG method and genetic matrix in order to write more accurate phylogenetic trees, about which we are now on the working bench.

7. Phylogenetic Tree by Cytochrome C

A phylogenetic tree written by using Cytochrome C is shown in Fig. 6. The genetic matrix for this phylogenetic tree is due to the distance ρ_1. We here consider the following species: Vertebrate (human, horse), Invertebrate (locust, garden snail), Higher plats (wheat, ginkgo biloba), Algae (enteromorpha intestinalis), Fungi (yeast, debaryomyces kloeckeri), Protozoa (euglena gracilis, tetrahymena pyriformis). Cytochrome C is suited for estimating phyletic lines among organisms being far from each other because the substi-

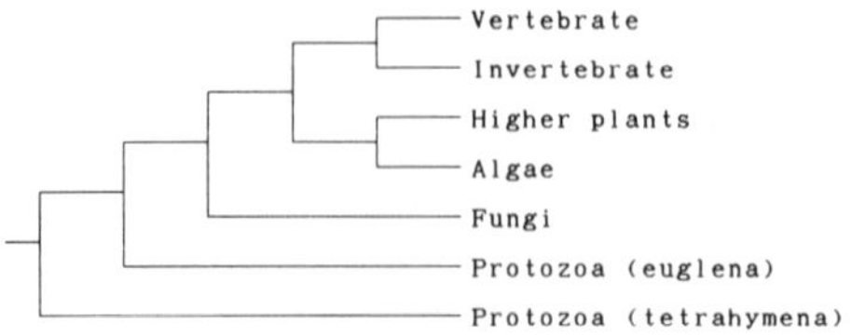

Fig. 6 Tree constructed by entropy ratio with C.

tution of amino acids in Cytochrome C may take a longer time than that in other protein. This phylogenetic tree well matches to some known results, for example, by Whittaker's[(8)].

8. Consequences

In this letter, we try to show that the information theoretical treatments will be important for the study of genes such as molecular evolution. We introduced the symmetrized entropy ratio as an application of the mutual entropy, and we wrote phylogenetic trees of some species by the following genetic matrix (GM for short) with the UPG method: (1) GM constructed from the symmetrized entropy ratio for hemoglobin α; (2) GM constructed by a conventional method, namely, the substitution rate of amino acids, for hemoglobin α; (3) (1) for cytochrome C. We compared the phylogenetic trees written by the above methods with that by the fossils of the same species. As a consequence based on the RF-criterion, the tree by (1) is better than that by (2). Namely, the information theoretical treatment gives us better description for biological evolution than the conventional treatment with the substitution rate. Moreover, the phylogenetic tree (Fig. 6) for cytochrome C by our measure turns out to be well-matched to an experimental result as explained in Sect. 7. In the case when the fossils of some organisms have not been found, we can use the entropy ratio and the UPG method to construct the phylogenetic tree for these organisms. Therefore we can conclude that the information theoretical approach to genetics might give us a clue to understand some aspects of biological evolution.

Further development of the alignment and more detail examinations for phylogenetic trees with an extension of the UPG method will be discussed elsewhere.

References

(1) S. B. Needleman and C. D. Wunsch: "A general method applicable to search for similarities in the amino acid sequences of two proteins", J. Mol. Biol., **48**, 443-453 (1970).

(2) P. H. Sellers: "On the theory and computation of evolutionary distances", SIAM J. Math., **26**, pp. 787-793 (1974).

(3) M. Ohya and Y. Uesaka: "Amino acid sequences and DP matching", Amino Acid Sequences and DP Matching", Sci. Univ. of Tokyo, Res. Rep. (1986).

(4) M. Ohya and Y. Kitagawa: "A mathematical analysis of DNA sequences", Symp. Appl. Fuct. Anal., **8**, pp. 36-47 (1985).

(5) P. H. A. Sneath and R. R. Sokal: "Numerical taxonomy", WH Freeman, San Francisco (1973).

(6) W. C. Barker , et al.: "Protein sequence database of the protein identification resource (PIR)", NBRF (1985).

(7) D. F. Robinson and L. R. Foulds: "Comparison of phylogenetic trees", Math. Biosci., **53**, pp. 131-147 (1981).

(8) L. Margulis and K. V. Schwartz: "Five Kingdoms", W. H. Freeman (1982).

(9) H. Umegaki and M. Ohya: "Entropies in Probablistic Systems", Kyoritsu Pub. Company (1983).

Vol. 27 (1989) *REPORTS ON MATHEMATICAL PHYSICS* No. 2

SOME ASPECTS OF QUANTUM INFORMATION THEORY AND THEIR APPLICATIONS TO IRREVERSIBLE PROCESSES*

MASANORI OHYA

Department of Information Sciences,
Science University of Tokyo, Noda City, Chiba 278, Japan

(Received October 12, 1986 — Revised May 16, 1988)

Several quantum entropies are systematically studied and the mathematical structure of a channel in optical communication processes is presented. As applications of these entropies and channel, general formulas of error probability in some communication processes using, for instance, coherent or squeezed states, are obtained and the irreversibility for some dynamical processes is discussed.

Introduction

The notion of entropy was introduced by Clausius around 1865 in order to discuss the thermal behaviour of physical systems on the basis of Carnot's work. Since then, the irreversibility of physical systems such as the second law of thermodynamics has been understood in terms of the entropy increase. It was Boltzmann who first tried to explain the entropy increase from the microscopic dynamics, that is, from the dynamics of large numbers of atoms. As every fundamental equation of motion such as Newton's equation, Schrödinger equation or Liouville equation, is invariant under the time reflection, it is almost impossible to show the entropy increase, hence to explain the irreversibility, by a direct application of such a fundamental equation of motion.

In this paper we consider quantum mechanical systems but our formulation mathematically contains any classical system as a special case. A quantum physical system is usually described by a density operator ϱ, and the entropy for the state ϱ is given by

$$S(\varrho) = -\mathrm{tr}\,\varrho\log\varrho$$

according to von Neumann [1]. For a Hamiltonian dynamics, the state ϱ changes in time due to the unitary time evolution U_t generated by the Hamiltonian H of the

* An invited review paper.

system: $U_t = \exp(itH)$ and $\varrho_t = U_t \cdot \varrho U_t$, so that we have

$$S(\varrho_t) = S(\varrho),$$

that is, the entropy is invariant under the time evolution of the system. Therefore, in order to explain the irreversible phenomena rigorously, we have to (1) modify the fundamental equation of motion in quantum mechanics, for instance, by adding some external effects (noise, fluctuation, etc.) to the reversible equation, (2) introduce new concept or criteria, besides the entropy, interpreting the irreversibility, or (3) construct a new theory containing quantum mechanics as a special case. There are many trials along the line of (1), but most of such trials are not so satisfactory, namely, some modifications do not involve the entropy increase in themselves and others are not mathematically well controlled. As for (2), there are a few different directions introducing new criteria, one of which is to develop von Neumann's quantum mechanical entropy and formulate the so-called quantum information theory along the ideas of Shannon [2], in which we might be able to find a useful expression for an irreversible process. Apart from comprehension of the irreversibility, rigorous formulation of quantum information (communication) theory is very important from both mathematical and physical points of view because of the following reasons: (i) Optical communication of information is an indispensable technique today, and photon is a typical object of quantum mechanics, so that we like to have a rigorous theory describing quantum communication processes. Concerning this in terms of quantum control theory and quantum statistics, there have been several interesting investigations initiated by Helstrom, Liu, Gordon, Holevo and others [68, 69, 72, 73, 74, 75, 76, 77], although I will not review them here but present them as the Part II of this paper on some other occasion. (ii) An aspect of Shannon's information theory is the theory of entropy, so the formulation of quantum information theory agrees to some extent with the development of the theory of quantum mechanical entropy initiated by von Neumann. And such a development will be useful in physics, for example, to study some cooperative behaviours [3, 4] of physical systems.

In this paper, we review and systematically reformulate, with a few new concepts and extensions, some of *our works* concerning (1) and (2) above. More precisely, the following topics are considered here:

⟨1⟩ Formulation of several entropies in general quantum systems.

⟨2⟩ Mathematical construction of a communication channel and its application to optical communication processes.

⟨3⟩ Reconsideration of irreversibility with some entropies.

§1. Preliminaries

In this section, we briefly review the gist of information theory in classical (commutative) dynamical systems (CDS for short) and fix the notations used throughout this paper both for CDS and quantum (noncommutative) dynamical systems (QDS for short).

Once we make clear all mechanisms of the state change, we can know almost all properties of a physical system because each aspect of the physical system is described by a certain state. One of the most general descriptions of the state change for CDS is suggested in communication theory of Shannon and its measure theoretic extension by Kullback, Leibler, Kolmogorov, Gelfand, Yaglom and others [5, 6, 7]. Therefore it is interesting to study the state change for QDS following Shannon's philosophy, which is a motivation of the present work.

Every CDS is generally described by measure theoretic terminology, that is, a state in CDS is expressed by a probability measure μ on a measurable space $(\Omega, \mathscr{F})$ and an observable of the system corresponds to a real random variable on Ω. The state change in CDS is described by a mapping from $P(\Omega)$, the set of all probability measures on Ω, into itself (or $P(\bar{\Omega})$ on another measurable space $(\bar{\Omega}, \bar{\mathscr{F}})$). Generally, the state change in CDS is given by a mapping between two dynamical systems, namely, an input system $(\Omega, \mathscr{F}, P(\Omega))$ and an output system $(\bar{\Omega}, \bar{\mathscr{F}}, P(\bar{\Omega}))$. The following linear mapping Λ^* from $P(\Omega)$ to $P(\bar{\Omega})$ is important and called a *channel* (or channeling transformation):

$$\bar{\varphi}(Q) = \Lambda^* \varphi(Q) = \int_\Omega \lambda(\omega, Q) d\varphi(\omega), \qquad \varphi \in P(\Omega), \tag{1.1}$$

where λ is a mapping from $\Omega \times \bar{\mathscr{F}}$ to $[0, 1]$ satisfying the following conditions: (i) $\lambda(\cdot, Q)$ is a measurable function on Ω for each $Q \in \bar{\mathscr{F}}$ and (ii) $\lambda(\omega, \cdot) \in P(\bar{\Omega})$ for each $\omega \in \Omega$. This λ is often called the Markov kernel on $\Omega \times \bar{\mathscr{F}}$. It was shown by Umegaki [8, 9] that for the above λ, there exists a unique bounded linear map Λ from $B(\bar{\Omega})$, the set of all bounded measurable functions on $\bar{\Omega}$, to $B(\Omega)$ satisfying (i) $f \geqslant 0 \rightarrow \Lambda(f) \geqslant 0$; (ii) $f_n \downarrow 0 \rightarrow \Lambda(f_n) \downarrow 0$; (iii) $\bar{\varphi}(f)$ $(\equiv \int_{\bar{\Omega}} f d\bar{\varphi}) = \int_\Omega \Lambda(f) d\varphi$.

Shannon's definition of the entropy of a state (probability distribution) $p = \{p_k\}$ is

$$S(p) = -\sum_k p_k \log p_k. \tag{1.2}$$

When Ω is a discrete set, say $\Omega = \{\omega_1, \omega_2, \omega_3, \ldots, \omega_n\}$, the occurrence probability of an event $\omega \in \Omega$ is denoted by $p_k = p(\omega_k)$ and the probability distribution of Ω is denoted simply by p, namely, $p = \{p_1, p_2, p_3, \ldots, p_n\}$. For two states p, q on Ω, the relative uncertainty between p and q is expressed by

$$S(p|q) = \sum_k p_k \log(p_k/q_k), \tag{1.3}$$

which is called the relative entropy of p and q, and it was extended in a general probability space $(\Omega, \mathscr{F}, P(\Omega))$ by Kullback, Leibler and others. Their definition of the relative entropy is

$$S(\varphi|\psi) = \begin{cases} \int_\Omega \log(d\varphi/d\psi) d\varphi & \text{if } \varphi \ll \psi, \\ +\infty & \text{otherwise} \end{cases} \tag{1.4}$$

for φ and ψ in $P(\Omega)$.

When a state φ of an input system dynamically changes to a state $\bar{\varphi}(\equiv \Lambda^*\varphi)$ of an output system under a channel Λ^*, we ask how much information carried by φ can be transmitted to the output system. It is the mutual entropy (information) that represents this amount of information transmitted from φ to $\bar{\varphi}$, which is defined in terms of a compound state of φ and $\bar{\varphi}$ and the relative entropy. The compound state Φ of φ and $\bar{\varphi}$ is a measure (state) expressing the correlation between φ and $\bar{\varphi}$, and it is given by

$$\Phi(Q_1, Q_2) = \int_{Q_1} \lambda(\omega, Q_2) d\varphi \tag{1.5}$$

for any $Q_1 \in \mathscr{F}$, $Q_2 \in \bar{\mathscr{F}}$. The mutual entropy in CDS is defined by the relative entropy $S(\cdot|\cdot)$ such as

$$I(\varphi; \Lambda^*) = S(\Phi|\Phi_0), \tag{1.6}$$

where Φ_0 is the direct product state (measure) $\varphi \otimes \bar{\varphi}$. In particular, if φ is a probability distribution $p = \{p_k\}$ and the channel Λ^* is a transition probability (p_{ij}), then the compound state Φ is the joint probability distribution: $\Phi = \{p(i, j)\}$ with $p(i, j) = p_{ij} p_j$ and the mutual entropy becomes

$$I(p; \Lambda^*) = \sum_{i,j} p(i, j) \log(p(i, j)/p_j q_i) \tag{1.7}$$

with $q_i = \sum_j p_{ij} p_j$. The following inequality is a fundamental inequality for communication theory.

THEOREM 1.1 (Shannon). $0 \leqslant I(p; \Lambda^*) \leqslant S(p)$.

See [9–12] for the classical information theory.

Around 1950 von Neumann reformulated quantum mechanics on Hilbert space and demonstrated the theory of operator algebras [1] and Haag and Kastler [13] found that the C^*-algebraic method is important for studying some physical systems with infinite degrees of freedom. That is, the study of such a physical system without using a Hilbert space is essential when the system involves, for instance, a kind of symmetry breaking. The operator algebraic method essentially starts from a set $\mathscr{A}$ containing all physical observables of interest and the set $\mathfrak{S}$ (or $\mathfrak{S}(\mathscr{A})$) of all states on $\mathscr{A}$. The most fundamental set $\mathscr{A}$ is a C^*-algebra. A symmetry of a physical system is defined by a * preserving automorphism α of $\mathscr{A}$ and the action of a symmetry group G is a homomorphism α: $G \to \mathrm{Aut}(\mathscr{A})$, the set of all automorphisms of $\mathscr{A}$ satisfying (i) $\alpha_g(\alpha_k(A)) = \alpha_{gk}(A)$ for any $g, k \in G$ and $A \in \mathscr{A}$; (ii) $\alpha_g(\cdot)$ is continuous in some topology. A concrete and important C^*-algebra is a von Neumann algebra although it is defined on a Hilbert space $\mathscr{H}$. A subset $\mathfrak{N}$ of $B(\mathscr{H})$ is said to be a von Neumann algebra acting on a Hilbert space $\mathscr{H}$ if $\mathfrak{N}'' = \mathfrak{N}$, where $\mathfrak{N}'' = (\mathfrak{N}')'$ with $\mathfrak{N}' = \{A \in B(\mathscr{H}); AB = BA, B \in \mathfrak{N}\}$.

Before closing this section, we show the correspondences between CDS and QDS in Table 1.

Table 1

	CDS	QDS
Observable	real random variable on measurable space $(\Omega, \mathscr{F})$	self-adjoint element of $\mathscr{A}$ (C^*-alg.) or $\mathfrak{N}$ (v. N. alg.)
State	probability measure $\varphi \in P(\Omega)$	p. l. fnal $\varphi \in \mathfrak{S}(\mathscr{A})$ with $\varphi(I) = 1$
Expectation	$\langle A \rangle = \int_\Omega A d\varphi$	$\langle A \rangle = \varphi(A)$

Consult Refs. [14–21] for the details of operator algebra and noncommutative probability theory and their physical applications.

§2. Entropy in C^*-systems

Let us discuss the entropy in C^*-dynamical systems introduced in [22]. The formulation of quantum mechanical entropy was presented by von Neumann about 1930, 20 years ahead of Shannon, and it now becomes a fundamental tool in analysing physical phenomena. His entropy is mentioned in Introduction; namely, for a density operator $\varrho \in \mathfrak{S}(\mathscr{H})$, the set of all density operators in a Hilbert space $\mathscr{H}$, the entropy is

$$S(\varrho) = -\operatorname{tr}\varrho\log\varrho, \tag{2.1}$$

which, in terms of any CONS $\{x_k\}$ in $\mathscr{H}$, equals

$$S(\varrho) = -\sum_k \langle x_k, \varrho\log\varrho\, x_k\rangle. \tag{2.2}$$

This does not depend on the choice of the CONS $\{x_k\}$.

Now, the spectral set of ϱ is discrete, so that we write the spectral decomposition of ϱ as

$$\varrho = \sum_n \lambda_n P_n,$$

where λ_n is an eigenvalue of ϱ and P_n is the projection from $\mathscr{H}$ onto the eigenspace associated with λ_n. Therefore, if every eigenvalue λ_n is non-degenerate, then the dimension of the range of P_n is one (we denote this by $\dim P_n = 1$). If a certain eigenvalue, say λ_n, is degenerate, then P_n can be further decomposed into one-dimensional projections:

$$P_n = \sum_{j=1}^{\dim P_n} E_j^{(n)}. \tag{2.3}$$

where $E_j^{(n)}$ is a one-dimensional projection expressed by $E_j^{(n)} = |x_j^{(n)}\rangle\langle x_j^{(n)}|$ with the eigenvector $x_j^{(n)}$ ($j = 1, 2, \ldots, \dim P_n$) for λ_n. By relabelling the indices j, n of $\{E_j^{(n)}\}$, we write

$$\varrho = \sum_n \lambda_n E_n \tag{2.4}$$

with

$$\lambda_1 \geqslant \lambda_2 \geqslant \ldots \geqslant \lambda_n \geqslant \ldots, \tag{2.5}$$

$$E_n \perp E_m \quad (n \neq m). \tag{2.6}$$

We call this decomposition the Schatten decomposition [23]. Now, in (2.5), the eigenvalue of multiplicity n is repeated precisely n times. For example, if the multiplicity of λ_1 is 2, then $\lambda_1 = \lambda_2$. Moreover, this decomposition is unique if and only if no eigenvalue is degenerate. *In the sequel, when we write* $\varrho = \sum_n \lambda_n E_n$, *it is the Schatten decomposition of* ϱ, *otherwise stated.*

For two Hilbert spaces $\mathscr{H}_1$ and $\mathscr{H}_2$, let $\mathscr{H} = \mathscr{H}_1 \otimes \mathscr{H}_2$ be the tensor product Hiblert space of $\mathscr{H}_1$ and $\mathscr{H}_2$ and let us denote the tensor product of two operators A and B acting on $\mathscr{H}_1$ and $\mathscr{H}_2$, respectively, by $A \otimes B$. The reduced states ϱ_1 in $\mathscr{H}_1$ and ϱ_2 in $\mathscr{H}_2$ for a state ϱ in $\mathscr{H}$ are given by the partial traces, which are denoted by $\varrho_k = \mathrm{tr}_{\mathscr{H}_j} \varrho$ ($j \neq k$; $j, k = 1, 2$). The properties of $S(\varrho)$ are summarized in

THEOREM 2.1. *For any density operator* $\varrho \in \mathfrak{S}(\mathscr{H})$, *the followings hold:*

(1) *Positivity:* $S(\varrho) \geqslant 0$.

(2) *Symmetry: Let* $\varrho' = U^{-1} \varrho U$ *for an invertible operator* U. *Then*

$$S(\varrho') = S(\varrho).$$

(3) *Concavity:* $S(\lambda \varrho_1 + (1-\lambda)\varrho_2) \geqslant \lambda S(\varrho_1) + (1-\lambda) S(\varrho_2)$ *for any* $\varrho_1, \varrho_2 \in \mathfrak{S}(\mathscr{H})$ *and any* $\lambda \in [0, 1]$.

(4) *Additivity:* $S(\varrho_1 \otimes \varrho_2) = S(\varrho_1) + S(\varrho_2)$ *for any* $\varrho_k \in \mathfrak{S}(\mathscr{H}_k)$.

(5) *Subadditivity: For the reduced states* ϱ_1, ϱ_2 *of* $\varrho \in \mathfrak{S}(\mathscr{H}_1 \otimes \mathscr{H}_2)$,

$$S(\varrho) \leqslant S(\varrho_1) + S(\varrho_2).$$

(6) *Lower Semicontinuity: If* $\|\varrho_n - \varrho\|_1$ $(\equiv \mathrm{tr}|\varrho_n - \varrho|) \to 0$, *then*

$$S(\varrho) \leqslant \liminf S(\varrho_n).$$

(7) *Continuity: Let* ϱ_n, ϱ *be elements in* $\mathfrak{S}(\mathscr{H})$ *which satisfy the following conditions:* (i) $\varrho_n \to \varrho$ *weakly as* $n \to \infty$, (ii) $\varrho_n \leqslant A$ $(\forall n)$ *for some compact operator* A, *and* (iii) $-\sum_k a_k \log a_k < +\infty$ *for the eigenvalues* $\{a_k\}$ *of* A. *Then* $S(\varrho_n) \to S(\varrho)$.

(8) *Strong Subadditivity: Let* $\mathscr{H} = \mathscr{H}_1 \otimes \mathscr{H}_2 \otimes \mathscr{H}_3$ *and denote the reduced states* $\mathrm{tr}_{\mathscr{H}_k} \varrho$ *and* $\mathrm{tr}_{\mathscr{H}_i \otimes \mathscr{H}_j} \varrho$ *by* ϱ_{ij} *and* ϱ_k, *respectively. Then* $S(\varrho) + S(\varrho_2) \leqslant S(\varrho_{12}) + S(\varrho_{23})$ *and* $S(\varrho_1) + S(\varrho_2) \leqslant S(\varrho_{13}) + S(\varrho_{23})$.

The proofs of the propositions of this theorem can be seen in [24–29]. There exists some difference between in CDS and in QDS, for instançe, the monotonicity is satisfied in CDS but not in QDS.

In order to discuss some physical phenomena, for instance, phase transitions, we had better start without Hilbert space. *Therefore we here formulate the entropy of a state in a C*-dynamical system.* Let $(\mathscr{A}, \mathfrak{S}, \alpha(R))$ be a C^*-dynamical system and $\mathscr{S}$ be a weak* compact and convex subset of $\mathfrak{S}$, ex$\mathscr{S}$ be the set of all extreme points of $\mathscr{S}$. From the Krein–Milman theorem, $\mathscr{S}$ is equal to the weak* closure of convex hull of ex$\mathscr{S}$. We are interested in the following three cases for the set $\mathscr{S}$: (1) $\mathscr{S} = \mathfrak{S}$, (2) $\mathscr{S} = I(\alpha)$, the set of all α-invariant states (i.e., $\varphi(\alpha_t(\cdot)) = \varphi(\cdot)$), (3) $\mathscr{S} = K(\alpha)$, the set of all KMS states at an inverse temperature β (i.e., for any A, B in $\mathscr{A}$, there exists a bounded function $F_{A,B}(z)$ of a complex value z continuous on and holomorphic in the strip $-\beta \leqslant \operatorname{Im} z \leqslant 0$ with boundary values: $F_{A,B}(t) = \varphi(\alpha_t(A)B)$ and $F_{A,B}(t - i\beta) = \varphi(B\alpha_t(A))$ for any $t \in R$ [30]). Note that $K(\alpha) \subset I(\alpha)$.

Every state $\varphi \in \mathscr{S}$ has a maximal measure μ pseudosupported [19, 20] on ex$\mathscr{S}$ such that

$$\varphi = \int_{\mathscr{S}} \omega d\mu. \tag{2.7}$$

The measure μ giving the above decomposition is not unique unless $\mathscr{S}$ is a Choquet simplex [20, 31], so that we denote the set of all such measures by $M_\varphi(\mathscr{S})$. Take

$$D_\varphi(\mathscr{S}) = \{M_\varphi(\mathscr{S}); \exists\{\mu_k\} \subset R^+ \text{ and } \{\varphi_k\} \subset \text{ex}\,\mathscr{S} \text{ s.t. } \sum_k \mu_k = 1,\ \mu = \sum_k \mu_k \delta(\varphi_k)\},$$

where $\delta(\varphi)$ is the Dirac measure concentrated on $\{\varphi\}$, and put

$$H(\mu) = -\sum_k \mu_k \log \mu_k$$

for a measure $\mu \in D_\varphi(\mathscr{S})$. Then the entropy of a state $\varphi \in \mathscr{S}$ w.r.t. $\mathscr{S}$ is defined by

$$S^{\mathscr{S}}(\varphi) = \begin{cases} \inf\{H(\mu);\ \mu \in D_\varphi(\mathscr{S})\}, \\ +\infty \quad \text{if } D_\varphi(\mathscr{S}) = \emptyset. \end{cases} \tag{2.8}$$

This entropy is an extension of von Neumann's entropy as shown below, and it depends on the set $\mathscr{S}$ chosen. Hence it represents *the uncertainty of the state measured in the reference system $\mathscr{S}$*. Three interesting entropies $S^{\mathfrak{S}}(\varphi)$ $(= S(\varphi)$ for short), $S^{I(\alpha)}(\varphi)$ $(= S^I(\varphi)$ for short) and $S^{K(\alpha)}(\varphi)$ $(= S^K(\varphi)$ for short) are generally different even for $\varphi \in K(\alpha)$.

THEOREM 2.2. *When $\mathscr{A} = B(\mathscr{H})$ and $\alpha_t = \operatorname{Ad}(U_t)$ with a unitary operator U_t, for any state φ given by $\varphi(\cdot) = \operatorname{tr}\varrho\,\cdot$ with a density operator ϱ, the followings hold:*

(1) $S(\varphi) = -\mathrm{tr}\,\varrho \log \varrho$.

(2) *If φ is an α-invariant faithful state and every eigenvalue of ϱ is non-degenerate, then* $S^I(\varphi) = S(\varphi)$.

(3) *If* $\varphi \in K(\alpha)$, *then* $S^K(\varphi) = 0$.

Sketch of proof: (1) Let $\varrho = \sum_k \lambda_k \varrho_k$ be a decomposition of ϱ into extremal states mϱ_k (i.e., $\varrho_k^2 = \varrho_k$). It is easily seen that $-\sum_k \lambda_k \log \lambda_k$ attains the minimum value when the above extremal decomposition is the Schatten decomposition of ϱ. Hence $S(\varphi) = -\mathrm{tr}\,\varrho \log \varrho$.

(2) Since φ is α-invariant, the equality $[U_t, \varrho] = 0$ holds for all $t \in R$. From the assumptions, we have $[U_t, E_k] = 0$ for each E_k of the Schatten decomposition $\varrho = \sum_k \lambda_k E_k$. Thus E_k is α-invariant for every k, by which we obtain $S(\varphi) \geqslant S^I(\varphi)$. The converse inequality is shown by using the ergodic decomposition of φ.

(3) The KMS state is unique for $\mathscr{A} = B(\mathscr{H})$, so $S^K(\varphi) = 0$, Q.E.D.

There are some relations among $S(\varphi)$, $S^I(\varphi)$ and $S^K(\varphi)$, for instance, we have

THEOREM 2.3. *For any $\varphi \in K(\alpha)$, the followings hold:*

(1)$S^K(\varphi) \leqslant S^I(\varphi)$.

(2) $S^K(\varphi) \leqslant S(\varphi)$.

(3) *If our dynamical system $(\mathscr{A}, \alpha(R))$ is G-abelian on φ, then*

$$S^K(\varphi) \leqslant S^I(\varphi) \leqslant S(\varphi).$$

($(\mathscr{A}, \alpha(R))$ is called G-Abelian if $E_\varphi \pi_\varphi(\mathscr{A})'' E_\varphi$ is an Abelian von Neumann algebra, where E_φ is the projection from $\mathscr{H}$ onto the set of all $U_\varphi(t)$-invariant vectors.)

(4) *If our dynamical system $(\mathscr{A}, \alpha(R))$ is η-Abelian, then*

$$S^K(\varphi) \leqslant S^I(\varphi).$$

($(\mathscr{A}, \alpha(R))$ is called η-Abelian if the equality

$$\lim_{T\to\infty} \frac{1}{T}\int_0^T \varphi(C^*[\alpha_t(A), B]C)dt = 0$$

holds for any $A, B, C \in \mathscr{A}$.)

Sketch of proof: It is enough to prove the case when the decomposition of φ is discrete: $\varphi = \sum_k \lambda_k \varphi_k$. The set of such states is denoted by $\mathscr{S}_d$.

(1) The extremal decomposition $\varphi = \sum_k \lambda_k \varphi_k$ of $\varphi \in K(\alpha)_d$ into ex$K(\alpha)$ is unique and orthogonal (i.e., $\varphi_n \perp \varphi_m$ for $n \neq m$; see Section 5 for the definition of the orthogonality), so the set inclusion ex $K(\alpha) \subset I(\alpha)$ implies that each φ_n can be further decomposed into the ergodic states. We denote this ergodic decomposition by

$\varphi_n = \sum_k \mu_k^n \psi_k$, $\psi_k \in \mathrm{ex}\, I(\alpha)$. Then

$$S^I(\varphi) = \inf\{-\sum_{k,n} \lambda_n \mu_k^n \log \lambda_n \mu_k^n;\ \{\mu_k^n\}\} = \sum_n \lambda_n S^I(\varphi_n) + S^K(\varphi) \geqslant S^K(\varphi).$$

(2) is proved similarly as (1) and (3) is a consequence of the uniqueness of the ergodic decomposition of φ by the G-Abelianness, and (4) is due to the set inclusion $\mathrm{ex}\, K(\alpha) \subset \mathrm{ex}\, I(\alpha)$ obtained by the η-Abelianness, Q.E.D.

This theorem tells us that even if the entropy $S(\varphi)$ of φ is infinite, the entropy $S^{\mathscr{S}}(\varphi)$ measuring in a proper reference system $\mathscr{S}$ becomes finite. Therefore our entropy can be applied in continuous systems for CDS. Moreover, when a physical system has a symmetry breaking, the entropy $S^K(\varphi)$ might change w.r.t. some parameters such as temperature, so that our entropy can be used to study some phase transitions in physical systems, which will be discussed elsewhere [32].

By the way, most properties of von Neumann entropy $S(\varrho)$ also hold for our entropy $S^{\mathscr{S}}(\varphi)$ under some conditions, for instance:

THEOREM 2.4. (1) *Positivity: For any* $\varphi \in \mathscr{S}$, $S^{\mathscr{S}}(\varphi) \geqslant 0$ *and* $= 0$ *iff* $\varphi \in \mathrm{ex}\, \mathscr{S}$.

(2) *Symmetry: For the dual map* ε^* *from* $\mathrm{ex}\, \mathscr{S}$ *to* $\mathrm{ex}\, \mathscr{S}$ *of a *-automorphism* ε *from* $\mathscr{A}$ *to* $\mathscr{A}$, *put* $\varphi' = \varepsilon^*(\varphi)$. *Then* $S^{\mathscr{S}}(\varphi') = S^{\mathscr{S}}(\varphi)$.

(3) *Concavity: For any* φ, $\psi \in \mathfrak{S}$ *and* $\lambda \in [0, 1]$, *put* $\omega = \lambda\varphi + (1-\lambda)\psi$. *Then*

(i) $S(\omega) \geqslant \lambda S(\varphi) + (1-\lambda) S(\psi)$;

(ii) *when* $\varphi, \psi \in K(\alpha)$, $S^K(\omega) \geqslant \lambda S^K(\varphi) + (1-\lambda) S^K(\psi)$ *and* $S^I(\omega) \geqslant \lambda S^I(\varphi) + (1-\lambda) S^I(\psi)$.

(iii) *when* $\varphi, \psi \in I(\alpha)$ *and if one of the following conditions is satisfied:*

(a) *every element of the centres* Z_φ *for* φ *and* Z_φ *for* ψ *is invariant under* $\alpha_t^\sim$ $(t \in R)$, *the canonical extension of* α_t;

(b) $(\mathscr{A}, \alpha(R))$ *is G-Abelian for* φ *and* ψ, *then we have* $S^I(\omega) \geqslant \lambda S^I(\varphi) + (1-\lambda) S^I(\psi)$.

(4) *Additivity: Let* $\mathscr{B} = \mathscr{A} \otimes \bar{\mathscr{A}}$ *and* $\gamma_t = \alpha_t \otimes \bar{\alpha}_t$. *For a weak* compact convex subset* $\mathscr{S}(\gamma)$, *if the extremal decomposition of any state in* $\mathscr{S}(\gamma)$ *is unique, then* $S^{\mathscr{S}(\gamma)}(\varphi \otimes \psi) = S^{\mathscr{S}(\alpha)}(\varphi) + S^{\mathscr{S}(\bar{\alpha})}(\psi)$.

(5) *Lower Semicontinuity: Suppose that there exists a unique maximal orthogonal decomposition measure defining* (2.7) *for each* φ *in* $\mathscr{S}$ *and any two states* $\omega, \psi \in \mathrm{ex}\, \mathscr{S}$ *are orthogonal. When a sequence* $\{\varphi_n\} \subset \mathscr{S}$ *converges to a state* $\varphi \in \mathscr{S}$ *in norm* $\|\varphi_n - \varphi\| \to 0$ *as* $n \to \infty$, *we have* $S^{\mathscr{S}}(\varphi) \leqslant \liminf S^{\mathscr{S}}(\varphi_n)$.

We omit the proof of this theorem, which is essentially same as those given in [22, 33].

§3. Quantum mechanical channels

By a direct extension of the classical channel Λ^* given by (1.3) and its dual expression, we define quantum mechanical channels in this section on the basis of [34, 35, 72, 75]. In order to define a channel in QDS, we need two dynamical systems, an input system and an output system denoted by C^*-triples $(\mathscr{A}, \mathfrak{S}, \alpha)$ and $(\bar{\mathscr{A}}, \bar{\mathfrak{S}}, \bar{\alpha})$, respectively. A mapping Λ from $\bar{\mathscr{A}}$ to $\mathscr{A}$ satisfying

$$\sum_{i,j=1}^{n} B_i \Lambda(A_i^* A_j) B_j \geqslant 0 \tag{3.1}$$

for any $B_i \in \mathscr{A}$, $A_j \in \bar{\mathscr{A}}$ and every $n \in N$ is called a completely positive map. Remark that the usual completely positive map is a linear map with (3.1). The dual map Λ^* of Λ from $\mathscr{P}$, the set of all positive functionals on $\mathscr{A}$, to $\bar{\mathscr{P}}$ on $\bar{\mathscr{A}}$ is called a *quasichannel*, and the map Λ^* from $\mathfrak{S}$ to $\bar{\mathfrak{S}}$ is called a *channel*. Namely, a channel is the dual map of a linear complete positive map. In this paper, we mainly deal with channels, but a quasichannel is indispensable when we consider Gaussian measures and their transformation [36]. Most physical state changes are described by such channels and we here give some examples encountered in usual discussions in physics [28, 37–40]. Let ϱ be a density operator in a Hilbert space $\mathscr{H}$ of a physical system.

⟨1⟩ *Unitary evolution*: $\varrho \to \Lambda_t^* \varrho = \operatorname{Ad} U_t(\varrho) \equiv U_t^* \varrho U_t$, $t \in R$, where U_t is a unitary operator on $\mathscr{H}$ generated by the Hamiltonian of the system, i.e., $U_t = \exp(itH)$.

⟨2⟩ *Semigroup evolution*: $\varrho \to \Lambda_t^* \varrho = V_t^* \varrho V_t$, $t \in R^+$, where $\{V_t; t \in R^+\}$ is a one-parameter semigroup on $\mathscr{H}$.

⟨3⟩ *Measurement*: When we measure an observable $A = \sum_n a_n P_n$ (spectral decomposition) in a state ϱ, the state ϱ changes to a state $\Lambda^* \varrho$ by this measurement according to the rule $\varrho \to \Lambda^* \varrho = \sum_n P_n \varrho P_n$.

⟨4⟩ *Reduction*: If a system Σ_1 interacts with an external system Σ_2 described by another Hilbert space $\mathscr{K}$ and the initial states of Σ_1 and Σ_2 are ϱ and σ, respectively, then the combined state θ_t of Σ_1 and Σ_2 at time t after the interaction between two systems is given by

$$\theta_t = U_t^* (\varrho \otimes \sigma) U_t,$$

where $U_t = \exp(itH)$ with the total Hamiltonian H of Σ_1 and Σ_2. A channel is obtained by taking the partial trace w.r.t. $\mathscr{K}$, viz.,

$$\varrho \to \Lambda_t^* \varrho = \operatorname{tr}_{\mathscr{K}} \theta_t.$$

⟨5⟩ *Conditional expectation* [41]: Let $\mathfrak{N}$ be a von Neumann algebra, $\mathfrak{M}$ be its von Neumann subalgebra and $\mathscr{E}$ be the conditional expectation (norm one projection [42]) if it exists, or the generalized conditional expectation [43] from $\mathfrak{N}$ to $\mathfrak{M}$. Then the dual map $\mathscr{E}^*$ of $\mathscr{E}$ is a channel.

The following channels have been introduced in [35] to study in quantum communication processes. Let two C^*-triples $(\mathscr{A}, \mathfrak{S}, \alpha)$ and $(\bar{\mathscr{A}}, \bar{\mathfrak{S}}, \bar{\alpha})$ be an input and an output system, respectively, and Λ^* be a channel from $\mathfrak{S}$ to $\bar{\mathfrak{S}}$.

(C1) Λ^* is said to be *stationary* if $\Lambda \circ \bar{\alpha}_t = \alpha_t \circ \Lambda$ for any $t \in R$.

(C2) Λ^* is said to be *ergodic* if it is stationary and $\Lambda^*(\operatorname{ex} I(\alpha)) \subset \operatorname{ex} I(\bar{\alpha})$ holds.

(C3) Λ^* is said to be *orthogonal* if it maps any pair of orthogonal states into orthogonal states (i.e., $\varphi \perp \psi \to \Lambda^* \varphi \perp \Lambda^* \varphi$).

(C4) Λ^* is said to be *deterministic* if it is orthogonal and bijective.

(C5) Λ^* is said to be *chaotic for* $\mathcal{S}$ ($\subset \mathfrak{S}$) if $\Lambda^* \varphi = \Lambda^* \psi$ for any pair $\varphi, \psi \in \mathrm{ex}\,\mathcal{S}$, and it is said to be *chaotic* if $\mathcal{S} = \mathfrak{S}$.

§4. Relative entropy

The relative entropy of two states was first introduced by Umegaki [44] for σ-finite and semifinite von Neumann algebras. For two density operators ϱ and σ it is defined as

$$S(\varrho|\sigma) = \mathrm{tr}\,\varrho(\log\varrho - \log\sigma). \tag{4.1}$$

Lindblad [45, 46] studied some fundamental properties of this relative entropy corresponding to those of Shannon's type relative entropy in CDS. There were several trials to extend the relative entropy to more general quantum systems and to apply it to some other fields [35, 47–61]. Here we review Araki's [49, 50] and Uhlmann's [52] definitions of the relative entropy and state the fundamental properties of the relative entropy.

[Araki's definition] Let $\mathfrak{N}$ be σ-finite (this condition is easily removed [50]) von Neumann algebra acting on a Hilbert space $\mathcal{H}$ and φ, ψ normal states on $\mathfrak{N}$ given by $\varphi(\cdot) = \langle x, \cdot x\rangle$ and $\psi(\cdot) = \langle y, \cdot y\rangle$ with $x, y \in \mathcal{K}$ (a positive natural cone). The operator $S_{x,y}$ is defined by

$$S_{x,y}(Ay + z) = s^{\mathfrak{N}}(y)A^* x, \quad A \in \mathfrak{N}, \quad s^{\mathfrak{N}'}(y)z = 0, \tag{4.2}$$

on the domain $\mathfrak{N}y + (I - s^{\mathfrak{N}'}(y))\mathcal{H}$, where $s^{\mathfrak{N}}(y)$ is the projection from $\mathcal{H}$ to $\{\mathfrak{N}'y\}^-$, the $\mathfrak{N}$-support of y. Using this $S_{x,y}$, the relative modular operator $\Delta_{x,y}$ is defined as $\Delta_{x,y} = (S_{x,y})^* S_{x,y}^-$, with spectral decomposition denoted by $\int_0^\infty \lambda de_{x,y}(\lambda)$. Then the relative entropy is given by

$$S(\psi|\varphi) = \begin{cases} \int_0^\infty \log\lambda \, d\langle y, e_{x,y}(\lambda) y\rangle & \text{if } \psi \ll \varphi, \\ +\infty & \text{otherwise,} \end{cases} \tag{4.3}$$

where $\psi \ll \varphi$ means that $\varphi(A^*A) = 0$ implies $\psi(A^*A) = 0$ for $A \in \mathfrak{N}$.

[Uhlmann's definition] Let $\mathcal{L}$ be a complex linear space and p, q be two seminorms on $\mathcal{L}$. Moreover, let $H(\mathcal{L})$ be the set of all positive hermitian forms α on $\mathcal{L}$ satisfying $|\alpha(x, y)| \leqslant p(x)q(y)$ for all $x, y \in \mathcal{L}$. Then the quadratical mean $\mathrm{QM}(p, q)$ of p and q is defined by

$$\mathrm{QM}(p, q)(x) = \sup\{\alpha(x, x)^{1/2};\ \alpha \in H(\mathcal{L})\}, \quad x \in \mathcal{L}, \tag{4.4}$$

and there exists a function $p_t(x)$ of $t \in [0, 1]$ for each $x \in \mathscr{L}$ satisfying the following conditions [51, 52]:

(1) For any $x \in \mathscr{L}$, $p_t(x)$ is continuous in t,

(2) $p_{1/2} = \mathrm{QM}(p, q)$,

(3) $p_{t/2} = \mathrm{QM}(p, p_t)$,

(4) $p_{(t+1)/2} = \mathrm{QM}(p_t, q)$.

This seminorm p_t is denoted by $\mathrm{QI}_t(p, q)$ and is called the quadratical interpolation from p to q. It is shown [52] that for any positive hermitian forms α, β, there exists a unique function $\mathrm{QF}_t(\alpha, \beta)$ of $t \in [0, 1]$ with values in the set $H(\mathscr{L})$ such that $\mathrm{QF}_t(\alpha, \beta)(x, x)^{1/2}$ is the quadratical interpolation from $\alpha(x, x)^{1/2}$ to $\beta(x, x)^{1/2}$. The relative entropy functional $S(\alpha|\beta)(x)$ of α and β is defined as

$$S(\alpha|\beta)(x) = -\liminf_{t\to 0}(1/t)\{\mathrm{QF}_t(\alpha, \beta)(x, x) - \alpha(x, x)\} \tag{4.5}$$

for $x \in \mathscr{L}$. Let $\mathscr{L}$ be a $*$-algebra $\mathscr{A}$ and φ, ψ be positive linear functionals on $\mathscr{A}$ defining two hermitian forms φ^{L}, ψ^{R} such as $\varphi^{\mathrm{L}}(A, B) = \varphi(A^*B)$ and $\psi^{\mathrm{R}}(A, B) = \psi(BA^*)$. Then the relative entropy of φ and ψ is defined by

$$S(\psi|\varphi) = S(\psi^{\mathrm{R}}|\varphi^{\mathrm{L}})(I). \tag{4.6}$$

If $\mathscr{L}$ is a von Neumann algebra $\mathfrak{N}$ and φ, ψ are normal positive linear functionals on $\mathfrak{N}$, then the Uhlmann relative entropy is shown [57] to be equal to the Araki relative entropy. For a C^*-algebra $\mathscr{A}$ and two positive linear functionals φ, ψ on $\mathscr{A}$, Uhlmann's definition can be directly applied. Further, by considering the GNS representation $\pi(=\pi_{\varphi+\varphi})$ of the functional $\varphi+\psi$ and the canonical extensions $\varphi^{\sim}$, $\psi^{\sim}$ of φ, ψ to $\pi(\mathscr{A})''$, we have the following [57]:

THEOREM 4.1. *In the above notations, the relative entropy $S(\varphi|\psi)$ in a C^*-system is equal to $S(\varphi^{\sim}|\psi^{\sim})$ for its canonically extended von Neumann system.*

Both definitions can be thus used for states in C^*-systems. Let us show that the expression for the relative entropy of two density operators ϱ and σ can be derived from the Uhlmann expression. For normal states φ, ψ on a von Neumann algebra $B(\mathscr{H})$ such that $\varphi(\cdot) = \mathrm{tr}\varrho\cdot$ and $\psi(\cdot) = \mathrm{tr}\sigma\cdot$ with density operators ϱ and σ, we get

$$\mathrm{QF}_t(\psi^{\mathrm{R}}, \varphi^{\mathrm{L}})(I, I) = \mathrm{tr}\varrho^{1-t}\sigma^t, \tag{4.7}$$

hence

$$\begin{aligned} S(\psi|\varphi) = S(\psi^{\mathrm{R}}|\varphi^{\mathrm{L}})(I) &= -\liminf_{t\to\theta}(1|t)\{\mathrm{QF}_t(\psi^{\mathrm{R}}, \varphi^{\mathrm{L}})(I, I) - \psi^{\mathrm{R}}(I, I)\} \\ &= -\liminf_{t\to\theta}(1/t)\mathrm{tr}(\varrho^{1-t}\sigma^t - \varrho) = \mathrm{tr}\varrho(\log\varrho - \log\sigma). \end{aligned} \tag{4.8}$$

Here we summarize the fundamental properties of the relative entropy. For notational simplicity, we write a theorem in the von Neumann algebraic terminology. Namely, let φ, ψ be normal states and $\{\varphi_n\}$, $\{\psi_n\}$ sequences of normal states on a von Neumann algebra $\mathfrak{N}$.

THEOREM 4.2. (1) *Positivity*: $S(\varphi|\psi) \geqslant 0$.

(2) *Joint Convexity*: $S(\lambda\psi_1+(1-\lambda)\psi_2|\lambda\varphi_1+(1-\lambda)\varphi_2) \leqslant \lambda S(\psi_1|\varphi_1)+(1-\lambda)S(\psi_2|\varphi_2)$ *for any* $\lambda\in[0, 1]$.

(3) *Additivity*: $S(\psi_1\otimes\psi_2|\varphi_1\otimes\varphi_2) = S(\psi_1|\varphi_1)+S(\psi_2|\varphi_2)$.

(4) *Lower Semicontinuity*: *If* $\lim_{n\to\infty} \|\psi_n-\psi\| = 0$ *and* $\lim_{n\to\infty} \|\varphi_n-\varphi\| = 0$, *then* $S(\psi|\varphi) \leqslant \liminf_{n\to\infty} S(\psi_n|\varphi_n)$. *Moreover, if there exists a positive number* λ *satisfying* $\psi_n \leqslant \lambda\varphi_n$, *then* $\lim_{n\to\infty} S(\psi_n|\varphi_n) = S(\psi|\varphi)$.

(5) *Monotonicity*: *For a channel* Λ^* *from* $\mathfrak{S}$ *to* $\bar{\mathfrak{S}}$,

$$S(\Lambda^*\psi|\Lambda^*\varphi) \leqslant S(\psi|\varphi).$$

(6) *Lower Bound*: $\|\psi-\varphi\|^2/4 \leqslant S(\psi|\varphi)$.

The proofs of (1)–(4) are given in [49, 50], and the proof of (5) is essentially given in [52], that of (6) in [54]. The relative entropy is related to the concept of sufficiency, and it can be used to classify some equilibrium states and stationary states [54, 57, 60, 62, 63].

§5. Compound state and mutual entropy

As discussed in Introduction, when a state φ changes to another state $\bar{\varphi}$ under a physical transformation, we ask how much information of φ is correctly transmitted to $\bar{\varphi}$, and the amount of this information is expressed by the mutual entropy (information) in CDS. We like to formulate this mutual entropy in QDS for two states φ and $\bar{\varphi} = \Lambda^*\varphi$ with a channel Λ^*, so that we first set the compound state of the initial state φ and the final state $\bar{\varphi}$ expressing the correlation existing between these two states as an extension of the compound measure given by (1.7).

The *compound state* Φ on the tensor product C^*-algebra $\mathscr{A}\otimes\bar{\mathscr{A}}$ of two states φ on $\mathscr{A}$ and $\bar{\varphi}$ on $\bar{\mathscr{A}}$ should satisfy the following properties:

(c. 1) $\Phi(A\otimes I) = \varphi(A)$ for any $A\in\mathscr{A}$;

(c. 2) $\Phi(I\otimes B) = \bar{\varphi}(B)$ for any $B\in\bar{\mathscr{A}}$;

(c. 3) the expression for Φ contains the classical expression as a special case;

(c. 4) Φ indicates the correspondence between each elementary component (pure state) of φ and that of $\bar{\varphi}$.

There are several states satisfying the above two conditions (c. 1) and (c. 2). For instance, the direct product state Φ_0 of φ and $\bar{\varphi}$ given by

$$\Phi_0 = \varphi\otimes\bar{\varphi} \tag{5.1}$$

is such a state, which corresponds to the direct product measure in CDS. We call a state satisfying the conditions (c. 1) and (c. 2) a *quasicompound state*. Let us define the "true" compound state having all the above conditions. Such a compound state is given through the decomposition (2.7) of the state φ. For a state φ in a weak* compact convex subset $\mathscr{S}$ of $\mathfrak{S}$ and a channel Λ^*, let μ be an extremal decomposition measure of φ. A compound state Φ_μ of φ and $\Lambda^*\varphi$ with respect to $\mathscr{S}$ and μ was introduced in [58, 64]

$$\Phi_\mu^{\mathscr{S}} = \int_{\mathscr{S}} \omega \otimes \Lambda^* \omega \, d\mu. \tag{5.2}$$

This state obviously satisfies (c. 1), (c. 2) and (c. 4) because the measure μ is pseudosupported by ex$\mathscr{S}$. The condition (c. 3) is indeed satisfied: When $\mathscr{A}$ and $\bar{\mathscr{A}}$ are Abelian algebras with measurable spaces $(\Omega, \mathscr{F})$ and $(\bar{\Omega}, \bar{\mathscr{F}})$, respectively, and φ is a probability measure on Ω, the extremal decomposition of φ is unique and given by

$$\varphi = \int_{\Omega} \delta_x \, d\varphi, \tag{5.3}$$

where δ_x is the Dirac measure concentrated at a point $x \in \Omega$. Put $\lambda(x, Q) = \Lambda^* \delta_x(Q)$ for any $x \in \Omega$ and $Q \in \bar{\mathscr{F}}$. Then λ is the Markov kernel defining the classical channel, and we have

$$\begin{aligned} \Phi_\varphi(P, Q) &= \int_{\Omega} \delta_x(P) \Lambda^* \delta_x(Q) \, d\varphi \\ &= \int_{\Omega} 1_P(x) \lambda(x, Q) \, d\varphi = \int_P \lambda(x, Q) \, d\varphi \end{aligned} \tag{5.4}$$

for any $P \in \mathscr{F}$, $Q \in \bar{\mathscr{F}}$. Thus our compound state defined by (5.2) is the desired one, but the uniqueness condition of the compound state is an open question. This compound state might play a similar role as the joint probability in CDS although the joint probability does not exist in QDS [65].

Now let us formulate the mutual entropy representing the information transmitted from an initial state $\varphi \in \mathfrak{S}$ to the final state $\Lambda^*\varphi \in \mathfrak{S}$. The *mutual entropy w.r.t. an initial state* $\varphi \in \mathscr{S}$, the *decomposition measure* μ and a *channel* Λ^* are defined by

$$I_\mu^{\mathscr{S}}(\varphi; \Lambda^*) = S(\Phi_\mu^{\mathscr{S}} | \Phi_\theta), \tag{5.5}$$

where $S(\cdot|\cdot)$ is the relative entropy for two states in a C^*-algebra. The *mutual entropy w.r.t. an initial state* $\varphi \in \mathscr{S}$ and a *channel* Λ^* is now defined by

$$I^{\mathscr{S}}(\varphi; \Lambda^*) = \limsup \{ I_\mu^{\mathscr{S}}(\varphi; \Lambda^*); \mu \in F_\varphi(\mathscr{S}; \varepsilon) \}, \tag{5.6}$$

where $\varepsilon \geqq 0$ and $F_\varphi(\mathscr{S}; \varepsilon)$ is the subset of the set $M_\varphi(\mathscr{S})$ such that $F_\varphi(\mathscr{S}; \varepsilon) = \{\mu \in D_\varphi(\mathscr{S});\ S^{\mathscr{S}}(\varphi) \leqslant H(\mu) < S^{\mathscr{S}}(\varphi) + \varepsilon < +\infty\}$ $(F_\varphi(\mathscr{S}; 0) = \{\mu \in D_\varphi(\mathscr{S});\ S^{\mathscr{S}}(\varphi) = H(\mu)\})$ or $F_\varphi(\mathscr{S}; \varepsilon) = M_\varphi(\mathscr{S})$ when $S^{\mathscr{S}}(\varphi) = \infty$. The above sets $M_\varphi(\mathscr{S})$, $D_\varphi(\mathscr{S})$ and the functional $H(\mu)$ are those introduced in Section 2. Note that the mutual entropy (5.5) should be used when the decomposition measure is fixed. In the sequel we use the simple notations Φ_μ, $I_\mu(\varphi; \Lambda^*)$ and $I(\varphi; \Lambda^*)$ when $\mathscr{S} = \mathfrak{S}$. Before discussing the fundamental properties of the mutual entropy, we introduce another mutual type entropy for an initial state φ and a final state ψ. We call it the *quasimutual entropy* and denote it by $I^0(\varphi, \psi)$. Let $\mathfrak{S}_{qc}$ be the set of all quasicompound states in $\mathfrak{S}$ for φ and ψ, and let Ψ_0 be $\varphi\otimes\psi$. Here it is not necessary that ψ is connected to φ through a channel. We define the quasimutual entropy for φ and ψ by

$$I^0(\varphi, \psi) = \sup\{S(\Psi|\Psi_0);\ \Psi\in\mathfrak{S}_{qc},\ \Psi\leqslant\Psi_0\}, \tag{5.7}$$

which will be used to define the ε-entropy for QDS.

In the remainder of this section, we assume that $\mathcal{A}$ and $\bar{\mathcal{A}}$ are von Neumann algebras acting on Hilbert spaces $\mathcal{H}$ and $\bar{\mathcal{H}}$, respectively, and the states denoted by φ, φ_k and ψ are normal states on a von Neumann algebra $\mathcal{A}$. Furthermore, let $\mathcal{K}$ and $\bar{\mathcal{K}}$ be positive natural cones for $\mathcal{A}$ and $\bar{\mathcal{A}}$, respectively.

Two states φ_1 and φ_2 are said to be orthogonal to each other (denoted by $\varphi_1\perp\varphi_2$) if their supports $s(\varphi_1)$ and $s(\varphi_2)$ are orthogonal, where the support $s(\varphi)$ of φ means the smallest projection E satisfying $\varphi(I-E)=0$. The measure $\mu\in M_\varphi(\mathcal{S})$ is said to be orthogonal if $(\int_Q \omega d\mu)\perp(\int_{\mathcal{S}/Q}\omega d\mu)$ is satisfied for every Borel set Q in $\mathcal{S}$. A channel is called normal if it sends a normal state to a normal state. The following lemmas are easily proved.

LEMMA 5.1. *For any normal channel Λ^*, $\varphi_1\perp\varphi_2$ implies $\varphi_1\otimes\Lambda^*\varphi_1\perp\varphi_2\otimes\Lambda^*\varphi_2$.*

LEMMA 5.2. *For x, z in $\mathcal{K}$ and y, w in $\bar{\mathcal{K}}$, we have*

$$\Delta_{x\otimes y, z\otimes w} = \Delta_{x,z}\otimes\Delta_{y,w}.$$

THEOREM 5.3. *For a normal channel Λ^* and a normal state φ, if a measure μ is in the set $F_\varphi(\mathfrak{S};\varepsilon)\cap D_\varphi(\mathfrak{S})$ and is orthogonal, then*

$$I_\mu(\varphi;\Lambda^*) = \int_{\mathfrak{S}} S(\Lambda^*\omega|\Lambda^*\varphi)d\mu < S(\varphi)+\varepsilon.$$

Proof: It suffices to prove the theorem for the case $\varphi=\mu_1\varphi_1+\mu_2\varphi_2$, $\varphi_1\perp\varphi_2$. Let x, x_k, y, y_k $(k=1, 2)$ be the vectors in positive cones such that $\varphi(A)=\langle x, Ax\rangle$, $\varphi_k(A)=\langle x_k, Ax_k\rangle$, $\Lambda^*\varphi(A)=\langle y, Ay\rangle$ and $\Lambda^*\varphi_k(A)=\langle y_k, Ay_k\rangle$ for any $A\in\mathcal{A}$. According to Theorem 3.6 of [50] (i.e., if $\psi_1\perp\psi_2$, then $S(\psi_1+\psi_2|\psi)=S(\psi_1|\psi)+$ $+S(\psi_1|\psi)$ for any ψ) and the above Lemmas 5.1 and 5.2, we obtain

$$\begin{aligned}
I_\mu(\varphi;\Lambda^*) &= S(\Phi_\mu|\Phi_0)\\
&= S(\mu_1\varphi_1\otimes\Lambda^*\varphi_1|\Phi_0)+S(\mu_2\varphi_2\otimes\Lambda^*\varphi_2|\Phi_0)\\
&= \mu_1 S(\varphi_1\otimes\Lambda^*\varphi_1|\Phi_0)+\mu_2 S(\varphi_2\otimes\Lambda^*\varphi_2|\Phi_0)+\\
&\quad+\mu_1\log\mu_1+\mu_2\log\mu_2\\
&= \mu_1\langle x_1\otimes y_1, (\log\Delta_{x_1\otimes y_1, x\otimes y})x_1\otimes y_1\rangle+\\
&\quad+\mu_2\langle x_2\otimes y_2, (\log\Delta_{x_2\otimes y_2, x\otimes y})x_2\otimes y_2\rangle+
\end{aligned}$$

$$
\begin{aligned}
&+\mu_1\log\mu_1+\mu_2\log\mu_2\\
&=\mu_1\langle x_1,(\log\Delta_{x_1,x})x_1\rangle+\mu_1\langle y_1,(\log\Delta_{y_1,y})y_1\rangle+\\
&\quad+\mu_2\langle x_2,(\log\Delta_{x_2,x})x_2\rangle+\mu_2\langle y_2,(\log\Delta_{y_2,y})y_2\rangle+\\
&\quad+\mu_1\log\mu_1+\mu_2\log\mu_2\\
&=\mu_1 S(\varphi_1|\varphi)+\mu_1\log\mu_1+\mu_2 S(\varphi_2|\varphi)+\mu_2\log\mu_2+\\
&\quad+\mu_1 S(\Lambda^*\varphi_1|\Lambda^*\varphi)+\mu_2 S(\Lambda^*\varphi_2|\Lambda^*\varphi)\\
&=S(\mu_1\varphi_1+\mu_2\varphi_2|\varphi)+\mu_1 S(\Lambda^*\varphi_1|\Lambda^*\varphi)+\\
&\quad+\mu_2 S(\Lambda^*\varphi_2|\Lambda^*\varphi)\\
&=\mu_1 S(\Lambda^*\varphi_1|\Lambda^*\varphi)+\mu_2 S(\Lambda^*\varphi_2|\Lambda^*\varphi).
\end{aligned}
$$

The second inequality is obvious for the case of $S(\varphi)=+\infty$. When $S(\varphi)<+\infty$, we have

$$
\begin{aligned}
I_\mu(\varphi;\Lambda^*)&=\sum_n\mu_n S(\Lambda^*\varphi_n|\Lambda^*\varphi)\ \left(=\int S(\Lambda^*\omega|\Lambda^*\varphi)d\mu\right)\\
&\leqslant\sum_n\mu_n S(\varphi_n|\varphi)\\
&=S\Big(\sum_n\mu_n\varphi_n|\varphi\Big)+\Big(-\sum_n\mu_n\log\mu_n\Big)\\
&=-\sum_n\mu_n\log\mu_n<S(\varphi)+\varepsilon,
\end{aligned}
$$

where we used the monotonicity $S(\Lambda^*\varphi_n|\Lambda^*\varphi)\leqslant S(\varphi_n|\varphi)$ and the assumption $\mu\in F_\varphi(\mathfrak{S};\varepsilon)$, Q.E.D.

When $\mathcal{A}$ is the full algebra $B(\mathcal{H})$, any normal state φ is described by a density operator ϱ such as $\varphi(A)=\mathrm{tr}\varrho A$ for any $A\in\mathcal{A}$. Then our entropy $S(\varphi)$ defined by (2.8) is equal to that of von Neumann: $S(\varphi)\ (=S(\varrho))=-\mathrm{tr}\varrho\log\varrho$. Every Schatten decomposition $\varrho=\sum_n\lambda_n E_n$, $E_n=|x_n\rangle\langle x_n|$ (i.e., λ_n is the eigenvalue of ϱ and x_n is its associated eigenvector) provides every orthogonal measure in $D_\varphi(\mathfrak{S})$ defining the entropy $S(\varphi)$. Since the Schatten decomposition of ϱ is not unique unless every eigenvalue λ_n is nondegenerate, the compound state Φ given by (5.2) is expressed as

$$
\Phi_E(Q)=\mathrm{tr}\sigma_E Q,\qquad Q\in\mathcal{A}\otimes\bar{\mathcal{A}},
$$

with

$$
\sigma_E=\sum_n\lambda_n E_n\otimes\Lambda^* E_n,
$$

where E represents a Schatten decomposition $\{E_n\}$. Then the mutual entropy for φ and the channel Λ^* is given by

$$
I(\varphi;\Lambda^*)=\sup\{I_E(\varphi;\Lambda^*);\ E=\{E_n\}\ \text{of}\ \varrho\},
$$

with

$$I_E(\varphi;\ \Lambda^*) = S(\sigma_E|\sigma_0) = \operatorname{tr}\sigma_E(\log\sigma_E - \log\sigma_0),$$

where $\sigma_0 = \varrho \otimes \Lambda^* \varrho$.

Since every Schaten decomposition is discrete and orthogonal, for a state φ given by a density operator ϱ such as $\varphi(A) = \operatorname{tr}\varrho A$, we have the following fundamental inequality.

THEOREM 5.4. $0 \leqslant I(\varphi; \Lambda^*) \leqslant \min\{S(\varphi),\ S(\Lambda^* \varphi)\}$.

Sketch of the proof: According to Theorem 5.3, we have $I_E(\varphi;\ \Lambda^*) \leqslant \min\{S(\varphi),\ S(\Lambda^* \varphi)\}$ for every Schatten decomposition E. This theorem follows by taking the supremum over E, Q.E.D.

This theorem means that the information correctly transmitted from the input system to the output system is less than that carried by the initial state. When we send an information through a channel, we have to consider the efficiency of the communication. This efficiency is measured by the mutual entropy; namely, we ask for which channel is the mutual entropy larger.

THEOREM 5.5. *For a state φ given by $\varphi(\cdot) = \operatorname{tr}\varrho\cdot$ and a channel Λ^*, we have*

(1) *if Λ^* is deterministic, then $I(\varphi;\ \Lambda^*) = S(\varphi)$;*

(2) *if Λ^* is chaotic, then $I(\varphi;\ \Lambda^*) = 0$;*

(3) *if Λ^* is ergodic and φ is stationary for a time evolution $\alpha_t = A\,dU_t$, and if every eigenvalue of ϱ is nonzero and nondegenerate, then $I(\varphi;\ \Lambda^*) = S(\Lambda^* \varphi)$.*

Sketch of the proof: (1) For the Schatten decomposition $\sum_k \lambda_k E_k$ of ϱ, $\Lambda^* E_k$ is a pure state because Λ^* is deterministic. This fact and the equality of Theorem 5.3 conclude (1).

(2) Since Λ^* is chaotic, the compound state σ_E is equal to σ_0. Thus $I(\varphi;\ \Lambda^*) = S(\sigma_0|\sigma_0) = 0$.

(3) As ϱ is α-invariant, every pure state E_k appearing in the Schatten decomposition becomes α-invariant, hence ergodic. The ergodicity of Λ^* implies that $\Lambda^* E_k$ is ergodic and pure. Therefore the conclusion follows because of $S(\Lambda^* E_k) = 0$, Q.E.D.

Before closing this section, we provide the definition of the quantum mechanical ε-entropy as a generalization of the ε-entropy introduced by Kolmogorov [66] for random variables. Let $(\Omega, \mathscr{F}, \mu)$ be a probability space and $M(\Omega)$ be the set of all random variables. Then for any pair of random variables f and g, there exist the joint probability measure μ_{fg} and the direct probability measure $\mu_f \otimes \mu_g$, by which we define a functional $I^0(f, g)$ like the mutual entropy in CDS by

$$I^0(f, g) = S(\mu_{fg}|\mu_f \otimes \mu_g). \tag{5.8}$$

The ε-entropy $S(f;\varepsilon)$ for $f\in M(\Omega)$ and any $\varepsilon>0$ is defined as

$$S(f;\varepsilon)=\inf\{I^0(f,g);\ g\in M_d(f;\varepsilon)\},\tag{5.9}$$

where d is a metric on the set of random variables in the probability space and $M_d(f;\varepsilon)$ is the set of all random variables h satisfying the following inequality: $\{\int d(f,h)^2 d\mu\}^{1/2}\leqslant\varepsilon$.

Analogously, we define the ε-entropy $S(\varphi;\varepsilon)$ of a state φ in QDS as follows: Let $\mathfrak{S}(\varphi;\varepsilon)$ be the set of all states $\psi\in\mathfrak{S}$ satisfying $\|\varphi-\psi\|\leqslant\varepsilon$. Then

$$S(\varphi;\varepsilon)=\inf\{I^0(\varphi,\psi);\ \psi\in\mathfrak{S}(\varphi;\varepsilon)\},\tag{5.10}$$

where $I^0(\varphi,\psi)$ is the quasimutual entropy defined by (5.7). This ε-entropy and a bit more restricted ε-entropy in QDS can be used to define fractal dimensions for quantum systems [67].

§6. Construction of channel for optical communication processes

In this section we present the mathematical construction of a channel describing some optical communication processes [35].

Let $\mathscr{A}=B(\mathscr{H})$ be the set of all bounded linear operators on a separable Hilbert space $\mathscr{H}$, $\mathfrak{S}=\mathfrak{S}(\mathscr{H})$ being the set of all normal states (density operators) on $\mathscr{A}$. In order to discuss communication processes, we need two dynamical systems: an input system $(\mathscr{A}_1,\mathfrak{S}_1)$ and an output system $(\mathscr{A}_2,\mathfrak{S}_2)$ acting on the Hilbert spaces $\mathscr{H}_1$ and $\mathscr{H}_2$, respectively.

Here we determine the general form of a channel for an optical communication process by taking account of direct effects of noise and loss existing in the course of the communication process. In addition to the Hilbert spaces $\mathscr{H}_1$ and $\mathscr{H}_2$, we need two more Hilbert spaces $\mathscr{K}_1$ and $\mathscr{K}_2$ describing the effect of noise and loss, respectively.

(Noise) $\mathfrak{S}(\mathscr{K}_1)$

$\downarrow$

$\mathfrak{S}(\mathscr{H}_1)\ \varrho \longrightarrow\longrightarrow \bar{\varrho}=\Lambda^*\varrho\in\mathfrak{S}(\mathscr{H}_2)$

(Input) $\qquad\uparrow\qquad$ (Output)

(Loss) $\mathfrak{S}(\mathscr{K}_2)$

Let $\nu\in\mathfrak{S}(\mathscr{K}_1)$ be a state representing the noise and a,π,γ the following maps:

(1) the map a is an amplification from $B(\mathscr{H}_2)$ to $B(\mathscr{H}_2\otimes\mathscr{K}_2)$ given by $a(A)=A\otimes I$ for any $A\in B(\mathscr{H}_2)$,

(2) the map π is a completely positive map from $\mathscr{B}(\mathscr{H}_2\otimes\mathscr{K}_2)$ to $\mathscr{B}(\mathscr{H}_1\otimes\mathscr{K}_1)$ with $\pi(I)=I$ describing the physical mechanism of the channel,

(3) the map γ is given by $\gamma(Q)=\mathrm{tr}_{\mathscr{K}_1}\nu Q$ for any $Q\in B(\mathscr{H}_1\otimes\mathscr{K}_1)$.

Then we define a mapping Λ from $B(\mathscr{H}_2)$ to $B(\mathscr{H}_1)$ such that

$$\Lambda=\gamma\circ\pi\circ a.$$

We next consider the dual maps of a, π, γ:

(1′) the dual map a^* of a is a map from $\mathfrak{S}(\mathscr{H}_2 \otimes \mathscr{K}_2)$ to $\mathfrak{S}(\mathscr{H}_2)$ such that $a^*(\theta) = \mathrm{tr}_{\mathscr{K}_2}\theta$,

(2′) the dual map π^*: $\mathfrak{S}(\mathscr{H}_1 \otimes \mathscr{K}_1) \rightarrow \mathfrak{S}(\mathscr{H}_2 \otimes \mathscr{K}_2)$ is given by $\mathrm{tr}\,\pi^*(\theta) W = \mathrm{tr}\,\theta\pi(W)$ for any $\theta \in \mathfrak{S}(\mathscr{H}_1 \otimes \mathscr{K}_1)$ and any $W \in B(\mathscr{H}_2 \otimes \mathscr{K}_2)$,

(3′) the dual map γ^*: $\mathfrak{S}(\mathscr{H}_1) \rightarrow \mathfrak{S}(\mathscr{H}_1 \otimes \mathscr{K}_1)$ is given by $\gamma^*(\varrho) = \varrho \otimes \nu$.

Therefore, once we know the noise ν and the mechanism of the transformation π, we can write down a channel explicitly as

$$\Lambda^* = a^* \circ \pi^* \circ \gamma^*$$

or equivalently,

$$\Lambda^*(\varrho) = \mathrm{tr}_{\mathscr{K}_2} \pi^*(\varrho \otimes \nu) \tag{6.1}$$

for any $\varrho \in \mathfrak{S}(\mathscr{H}_1)$. (6.1) is a mathematical expression of the channel for an optical communication process introduced [35].

We now construct a more concrete model of the channel for an attenuation process. A quantum system composed of photons is described by the Hamiltonian $H = a^* a + 1/2$, where a^* and a are creation and annihilation operators of a photon, respectively. The Schrödinger equation $Hx(q) = Ex(q)$ is easily solved. The eigenvalue E_n is $E_n = n + 1/2$ $(n \geqslant 0)$ and the eigenvector $x_n(q)$ to E_n is $\left(1/(\pi^{1/2} n!)^{1/2}\right) H_n \times \times (2^{1/2} q) \exp(-q^2/2)$, where $H_n(q)$ is the nth Hermite function. The Hilbert space of this system is the closed linear span of the linear combinations of $x_n(q)$ $(n = 0, 1, 2, \ldots)$. Our model for an optical communication process is considered as follows: When n_1 photons are transmitted from the input system, m_1 photons from the noise system add to the signal. Then m_2 photons are lost to the loss system through the channel, and n_2 photons are detected in the output system. In this model, the Hilbert spaces $\mathscr{H}_1$, $\mathscr{H}_2$, $\mathscr{K}_1$, $\mathscr{K}_2$ and their coordinates are denoted in Table 2.

Table 2. Quantum systems

System	Hilbert sp.	CONS	Coordinate
Input	$\mathscr{K}_1$	$x_n^{(1)}(q_1)$	q_1
Noise	$\mathscr{K}_1$	$y_m^{(1)}(t_1)$	t_1
Output	$\mathscr{H}_2$	$x_n^{(2)}(q_2)$	q_2
Loss	$\mathscr{K}_2$	$y_m^{(2)}(t_2)$	t_2

According to the conservation of energy $(n_1 + m_1 = n_2 + m_2)$, we assume the following linear transformation among the coordinates q_1, t_1, q_2, t_2 of the input; noise, output and loss systems, respectively:

$$q_2 = \alpha q_1 + \beta t_1, \qquad \alpha^2 + \beta^2 = 1.$$
$$t_2 = -\beta q_1 + \alpha t_1,$$

For simplicity, we put $m_1 = 0$. By using this linear transformation, we define the mapping $\pi^* = V(\cdot)V^*$ by

$$V(x_n^{(1)} \otimes y_0^{(1)})(q_2, t_2) = x_n^{(2)} \otimes y_0^{(2)}(\alpha q_2 - \beta t_2, \beta q_2 + \alpha t_2)$$

$$= \sum_{j=0}^{n} c_j^{(n)} x_j^{(2)} \otimes y_{n-j}^{(2)}(q_2, t_2),$$

where $c_j^{(n)}$, after performing calculations, is found to be

$$c_j^{(n)} = \sqrt{n!/\{(n-j)!\, j!\}} \cdot \alpha^j (-\beta)^{n-j},$$

so that the channel Λ^* is expressed by

$$\Lambda^*(\varrho) = \mathrm{tr}_{\mathscr{K}_2} V^*(\varrho \otimes \nu) V, \tag{6.2}$$

with the noise $\nu = |y_0^{(1)}\rangle\langle y_0^{(1)}| \in \mathfrak{S}\mathscr{K}_1)$ due to the "zero point fluctuation" of electromagnetic field ($y_0^{(1)}$ is a vacuum state vector in $\mathscr{K}_1$). Now α^2 can be regarded as the transmition efficiency η for the channel.

This expression (6.2) provides us a concrete expression for the error probability of some optical communication processes as seen in the next section.

§7. Rigorous derivation of error probability

The mathematical formulation of quantum control theory was initiated by Helstrom [68] on the basis of the work [69], and many works related to this topics, for instance, quantum estimation and quantum decision, have appeared [70–78].

Here we derive a general formula for the error probability in quantum control communication processes. In particular, the error probabilities for a coherent state or a squeezed state taken as an input state are given.

Let us first explain quantum coding. Suppose that we encode an information by some procedure and we represent the information by a sequence of alphabets $\Pi c^{(k)}$, where $c^{(k)}$ is an element in the set $\mathscr{C}$ of some proper symbols. Then it is necessary for us to express this symbol or the sequence of symbols by some quantum code representing optical signals. This expression is called the quantum mechanical coding. Let ξ_i be the quantum code corresponding to a symbol $c_i \in \mathscr{C}$. In the sequel we take

$$\mathscr{C} = \{0, 1\}, \qquad \Xi = \{\xi_0, \xi_1\}$$

for simplicity. One expression of the quantum code represents the state of a photon; for instance, ξ_0 is the vacuum state and ξ_1 is another state such as coherent or squeezed state.

Two states (quantum mechanical codes) $\xi_0^{(1)}$ and $\xi_1^{(1)}$ in the input system are transmitted to the output system through a channel Λ^*. We here introduce a general formula for the error probability as an application of the construction of a channel Λ^* given by (6.1) in Section 6. The signal "0" represented by the vaccum state

$\xi_0^{(1)}$ goes always to "0" because we have only disturbance due to the zero point fluctuation. However the signal "1" represented by some other state $\xi_1^{(1)}$ can be read either "0" or "1" at the output system due to the loss of information. Therefore the error probability q_e that the signal "1" is recognized as the signal "0" is given by

$$q_e = \operatorname{tr} \Lambda^*(\xi_1^{(1)})\xi_0^{(2)} = \operatorname{tr}_{\mathscr{H}_2}(\operatorname{tr}_{\mathscr{K}_2}\pi^*(\xi_1^{(1)} \otimes \nu))\xi_0^{(2)}, \tag{7.1}$$

where $\xi_0^{(2)}$ is the vacuum state for the output system. Based on this error probability q_e above, the error probability of PCM modulation with the t_0-tuple error correcting code with the weight N and that for PPM modulation are given by the following formulae (7.2) and (7.3), respectively:

$$P_e^{\mathrm{PCM}} = \sum_{j=t_0+1}^{N} {}_NC_j q_e^j (1-q_e)^{N-j}, \tag{7.2}$$

$$P_e^{\mathrm{PPM}} = q_e, \tag{7.3}$$

where ${}_NC_j = N!/\{N-j)!j!\}$.

Now, using the concrete channel (6.2) constructed in the previous section, we can calculate the error probability when we take the coherent state or the squeezed state as an input state. When the input state is a coherent state $\varrho = |\theta\rangle\langle\theta|$ with the eigenvector $|\theta\rangle$ associated to the eigenvalue θ for the annihilation operator a of photon [38, 79], the error probabilities q_e, P_e^{PCM} and P_e^{PPM} are given by

$$q_e = \exp(-\eta|\theta|^2),$$

$$P_e^{\mathrm{PCM}} = \sum_{j=t_0+1}^{N} {}_NC_j \{\exp(-\eta|\theta|^2)\}^j \{1-\exp(-\eta|\theta|^2)\}^{N-j},$$

$$P_e^{\mathrm{PPM}} = q_e$$

according to the general formulae (7.1)–(7.3). These formulas are exactly the same as those used in usual discussions of optical communication processes [76, 77].

When the input state is a squeezed state [80–82] denoted by $|\theta; \lambda, \mu\rangle$ which is the eigenvector of the annihilation operator $b = \lambda a + \mu a^*$: $b|\theta; \lambda, \mu\rangle = (\lambda\theta + \mu\bar{\theta})|\theta; \lambda, \mu\rangle$, the error probability q_e is given as

$$q_e = \sqrt{\tau}\exp[\{(1-\eta)\tau - 1\}|\gamma|^2 + [1-(1-\eta)^2\tau]\{\bar{\mu}\gamma^2/(2\lambda) + \mu\bar{\gamma}^2/(2\bar{\lambda})\}],$$

where $\tau = \{|\lambda|^2 - (1-\eta)^2|\mu|^2\}^{-1}$. Other error probabilities P_e^{PCM} and P_e^{PPM} are obtained in terms of q_e from (7.2) and (7.3), respectively.

§8. Dynamical change of entropy

As discussed in Introduction, the irreversibility of a physical system is understood in terms of the entropy change, in particular, the entropy increase. However, this entropy change can not be achieved from fundamental equations of quantum mechanics. We consider the dynamical change of the entropy in this section and that of the mutual entropy in the next section. Concerning the change of entropy, we address ourselves to the following questions:

(E1) For a dynamical change of a state φ to $\bar{\varphi}$ under some external effects, under what conditions does the entropy $S(\varphi)$ increase: $S(\varphi) \leqslant S(\bar{\varphi})$?

(E2) When a state depends on time, for which time evolution does the entropy $S(\varphi_t)$ converge to some definite value: the existence of a state ψ and the limit $\lim_{t\to\infty} S(\varphi_t) = S(\psi)$?

(E3) Under what conditions are both (E1) and (E2) satisfied: $S(\varphi_t)\uparrow S(\psi)$?

We study these problems in this section, but it is not so easy to get complete answers to them, so we also consider the following converse question to the above:

(E4) Under what conditions is the entropy invariant in the course of dynamics ?

Here we discuss the above problems from a general standing and in the linear response framework.

Let us consider the case when the state change is caused by a time independent or time dependent channel; namely, $\varphi \to \Lambda^*\varphi$ or $\Lambda_t^*\varphi$. The following theorem might be one of the most general results concerning the question (E1). Let $\mathscr{H}$ and $\mathscr{L}$ be the Hilbert spaces corresponding to the input and the output systems, respectively, and Λ^* be a channel from $\mathfrak{S}(\mathscr{H})$, the set of all density operators on $\mathscr{H}$, to $\mathfrak{S}(\mathscr{L})$.

THEOREM 8.1. *If the dual map Λ of the above channel Λ^*: $\mathfrak{S}(\mathscr{H}) \to \mathfrak{S}(\mathscr{L})$ satisfies the equality* $\operatorname{tr}\Lambda(\varrho) = \operatorname{tr}\varrho$ *for any* $\varrho \in \mathfrak{S}(\mathscr{H})$, *then* $S(\Lambda^*\varrho) \geqslant S(\varrho)$.

The proof of this theorem is based on the convexity of the entropy and the convex analysis of operators (see [83–85] for the details).

Some special cases of this theorem have been discussed by several authors such as Nakamura and Umegaki [37], Davis [87] and Ingarden and Urbanik [88]. When we measure an operator Q having a discrete spectral decomposition such as $Q = \sum_n q_n P_n$, the state ϱ describing the system suffers change after the measurement of Q, viz. $\varrho \to \bar{\varrho} \equiv \Gamma_Q \varrho \equiv \sum_n P_n \varrho P_n$. The transformation Γ_Q is a channel and satisfies all conditions of Theorem 8.1. Moreover, this Γ_Q is nothing but a conditional expectation from $B(\mathscr{H})$ to the von Neumann algebra $\mathfrak{M}$ generated by the operators Q and I: $\mathfrak{M} = \{Q\}''$. For this Γ_Q, not only the inequality $S(\Gamma_Q\varrho) \geqslant S(\varrho)$ but also the necessary and sufficient condition satisfying the equality $S(\Gamma_Q\varrho) = S(\varrho)$ are shown in [37]. That is $[S(\Gamma_Q\varrho) = S(\varrho)$ iff $\varrho \in \{Q\}']$. We here give a simpler proof of this fact: If $\varrho \in \{Q\}'$, then $\Gamma_Q\varrho = \varrho$, hence the equality holds. Conversely, if the equality $S(\Gamma_Q\varrho) = S(\varrho)$ holds, then

$$\begin{aligned} 0 = S(\Gamma_Q\varrho) - S(\varrho) &= -\operatorname{tr}\Gamma_Q(\varrho)\log\Gamma_Q(\varrho) + \operatorname{tr}\varrho\log\varrho \\ &= -\operatorname{tr}\Gamma_Q(\varrho\log\Gamma_Q(\varrho)) + \operatorname{tr}\varrho\log\varrho = -\operatorname{tr}\varrho\log\Gamma_Q(\varrho) + \\ &\quad + \operatorname{tr}\varrho\log\varrho = S(\varrho|\Gamma_Q(\varrho)). \end{aligned}$$

According to the inequality (6) of Theorem 4.2, we have $\Gamma_Q(\varrho) = \varrho$, so that $\varrho \in \{Q\}'$. Ingarden and Urbanik called the entropy $S(\Gamma_Q \varrho)$ the Q-entropy and showed [88] that it has similar properties as the von Neumann entropy such as positivity, concavity, additivity, etc.

From Theorem 8.1 we can conclude that a time dependent channel Λ_t^* satisfying all conditions of the theorem yields a monotone increase of the entropy $S(\Lambda_t^* \varrho)$. Thus it is interesting to find physical systems having these properties or to find some conditions under which a suitable time dependent channel can be constructed from the first principle of quantum mechanics.

Now we discuss the problem (E2) in a more concrete dynamics having irreversibility in itself, namely, in the linear response dynamics. A mathematical rigorous treatment of the linear response theory (Kubo theory) has been discussed by several authors [89–91]. Here we introduce the entropy for the linear response theory and study the dynamical change of the entropy in the linear response dynamics.

Let a physical system be described by a triple $(\mathscr{A}, \mathfrak{S}, \alpha(R))$, where $\mathscr{A} = B(\mathscr{H})$, $\mathfrak{S}$ is the set of all density operators on $\mathscr{H}$ and $\alpha_t(\cdot) = U_t \cdot U_t^*$, $U_t = \exp(itH)$ with a Hamiltonian H. Take a state $\varphi \in \mathfrak{S}$ such that $\varphi(\cdot) = \mathrm{tr}\,\varrho\,\cdot$. We suppose that φ is a faithful (i.e., $\varphi(A^*A) = 0 \to A = 0$) KMS state at $\beta = 1$ w.r.t. α_t and the KMS Hamiltonian H is lower bounded.

When the physical system described by the above state φ is perturbed by an α-analytic self-adjoint operator $\lambda V (\lambda \in (0, 1))$ in $\mathscr{A}$, the perturbed time evolution α_t^V and the perturbed state φ^V are expressed as follows:

$$\alpha_t^V(A) = \sum_{n \geqslant 0} (i\lambda)^n \int_{0 \leqslant t_1 \leqslant \ldots \leqslant t_n \leqslant t} dt_1 \ldots \int dt_n \, [\alpha_{t_1}(V), [\ldots[\alpha_{t_n}(V), \alpha_t(A)]\ldots]] \tag{8.1}$$

for $t \geqslant 0$ (the case $t < 0$ is due to exchange of 0 and t in the integral domain) and

$$\varphi^V(A) = \varphi(W^*AW)/\varphi(W^*W), \tag{8.2}$$

$$W = \sum_{n \geqslant 0} (-\lambda)^n \int_{0 \leqslant t_1 \leqslant \ldots \leqslant t_n \leqslant 1/2} dt_1 \ldots \int dt_n \; \alpha_{it_1}(V) \ldots \alpha_{it_n}(V). \tag{8.3}$$

Taking the first order approximation w.r.t. λ in (8.1) and (8.2), we have

$$\alpha_t^{V,1}(A) = \alpha_t(A) + i\lambda \int_0^t ds [\alpha_s(V), \alpha_t(A)], \tag{8.4}$$

$$\varphi^{V,1}(A) = \varphi(A) - \lambda \int_0^1 ds \varphi(A\alpha_{is}(V)) + \lambda\varphi(A)\varphi(V), \tag{8.5}$$

where we need some computation to obtain (8.5) from (8.3) [91]. In the sequel we call $\varphi^{V,1}$ and $(\alpha_t^{V,1})^* \varphi$ a "linear response perturbed state" and a "linear response state", respectively, although they do not exhibit the positivity of a state and the linear response time evolution $\alpha_t^{V,1}$ is not an automorphism of $\mathscr{A}$.

It is known [92] that φ^V satisfies the KMS condition at $\beta = 1$ w.r.t α_t^V, hence φ^V is an equilibrium state for the Hamiltonian $H + \lambda V$. Therefore it is natural to expect the state change under α_t^V such that

$$w^* - \lim_{t \to \infty} (\alpha_t^V)^* \varphi = \varphi^V. \tag{8.6}$$

That is, when a perturbation λV acts on the system in equilibrium described by φ and H, we should observe that the state φ gradually approaches to new equilibrium state for the Hamiltonian $H + \lambda V$. Regardless of (8.6), the entropy of the time dependent state $(\alpha_t^V)^* \varphi$ is always equal to that of φ (i.e., $S((\alpha_t^V)^* \varphi) = S(\varphi)$) because α_t^V is implemented by a unitary operator generated by $H + \lambda V$. This means that the entropy of a state is a dynamical invariant even when the state changes to a different equilibrium state like (8.6), which contradicts the macroscopic behavior of physical systems in our real world. This result is not surprising because quantum mechanics does not contain the irreversibility in itself as discussed before. The linear response theory succeeded to explain some irreversible phenomena, so we hope that the entropy of the linear response state might change in time. From the assumptions imposed on our dynamical system and φ, we obtain by a simple calculation

$$\varphi(\alpha_t^{V,1}(A)) = \mathrm{tr}\,\varrho^{V,1}(t) A,$$

$$\varphi^{V,1}(A) = \mathrm{tr}\,\varrho^{V,1} A$$

with

$$\varrho^{V,1} = \Big(I - \lambda \int_0^1 \alpha_{is}(V) ds + \mathrm{tr}\,\varrho V\Big)\varrho,$$

$$\varrho^{V,1}(t) = \Big(I - i\lambda \int_0^t \alpha_{-s}(V) ds + i\lambda \int_0^t \alpha_{-s+i}(V) ds\Big)\varrho.$$

Since the set of all trace class operators is an ideal for $\mathscr{A}$, both $\varrho^{V,1}$ and $\varrho^{V,1}(t)$ are trace class operators, so we define normal states $\theta(t)$ and θ by

$$\theta(t) = |\varrho^{V,1}(t)| / \mathrm{tr} |\varrho^{V,1}(t)|,$$

$$\theta = |\varrho^{V,1}| / \mathrm{tr} |\varrho^{V,1}|,$$

where $|A| \equiv (A^* A)^{1/2}$. The entropy $S(\varrho^{V,1}(t))$ of the linear response state $\varrho^{V,1}(t)$ and the entropy $S(\varrho^{V,1})$ of the linear response perturbed state $\varrho^{V,1}$ are defined as

$$S(\varrho^{V,1}(t)) = -\mathrm{tr}\,\theta(t) \log \theta(t),$$

$$S(\varrho^{V,1}) = -\mathrm{tr}\,\theta \log \theta.$$

If $\varrho^{V,1}(t) \geqslant 0$, then our entropy is identical to that of von Neumann because of $\mathrm{tr}\,\varrho^{V,1}(t) = 1$.

Now, our problem is as follows: *When the linear response state* $\varrho^{V,1}(t)$ *approaches to the linear response perturbed state* $\varrho^{V,1}$ *in the weak topology* (*i.e.*, $\lim_{t\to\infty} \mathrm{tr}\,\varrho^{V,1}(t)A = \mathrm{tr}\,\varrho^{V,1}A$ *for any* A *in* $\mathscr{A}$), *under what conditions does the entropy* $S(\varrho^{V,1}(t))$ *dynamically change to* $S(\varrho^{V,1})$? Here we assume the existence of the limit $w-\lim_{t\to\infty} \varrho^{V,1}(t) = \varrho^{V,1}$, but this existence is realized under some ergodic conditions [91]. Then we can prove the following lemma.

LEMMA 8.2. *If* $\varrho^{V,1}(t)$ *weakly converges to* $\varrho^{V,1}$ *as* $t\to\infty$, *then we have*

(1) $\|\theta(t)-\theta\|_1 \to 0$ *as* $t\to\infty$, *where* $\|\cdot\|_1 \equiv \mathrm{tr}|\cdot|$,

(2) *there exist positive constants a, b such that* $a\varrho \leqslant |\varrho^{V,1}(t)| \leqslant b\varrho$ *for a sufficiently small* λ.

Sketch of the proof: (1) Since $\mathrm{tr}\,\varrho^{V,1}(t) = \mathrm{tr}\,\varrho^{V,1} = 1$, the following implications hold: $\varrho^{V,1}(t) \to \varrho^{V,1}(\text{weak}) \Rightarrow \|\varrho^{V,1}(t)-\varrho^{V,1}\|_1 \to 0 \Rightarrow \|\varrho^{V,1}(t)-\varrho^{V,1}\| \to 0 \Rightarrow \| |\varrho^{V,1}(t)| - |\varrho^{V,1}| \| \to 0$ and $|\mathrm{tr}|\varrho^{V,1}(t)| - \mathrm{tr}|\varrho^{V,1}|| \to 0 \Rightarrow \theta(t) \to \theta$ (weak) $\Rightarrow \|\theta(t)-\theta\|_1 \to 0$.

(2) Since φ is faithful and normal, $\varphi(\alpha_t^{V,1}(A))$ can be expressed as [90, 91]

$$\varphi(\alpha_t^{V,1}(A)) = \varphi(A) - \langle TA^*x,\ T(I-\exp(-itH))Vx\rangle,$$

where $T = \{1-\exp(-H)\}/H$. This fact with the lower boundness of H implies the conclusion, Q.E.D.

Applying (7) of Theorem 2.1, we get

THEOREM 8.3. *If* $\varrho^{V,1}(t)$ *weakly converges to* $\varrho^{V,1}$ *as* $t\to\infty$ *and* $S(\varrho)$ *is finite, then* $S(\varrho^{V,1}(t))$ *converges to* $S(\varrho^{V,1})$ *as* $t\to\infty$.

This theorem might provide a reason why the linear response theory has been useful for the description of some irreversible phenomena. Finally, we discuss the problem (E5). For two dynamical systems $(\mathscr{A}, \mathfrak{S}, \alpha)$ and $(\bar{\mathscr{A}}, \bar{\mathfrak{S}}, \bar{\alpha})$ and a channel from $\mathfrak{S}$ to $\bar{\mathfrak{S}}$, let $\mathscr{S}$ be a weak* compact convex subset of $\mathfrak{S}$ and $\bar{\mathscr{S}}$ be that of $\bar{\mathfrak{S}}$. If we have a certain one-to-one correspondence between $\mathscr{S}$ and $\bar{\mathscr{S}}$ by the channel Λ^*, then the equality $S^{\mathscr{S}}(\varphi) = S^{\bar{\mathscr{S}}}(\Lambda^*\varphi)$ is expected. We can prove the following intuitively trivial facts about the entropy invariance [22].

THEOREM 8.4. *For a stationary channel* Λ^*, *the followings hold*:

(1) *If* Λ^* *is injective and a* $*$*-homomorphism, then* $S^K(\varphi) = S^K(\Lambda^*\varphi)$ *for any* $\varphi \in K(\alpha)$.

(2) *If* Λ^* *is a bijection from* $exI(\alpha)$ *to* $exI(\alpha)$, *then* $S^I(\varphi) = S^I(\Lambda^*\varphi)$ *for any* $\varphi \in I(\alpha)$.

(3) *If* Λ^* *is a bijection on* $\mathfrak{S}$, *then* $S(\varphi) = S(\Lambda^*\varphi)$ *for any* $\varphi \in \mathfrak{S}$.

The entropy change in the linear response dynamics for more general quantum systems can be similarly discussed, and the entropy change for some open systems [39, 93] and for some quantum stochastic processes [95] might be worth considering for the study of the irreversibility, which will be discussed elsewhere.

§9. Dynamical change of mutual entropy

As discussed in Introduction, we study the time development of the mutual entropy defined by (5.6) when the state is changed by a time dependent channel Λ_t^*. Here we assume that $\mathscr{A}$ is a von Neumann algebra acting on a Hilbert space $\mathscr{H}$ with $\mathscr{A} = \bar{\mathscr{A}}$ and $\Lambda(R^+) = \{\Lambda_t;\ t \in R^+\}$ is a dynamical semigroup on a von Neumann algebra $\mathscr{A}$ (i.e., $\Lambda(R^+)$ is a weakly* continuous semigroup and Λ_t^* is a normal channel for each $t \in R^+$) and there exists at least one faithful normal stationary state ψ w.r.t. Λ_t^* (i.e., $\Lambda_t^*\psi = \psi$ for all $t \in R^+$).

Define two subsets $\mathscr{A}(\Lambda)$ and $\mathscr{A}(C)$ of $\mathscr{A}$ such as

$$\mathscr{A}(\Lambda) = \{A \in \mathscr{A} |\ \Lambda_t(A) = A,\ t \in R^+\},$$

$$\mathscr{A}(C) = \{A \in \mathscr{A} |\ \Lambda_t(A^*A) = \Lambda_t(A^*)\Lambda_t(A),\ t \in R^+\}.$$

Then $\mathscr{A}(\Lambda)$ is a von Neumann subalgebra of $\mathscr{A}$ and there exists a unique conditional expectation $\mathscr{E}$ from $\mathscr{A}$ to $\mathscr{A}(\Lambda)$ [15, 95].

THEOREM 9.1. *Under the same conditions as those of Theorem 5.3, if $\mathscr{A}(\Lambda) = \mathscr{A}(C)$ holds and $\mathscr{A}$ is type I, then $I_\mu(\varphi;\ \Lambda_t^*)$ decreases in time and approaches $I_\mu(\varphi;\ \mathscr{E}^*)$ as $t \to \infty$.*

Proof: Since $\mathscr{A}(\Lambda) = \mathscr{A}(C)$, it is known [95] that $\Lambda_t^*\omega$ weakly converges to $\mathscr{E}^*\omega$ for any normal state ω. The assumption that $\mathscr{A}$ is σ-finite and type I implies the norm convergence of $\Lambda_t^*\omega$ [96]:

$$\|\Lambda_t^*\varphi - \mathscr{E}^*\varphi\| \to 0 \quad \text{and} \quad \|\Lambda_t^*\varphi_n - \mathscr{E}^*\varphi_n\| \to 0, \quad t \to \infty,$$

where $\varphi = \sum_n \mu_n \varphi_n$ with $\varphi_n \perp \varphi_m$ $(n \neq m)$. As there exists a constant $\lambda_n \in R^+$ satisfying $\varphi_n \leqslant \lambda_n \varphi$ for each n, the inequality $\Lambda_t^*\varphi_n \leqslant \lambda_n \Lambda_t^*\varphi$ holds for all $t \in R^+$. Therefore (4) of Theorem 4.2 applies and we obtain

$$\lim_{t \to \infty} S(\Lambda_t^*\varphi_n | \Lambda_t^*\varphi) = S(\mathscr{E}^*\varphi_n | \mathscr{E}^*\varphi).$$

This equality and the equality of Theorem 5.3 proves the existence of $\lim_{t\to\infty} I_\mu(\varphi;\ \Lambda_t^*)$. This limit is decreasing in time because $S(\Lambda_{t+s}^*\varphi_n | \Lambda_{t+s}^*\varphi) \leqslant S(\Lambda_t^*\varphi_n | \Lambda_t^*\varphi)$ for all $s \in R^+$, Q.E.D.

THEOREM 9.2. *If $\mathscr{A} = B(\mathscr{H})$ and $\mathscr{A}(\Lambda) = \mathscr{A}(C)$ holds, then for any normal state ϱ*
(1) *$I(\varrho;\ \Lambda_t^*)$ decreases to $I(\varrho;\ \mathscr{E}^*)$ as $t \to +\infty$, and*
(2) *there exists only one stationary state w.r.t. Λ_t^* iff $I(\varrho;\ \mathscr{E}^*) = 0$ for all ϱ.*

Proof: The convergence of $I_E(\varrho;\ \Lambda_t^*)$ to $I_E(\varrho;\ \mathscr{E}^*)$ as $t \to +\infty$ is proved in Theorem 9.1. Hence we obtain (1) by taking the supremum over E. The statement (2) is easy to prove, Q.E.D.

THEOREM 9.3. *When $\Lambda(R)$ is unitarily implemented, the mutual entropy $I(\varrho;\ \Lambda_t^*)$ is equal to the entropy $S(\varrho)$.*

Theorem 9.1 and item (1) of Theorem 9.2 show that the information of the state φ is lost in time when Λ_t^* is dissipative. Item (2) of Theorem 9.2 tells that the information of an initial state is completely lost in equilibrium if there exists only one stationary state for Λ_t^*. These results might be quite natural for dissipative processes from physical points of view.

About ten years ago, Toyoda conjectured [97] that the quantum measurement process should be treated similarly as a communication process of information. This program is now progressing [32] by using a communication channel and the mutual entropy.

Acknowledgements

The author thanks Professors H. Umegaki and R. S. Ingarden for their hospitality and encouragement.

He also thanks Professor Hiai for his critical reading and useful comments.

REFERENCES

[1] J. von Neumann, *Die Mathematischen Grundlagen der Quantenmechanik*, Springer, Berlin, 1932.
[2] C. R. Shannon, *Bell System Tech. J.*, **27** (1948), 379 and 623.
[3] G. Nicolis and I. Prigogine, *Self-organization in Nonequilibrium Systems*, J. Wiley & Sons, 1977.
[4] H. Haken, *Synergetics—An Introduction*, Springer-Verlag, Berlin, 1978.
[5] S. Kullback and R. A. Leibler, *Ann. Math. Stat.*, **22** (1951), 79.
[6] A. N. Kolmogorov, *IEEE on Information Theory*, **2** (1956), 102.
[7] I. M. Gelfand and A. M. Yaglom, *Amer. Math. Soc. Transl.*, **12** (1959), 199.
[8] H. Umegaki, *J. Math. Anal. Appl.*, **25** (1969), 41.
[9] H. Umegaki and M. Ohya, *Entropies in Probabilistic Systems* (in Japanese), Kyoritsu Publishing Company, 1983.
[10] M. S. Pinsker, *Information and Information Stability of Random Variables and Processes*, Holden-Day, Inc., 1964.
[11] R. Ash, *Information Theory*, Interscience, 1965.
[12] S. Guiasu, *Information Theory with Applications*, McGraw-Hill, 1977.
[13] R. Haag and D. Kastler, *J. Math. Phys.*, **5** (1964), 848.
[14] M. Tomita, *Quasi-standard von Neumann Algebras*, Mimeographed Notes, Kyushu Univ., 1967.
[15] M. Takesaki, *Tomita's Theory of Modular Hilbert Algebras and its Applications*, Lecture Notes Math., **128**, Springer, 1970.
[16] S. Sakai, *C*-algebras and W*-algebras*, Springer, Berlin, 1971.
[17] H. Ezawa, Structure of quantum mechanics (in Japanese), in: *Quantum Mechanics II*, ed. by H. Yukawa and T. Toyoda, Iwanami Publishing Company, 1978.
[18] O. Bratteli and D. W. Robinson, *Operator Algebras and Quantum Statistical Mechanics I*, Springer, New York, 1981.
[19] O. Bratteli and D. W. Robinson, *Operator Algebras and Quantum Statistical Mechanics II*, Springer, New York, 1981.
[20] M. Takesaki, *Theory of Operator Algebras I*, Springer, New York, 1983.

[21] H. Umegaki, M. Ohya and F. Hiai, *Introduction to Operator Algebras* (in Japanese), Kyoritsu Publishing Company, 1985.
[22] M. Ohya, *J. Math. Anal. Appl.*, **100** (1984), 222.
[23] R. Schatten, *Norm Ideals of Completely Continuous Operators*, Springer-Verlag, 1970.
[24] E. H. Lieb and M. B. Ruskai, *J. Math. Phys.*, **14** (1973), 1938.
[25] A. Wehrl, *Rep. Math. Phys.*, **6** (1976), 15.
[26] A. Wehrl, *Rev. Mod. Phys.*, **50** (1978), 221.
[27] G. Lassner and G. A. Lassner, *Rep. Math. Phys.*, **15** (1979), 41.
[28] W. Thirring, *Quantum Mechanics of Large Systems*, Springer-Verlag, 1983.
[29] H. Umegaki and M. Ohya, *Quantum Mechanical Entropies* (in Japanese), Kyoritsu Publishing Company, 1984.
[30] R. Haag, N. M. Hugenholtz and M. Winnink, *Commun. Math. Phys.*, **5** (1967), 215.
[31] G. Choquet, *Lecture Analysis I, II, III*, Benjamin, New York, 1969.
[32] M. Ohya, Symmetry breaking and quantum measurement (in preparation).
[33] M. Ohya and T. Matsuoka, *J. Math. Phys.*, **27**, No. 8 (1986), 2076.
[34] M. Ohya, *J. Math. Anal., Appl.*, **84** (1981), 318.
[35] M. Ohya, *IEEE on Information Theory*, **29** (1983), 770.
[36] M. Ohya and N. Watanabe, *Japan J. Appl. Math.*, **3**, No. 1 (1986), 197.
[37] M. Nakamura and H. Umegaki, *Math. Jap.*, **7** (1962), 151.
[38] J. R. Klauder and E. C. G. Sudershan, *Fundamentals of Quantum Optics*, Benjamin, New York, 1968.
[39] K. Kraus, *Ann. Phys.*, **64** (1970), 311.
[40] M. Ohya, *Kodai. Math. J.*, **3** (1980), 287.
[41] H. Umegaki, *Tohoku Math. J.*, **6** (1954), 177.
[42] J. Tomiyama, *Proc. Japan Acad.*, **33** (1957), 608.
[43] L. Accardi and C. Cecchini, *J. Func. Anal.*, **45** (1982), 275.
[44] H. Umegaki, *Kodai Sem. Rep.*, **14** (1962), 59.
[45] G. Lindblad, *Commun. Math. Phys.*, **33** (1973), 305.
[46] G. Lindblad, *Commun. Math. Phys.*, **40** (1975), 147.
[47] F. Schlogl, *Ann. Phys.*, **45** (1967), 155.
[48] A. Connes abd E. Størmer, *Acta. Math.*, **134** (1975), 289.
[49] H. Araki, *Publ. RIMS Kyoto Univ.*, **11** (1976), 809.
[50] H. Araki, *Publ. RIMS Kyoto Univ.*, **13** (1977), 173.
[51] W. Pusz and S. L. Woronowicz, *Rep. Math. Phys.*, **8** (1975), 159.
[52] A. Uhlmann, *Commun. Mat. Phys.*, **54** (1977), 21.
[53] H. Spon, *J. Math. Phys.*, **19** (1978), 1227.
[54] F. Hiai, M. Ohya and M. Tsukada, *Pacific J. Math.*, **96** (1981), 99.
[55] R. S. Ingarden, *Int. J. Engng. Sci.*, **19** No. 12 (1981), 1609.
[56] R. S. Ingarden, H. Janyszek, A. Kossakowski and T. Kawaguchi, *Tensor, N. S.*, **37** (1982), 105.
[57] F. Hiai, M. Ohya and M. Tsukada, *Pacific J. Math.*, **107** (1983), 117.
[58] M. Ohya, *Lecture Note in Math.*, **1136** (1985), 397.
[59] M. Donald, *Commun. Math. Phys.*, **105** (1986), 13.
[60] D. Petz, *Commun. Math. Phys.*, **105** (1986), 123.
[61] R. S. Ingarden, *Geometry of Thermodynamics*, preprint.
[62] H. Umegaki, *Kodai Math. Sem. Rep.*, **11** (1959), 51.
[63] S. Gudder and J. P. Marchand, *J. Math. Phys.*, **13** (1972), 799.
[64] M. Ohya, *L. Nuovo Cimento*, **38** (1983), 402.
[65] K. Urbanik, *Stud. Math.*, **21** (1961), 317.
[66] A. N. Kolmogorov and V. M. Tikhomirov, *Amer. Math. Soc. Translation, Ser. 2*, **17** (1961), 277.

[67] M. Ohya, *Fractal dimensions for general quantum states*, to appear.
[68] C. W. Helstrom, *Quantum Detection and Estimation Theory*, Academic Press, New York, 1976.
[69] C. W. Helstrom, J. W. S. Liu and J. P. Gordon, *Proc. IEEE*, **58** (1970), 1578.
[70] R. S. Ingarden, *Fortschritte der Physik*, **12** (1964), 567.
[71] R. S. Ingarden, *Fortschritte der Physik*, **13** (1965), 755.
[72] A. S. Holevo, *Problemy Peredachi Informatsii* **8** (1972), 62.
[73] A. S. Holevo, *Problemy Peredachi Informatsii* **9** (1973), 31.
[74] R. S. Ingarden, *Rep. Math. Phys.*, **10** (1976), 43.
[75] A. S. Holevo, *Rep. Math. Phys.*, **2** (1977), 273.
[76] R. J. McEliece, *IEEE. Trans. Information Theory*, **27** (1981), 393.
[77] O. Hirota, K. Yamazaki, M. Nakagawa and M. Ohya, *Trans. IECE Japan*, **E69** (1986), 917.
[78] M. Ohya and N. Watanabe, *Efficiency of modulations in optical communication processes*, to appear.
[79] R. J. Glauber, *Phys. Rev.*, **131** (1963), 2766.
[80] D. Stoler, *Phys. Rev.*, **D1** (1970), 3217, **4** (1971), 1925.
[81] J. N. Hollenhorst, *Phys. Rev.*, **D19** (1979), 1669.
[82] H. P. Yuen, *Phys. Rev.*, **A13** (1976), 2226.
[83] J. Bendat and S. Sherman, *Trans. Amer. Math. Soc.*, **79** (1955), 58.
[84] M. Nakamura and H. Umegaki, *Proc. Japan Acad.*, **37** (1961), 149.
[85] W. Donogue, *Monotone Matrix Functions and Analytic Continuation*, Springer, 1974.
[86] T. Ando, *Topics on Operator Inequality*, Res. Rep. of Hokkaido Univ.
[87] C. Davis, *Proc. Japan Acad.*, **37** (1961), 533.
[88] R. S. Ingarden and K. Urbanik, *Acta Phys. Polon.*, **21** (1962), 281.
[89] J. Naudts, A. Verbeure and R. Weder, *Commun. Math. Phys.*, **44** (1975), 87.
[90] A. Verbeure and R. Weder, *Commun. Math. Phys.*, **44** (1975), 101.
[91] M. Ohya, *Rep. Math. Phys.*, **16** (1979), 305.
[92] H. Araki, *Publ. RIMS Kyoto Univ.*, **9** (1973), 165.
[93] E. B. Davies, *Quantum Theory of Open System*, Academic Press, New York, 1976.
[94] L. Accardi, A. Frigerio and J. T. Lewis, *Publ. RIMS Kyoto Univ.*, **18** (1982), 97.
[95] A. Frigerio, *Commun. Math. Phys.*, **63** (1978), 269.
[96] G. F. Dell'Antonio, *Commun. Pure Appl. Math.*, **20** (1967), 413.
[97] R. Toyoda, Quantum mechanics and physics for information (in Japanese), in: *Quantum Mechanics II*, ed. by H. Yukawa and T. Toyoda, Iwanami Publishing Company, 1978.

© 1983 IEEE. Reprinted, with permission, IEEE TRANSACTIONS ON INFORMATION THEORY, VOL. IT - 29, NO. 5, SEPTEMBER 1983

On Compound State and Mutual Information in Quantum Information Theory

MASANORI OHYA

Abstract—**A quantum mechanical compound state of an input state and its output state generated through a communication channel is constructed. The mutual information of quantum communication theory is defined by using the compound state, and its fundamental properties are studied.**

I. Introduction

The communication of information by using a laser is an important technique. Laser action is a typical quantum effect, and therefore we should reconstruct information (communication) theory in terms of quantum mechanical language [3], [6], [7], [8], [10], [13], [14]. This program is not so easy to complete. One of the main difficulties is to formulate *in a systematic way* the so-called mutual information which plays an essential role in information theory. In order to formulate it rigorously, we have to know proper relations between input states and output states connected with each other by a communication channel. There have been some attempts [7], [10] to proceed with this program. This short note is one more attempt to formulate the communication process with the following aims: 1) construct a suitable compound state of an input state and its output state generated through a communication channel; 2) formulate the mutual information by using the compound state constructed and study its fundamental properties; 3) examine a simple process of photon-transmission in our formulation in order to show that our formulation is indeed an extension of the usual one.

Manuscript received February 3, 1982; revised November 10, 1982.

The author is with the Department of Information Sciences, Science University of Tokyo, Noda City, Chiba 278, Japan.

II. Quantum Dynamical System

We briefly review the density operator method to describe a quantum dynamical system. Since von Neumann's famous book [25] appeared, a quantum mechanical system has been described by the method of Hilbert space. Let $\mathcal{H}$ be a separable complex Hilbert space and $B(\mathcal{H})$ be the set of all bounded linear operators on $\mathcal{H}$. A self-adjoint operator of $B(\mathcal{H})$ might be called a physical observable. We consider only bounded operators in this correspondence.

The expected value of $A \in B(\mathcal{H})$ in a state vector $\Omega \in \mathcal{H}$ is given by $\langle \Omega, A\Omega \rangle$, where $\langle \cdot , \cdot \rangle$ is the inner product of $\mathcal{H}$. More generally, let $T(\mathcal{H})_{+,1}$ be the set of all positive trace class operators (density operators) on $\mathcal{H}$ with $\operatorname{tr}\rho = 1$. That is, $T(\mathcal{H})_{+,1} = \{\rho \in B(\mathcal{H}) | \rho \geqslant 0,\ \operatorname{tr}\rho \equiv \Sigma_n \langle \Phi_n, \rho\Phi_n \rangle = 1$ for any complete orthonormal system (CONS) $\{\Phi_n\}$ of $\mathcal{H}\}$. We call a density operator a (normal) state on $B(\mathcal{H})$ as usual [21].

A dynamical change of a quantum system is described by either the Heisenberg picture or the Schrödinger picture. The Heisenberg picture provides the dynamics of observables and the Schrödinger picture provides that of states. These pictures are completely equivalent for Hamiltonian dynamics. In our notation, the time evolution dynamics of a quantum system is given by a one-parameter group $\alpha(R)$ of automorphisms of $B(\mathcal{H})$ such that $\alpha_t(A) = u_t A u_{-t}$ for any $A \in B(\mathcal{H})$ and all $t \in R$, the set of real numbers, where u_t is a unitary operator on $\mathcal{H}$ generated by the Hamiltonian H of the system: $u_t = \exp(itH)$. Thus our quantum dynamical system is described by a triple $(B(\mathcal{H}), T(\mathcal{H})_{+,1}, \alpha(R))$.

III. Quantum Mechanical Channel

In information theory we have to consider two dynamical systems: an input system and an output system. An input system is described by a triple $(B(\mathcal{H}_1), T(\mathcal{H})_{+,1}, \alpha(R))$ and an output system is described by another triple $(B(\mathcal{H}_2), T(\mathcal{H}_2), \tau(R))$, where $\tau_t(A) = v_t A v_{-t}$ for a unitary operator v_t on $\mathcal{H}_2$. These time evolutions α_t, τ_t characterize the input states and the output states respectively. They have nothing to do with the dynamics of a channel. As in classical case (cf. [23]), we often need to prepare a stationary state as an input state in quantum case.

Information (entropy) of the input system is transmitted to the output system. A channel exhibits all dynamical effects for the information transmission. In Shannon's theory [2], [18], the information of a system is carried by a probability distribution of events, and a channel induces a change of this probability distribution. The concept of state in a quantum dynamical system can be regarded as an extension of that of probability distribution. Therefore *the information of quantum system is carried by a state, and a channel provides a dynamical change of states.*

It is shown [23] that the dual map of a channel in classical theory transforms any positive bounded function of an input system to that of an output system. By analogy with this fact, a

IEEE TRANSACTIONS ON INFORMATION THEORY, VOL. IT-29, NO. 5, SEPTEMBER 1983

quantum mechanical channel might be formulated as follows [7], [13]: *A mapping* Λ^* *from* $T(\mathcal{H}_1)_{+,1}$ *to* $T(\mathcal{H}_2)_{+,1}$ *is said to be a channel if its dual map* Λ *from* $B(\mathcal{H}_2)$ *to* $B(\mathcal{H}_1)$ *satisfies the following two conditions*: a) Λ *is completely positive* (i.e., *for any positive* $n \times n$*-matrix* $(Q_{i,j})$ *with* $Q_{i,j} \in B(\mathcal{H}_2)$, *the matrix* $(\Lambda Q_{i,j})$ *is positive for all* $n \in N$), *and* b) $\Lambda I_2 = I_1$ *for the identity operator* I_k *on* $\mathcal{H}_k$ $(k = 1,2)$ (we denote every identity operator by I in the sequel when no confusion occurs).

The physical significance of the above conditions is the following. 1) We need "complete positivity" instead of "positivity" because the tensor product $\Lambda_1^* \otimes \Lambda_2^*$ of two channels Λ_1^* and Λ_2^* should also be a channel, and this is not the case in quantum systems unless both Λ_1 and Λ_2 are completely positive [21, p. 218]. 2) The condition b) is necessary for a channel to transfer a state to another state. These conditions are very weak so that most of physical transformations satisfy them [14], [27]. Therefore our definition of a channel might be general enough in order to construct a mathematical framework of quantum communication.

By Stinespring's theorem [19], [26] a completely positive normal map Λ is expressed as $\Lambda Q = \Sigma_k V_k^* Q V_k$ for any $Q \in B(\mathcal{H}_2)$, where V_k is a bounded operator from $\mathcal{H}_1$ to $\mathcal{H}_2$ (i.e., $V_k \in B(\mathcal{H}_1, \mathcal{H}_2)$) with $\Sigma_k V_k^* V_k = I$. Since $\operatorname{tr} \Lambda^* \rho Q = \operatorname{tr} \rho \Lambda Q$, we have $\Lambda^* \rho = \Sigma_k V_k \rho V_k^*$ for any $\rho \in T(\mathcal{H}_1)_{+,1}$. In particular when $\mathcal{H}_1 = \mathcal{H}_2$ and V_k is a projection for each k, the change of ρ to $\Lambda^*\rho$ is often called a quantum measurement [4, p. 15].

Finally we note that a channel considered in this note is supposed to be memoryless. In particular, we do not deal with the transmission of sequences of states.

IV. Compound State

An input state ρ is transmitted through a channel is Λ^* to an output system and the resulting state $\Lambda^*\rho$. The state ρ carries all information of the input system in a certain physical situation. We should like to know how much information is correctly transferred to the output system. For this purpose, we have to find some relation between constituents of ρ and those of $\Lambda^*\rho$.

The compound system of the input and output systems is described by the pair $(B(\mathcal{H}), T(\mathcal{H})_{+,1})$, where $\mathcal{H}$ is the tensor product Hilbert space of $\mathcal{H}_1$ and $\mathcal{H}_2$: $\mathcal{H} = \mathcal{H}_1 \otimes \mathcal{H}_2$.

If an input state ρ is given and its associated output state $\Lambda^*\rho$ is observed, then a compound state σ of ρ and $\Lambda^*\rho$ should satisfy the following conditions: a) $\sigma \in T(\mathcal{H})_{+,1}$, b) $\operatorname{tr} \sigma(A \otimes I) = \operatorname{tr} \rho A$ for any $A \in B(\mathcal{H}_1)$, and c) $\operatorname{tr} \sigma(I \otimes B) = \operatorname{tr} \Lambda^* \rho B$ for any $B \in B(\mathcal{H}_2)$. There are many such states, for example, a state defined by

$$\sigma_0 = \rho \otimes \Lambda^*\rho \tag{3.1}$$

is a typical state satisfying the above conditions. We call this state σ_0 *a trivial compound state*. However, it does not provide for any correlation between ρ and $\Lambda^*\rho$. Not every state satisfying a), b), and c) is suitable for expressing the correlation. Thus we have to construct a proper compound state.

Once an input state ρ is given, we can obtain the eigenvalues $\{\lambda_k\}$ of ρ and have a spectral decomposition of ρ such that

$$\rho = \Sigma_k \lambda_k P_k, \tag{3.2}$$

where P_k is the projection from $\mathcal{H}_1$ to the eigenspace associated to the eigenvalue λ_k. The projection P_k is further decomposed into one-dimensional projections [19], though this decomposition is not always unique. That is,

$$P_k = \Sigma_{j=1}^{n_k} E_{k,j}, \tag{3.3}$$

where n_k is the dimension of P_k and $E_{k,j}$ is the projection to the one-dimensional subspace of $\mathcal{H}$ generated by an eigenvector $\Phi_{k,j}$ associated with λ_k: $E_{k,j} = |\Phi_{k,j}\rangle\langle\Phi_{k,j}|$ in the Dirac notation. By relabeling the indices of $\{E_{k,j}\}$, ρ can be written as

$$\rho = \Sigma_k \lambda_k E_k, \tag{3.4}$$

with

$$E_k = |\Phi_k\rangle\langle\Phi_k|, \tag{3.5}$$

where Φ_k $(k = 1,2,\cdots)$ are mutually orthogonal normalized eigenvectors of ρ. In the decomposition (3.4), the eigenvalue of multiplicity n is repeated precisely n times. Furthermore, this decomposition is unique if and only if no eigenvalue is degenerate. For instance, if the multiplicity of an eigenvalue μ of ρ is two and Ψ_1 and Ψ_2 are its orthonormal eigenvectors, then two vectors defined by $\Omega_1 = 2^{-1/2}(\Psi_1 + \Psi_2)$ and $\Omega_2 = 2^{-1/2}(\Psi_1 - \Psi_2)$ are also eigenvectors of ρ associated with μ. Hence ρ has the following two different expressions:

$$\rho = \mu(|\Psi_1\rangle\langle\Psi_1| + |\Psi_2\rangle\langle\Psi_2|) + \text{other terms}, \tag{3.6}$$

and

$$\rho = \mu(|\Omega_1\rangle\langle\Omega_1| + |\Omega_2\rangle\langle\Omega_2|) + \text{other terms}. \tag{3.7}$$

Each E_k appearing in the decomposition (3.4) is a pure state (i.e., $E_k^2 = E_k$) and is here called an elementary event. Thus the input state ρ can be considered as an event composed of elementary events E_k occurring with probability λ_k. These elementary events are transmitted to the output system through a communication channel Λ^*. Then we should find a connection among the input constituents $\{E_k\}$ and the output constituents $\{\Lambda^* E_k\}$ in order to express the correlation existing between ρ and $\Lambda^*\rho$. We introduce a compound state representing this connection:

$$\sigma_E = \Sigma_k \lambda_k E_k \otimes \Lambda^* E_k. \tag{3.8}$$

This compound state does depend on the decomposition (3.4) of ρ, that is, on the choice of $E = \{E_k\}$. It is clear that σ_E satisfies the conditions a), b), and c) mentioned previously.

The compound state introduced here plays the role of the joint probability in classical theory. In the following sections, we shall see that our construction of the compound state is appropriate for discussing quantum communication.

V. Mutual Information

In this section, we define the mutual information by applying the compound state constructed in the previous section.

The information (entropy) of an input system described by a state ρ is given by von Neumann [25] as

$$S(\rho) = -\operatorname{tr} \rho \log \rho. \tag{4.1}$$

This quantity represents the randomness or uncertainty of ρ.

We now have the following question: *when we send information carried by a state* ρ *to an output system through a channel* Λ^*, *how much information can be transmitted to a receiver observing the output state?*

In Shannon's communication theory, the mutual information gives this amount. It is a partial information of the input state gained by observing the output state. We have to define the mutual information in our quantum dynamical systems. In the sequel, we denote the mutual information by $I(\rho; \Lambda^*)$ for an input state ρ and a channel Λ^*.

When the mutual information $I(\rho; \Lambda^*)$ is properly defined, it should have the following properties: if there exists a one-to-one correspondence between constituents of ρ and those of $\Lambda^*\rho$, then we can expect that a receiver gets all information of ρ, namely, $I(\rho; \Lambda^*) = S(\rho)$. On the other hand, if there does not exist any relation among constituents of ρ and those of $\Lambda^*\rho$, then $I(\rho; \Lambda^*)$ should be zero. Except for these two extreme cases, $I(\rho; \Lambda^*)$ should be larger than 0 and less than $S(\rho)$.

The Kullback–Leibler information [9] is applied when we formulate the mutual information in classical information theory. Umegaki first extended the Kullback–Leibler information to noncommutative dynamical systems and called it the relative information (entropy) [24]. The relative entropy has been further generalized and studied by Araki [1], Lindblad [11], Uhlmann [22], Hiai, Ohya, and Tsukada [5], and others. Using the relative entropy we formulate the mutual information with respect to an input state ρ and a communication channel Λ^* as follows:

$$I(\rho;\Lambda^*) = \sup_E\{S(\sigma_E|\sigma_0)|E = \{E_k\}\}, \tag{4.2}$$

where σ_E and σ_0 are compound states defined in Section IV and $S(\sigma_E|\sigma_0)$ is the relative entropy of σ_E from σ_0 given by

$$S(\sigma_E|\sigma_0) = \operatorname{tr}\sigma_E(\log\sigma_E - \log\sigma_0). \tag{4.3}$$

In order to discuss the properties of the mutual information $I(\rho;\Lambda^*)$, we need a few definitions concerning the channel Λ^*.

1) A channel Λ^* is said to be deterministic if Λ^* sends a pure state in $T(\mathcal{H}_1)_{+,1}$ to a pure state in $T(\mathcal{H}_2)_{+,1}$ and this correspondence is one-to-one and orthogonal (i.e., $E_k \perp E_j \to \Lambda^*E_k \perp \Lambda^*E_j$).
2) A channel Λ^* is said to be chaotic if $\Lambda^*E_k = \Lambda^*E_j$ for any pair E_k, E_j appearing in the decomposition (3.4). (This means that Λ^* cannot distinguish elementary events of ρ.)
3) A channel Λ^* is said to be stationary if $\Lambda\circ\tau_t = \alpha_t\circ\Lambda$ holds (i.e., $\Lambda(\tau_t(A)) = \alpha_t(\Lambda(A))$ for any $A\in B(\mathcal{H}_2)$) for all $t\in R$.

The ergodicity of a channel is an important concept in classical information theory [23]. In [13] we introduced ergodic channels in quantum systems and studied their dynamical properties.

4) A channel Λ^* is said to be ergodic if Λ^* is stationary and Λ^* sends any extremal α-invariant state in $T(\mathcal{H}_1)_{+,1}$ to an extremal τ-invariant state in $T(\mathcal{H}_2)_{+,1}$.

Recall that a state ρ is α-invariant if $\operatorname{tr}\rho\alpha_t(A) = \operatorname{tr}\rho A$ holds for any $A\in B(\mathcal{H}_1)$ and all $t\in R$. An α-invariant state is a state describing a thermodynamical equilibrium of a physical system with a time evolution α_t. Furthermore, an α-invariant state is extremal if it cannot be expressed by a convex combination of other α-invariant states. An extremal α-invariant state is often called an ergodic state α_t, and it describes a pure phase in thermodynamical equilibrium. For more detailed physical meaning of such states, consult [15, ch. 6].

We have the following results.

Theorem 1: For an input state ρ and a channel Λ^*, the inequalities $0 \leqslant I(\rho;\Lambda^*) \leqslant \min\{S(\rho), S(\Lambda^*\rho)\}$ hold.

Theorem 2: For an input state ρ and a channel Λ^*, we have 1) if Λ^* is deterministic, then $I(\rho;\Lambda^*) = S(\rho)$; 2) if Λ^* is chaotic, then $I(\rho;\Lambda^*) = 0$; 3) if ρ is an α-invariant state whose eigenvalues are nonzero and nondegenerate, and if Λ^* is ergodic, then $I(\rho;\Lambda^*) = S(\Lambda^*\rho)$.

Let us prove these theorems. First of all, we prove the following lemma.

Lemma 3: For a decomposition (3.4) of an input state ρ and the compound state σ_E defined by (3.8), we have

$$S(\sigma_E|\sigma_0) = \Sigma_k\lambda_k S(\Lambda^*E_k|\Lambda^*\rho). \tag{4.4}$$

Proof: Let $\{\Phi_n\}$ be a complete orthonormal system of $\mathcal{H}_1$ containing all eigenvectors of ρ and $\{\Psi_m\}$ be a CONS of $\mathcal{H}_2$ containing all eigenvectors of $\Lambda^*\rho$. Then $\{\Phi_n\otimes\Psi_m\}$ becomes a CONS of the tensor product Hilbert space $\mathcal{H} = \mathcal{H}_1\otimes\mathcal{H}_2$. The relative entropy of the compound states σ_E and σ_0 is given by $S(\sigma_E|\sigma_0) = \operatorname{tr}(\sigma_E\log\sigma_E - \sigma_E\log\sigma_0)$.

Put $E_n = |\Phi_n\rangle\langle\Phi_n|$ for every n. Then $\rho E_n = \lambda_n E_n$ if Φ_n is an eigenvector of ρ and $\rho E_n = 0$ otherwise. According to this fact and the orthonormality of Φ_n, we have

$$\begin{aligned}
\operatorname{tr}\sigma_E\log\sigma_E &= \Sigma_{n,m}\Sigma_{n',m'}\langle\Phi_n\otimes\Psi_m, \sigma_E\Phi_{n'}\otimes\Psi_{m'}\rangle\\
&\quad\cdot\log\langle\Phi_{n'}\otimes\Psi_{m'}, \sigma_E\Phi_n\otimes\Psi_m\rangle\\
&= \Sigma_{n,m}\Sigma_{n',m'}\Sigma_k\lambda_k\langle\Phi_n, E_k\Phi_{n'}\rangle\langle\Psi_m, \Lambda^*E_k\Psi_{m'}\rangle\\
&\quad\cdot\log\Sigma_j\lambda_j\langle\Phi_{n'}, E_j\Phi_n\rangle\langle\Psi_{m'}, \Lambda^*E_j\Psi_m\rangle\\
&= \Sigma_{n,m}\Sigma_{m'}\lambda_n\langle\Psi_m, \Lambda^*E_n\Psi_{m'}\rangle\log\lambda_n\langle\Psi_{m'}, \Lambda^*E_n\Psi_m\rangle\\
&= \Sigma_{n,m}\langle\Psi_m, \lambda_n\Lambda^*E_n\log\lambda_n\Lambda^*E_n\Psi_m\rangle\\
&= \Sigma_{n,m}\lambda_n\log\lambda_n\langle\Psi_m, \Lambda^*E_n\Psi_m\rangle\\
&\quad+\Sigma_{n,m}\lambda_n\langle\Psi_m, \Lambda^*E_n\log\Lambda^*E_n\Psi_m\rangle\\
&= \Sigma_n\lambda_n\log\lambda_n\operatorname{tr}\Lambda^*E_n + \Sigma_n\lambda_n\operatorname{tr}\Lambda^*E_n\log\Lambda^*E_n\\
&= \Sigma_n\lambda_n\log\lambda_n + \Sigma_n\lambda_n(-S(\Lambda^*E_n)).
\end{aligned}$$

In the course of the above calculation, we used the fact that Λ^*E_n is a state, hence $\operatorname{tr}\Lambda^*E_n = 1$. In the same manner, we obtain

$$-\operatorname{tr}\sigma_E\log\sigma_0 = -\Sigma_n\lambda_n\log\lambda_n + S(\Lambda^*\rho).$$

Thus,

$$\begin{aligned}
S(\sigma_E|\sigma_0) &= S(\Lambda^*\rho) - \Sigma_n\lambda_n S(\Lambda^*E_n)\\
&= \Sigma_n\lambda_n S(\Lambda^*E_n|\Lambda^*\rho).
\end{aligned}$$

Q.E.D.

Proof of Theorem 1: The positivity of $S(\sigma_E|\sigma_0)$ implies that of $I(\rho;\Lambda^*)$. As shown in the proof of Lemma 3, $S(\sigma_E|\sigma_0) \leqslant S(\Lambda^*\rho)$, from which the inequality $I(\rho;\Lambda^*) \leqslant S(\Lambda^*\rho)$ follows. Let us show another inequality $I(\rho;\Lambda^*) \leqslant S(\rho)$. Since Λ is a completely positive and identity preserving map, it is known [11], [22] that $S(\Lambda^*\rho_1|\Lambda^*\rho_2) \leqslant S(\rho_1|\rho_2)$ for any ρ_1 and ρ_2 in $T(\mathcal{H}_1)_{+,1}$. Hence from Lemma 3, we have

$$\begin{aligned}
S(\sigma_E|\sigma_0) &\leqslant \Sigma_n\lambda_n S(E_n|\rho)\\
&= \Sigma_n\lambda_n(\operatorname{tr}E_n\log E_n - \operatorname{tr}E_n\log\rho)\\
&= -\operatorname{tr}\Sigma_n\lambda_n E_n\log\rho = -\operatorname{tr}\rho\log\rho = S(\rho).
\end{aligned}$$

Taking the supremum over E, we get $I(\rho;\Lambda^*) \leqslant S(\rho)$. Q.E.D.

Proof of Theorem 2: 1) Since Λ^* is deterministic, for each elementary event E_k of ρ, Λ^*E_k is a pure state and $\Lambda^*\rho = \Sigma_k\lambda_k\Lambda^*E_k$ is an extremal decomposition of $\Lambda^*\rho$ into pure states (i.e., a decomposition of type (3.4)). Hence, $S(\Lambda^*\rho) = S(\rho) = -\Sigma_k\lambda_k\log\lambda_k$. From Lemma 3, we have $S(\sigma_E|\sigma_0) = S(\Lambda^*\rho)$ because Λ^*E_k is pure (so $S(\Lambda^*E_k) = 0$). Therefore we get $I(\rho;\Lambda^*) = S(\rho)$.

2) Since Λ^* is chaotic, the compound state σ_E is equivalent to σ_0. Indeed

$$\sigma_E = \Sigma_k\lambda_k E_k\otimes\Lambda^*E_k = \Sigma_k\lambda_k E_k\otimes\Lambda^*\rho = \rho\otimes\Lambda^*\rho = \sigma_0.$$

Thus we get

$$I(\rho;\Lambda^*) = \sup_E S(\sigma_E|\sigma_0) = S(\sigma_0|\sigma_0) = 0.$$

3) When ρ is α-invariant, every pure state E_k appearing in the decomposition (3.4) is also α-invariant. Let us prove this fact. Since ρ is α-invariant, $[u_t,\rho] = 0$ holds for all $t\in R$. From the assumptions for the eigenvalues of the state ρ, the equality $[u_t, E_k] = 0$ holds for all t and k. Thus E_k is a α-invariant state, hence ergodic. Since it is assumed that Λ^* is ergodic, Λ^*E_k is an ergodic state for τ_t. We now show that this ergodic state Λ^*E_k is also pure. Suppose that Λ^*E_k is not pure; then we have an extremal decomposition of Λ^*E_k into pure states such as $\Lambda^*E_k = \Sigma_n\mu_n^k\theta_n^k$. The τ-invariance of Λ^*E_k reduces that of θ_n^k as discussed above. This contradicts the fact that Λ^*E_k is an ergodic state.

Thus $\Lambda^* E_k$ is a pure state. From this result we have

$$I(\rho;\Lambda^*) = \sup_E S(\sigma_E|\sigma_0) = S(\Lambda^*\rho)$$

because of $S(\Lambda^* E_k) = 0$. Q.E.D.

Part 3) of Theorem 2 tells that for an ergodic channel the mutual information is equal to the information carried by the output state $\Lambda^*\rho$; hence an ergodic channel is the most effective channel for α-invariant states having nondegenerate eigenvalues.

It is seen from Theorems 1 and 2 that the mutual information defined by (4.2) satisfies most of the properties desired of an information gain. The mutual information is given by a compound state; therefore our construction of a compound state will be on a right track for quantum communication theory.

Before closing this section, we make a few more remarks.

Remark 1: Let us briefly discuss the correspondence between our formulation and that of Shannon. Suppose for simplicity that both input and output classical systems have n events. In classical theory an input state ρ is a probability distribution $(q_j)_{j=1}^n$ of events, and a channel Λ^* is given by a $n \times n$ transition matrix $[p_{ij}]_{i,j=1}^n$. Thus we have the following correspondences:

a) output state $\Leftrightarrow$ probability distribution $\Lambda^*\rho = (\Sigma_{j=1}^n p_{ij}q_j)_{i=1}^n\ (\equiv (r_i))$;
b) compound state $\Leftrightarrow$ joint probability distribution $(p(i,j))_{i,j=1}^n$ with $p(i,j) \equiv p_{ij}q_j$;
c) mutual information $I = \Sigma_{i,j} p(i,j)(\log p(i,j) - \log r_i q_j)$.

Remark 2: The capacity of the channel is an important quantity for analyzing the transmission of information. In quantum communication theory, it might be given by taking the supremum of $I(\rho;\Lambda^*)$ over ρ in a certain subset $\mathbb{S}$ of $T(\mathcal{H}_1)_{+,1}$. As we need more mathematical preparation to discuss the capacity and its properties and we are still working on this subject, we do not treat it here.

VI. A Construction of Channel and an Application

We have two aims in this section. One of them is to determine the form of a channel Λ^* by taking account of direct effects of noise and loss existing in the channel, and another is to examine the validity of our formulation by considering a simple process of photon transmission.

In addition to the Hilbert spaces $\mathcal{H}_1$ and $\mathcal{H}_2$, we here use two more Hilbert spaces $\mathcal{K}_1$ and $\mathcal{K}_2$ in order explicitly to describe some external effects to the input and output states. For instance a state in $\mathcal{K}_1$ induces a noise into the channel and a state in $\mathcal{K}_2$ indicates a loss of information at the output system.

Let ξ be a state in $\mathcal{K}_1$ (i.e., $\xi \in T(\mathcal{K}_1)_{+,1}$) describing the noise in the channel. We consider the following maps:

$$B(\mathcal{H}_2) \xrightarrow{a} B(\mathcal{H}_2 \otimes \mathcal{K}_2) \xrightarrow{\Pi} B(\mathcal{H}_1 \otimes \mathcal{K}_1) \xrightarrow{\Gamma} B(\mathcal{H}_1). \tag{5.1}$$

These maps are defined as follows. The map a is an amplification from $B(\mathcal{H}_2)$ to $B(\mathcal{H}_2 \otimes \mathcal{K}_2)$ given by $a(A) = A \otimes I$ for any $A \in B(\mathcal{H}_2)$. The map Π is a completely positive map from $B(\mathcal{H}_2 \otimes \mathcal{K}_2)$ to $B(\mathcal{H}_1 \otimes \mathcal{K}_1)$ with $\Pi(I) = I$. The map Γ is given by $\Gamma(Q) = \mathrm{tr}_{\mathcal{K}_1} \xi Q$ for any $Q \in B(\mathcal{H}_1 \otimes \mathcal{K}_1)$, where $\mathrm{tr}_{\mathcal{K}_1}$ is a partial trace with respect to the Hilbert space $\mathcal{K}_1$ [4, p. 150].

It is easy to show [12] that these maps are completely positive, hence $\Lambda = \Gamma \circ \Pi \circ a$ is also completely positive.

We next consider the dual maps of a, Π, and Γ. The dual map a^* of a is a map from $T(\mathcal{H}_2 \otimes \mathcal{K}_2)_{+,1}$ to $T(\mathcal{H}_2)_{+,1}$ such that $\mathrm{tr}\, a^*(\theta)A = \mathrm{tr}\, \theta a(A)$ for any $\theta \in T(\mathcal{H}_2 \otimes \mathcal{K}_2)_{+,1}$ and any $A \in B(\mathcal{H}_2)$. Thus, $a^*(\theta) = \mathrm{tr}_{\mathcal{K}_2} \theta$. The dual map Π^*: $T(\mathcal{H}_1 \otimes \mathcal{K}_1)_{+,1} \to T(\mathcal{H}_2 \otimes \mathcal{K}_2)_{+,1}$ is given by $\mathrm{tr}\, \Pi^*(\sigma)W = \mathrm{tr}\, \sigma\Pi(W)$ for any $\sigma \in T(\mathcal{H}_1 \otimes \mathcal{K}_1)_{+,1}$ and any $W \in B(\mathcal{H}_2 \otimes \mathcal{K}_2)$. The dual map Γ^*: $T(\mathcal{H}_1)_{+,1} \to T(\mathcal{H}_1 \otimes \mathcal{K}_1)_{+,1}$ is given by $\mathrm{tr}\, \Gamma^*(\rho)Q = \mathrm{tr}\, \rho\Gamma(Q)$ for any $\rho \in T(\mathcal{H}_1)_{+,1}$ and any $Q \in B(\mathcal{H}_1 \otimes \mathcal{K}_1)$. It is easily seen that Γ^* is expressed as $\Gamma^*(\rho) = \rho \otimes \xi$.

Therefore, once we know the noise ξ and the mechanism of the transformation Π, we can write down a channel explicitly such that

$$\Lambda^* = a^* \circ \Pi^* \circ \Gamma^* \tag{5.2}$$

or equivalently,

$$\Lambda^*\rho = \mathrm{tr}_{\mathcal{K}_2} \Pi^*(\rho \otimes \xi) \tag{5.3}$$

for any $\rho \in T(\mathcal{H}_1)_{+,1}$.

If neither noise nor loss exists in the course of information transmission, then $\Lambda^* = \Pi^*$ because $\mathcal{K}_1 = C$ and $\mathcal{K}_2 = C$, the set of all complex numbers. In this case the information of the input system should be completely transmitted: $I(\rho;\Lambda^*) = S(\rho)$. Therefore Π^* should be deterministic.

We apply our formulation to a very simple model of information transmission. This model has been discussed in [20], and we briefly sketch it. A transmitter sends photons with frequency in the laser range, and a receiver detects them. The system composed of photons is described by a Hamiltonian $H = a^*a + 1/2$, where a^* and a are creation and annihilation operators of photon, respectively. Here we take $h\nu = 1$ for simplicity. The Schrödinger equation $H\Phi(x) = \epsilon\Phi(x)$ is easily solved [16, p. 182]: the eigenvalue ϵ_n is given by $\epsilon_n = n + 1/2$ ($n \in N$, the set of all natural numbers) and the associated eigenvector is given by $\Phi_n(x) = (1/\sqrt{\pi^{1/2}n!})H_n(\sqrt{2}x)\exp(-x^2/2)$ for each n, where $H_n(x)$ is the nth Hermite function. The Hilbert space $\mathcal{H}$ of this system is the closed linear span of linear combinations of Φ_n ($n = 1,2,3,\cdots$). In this model the Hilbert spaces $\mathcal{H}_1$, $\mathcal{H}_2$, $\mathcal{K}_1$, and $\mathcal{K}_2$ considered above and their related quantities are denoted in a Table I.

TABLE I
SIMPLE MODEL

Hilbert Space	Annihilation Creation Op.	Hamiltonian	Eigenvalue	CONS
$\mathcal{H}_1$	a_1, a_1^*	$a_1^*a_1 + 1/2$	$n_1 + 1/2$	$\Phi_{n_1}(x_1)$
$\mathcal{K}_1$	a_2, a_2^*	$a_2^*a_2 + 1/2$	$n_2 + 1/2$	$\Phi_{n_2}(x_2)$
$\mathcal{H}_2$	b_1, b_1^*	$b_1^*b_1 + 1/2$	$m_1 + 1/2$	$\Phi_{m_1}(y_1)$
$\mathcal{K}_2$	b_2, b_2^*	$b_2^*b_2 + 1/2$	$m_2 + 1/2$	$\Phi_{m_2}(y_2)$

When n_1 photons are sent through a channel having n_2 photons as its noise, let m_1 photons be detected and m_2 photons be lost. Then a relation a) $n_1 + n_2 = m_1 + m_2$ should hold because of conservation of energy (photon number). Since $a = 1/\sqrt{2}(x + \partial/\partial x)$ and $a^* = 1/\sqrt{2}(x - \partial/\partial x)$, the relation a) is satisfied under, for example, a linear transformation of coordinates x_1, x_2, y_1, y_2 such that b) $x_1 = \alpha y_1 - \beta y_2$ and $x_2 = \beta y_1 + \alpha y_2$ with $\alpha^2 + \beta^2 = 1$. We now suppose for simplicity that the noise is due to the "zero point fluctuation" of electromagnetic field, that is, $n_2 = 0$. Hence the noise source is described by a state $\xi = |\Phi_0(x_2)\rangle\langle\Phi_0(x_2)|$. For an input state $E_n = |\Phi_n(x_1)\rangle\langle\Phi_n(x_1)|$, the change of the state $E_n \otimes \xi$ under the transformation b) follows from the change of the vector $\Phi_n(x_1) \otimes \Phi_0(x_2)$ such that

$$\Phi_n(x_1) \otimes \Phi_0(x_2) \xrightarrow{\text{b)}} V\Phi_n(x_1) \otimes \Phi_0(x_2) (= \Psi_n(y_1, y_2))$$
$$= \Sigma_{j=0}^n c_j^n \Phi_j(y_1) \otimes \Phi_{n-j}(y_2),$$

where $c_j^n = \sqrt{n!/j!(n-j)!}(-\beta)^{n-j}\alpha^j$. By (5.3) we have

$$\Lambda^* E_n = \mathrm{tr}_{\mathcal{K}_2} V(\rho \otimes \xi)V^* = \mathrm{tr}_{\mathcal{K}_2}|\Psi_n\rangle\langle\Psi_n| = \Sigma_{j=0}^n |c_j^n|^2 |\Phi_j\rangle\langle\Phi_j|.$$

Therefore for an input state $\rho = \Sigma_{n=0}^{N} \lambda_n E_n$ $(0 \leqslant N \leqslant \infty)$ with $\Sigma_n \lambda_n = 1$ and $\lambda_k \neq \lambda_j$ $(k \neq j)$, the compound state and the trivial compound state introduced in Section IV are given by

$$\sigma_E = \Sigma_n \lambda_n E_n \otimes \Lambda^* E_n = \Sigma_{n=0}^{N} \Sigma_{j=0}^{n} \lambda_n |c_j^n|^2 E_n \otimes \theta_j,$$

$$\sigma_0 = \Sigma_n \lambda_n E_n \otimes \Sigma_k \lambda_k \Lambda^* E_k = \Sigma_{n,k} \Sigma_{j=0}^{k} \lambda_n \lambda_k |c_j^k|^2 E_n \otimes \theta_j,$$

where θ_j is defined by $\theta_j = |\Phi_j\rangle\langle\Phi_j| \in T(\mathcal{H}_2)_{+,1}$. Thus we can calculate the mutual information as follows:

$$\begin{aligned} I(\rho; \Lambda^*) = & -\Sigma_n \lambda_n \log \lambda_n - \Sigma_{j=0}^{N} \left(\Sigma_{n=j}^{N} \lambda_n |c_j^n|^2 \right) \\ & \cdot \log \left(\Sigma_{n=j}^{N} \lambda_n |c_j^n|^2 \right) + \Sigma_n \Sigma_{j=0}^{n} \left(\lambda_n |c_j^n|^2 \right) \log \left(\lambda_n |c_j^n|^2 \right) \\ = & \Sigma_{j=0}^{N} \Sigma_{n=j}^{N} \lambda_n |c_j^n|^2 \log \left(|c_j^n|^2 / \left(\Sigma_{n=j}^{N} \lambda_n |c_j^n|^2 \right) \right). \end{aligned}$$

This result is also obtained from the Shannon's formula with an initial probability $\langle \lambda_n \rangle$ and the conditional probability $\langle |c_j^n|^2 \rangle$ given above. As the above model is very simple, this conditional probability can be calculated [20] without introducing a channel Λ^* explicitly. Once the conditional probability is known (e.g., when every eigenvalue of ρ is nondegenerate and Λ^* is given explicitly, it can be obtained by taking proper CONS's at the input and output systems), Shannon's expression for the mutual information is equivalent to ours. Without formulating a channel and a compound state precisely, neither have we a general rule to compute the mutual information (the joint probability) nor can we study quantum communication processes in general or the properties of a channel and the mutual information in particular.

Acknowledgment

The author wishes to express his gratitude to Professor H. Umegaki for his valuable comments and constant encouragement. He also thanks the referees for their valuable suggestions and advice, which enabled him to revise his correspondence in the present form.

References

[1] H. Araki, "Relative entropy of states of von Neumann algebras," *Publ. RIMS*, Kyoto Univ., vol. 11, pp. 809–833, 1976.

[2] R. Ash, *Information Theory*. New York: Interscience, 1965.

[3] E. B. Davies, "Information and quantum measurement," *IEEE Trans. Inform. Theory*, pp. 530–534, 1977.

[4] ——, *Quantum Theory of Open System*. New York: Academic, 1976.

[5] F. Hiai, M. Ohya, and M. Tsukada, "Sufficiency, KMS condition and relative entropy in von Neumann algebras," *Pacific J. Math.*, vol. 96, pp. 99–109, 1981.

[6] C. W. Helstrom, J. W. S. Liu, and J. P. Gordon, "Quantum mechanical communication theory," in *Proc. IEEE*, vol. 58, pp. 1578–1598, 1970.

[7] A. S. Holevo, "Problems in the mathematical theory of quantum communication channels," *Rep. Math. Phys.*, vol. 12, pp. 273–278, 1977.

[8] ——, "Information theoretical aspects of quantum measurement," *Probl. Peredači Inform.*, vol. 9, no. 2, pp. 31–42, 1973.

[9] S. Kullback and R. A. Leibler, "On information and sufficiency," *Ann. Math. Statist.*, vol. 22, pp. 79–86, 1951.

[10] R. S. Ingarden, "Quantum information theory," *Rep. Math. Phys.*, vol. 10, pp. 43–73, 1976.

[11] G. Lindblad, "Completely positive maps and entropy inequalities," *Commun. Math. Phys.*, vol. 40, pp. 147–151, 1975.

[12] ——, "On the generators of quantum dynamical semigroups," *Commun. Math. Phys.*, vol. 48, pp. 119–130, 1975.

[13] M. Ohya, "Quantum ergodic channels in operator algebras," *J. Math. Anal. Appl.*, vol. 84, pp. 318–328, 1981.

[14] ——, "Entropy transmission in C^*-dynamical systems," preprint.

[15] D. Ruelle, *Statistical Mechanics*. New York: Benjamin, 1965.

[16] L. I. Schiff, *Quantum Mechanics*, 2nd ed. New York: Wiley, 1968.

[17] R. Schatten, *Norm Ideals of Completely Continuous Operators*. New York: Springer-Verlag, 1970.

[18] C. E. Shannon, *A Mathematical Theory of Communication*. Urbana, IL: Univ. Illinois, 1949.

[19] N. F. Stinespring, "Positive functions on C^*-algebras," in *Proc. Amer. Math. Soc.*, vol. 6, pp. 211–216, 1955.

[20] H. Takahasi, *Information Theory of Quantum Mechanical Channels, Advances in Communication Systems*, Vol. 1. New York: Academic, 1966, pp. 227–310.

[21] M. Takesaki, *Theory of Operator Algebra I*. New York: Springer-Verlag, 1981.

[22] A. Uhlmann, "Relative entropy and the Wigner–Yanase–Dyson–Lieb concavity in an interpolation theory," *Commun. Math. Phys.*, vol. 54, pp. 21–32, 1977.

[23] H. Umegaki, "Representation and extremal properties of averaging operators and their applications to information channels," *J. Math. Anal. Appl.*, vol. 25, pp. 41–73, 1969.

[24] ——, "Conditional expectation in an operator algebras IV," *Kodai Math. Sem. Rep.*, vol. 14, pp. 59–85, 1962.

[25] J. von Neumann, *Die Mathematischen Grundlagen der Quantenmechanik*. Berlin: Springer, 1932.

[26] K. Kraus, "General state changes in quantum theory," *Ann. Phys.*, vol. 64, pp. 311–335, 1970.

[27] V. Gorini, A. Frigerio, M. Verri, A. Kossakowski, and E. L. G. Sudarshan, "Properties of quantum Markovian master equations," *Rep. Math. Phys.*, vol. 13, pp. 149–173, 1982.

LETTERE AL NUOVO CIMENTO VOL. 38, N. 11 12 Novembre 1983

Note on Quantum Probability.

M. OHYA

Department of Information Sciences, Science University of Tokyo
Noda City, Chiba 278, Japan

(ricevuto il 3 Agosto 1983)

PACS. 03.65. – Quantum theory; quantum mechanics.

Summary. – When a state of a physical system dynamically changes to another state, it is important to know the correlation existing between the initial state and the final state. This correlation is described by a compound state (measure) in classical systems. In this note, we show a way how to construct such a compound state in quantum systems which is an extension of the classical compound state.

It is rather important in many physical sciences to study the dynamical change of states of a system. One of the most general description of this state change for classical systems is suggested in the communication theory of Shannon.

A state of a classical dynamical system is expressed by a probability measure on that system and its dynamical change is generally considered as follows [1]: let X, Y be compact Hausdorff spaces and $\mathscr{F}_X$, $\mathscr{F}_Y$ be their Borel fields, respectively. We denote the set of all regular probability measures (states) on $(X, \mathscr{F}_X)$ by $P(X)$ and on $(Y, \mathscr{F}_Y)$ by $P(Y)$. A mapping $\lambda\colon X \times \mathscr{F}_Y \to R^+$ satisfying the following two conditions is called a channel: i) $\lambda(x, \cdot) \in P(Y)$ for each fixed $x \in X$ and ii) $\lambda(\cdot, Q)$ is a continuous measurable function on X for each fixed $Q \in \mathscr{F}_Y$. This mapping is often called a transition (or Markov) kernel and is useful to study, for instance, information transmission and stochastic processes. A channel so defined provides a mechanism of state change. Namely, a state $\varphi \in P(X)$ is transferred to a state $\psi \in P(Y)$ under a channel λ such as

$$\psi(Q) = \int_X \lambda(x, Q)\,\varphi(\mathrm{d}x)\,, \qquad Q \in \mathscr{F}_Y\,. \tag{1}$$

Moreover, in order to study the process of state change and the property of a channel itself, we need a compound state (joint probability) indicating the correlation existing

(1) H. UMEGAKI: *J. Math. Anal. Appl.*, **25**, 41 (1969).

between the initial state φ and the final state ψ. The compound state Φ is given by

$$\Phi(Q_1, Q_2) = \int_{Q_1} \lambda(x, Q_2)\,\varphi(\mathrm{d}x) \tag{2}$$

for any $Q_1 \in \mathscr{F}_X$ and $Q_2 \in \mathscr{F}_Y$.

In quantum dynamics, we take two C^*-systems $(\mathscr{A}, \mathfrak{S}(\mathscr{A}))$ and $(\mathscr{B}, \mathfrak{S}(\Lambda))$, one of which describes an initial (input) system corresponding to $(X, \mathscr{F}_X, P(X))$ above and another describes a final (output) system corresponding to $(Y, \mathscr{F}_Y, P(Y))$. Here $\mathscr{A}$ (respectively, $\mathscr{B}$) is a C^*-algebra with unity $I_{\mathscr{A}}$ (respectively, $I_{\mathscr{B}}$) and $\mathfrak{S}(\mathscr{A})$ (respectively, $\mathfrak{S}(\mathscr{B})$) is the set of all states (*i.e.* normalized positive linear functionals) on $\mathscr{A}$ (respectively, $\mathscr{B}$) ([2]). Then let us consider a mapping Λ^* from $\mathfrak{S}(\mathscr{A})$ to $\mathfrak{S}(\mathscr{B})$ such that its dual map Λ: $:\mathscr{B} \to \mathscr{A}$ is completely positive ([2]) with $\Lambda I_{\mathscr{B}} = I_{\mathscr{A}}$. This mapping Λ^* is called a channel between two quantum-dynamical systems ([3]). In particular, when $\mathscr{A} = C(X)$, the set of all continuous functions on X, and $\mathscr{B} = C(Y)$, the formula (1) defines a channel Λ^* from $P(X)$ to $P(Y)$ (*i.e.* $\psi = \Lambda^*\varphi$) because every probability measure φ on $(X, \mathscr{F}_X)$ can be regarded as a state on $C(X)$ by the Riesz-Markov-Kakutani theorem. We meet several channels in several fields of physics. For example, time evolution automorphism group, dynamical semi-group and conditional expectation on a certain algebra are typical channels.

Now it is well known ([4]) that the joint probability measure does not generally exist in quantum systems. Hence it has been difficult to define a compound state describing the correlation existing between an initial state $\varphi \in \mathfrak{S}(\mathscr{A})$ and its final state $\Lambda^*\varphi \in \mathfrak{S}(\mathscr{B})$. The aim of this note is to construct such a compound state and to show that our compound state is an extension of the classical one given by (2).

For an initial state φ and the final state $\Lambda^*\varphi$, a compound state Φ on $\mathscr{A} \otimes \mathscr{B}$ of φ and $\Lambda^*\varphi$ should satisfy the following two conditions: i) $\Phi(A \otimes I_{\mathscr{B}}) = \varphi(A)$ for any $A \in \mathscr{A}$ and ii) $\varphi(I_{\mathscr{A}} \otimes B) = \Lambda^*\varphi(B)$ for any $B \in \mathscr{B}$. There exist many states satisfying these conditions, for instance, $\Phi_0 = \varphi \otimes \Lambda^*\varphi$ is such a state. But this state does not carry any correlation between φ and $\Lambda^*\varphi$.

For any weak *-compact convex subset $\mathscr{S}$ of $\mathfrak{S}(\mathscr{A})$, there exists a maximal measure μ such that φ is the barycentre of μ and μ is pseudosupported by the set $e \times \mathscr{S}$ of all extreme points of $\mathscr{S}$ in the sense that $\mu(Q) = 1$ for every Baire subset Q of $\mathscr{S}$ with $Q \supset e \times \mathscr{S}$ ([5]). In this case, we write

$$\varphi = \int_{(e\times\mathscr{S})} \omega \,\mathrm{d}\mu\,. \tag{3}$$

Note that the above maximal measure μ is not always unique, and we denote the set of all such measures by $M_\varphi(\mathscr{S})$.

We now construct a « true » compound state of φ and $\Lambda^*\varphi$. For each $\mu \in M_\varphi(\mathscr{S})$, define

$$\varphi_\mu = \int_{(e\times\mathscr{S})} \omega \otimes \Lambda^*\omega \,\mathrm{d}\mu\,. \tag{4}$$

It is easy to see that this state Φ_μ satisfies the conditions i) and ii) mentioned above. Let us now show that the compound state defined by (4) is indeed an extension of the

([2]) M. Takesaki: *Theory of operator algebra I* (Berlin, 1981).
([3]) M. Ohya: *J. Math. Anal. Appl.*, **84**, No. 2, 318 (1981).
([4]) K. Urbanik: *Studia Math.*, **21**, 117 (1961).

classical one. When $\mathscr{S} = P(X)$, the extremal decomposition of a state $\varphi \in P(X)$ is unique and given by

$$\varphi = \int_X \delta_x \varphi(\mathrm{d}x) , \tag{5}$$

where δ_x is the Dirac measure concentrated at a point $x \in X$. Since $(\Lambda^* \delta_x)(Q_2) = \lambda(x, Q_2)$ for any $Q_2 \in \mathscr{F}_Y$, we have

$$\Phi_\varphi(Q_1 \times Q_2) = \int_X \delta_x(Q_1) \Lambda^* \delta_x(Q_2) \varphi(\mathrm{d}x) = \int_X 1_{Q_1}(x) \lambda(x, Q_2) \varphi(\mathrm{d}x) = \int_{Q_1} \lambda(x, Q_2) \varphi(\mathrm{d}x) ,$$

for any $Q_1 \in \mathscr{F}_X$ and $Q_2 \in \mathscr{F}_Y$.

We finally consider the case of $\mathscr{A} = C(\mathscr{H}_1) = C(\mathscr{H}_1) + CI$, where $C(\mathscr{H})$ is the set of all compact operators on a separable Hilbert space $\mathscr{H}$. Then $\mathfrak{S}(\mathscr{A})$ contains the set $T(\mathscr{H}_1)_{+,1}$ of all positive trace class operators on $\mathscr{H}_1$ with unit trace and so does $\mathfrak{S}(\mathscr{B})$. Moreover, a channel Λ^* is a trace-preserving completely positive map from $T(\mathscr{H}_1)_{+,1}$ to $T(\mathscr{H}_2)_{+,1}$. In this case, if an extremal decomposition of a state $\varrho \in T(\mathscr{H}_1)_{+,1}$ is given by $\varrho = \sum_n \lambda_n \varrho_n$, then our compound state is

$$\sigma = \sum_n \lambda_n \varrho_n \otimes \Lambda^* \varrho_n .$$

Among these compound states, the following is the most important:

$$\sigma_E = \sum_n \lambda_n E_n \otimes \Lambda^* E_n . \tag{6}$$

The symbols appearing in (6) mean the following: *a*) λ_n is the eigenvalue of ϱ, and the eigenvalue of multiplicity m is repeated precisely m times. *b*) $E_n = |x_n\rangle\langle x_n|$ in Dirac's notation, where x_n is an eigenvector associated with λ_n and it is unique if and only if λ_n is nondegenerate (5). *c*) E represents $\{E_n\}$. The compound state σ_E depends on E, hence on the choice of the eigenvectors $\{x_n\}$. This compound state is first introduced in (2), but it is a special case of our compound state Φ_μ as just discussed.

The compound state constructed for quantum systems plays a similar role as the joint probability measure for classical systems, so that it is useful to formulate quantum communication theory (6) and to study the dynamics of state change (7).

* * *

The author wishes to express his gratitude to Prof. H. UMEGAKI for his interest in and valuable comments to the present work.

(5) R. SCHATTEN: *Norm ideals of completely continuous operators* (Berlin, 1970).

(6) M. OHYA: *On compound state and mutual information in quantum information theory*, to appear in *IEEE Information Theory*.

(7) M. OHYA: *Remarks on dynamical change of states*, preprint.

Reprinted from JOURNAL OF MATHEMATICAL ANALYSIS AND APPLICATIONS Vol. 84, No. 2, December 1981

Quantum Ergodic Channels in Operator Algebras

MASANORI OHYA

Department of Information Sciences,
Science University of Tokyo, 278 Noda City, Chiba Japan

Submitted by G.-C. Rota

The quantum ergodic channel is studied by operator algebraic methods. The ergodic and KMS channels are introduced and their dynamical properties are discussed.

INTRODUCTION

After Shannon's work [19], information theory deals not only with entropy (information) itself but also with its transmission. This mechanism of transmission has been revealed by the so-called channel between input sources and output receivers.

Laser physics has been much developed and its usefulness to information theory has recently been realized. It is therefore necessary to formulate communication theory in terms of quantum mechanical language. Many authors have tried to find a mathematical basis of information theory [2, 9–12, 23]. The present paper is one trial along this line.

In Section 1, we introduce several channels and formulate the problems to be investigated in this paper. In Section 2, we discuss the existence of ergodic channels introduced in Section 1. In Section 3, we characterize the ergodic channel and prove a theorem of the Radon–Nikodym type for a stationary channel. In Section 4, we study the KMS and weakly mixing channels and consider some relations between such channels.

1. PRELIMINARIES

The classical communication channel has been described in the following way: Let X, Y be compact Hausdorff spaces and $\mathscr{B}_X$, $\mathscr{B}_Y$ be their σ-fields, respectively. In information theory we call $(X, \mathscr{B}_X)$ an input (message) space and $(Y, \mathscr{B}_Y)$ an output (message) space. We denote the sets of probability measures by $P(X)$ on $(X, \mathscr{B}_X)$ and by $P(Y)$ on $(Y, \mathscr{B}_Y)$. A mapping

$\nu: X \otimes \mathscr{B}_Y \to R^+$ is said to be a channel from the input space to the output space when the following conditions are met:

(1) $\nu(x, \cdot) \in P(X)$ for each fixed $x \in X$,

(2) $\nu(\cdot, Q)$ is measurable on $(X, \mathscr{B}_X)$ for each fixed Q in $\mathscr{B}_Y$.

A probability measure μ is transferred to a probability measure μ' through a channel ν in the manner of

$$\mu'(Q) = \int_X \nu(x, Q)\, d\mu(x),$$

for any $Q \in \mathscr{B}_Y$.

Let $B(X)$ and $B(Y)$ be C^*-algebras of all bounded Baire measurable functions on X and Y, respectively. Let $K(B(Y), B(X))$ be the set of all bounded linear transformations from $B(Y)$ to $B(X)$ satisfying for any $K \in K(B(Y), B(X))$, (1) $K1 = 1$, (2) $Kf \geqslant 0$ for any $f \geqslant 0$ in $B(Y)$, and (3) $Kf_n \downarrow 0$ for any $f_n \downarrow 0$ in $B(Y)$.

Umegaki has shown [23] that a one to one affine correspondence exists between the set of channels and the set $K(B(Y), B(X))$ above such that

$$(Kf)(x) = \int_Y f(y)\, \nu(x, dy)$$

for any $f \in B(Y)$.

A quantum mechanical dynamical system is described by a triple $(\mathscr{A}, \mathfrak{S}(\mathscr{A}), \alpha(R))$, where $\mathscr{A}$ is a von Neumann algebra with unity I acting on a Hilbert space $\mathscr{H}$, $\mathfrak{S}(\mathscr{A})$ is the set of all normal states on $\mathscr{A}$, and $\alpha(R)$ is a strongly continuous one-parameter group of automorphism of $\mathscr{A}$.

We need two such dynamical systems in the information theory, namely, an input system and an output system. An input system is described by a triple $(\mathscr{A}, \mathfrak{S}(\mathscr{A}), \alpha(R))$ above and an output system is described by another von Neumann algebraic triple $(\mathscr{B}, \mathfrak{S}(\mathscr{B}), \tau(R))$ acting on another Hilbert space $\mathscr{K}$.

A message of the input system is sent to the output system through a channel. By analogy with Umegaki's result discussed above, it is natural for us to take a channel for the quantum case as a mapping from $\mathfrak{S}(\mathscr{A})$ to $\mathfrak{S}(\mathscr{B})$. More precisely, a mapping, denoted by Λ^*, from $\mathfrak{S}(\mathscr{A})$ to $\mathfrak{S}(\mathscr{B})$ is said to be a channel if it is a dual map of Λ from $\mathscr{B}$ to $\mathscr{A}$ such that (1) Λ is completely positive [1, 4, 14, 15, 20, 21] (i.e., it sends every positive $n \times n$ matrix (B_{ij}) with entries B_{ij} in $\mathscr{B}$ to a positive $n \times n$ matrix (ΛB_{ij}) for al $n \in N$), (2) Λ is ultraweakly continuous, and (3) $\Lambda(I) = I$ (cf. Ref. [10]).

The set of all invariant states in $\mathfrak{S}(\mathscr{A})$ under α_t $(t \in R)$ is denoted by $I(\alpha)$ and the extreme points of $I(\alpha)$ are denoted by $exI(\alpha)$. We similarly define $I(\tau)$ and $exI(\tau)$.

A channel Λ^* is said to be stationary (or covariant) if $\Lambda \circ \tau_t = \alpha_t \circ \Lambda$ holds for all $t \in R$. The set of all stationary channels is denoted by $SC(\mathscr{A}, \mathscr{B})$ or simply by SC.

A channel Λ^* is said to be ergodic if $\Lambda^* \in SC$ and $\varphi \in exI(\alpha)$ implies $\Lambda^*\varphi \in exI(\tau)$. The set of all ergodic channels is denoted by $EC(\mathscr{A}, \mathscr{B})$ or EC.

These channels play important roles in usual (classical) information theory (cf. Ref. [23]).

In quantum statistical mechanics, the equilibrium state of a dynamical system is an important tool in studying the property of such a system [8, 17]. The most general formulation of an equilibrium state is due to the KMS condition [6, 8, 22]. A state $\varphi \in \mathfrak{S}(\mathscr{A})$ satisfies this condition with respect to α_t if for any pair $A, B \in \mathscr{A}$, there exists a bounded function $F_{A,B}(z)$ of $z \in C$ analytic in,and continuous on, the strip $0 \leqslant \operatorname{Im} z \leqslant 1$ with boundary values: $F_{A,B}(t) = \varphi(\alpha_t(A)\,B)$ and $F_{A,B}(t+i) = \varphi(B\alpha_t(A))$. Let us denote the set of all KMS states w.r.t. α_t by $K(\alpha)$ and the set of all extreme points of $K(\alpha)$ by $exK(\alpha)$. We also define $K(\tau)$ and $exK(\tau)$ for τ_t.

A channel Λ^* is said to be KMS if $\Lambda^* \in SC$ and $\varphi \in K(\alpha)$ implies $\Lambda^*\varphi \in K(\tau)$. We denote the set of all KMS channels by $KC(\mathscr{A}, \mathscr{B})$ or KC.

A channel Λ^* is said to be extremal in SC if $\Lambda^* = \lambda\Lambda_1^* + (1-\lambda)\,\Lambda_2^*$ for some $\Lambda_1^*, \Lambda_2^* \in SC$ and $\lambda \in (0, 1)$ implies $\Lambda_1^* = \Lambda_2^* = \Lambda^*$.

We introduce one more concept in this section: This concept is somewhat related to that of ergodic channel.

We call a state φ weakly mixing (or η-clustering) for α if $\varphi(\eta_\alpha(A)\,B) = \varphi(\eta_\alpha(A))\,\varphi(B)$ holds for any A, B in $\mathscr{A}$, where η_α is defined as

$$\eta_\alpha(A) = \lim_{T\to\infty} \frac{1}{T}\int_0^T dt\,\alpha_t(A)$$

for any $A \in \mathscr{A}$. We denote the set of all weakly mixing states for α by $WM(\alpha)$. For τ, we define $WM(\tau)$ in the same manner.

A channel Λ^* is said to be weakly mixing if $\varphi \in WM(\alpha)$ implies $\Lambda^*\varphi \in WM(\tau)$. The set of all weakly mixing channels is denoted by $WMC(\mathscr{A}, \mathscr{B})$ or WMC.

The problems to which we address ourselves are as follows:

(1) Do the channels defined above exist?

(2) What kinds of characterizations do we have for such ergodic channels?

Remark. Some authors define a weakly mixing state ϕ by $\lim_{T\to\infty} (1/T)$

$\int_0^T dt\,|\phi(\alpha_t(A)\,B)-\phi(\alpha_t(A))\,\phi(B)|=0$. This definition is slightly general, but we take the above definition here.

2. Existence of Ergodic Channels

Let $\mathfrak{N}$ be the von Neumann algebraic tensor product of $\mathscr{A}$ and $\mathscr{B}$. Let $\mathrm{Aut}(\mathfrak{N})$ be the set of all automorphisms of $\mathfrak{N}=\mathscr{A}\otimes\mathscr{B}$. We denote an embedding from $\mathscr{B}$ into $\mathfrak{N}$ by j; that is, $j(B)=1\otimes B$ for any $B\in\mathscr{B}$.

Suppose that $exI(\tau)\neq\varnothing$ (resp. $K(\tau)\neq\varnothing$ or $WM(\tau)\neq\varnothing$). We take a state $\bar{\psi}\in WM(\tau))$ and define $\Lambda=(1\otimes\bar{\psi})\circ j$; then it is readily seen that the dual map Λ^* of Λ is an ergodic (resp. KMS or weakly mixing) channel.

More sophisticated channels can be constructed. For example, let us take $\sigma\in\mathrm{Aut}(\mathfrak{N})$ satisfying the intertwining property $\sigma\circ\gamma_t=\gamma_t\circ\sigma$ $(\gamma_t=\alpha_t\otimes\tau_t,$ $t\in R)$ and define $\Lambda=(1\otimes\bar{\psi})\circ\sigma$ for some $\bar{\psi}\in\mathfrak{S}(\mathscr{B})$.

When $\bar{\psi}\in WM(\tau)$, we can show that Λ^* is a weakly mixing channel. Let us sketch the proof: For any $\varphi\in WM(\alpha)$, we have

$$\begin{aligned}\Lambda^*\varphi(\eta_\tau(B_1)\,B_2)&=\varphi\otimes\bar{\psi}(\sigma(\eta_\tau(B_1))\,\sigma(B_2))\\&=\varphi\otimes\bar{\psi}(\eta_\gamma(\sigma(B_1))\,\sigma(B_2))\end{aligned}$$

for any $B_1,B_2\in\mathscr{B}$. It is easy to see that

$$\varphi\otimes\bar{\psi}(\eta_\gamma(A_1\otimes B_1)\,A_2\otimes B_2)=\varphi(\eta_\alpha(A_1))\,\varphi(A_2)\,\bar{\psi}(\eta_\tau(B_1))\,\bar{\psi}(B_2),$$

for any $A_1,A_2\in\mathscr{A}$ and $B_1,B_2\in\mathscr{B}$. Since $\mathfrak{N}=\{\sum_i A_i\otimes B_i : A_i\in\mathscr{A},$ $B_i\in\mathscr{B}\}^{-W}$, we obtain

$$\begin{aligned}\Lambda^*\varphi(\eta_\tau(B_1)\,B_2)&=\varphi\otimes\bar{\psi}(\eta_\gamma(\sigma(B_1)))\,\varphi\otimes\bar{\psi}(\sigma(B_2))\\&=\Lambda^*\varphi(\eta_\tau(B_1))\,\Lambda^*\varphi(B_2).\end{aligned}$$

Hence $\Lambda^*\varphi\in WMC$.

When $\bar{\psi}\in K(\tau)$, Λ^* becomes a KMS channel: For any $\varphi\in K(\alpha)$, the state $\varphi\otimes\bar{\psi}$ on $\mathfrak{N}$ satisfies the KMS condition w.r.t. γ_t. There then exists a bounded function $F_{Q,R}(z)$ of $z\in C$ for any pair $Q,R\in\mathfrak{N}$ analytic in and continuous on the strip $0\leqslant \mathrm{Im}\,z\leqslant 1$ with boundary values: $F_{Q,R}(t)=\varphi\otimes\bar{\psi}(\gamma_t(Q)\,R)$ and $F_{Q,R}(t+i)=\varphi\otimes\bar{\psi}(R\gamma_t(Q))$. In particular, put $Q=\sigma(B_1)$ and $R=\sigma(B_2)$ for any $B_1,B_2\in\mathscr{B}$. Then

$$\begin{aligned}F_{B_1,B_2}(t)&=\Lambda^*\varphi(\tau_t(B_1)\,B_2),\\F_{B_1,B_2}(t+i)&=\Lambda^*\varphi(B_2\,\tau_t(B_1)).\end{aligned}$$

Thus, Λ^* is a KMS channel.

By these simple arguments, we can answer question (1) in Section 1 affirmatively under rather plausible conditions (e.g., $exI(\tau) \neq \emptyset$).

We remark here that the condition $\sigma \circ \gamma_t = \gamma_t \circ \sigma$ is indeed satisfied for several physical models.

3. Ergodic Channel

We give some characterizations of an ergodic channel.

Before stating and proving the results, we note one point of terminology; that is, for a subset $\mathscr{S}$ of $\mathfrak{S}(\mathscr{A})$, when $\varphi(Q) = \varphi(R)$ is satisfied for every $\varphi \in \mathscr{S}$, we say that Q is equal to R mod $\mathscr{S}$.

THEOREM 1. *If a channel Λ^* is ergodic, then it is extremal in SC* mod $I(\alpha)$ *(i.e., $\Lambda_1^* = \Lambda_2^*$ iff $\varphi(\Lambda_1(B)) = \varphi(\Lambda_2(B))$ for any $\varphi \in I(\alpha)$ and any $B \in \mathscr{B}$).*

Proof. Suppose that there exist $\Lambda_1^*, \Lambda_2^* \in SC$ and $\lambda \in (0, 1)$ such that $\Lambda^* = \lambda \Lambda_1^* + (1 - \lambda) \Lambda_2^*$. We then have

$$\Lambda^* \varphi(B) = \lambda \Lambda_1^* \varphi(B) + (1 - \lambda) \Lambda_2^* \varphi(B)$$

for any $\varphi \in exI(\alpha)$ and any $B \in \mathscr{B}$. Now, it is easy to see that $\Lambda_k^* \varphi(\tau_t(B)) = \Lambda_k^* \varphi(B)$ $(k = 1, 2)$ for any $B \in \mathscr{B}$, so $\Lambda_k^* \varphi \in I(\tau)$. Since $\Lambda^* \in EC$, the state $\Lambda^* \varphi$ is ergodic, which implies $\Lambda_1^* \varphi = \Lambda_2^* \varphi = \Lambda^* \varphi$. The set $I(\alpha)$ is the closure of the convex hull of $exI(\alpha)$ in the weak*-topology, hence the equality $\Lambda_1^* \varphi = \Lambda_2^* \varphi = \Lambda^* \varphi$ holds for every $\varphi \in I(\alpha)$. Therefore, $\Lambda_1^* = \Lambda_2^* = \Lambda^*$ mod $I(\alpha)$. Q.E.D.

THEOREM 2. *When $\mathscr{B}$ is simple and $\mathscr{A} = B(\mathscr{H})$, a channel Λ^* is ergodic if and only if Λ^* is extremal in SC* mod $I(\alpha)$.

Proof. The "only if" part has been proved in Theorem 1. We show the "if" part here. Suppose that a channel Λ^* is not ergodic; then there exists an ergodic state φ in $I(\alpha)$ such that $\Lambda^* \varphi$ is not in $exI(\tau)$. Put $\psi = \Lambda^* \varphi$. The state ψ can be decomposed into two other invariant states ψ_1, ψ_2 such that $\psi = \lambda \psi_1 + (1 - \lambda) \psi_2$ for some $\lambda \in (0, 1)$. Thus the states ψ_k $(k = 1, 2)$ are dominated by ψ. Hence there exist Q_k^ψ $(k = 1, 2)$ in $\pi_\psi(\mathscr{B})' \cap U^\psi(R)'$ [5, 13, 17, 18] such that $Q_k^\psi \geq 0$,

$$\psi_k(B) = \langle \Psi, Q_k^\psi \pi_\psi(B) \Psi \rangle$$

and

$$\langle \Psi, Q_k^\psi \Psi \rangle = 1,$$

in which π_ψ, $\{U_t^\psi : t \in R\}$, and Ψ are, respectively, the representation, one-parameter unitary group, and cyclic vector obtained by the GNS construction theorem [6, 17, 18]. Since the von Neumann algebra $\mathscr{B}$ is assumed to be simple, the representation π_ψ is faithful. Define $\Xi^\psi = \Lambda \circ \pi_\psi^{-1}$. This map Ξ^ψ can be extended to a complete positive map from $B(\mathscr{H}_\psi)$ to $\mathscr{A}$ because $\mathscr{A} = B(\mathscr{H})$. We use the same notation for this extended map. We therefore have

$$\Lambda^*\varphi = (\Xi^\psi \circ \pi_\psi)^* \varphi,$$

$$\psi_k(B) = \Xi^{\psi^*}\varphi(Q_k^\psi \pi_\psi(B)),$$

for any $B \in \mathscr{B}$. It is easily shown that π_ψ, Q_k^ψ are completely positive, so the map $\Lambda_k^\psi = \Xi^\psi \circ Q_k^\psi \circ \pi_\psi$ is completely positive from $\mathscr{B}$ to $\mathscr{A}$. Consider a map $\bigoplus_{\varphi \in \mathfrak{S}(\mathscr{A})} \Lambda_k^\psi$ (remember $\psi = \Lambda^*\varphi$), where $\Lambda_k^\psi = \Lambda$ when $\psi \notin I(\tau)$. This map is completely positive from $\mathscr{B}$ to $\bigoplus_{\varphi \in I(\alpha)} \mathscr{A}_\varphi$ with $\mathscr{A}_\varphi = \mathscr{A}$ for all $\varphi \in \mathfrak{S}(\mathscr{A})$. We define a projection E from $\bigoplus \mathscr{A}_\varphi$ to $\mathscr{A}$ such as $\omega(E(\bigoplus \Lambda_k^\psi(B))) = \omega(\Lambda_k^\omega(B))$ for every state $\omega \in \mathfrak{S}(\mathscr{A})$.

Let us define a map

$$\Lambda_k = E \circ \bigoplus \Lambda_k^\psi.$$

Then Λ_k is a completely positive stationary map and $\Lambda_k(I) = I$. According to the decomposition of ψ, we have $\Lambda^* = \lambda \Lambda_1^* + (1-\lambda)\Lambda_2^*$, which contradicts the extremality of Λ^*. Q.E.D.

Remark. In the proof of this theorem, if the state $\Lambda^*\varphi$ is ergodic (i.e., $\Lambda^*\varphi \in exI(\tau)$), then Q_k^ψ $(k = 1, 2)$ are equal to the unity I.

A completely positive map $\Lambda: \mathscr{B} \to \mathscr{A}$ can be represented by the Stinespring theorem [3, 20] as follows: There exist a representation ρ from $\mathscr{B}$ into the set of bounded linear operators $\mathbb{B}(\mathscr{L})$ on a Hilbert space $\mathscr{L}$ and a bounded linear transformation V on $\mathscr{H}$ into $\mathscr{L}$, by which we obtain $\Lambda(B) = V^*\rho(B)V$ for any $B \in \mathscr{B}$.

As in the GNS theorem for an invariant state, when a channel is a stationary, we expect that there exists a strongly continuous unitary operator T_t $(t \in R)$ such as $\rho(\tau_t(B)) = T_t \rho(B) T_{-t}$. We can indeed construct this unitary operator. According to the Stinespring theorem, the Hilbert space $\mathscr{L}$ above is the completion of $\mathscr{B} \otimes \mathscr{H}/\mathscr{N}$ w.r.t. an inner product $\langle \cdot, \cdot \rangle$ defined below. Here, $\mathscr{B} \otimes \mathscr{H}$ is the algebraic tensor product of $\mathscr{B}$ and $\mathscr{H}$, the inner product $\langle \cdot, \cdot \rangle$ is given by

$$\langle \Omega, \Theta \rangle = \Sigma \langle \Phi_i, \Lambda(B_i^* C_k) \Psi_k \rangle$$

for any $\Omega = \sum_{k=1}^n B_k \otimes \Phi_k$, $\Theta = \sum_{k=1}^n C_k \otimes \Psi_K$ in $\mathscr{B} \otimes \mathscr{H}$, and $\mathscr{N}$ is the set $\{\Omega \in \mathscr{B} \otimes \mathscr{H} \mid \langle \Omega, \Omega \rangle = 0\}$. Let us denote the equivalent class of Ω by $[\Omega]$. Then we have

$$\langle \Omega, \Theta \rangle = \langle [\Omega], [\Theta] \rangle,$$
$$\rho(B)[\sum B_k \otimes \Phi_k] = [\sum BB_k \otimes \Phi_k],$$
$$V\Phi = [I \otimes \Phi].$$

For simplicity of discussion, we assume that α_t is spatial (i.e., there exists a strongly continuous unitary operator U_t such as $\alpha_t(A) = U_t A U_{-t}$ for any $A \in \mathscr{A}$ and $t \in R$). By analogy with the GNS theorem for an invariant state, define an operator S_t by

$$S_t(\sum B_k \otimes \Phi_k) = [\sum \tau_t(B_k) \otimes U_t \Phi_k].$$

It is then easy to see that

$$\langle S_t \Omega, S_t \Theta \rangle = \langle \Omega, \Theta \rangle$$

for any Ω, $\Theta \in \mathscr{B} \otimes \mathscr{H}$, and

$$\langle S_t \Theta, S_t \Theta \rangle = 0$$

for any $\Theta \in \mathscr{N}$. S_t can be extended to an unitary operator T_t on $\mathscr{L} = \overline{\mathscr{B} \otimes \mathscr{H}/\mathscr{N}}$. This operator T_t is given by

$$T_t[\Omega] = [S_t \Omega].$$

Now, for any $B, B_k \in \mathscr{B}$, $\Phi_k \in \mathscr{H}$, we obtain

$$\begin{aligned} T_t \rho(B) T_{-t}[\sum B_k \otimes \Phi_k] &= T_t[\sum B\tau_{-t}(B_k) \otimes U_{-t}\Phi_k] \\ &= [\sum \tau_t(B\tau_{-t}(B_k)) \otimes \Phi_k] \\ &= \rho(\tau_t(B))[\sum B_k \otimes \Phi_k]. \end{aligned}$$

Hence, $\rho(\tau_t(B)) = T_t \rho(B) T_{-t}$ holds for any $B \in \mathscr{B}$.

Furthermore, $T_t V\Phi = T_t[I \otimes \Phi] = [I \otimes U_t \Phi] = VU_t \Phi$ is satisfied for any $\Phi \in \mathscr{H}$. If Φ is an invariant vector under U_t, then $T_t V\Phi = V\Phi$. We therefore obtain

THEOREM 3. *If a channel Λ^* is stationary and α_t is spatial such that $\alpha_t(A) = U_t A U_{-t}$ for any $A \in \mathscr{A}$, then there exists a strongly continuous one-parameter unitary group $\{T_t \mid t \in R\}$ such that $\rho(\tau_t(B)) = T_t \rho(B) T_{-t}$ for any $B \in \mathscr{B}$ and $T_t V\Phi = V\Phi$ for any invariant vector Φ of $\mathscr{H}$ under U_t.*

By using this theorem, we have the following Radon–Nikodym theorem for a stationary channel.

THEOREM 4. *Let* Λ^* *be a stationary channel. Then there exists a completely positive stationary map* Ξ *from* $\mathscr{B}$ *to* $\mathscr{A}$ *dominated by* Λ *if and only if* $\rho(\mathscr{B})' \cap T(R)' \neq CI$.

Proof. Suppose that such a completely positive map Ξ exists. According to Størmer [21], there exists an positive operator Q in $\rho(\mathscr{B})'$ such that $Q \notin CI$ and

$$\Xi(B) = V^* Q\rho(B) V$$

for any $B \in \mathscr{B}$. It has been shown that $T_t V\Phi = [I \otimes U_t \Phi] = VU_t \Phi$ for any $\Phi \in \mathscr{H}$ and all $t \in R$. Hence $T_t V = VU_t$ holds for all $t \in R$. Since Ξ is stationary, we obtain

$$\begin{aligned} V^* Q T_t \rho(B) T_{-t} V &= V^* Q \rho(\tau_t(B)) V = \Xi(\tau_t(B)) \\ &= U_t V^* Q \rho(B) V U_{-t} = V^* T_t Q \rho(B) T_{-t} V \end{aligned}$$

for any $B \in \mathscr{B}$. Thus, $QT_t = T_t Q$ for all $t \in R$, which means that Q is an operator in $\rho(\mathscr{B})' \cap T(R)'$. This concludes $\rho(\mathscr{B})' \cap T(R)' \neq CI$.

Conversely, if we take an operator Q from $\rho(\mathscr{B})' \cap T(R)'$ such as $Q \notin CI$ and $0 \leqslant Q \leqslant I$. Let us define a map Ξ as $\Xi(B) = V^* Q\rho(B) V$ for any $B \in \mathscr{B}$. Then it is readily seen that Ξ is a completely positive stationary map from $\mathscr{B}$ to $\mathscr{A}$ and dominated by Λ. Q.E.D.

4. WEAKLY MIXING AND KMS CHANNELS

We study the property of weakly mixing and KMS channels and some relations to ergodic channel.

When a channel Λ^* is stationary and weakly mixing, we have

$$\Lambda^*\phi(\eta_\tau(B_1) B_2) = \Lambda^*\phi(\eta_\tau(B_1)) \Lambda^*\phi(B_2)$$

for any $B_1, B_2 \in \mathscr{B}$ and $\phi \in WM(\alpha)$. Since Λ^* is stationary, the right hand side of the above equality becomes

$$\begin{aligned} \phi(\eta_\alpha(\Lambda(B_1)) \phi(\Lambda(B_2)) &= \phi(\eta_\alpha(\Lambda(B_1)) \Lambda(B_2)) \\ &= \phi(\Lambda\eta_\tau(B_1) \Lambda(B_2)). \end{aligned}$$

Hence, $\Lambda(\eta_\tau(B_1) B_2) = \Lambda(\eta_\tau(B_1)) \Lambda(B_2)$ mod $WM(\alpha)$.

Conversely, for any $\phi \in WM(\alpha)$, we obtain

$$\Lambda^*\phi(\eta_\tau(B_1)\,B_2) = \phi(\Lambda(\eta_\tau(B_1)\,B_2)) = \phi(\Lambda(\eta_\tau(B_1))\,\Lambda(B_2))$$
$$= \phi(\eta_\alpha(\Lambda(B_1))\,\Lambda(B_2)) = \phi(\eta_\alpha(\Lambda(B_1)))\,\phi(\Lambda(B_2))$$

for any $B_1, B_2 \in \mathscr{B}$. Thus $\Lambda^*\phi \in WM(\alpha)$.

Let E_ϕ be a projection associated with a state $\phi \in \mathfrak{S}(\mathscr{A})$ from the GNS Hilbert space $\mathscr{H}_\phi$ to the set of invariant vectors under the unitary group $\{U_t^\phi \Phi \mid t \in R\}$ of ϕ (i.e., $\pi_\phi(\alpha_t(A)) = U_t^\phi \pi_\phi(A)\,U_{-t}^\phi$ and $U_t^\phi \Phi = \Phi$ for the GNS representation π_ϕ and the cyclic vector Φ).

We call a dynamical system $(\mathscr{A}, \alpha)$ G-abelian on an invariant state ϕ if for any $A, B \in \mathscr{A}$, $\tilde{\phi}([\eta_\alpha(A), B]) = 0$ is satisfied, where $\tilde{\phi}$ is any vector state of $E_\phi \mathscr{H}_\phi$. In particular, when $(\mathscr{A}, \alpha)$ is G-abelian on every invariant state for α, we merely say that $(\mathscr{A}, \alpha)$ is G-abelian.

Suppose that the input system $(\mathscr{A}, \alpha)$ is G-abelian and a channel Λ^* is stationary and weakly mixing. Then $exI(\alpha) = WM(\alpha) \cap I(\alpha)$. Since $\Lambda^* \in WMC \cap SC$, $\Lambda^*\phi \in WM(\tau)$, which implies $\Lambda^*\phi \in exI(\tau)$. Hence Λ^* is ergodic.

For the G-abelian input system $(\mathscr{A}, \alpha)$ and a stationary channel Λ^*, if Λ is a *-homomorphism, we can readily show that $EC = WMC \cap SC$.

Let $\tilde{\mathscr{B}}$ be the set of τ-analytic elements of $\mathscr{B}$. Then it is an easy exercise to show that $\Lambda^* \in KC$ if and only if $\Lambda(\tau_t(B_1)\,B_2) = \Lambda(B_2\tau_{t+i}(B_1)) \bmod K(\alpha)$ for any $B_1 \in \tilde{\mathscr{B}}$ and $B_2 \in \mathscr{B}$. Moreover, if the input system $(\mathscr{A}, \alpha)$ is η-abelian (i.e., $\phi(D^*[\eta_\alpha(A), B]\,D) = 0$ for every invariant state ϕ and any $A, B, D \in \mathscr{A}$) and Λ is a *-holomorphism, then we can easily see that Λ^* is KMS if and ony only if Λ^* is ergodic.

We summarize these facts as

THEOREM 5. *For a stationary channel Λ^*, we have:*

(1) $\Lambda^* \in WMC$ *if and only if* $\Lambda(\eta_\tau(B_1)\,B_2) = \Lambda(\eta_\tau(B_1))\,\Lambda(B_2)$ mod $WM(\alpha)$ *for any* $B_1, B_2 \in \mathscr{B}$.

(2) *If the input system $(\mathscr{A}, \alpha)$ is G-abelian and $\Lambda^* \in WMC$, then Λ^* is ergodic.*

(3) *If $(\mathscr{A}, \alpha)$ is G-abelian and Λ is a *-homomorphism, then $EC = WMC \cap SC$.*

(4) $\Lambda^* \in KC$ *if and only if* $\Lambda(\tau_t(B_1)\,B_2) = \Lambda(B_2\tau_{t+1}(B_1))$ mod $K(\alpha)$ *for any* $B_1 \in \tilde{\mathscr{B}}$ *and* $B_2 \in \mathscr{B}$.

(5) *If $(\mathscr{A}, \alpha)$ is η-abelian and Λ is a *-homomorphism, then $KC = EC$.*

Conclusion

We introduced and studied several ergodic channels (ergodic, KMS, weakly mixing) in this paper. We hope that these concepts are useful to analyze the information theory for quantum systems. One of the applications is the transmission of entropy (information), which is discussed in Ref. [16].

Acknowledgments

The author thanks Professor H. Umegaki for valuable discussion and useful comments.

References

1. M. D. Choi, *Ill. J. Math.* **18** (1974), 565.
2. E. B. Davies, *IEEE Trans. Inform. Theory* (1977), 530.
3. E. B. Davies, "Quantum Theory of Open System," Academic Press, New York, 1976.
4. C. Davis, *Proc. Amer. Math. Soc.* **8** (1957), 42.
5. S. Doplicher, D. Kastler, and E. Størmer, *J. Funct. Anal.* **5** (1969), 419.
6. G. G. Emch, "Algebraic Methods in Statistical Mechanics and Quantum Field Theory," Wiley, New York, 1971.
7. R. Haag and D. Kastler, *J. Math. Phys.* **5** (1964), 848.
8. R. Haag, N. M. Hugenholtz, and W. Winnink, *Commun. Math. Phys.* **5** (1967), 215.
9. C. W. Helstrom, J. W. S. Liu, and J. P. Gordon, *Proc. IEEE* **58** (1970), 1578.
10. A. S. Holevo, *Rep. Math. Phys.* **12** (1977), 273.
11. A. S. Holevo, *Problemy Peredači Informacii* **9**, No. 2 (1973), 31.
12. R. S. Ingarden, *Rep. Math. Phys.* **10** (1976), 43.
13. I. Kovacs and J. Szucs, *Acta Sci. Math. (Szeged)* **27** (1966), 233,
14. K. Kraus, *Ann. Phys.* **64** (1971), 311.
15. G. Lindblad, *Commun. Math. Phys.* **48** (1976), 119.
16. M. Ohya, preprint.
17. D. Ruelle, "Statistical Mechanics," Benjamin, New York, 1969.
18. S. Sakai, "C^*-Algebras and W^*-Algebras," Springer-Verlag, New York/Berlin, 1971.
19. C. E. Shannon, *Bell System Tech. J.* **27** (1948), 379.
20. N. F. Stinespring, *Proc. Amer. Math. Soc.* **6** (1965), 211.
21. E. Størmer, *Acta Math.* **110** (1971), 233.
22. M. Takesaki, "Tomita's Theory of Modular Hilbert Algebras and its Applications," Lecture Notes in Mathematics, No. 128, Springer-Verlag, Berlin, 1970.
23. H. Umegaki, *J. Math. Anal. Appl.* **25**, No. 1 (1969), 41.

PACIFIC JOURNAL OF MATHEMATICS
Vol. 96, No. 1, 1981

SUFFICIENCY, KMS CONDITION AND RELATIVE ENTROPY IN VON NEUMANN ALGEBRAS

FUMIO HIAI, MASANORI OHYA AND MAKOTO TSUKADA

The sufficiency in von Neumann algebras is discussed with some applications to classification of normal states. It is shown that the concept of sufficiency characterizes the KMS-states and the invariant states with respect to a modular automorphism group. The relations between the sufficiency and the relative entropy are established.

Introduction. Since the investigation of sufficient statistics in abstract measure theoretic terms was initiated by Halmos and Savage [10], the concept of sufficiency has been developed by many mathematical statisticians in terms of various relations given by comparison of experiments, risk functions within the framework of statistical decision problems and so on. A characterization of sufficiency was given in [12] through the measure of Kullback-Leibler information.

The concept of sufficiency was first generalized by Umegaki [22, 23] to the noncommutative case of semi-finite von Neumann algebras with some extension of the Kullback-Leibler information (usually called the relative entropy). Later the related discussions especially concerning the relative entropy for quantum systems have been made by several authors, e.g., Araki [2, 3], Gudder and Marchand [7], and Lindblad [13].

As defined precisely and explained in §§1 and 4 of this paper, the concept of sufficiency is more or less considered through the informativity of a certain subalgebra with respect to a given algebra for a dynamical system of interest. Namely, in the case that such a subalgebra is sufficient, the relative entropy on the subalgebra is equal to that on the given algebra. This fact may or may not be a reason why the concept of sufficiency has not been entered into analysis of physical systems, in which the change of entropy is thought of more relevant.

The Kubo-Martin-Schwinger (KMS) condition was introduced by these three authors [11, 14] as a boundary condition of the thermal Green function. Haag, Hugenholtz and Winnink [8] showed that in the operator algebraic framework this condition is a fundamental one describing thermal equilibrium of quantum systems. The KMS condition through the Tomita-Takesaki theory now becomes a core of studying von Neumann algebras.

Under the above historical basis, our main motivation of this work is as follows: How useful for quantum systems is the concept of sufficiency? How much of the related topics of sufficiency, mostly done for the commutative case, can be generalized to the noncommutative case?

Having these questions in our mind, we discuss the sufficiency with some applications to classification of normal states on the basis of recent development of von Neumann algebras.

In § 1 of this paper, we establish definitions and notations used throughout and also give some simple facts.

In § 2, it is shown that the concept of sufficiency characterizes the invariant states and the KMS-states with respect to a modular automorphism group.

In § 3, we prove some formulas on the relative entropy using Araki's definition of relative entropy.

In § 4, combining several theorems obtained in the previous sections, we establish some results which indicate the relations between the sufficiency and the relative entropy.

As a whole, we like to claim that the concept of sufficiency might be very useful for analysing von Neumann algebras and hence some quantum systems.

1. Definition and preliminaries. Throughout this paper, let $\mathfrak{N}$ be a von Neumann algebra with unity I acting on a Hilbert space $\mathscr{H}$, and $\mathfrak{S}$ be the set of all normal states of $\mathfrak{N}$. A dynamical system of physically interest is described by a triple $(\mathfrak{N}, \mathfrak{S}, \alpha)$, where α_t, $t \in \boldsymbol{R}$, is a strongly continuous one-parameter automorphism group of $\mathfrak{N}$. A state $\varphi \in \mathfrak{S}$ is said to satisfy the *Kubo-Martin-Schwinger* (KMS) *boundary condition* at a certain constant $\beta > 0$ with respect to α_t if for every pair $A, B \in \mathfrak{N}$ there exists a bounded function $F_{A,B}(z)$ continuous on and holomorphic in the strip $0 \leqq \mathrm{Im}\, z \leqq \beta$ with boundary values:

$$F_{A,B}(t) = \varphi(\alpha_t(A)B) \quad \text{and} \quad F_{A,B}(t + i\beta) = \varphi(B\alpha_t(A)) .$$

If φ satisfies the KMS condition with respect to α_t, then φ is proved to be α_t-invariant, i.e., $\varphi \circ \alpha_t = \varphi$. Considering $\alpha_{\beta t}$, we may take $\beta = 1$ in the sequel discussions. Takesaki showed in [17] using Tomita's theory that to every faithful state $\varphi \in \mathfrak{S}$ there exists a unique one-parameter automorphism group (i.e., the so-called *modular automorphism group*) σ_t^φ with respect to which φ satisfies the KMS condition at $\beta = 1$.

In this paper, a subalgebra $\mathfrak{M}$ always means a von Neumann subalgebra of $\mathfrak{N}$ with I. For a subalgebra $\mathfrak{M}$ and a state $\varphi \in \mathfrak{S}$, let $E_\varphi(\cdot|\mathfrak{M})$ denote the conditional expectation with respect to $\mathfrak{M}$ and φ (if it exists), which is characterized as a norm one normal projection from $\mathfrak{N}$ onto $\mathfrak{M}$ satisfying $\varphi(A) = \varphi(E_\varphi(A|\mathfrak{M}))$ for all $A \in \mathfrak{N}$ (cf. [19, 21]). It was shown by Takesaki [18] that for a faithful state $\varphi \in \mathfrak{S}$ the conditional expectation $E_\varphi(\cdot|\mathfrak{M})$ exists if and only if $\mathfrak{M}$ is invariant under the modular automorphism group σ_t^φ.

According to [5], for any two faithful states $\varphi, \psi \in \mathfrak{S}$ there exists a strongly continuous function $t \mapsto u_t$ of $\boldsymbol{R}$ into the unitary group of $\mathfrak{N}$ which is a φ-cocycle, i.e.,

$$u_{s+t} = u_s \sigma_s^\varphi(u_t)\,, \quad s, t \in \boldsymbol{R}\,,$$

and which satisfies

$$\sigma_t^\psi(A) = u_t \sigma_t^\varphi(A) u_t^*\,, \quad t \in \boldsymbol{R}\,, \quad A \in \mathfrak{N}\,.$$

This φ-cocycle u_t is denoted by $u_t = (D\psi : D\varphi)_t$ and is called the *Connes Radon-Nikodym derivative* of ψ with respect to φ. Some discussions are found in [4, 9] concerning Connes Radon-Nikodym derivatives and conditional expectations.

Let S be a subset of $\mathfrak{S}$. A subalgebra $\mathfrak{M}$ is said to be *sufficient* for S if $E_\varphi(\cdot|\mathfrak{M})$ exists for each $\varphi \in S$ and for every $A \in \mathfrak{N}$ there exists an $A_0 \in \mathfrak{M}$ such that

$$A_0 = E_\varphi(A|\mathfrak{M}) \quad \text{a.e. } [\varphi]\,, \quad \varphi \in S\,,$$

where $A = B$ a.e. $[\varphi]$ means $\varphi(|A - B|) = 0$. This definition of sufficient subalgebras is somewhat weaker than that in [22]. Also we call $\mathfrak{M}$ to be *minimal sufficient* for S if $\mathfrak{M}$ is sufficient for S and any subalgebra being sufficient for S includes $\mathfrak{M}$.

For $\varphi, \psi \in \mathfrak{S}$, it is said that ψ is absolutely continuous with respect to φ (we write $\psi \ll \varphi$) if for each $A \in \mathfrak{N}$, $\varphi(A^*A) = 0$ implies $\psi(A^*A) = 0$; that is, $\psi \ll \varphi$ if and only if $s(\psi) \leqq s(\varphi)$ where $s(\varphi)$ is the support projection of φ. We give here the elementary facts of sufficiency which are readily seen from the definition.

(1°) Let $\varphi, \psi \in \mathfrak{S}$ with $\psi \ll \varphi$. Then a subalgebra $\mathfrak{M}$ is sufficient for $\{\varphi, \psi\}$ if and only if $E_\varphi(\cdot|\mathfrak{M})$ exists and $\psi(A) = \psi(E_\varphi(A|\mathfrak{M}))$ for all $A \in \mathfrak{N}$.

(2°) If a subalgebra $\mathfrak{M}$ is sufficient for $\{\varphi, \psi\}$, then $\varphi = \psi$ on $\mathfrak{N}$ if and only if $\varphi = \psi$ on $\mathfrak{M}$.

When $S(\subset \mathfrak{S})$ contains a faithful state φ, then:

(3°) A subalgebra $\mathfrak{M}$ is sufficient for S if and only if $\mathfrak{M}$ is sufficient for every pair $\{\varphi, \psi\}$ with $\psi \in S$.

(4°) If $\mathfrak{M}$ is sufficient for S, then any subalgebra $\mathfrak{M}_1$ including $\mathfrak{M}$ is sufficient for S whenever $E_\varphi(\cdot|\mathfrak{M}_1)$ exists.

2. Sufficiency and characterization of states. The following lemma is a restatement of [4, Lemma 1.6] in our terminology. We give the proof for completeness.

LEMMA 2.1. *For each subalgebra* $\mathfrak{M}$ *and two faithful states* $\varphi, \psi \in \mathfrak{S}$, *the following conditions are equivalent*:

(i) $\mathfrak{M}$ *is sufficient for* $\{\varphi, \psi\}$;

(ii) $E_\varphi(\cdot|\mathfrak{M})$ *exists and* $(D\psi: D\varphi)_t \in \mathfrak{M}$ *for every* $t \in \boldsymbol{R}$.

Proof. Let $\hat{\varphi} = \varphi \upharpoonright \mathfrak{M}$ and $\hat{\psi} = \psi \upharpoonright \mathfrak{M}$. Assume that $\mathfrak{M}$ is sufficient for $\{\varphi, \psi\}$. Then the conditional expectation $E_\varphi(\cdot|\mathfrak{M})$ exists and $\psi(A) = \psi(E_\varphi(A|\mathfrak{M}))$ for $A \in \mathfrak{N}$. By [5, Lemma 1.4.4], we have

$$(D\psi: D\varphi)_t = (D(\hat{\psi}\circ E): D(\hat{\varphi}\circ E))_t = (D\hat{\psi}: D\hat{\varphi})_t \in \mathfrak{M}$$

for every $t \in \boldsymbol{R}$, where $E(\cdot) = E_\varphi(\cdot|\mathfrak{M})$. Conversely assume that $E_\varphi(\cdot|\mathfrak{M})$ exists and $(D\psi: D\varphi)_t \in \mathfrak{M}$ for all $t \in \boldsymbol{R}$. Since $\sigma_t^{\hat{\varphi}} = \sigma_t^\varphi \upharpoonright \mathfrak{M}$, it follows that $u_t = (D\psi: D\varphi)_t$ is a $\hat{\varphi}$-cocycle. By [5, Theorem 1.2.4], there exists a unique faithful normal semi-finite weight $\tilde{\psi}$ on $\mathfrak{M}$ such that $(D\tilde{\psi}: D\hat{\varphi})_t = u_t$. Define a faithful normal semi-finite weight ψ' on $\mathfrak{N}$ by $\psi'(A) = \tilde{\psi}(E_\varphi(A|\mathfrak{M}))$ for $A \in \mathfrak{N}$. Then it follows that

$$(D\psi': D\varphi)_t = (D\tilde{\psi}: D\hat{\varphi})_t = (D\psi: D\varphi)_t\,, \quad t \in \boldsymbol{R}\,.$$

Hence we have $\psi' = \psi$, so that $\psi(A) = \psi(E_\varphi(A|\mathfrak{M}))$ for every $A \in \mathfrak{N}$. This shows that $\mathfrak{M}$ is sufficient for $\{\varphi, \psi\}$. □

In this section, let φ be a fixed faithful normal state of $\mathfrak{N}$ and σ_t^φ its modular automorphism group. Let Z_φ be the subalgebra consisting of all $A \in \mathfrak{N}$ such that $\varphi(AB) = \varphi(BA)$ for every $B \in \mathfrak{N}$. The subalgebra Z_φ is called the *centralizer* of φ and is exactly the fixed point algebra of σ_t^φ (cf. [17, Lemma 15.8]), i.e.,

$$Z_\varphi = \{A \in \mathfrak{N}: \sigma_t^\varphi(A) = A,\, t \in \boldsymbol{R}\}\,.$$

Let $\mathfrak{Z}$ be the center of $\mathfrak{N}$, i.e., $\mathfrak{Z} = \mathfrak{N} \cap \mathfrak{N}'$. Clearly $\mathfrak{Z} \subset Z_\varphi$. Let $I(\varphi)$ be the set of all σ_t^φ-invariant states in $\mathfrak{S}$, and $K(\varphi)$ be the set of all states in $\mathfrak{S}$ satisfying the KMS condition with respect to σ_t^φ at $\beta = 1$. Then we have:

THEOREM 2.2. (1) *For each* $\psi \in \mathfrak{S}$, $\psi \in I(\varphi)$ *if and only if* Z_φ *is sufficient for* $\{\varphi, \psi\}$.

(2) *The centralizer* Z_φ *is minimal sufficient for* $I(\varphi)$.

Proof. (1) Let $\psi \in \mathfrak{S}$ and take $\psi_1 = (\psi + \varphi)/2$. Then we easily see that $\psi \in I(\varphi)$ is equivalent to $\psi_1 \in I(\varphi)$, and the sufficiency of Z_φ for $\{\varphi, \psi\}$ is equivalent to that for $\{\varphi, \psi_1\}$. Therefore we can assume that ψ is faithful. Since Z_φ is elementwise invariant under σ_t^φ, there exists the conditional expectation $E_\varphi(\cdot|Z_\varphi)$ from $\mathfrak{N}$ onto Z_φ. Hence, in view of Lemma 2.1, it suffices to show that $\psi \in I(\varphi)$ if and only if $(D\psi: D\varphi)_t \in Z_\varphi$ for every $t \in \boldsymbol{R}$. If $\psi \in I(\varphi)$, then by [5, Lemma 1.2.3] there exists a positive self-adjoint operator h affiliated with Z_φ such that $(D\psi: D\varphi)_t = h^{it} \in Z_\varphi$ for all $t \in \boldsymbol{R}$. Conversely suppose that $(D\psi: D\varphi)_t \in Z_\varphi$ for every $t \in \boldsymbol{R}$. Since

$$\sigma_t^\psi(A) = (D\psi: D\varphi)_t \sigma_t^\varphi(A)(D\psi: D\varphi)_t^* , \tag{2.1}$$

we have

$$\varphi(\sigma_t^\psi(A)) = \varphi(\sigma_t^\varphi(A)) = \varphi(A) , \quad A \in \mathfrak{N} .$$

Hence it follows that φ is σ_t^ψ-invariant, and thus ψ is σ_t^φ-invariant (cf. [17, Theorem 15.2]).

(2) It follows from (1) that Z_φ is sufficient for every pair $\{\varphi, \psi\}$ with $\psi \in I(\varphi)$. Hence Z_φ is sufficient for $I(\varphi)$. To show the minimality of Z_φ, let $\mathfrak{M}$ be any subalgebra which is sufficient for $I(\varphi)$. We now prove that $Z_\varphi \subset \mathfrak{M}$. Take any positive invertible operator $h \in Z_\varphi$ with $\varphi(h) = 1$, and define a faithful state $\psi \in \mathfrak{S}$ by $\psi(A) = \varphi(hA)$ for $A \in \mathfrak{N}$. Then we have $\psi \in I(\varphi)$ and $(D\psi: D\varphi)_t = h^{it}$. Since $(D\psi: D\varphi)_t \in \mathfrak{M}$ for every $t \in \boldsymbol{R}$ by Lemma 2.1, it follows that $h \in \mathfrak{M}$. Thus $Z_\varphi \subset \mathfrak{M}$. □

THEOREM 2.3. (1) *For each $\psi \in \mathfrak{S}$, $\psi \in K(\varphi)$ if and only if $\mathfrak{Z}$ is sufficient for $\{\varphi, \psi\}$.*

(2) *The center $\mathfrak{Z}$ is minimal sufficient for $K(\varphi)$.*

Proof. As in the proof of Theorem 2.2, we can assume that ψ is faithful. If $\psi \in K(\varphi)$, then by [15, Theorem 5.4] there exists a positive self-adjoint operator h affiliated with $\mathfrak{Z}$ such that $\psi(A) = \varphi(hA)$ for $A \in \mathfrak{N}$, so that $(D\psi: D\varphi)_t = h^{it} \in \mathfrak{Z}$ for every $t \in \boldsymbol{R}$. Conversely if $(D\psi: D\varphi)_t \in \mathfrak{Z}$ for every $t \in \boldsymbol{R}$, then by (2.1) we have $\sigma_t^\psi = \sigma_t^\varphi$ and hence $\psi \in K(\varphi)$. Thus (1) is proved. The proof of (2) is analogous to that of Theorem 2.2. □

3. Relative entropy. When $\mathfrak{N}$ is finite dimensional, for each φ and ψ in $\mathfrak{S}$ the relative entropy $S(\varphi|\psi)$ is defined by

$$S(\varphi|\psi) = \mathrm{tr}(\rho_\psi \log \rho_\psi - \rho_\psi \log \rho_\varphi) ,$$

where ρ_φ and ρ_ψ are density matrices for φ and ψ. Araki [2, 3] extended the relative entropy to the case for normal positive linear functionals of general von Neumann algebras, and studied its several properties such as joint convexity, lower semiconitinuity and monotonicity.

In this section, we assume as in [3] that $\mathfrak{N}$ has a cyclic and separating vector. Let V be a natural positive cone (cf. [1]) for $\mathfrak{N}$ and let φ and ψ be states in $\mathfrak{S}$. By [1, Theorem 6], there exist unique vector representatives Φ and Ψ of φ and ψ in V such that $\varphi(A) = \langle\Phi, A\Phi\rangle$ and $\psi(A) = \langle\Psi, A\Psi\rangle$ for all $A \in \mathfrak{N}$. The operator $S_{\Phi,\Psi}$ with the domain

$$D(S_{\Phi,\Psi}) = \mathfrak{N}\Psi + (I - s^{\mathfrak{N}'}(\Psi))\mathscr{H}$$

is defined by

$$S_{\Phi,\Psi}(A\Psi + \Omega) = s^{\mathfrak{N}}(\Psi)A^*\Phi\ , \quad A \in \mathfrak{N}\ , \quad s^{\mathfrak{N}'}(\Psi)\Omega = 0\ ,$$

where $s^{\mathfrak{N}}(\Psi)(\in \mathfrak{N})$ denotes the $\mathfrak{N}$-support of the vector Ψ. Then $S_{\Phi,\Psi}$ is a closable conjugate-linear operator (cf. [3]) and the relative modular operator $\Delta_{\Phi,\Psi}$ is defined by

$$\Delta_{\Phi,\Psi} = (S_{\Phi,\Psi})^*\overline{S_{\Phi,\Psi}}\ .$$

The relative entropy $S(\varphi|\psi)$ is now given by

$$S(\varphi|\psi) = \begin{cases} -\langle\Psi, (\log \Delta_{\Phi,\Psi})\Psi\rangle & \text{if} \quad \psi \ll \varphi\ , \\ +\infty & \text{otherwise}\ . \end{cases}$$

Then $S(\varphi|\psi) \geqq 0$, and $S(\varphi|\psi) = 0$ if and only if $\varphi = \psi$. For each subalgebra $\mathfrak{M}$, let $S_{\mathfrak{M}}(\varphi|\psi)$ denote the relative entropy of the restrictions of φ and ψ to $\mathfrak{M}$. By the monotonicity of relative entropy generally proved in [20], it holds that

$$S_{\mathfrak{M}}(\varphi|\psi) \leqq S(\varphi|\psi) \tag{3.1}$$

for every subalgebra $\mathfrak{M}$ (also see [2, 3, 13, 23]).

THEOREM 3.1. *For each* $\varphi, \psi \in \mathfrak{S}$,

$$\|\varphi - \psi\| \leqq \{2S(\varphi|\psi)\}^{1/2}\ .$$

Proof. By [16, p.31], we can take two normal positive linear functionals φ_1 and φ_2 such that $\varphi - \psi = \varphi_1 - \varphi_2$, $\|\varphi - \psi\| = \|\varphi_1\| + \|\varphi_2\|$ and $s(\varphi_1) \perp s(\varphi_2)$. Let $e = s(\varphi_1)$. Then it follows that

$$\begin{aligned}\|\varphi - \psi\| &= (\varphi - \psi)(e) - (\varphi - \psi)(I - e) \\ &= 2(\varphi(e) - \psi(e))\ .\end{aligned}$$

Let $\mathfrak{M}$ be the subalgebra generated by e and $I - e$. By using the monotonicity, we have

$$\begin{aligned}S(\varphi|\psi) &\geqq S_{\mathfrak{M}}(\varphi|\psi) \\ &= \psi(e)\log\frac{\psi(e)}{\varphi(e)} + \psi(I - e)\log\frac{\psi(I - e)}{\varphi(I - e)}\ .\end{aligned}$$

It was shown in [6] that

$$2|\alpha-\beta| \leqq \left\{2\left(\beta\log\frac{\beta}{\alpha}+(1-\beta)\log\frac{1-\beta}{1-\alpha}\right)\right\}^{1/2}$$

for $0\leqq\alpha,\beta\leqq 1$. Taking $\alpha=\varphi(e)$ and $\beta=\psi(e)$, we deduce the desired inequality. □

THEOREM 3.2. *Let $\varphi,\psi\in\mathfrak{S}$ be faithful and $\mathfrak{M}$ be a subalgebra such that $\mathfrak{M}\subset Z_\varphi$. Define $\psi'\in\mathfrak{S}$ by $\psi'(A)=\psi(E_\varphi(A|\mathfrak{M}))$ for $A\in\mathfrak{N}$. If $S_{\mathfrak{M}}(\varphi|\psi)<+\infty$, then*

$$S(\psi'|\psi)=S(\varphi|\psi)-S_{\mathfrak{M}}(\varphi|\psi)\,.$$

Proof. First note that ψ' is well defined from $\mathfrak{M}\subset Z_\varphi$. Since $\varphi\restriction\mathfrak{M}$ is a faithful normal trace, there exists a positive self-adjoint operator h affiliated with $\mathfrak{M}$ such that $\psi(A)=\varphi(hA)$ for all $A\in\mathfrak{M}$. Take the spectral decomposition $h=\int_0^\infty\lambda\,de(\lambda)$ and $h_n=\int_0^n\lambda\,de(\lambda)$. Since $h_n\in\mathfrak{M}$, we have for every $A\in\mathfrak{N}$

$$\begin{aligned}\psi'(A)&=\psi(E_\varphi(A|\mathfrak{M}))=\lim_{n\to\infty}\varphi(h_nE_\varphi(A|\mathfrak{M}))\\&=\lim_{n\to\infty}\varphi(E_\varphi(h_nA|\mathfrak{M}))=\lim_{n\to\infty}\varphi(h_nA)=\varphi(hA)\,.\end{aligned}$$

Hence it follows (cf. [5, Lemma 1.2.3]) that $(D\psi':D\varphi)_t=h^{it}$ for all $t\in\boldsymbol{R}$. By the relations

$$\begin{aligned}(D\psi':D\psi)_t&=(D\psi':D\varphi)_t(D\varphi:D\psi)_t=h^{it}(D\varphi:D\psi)_t\,,\\(D\psi':D\psi)_t&=(\Delta_{\Psi',\Psi})^{it}\Delta_\Psi^{-it}\,,\\(D\varphi;D\psi)_t&=(\Delta_{\Phi,\Psi})^{it}\Delta_\Psi^{-it}\,,\end{aligned}$$

where $\Delta_\Psi=\Delta_{\Psi,\Psi}$, we deduce that $(\Delta_{\Psi',\Psi})^{it}=h^{it}(\Delta_{\Phi,\Psi})^{it}$ for all $t\in\boldsymbol{R}$. Moreover since $h^{it}\in\mathfrak{M}\subset Z_\varphi$ and

$$\sigma_t^\varphi(A)=(\Delta_{\Phi,\Psi})^{it}A(\Delta_{\Phi,\Psi})^{-it}\,,\quad A\in\mathfrak{N}\,,$$

it follows that h^{it} and $(\Delta_{\Phi,\Psi})^{it}$ commute. Now let $\hat{\Phi}$ and $\hat{\Psi}$ be vector representatives of $\varphi\restriction\mathfrak{M}$ and $\psi\restriction\mathfrak{M}$ in a natural positive cone $\hat{V}$ for $\mathfrak{M}$. Since $\varphi\restriction\mathfrak{M}$ is a trace, it follows that $\Delta_{\hat{\Psi},\hat{\Phi}}=h$. By $\log\Delta_{\hat{\Phi},\hat{\Psi}}=-J(\log\Delta_{\hat{\Psi},\hat{\Phi}})J$ where J is the modular conjugation operator associated with $\hat{V}$ (cf. [3, Remark 3.4]), we have

$$\begin{aligned}S_{\mathfrak{M}}(\varphi|\psi)&=-\langle\hat{\Psi},(\log\Delta_{\hat{\Phi},\hat{\Psi}})\hat{\Psi}\rangle\\&=\langle\hat{\Psi},(\log h)\hat{\Psi}\rangle=\langle\Psi,(\log h)\Psi\rangle\,,\end{aligned}\tag{3.2}$$

which is finite from the assumption. Therefore we obtain

$$S(\psi'|\psi)=-\langle\Psi,(\log\Delta_{\Psi',\Psi})\Psi\rangle$$

$$= -\langle \Psi, (\log \Delta_{\Phi,\Psi})\Psi \rangle - \langle \Psi, (\log h)\Psi \rangle$$
$$= S(\varphi|\psi) - S_{\mathfrak{M}}(\varphi|\psi) . \qquad \square$$

THEOREM 3.3. *Let* $\varphi \in \mathfrak{S}$ *be faithful and* $\mathfrak{M}$ *be a subalgebra such that* $\mathfrak{M} \subset Z_\varphi$. *Let* $\psi \in \mathfrak{S}$ *and define* $\psi' \in \mathfrak{S}$ *by* $\psi'(A) = \psi(E_\varphi(A|\mathfrak{M}))$ *for* $A \in \mathfrak{N}$. *Suppose either* (a) ψ *is faithful or* (b) $\psi \leqq \lambda\varphi$ *for some* $\lambda > 0$. *If* $S_{\mathfrak{M}}(\varphi|\psi) < +\infty$, *then*

$$\|\psi' - \psi\| \leqq \{2(S(\varphi|\psi) - S_{\mathfrak{M}}(\varphi|\psi))\}^{1/2} .$$

Proof. For the case (a), the desired inequality is immediate from Theorems 3.1 and 3.2. Now suppose that $\psi \leqq \lambda\varphi$ for some $\lambda > 0$. For each $\varepsilon > 0$, let $\psi_\varepsilon = (1+\varepsilon)^{-1}(\psi + \varepsilon\varphi) \in \mathfrak{S}$ and define $\psi'_\varepsilon \in \mathfrak{S}$ by $\psi'_\varepsilon(A) = \psi_\varepsilon(E_\varphi(A|\mathfrak{M}))$. By the convexity of relative entropy (cf. [3, Theorem 3.8]), we have

$$S_{\mathfrak{M}}(\varphi|\psi_\varepsilon) \leqq (1+\varepsilon)^{-1} S_{\mathfrak{M}}(\varphi|\psi) < +\infty .$$

Hence it follows from the case (a) that

$$\|\psi'_\varepsilon - \psi_\varepsilon\| \leqq \{2(S(\varphi|\psi_\varepsilon) - S_{\mathfrak{M}}(\varphi|\psi_\varepsilon))\}^{1/2} . \tag{3.3}$$

Since $\psi_\varepsilon \leqq \lambda\varphi$ for each $\varepsilon > 0$, by [3, Theorem 3.7] we have

$$\lim_{\varepsilon \to +0} S(\varphi|\psi_\varepsilon) = S(\varphi|\psi) ,$$
$$\lim_{\varepsilon \to +0} S_{\mathfrak{M}}(\varphi|\psi_\varepsilon) = S_{\mathfrak{M}}(\varphi|\psi) .$$

Since $\psi'_\varepsilon = (1+\varepsilon)^{-1}(\psi' + \varepsilon\varphi)$, we obtain the desired inequality by letting $\varepsilon \to +0$ in (3.3). $\square$

Before closing this section, we have to note that Professor Araki gave us very important comments to some results of our first version of this paper, which make us enable to write them in the above form.

4. Sufficiency and relative entropy. In this section, let a faithful state $\varphi \in \mathfrak{S}$ be fixed as in § 2.

THEOREM 4.1. *For each subalgebra* $\mathfrak{M} \subset Z_\varphi$ *and each* $\psi \in \mathfrak{S}$, *the following statements hold:*

(1) *Suppose the condition* (a) *or* (b) *in Theorem* 3.3. *If* $\mathfrak{M}$ *is sufficient for* $\{\varphi, \psi\}$, *then* $S_{\mathfrak{M}}(\varphi|\psi) = S(\varphi|\psi)$, *and conversely if* $S_{\mathfrak{M}}(\varphi|\psi) = S(\varphi|\psi) < +\infty$, *then* $\mathfrak{M}$ *is sufficient for* $\{\varphi, \psi\}$.

(2) *If* $\mathfrak{M}$ *is sufficient for* $\{\varphi, \psi\}$, *then* $S_{\mathfrak{M}}(\varphi|(\psi+\varphi)/2) = S(\varphi|(\psi+\varphi)/2)$, *and conversely if* $S_{\mathfrak{M}}(\varphi|(\psi+\varphi)/2) = S(\varphi|(\psi+\varphi)/2) < +\infty$, *then* $\mathfrak{M}$ *is sufficient for* $\{\varphi, \psi\}$.

Proof. (1) We use the notations in the proof of Theorem 3.2. Let $\mathfrak{M}$ be sufficient for $\{\varphi, \psi\}$, and suppose the condition (a). There exists a positive self-adjoint operator h affiliated with $\mathfrak{M}$ such that $\psi(A) = \varphi(hA)$ for all $A \in \mathfrak{M}$. Then we have

$$\psi(A) = \psi(E_\varphi(A \mid \mathfrak{M})) = \varphi(hA) , \quad A \in \mathfrak{N} ,$$

as in the proof of Theorem 3.2. Hence it follows that $(D\psi: D\varphi)_t = h^{it}$, and we have

$$(\Delta_{\Phi,\Psi})^{it} = (D\varphi: D\psi)_t \Delta_\Psi^{it} = h^{-it}\Delta_\Psi^{it} .$$

Since h^{it} and $(\Delta_{\Phi,\Psi})^{it}$ commute, it follows that h^{-it} and Δ_Ψ^{it} commute. We thus have

$$\begin{aligned} S(\varphi \mid \psi) &= -\langle \Psi, (-\log h + \log \Delta_\Psi)\Psi \rangle \\ &= \langle \Psi, (\log h)\Psi \rangle , \end{aligned} \tag{4.1}$$

by $\Delta_\Psi \Psi = \Psi$. From (3.2) and (4.1), we obtain $S_{\mathfrak{M}}(\varphi \mid \psi) = S(\varphi \mid \psi)$. The case (b) is proved from the case (a) by taking limits as in the proof of Theorem 3.3.

Assume conversely that $S_{\mathfrak{M}}(\varphi \mid \psi) = S(\varphi \mid \psi) < +\infty$. Then it follows from Theorem 3.3 that $\psi' = \psi$, which implies that $\mathfrak{M}$ is sufficient for $\{\varphi, \psi\}$.

(2) is immediate from (1), since $\mathfrak{M}$ is sufficient for $\{\varphi, \psi\}$ if and only if $\mathfrak{M}$ is sufficient for $\{\varphi, (\psi + \varphi)/2\}$. □

The above fact (1) extends the result [23, Theorem 5] which was proved under some strong assumptions. Combining Theorem 4.1 with Theorems 2.2 and 2.3, we have the following:

COROLLARY 4.2. (1) *Suppose the condition* (a) *or* (b) *in Theorem* 3.3. *If* $\psi \in I(\varphi)$, *then* $S_{Z_\varphi}(\varphi \mid \psi) = S(\varphi \mid \psi)$, *and conversely if* $S_{Z_\varphi}(\varphi \mid \psi) = S(\varphi \mid \psi) < +\infty$, *then* $\psi \in I(\varphi)$.

(2) *If* $\psi \in I(\varphi)$, *then* $S_{Z_\varphi}(\varphi \mid (\psi + \varphi)/2) = S(\varphi \mid (\psi + \varphi)/2)$, *and conversely if* $S_{Z_\varphi}(\varphi \mid (\psi + \varphi)/2) = S(\varphi \mid (\psi + \varphi)/2) < +\infty$, *then* $\psi \in I(\varphi)$.

COROLLARY 4.3. (1) *Suppose the condition* (a) *or* (b) *in Theorem* 3.3. *If* $\psi \in K(\varphi)$, *then* $S_3(\varphi \mid \psi) = S(\varphi \mid \psi)$, *and conversely if* $S_3(\varphi \mid \psi) = S(\varphi \mid \psi) < +\infty$, *then* $\psi \in K(\varphi)$.

(2) *If* $\psi \in K(\varphi)$, *then* $S_3(\varphi \mid (\psi + \varphi)/2) = S(\varphi \mid (\psi + \varphi)/2)$, *and conversely if* $S_3(\varphi \mid (\psi + \varphi)/2) = S(\varphi \mid (\psi + \varphi)/2) < +\infty$, *then* $\psi \in K(\varphi)$.

The monotonicity (3.1) says that the restriction of measurement to a subalgebra $\mathfrak{M}$ usually makes it more difficult to discriminate between two states. From Theorem 4.1, the physical meaning of sufficiency might be explained as follows: If a subalgebra $\mathfrak{M}$ is sufficient for $\{\varphi, \psi\}$, then we obtain from the measurement of $\mathfrak{M}$ as much information as from that of $\mathfrak{N}$ to discriminate between φ and ψ. In particular, to distinguish $\psi \in I(\varphi)$ (resp. $\psi \in K(\varphi)$) from φ, the measurement of Z_φ (resp. $\mathfrak{Z}$) gives as much information as $\mathfrak{N}$.

ACKNOWLEDGMENT. The authors wish to express their gratitude to Professor H. Umegaki for his valuable advice and constant encouragement. They thank Professor H. Araki for his critical reading and essential comments for this work. Professor J. Tomiyama and the referee kindly indicated the useful paper [4] to the authors.

REFERENCES

1. H. Araki, *Some properties of modular conjugation operator of von Neumann algebras and a non-commutative Radon-Nikodym theorem with a chain rule*, Pacific J. Math., **50** (1974), 309-354.

2. ———, *Relative entropy of states of von Neumann algebras*, Publ. RIMS, Kyoto Univ., **11** (1976), 809-833.

3. ———, *Relative entropy for states of von Neumann algebras II*, Publ. RIMS, Kyoto Univ., **13** (1977), 173-192.

4. F. Combes and C. Delaroche, *Groupe modulaire d'une espérance conditionnelle dans une algébre de von Neumann*, Bull. Soc. Math. France, **103** (1975), 385-426.

5. A. Connes, *Une classification des facteurs de type III*, Ann. Sci. École Norm. Sup. Sér. 4, **6** (1973), 133-252.

6. I. Csiszár, *Information-type measures of difference of probability distributions and indirect observations*, Studia Sci. Math. Hungar., **2** (1967), 299-318.

7. S. Gudder and J.-P. Marchand, *Noncommutative probability on von Neumann algebras*, J. Math. Phys., **13** (1972), 799-806.

8. R. Haag, N. M. Hugenholtz and M. Winnink, *On the equilibrium states in quantum statistical systems*, Commun. Math. Phys., **5** (1967), 215-236.

9. U. Haagerup, *Operator valued weights in von Neumann algebras, I*, J. Functional Analysis, **32** (1979), 175-206.

10. P. R. Halmos and L. J. Savage, *Application of the Radon-Nikodym theorem to the theory of sufficient statistics*, Ann. Math. Statistics, **20** (1949), 225-241.

11. R. Kubo, *Statistical mechanical theory of irreversible processes I*, J. Phys. Soc. Japan, **12** (1957), 570-586.

12. S. Kullback and R. A. Leibler, *On information and sufficiency*, Ann. Math. Statistics, **22** (1951), 79-86.

13. G. Lindblad, *Expectations and entropy inequalities for finite quantum systems*, Commun. Math. Phys., **39** (1974), 111-119.

14. P. C. Martin and J. Schwinger, *Theory of many-particle systems I*, Phys. Rev., **115** (1959), 1342-1373.

15. G. K. Pedersen and M. Takesaki, *The Radon-Nikodym theorems for von Neumann algebras*, Acta Math., **130** (1973), 53-87.

16. S. Sakai, *C*-Algebras and W*-Algebras*, Springer, Berlin, 1971.

17. M. Takesaki, *Tomita's Theory of Modular Hilbert Algebras and its Applications*, Springer, Lecture notes in math. Vol. 128, 1970.

18. ———, *Conditional expectations in von Neumann algebras*, J. Functional Analysis, **9** (1972), 306-321.

19. J. Tomiyama, *On the projection of norm one in W^*-algebras,* Proc. Japan Acad., **33** (1957), 608–612.
20. A. Uhlmann, *Relative entropy and the Wigner-Yanase-Dyson-Lieb concavity in an interpolation theory,* Commun. Math. Phys., **54** (1977), 21–32.
21. H. Umegaki, *Conditional expectation in an operator algebra,* Tôhoku Math. J., **6** (1954), 177–181.
22. ———, *Conditional expectation in an operator algebra, III,* Kōdai Math. Sem. Rep., **11** (1959), 51–64.
23. ———, *Conditional expectation in an operator algebra, IV (entropy and information),* Kōdai Math. Sem. Rep., **14** (1962), 59–85.

Received March 13, 1979 and in revised form January 4, 1980.

SCIENCE UNIVERSITY OF TOKYO
NODA CITY, CHIBA 278, JAPAN

M. OHYA
KODAI MATH. J.
3 (1980), 287—294

ON OPEN SYSTEM DYNAMICS —AN OPERATOR ALGEBRAIC STUDY—

BY MASANORI OHYA

Abstract

The open system dynamics is rigorously studied within the C^*-algebraic framework in terms of the approach to equilibrium. It is pointed out that every combined state of every state of a finite system and an equilibrium state describing an infinite reservoir relaxes to equilibrium through an interaction between both systems when the total Hamiltonian of the combined system satisfies some spectral properties.

Sec. I: Introduction.

The approach to equilibrium of a dynamical system is one of the most important problems to be solved in quantum statistical mechanics [1, 2, 3, 4]. The principle aim of this paper is to study the problem of this type for a finite system interacting with an infinite reservoir in equilibrium.

The motivation of this work is as follows: Some physicists think that the system to be measured or in which some experiments are performed should be finite even if it is large compared with the radius of an atom. However, if a system is finite and isolated, any time dependent state of the system will not relax to equilibrium because the basic Schrödinger equation of motion is reversible under time reflection. From evidence accumulated by experiments, most of physical systems relax to some equilibrium after long time. This fact tells that we had better treat such finite system as open system, namely, interactions between the finite system and the outside of the system (the so-called reservoir) should be taken into account. We then expect that physically interesting combined states of the system and the reservoir will relax to equilibrium through an interaction between them.

In this paper, we obtain conditions under which such relaxation occurs. We here take the Kubo-Martin-Schwinger (K. M. S.) condition [5, 6, 7] as that of equilibrium of our systems considered since any Gibbs state satisfies this condition and the K. M. S. condition seems most appropriate [8, 9] to discuss thermal equilibrium in quantum statistical mechanics.

Received April 18, 1979

Sec. II: Formulation of the Problems.

Let $\mathcal{H}^S$ be a Hilbert space of a system S and H^S denote a self-adjoint lower bounded Hamiltonian of S. We often call the system S finite when the volume of S is finite or the degrees of freedom of S is finite. Our system S is assumed to have finite volume, so the spectrum of H^S becomes discrete. Let our finite system S be described by a triple $(\mathcal{A}^S, \mathfrak{S}^S, \alpha_t^S)$, where $\mathcal{A}^S$ is the C^*-algebra $B(\mathcal{H}^S)$ of all bounded linear operators on the Hilbert space $\mathcal{H}^S$, $\mathcal{S}^S$ is the set of all normal states on $\mathcal{A}^S$ (i.e., the set of all linear functionals ϕ^S on $\mathcal{A}^S$ such that $\phi^S(A^*A)\geqq 0$ for any $A\in\mathcal{A}^S$, $\phi^S(I^S)=1$ for unity I^S of $\mathcal{A}^S$ and $\phi^S(A_\alpha)\uparrow\phi(A)$ for $A_\alpha\uparrow A$, filtering upwards, in $\mathcal{A}^S$), and α_t^S $(t\in\boldsymbol{R})$ is the time evolution automorphism of $\mathcal{A}^S$ generated by the Hamiltonian H^S.

On the other hand, an infinite reservoir R is described by another triple $(\mathcal{A}^R, \mathcal{S}^R, \alpha_t^R)$, where $\mathcal{A}^R$ is a C^*-algebra with unity I^R, $\mathcal{S}^R$ is the set of all states on $\mathcal{A}^R$ and α_t^R $(t\in\boldsymbol{R})$ is a strongly continuous one-parameter automorphism of $\mathcal{A}^R$. We assume that the infinite reservoir is initially in equilibrium described by a faithful K. M. S. state φ^R at the inverse temperature β with respect to α_t^R. It is said that the state φ^R satisfies the K. M. S. condition at β w. r. t. the automorphism α_t^R if for any pair A, B in $\mathcal{A}^R$, there exists a bounded function $F_{A,B}(z)$ of the complex number z holomorphic in and continuous on the strip $-\beta\leqq \mathrm{Im}\, z \leqq 0$ such that $F_{A,B}(t)=\varphi^R(\alpha_t^R(A)B)$ and $F_{A,B}(t-i)=\varphi^R(B\alpha_t^R(A))$ for any $t\in\boldsymbol{R}$. By the Gelfand-Naimark-Segal (G. N. S.) construction theorem, to the state φ^R there correspond a Hilbert space $\mathcal{H}^R$, a cyclic vector Φ^R, a representation π^R being a *-homomorphism from the C^*-algebra $\mathcal{A}^R$ to the set $B(\mathcal{H}^R)$ of all bounded linear operators on the Hilbert space $\mathcal{H}^R$ and a strongly continuous one-parameter unitary group U_t^R such that $\pi^R(\alpha_t^R(A))=U_t^R\pi^R(A)U_t^{R*}$ for any $A\in\mathcal{A}^R$ and $U_t^R\Phi^R=\Phi^R$.

Let us take any faithful state $\phi^S\in\mathcal{S}^S$ of S and consider its time development. When the system S is isolated and the state ϕ^S is not α_t^S-invariant (i. e. $\phi^S(\alpha_t^S(A))\neq\phi^S(A)$ for some $A\in\mathcal{A}^S$), the expectation value $\phi^S(\alpha_t^S(A))$ is periodic or at least almost periodic function in t because the system S is finite. Thus $\phi^S(\alpha_t^S(A))$ does not relax to equilibrium for all $A\in\mathcal{A}^S$ when time t tends to infinite. That is, the infinite time limit of $\phi^S\circ\alpha_t^S$ in the weak*-topology does not exist. We hence need to take account of the effect of an infinite reservoir on the finite system S in order to explain such relaxation behavior.

The initial (non-interacting) combined system of the system S and the reservoir R is described by the following:

(1-1) Algebra: $\mathcal{A}=\mathcal{A}^S\otimes\mathcal{A}^R$,

(1-2) State: $\phi=\phi^S\otimes\varphi^R\in\mathcal{S}=\mathcal{S}^S\otimes\mathcal{S}^R$,

(1-3) Time evolution: $\alpha_t^0=\alpha_t^S\otimes\alpha_t^R$,

(1-4) Hilbert space: $\mathcal{H}=\mathcal{H}^S\otimes\mathcal{H}^R$,

(1-5) Representation: $\pi=i^S\otimes\pi^R$ (i^S is the identity map onto $\mathcal{A}^S$).

Let us introduce an interaction between the systems S and R. The interaction will be a bounded self-adjoint element $V=V^*\in\mathcal{A}$. By Stone's theorem, there exists a self-adjoint operator H^R which generates the unitary one parameter group U_t^R, i. e., $U_t^R = \exp(iH^R t)$. We may call $H = H^S + H^R + \pi(V)$ the total Hamiltonian of the combined system, which generates the so-called perturbed time evolution automorphism α_t of $\mathcal{A}$ [10]:

$$\alpha_t(A) = \sum_{n \geq 0} i^n \int_{0 \leq t_1 \leq \cdots \leq t_n \leq t} dt_1 \cdots \int dt_n [\alpha_{t_1}^0(V), \cdots [\alpha_{t_n}^0(V), \alpha_t^0(A)] \cdots] \tag{1-6}$$

for $t \geqq 0$ (the case $t \leqq 0$ is due to exchange of 0 and t in the above integral domain).

Considering the combined system and any faithful state on $\mathcal{A}$ denoted by $\phi = \phi^S \otimes \varphi^R$, it is natural for us to ask the following question:

> " *Under what conditions on the dynamical system does the limit* $w^* - \lim_{|t| \to \infty} \phi \circ \alpha_t$ *exist and is it identical to a proper eqilibrium (K. M. S.) state at the inverse temperature* β *with respect to the automorphism* α_t*?*"

This problem concerns the relaxation process of the combined system. In conventional discussions of physics, one does not worry about such question but takes it for granted. By answering this question, we expect that the restriction of the limiting state to the algebra $\mathcal{A}^S$ of the system S might be close, in some sense, to the K. M. S. state of S with the same temperature of the reservoir R, and we can also explain some physical phenomena of the so-called relaxation processes; for example, if the temperature of the system is initially different from that of the reservoir, our experience tells that if the system is in contact with the reservoir, then the temperature of the system goes to that of the reservoir. We finally note that our investigation here is concerned with the time development of the combined system but not of the system itself. It is really important to study directly the time development of a state of the system under the effect of some interaction with an infinite reservoir. For this purpose, the technique of conditional expectation invented by H. Umegaki [11] will be essential. This aspect will be discussed elsewhere [12].

Sec. III: Relaxation Process.

In this section, we study the problem presented in the previous section.

As mentioned before, the initial temperature of the system S might be different from that of the reservoir R, or the initial state ϕ^S of S might not be α_t^S-invariant. In any case, there exists a trace class operator $\rho^S = \exp(-\beta H^S)/\mathrm{Tr}\exp(-\beta H^S)$ so that the state φ^S defined by $\varphi^S(A) = \mathrm{Tr}\,\rho^S A$ for any $A \in \mathcal{A}^S$ satisfies the K. M. S. condition at the inverse temperature β of the reservoir with respect to the automorphism α_t^S of the system. Thus the state $\varphi^S \otimes \varphi^R$ on $\mathcal{A}$ satisfies this condition at β w. r. t. $\alpha_t^0 = \alpha_t^S \otimes \alpha_t^R$. Let us denote this state by φ in the sequel discussions. Moreover, for a faithful normal state ϕ^S on $\mathcal{A}^S$ which may not be α_t^S-invariant, we denote the combined state with the equilibrium state φ^R of the reservoir by $\phi=\phi^S\otimes\varphi^R$ as in (1-2).

Let us start by proving several lemmas.

LEMMA 1. *Let ω be a state on $\mathcal{A}$ and χ be another state on $\mathcal{A}$ dominated by ω, i.e., $\chi\leqq\lambda\omega$ for some $\lambda>0$. Then for any $\varepsilon>0$, there exists an element W in $\mathcal{A}$ such that*

$$|\chi(A)-\omega(AW)|<\varepsilon\|A\|$$

for any $A\in\mathcal{A}$.

Proof. It is well-known that for any state χ dominated by ω, there exists a positive operator B in $\pi_\omega(\mathcal{A})'$ such as

$$\chi(A)=(\Omega,\ B\pi_\omega(A)B\Omega),$$

where Ω is the G.N.S. cyclic vector induced by the state ω. The cyclicity of Ω implies that for any $\varepsilon>0$, there exists an element W in $\mathcal{A}$ such that $\|B^2\Omega-\pi_\omega(W)\Omega\|<\varepsilon$ holds. We hence have $|\chi(A)-\omega(AW)|<\varepsilon\|A\|$ for any $A\in\mathcal{A}$. (q.e.d.)

LEMMA. 2. *For the state $\phi=\phi^S\otimes\varphi^R$ introduced above, there exists an element K in $\mathcal{A}$ such that for any $\varepsilon>0$,*

$$|\phi(A)-\varphi(AK)|<\varepsilon\|A\|$$

for any $A\in\mathcal{A}$.

Proof. Let us consider two following subsets $\mathcal{V}^S$ and $\mathcal{D}^S$ of $\mathcal{S}^s$:

$$\mathcal{V}^S=\{\psi^S\in\mathcal{S}^S:\ \psi^S(A)=(\Psi^S,\ A\Psi^S)\text{ for any }A\in\mathcal{A}^S\text{ and }\Psi^S\in\mathcal{H}^S\}$$

and

$$\mathcal{D}^S=\{\psi^S\in\mathcal{S}^S:\ \psi^S\leqq\lambda\varphi^S\text{ for some }\lambda\in R^+\}.$$

As is known [13, 14], the set $\mathcal{V}^S$ is dense in $\mathcal{S}^S$ and the set $\mathcal{D}^S$ is dense in $\mathcal{V}^S$ because φ^S is a K.M.S. state. Hence for the state ϕ^S and any $\varepsilon>0$, there exists a state ω in $\mathcal{V}^S$ such that $|\phi^S(A)-\omega(A)|<(\varepsilon/3)\|A\|$ holds for any $A\in\mathcal{A}^S$. Moreover, for the above state ω, there exists a state χ in $\mathcal{D}^S$ such that $|\omega(A)-\chi(A)|<(\varepsilon/3)\|A\|$ for any $A\in\mathcal{A}^S$. As χ is dominated by φ^S, according to Lemma 1, there exists an element W in $\mathcal{A}^S$ such as $|\chi(A)-\varphi^S(AW)|<(\varepsilon/3)\|A\|$. We therefore obtain $|\phi(A)-\varphi(AK)|<\varepsilon\|A\|$ for any $A\in\mathcal{A}$, where K is taken as $K=W\otimes I^R$. (q.e.d.)

LEMMA 3. *For the state $\phi=\phi^S\otimes\varphi^R$, there exist a state ψ on $\mathcal{A}$ and an element R in $\mathcal{A}$ such that*

(i) *the state ψ satisfies the K.M.S. condition at β w.r.t. α_t, and*

(ii) *for any $\varepsilon>0$, $|\phi(A)-\psi(AR)|<\varepsilon\|A\|$ holds for any $A\in\mathcal{A}$.*

Proof. Let us introduce a vector as

$$\Psi=D^V\Phi/\|D^V\Phi\|,$$

where Φ is the cyclic vector associated with φ such that $\Phi=\Phi^S\otimes\Phi^R$ and $\varphi^S(A)=(\Phi^S, A\Phi^S)$ for any $A\in\mathcal{A}^S$, and D^V is given by

$$D^V=\sum_{n\geq 0}(-1)^n\int_{0\leq t_1\leq\cdots\leq t_n\leq\beta/2}dt_1\cdots\int dt_n\tilde{\alpha}^0_{it_1}(\pi(V))\cdots\tilde{\alpha}^0_{it_n}(\pi(V)), \tag{2-1}$$

where $\tilde{\alpha}^0$ is the canonical extension of α^0 to the von Neumann algebra $\pi(\mathcal{A})''$. The above vector Ψ is always in the Hilbert space $\mathcal{H}$, although D^V is not in the C^*-algebra $\pi(\mathcal{A})$ except when V is a α^0-analytic element of $\mathcal{A}$. Define a state ψ by $\psi(A)=(\Psi, \pi(A)\Psi)$ for any $A\in\mathcal{A}$. This state ψ satisfies the K.M.S. condition at β w.r.t. α_t [10]. The state ψ and the time evolution automorphism α_t are constructed by φ and α^0_t through (2-1) and (1-6) respectively. Conversely, it is easily seen that φ and α^0_t can be constructed back from ψ and α_t with the interaction—V. Namely, putting $\Phi'=Q^V\Psi/\|Q^V\Psi\|$ with

$$Q^V=\sum_{n\geq 0}\int_{0\leq t_1\leq\cdots\leq t_n\leq\beta/2}dt_1\cdots\int dt_n\tilde{\alpha}_{it_1}(\pi(V))\cdots\tilde{\alpha}_{it_n}(\pi(V)).$$

We can then readily show that $\Phi'=\Phi$. Moreover, by the simple but rather tedious computations using the boundary properties of the K.M.S. state, we obtain

$$\varphi(A)=(Q^V\Psi, \pi(A)Q^V\Psi)/\|Q^V\Psi\|^2=(\Psi, \pi(A)S^V_0\Psi),$$

where S^V_0 is given by $S^V/|(\Psi, S^V\Psi)|$ and

$$S^V=\sum_{n\geq 0}\int_{0\leq t_1\leq\cdots\leq t_n\leq\beta}dt_1\cdots\int dt_n\tilde{\alpha}_{it_1}(\pi(V))\cdots\tilde{\alpha}_{it_n}(\pi(V)).$$

If the interaction V is a α^0-analytic (hence α-analytic) element of $\mathcal{A}$, the above S^V_0 is given by $\pi(\tilde{S}^V_0)$, where $\tilde{S}^V_0=\tilde{S}^V/|\psi(\tilde{S}^V)|$ and

$$\tilde{S}^V=\sum_{n\geq 0}\int_{0\leq t_1\leq\cdots\leq t_n\leq\beta}dt_1\cdots\int dt_n\alpha_{it_1}(V)\cdots\alpha_{it_n}(V).$$

Since the set of all α-analytic element of $\mathcal{A}$ is dense in $\mathcal{A}$ in the norm topology, for any $V=V^*\in\mathcal{A}$ and any $\varepsilon>0$, there exists a α-analytic element $V_0=V_0{}^*\in\mathcal{A}$ such that $\|V-V_0\|<\varepsilon$. According to Theorem 3.1 of [10], we easily obtain

$$\|Q^V\Psi-Q^{V_0}\Psi\|<\frac{1}{2}\beta\|V-V_0\|\exp\Big(\frac{1}{2}\beta\|V-V_0\|\Big).$$

Hence for any $\varepsilon>0$, $\|Q^V\Psi/\|Q^V\Psi\|-Q^{V_0}\Psi/\|Q^{V_0}\Psi\|\|<\varepsilon$ holds. It is thus a easy exercise to show that for any $\varepsilon>0$, the inequality $|\varphi(A)-\psi(A\tilde{S}^{V_0}_0)|<\frac{1}{2}\varepsilon\|A\|$ is satisfied for any $A\in\mathcal{A}$. Now, by Lemma 2, there exists an element K in $\mathcal{A}$ such that $|\phi(A)-\varphi(AK)|<\frac{1}{2}\varepsilon\|A\|$ holds for any $A\in\mathcal{A}$. Taking $K\tilde{S}^{V_0}_0=R$, this R is an element of $\mathcal{A}$ and the inequality $|\phi(A)-\psi(AR)|<\varepsilon\|A\|$ holds for any $A\in\mathcal{A}$. (q.e.d.)

Let us find conditions under which every state $\phi=\phi^S\otimes\varphi^R$ ($\phi^S\in\mathcal{S}^S$) relaxes, under the time evolution α_t, to an equilibrium (K. M. S.) state at β w. r. t. α_t, as physically expected.

The spectrum of a Hamiltonian is one of the most important quantities of physics, and most of physicists are interested in the property of it; for example, how dense it is in R. In our case, the system considered is finite, so the spectrum of the Hamiltonian H^S is discrete, which forbids the approach to equilibrium. Therefore we needed some interaction with suitable reservoir. The Hamiltonian to be studied was the so-called total one $H=H^S+H^R+\pi(V)$. What we ask is the following: Which conditions do we have to impose on the total Hamiltonian H? In other words, which interaction V do we have to choose?

THEOREM 4. *If the rank of the projection E to the null space of H is one and the spectrum of H consists of* $\{0\}$ *and absolutely continuous one, then the time evolution automorphism α_t admits unique* K. M. S. *state ψ in the representation space $\mathcal{H}$ and the limit w^*-lim $\phi\circ\alpha_t$ is equal to ψ.*

Proof. In Lemma 3, we constructed the state ψ satisfying the K. M. S. condition w. r. t. α_t. For any $A, B\in\mathcal{A}$, let us consider $\psi(A\alpha_t(B))$, which is equal to $(\Psi, \pi(A)\exp(+itH)\pi(B)\Psi)$ because the K. M. S. state ψ is α_t-invariant. According to the spectral decomposition of H, we have

$$\psi(A\alpha_t(B))=(\pi(A)^*\Psi, \int\exp(+itr)de(r)\pi(B)\Psi).$$

When t tends to infinite, the above expression becomes

$$(\pi(A)^*\Psi, E\pi(B)\Psi)$$

because of the spectrum properties of H. Since the rank of E is one, $E\pi(B)\Psi=(\Psi, \pi(B)\Psi)\Psi$ for any $B\in\mathcal{A}$. We thus obtain

$$\lim_{|t|\to\infty}\psi(A\alpha_t(B))=\psi(A)\psi(B)$$

for any $A, B\in\mathcal{A}$. Namely the state ψ is clustering for α_t. This fact tells [13] that ψ is the unique K. M. S. state of $\mathcal{H}$. Let us now consider

$$|\phi(\alpha_t(A))-\psi(A)|,$$

which is less than

$$|\phi(\alpha_t(A))-\psi(\alpha_t(A)R)|+|\psi(\alpha_t(A)R)-\psi(A)|,$$

where R is an element in $\mathcal{A}$ obtained in Lemma 3. The first term of the above expression is again less than $\varepsilon\|A\|$ because of Lemma 3. We now estimate the second term:

$$I(t)=|\psi(\alpha_t(A)R)-\psi(A)|.$$

As shown that the state ψ is clustering for α_t and R is in $\mathcal{A}$,

$$\lim_{|t|\to\infty} I(t)=0$$

because of $\psi(R)=1$. We hence have

$$w^*\text{-}\lim_{|t|\to\infty} \phi\circ\alpha_t=\psi. \qquad \text{(q. e. d.)}$$

We finally see what happens to the state ψ when the strength of the interaction V becomes very weak but the interaction still has the properties of Theorem 4.

THEOREM 5. *Under the conditions of Theorem* 4, *we have*

$$w^*\text{-}\lim_{\|V\|\to\infty}\ \lim_{|t|\to\infty} \phi\circ\alpha_t=\varphi^S \text{ on } \mathcal{A}^S.$$

Proof. From Theorem 4, we have only to show

$$w^*\text{-}\lim_{\|V\|\to 0} \psi=\varphi^S \text{ on } \mathcal{A}^S.$$

As discussed in the proof of Lemma 3, the unique K. M. S. state ψ is given through $\psi(A)=(T^V\Phi, \pi(A)T^V\Phi)$ for any $A\in\mathcal{A}$, where T^V is given by

$$T^V=D^V/\|D^V\Phi\| \text{ with } D^V \text{ defined in the proof of Lemma 3.}$$

Therefore the following inequality holds:

$$|\psi(A)-\varphi(A)|\leqq 2\|(I-T^V)\Phi\|\|A\|.$$

As mentioned in Lemma 3, for any $V=V^*\in\mathcal{A}$ and any $\varepsilon>0$, there exists a α^0-analytic element $V_0=V_0^*\in\mathcal{A}$ such that $\|T^V\Phi-T^{V_0}\Phi\|<\varepsilon$. We hence have

$$|\psi(A)-\varphi(A)|\leqq 2\|(I-T^{V_0})\Phi\|\|A\|.$$

The above inequality together with the facts that $T^{V_0=0}=I$ implies that $w^*\text{-}\lim_{\|V\|\to 0}\psi=\varphi$ on $\mathcal{A}$. As the restriction of the state φ to $\mathcal{A}^S$ is φ^S, we have the conclusion. (q. e. d.)

This theorem shows that if we can choose the interaction so that its strength is sufficiently weak but the total Hamiltonian $H=H^S+H^R+\pi(V)$ still satisfies the condition of Theorem 4, then the limiting state of ψ under $\|V\|\to 0$ is enough close to the equilibrium state φ, that is, any state ϕ^S on $\mathcal{A}^S$ approaches to the equilibrium state φ^S in the above sense.

The theorem 4 will be somewhat related to the derivation of equilibrium state [15, 16], about which we will discuss elsewhere. The conditions of Theorems 4 and 5 might be realized in some physical models.

Acknowledgement

The author thanks Professors H. Umegaki, D. Kastler, H. Ezawa, K. Nakamura and F. Hiai for fruitful discussions and useful comments to the related topics of this work. He also thanks Professor K. Kunisawa for his kindness.

REFERENCES

[1] F. HAAKE, "Statistical Treatment of Open System by Generalized Master Equation", Springer-Verlag, 1972.
[2] E.B. DAVIES, "Quantum Theory of Open Systems", Academic Press, 1976.
[3] K. HEPP AND E.H. LIEB, Helv. Phys. Acta. **46**, 573, 1975.
[4] G.G. EMCH AND C. RADIN, J. Math. Phys. **12**, 2043, 1971.
[5] R. KUBO, J. Phys. Soc. Japan, **21**, 570, 1957.
[6] P.C. MARTIN AND J. SCHWINGER, Phys. Rev. **115**, 1342, 1959.
[7] R. HAAG, N. HUGENHOLTZ AND M. WINNIK, Commun. Math. Phys. **5**, 215, 1967.
[8] G.L.SEWELL, "The Description of Thermodynamical Phases in Statistical Mechanics", 1974.
[9] G.G. EMCH AND J.F. KNOPS, J. Math. Phys. **12**, 2043, 1971.
[10] H. ARAKI, R.I.M.S. Kyoto, **9**, 165, 1973.
[11] H. UMEGAKI, Tohoku J. Math. **6**, 177, 1954 and **8**, 86, 1956., Kodai Math. Sem. Rep. **11**, 51, 1959 and **14**, 59, 1962.
[12] M. OHYA, in preparation.
[13] M. TAKESAKI, "Tomita's Theory of Modular Hilbert Algebras and its Applications", Springer-Verlag, 1970.
[14] J. DIXMIER, "Les algèbres d'opérateurs dans l'espace Hilbertien", Gauthier-Villars, Paris, 1969.
[15] R. HAAG, D. KASTLER AND E. TRYCH-POHLMEYER, Commun. Math. Phys. **38**, 173, 1974.
[16] D. KASTLER, Private Communication.

DEPARTMENT OF INFORMATION SCIENCES,
THE SCIENCE UNIVERSITY OF TOKYO,
278, NODA CITY, CHIBA, JAPAN.

Vol. 16 (1979) *REPORTS ON MATHEMATICAL PHYSICS* No. 3

DYNAMICAL PROCESS IN LINEAR RESPONSE THEORY

MASANORI OHYA

Department of Information Sciences, The Science University of Tokyo, Noda City, Chiba, Japan*

and

Institute of Fundamental Studies, Department of Physics and Astronomy, University of Rochester, Rochester, New York, U.S.A.

(*Received June 15, 1977—Revised February 28, 1978*)

The problems of a dynamical process in the linear response theory are studied in the C^*-algebraic framework. We outline the domain of validity of the linear response approximation by considering the similarity and difference between the linear response dynamics and the exact dynamics. It is shown that the linear response method is identical to the exact method when we consider the return to equilibrium of all locally perturbed states. As far as the stability of a dynamical system is concerned, the linear response theory is shown to differ from the exact theory. It is pointed out that the phase transition in a dynamical process does not occur in the linear response theory.

In Section II, we define the perturbed dynamical systems under a perturbation $V = V^* \in \mathfrak{A}$: $(\mathfrak{A}, \varphi^V, \alpha^V)$ in the "exact" sense (E.S. for brevity) and $(\mathfrak{A}, \varphi^{V,1}, \alpha^{V,1})$ in the "linear response" sense (L.R.S. for brevity). In that section, we formulate the problems to be investigated in this paper. In Section III, it is shown that the linear response method is identical to the exact method when we consider the return to equilibrium of all perturbed states. In Section IV, we study the stability of the dynamical system $(\mathfrak{A}, \varphi, \alpha)$, and it is pointed out that the linear response theory differs from the exact theory. In Section V, the phase transition in a dynamical process is discussed and it is shown that we do not have such transition in the linear response dynamics. In Section VI, we illustrate general results obtained in the previous sections by the $X-Y$ model.

I. Introduction

Whereas the linear response theory [1, 2] has succeeded to explain many physical phenomena, it is nevertheless true that one does not know how to control this theory rigorously. Recently some work [3, 4] has been done to formulate the linear response theory in the von Neumann algebraic framework. Our work is an attempt to study further

* Mailing address.

the linear response theory in the C^*-algebraic framework with the purpose of determining the domain of validity of the linear response approximation.

An infinite dynamical system is described by a triple $(\mathfrak{A}, \varphi, \alpha)$, where $\mathfrak{A}$ is a C^*-algebra with unity I, φ is a state on $\mathfrak{A}$ and α is a continuous homomorphism from the set of real numbers R to Aut($\mathfrak{A}$), the set of automorphisms of $\mathfrak{A}$. We call α_t, $t \in R$, a time evolution (automorphism) of $\mathfrak{A}$. It is assumed that α_t satisfies the following continuity: $\lim_{t\to 0} \|\alpha_t(A)-A\| = 0$ for any $A \in \mathfrak{A}$. We further assume that the above state φ satisfies the K.M.S. condition [5] with respect to α_t at an inverse temperature β (we take $\beta = 1$ for simplicity). So our dynamical system in the sequel is $(\mathfrak{A}, \varphi, \alpha)$ with the K.M.S. condition. Since the K.M.S. state φ is time invariant under α_t, there exists a strongly continuous one parameter group of unitary operators U_t implementing α_t: $\Pi_\varphi(\alpha_t(A)) = U_t \Pi_\varphi(A) U_{-t}$, defined through $U_t \Phi = \Phi$, where Π_φ is the G.N.S. representation and Φ is the cyclic vector induced by φ. Moreover, by Stone's theorem [7], there exists a self-adjoint operator H such that $U_t = \exp(itH)$.

In the following sections, we denote by $\tilde{\varphi}$ and $\tilde{\alpha}_t$ the canonical extensions of φ and α_t from the C^*-algebra to the von Neumann algebra $\Pi_\varphi(\mathfrak{A})''$, the weak closure of $\Pi_\varphi(\mathfrak{A})$ [8].

II. Formulation of the problems

In this section, we consider the effect of a perturbation on the dynamical system $(\mathfrak{A}, \varphi, \alpha)$. We then formulate the problems to be investigated in the following sections.

If our dynamical system $(\mathfrak{A}, \varphi, \alpha)$ is disturbed by a perturbation λV, where V is a self-adjoint element of $\mathfrak{A}$ and λ is a positive real coupling constant, then the exact perturbed system is denoted by a new triple $(\mathfrak{A}, \varphi^V, \alpha^V)$, which is precisely defined below [9, 10]: The exact perturbed state φ^V is given by, for any $A \in \mathfrak{A}$,

$$\langle \varphi^V; A\rangle = (\Phi^V, \Pi_\varphi(A)\Phi^V), \tag{2.1}$$

where

$$\Phi^V \equiv \Pi_\varphi(W^V_{i/2})\Phi / \|\Pi_\varphi(W^V_{i/2})\Phi\| \equiv \Pi_\varphi(W^V)\Phi, \tag{2.2}$$

$$W^V_{i/2} = \sum_{n\geq 0} (-\lambda)^n \int\limits_{0\leq t_1 \leq \ldots \leq t_n \leq 1/2} dt_1 \ldots \int dt_n \, \alpha_{it_1}(V) \ldots \alpha_{it_n}(V). \tag{2.3}$$

This perturbed vector Φ^V is also cyclic and separating for $\Pi_\varphi(\mathfrak{A})''$ [9]. The above $W^V_{i/2}$ (hence W^V) is actually an element of $\mathfrak{A}$ [9, 10], provided that the perturbation V is an α-analytic element of $\mathfrak{A}$. However, $\Pi_\varphi(W^V_{i/2})\Phi$ is an element of the G.N.S. Hilbert space $\mathscr{H}_\varphi$ for every $V = V^* \in \mathfrak{A}$. The exact perturbed time evolution automorphism α^V_t of $\mathfrak{A}$ is given by the uniformly convergent Dyson series: For any $A \in \mathfrak{A}$,

$$\alpha^V_t(A) = \sum_{n\geq 0} (i\lambda)^n \int\limits_{0\leq t_1 \leq \ldots \leq t_n \leq t} dt_1 \ldots \int dt_n [\alpha_{t_1}(V), [\ldots, [\alpha_{t_n}(V), \alpha_t(A)] \ldots]] \tag{2.4}$$

for $t \geq 0$ (the case $t < 0$ results by exchange of 0 and t in the above integral).

We here note that the above perturbed state φ^V satisfies the K.M.S. condition with respect to the perturbed time evolution α_t^V [9, 10].

We now consider the linear response dynamical system denoted by a triple $(\mathfrak{A}, \varphi, \alpha)$, which is obtained by linearizing the exact perturbed dynamical system w.r.t. λ. The linear response functional $\varphi^{V,1}$ is given by

$$\langle\varphi^{V,1}; A\rangle = \langle\varphi; A\rangle - \lambda\Big\langle\varphi; \int_0^{1/2} ds\alpha_{-is}(V)A\Big\rangle - $$

$$-\lambda\Big\langle\varphi; \int_0^{1/2} dsA\alpha_{is}(V)\Big\rangle + \lambda\langle\varphi; A\rangle\,\langle\varphi; V\rangle \tag{2.5}$$

for any $A \in \mathfrak{A}$. The linear response time evolution $\alpha_t^{V,1}$ is given by

$$\alpha_t^{V,1}(A) = \alpha_t(A) + i\lambda\int_0^t ds[\alpha_s(V), \alpha_t(A)] \tag{2.6}$$

for any $A \in \mathfrak{A}$ and $t \geqslant 0$ (the case $t < 0$ results by exchange of 0 and t in the above integral).

We study, in the exact dynamics (see, e.g., [10, 11, 12]), the existence of the limits $w^*\text{-}\lim_{t\to\pm\infty} \varphi^V \circ \alpha_t \equiv \varphi_\pm^V$, $w^*\text{-}\lim_{t\to\pm\infty} \varphi \circ \alpha_t^V \equiv \varphi_\pm$ and discuss the properties of the limits $\varphi_\pm^V$ and $\varphi_\pm$, provided that they exist. In the linear response dynamics, the following questions are pertinent to our investigations:

(1) Under which conditions on V do the limits $\varphi_\pm^{V,1} \equiv w^*\text{-}\lim_{t\to\pm\infty} \varphi^{V,1} \circ \alpha_t$ exist?

(2) Under which conditions on V are $\varphi_\pm^{V,1}$ equal to φ?

(3) Under which conditions on V do the limits $\varphi_{\pm,1} \equiv w^*\text{-}\lim_{t\to\pm\infty} \varphi \circ \alpha_t^{V,1}$ exist? (If they exist, the system is said to be stable in L.R.S. under the perturbation V [4].)

(4) Under which conditions on V are $\varphi_{\pm,1}$ equal to $\varphi^{V,1}$?

Comparing the results for the above questions with those for the exact dynamics, we ask

(5) Can we outline the domain of validity of the linear response approximation?

By analogies to the exact dynamics [10, 11, 12, 13, 15, 21], the question (1) may be regarded as the problem of the approach to equilibrium, the question (2) is of the return to equilibrium and the questions (3), (4) are of the stability of the dynamical system, all in L.R.S. under a perturbation V. The question (3) has been studied by Verbeure and Weder [4], whose C^*-algebraic version is as follows:

THEOREM II.1. *For the dynamical system* $(\mathfrak{A}, \varphi, \alpha)$ *and a perturbation* $V = V^* \in \mathfrak{A}$, *the dynamical system is stable in L.R.S. under* V *if and only if* $\exp(itH)\Pi_\varphi(V)\Phi$ *weakly converges in the G.N.S. Hilbert space* $\mathscr{H}_\varphi$ *as* $t \to \pm\infty$. *Moreover, in this case, we have*

$$\langle\varphi_{\pm,1}; A\rangle = \langle\varphi; A\rangle \mp \lambda\big(\Phi, \Pi_\varphi(A)T^2(1-E_\varphi)\Pi_\varphi(V)\Phi\big)$$

for any $A \in \mathfrak{A}$.

In the above theorem, T is a self-adjoint positive operator defined by

$$T=(\Delta-1)^{1/2}/(\ln\Delta) \quad \text{and} \quad \Delta=\exp(-H),$$

and E_φ is the projection on the null space of H. This theorem is the C^*-algebraic version of the Theorem 3 of [4], and it can be readily proved.

Before closing this section, we remark that the linear response functional $\varphi^{V,1}$ may not be a state on $\mathfrak{A}$ in general unless certain conditions are satisfied. The linear response time evolution $\alpha_t^{V,1}$ introduced by Verbeure and Weder [4] in the algebraic framework is not an automorphism of $\mathfrak{A}$.

III. Return to equilibrium

In this section, we consider the time development of $\varphi^{V,1}$ after removing the perturbation V dynamically. Namely, we study the questions (1) and (2) given in the previous section. The question (1) of the approach to equilibrium can be solved as follows.

THEOREM III.1. *For the dynamical system* $(\mathfrak{A},\varphi,\alpha)$ *and a perturbation* $V=V^*\in\mathfrak{A}$, *the limits* $\varphi_\pm^{V,1}=w^*\text{-}\lim_{t\to\pm\infty}\varphi^{V,1}\circ\alpha_t$ *exist if and only if* $\exp(-itH)\Pi_\varphi(V)\Phi$ *weakly converges in the Hilbert space* $\mathscr{H}_\varphi$ *as* $t\to\pm\infty$. *Moreover, in this case, we have*

$$\langle\varphi_\pm^{V,1};A\rangle=\langle\varphi;A\rangle-\lambda\big(\Phi,\Pi_\varphi(A)E_\varphi\Pi_\varphi(V)\Phi\big)+\lambda\langle\varphi;A\rangle\langle\varphi;V\rangle$$

for any $A\in\mathfrak{A}$.

We need a lemma to prove this theorem.

LEMMA III.1. *For the dynamical system* $(\mathfrak{A},\varphi,\alpha)$ *and a perturbation* $V=V^*\in\mathfrak{A}$, *we have*

$$\langle\varphi^{V,1};A\rangle=\langle\varphi;A\rangle-\lambda\int_0^1 ds\langle\varphi;A\alpha_{is}(V)\rangle+\lambda\langle\varphi;A\rangle\langle\varphi;V\rangle$$

for any $A\in\mathfrak{A}$.

Proof: According to the definition (2.5) of $\varphi^{V,1}$, we have only to show that $\int_0^{1/2}ds\langle\varphi;\alpha_{-is}(V)A\rangle=\int_{1/2}^1 ds\langle\varphi;A\alpha_{is}(V)\rangle$ for any $A\in\mathfrak{A}$. Let us first prove the above equality for any α-analytic element $\tilde V=\tilde V^*\in\mathfrak{A}$. Since φ is a K.M.S. state w.r.t. α_t, the equality $\langle\varphi;\alpha_{-is}(\tilde V)A\rangle=\langle\varphi;A\alpha_{-is+i}(\tilde V)\rangle$ holds for any $A\in\mathfrak{A}$. This immediately yields the desired equality. We now consider the case of a general perturbation $V=V^*\in\mathfrak{A}$. Then it suffices for us to show that for any $V=V^*\in\mathfrak{A}$ and any $\varepsilon>0$, there exists an α-analytic element $\tilde V=\tilde V^*\in\mathfrak{A}$ such that

$$\left|\int_0^1 ds\langle\varphi;A\alpha_{is}(\tilde V)\rangle-\int_0^1 ds\langle\varphi;A\alpha_{is}(V)\rangle\right|<\varepsilon.$$

It is easily seen that for any $A \in \mathfrak{A}$, we have

$$\int_0^1 ds \Delta^s \Pi_\varphi(A)\Phi = T^2 \Pi_\varphi(A)\Phi.$$

Therefore,

$$\left|\int_0^1 ds\langle\varphi;\, A\alpha_{is}(\tilde{V})\rangle - \int_0^1 ds\langle\varphi;\, A\alpha_{is}(V)\rangle\right| = \left|\int_0^1 ds\big(\Delta^s\Pi_\varphi(A)^*\Phi, \Pi_\varphi(\tilde{V}-V)\Phi\big)\right|$$

$$= \left|\big(T^2\Pi_\varphi(A)^*\Phi, \Pi_\varphi(\tilde{V}-V)\Phi\big)\right| = \left|\big(T\Pi_\varphi(A)^*\Phi,\, T\Pi_\varphi(\tilde{V}-V)\Phi\big)\right|$$

$$\leqslant \|T\Pi_\varphi(A)^*\Phi\|\, \|T\big(\Pi_\varphi(\tilde{V})-\Pi_\varphi(V)\big)\Phi\|. \tag{3.1}$$

Let $\tilde{\mathfrak{N}}$ be a subset of the von Neumann algebra $\mathfrak{N} \equiv \Pi_\varphi(\mathfrak{A})''$ defined by $\tilde{\mathfrak{N}} = \{\int dt\,\tilde{\alpha}_t(Q)\times \times f(t)\colon Q \in \mathfrak{N},\, f \in \hat{\mathfrak{D}}\}$, where $\hat{\mathfrak{D}}$ is the Fourier transform of functions in the space $\mathfrak{D}$ of infinitely differentiable functions with compact supports. Any element of $\tilde{\mathfrak{N}}$ is $\tilde{\alpha}$-analytic. We introduce a sesquilinear form on the von Neumann algebra $\mathfrak{N}$ in such a manner that $(R, Q)_\# = (TR\Phi, TQ\Phi)$ for any R, Q in $\mathfrak{N}$. We then obtain a Hilbert space $\mathfrak{N}_\#$ by closing $\mathfrak{N}$ w.r.t. the above sesquilinear form. It is known [3] that $\tilde{\mathfrak{N}}$ is dense in $\mathfrak{N}_\#$. It is an easy exercise to show that for any $Q \in \tilde{\mathfrak{N}}$ and any $\varepsilon > 0$, there exists an α-analytic element $\tilde{V} = \tilde{V}^*$ such that $\|Q - \Pi_\varphi(\tilde{V})\|_\# < \varepsilon$. Hence, for any $V = V^* \in \mathfrak{A}$ and any $\varepsilon > 0$, there exists an α-analytic element $\tilde{V} = \tilde{V}^* \in \mathfrak{A}$ such that (3.1) is less than ε. (Q.E.D.)

Proof of Theorem III.1: According to the above lemma and the time translational invariance of the state φ under α_t, we have

$$\langle\varphi^{V,1};\, \alpha_t(A)\rangle = \langle\varphi;\, A\rangle - \lambda\int_0^1 ds\langle\varphi;\, \alpha_t(A)\,\alpha_{is}(V)\rangle + \lambda\langle\varphi;\, A\rangle\,\langle\varphi;\, V\rangle.$$

Since $U_t = \exp(itH)$ and $U_t\Phi = \Phi$, $\langle\varphi;\, \alpha_t(A)\,\alpha_{is}(V)\rangle$ is equal to $\big(\Pi_\varphi(A^*)\Phi, \exp(-itH- -sH)\Pi_\varphi(V)\Phi\big)$. Hence, if $\exp(-itH)\Pi_\varphi(V)\Phi$ weakly converges in the Hilbert space $\mathscr{H}_\varphi$ as $t \to \pm\infty$, then the limits $\lim_{t\to\pm\infty} \langle\varphi^{V,1};\, \alpha_t(A)\rangle$ exist for any $A \in \mathfrak{A}$. Conversely, if the above limits exist for any $A \in \mathfrak{A}$, then $\exp(-itH)\Pi_\varphi(V)\Phi$ weakly converges in $\mathscr{H}_\varphi$ because the set $\{\Pi_\varphi(A^*)\Phi\colon\ A \in \mathfrak{A}\}$ is dense in $\mathscr{H}_\varphi$. Finally, if the weak limits $w\text{-}\lim_{t\to\pm\infty} \exp(-itH)\Pi_\varphi(V)\Phi$ exist in $\mathscr{H}_\varphi$, then they are equal to $E_\varphi\Pi_\varphi(V)\Phi$ [6]. We thus conclude

$$\langle\varphi^{V,1}_\pm;\, A\rangle = \langle\varphi;\, A\rangle - \lambda\big(\Phi, \Pi_\varphi(A)\,E_\varphi\Pi_\varphi(V)\Phi\big) + \lambda\langle\varphi;\, A\rangle\,\langle\varphi;\, V\rangle.$$

(Q.E.D.)

A remarkable feature is that the above theorem can be proved under the same condition of the Theorem II.1. Thus we have an interesting consequence: For the dynamical system $(\mathfrak{A}, \varphi, \alpha)$, the approach to equilibrium occurs in L.R.S. under a perturbation V if and only if the dynamical system is stable in L.R.S. under V. We might not have this situation in the exact dynamics.

Now let us answer the question (2). Before stating the result, we introduce one concept.

DEFINITION III.1. The projection E_φ is said to be *one-dimensional* for $A \in \mathfrak{A}$ if $E_\varphi \Pi_\varphi(A)\Phi = (\Phi, \Pi_\varphi(A)\Phi)\Phi$.

We can answer the question (2) as

THEOREM III.2. *For the dynamical system* $(\mathfrak{A}, \varphi, \alpha)$ *and a perturbation* $V = V^* \in \mathfrak{A}$, *the return to equilibrium in L.R.S. occurs under* V *if and only if* $\varphi_\pm^{V,1}$ *exist and* E_φ *is one-dimensional for* V.

Remark. The existence of $\varphi_\pm^{V,1}$ is always in the weak sense in our forthcoming discussions, so we merely say that $\varphi_\pm^{V,1}$ exist as in the above theorem.

Proof: The "if" part can be proved as follows: When $\varphi_\pm^{V,1}$ exist, they are given by

$$\langle \varphi_\pm^{V,1}; A\rangle = \langle \varphi; A\rangle - \lambda(\Phi, \Pi_\varphi(A)E_\varphi \Pi_\varphi(V)\Phi) + \lambda\langle \varphi; A\rangle\langle \varphi; V\rangle$$

for any $A \in \mathfrak{A}$. E_φ is assumed to be one-dimensional for V, we thus obtain the equality $(\Phi, \Pi_\varphi(A)E_\varphi\Pi_\varphi(V)\Phi) = \langle\varphi; A\rangle\langle\varphi; V\rangle$, which immediately implies $\langle \varphi_\pm^{V,1}; A\rangle = \langle\varphi; A\rangle$ for any $A \in \mathfrak{A}$, i.e. return to equilibrium.

Conversely, if the return to equilibrium in L.R.S. occurs under a perturbation $V = V^* \in \mathfrak{A}$, then for any $A \in \mathfrak{A}$, $\langle \varphi_\pm^{V,1}; A\rangle = \langle \varphi; A\rangle$, which implies

$$(\Phi, \Pi_\varphi(A)E_\varphi\Pi_\varphi(V)\Phi) = (\Phi, \Pi_\varphi(A)\Phi)(\Phi, \Pi_\varphi(V)\Phi).$$

Hence we obtain

$$\left(\Pi_\varphi(A^*)\Phi, \left(E_\varphi\Pi_\varphi(V) - (\Phi, \Pi_\varphi(V)\Phi)\right)\Phi\right) = 0.$$

As $\Pi_\varphi(\mathfrak{A})\Phi$ is dense in $\mathscr{H}_\varphi$, the above equality shows that

$$\left(E_\varphi\Pi_\varphi(V) - (\Phi, \Pi_\varphi(V)\Phi)\right)\Phi = 0,$$

which means that E_φ is one-dimensional for V. (Q.E.D.)

Using this result, we have a remarkable property which tells us the similarity between the linear response dynamics and the exact dynamics.

THEOREM III. 3. *For the dynamical system* $(\mathfrak{A}, \varphi, \alpha)$, *the following two conditions are equivalent*:

(1) *The return to equilibrium in E.S. occurs under all perturbations* $V = V^* \in \mathfrak{A}$.

(2) *The return to equilibrium in L.R.S. occurs under all perturbations* $V = V^* \in \mathfrak{A}$.

Proof: Since φ is a K.M.S. state with respect to α_t, the return to equilibrium in E.S. occurs under all perturbations $V = V^* \in \mathfrak{A}$ if and only if the dynamical system $(\mathfrak{A}, \varphi, \alpha)$ is clustering (i.e., $\lim_{t\to\pm\infty} \langle\varphi; \alpha_t(A)B\rangle = \langle\varphi; A\rangle\langle\varphi; B\rangle$ for any $A, B \in \mathfrak{A}$) [10]. Moreover, the dynamical system is clustering if and only if it is stable in L.R.S. under all $V = V^* \in \mathfrak{A}$ and η-clustering (i.e., $\eta\langle\varphi; \alpha_t(A)B\rangle = \langle\varphi; A\rangle\langle\varphi; B\rangle$ for any $A, B \in \mathfrak{A}$, where η is an invariant mean on R). This fact is the C^*-algebraic version of the Theorem 6 of Ref. [4]. By the Theorems II.1 and III.1, the weak* limits of $\varphi^{V,1} \circ \alpha_t$ as $t \to \pm\infty$ exist. Furthermore, as the state φ is time invariant under α_t, the dynamical system is η-

clustering if and only if the projection E_φ is one-dimensional [6]. Any element of $\mathfrak{A}$ is expressible as a sum of two self-adjoint elements of $\mathfrak{A}$. We therefore arrive at the assertion of this theorem by virtue of Theorem III.2. (Q.E.D.)

This theorem can be realized [12] by some dynamical systems; for example, the $X-Y$ model [15] and the free Fermi gas [16]. We will discuss the $X-Y$ model in Section VI.

IV. Stability of a dynamical system

In this section, we investigate the question (4) and discuss the difference between the linear response dynamics and the exact dynamics. We will show the conditions on the perturbation V under which the linear response functional $\varphi^{V,1}$ is achieved by a purely dynamical process from the initial equilibrium state φ.

The answer of this question is

THEOREM IV.1. *The linear response functional $\varphi^{V,1}$ is dynamically achieved by the linear response time evolution $\alpha_t^{V,1}$ as $t \to \infty$ (i.e., $\varphi^{V,1} = w^*\text{-}\lim_{t\to\infty} \varphi \circ \alpha_t^{V,1}$) if and only if the dynamical system $(\mathfrak{A}, \varphi, \alpha)$ is stable in L.R.S. under the perturbation $V = V^* \in \mathfrak{A}$ and the projection E_φ is one-dimensional for V.*

Proof: The "if" part can be shown as follows: According to Lemma III.1, $\varphi^{V,1}$ is given by

$$\langle \varphi^{V,1}; A\rangle = \langle \varphi; A\rangle - \lambda \int_0^1 ds \langle \varphi; A\alpha_{is}(V)\rangle + \lambda \langle \varphi; A\rangle \langle \varphi; V\rangle$$

for any $A \in \mathfrak{A}$. It is easily seen that

$$\int_0^1 ds \langle \varphi; A\alpha_{is}(V)\rangle = (\Phi, \Pi_\varphi(A) T^2 \Pi_\varphi(V)\Phi).$$

Since the dynamical system $(\mathfrak{A}, \varphi, \alpha)$ is stable in L.R.S. under V, we have from Theorem II.1

$$\lim_{t\to\infty} \langle \varphi; \alpha_t^{V,1}(A)\rangle = \langle \varphi; A\rangle - \lambda(\Phi, \Pi_\varphi(A) T^2 (1-E_\varphi)\Pi_\varphi(V)\Phi),$$

which turns out to be equal to

$$\langle \varphi; A\rangle - \lambda(\Phi, \Pi_\varphi(A) T^2 \Pi_\varphi(V)\Phi) + (\Phi, \Pi_\varphi(A) E_\varphi \Pi_\varphi(V)\Phi)$$

because $T^2 E_\varphi \Pi_\varphi(V)\Phi = E_\varphi \Pi_\varphi(V)\Phi$.

The projection E_φ is assumed to be one-dimensional for V, so we obtain

$$(\Phi, \Pi_\varphi(A) E_\varphi \Pi_\varphi(V)\Phi) = \langle \varphi; A\rangle \langle \varphi; V\rangle. \tag{4.1}$$

Hence we conclude

$$\varphi^{V,1} = w^*\text{-}\lim_{t\to\infty} \varphi \circ \alpha_t^{V,1}. \tag{4.2}$$

Conversely, the "only if" part is shown as follows: If the equality (4.2) is satisfied, then the dynamical system is stable in L.R.S. under the perturbation V and the equality (4.1) holds. Hence, as proved in Theorem III.2, the projection E_φ is one-dimensional for V. (Q.E.D.)

According to the above theorem and Theorem III.2, we can readily prove

COROLLARY IV.1. *If the dynamical system* $(\mathfrak{A}, \varphi, \alpha)$ *is stable in L.R.S. under a perturbation* $V = V^* \in \mathfrak{A}$ *and the projection* E_φ *is one-dimensional for* V, *then* $w^*\text{-}\lim_{s\to\pm\infty} \lim_{t\to\infty} \varphi \circ \alpha_t^{V,1} \circ \alpha_s = \varphi$.

When the dynamical system $(\mathfrak{A}, \varphi, \alpha)$ is clustering, it is stable in L.R.S. under all perturbations and η-clustering [4]. Applying the above theorem and corollary, we obtain

THEOREM IV.2. *If the dynamical system* $(\mathfrak{A}, \varphi, \alpha)$ *is clustering, then*

(1) $\varphi^{V,1} = w^*\text{-}\lim_{t\to\infty} \varphi \circ \alpha_t^{V,1}$ *for all perturbations* $V = V^* \in \mathfrak{A}$, *and*

(2) $\varphi = w^*\text{-}\lim_{s\to\pm\infty} \lim_{t\to\infty} \varphi \circ \alpha_t^{V,1} \circ \alpha_s$ *for all* $V = V^* \in \mathfrak{A}$.

This result shows a sharp contrast between the exact theory and the linear response theory. Namely, we can not assert the same fact for the exact dynamics. This aspect will be illustrated in Section VI by an exactly solvable model: $X-Y$ model.

V. Phase transition in dynamical process

When there exist several different thermodynamical phases at an inverse temperature $\beta\ (=1)$, it is interesting to consider the change in a dynamical process from one state to another. For instance, we have metastable states and stable states for some physical systems [17, 18, 19], we then ask when a metastable state goes over to a stable state by the effect of a perturbation. In our terminology, if we have two different states φ and ψ of an infinite dynamical system, which satisfy the K.M.S. condition with respect to the time evolution α_t of the system, the question of interest is whether the transition from φ to ψ can occur dynamically through a perturbation $V = V^* \in \mathfrak{A}$; that is, $\psi = w^*\text{-}\lim_{t\to\infty} \varphi^V \circ \alpha_t$ in E.S. and $\psi = w^*\text{-}\lim_{t\to\infty} \varphi^{V,1} \circ \alpha_t$ in L.R.S. We assume that the state φ is an extremal K.M.S. state and ψ is another K.M.S. state, both w.r.t. α_t.

In exact dynamics, such transition possibly occurs [20] under some perturbation. However, in the linear response dynamics, we have a negative result as discussed below.

LEMMA V.1. *For the dynamical system* $(\mathfrak{A}, \varphi, \alpha)$, *if* φ *is a K.M.S. state w.r.t.* α_t *and* $\varphi_\pm^{V,1}$ *exist as states on* $\mathfrak{A}$ *under a perturbation* $V = V^* \in \mathfrak{A}$, *then* $\varphi_\pm^{V,1}$ *are vector states of* φ.

Proof: When $\varphi_\pm^{V,1}$ exist, they are given by

$$\langle \varphi_\pm^{V,1}; A\rangle = \langle \tilde{\varphi}; \Pi_\varphi(A)\rangle - \lambda(\Phi, \Pi_\varphi(A) E_\varphi \Pi_\varphi(V)\Phi) + \\ + \lambda\langle\tilde{\varphi}; \Pi_\varphi(A)\rangle \langle\tilde{\varphi}; \Pi_\varphi(V)\rangle = \langle \tilde{\varphi}_\pm^{V,1}; \Pi_\varphi(A)\rangle,$$

where $\tilde{\varphi}_\pm^{V,1}$ and $\tilde{\varphi}$ are the canonical extensions of $\varphi_\pm^{V,1}$ and φ to the von Neumann algebra

$\Pi_\varphi(\mathfrak{A})''$. Let $\{P_\gamma\}$ be an increasing net of projections in $\Pi_\varphi(\mathfrak{A})''$ converging to P (a projection on the closure of $\bigcup_\gamma P_\gamma \mathscr{H}_\varphi$). Then

$$|\langle\tilde{\varphi}_\pm^{V,1}; P_\gamma - P\rangle| \leqslant |\langle\tilde{\varphi}; P_\gamma - P\rangle| \cdot \left(1+\lambda\,|\langle\tilde{\varphi}; \Pi_\varphi(V)\rangle|\right) + \lambda \cdot \|(P_\gamma - P)\Phi\|\,\|E_\varphi \Pi_\varphi(V)\Phi\|,$$

which goes to zero as $\gamma \to \infty$ because $\tilde{\varphi}$ is a normal state on $\Pi_\varphi(\mathfrak{A})''$. This means that $\tilde{\varphi}_\pm^{V,1}$ are normal states on $\Pi_\varphi(\mathfrak{A})''$. Since φ is a K.M.S. state w.r.t. α_t, $\varphi_\pm^{V,1}$ are vector states of φ (see Theorem 4, p. 233 of Ref. [24]). (Q.E.D.)

Let φ and ψ be two K.M.S. states as assumed above. We now suppose that the desired transition occurs in L.R.S. under $V = V^* \in \mathfrak{A}$ (i.e., $\psi = w^*\text{-}\lim_{t\to\infty} \varphi^{V,1} \circ \alpha_t = \varphi_+^{V,1}$). Since the canonical extension $\tilde{\varphi}_+^{V,1}$ of $\varphi_+^{V,1}$ is a faithful normal state on $\Pi_\varphi(\mathfrak{A})''$, we can thus express $\varphi_+^{V,1}$ as [14]

$$\langle\varphi_+^{V,1}; A\rangle = \langle\tilde{\varphi}_+^{V,1}; \Pi_\varphi(A)\rangle = \left(\Phi, R^*\Pi_\varphi(A)R\Phi\right),$$

where R is a self-adjoint positive operator affiliated with the center $Z = \Pi_\varphi(\mathfrak{A})'' \cap \Pi_\varphi(\mathfrak{A})'$. Since φ is assumed to be extremal, the von Neumann algebra $\Pi_\varphi(\mathfrak{A})''$ is a factor: $Z = CI$. Thus there exists a constant $\mu \in C$ such that $R = \mu I$. As $\langle\varphi_+^{V,1}; I\rangle = \langle\varphi^{V,1}; I\rangle = 1$ is satisfied, $|\mu|^2 = 1$. This implies $\varphi_+^{V,1} = \varphi$, which means that we do not have the phase change expected. Hence

THEOREM V.1. *For the dynamical system $(\mathfrak{A}, \varphi, \alpha)$, if φ is extremal among the K.M.S. states w.r.t. α_t, then the phase transition in the dynamical process does not occur in L.R.S. under any perturbation $V = V^* \in \mathfrak{A}$.*

This theorem might tell us that we need at least some non-linear terms to discuss the phase transition effected by perturbations.

VI. $X-Y$ model

We illustrate some of general results obtained in the previous sections by the $X-Y$ model. The $X-Y$ model has been rigorously studied by several authors [15, 21, 22]. We mainly follow the paper of Emch and Radin [21] here.

Let us consider a one dimensional lattice Z. Every site of Z is occupied by a particle with spin 1/2. Thus to every site $k \in Z$, an algebra $\mathfrak{A}_k$ is associated, and to every finite region $\Lambda \subset Z$, an algebra $\mathfrak{A}_\Lambda$ is associated. These algebras are given by

$$\mathfrak{A}_k = \{\lambda_1 I_k + \lambda_2 \sigma_k^x + \lambda_3 \sigma_k^y + \lambda_4 \sigma_k^z : \lambda_j \in C\ (j = 1, 2, 3, 4)\},$$
$$\mathfrak{A}_\Lambda = \bigotimes_{k\in\Lambda} \mathfrak{A}_k,$$

where $\sigma_k^{x,y,z}$ are the Pauli matrices and I_k is the identity matrix at the site $k \in Z$. We then define an algebra of local elements: $\mathfrak{A}_0 = \bigcup_{\Lambda\subset Z} \mathfrak{A}_\Lambda$. The quasi-local C^*-algebra of our lattice system is the inductive limit of the above algebra: $\mathfrak{A} = \overline{\bigcup_{\Lambda\subset Z} \mathfrak{A}_\Lambda}^{n}$.

The local Hamiltonian H_n (n is a positive integer associated with a finite region $\Lambda_n = [-n, n] \subset Z$) of the $X-Y$ model is

$$H_n = -\sum_{k=-n}^{n-1} \{(1+\xi)\sigma_k^x\sigma_{k+1}^x + (1-\xi)\sigma_k^y\sigma_{k+1}^y\},$$

where ξ is some real constant in (0, 1). The time evolution generated by H_n is $\alpha_t^n(A) = \exp(itH_n)A\exp(-itH_n)$ for any $A \in \mathfrak{A}$ and $t \in R$. The canonical equilibrium state φ_n on $\mathfrak{A}$ with an inverse temperature $\beta = 1$ is given by $\varphi_n = \exp(-H_n)/\mathrm{Tr}\exp(-H_n)$. It is then shown [22] that there exists a one parameter group of automorphisms α_t of $\mathfrak{A}$ and a state φ on $\mathfrak{A}$ such that $\lim_{n\to\infty} \|\alpha_t^n(A)-\alpha_t(A)\| = 0$ and $\lim_{n\to\infty} |\langle\varphi_n; A\rangle - \langle\varphi; A\rangle| = 0$. Then φ becomes an extremal K.M.S. state w.r.t. α_t.

Let $\mathscr{A}$ and ω, ν_t be the even C^*-subalgebra of $\mathfrak{A}$ and the restrictions of φ and α_t to this subalgebra respectively (cf. [21]). The state ω turns out to be an extremal K.M.S. state w.r.t. ν_t. The dynamical system of the $X-Y$ model is now set by the triple $(\mathscr{A}, \omega, \nu)$. It can be shown [21, 23] that $(\mathscr{A}, \nu)$ is asymptotic Abelian in norm (i.e., $\lim_{t\to\pm\infty} \|[A, \alpha_t(B)]\| = 0$ for any $A, B \in \mathscr{A}$), hence the dynamical system $(\mathscr{A}, \omega, \nu)$ is clustering. As discussed in the proof of Theorem III.3, the return to equilibrium in E.S. occurs under all perturbations $V = V^* \in \mathscr{A}$ (i.e., $w^*\text{-}\lim_{t\to\pm\infty} \omega^V \circ \nu_t = \omega$). On the other hand, the perturbed dynamical system in E.S. $(\mathscr{A}, \omega^V, \nu^V)$ is not always asymptotic Abelian. For example, consider a perturbation $V = V^* = \sum_{k=-p}^{p-1} \{(1+\xi)\sigma_k^x\sigma_{k+1}^x + (1-\xi)\sigma_k^y\sigma_{k+1}^y\}$ (p: integer $\geqslant 1$), then the local algebra $\mathscr{A}_{[-p+1, p-1]}$ is left pointwise invariant under the perturbed time evolution ν_t^V; namely, for any $A \in \mathscr{A}_{[-p+1, p-1]}$, there exists $B \in \mathscr{A}$ such that $\|[B, \nu_t^V(A)]\| = \|[B, A]\| \neq 0$. We therefore know that for such perturbation $V = V^* \in \mathscr{A}$, the equality $\omega^V = w^*\text{-}\lim_{t\to\infty} \omega \circ \nu_t^V$ is not satisfied. According to the Theorems III.3 and IV.1, we however have the following situation in the linear response dynamics.

THEOREM VI.1. *For the dynamical system $(\mathscr{A}, \omega, \nu)$ of the X–Y model,*

(1) *the return to equilibrium in L.R.S. (i.e., $w^*\text{-}\lim_{t\to\pm\infty} \omega^{V,1} \circ \nu_t = \omega$) occurs under all perturbations $V = V^* \in \mathscr{A}$, and*

(2) *the linear response functional $\omega^{V,1}$ is dynamically achieved (i.e., $w^*\text{-}\lim_{t\to\infty} \omega \circ \nu_t^{V,1} = \omega^{V,1}$) for all perturbations $V = V^* \in \mathfrak{A}$.*

Acknowledgment

The author thanks Professor G. G. Emch for valuable discussions and critical comments. He also thanks Professor H. Umegaki for valuable suggestions and kindness at Tokyo Institute of Technology. He finally thanks the referee for critical comments.

REFERENCES

[1] R. Kubo: *J. Phys. Soc. Japan* **21** (1957), 570.
[2] H. Mori: *Progr. Theor. Phys.* **33** (1965), 423.
[3] J. Naudts, A. Verbeure and R. Weder: *Commun. Math. Phys.* **44** (1975), 87.
[4] A. Verbeure and R. Weder: *Commun. Math. Phys.* **44** (1975), 101.
[5] R. Haag, N. Hugenholtz and M. Winnik: *Commun. Math. Phys.* **5** (1967), 215.
[6] G. G. Emch: *Algebraic methods in statistical mechanics and quantum field theory*, Wiley, 1972.
[7] See, for example, M. Reed and B. Simon: *Functional analysis*, Acad. Press, 1972.
[8] See, for example, page 200 of [6].
[9] H. Araki: *R.I.M.S., Kyoto* **9** (1973), 165.
[10] D. W. Robinson: *C*-algebras and quantum statistical mechanics*, Lecture in Varenna School, 1973.
[11] G. G. Emch: *J. Math. Phys.* **7** (1966), 1198.
[12] M. Ohya: *Thesis* at University of Rochester, 1976, and references quoted therein.
[13] See, for example, O. E. Lanford III and D. W. Robinson: *Commun. Math. Phys.* **24** (1972), 193.
[14] M. Takesaki: *Tomita's theory of modular Hilbert algebras and its applications*, Springer-Verlag, 1970.
[15] D. B. Abraham, E. Barouch, G. Gallavotti and A. Martin-Løf: *Phys. Rev. Letters* **25 II** (1970), 1449.
[16] See page 305 of [6].
[17] G. G. Emch and J. F. Knops: *J. Math. Phys.* **11** (1970), 3008.
[18] O. Penrose and J. L. Lebowitz: *J. Stat. Mech.* **3** (1971), 211.
[19] G. L. Sewell: *Letteu al Nuovo Cimento* **10** (1974), 430.
[20] See Chapters III and V of [12].
[21] G. G. Emch and C. Radin: *J. Math. Phys.* **12** (1971), 2043.
[22] H. Araki: *Commun. Math. Phys.* **14** (1969), 120.
[23] H. Narnhofer: *Acta Phys. Austrica* **31** (1970), 349.
[24] J. Dixmier: *Les algèbres d'opérateures dans l'espace Hilbertien*, Paris 1957.

Stability of Weiss Ising model

Masanori Ohya

The Science University of Tokyo, Department of Information Science, 278, Noda, Chiba, Japan [a)]
and Institute of Fundamental Studies, Department of Physics and Astronomy, University of Rochester, Rochester, New York 14627
(Received 21 June 1977)

The problems of stability and approach to equilibrium of the Weiss Ising model are studied. Our investigations are performed in the exact and linear response senses in order to compare both theories. The change of a metastable state of the Weiss Ising model is discussed under local perturbations.

I. INTRODUCTION

The problem of stability of a dynamical system with infinitely many degrees of freedom encountered in quantum statistical mechanics has been recently paid much attention by several authors.[1–4] Verbeure and Weder[3] have studied this problem in the linear response theory. Ohya[4] has considered similar problems in the linear response and exact theories. In this paper, we apply some of the general results obtained in Refs. 3, 4 to the Weiss Ising model (WIM for brevity in the following) formulated in the operator algebraic framework.[5]

In Sec. II, we review the WIM briefly. In Sec. III, we show that the WIM provides us a nontrivial example of the work by Verbeure and Weder[3] concerning the stability of a dynamical system in the linear response sense (LRS for brevity). In Sec. IV, we study the dynamical behavior of locally perturbed states of the WIM by bounded self-adjoint operators belonging to the quasilocal C^*-algebra of this model. Namely, we look at the time development of the perturbed states after removing the perturbations dynamically. The discussion of that section is of the approach to equilibrium in both linear response and exact dynamics. We also discuss the dynamical change of metastable states in the WIM in Sec. IV. The change of a metastable state to a stable state under some local perturbation is extremely interesting to be investigated in quantum statistical mechanics and quantum measuring processes.[6–8] Although we first have to show the existence of metastable states in a dynamical system considered, we fortunately know[5] that there exist metastable states in the WIM, provided that the temperature is below a critical point. We are thus at the stage in the WIM to consider the dynamical change of the metastable states. We want to know whether there exists a local perturbation under which a metastable state goes to a stable state or not.

We take the inverse temperature of the WIM $\beta=1$ throughout this paper.

II. WEISS ISING MODEL

When we treat a physical system composed of many interacting particles, we have a powerful approximation, the so-called mean field method, which will be considered as follows: Observe a particle in the system, then we can regard the effect to this representative particle from other particles surrounding it as the field determined by an averaging procedure over the system of interest. Thus the representative particle can be statistically treated in the mean field. Conversely, we may determine the average field which the representative particle exerts on its neighbors. If all particles are identical, then every mean field should be coincident (requirement of self-consistency of mean field). This self-consistency requirement determines the mean field of the system precisely. We can study statistical properties of the system by this mean field. In this section, we briefly review one of the mean fields: the Weiss Ising model, in the algebraic framework.

Following Emch and Knops,[5] we consider a one-dimensional lattice Z (this restriction can be lifted, but we take it for simplicity). The quasilocal C^*-algebra of this lattice system Z is given by

$$\mathfrak{A}=\bar{\mathfrak{A}}_0^n$$

$$\mathfrak{A}_0=\bigcup_{\Lambda\subset Z}\mathfrak{A}_\Lambda$$

$$\mathfrak{A}_\Lambda=\bigotimes_{k\in\Lambda}\mathfrak{A}_k \quad (\Lambda \text{ is a finite region of } Z)$$

$$\mathfrak{A}_k=\{\lambda_1 I_k+\lambda_2\sigma_k^x+\lambda_3\sigma_k^y+\lambda_4\sigma_k^z:\lambda_j\in C\ (j=1,2,3,4)\},$$

where $\sigma_k^{x,y,z}$ are the Pauli matrices and I_k is the identity matrix at the site k of the lattice Z. The local C^*-algebra $\mathfrak{A}_\Lambda$ enjoys two properties, named "isotony" and "locality."[9]

The local Hamiltonian of a ferromagnetic system considered here is given by

$$H(\Lambda)=-B\sum_{k\in\Lambda}\sigma_k^z-\sum_{k,i\in\Lambda}J_{k,i}(\Lambda)\sigma_k^z\sigma_i^z,$$

where B is an external magnetic field on the system along the z axis and $J_{k,i}(\Lambda)$ are real coupling constants on Λ with $J_{k,i}(\Lambda)=J_{i,k}(\Lambda)=J_{|k-i|}(\Lambda)$, $J_{k,k}(\Lambda)=0$. In order to study more about our system, let us introduce the following notations:

$$H(\Lambda)=\sum_{k\in\Lambda}H_k,$$

$$H_k=-[B+B_k(\Lambda)]\sigma_k^z,$$

$$B_k(\Lambda)=2\sum_{i\in\Lambda}J_{k,i}(\Lambda).$$

We can assume $B=0$ in the sequel discussions without loss of generality.

The idea of the mean field approximation of the ferromagnetic system comes from that we replace the above volume dependent $B_k(\Lambda)$ with some averaged volume in-

[a)]Mailing address.

dependent one. Since the above local Hamiltonian $H(\Lambda)$ is bounded, it defines a time evolution automorphism of $\mathfrak{A}: \tau_t^\Lambda(A)=\exp[itH(\Lambda)]A\exp[-itH(\Lambda)]$ for any $A\in\mathfrak{A}$. We then ask whether and in which sense the time evolution τ_t after taking the limit $\Lambda\to\infty$ exists. Emch and Knops[5] have studied this question under the following conditions: (i) Stability condition: $\sum_{i\in\Lambda}|J_{k,i}(\Lambda)|\leq\frac{1}{2}c_k$ for any finite $\Lambda\subset Z$ containing k, where c_k is a Λ-independent constant. (ii) Van der Waals condition: $\lim_{\Lambda\to\infty}J_{k,i}(\Lambda)=0$ for all $k,i\in\Lambda$. (iii) Periodic condition: $J_{|k-i|}(\Lambda)=J_{|k-i|+p}(\Lambda)$, where p is a period of the lattice.

It is then shown that the volume infinite limit of $\tau_t^\Lambda(A)$ for any $A\in\mathfrak{A}$ exists for each representation Π such that its domain is restricted to the quasilocal C^*-algebra $\mathfrak{A}$, but its range cannot be in general the representation space $\Pi(\mathfrak{A})$ but the weak closure $\Pi(\mathfrak{A})''$. After taking the limit $\Lambda\to\infty$, the representation dependent time evolution τ_t can be extended to the automorphism $\tilde{\tau}_t$, the canonical extension of τ_t (see the remark at the end of this section), on the von Neumann algebra $\Pi(\mathfrak{A})''$. Here the time evolutions τ_t and $\tilde{\tau}_t$ should be written as τ_t^Π and $\tilde{\tau}_t^\Pi$ if we want to show their representation dependence explicitly, but we take the former expressions for simplicitly.

We now notice that if the von Neumann algebra $\Pi(\mathfrak{A})''$ is a factor, then the operator $\Pi(B_k)$, in the appropriate limit[5] of $\Pi(B_k(\Lambda))$ as $\Lambda\to\infty$, becomes a multiple of identity because of $\Pi(B_k)\in\Pi(\mathfrak{A})''\cap\Pi(\mathfrak{A})'$. We next have to construct an extremal KMS state φ in the following sense[5,10]:

(a) $\Pi_\varphi(\mathfrak{A})''\cap\Pi_\varphi(\mathfrak{A})'=CI$,

(b) $\int dt\langle\tilde{\varphi};\tilde{\tau}_t(\Pi_\varphi(A))\Pi_\varphi(B)\rangle f(t)$

$=\int dt\langle\tilde{\varphi};\Pi_\varphi(B)\tilde{\tau}_t(\Pi_\varphi(A))\rangle f(t-i)$

for any finite regions Λ, Λ' of Z, any $A\in\mathfrak{A}_\Lambda$, $B\in\mathfrak{A}_{\Lambda'}$, and any $f\in\hat{\mathcal{D}}$, the set of the Fourier-transforms of functions in the set $\mathcal{D}$ of infinitely differentiable functions with compact supports. The above $\tilde{\varphi}$ in (b) is the canonical extension of φ to $\Pi_\varphi(\mathfrak{A})''$ (see the remark below) and Π_φ is the GNS representation associated with φ. This construction can be carried out by the following facts:

(1) φ is a locally normal state; that is, for any finite region Λ of Z

$$\varphi\restriction\mathfrak{A}_\Lambda=\exp[-H(\Lambda)]/\mathrm{Tr}[-\exp H(\Lambda)], \tag{2.1}$$

where, from $\Pi_\varphi(B_k)=B_{k,\varphi}I_k(B_{k,\varphi}\in C)$, $H(\Lambda)$ is redefined as

$$H(\Lambda)=-\sum_{k\in\Lambda}B_{k,\varphi}\sigma_k^z.$$

(2) φ is a product state; that is,

$$\langle\varphi;AB\rangle=\langle\varphi;A\rangle\langle\varphi;B\rangle \tag{2.2}$$

for any A,B in $\mathfrak{A}_\Lambda$ and $\mathfrak{A}_{\Lambda'}$ respectively with $\Lambda\cap\Lambda'=\phi$.

(3) We have the self-consistency equations such that

$$\langle\varphi;\sigma_k^x\rangle=\langle\varphi;\sigma_k^y\rangle=0, \tag{2.3}$$

$$\langle\varphi;\sigma_k^z\rangle=\tanh B_{k,\varphi}, \tag{2.4}$$

$$B_{k,\varphi}=2\tilde{J}_k[\langle\varphi;\sigma_k^z\rangle], \tag{2.5}$$

where $\tilde{J}_k$ is the extension of a positive bounded linear functional J_k on a C^*-algebra $\mathcal{C}(Z)$ of all bounded functions on Z. Here J_k is defined on a subspace $\mathcal{C}_J(Z)$, for which $J_k[f]=\lim_{\Lambda\to\infty}\sum_{i\in\Lambda}J_{k,i}(\Lambda)f(i)$ exists. The extension from J_k to $\tilde{J}_k$ is always possible by the Hahn–Banach theorem. We can thus construct an extremal KMS locally normal state φ. For this state φ, the following important properties hold:

(4) $\tilde{\tau}_t$ is an automorphism of $\Pi_\varphi(\mathfrak{A})$. (2.6)

(5) $\tilde{\tau}_t$ is an automorphism of $\Pi_\varphi(\mathfrak{A}_\Lambda)$ for any finite $\Lambda\subset Z$. (2.7)

Let us finally notice that the canonical extension $\tilde{\varphi}$ of φ from $\mathfrak{A}$ to $\Pi_\varphi(\mathfrak{A})''$ is a faithful normal extremal KMS state with respect to $\tilde{\tau}_t$.[11]

Remark: (1) The canonical extension $\tilde{\tau}_t$ of τ_t to $\Pi_\varphi(\mathfrak{A})''$ is defined by $\tilde{\tau}_t(Q)=U_tQU_{-t}$ for any $Q\in\Pi_\varphi(\mathfrak{A})''$, where U_t is a strongly continuous one-parameter unitary group implementing $\tilde{\tau}_t$ on $\Pi_\varphi(\mathfrak{A})''$.

(2) The canonical extension $\tilde{\varphi}$ of φ to $\Pi_\varphi(\mathfrak{A})''$ is defined by $\langle\varphi;Q\rangle=(\Phi,Q\Phi)$ for any $Q\in\Pi_\varphi(\mathfrak{A})''$, where Φ is the GNS cyclic vector induced by φ.

III. STABILITY OF WEISS ISING MODEL

Verbeure and Weder[3] have studied the stability of dynamical systems in the linear response theory. Let a dynamical system be described by $(\mathcal{A},\phi,\alpha)$, where $\mathcal{A}$ is a C^*-algebra with an identity operator I, ϕ is a KMS state with respect to the time evolution α_t $(t\in R)$, a strongly continuous one-parameter group of automorphisms of $\mathcal{A}$. Verbeure and Weder introduced the linear response time evolution as follows: For any $A\in\mathcal{A}$ and a perturbation $\lambda V=\lambda V^*\in\mathcal{A}$ with $\lambda\in[0,1]$,

$$\alpha_t^{V,1}(A)=\alpha_t(A)+i\lambda\int_0^t ds[\alpha_s(V),\alpha_t(A)] \tag{3.1}$$

for $t\geq 0$, and

$$\alpha_t^{V,1}(A)=\alpha_t(A)+i\lambda\int_t^0 ds[\alpha_s(V),\alpha_t(A)] \tag{3.2}$$

for $t<0$. These are obtained by linearizations w.r.t. λ of the exact time evolution defined by

$$\alpha_t^V(A)=\sum_{n\geq 0}(i\lambda)^n\int\limits_{0\leq t_1\leq\cdots\leq t_n\leq t} dt_1\cdots\int dt_n[\alpha_{t_1}(V),$$

$$[\cdots[\alpha_{t_n}(V),\alpha_t(A)]\cdots]] \tag{3.3}$$

for $t\geq 0$, and

$$\alpha_t^V(A)=\sum_{n\geq 0}(i\lambda)^n\int\limits_{t\leq t_1\leq\cdots\leq t_n\leq 0} dt_1\cdots\int dt_n$$

$$\times[\alpha_{t_1}(V),[\cdots[\alpha_{t_n}(V),\alpha_t(A)]\cdots]] \tag{3.4}$$

for $t<0$.

The C^*-algebraic versions of the definitions of stability in LRS of the dynamical system $(\mathcal{A},\phi,\alpha)$ due to Verbeure and Weder are as followings.

Definition III.1: A dynamical system $(\mathcal{A},\phi,\alpha)$ is said to be stable in LRS under a perturbation $V=V^*\in\mathcal{A}$ if the weak* limits $\phi_{\pm,1}=\mathrm{w}^*\text{-}\lim_{t\to\pm\infty}\phi\circ\alpha_t^{V,1}$ exist under V.

Definition III.2: A dynamical system $(\mathcal{A}, \phi, \alpha)$ is said to be stable in LRS if it is stable in LRS under all perturbations $V = V^* \in \mathcal{A}$.

As we have seen in the previous section, the WIM should be originally treated by the triple $(\mathfrak{A}, \varphi, \tau)$, but since τ_t does not exist as an automorphism of $\mathfrak{A}$ for a general state φ on $\mathfrak{A}$, the WIM has to be considered through the von Neumann algebraic triple $(\Pi_\varphi(\mathfrak{A})'', \tilde{\varphi}, \tilde{\tau})$. However, for an extremal KMS state φ, the time evolution $\tilde{\tau}_t$ can be an automorphism not only of $\Pi_\varphi(\mathfrak{A})''$ but also of $\Pi_\varphi(\mathfrak{A})$, and hence of $\mathfrak{A}$ (since all representations are faithful). We thus define the linear response time evolution $\tilde{\tau}_t^{\Pi_\varphi(V),1}$ by replacing $\alpha_t, A, \lambda V$ with $\tilde{\tau}_t$, $\Pi_\varphi(A)$, $\lambda\Pi_\varphi(V)$ in (3.1) and (3.2). It is readily shown by merely extending the von Neumann algebraic results obtained by Verbeure and Weder[3] to the C^*-algebraic dynamical system $(\mathcal{A}, \phi, \alpha)$ that the dynamical system $(\mathcal{A}, \phi, \alpha)$ is clustering [i.e., $\lim_{t\to\pm\infty}\langle\phi; \alpha_t(A)B\rangle = \langle\phi; A\rangle\langle\phi; A\rangle$ for any $A, B \in \mathcal{A}$] if and only if it is η_t clustering [i.e., $\eta_t\langle\phi;\ \alpha_t(A)B\rangle = \langle\phi;\ A\rangle\langle\phi;\ B\rangle$ for any $A, B \in \mathcal{A}$, where η_t is a mean over $t \in R$] and stable in LRS. This fact implies that if a dynamical system is clustering for its KMS state (e.g., $X-Y$ model, free Fermi gas), Definitions III.1 and III.2 become equivalent. Therefore, for such systems, the problem of stability in LRS is not so attractive. However, the WIM is even not η_t clustering, so that it may be interesting to study the problem of stability for this model. We ask a following question: Under which conditions on $V = V^* \in \mathcal{A}$ is the WIM stable in LRS? In other words, can we determine the set of perturbations under which the WIM is stable in LRS?

Before answering this question, let us introduce a C^*-subalgebra $\mathfrak{A}^z$ of $\mathfrak{A}$ such that

$$\mathfrak{A}^z = \overline{\mathfrak{A}_0^z}^n$$

$$\mathfrak{A}_0^z = \bigcup_{\Lambda\subset Z} \mathfrak{A}_\Lambda^z\ ,$$

$$\mathfrak{A}_\Lambda^z = \bigotimes_{k\in\Lambda} \mathfrak{A}^z{}_k,$$

$$\mathfrak{A}_k^z = \{\mu_1 I_k + \mu_2\sigma_k^z\colon\ \mu_1, \mu_2 \in C\}.$$

We then have

Theorem III.1: The WIM is stable in LRS under a perturbation $V = V^* \in \mathfrak{A}$ if and only if $V \in \mathfrak{A}^z$.

Proof: According to Verbeure and Weder,[3] the WIM is stable in LRS under a perturbation $V = V^* \in \mathfrak{A}$ if and only if $U_t\Pi_\varphi(V)\Phi = \exp(itH)\Pi_\varphi(V)\Phi$ weakly converges in the GNS Hilbert space $\mathcal{H}_\varphi$ as time tends to infinite, where H is the infinitesimal generator of the unitary one-parameter group U_t implementing $\tilde{\tau}_t$. Let us first consider the case when the perturbation $V = V^* \in \mathfrak{A}$ is local; that is, there exists a finite region Λ of Z such that $V = V^* \in \mathfrak{A}_\Lambda$. As we have discussed, the time evolution $\tilde{\tau}_t$ is an automorphism of $\Pi_\varphi(\mathfrak{A}_\Lambda)$, hence of $\mathfrak{A}_\Lambda$. The restriction $\tilde{\tau}_t^\Lambda$ of $\tilde{\tau}_t$ to $\mathfrak{A}_\Lambda$ is generated by the local Hamiltonian $H(\Lambda) = -\sum_{k\in\Lambda} B_{k,\varphi}\sigma_k^z$. The extremal KMS product state $\tilde{\varphi}$ is identical to $\exp[-H(\Lambda)]/\mathrm{Tr}\exp[-H(\Lambda)]$ on the local algebra $\mathfrak{A}_\Lambda$. Therefore, the restriction $\tilde{\varphi}^\Lambda$ of $\tilde{\varphi}$ to $\mathfrak{A}_\Lambda$ is a KMS state with respect to the time evolution $\tilde{\tau}_t^\Lambda$. If the WIM is stable in LRS under the local perturbation V, $(\chi, U_t\Pi_\varphi(V)\Phi)$ converges to a definite value for any $\chi \in \mathcal{H}_\varphi$ as time tends to infinite.

Moreover,

$$U_t\Pi_\varphi(V)\Phi = \exp(itH)\Pi_\varphi(V)\Phi$$
$$= \exp[itH(\Lambda)]\Pi_\varphi(V)\Phi = \tilde{\tau}_t\,(\Pi_\varphi(V))\Phi.$$

As φ is a product state, the Hilbert space $\mathcal{H}_\varphi$ can be an incomplete direct produce space denoted by $\otimes_{k\in Z}\mathcal{H}_k$,[9] where $\mathcal{H}_k$ is isomorphic to C^2 generated by four-dimensional vectors at the site k. Hence Φ can be written as $\otimes_{k\in Z}\Phi_k$ by cyclic vectors Φ_k for $B(\mathcal{H}_k)$. Choosing $\chi = \chi_\Lambda \otimes_{k\in Z\setminus\Lambda}\Phi_k (Z\setminus\Lambda \equiv \{k\in Z : k\notin\Lambda\})$ for any χ_Λ in $\mathcal{H}_\Lambda = \otimes_{k\in\Lambda}\mathcal{H}_k$, $(\chi, U_t\pi_\varphi(V)\Phi)$ is equal to $(\chi_\Lambda, \exp[itH(\Lambda)] \times \Pi_\varphi(V)\Phi_\Lambda)$, where $\Phi_\Lambda = \otimes_{k\in\Lambda}\Phi_k$. Since Λ is a finite region of Z and χ_Λ is arbitrary element in $\mathcal{H}_\Lambda$, the function $t \in R \to (\chi_\Lambda, \exp[itH(\Lambda)]\,\Pi_\varphi(V)\Phi_\Lambda)$ is almost periodic in t. This fact implies[12] that if the weak limit $\text{w-lim}_{t\to\pm\infty}U_t \times\Pi_\varphi(V)\Phi$ exists, then $U_t\Pi_\varphi(V)\Phi$ is identical to $\Pi_\varphi(V)\Phi$. Since $U_t\Pi_\varphi(V)\Phi = \tilde{\tau}_t^\Lambda(\Pi_\varphi(V))\Phi$ and Φ is separating for the von Neumann algebra $\Pi_\varphi(\mathfrak{A})''$, we have $\tilde{\tau}_t^\Lambda(\Pi_\varphi(V)) = \Pi_\varphi(V)$. As the quasilocal C^*-algebra $\mathfrak{A}$ is simple, the representation Π_φ is faithful; hence $[\exp[itH(\Lambda)],\ V] = 0$. Thus we immediately conclude that V is in $\mathfrak{A}_\Lambda^z$ according to the forms of $H(\Lambda)$ and our local C^*-subalgebra $\mathfrak{A}_\Lambda^z$. Conversely, if the local perturbation $V = V^* \in \mathfrak{A}_\Lambda$ is in $\mathfrak{A}_\Lambda^z$, then it is obvious that $U_t\Pi_\varphi(V)\Phi$ weakly converges in $\mathcal{H}_\varphi$ as time tends to infinite.

Next we have to consider for a general perturbation $V = V^* \in \mathfrak{A}$. Namely, we have to show that the weak limit $\text{w-lim}_{t\to\pm\infty}U_t\Pi_\varphi(V)\Phi$ exists in $\mathcal{H}_\varphi$ if and only if $V = V^* \in \mathfrak{A}^z$. The "if" part is shown as follows: For any $V \in \mathfrak{A}^z$ and any $\epsilon > 0$, there exists $V_0 = V_0^* \in \mathfrak{A}_0^z$ such that $\|V - V_0\| < \epsilon$. Hence we obtain

$$\|U_t\Pi_\varphi(V)\Phi - \Pi_\varphi(V)\Phi\| \le \|U_t\Pi_\varphi(V)\Phi - U_t\Pi_\varphi(V_0)\Phi\|$$
$$+ \|U_t\Pi_\varphi(V_0)\Phi - \Pi_\varphi(V_0)\Phi\| + \|\Pi_\varphi(V_0)\Phi - \Pi_\varphi(V)\Phi\|$$
$$= 2\,\|\Pi_\varphi(V) - \Pi_\varphi(V_0)\| = 2\,\|V - V_0\| < 2\epsilon.$$

Conversely, let us prove the "only if" part. If there exists $V = V^* \in \mathfrak{A}$ such that $\text{w-lim}_{t\to\pm\infty}U_t\Pi_\varphi(V)\Phi$ exists in $\mathcal{H}_\varphi$, then as the local algebra $\mathfrak{A}_0 = \cup_{\Lambda\subset Z}\mathfrak{A}_\Lambda$ is norm dense in $\mathfrak{A}$, for any $V = V^* \in \mathfrak{A}$ and any $\epsilon > 0$, there exists $V_0 = V_0^* \in \mathfrak{A}_0$ such that $\|V - V_0\| < \epsilon$; hence it is easily seen that

$$\|U_t\Pi_\varphi(V)\Phi - U_t\Pi_\varphi(V_0)\Phi\| < \epsilon.$$

Therefore, if $\text{w-lim}_{t\to\pm\infty}U_t\Pi_\varphi(V)\Phi$ exists in $\mathcal{H}_\varphi$, then $\text{w-lim}_{t\to\pm\infty}U_t\Pi_\varphi(V_0)\Phi$ exists in $\mathcal{H}_\varphi$ too. Since V_0 is a local element of $\mathfrak{A}$, this V_0 should be in the algebra $\mathfrak{A}_0^z$ as seen before. This fact implies that the perturbation $V = V^* \in \mathfrak{A}$ is in the algebra $\mathfrak{A}^z$ because $\mathfrak{A}_0^z$ is norm dense in $\mathfrak{A}^3$.

Q.E.D.

From this theorem, we can divide the perturbations into two classes. The first class is the set of perturbations under which the WIM is stable in LRS. Under the perturbations of the second class, the WIM is not stable in LRS.

The following theorem tells us that when $V = V^* \in \mathfrak{A}^z$, the linear response approximation is exact as far as the stability of the WIM is concerned.

Theorem III.2: For any perturbation $V = V^* \in \mathfrak{A}^z$, we have

(1) $\text{w*-}\lim_{t\to\pm\infty} \tilde{\varphi}\circ \tilde{\tau}_t^{\Pi_\varphi(V),1} = \varphi$ on ,

(2) $\text{w*-}\lim_{t\to\pm\infty} \tilde{\varphi}\circ \tilde{\tau}_t^{\Pi_\varphi(V)} = \varphi$ on $\mathfrak{A}$.

Remark: (a) The above exact perturbed time evolution $\tilde{\tau}_t^{\Pi_\varphi(V)}$ is given by (3.3) and (3.4) under the replacements of α_t and λV with $\tilde{\tau}_t$ and $\lambda\Pi_\varphi(V)$. (b) This theorem can be read that if the WIM is stable in LRS under a perturbation $V=V^*\in\mathfrak{A}$, then it is stable in exact sense under the same V.

Proof: According to Verbeure and Weder,[3] we have

$$\lim_{t\to\pm\infty} \langle\tilde{\varphi}; \tilde{\tau}_t^{\Pi_\varphi(V),1}(\Pi_\varphi(V))\rangle$$
$$= \langle\tilde{\varphi}; \Pi_\varphi(A)\rangle - \lambda \lim_{t\to\pm\infty} (T\Pi_\varphi(A^*)\Phi, T[1-\exp(itH)]\Pi_\varphi(V)\Phi),$$

where $T=[(\Delta-1)/\ln\Delta]^{1/2}$ and $\Delta=\exp(-H)$. We have seen in the proof of the theorem III.1

$$\exp(itH)\Pi_\varphi(V)\Phi = \Pi_\varphi(V)\Phi,$$

for any $V=V^*\in\mathfrak{A}^z$. We hence have

$$(T\Pi_\varphi(A^*)\Phi,\ T[1-\exp(itH)]\Pi_\varphi(V)\Phi)=0,$$

which implies $\lim_{t\to\pm\infty}\langle\tilde{\varphi}; \tilde{\tau}_t^{\Pi_\varphi(V),1}(\Pi_\varphi(A))\rangle = \langle\tilde{\varphi}; \pi_\varphi(A)\rangle = \langle\varphi; A\rangle$ for any $A\in\mathfrak{A}$. Let us prove (2). Simple computation tells us

$$\frac{d}{dt}\tilde{\tau}_{-t}\tilde{\tau}_t^{\Pi_\varphi(V)}(\Pi_\varphi(A)) = i\lambda\tilde{\tau}_{-t}([\Pi_\varphi(V), \tilde{\tau}_t^{\Pi_\varphi(V)}(\Pi_\varphi(A))]).$$

Hence

$$\langle\tilde{\varphi}; \tilde{\tau}_t^{\Pi_\varphi(V)}(\Pi_\varphi(A))\rangle = \langle\tilde{\varphi}; \tilde{\tau}_{-t}\tilde{\tau}_t^{\Pi_\varphi(V)}(\Pi_\varphi(A))\rangle$$
$$= \langle\varphi; A\rangle + i\lambda\int_0^t ds\langle\tilde{\varphi}; [\Pi_\varphi(V), \tilde{\tau}_t^{\Pi_\varphi(V)}(\Pi_\varphi(A))]\rangle.$$

Therefore, we have only to show that the limits $\lim_{t\to\pm\infty}\int_0^t ds\langle\tilde{\varphi}; [\Pi_\varphi(V), \tilde{\tau}_t^{\Pi_\varphi(V)}(\Pi_\varphi(A))]\rangle$ exist for any A in $\mathfrak{A}_0=\cup_{\Lambda\subset Z}\mathfrak{A}_\Lambda$. Since $\tilde{\tau}_t(\Pi_\varphi(V))=\Pi_\varphi(V)$ for any $V=V^*\in\mathfrak{A}^z$, the perturbed time evolution $\tilde{\tau}_t^{\Pi_\varphi(V)}$ is given by

$$\tilde{\tau}_t^{\Pi_\varphi(V)}(\Pi_\varphi(A)) = \exp[it\Pi_\varphi(V)]\ \tilde{\tau}_t(\Pi_\varphi(A))\exp[-it\Pi_\varphi(V)].$$

Let us take any $V_0=V_0^*$ from $\mathfrak{A}^z_{\Lambda_0}$, where Λ_0 is a finite region of Z. For a local element A of $\mathfrak{A}$, there exists a finite region Λ of Z such that $A\in\mathfrak{A}_\Lambda$. According to the locally normality of the state φ, we obtain

$$\langle\tilde{\varphi}; [\Pi_\varphi(V_0), \tilde{\tau}_t^{\Pi_\varphi(V)}(\Pi_\varphi(A))]\rangle$$
$$= \operatorname{Tr}\exp[-H(\Lambda\cup\Lambda_0)][V_0, \exp(itV_0)\exp[itH(\Lambda)]A \times\exp[-itH(\Lambda)]\exp(-itV_0)]/\operatorname{Tr}\exp[-H(\Lambda\cup\Lambda_0)], \tag{3.5}$$

where $H(\Omega)=-\sum_{k\in\Omega}B_{k,\varphi}\sigma_k^z$ with $\Omega=\Lambda$ or $\Lambda\cup\Lambda_0$. The above (3.5) is equal to zero due to the forms of $H(\Omega)$ and V_0. As $\mathfrak{A}_0^z$ is norm dense in $\mathfrak{A}^z$, we conclude (2) by continuity. Q.E.D.

IV. APPROACH TO EQUILIBRIUM

One way of analyzing the structure of statistical mechanics is to study the time development of states which have been perturbed from equilibrium. If a perturbed state relaxes to another state as time tends to infinite, we may say that the approach to equilibrium occurs under that perturbation. In this section, we will study the time development of such perturbed states in the WIM. As discussed in the previous section, the WIM is described by a triple $(\Pi_\varphi(\mathfrak{A}), \tilde{\varphi}, \tilde{\tau})$ for an extremal KMS state φ. For a perturbation $\lambda V=\lambda V^*\in\mathfrak{A}$ with $\lambda\in[0,1]$, we denote the exact perturbed WIM by a triple $(\Pi_\varphi(\mathfrak{A}), \tilde{\varphi}^{\Pi_\varphi(V)}, \tilde{\tau}^{\Pi_\varphi(V)})$ and the linear response WIM by $(\Pi_\varphi(\mathfrak{A}), \tilde{\varphi}^{\Pi_\varphi(V),1}, \tilde{\tau}^{\Pi_\varphi(V),1})$. They are given as follows[4,13]: The exact perturbed and linear response time evolutions $\tilde{\tau}_t^{\Pi_\varphi(V)}, \tilde{\tau}_t^{\Pi_\varphi(V),1}$ respectively are given in Sec. III. Here, let us introduce the exact perturbed state $\tilde{\varphi}^{\Pi_\varphi(V)}$ and the linear response functional $\tilde{\varphi}^{\Pi_\varphi(V),1}$. For any $A\in\mathfrak{A}$,

$$\langle\tilde{\varphi}^{\Pi_\varphi(V)}; \Pi_\varphi(A)\rangle = (\Phi^V, \Pi_\varphi(A)\Phi^V), \tag{4.1}$$

$$\Phi^V = \Pi_\varphi(W^V_{i/2})\Phi/\|\Pi_\varphi(W^V_{i/2})\Phi\|, \tag{4.2}$$

$$W^V_{i/2} = \sum_{n\ge 0}(-\lambda)^n\int_{0\le t_1\le\cdots\le t_n\le 1/2} dt_1\cdots\int dt_n\,\tilde{\tau}_{it_1}(\Pi_\varphi(V))\cdots \times\tilde{\tau}_{it_n}(\Pi_\varphi(V)), \tag{4.3}$$

$$\langle\tilde{\varphi}^{\Pi_\varphi(V),1}; \Pi_\varphi(A)\rangle = \langle\tilde{\varphi}; \Pi_\varphi(A)\rangle - \lambda\langle\tilde{\varphi}; \Pi_\varphi(A)\int_0^1 ds\,\tilde{\tau}_{is}(\Pi_\varphi(V))\rangle + \lambda\langle\tilde{\varphi}; \Pi_\varphi(A)\rangle\langle\tilde{\varphi}; \Pi_\varphi(V)\rangle, \tag{4.4}$$

where $\tilde{\varphi}^{\Pi_\varphi(V),1}$ is obtained by linearizing $\tilde{\varphi}^{\Pi_\varphi(V)}$ w.r.t. λ and some computations.[4]

The motivation of this section is the following:

(i) Under which conditions on $V=V^*\in\mathfrak{A}$, do the limits $\tilde{\varphi}_\pm^{\Pi_\varphi(V),1} = \text{w*-}\lim_{t\to\pm\infty}\tilde{\varphi}^{\Pi_\varphi(V),1}\circ\tilde{\tau}_t$ exist on $\Pi_\varphi(\mathfrak{A})$? (Approach to equilibrium in LRS under V.)

(ii) Under which conditions on $V=V^*\in\mathfrak{A}$, do the limits $\tilde{\varphi}_\pm^{\Pi_\varphi(V)} = \text{w*-}\lim_{t\to\pm\infty}\tilde{\varphi}^{\Pi_\varphi(V)}\circ\tilde{\tau}_t$ exist on $\Pi_\varphi(\mathfrak{A})$? [Approach to equilibrium in ES (exact sense) under V.]

(iii) Is the linear response approximation useful to study some dynamical processes?

We study these questions through the WIM. The answer concerning the first two question is

Theorem IV.1: For the WIM initially in equilibrium described by an extremal KMS state φ, the approach to equilibrium in LRS occurs under a perturbation $V=V^*\in\mathfrak{A}$ if and only if the approach to equilibrium in ES occurs under the same V.

Proof: As we have shown[4] that for a dynamical system satisfying the KMS condition, the approach to equilibrium in LRS occurs under a perturbation $V=V^*$ belonging to the C^*-algebra of the dynamical system if and only if the dynamical system is stable in LRS under the same V. Therefore, for the WIM, when the approach to equilibrium in LRS occurs under a perturbation $V=V^*\in\mathfrak{A}$, the WIM is stable in LRS under this perturbation V. According to Theorem III.1, this perturbation V is in the C^*-algebra $\mathfrak{A}^z$. As seen in the proof of Theorem III.1, the equality $\tilde{\tau}_t(\Pi_\varphi(V))=\Pi_\varphi(V)$ holds for all $t\in R$. It is easily checked that $\Pi_\varphi(W^V_{i/2})\Phi = \exp[-\frac{1}{2}\lambda\Pi_\varphi(V)]\Phi$ is satisfied for $V=V^*\in\mathfrak{A}^z$. Thus we obtain

$$\langle\tilde{\varphi}^{\Pi_\varphi(V)}; \tilde{\tau}_t(\Pi_\varphi(A))\rangle$$
$$= (\exp[-\tfrac{1}{2}\lambda\Pi_\varphi(V)]\Phi,\ \tilde{\tau}_t(\Pi_\varphi(A))\exp[-\tfrac{1}{2}\lambda\Pi_\varphi(V)]\Phi)/ \|\exp[-\tfrac{1}{2}\lambda\Pi_\varphi(V)]\Phi\|^2,$$

which is equal to $\langle\tilde{\varphi}^{\Pi_\varphi(V)};\Pi_\varphi(A)\rangle$ because of $[\Pi_\varphi(V)]^n \in \Pi_\varphi(\mathfrak{A}^e)$ for any positive integer n and $[U_t,R]=0$ for any $R\in\Pi_\varphi(\mathfrak{A}^e)$. Hence the approach to equilibrium in ES occurs under $V=V^*\in\mathfrak{A}^e$. Conversely, if the approach to equilibrium in ES occurs under a perturbation $V=V^*\in\mathfrak{A}$, then the limits $\lim_{t\to\pm\infty}\langle\tilde{\varphi}^{\Pi_\varphi(V)};\tilde{\tau}_t(\Pi_\varphi(A))\rangle$ exist uniformly in $\lambda\in[0,1]$ for any $A\in\mathfrak{A}$. According to the definition of $\tilde{\varphi}^{\Pi_\varphi(V)}$, $\langle\tilde{\varphi}^{\Pi_\varphi(V)};\tilde{\tau}_t(\Pi_\varphi(A))\rangle$ is expressed by a power series of λ. Moreover, it is readily seen that the function $\lambda\in[0,1]\to\langle\tilde{\varphi}^{\Pi_\varphi(V)};\tilde{\tau}_t(\Pi_\varphi(A))\rangle$ is differentiable at all λ for any $A\in\mathfrak{A}$ and each term of the power series of λ is uniformly bounded in t. Hence

$$\lim_{t\to\pm\infty}\langle\tilde{\varphi}^{\Pi_\varphi(V),1};\tilde{\tau}_t(\Pi_\varphi(A))\rangle-\langle\tilde{\varphi};\Pi_\varphi(A)\rangle$$

$$=\lambda\lim_{t\to\pm\infty}\left[\frac{d}{d\lambda}\langle\tilde{\varphi}^{\Pi_\varphi(V)};\tilde{\tau}_t(\Pi_\varphi(A))\rangle\right]_{\lambda=0}$$

$$=\lambda\left[\frac{d}{d\lambda}\lim_{t\to\pm\infty}\langle\tilde{\varphi}^{\Pi_\varphi(V)};\tilde{\tau}_t(\Pi_\varphi(A))\rangle\right]_{\lambda=0}$$

exists for any $A\in\mathfrak{A}$. This concludes that the approach to equilibrium in LRS occurs under V. Q.E.D.

The above theorem tells us that the linear response method is a good approximation of the exact method as far as the approach to equilibrium is concerned for the WIM.

Let us consider the dynamical change of metastable states in the WIM. In physical systems, we often encounter metastable states which satisfy the same equilibrium condition of stable states. A stable state gives an absolute minimum to the free energy density of the system. On the contrary, a metastable state gives a relative minimum to the free energy density. A metastable state however easily changes to a stable state (or mixture of stable states) by a local external disturbance. The rigorous interpretation of the existence of metastable states and the transition from a metastable state to a stable state has not much done yet. In the WIM, we, however, know[5] that the metastable states exist below some critical temperature. These metastable states are extremal KMS states as the stable state. Let us discuss the above transition of a metastable state appeared in the WIM under the effect of local perturbations. We ask whether there exists a perturbation which causes such transition. Let $\tilde{\varphi}$ be a metastable state satisfying the KMS condition with respect to $\tilde{\tau}_t$. The locally perturbed state $\tilde{\varphi}^{\Pi_\varphi(V)}$ of $\tilde{\varphi}$ under a perturbation $\lambda V=\lambda V^*\in\mathfrak{A}$ is given by (4.1). We then switch off the perturbation at $t=0$, that is, we look at the time development of the perturbed state $\tilde{\varphi}^{\Pi_\varphi(V)}$ under $\tilde{\tau}_t$. It is interesting to know when the limit $\tilde{\varphi}_+^{\Pi_\varphi(V)}=\text{w*-}\lim_{t\to\infty}\tilde{\varphi}^{\Pi_\varphi(V)}\circ\tilde{\tau}_t$ exists and whether $\tilde{\varphi}$ and $\tilde{\varphi}_+^{\Pi_\varphi(V)}$ are disjoint. The first question has been answered by Theorem IV.1. We are interested in the second question here. If these states $\tilde{\varphi}$ and $\tilde{\varphi}_+^{\Pi_\varphi(V)}$ are disjoint, then there is possibility of having the desired transition from the metastable state to a stable state. Due to Theorem IV.1, the limit $\tilde{\varphi}_+^{\Pi_\varphi(V)}=\text{w*-}\lim_{t\to\infty}\tilde{\varphi}^{\Pi_\varphi(V)}\circ\tilde{\tau}_t$ exists if and only if the perturbation V is in the C^*-algebra $\mathfrak{A}^e$. As shown in the proof of that theorem, $\tilde{\varphi}_+^{\Pi_\varphi(V)}$ is identical to $\tilde{\varphi}^{\Pi_\varphi(V)}$. It is easily seen that the state $\tilde{\varphi}^{\Pi_\varphi(V)}$ is faithful normal on the von Neumann algebra $\Pi_\varphi(\mathfrak{A})''$. Thus the states $\tilde{\varphi}$ and $\tilde{\varphi}^{\Pi_\varphi(V)}$ cannot be disjoint. This fact implies that there does not exist a perturbation $V=V^*\in\mathfrak{A}$ under which a metastable state of the WIM goes to a stable state.

ACKNOWLEDGMENTS

The author thanks Professor G. G. Emch for valuable discussions and suggestions of this work. He also thanks Professor H. Umegaki for his interest to the author's work and his kindness at Tokyo Institute of Technology.

[1] R. Haag, D. Kastler, and E. Trych-Pohlymeyer, Commun. Math. Phys. 38, 173 (1974).
[2] J. L. Lebowitz, M. Aizemann, and S. Goldstein, J. Math. Phys. 6, 1284 (1975).
[3] A. Verbeure and R. Weder, Commun. Math. Phys. 44, 101 (1975).
[4] M. Ohya, "Dynamical Process in Linear Response Theory," preprint.
[5] G. G. Emch and J. F. Knops, J. Math. Phys. 12, 2043 (1971).
[6] G. G. Emch, J. Math. Phys. 7, 1198 (1966).
[7] G. L. Sewell, Lett. Nuovo Cimento 10, 430 (1974).
[8] A. Daneri, A. Loinger, and G. M. Prosperi, Nucl. Phys. 33, 297 (1962).
[9] G. G. Emch, *Algebraic Methods in Statistical Mechanics and Quantum Field Theory* (Wiley, New York, 1972).
[10] R. Haag, N. Hugenholtz, and M. Winnik, Commun. Math. Phys. 5, 215 (1967).
[11] M. Takesaki, *Tomita's Theory of Modular Hilbert Algebra and Its Applications* (Springer-Verlag, Berlin, 1970).
[12] M. Winnik, in *Fundamental Poblems in Statistical Mechanics*, edited by E. D. G. Cohen (North-Holland, Amsterdam, 1975).
[13] D. W. Robinson, lecture at Varenna School, 1973.

List of Publications

1. Papers

1 M.Ohya (1974) Dynamical process in quantum statistical mechanics, U.R Tech. Rep., 25, No.2, 12-25.

2 M.Ohya (1975) Remarks on noncommutative ergodic properties, U.R Tech. Rep., 26, No.3, 30-38.

3 M.Ohya (1978) Stability of Weiss Ising model, J. Math. Phys., 19, No.5, 967-971.

4 M.Ohya (1978) On linear response dynamics, L. Nuovo Cimento, 21, No.16, 573-576.

5 M.Ohya (1979) Dynamical process in linear response theory, Rep. Math. Phys, 16, No.3, 305-315.

6 M.Ohya (1980) On open system dynamics -an operator algebraic study-, Kodai Math. J., 3, 287-294.

7 F.Hiai, M.Ohya and M.Tsukada (1981) Sufficiency, KMS condition and relative entropy in von Neumann algebras, Pacific J. Math., 96, No.1, 99-109.

8 F.Hiai, M.Ohya and M.Tsukada (1981) Relative entropy in quantum dynamical systems, Res. Rep. Inform. Sci. of TIT, A-77.

9 M.Ohya (1981) Quantum ergodic channels in operator algebras, J. Math. Anal. Appl., 84, No.2, 318-327.

10 M.Ohya (1983) Note on quantum probability, L. Nuovo Cimento, 38, No.11, 402-404.

11 F.Hiai, M.Ohya and M.Tsukada (1983) Sufficiency and relative entropy in *-algebras with applications in quantum systems, Pacific J. Math., 107, No.1, 117-140.

12 M.Ohya (1983) On compound state and mutual information in quantum information theory, IEEE Trans. Information Theory, 29, No.5, 770-774.

13 大矢雅則，渡邉　昇 (1984) 量子論的通信過程における数理モデルの構成とその解析, 電子通信学会誌, J87-A, No.6, 548-552.

14 M.Ohya (1984) Entropy Transmission in C*-dynamical systems, J. Math. Anal. Appl., 100, No.1, 222-235.

15 大矢雅則 (1984) DNA のエントロピー解析, Semi. Appl. Func. Anal., 6, 45-56.

16 M.Ohya (1985) State change and entropies in quantum dynamical systems, Springer Lecture Notes in Math., 1136, 397-408.

17 M.Ohya (1985) Construction and analysis of a mathematical model in quantum communication processes, Scripta Thechnica, Inc., Elect. Commun. Japan, 68, No.2, 29-34.

18 M.Ohya (1985) A Mathematical Analysis of DNA Sequences, Symp. Appl. Funct. Anal., 8, 36-47.

19 M.Ohya and Y.Fujii (1986) Entropy change in linear response dynamics, IL

Nuovo Cimento, 91B, No.1, 25-30.

20 M.Ohya and T.Matsuoka (1986) Continuity of entropy and mutual entropy in C*-dynamical systems, J. Math. Phys., 27(8), 2076-2079.

21 M.Ohya and N.Watanabe (1986) A new treatment of communication processes with Gaussian Channels, Japan J. Appl. Math., 3, No.1, 197-206.

22 O.Hirota, K.Yamazaki, M.Nakagawa and M.Ohya (1986) Properties of error correcting code using photon pulse, Trans. IEICE, E69, No.9, 917 − 919.

23 M.Ohya, M.Tsukada and H.Umegaki (1987) A formulation of noncommutative McMillan theorem, Proc. Japan Academy, 63, Ser.A, No.3, 50-53.

24 K.Yamazaki, O.Hirota, M.Nakagawa T.Umeyama and M.Ohya (1987) Effect of error correcting code in photon communications with energy loss, Trans. IEICE, E70, No.8, 689-692.

25 M.Ohya (1987) Information theoretical analysis of phylogenetic trees for molecular evolution, Symp. Appl. Func. Anal., 10, 87-101.

26 M.Ohya (1988) Fractal dimensions for general quantum states, Symp. Appl. Func. Anal., 11, 45-57.

27 大矢雅則，吉見英弥，広田　修 (1988) 量子制御通信過程における誤り確率の厳密な導出, 電子情報通信学会論文誌, J71-B, No.4, 533-539.

28 大矢雅則 (1988) 光通信の数学的基礎－非可換系のエントロピー論と光通信過程－, 素粒子論研究, 第 78 巻 2 号, 通巻 500 号, B124-B140.

29 大矢雅則 (1988) 遺伝子の情報理論的解析, 電子情報通信学会誌, 71, No.3, 295-297.

30 M.Ohya (1989) Some aspects of quantum information theory and their applications to irreversible processes, Rep. Math. Phys., 27, 19-47.

31 大矢雅則, 村木尚文 (1989) Breiman の定理への補足, 電子情報通信学会誌 (A), J72-A, 2057-2060.

32 M.Ohya (1989) Information theoretical treatments of genes, Trans. IEICE, E72, No.5, 556-560.

33 大矢雅則, 宮崎　智, 大島靖史 (1989) アミノ酸配列の整列化法, Viva Origino, 17, No.3, 141-151.

34 大矢雅則 (1990) 状態の ε エントロピーとフラクタル次元, 素粒子論研究, 80, D138-D149.

35 大矢雅則, 須鎗弘樹 (1990) 量子制御通信過程における誤り確率の最適化, 電子情報通信学会論文誌, J73-B-I, No.3, 200-207.

36 大矢雅則, 井ノ口育美 (1990) 情報量を用いた進化系統樹作成の新手法, 生物物理学会誌, 30, No.2, 30-37.

37 M.Ohya (1991) Fractal dimensions of states, Quantum Probability and Related Topics, 6, 359-369.

38 M.Ohya (1991) Information dynamics and its applications to optical communication processes, Springer Lecture Notes in Physics, 378, 81-92.

39 M.Ohya and H.Suyari (1991) Rigorous derivation of error probability in coherent optical communication, Springer Lecture Note in Physics. 378,

203-212.

40 M.Ohya (1991) On fuzzy relative entropy, Symp. Appl. Func. Anal., 13, 105-115.

41 大矢雅則, 松永慎一 (1991) 符号化法と遺伝子, 電子情報通信学会論文誌, J74A, No.7, 1075-1084.

42 N.Muraki, M.Ohya and D.Petz (1992) Entropies of general quantum states, Open Systems and Information Dynamics, 1, No.1, 43-56.

43 N.Muraki and M.Ohya (1992) Note on continuity of information rate, Illinois J. of Math., 36, 529-550.

44 大矢雅則, 水　正範 (1992) 最適値問題に対する情報論的アプローチとその応用, 電子情報通信学会論文誌, J75-A, No.12, 1859-1864.

45 M.Ohya and Y.Uesaka (1992) Amino acid sequences and DP matching: a new method of alignment, Information Sciences, 63, 139-151.

46 M.Ohya, S.Miyazaki and K.Ogata (1992) On multiple alignment of genome sequences, IEICE Trans. Commun. E75-B, No.6, 453-457.

47 M.Ohya and S.Naritsuka (1993) On fuzzy relative entropy, Open Systems and Information Dynamics, 1, No.3, 397-408.

48 M.Ohya (1993) Quantum entropies and thier maximizations, Maximum Entropy and Bayesian Methods, 12, 189-194.

49 M.Ohya and N.Watanabe (1993) Information dynamics and its application to Gaussian communication processes, Maximum Entropy and Bayesian Methods, 12, 195-203.

50 L.Accardi, M.Ohya and H.Suyari (1994) Computation of mutual entropy in quantum markov chains, Open Systems and Information Dynamics, 2, No.3, 337-354.

51 F.Hiai, M.Ohya and D.Petz (1995) McMillan type convergence for quantum Gibbs states, Archives of Mathematics, 64, 154-158.

52 T.Matsuoka and M.Ohya (1995) Fractal dimensions of states and their applications to Ising model, Rep. Math. Phys., 36, 365-379.

53 M.Ohya and T.Matsuoka (1995) Simulated annealing and its application to Cobb-Douglas economic model, Open Systems and Information Dynamics, 3, No.3, 357-368.

54 L.Accardi, M.Ohya and H.Suyari (1995) An application of lifting theory to optical communication processes, Rep. Math. Phys., 36, 403-420.

55 M.Ohya (1995) State change, complexity and fractal in quantum systems, Quantum Communications and Measurement, Plenum, 2, 309-320.

56 L.Accardi, M.Ohya and H.Suyari (1995) Mutual entropy in quantum Markov chains, Quantum Communications and Measurement, Plenum, 2, 351-358.

57 M.Ohya and N.Watanabe (1995) A mathematical study of information transmission in quantum communication processes, Quantum Communications and Measurement, Plenum, 2, 371-378.

58 大矢雅則, 松岡隆志 (1996) 状態のフラクタル次元を用いたクレーター及び河川の複雑さの解析, 電子情報通信学会論文誌, J79-A, No.9, 1590-1599.
59 N.Muraki and M.Ohya (1996) Entropy functionals of Kolmogorov-Sinai type and their limit theorems, Letter in Mathematical Physics, 36, 327-335.
60 L.Accardi, M.Ohya and N.Watanabe (1996) Note on quantum dynamical entropies, Rep. Math. Phys., 38, No.3, 457-469.
61 M.Ohya and D.Petz (1996), Notes on quantum entropy, Studia Scientiarum Mathematicarum Hungarica, 31, 423-430.
62 M.Ohya and N.Watanabe (1996) Note on irreversible dynamics and quantum information, Contributions in Probability, Undine, Forum, 205-220.
63 S.Miyazaki, H.Sugawara and M.Ohya (1996) The efficiency of entropy evolution rate for construction of phylogenetic trees, Genes Genet. Syst., 71, 323-327.
64 M.Ohya (1997) Complexity, fractal dimension for quantum states, Open Systems and Information Dynamics, 4, 141-157.
65 大矢雅則, 小坂　稔 (1997) 情報力学によるカオス現象の考察, 電子情報通信学会論文誌, J80-A, No.12, 2138-2144.
66 L.Accardi, M.Ohya and N.Watanabe (1997) Dynamical entropy through quantum Markov chain, Open Systems and Information Dynamics, 4, 71-87.
67 M.Ohya, D.Petz and N.Watanabe (1997) On capacities of quantum channels, Probability and Mathematical Statistics, 17, 179-196.
68 大矢雅則, 須鎗弘樹, 古市　茂 (1997) スクイズド真空状態による誤り確率の改善, 電子情報通信学会論文誌, J80-A, No.5, 809-817.
69 大矢雅則, 須鎗弘樹, 樋口　慶 (1997) 対称コンタクトプロセスによる二義錯視図形の解析, 電子情報通信学会論文誌, J80-A, No.6, 1022-1029.
70 大矢雅則, 緒方浩二 (1997) 動的計画法を応用したマルチプル・アライメントとその改良, 電子情報通信学会論文誌, J80-A, No.4, 684-691.
71 大矢雅則 (1997) 遺伝子の情報理論的アプローチ, Viva Origino, 25, 223-246.
72 S.Furuichi, M.Ohya and H.Suyari (1997) Computation of mutual entropy in quantum amplifier processes, Quantum Communication, Computing, and Measurement, 3, 147-155.
73 M.Ohya and N.Watanabe (1997) Quantum capacity of noisy quantum channel, Quantum Communication, Computing, and Measurement, Plenum, 3, 213-220.
74 M.Ohya and H.Ogata (1997) Multiple alignment for amino acid sequences by dynamics programing, Electronics and Communications in Japan, Part 3, 81, No.4, 684-691.
75 K.Sato and S.Miyazaki and M.Ohya (1998) Analysis of HIV by entropy evolution rate, Amino Acids, 14, 343-352.
76 M.Ohya (1998) Complexities and their applications to characterization of chaos, International Journal of Theoretical Physics, 37, No.1, 495-505.
77 井上　啓, 大矢雅則, 林　治 (1998) 状態のフラクタル次元を用いた地震現象の解析, 日本応用数理学会論文誌, 8, No.2, 187-197.

78 M.Ohya, D.Petz and N.Watanabe (1998) Numerical computation of quantum capacity, International Journal of Theoretical Physics, 38, No.1, 507-510.

79 M.Ohya and N.Watanabe (1998) On Mathematical treatment of Fredkin-Toffoli-Milburn gate, Physica D, 120, 206-213.

80 K.Inoue, M.Ohya and H.Suyari (1998) Characterization of quantum teleportation by nonlinear quantum channel and quantum mutual entropy, Physica D, 120, 117-124.

81 大矢雅則 (1998) 量子チャネルの伝送容量について, 電子情報通信学会論文誌, J81-A, No.12, 1638-1643.

82 菊池慶一, 渡邉　昇, 大矢雅則 (1998) 結合チャネルの情報伝送効率について, 電子情報通信学会論文誌, J81-A, No.12, 1707-1714.

83 H.Ogata, M.Ohya and H.Umeyama (1998) Amino Acid Similarity Matrix for Homology Modeling Derived from Structural Alignment And Optimized by the Monte Carlo Method, J. Mol. Graphics and Mod., 16, 178-189.

84 M.Ohya (1998) Foundation of entropy, complexity and fractal in quantum systems, Probability Towards the Year 2000, 263-286.

85 A.Kossakowski, M.Ohya and N.Watanabe (1999) Quantum dynamical entropy for completely positive map, Infinite Dimensional Analysis, Quantum Probability and Related Topics, 2, No.2, 267-282.

86 L.Accardi and M.Ohya (1999) Compound channels, transition expectations, and liftings, Appl. Math. Optim., 39, 33-59.

87 M.Ohya (1999) Fundamentals of quantum mutual entropy and capacity, Open System and Information Dynamics, 6, No.1, 69-78.

88 S.Furuichi and M.Ohya (1999) Quantum mutual entropy for Jaynes-Cummings model, Rep. Math. Phys., 44, 81-86.

89 S.Furuichi and M.Ohya (1999) Entanglement degree in the time development of the Jaynes-Cummings model, Letter in Math. Phys., 49, 279-285.

90 H. Hirano, K. Sato, T. Yamaki and M. Ohya (1999) Study of the relation between the variation of HIV and the disease, progression by entropy evolution rate, Viva Origino, 27, 91-106.

91 K.Inoue and T.Matsuoka and M.Ohya (2000) ε-entropy and fractal dimension of a state for a Gaussian measure, Open Systems and Information Dynamics, 7, No.1, 41-53.

92 井上　啓, 上川原一, 大矢雅則 (2000) 量子エントロピー型カオス尺度を用いたスピン 1/2 系のカオスの解析, 日本応用数理学会論文誌, 10, No.1, 51-69.

93 K.Inoue, M.Ohya and K.Sato (2000) Application of chaos degree to some dynamical systems, Chaos, Soliton & Fractals, 11, 1377-1385.

94 M.Ohya (2000) Complexity in dynamics and computation, Acta Applicandae Mathematicae, 63, 293-306.

95 M.Ohya and N.Masuda (2000) NP problem in quantum algorithm, Open Systems and Information Dynamics, 7, No.1, 33-39.

96 M.Ohya and K.Sato (2000) Use of information theory to study genome sequences, Rep. Math. Phys., 46, No.3, 419-428.

97 V.P.Belavkin and M.Ohya (2001) Quantum entropy and information in discrete entangled states, Infinite Dimensional Analysis, Quantum Probability and Related Topics, 4, No.2, 137-160.

98 K-H.Fichtner and M.Ohya (2001) Quantum teleportation with entangled states given by beam splittings, Commun. Math. Phys., 222, 229-247.

99 K.Sato and M.Ohya (2001) Analysis of the disease course of HIV-1 by entropic chaos degree, Amino Acids, 20, 155-162.

100 K.Sato and M.Ohya (2001) Can information measure be one of markers to estimate disease progression in HIV-1-infected patients?, Open Systems and Information Dynamics, 8, 125-136.

101 M.Ohya, I.V.Volovich and N.Watanabe (2001) Quantum Logical Gate Based on Electron Spin Resonance, Quantum Information, 3 , 143-156.

102 M.Ohya,T.Nishioka and K.Sato (2001) Phylogenetic Relation of HIV-1 in the V3 Region by Information Measure, Quantum Information, 3, 127-142.

103 K-H.Fichtner and M.Ohya (2002) Quantum teleportation and beam splitting, Commun. Math. Phys., 225, 67-89.

104 K.Inoue, M.Ohya and I.V.Volovich (2002) Semiclassical properties and chaos degree for the quantum baker's map, J. Math. Phys., 43-2, 734-755.

105 V.P.Belavkin and M.Ohya (2002) Entanglement, quantum entropy and mutual information, Proc. R. Soc. Lond. Ser. A Math. Phys. Eng. Sci. 458, No.2017, 209-231

106 K.Inoue, I.V.Volovich and M.Ohya (2002) A Treatment of Quantum Baker's Map by Chaos Degree, Quantum Information, 4, 87-102.

107 M.Ohya (2002) An Information Theoretic Approach to the Study of Genome Sequences: An Application to the Evolution of HIV, Springer Lecture Notes in Computer Science, 2509, 50-57.

108 M.Ohya and I.V.Volovich (2003) On Quantum Capacity and Its Bound, Infinite Dimensional Analysis and Quantum Probability, 6, No.2, 301-310.

109 A.Kossakowski, M.Ohya and Y.Togawa (2003) How can we observe and describe chaos?, Open System and Information Dynamics, 10, No.3, 221-233.

110 M.Ohya and I.V.Volovich (2003) New quantum algorithm for studying NP-complete problems, Rep. Math. Phys., 52, No.1, 25-33.

111 M.Ohya and I.V.Volovich (2003) Quantum computing and chaotic amplifier, J. Opt. B, 5, No.6, 639-642.

112 池正人，佐藤圭子，谷田貝甲児，大矢雅則 (2003) 相互エントロピーを用いたアライメントの改良, 情報知識学会誌掲載, 13, No.3, 49-57.

113 M.Ohya (2003) Information Dynamics and its Application to Recognition Process, A Garden of Quanta : Essays in Honor of Hiroshi Ezawa, World Scientific Pub. Co. Inc., 445-462.

114 K-H.Fichtner, W.Freudenberg, M.Ohya (2003) Recognition and teleportation, Quantum probability and infinite dimensional analysis, QP–PQ: Quantum Prob. White Noise Anal., Quantum Probability and Infinite-Dimensional Analysis, World Sci. Publishing, 15, 85-105.

115 K.Inoue, M.Ohya, I.V.Volovich (2003) On quantum-classical correspondence and chaos degree for baker's map. Fundamental aspects of quantum physics (Tokyo, 2001), QP–PQ: Quantum Prob. White Noise Anal., World Sci. Publishing, 17, 177–187.

116 L.Accardi and M.Ohya (2004) A Stochastic Limit Approach to the SAT Problem, Open Systems and Information dynamics, 11-3, 219-233.

117 M.Ohya (2004) Foundation of Chaos Through Observation, Quantum Information and Complexity edited by T.Hida, K.Saito and Si Si, 391-410.

118 T.Mastuoka and M.Ohya (2004) A New Measurement of Time Serial Correlations in Stock Price Movements and Its Application, Quantum Information and Complexity edited by T.Hida, K.Saito and Si Si, 341-361.

119 M.Ohya (2004), On quantum information and algorithm, Mathematical Modelling in Physics, Engineering and Cognitive Sciences, 10, 451-468.

120 M.Ohya (2005) Quantum Algorithm for SAT Problem and Quantum Mutual Entropy, Reports on Mathematical Physics, 55, No.1, 109-125.

121 T.Matsuoka and M.Ohya (2005) Quantum Entangled State and Its Characterization, Foundations of Probability and Physics-3, American Institute of Physics, 750, 298-306.

122 K-H Fichtner, W.Freudenberg and M.Ohya (2005) Teleportation Schemes In Infinite Dimensional Hilbert Spaces, J. Math. Phys. 46, 102103-14.

123 T.Miyadera and M.Ohya (2005) Quantum Dynamical Entropy of Spin Systems, Reports on Mathematical Physics, 56, No.1, 1-10.

124 W. Freudenberg, M. Ohya and N. Watanabe (2005), On quantum logical gates on a general Fock space, QP-PQ: Quantum Probability and White Noise Analysis, World Sci. Publishing, 18, 252-268.

125 K-H Fichtner, W.Freudenberg and M.Ohya (2006) Recognition and Teleportation, Quantum Information, 5, 1-17.

126 T.Miyadera and M.Ohya (2006) On Halting Process of Quantum Turing Machine, Open System and Information Dynamics, 12, No.3, 261-264.

127 F.-H.Fichtner, T.Miyadera, M.Ohya (2006) Fidelity of Qunatum Teleportation Model Using Beam Splittings, QP-PQ: Quantum Probability and White Noise Analysis, Quantum Information and Computing, World Sci. Publishing, 19, 113-130.

128 W.Freudenberg, M.Ohya, N.Turchina, N.Watanabe (2006) Quantum Logical Gates Realized by Beam Splittings, QP-PQ: Quantum Prob. White Noise Anal., Quantum Information and Computing, World Sci. Publishing, 19, 131-148.

129 S.Iriyama, M.Ohya and I.V.Volovich (2006) Generalized Quantum Turing

Machine and its Application to the SAT Chaos Algorithm, QP-PQ: Quantum Prob. White Noise Anal., Quantum Information and Computing, World Sci. Publishing, 19, 204-225.

130 A.Kossakowski and M.Ohya (2006) Can Non-Maximal Entangled State Achieve a Complete Quantum Teleportation?, Reconsideration of Foundation-3, American Institute of Physics, 810, 211-216.

131 L.Accardi, T.Matsuoka and M.Ohya (2006) Entangled Markov Chains are Indeed Entangled, Infinite Dimensional Analysis Quantum Probability and Related Topics, 9, No.3, 379-390.

132 A.Kossakowski and M.Ohya (2007) New Scheme of Quantum Teleportation, Infinite Dimensional Analysis, Quantum Probability and Related Topics, 10, 3, 411-420.

133 M.Ohya (2008) Adaptive Dynamics and Its Applications to Chaos and NPC Problem, Quantum Bio-Informatics: From Quantum Information to Bio-Informatics, World Sci. Publishing, 1, 181-216.

134 K.-H. Fichtner, L. Fichtner, W. Freudenberg and M.Ohya, Quantum Models of the Recognition Process - On a Convergence Theorem, submitted to Journal of Mathematical Physics.

2. Other publications

1 大矢雅則（1988-1995）情報と数理 1 -20，数理科学，No.305，78-83 ; No.307，46-53;No.309，7177 ; No.311，78-83 ; No.314，76-83;No.316，75-83 ; No.318，76-83 ; No.322，72-76;No.323，78-83;No.326，80-83;No.330，74-78;No.337，79-83;No.342，79-83;No.346，68-73;No.351，70-75;No.351，70-75;No.353，57-63;No.358，76-83;No.371，61-67;No.373，59-64;No.385，77-83.

3. Books

1 梅垣壽春，大矢雅則（1983）確率論的エントロピー，共立出版.

2 梅垣壽春，大矢雅則（1984）量子論的エントロピー，共立出版.

3 梅垣壽春，大矢雅則，日合文雄（1985）作用素代数入門，共立出版.

4 梅垣壽春，大矢雅則，塚田　真（1987）測度・積分・確率，共立出版.

5 梅垣壽春，大矢雅則，垣原裕一郎（1987）集合・位相・距離，共立出版.

6 大矢雅則（1988）光通信理論とその応用（第一章），森北出版.

7 M.Ohya and D.Petz (1993) Quantum Entropy and its Use, Springer-Verlag, TMP-series.

8 大矢雅則，今井秀樹，小嶋　泉，中村八束，廣田正義（編集）（1995）数理情報科学辞典，朝倉書店.

9 大矢雅則，小嶋　泉　（編著）（1996）量子情報と進化の力学－研究最前線－，牧野書店.

10 R.S.Ingarden, A.Kossakowski and M.Ohya (1997), Information Dynamics and Open Systems, Kluwer Academic Publishers.

11 大矢雅則，渡邉　昇（1998）量子通信理論の基礎－量子情報から光通信へ－，牧野書店.

12 大矢雅則（1999）量子コンピュータの数理，丸善株式会社.

13 井上　啓，佐藤圭子，大矢雅則（1999）情報数理入門，サイエンス社.

14 大矢雅則（2005）情報進化論，岩波書店.

16 大矢雅則, 渡邉　昇 (2006), 量子暗号と量子テレポーテーション-新たな情報通信プロトコル-, 共立出版

17 大矢雅則（編著）(2008),「量子情報を巡る数理と哲理」－エントロピー・ゆらぎ・ミクロとマクロ・アルゴリズム・生命情報－, 国際高等研究所　Academic Publication.

15 M.Ohya and I.V.Volovich, Mathematical Foundation of Quantum Computers, Teleportations and Cryptography, to appear.